Geophysical Monograph Series

Geophysical Monograph Series

A. F. Spilhaus, Jr., Managing Editor

Geophysical Monograph 27

The Tectonic and Geologic Evolution of Southeast Asian Seas and Islands: Part 2

Dennis E. Hayes
Editor

American Geophysical Union
Washington, D.C.
1983

The Tectonic and Geologic Evolution
of Southeast Asian Seas and Islands: Part 2

Library of Congress Cataloging in Publication Data

(Revised for part 2)
Main entry under title:

The Tectonic and geologic evolution of Southeast Asian
  seas and islands.

  (Geophysical monograph series; no. 23,   )
  Includes bibliographies and index.
  1. Geology—South Pacific Ocean. 2. Geology—Asia,
Southeastern. I. Hayes, Dennis E. II. Series:
Geophysical monograph; no. 23, etc.
QE350.42.S68T43   551.46′08′0947   80-27080
ISBN 0-87590-023-2 (v. 1)

ISBN 0-87590-053-4

ISSN 0065-8448
Printed in the United States of America.

CONTENTS

Preface

This volume is the second AGU monograph on Studies of East Asian Tectonics and Resources (SEATAR). Like its predecessor, Geophysical Monograph 23, *The Tectonic and Geologic Evolution of Southeast Asian Seas and Islands*, Geophysical Monograph 27 is primarily an outgrowth of a major international program of cooperative research between earth scientists in the United States and their counterparts in Southeast Asia. From a workshop held in Bangkok in October 1973, this research program grew into an international, multimillion dollar program. Most of the U.S. SEATAR research programs were funded by the Office of the International Decade of Ocean Exploration (IDOE) of the National Science Foundation (NSF), but several closely related projects were supported by other sections of the NSF, by the Office of Naval Research, and by others. The formal U.S. SEATAR program is now drawing to a close.

Many new field programs and laboratory studies have been completed since the publication of Geophysical Monograph 23, and, in large part, they have focused on new geographic locales and/or new approaches. Hence this volume expands on our previously published studies rather than 'replaces' them. The two volumes should be viewed as being complementary—collectively they provide the most comprehensive analyses, conclusions, and speculations available anywhere regarding selected, key regions within Southeast Asia.

The contributions of Geophysical Monograph 27 span an unusually wide spectrum of disciplines, ranging from detailed geochemistry, through seismotectonics, classical stratigraphy, paleomagnetics, offshore multichannel seismic programs, to seafloor magnetotelluric soundings and more. Furthermore, there has been a considerable attempt to utilize the best information from both onshore and offshore and to integrate these results successfully. As a consequence I believe we have made an important contribution toward the understanding of the complex time-space-process matrix that has controlled the evolution of this fascinating region. Monographs 27 and 23 will continue to serve as fundamental reference volumes for geological and geophysical researches for many years to come. Those with current interests in the hydrocarbon potential of the area, especially the South China Sea and environs, and longstanding interests in metalliferous ore deposits and regional and local tectonism and volcanism will find that these volumes represent a timely statement on the state of the art here.

I wish to acknowledge the support of the Office of IDOE of the National Science Foundation in the implementation of the United States SEATAR program. Without this support most of the research projects reported here would not have been possible. Similarly, the outstanding cooperation of our scientific colleagues in Southeast Asia, especially those in the Philippines and Indonesia, is gratefully acknowledged.

Mary Garland, my administrative assistant, has dealt very effectively with many aspects of the editing process. I am sure the contributors share my appreciation for her pleasant tenacity regarding deadline 'reminders.' Her role in the publication of the volume has been invaluable.

I wish to acknowledge the excellent cooperation of the managers, editors, and production staff of the American Geophysical Union in the production of this monograph. Last, I thank the authors, who, after all, did the real work! Fortunately for them, a few important problems remain unsolved.

DENNIS E. HAYES
Editor

September 1982

Seismotectonics of the Northern Philippine Island Arc

MICHAEL W. HAMBURGER, RICHARD K. CARDWELL, AND BRYAN L. ISACKS

Department of Geological Sciences, Cornell University, Ithaca, New York 14853

This study synthesizes available data on seismicity and focal mechanisms for the complex zones of plate convergence in the Philippine island arc from Luzon to Taiwan. Evidence for westward subduction at the East Luzon Trough includes (1) a series of large earthquakes located landward of the bathymetric low with aftershock zones elongated parallel to the trough; (2) focal mechanisms of large and moderate-sized events implying low-angle westward underthrusting; (3) a diffuse zone of seismicity extending to at least 60-km, and possibly to 80-km, depth; and (4) published seismic reflection profiles interpreted to show oceanic basement dipping westward beneath a wedge of deformed sediment. Although the formation of this incipient subduction zone appears to be in progress, the seismicity and focal mechanisms are similar to those found in well-developed subduction zones. This similarity may be explained by a reactivation of a preexisting convergent margin east of central Luzon. An east-trending belt of seismicity at 15°N connects the East Luzon Trough with the Philippine Trench. A large strike-slip event and several obliquely oriented reverse-faulting events may be explained by a zone of distributed sinistral shear connecting the two convergent margins. Along the western side of the arc, a continuous east-dipping subduction zone extends from southwestern Luzon to southern Taiwan. Major transitions in the mode of plate convergence along this margin coincide with concentrated deformation of the overlying plate and contortion of the underthrusting plate: (1) At the southern end of the Manila Trench, a vertically dipping nest of shallow and intermediate-depth seismicity coincides with the transition from arc-continent collision south of Luzon to subduction of oceanic lithosphere beneath western Luzon. (2) North of Luzon, at 19°N, a cluster of seismicity, including one large strike-slip event and smaller dip-slip events, may be interpreted as a zone of upper plate shearing. This cluster of seismicity coincides with (a) a marked bend in the strike of the trench, forearc basin, and subducted lithosphere; (b) a zone of anomalous breadth of the volcanic arc, and (c) the approach of the boundary between South China Sea oceanic lithosphere and transitional/continental crust of the Asian margin. This portion of the subduction zone may reflect a transition to restricted subduction of buoyant Asian margin lithosphere. Just north of this transition, a set of normal-faulting events located near the Manila Trench indicates extensional deformation of the subducting slab and may be related to the resistance of buoyant continental lithosphere to subduction. (3) South of Taiwan, a dense cluster of large shallow earthquakes overlies a vertically dipping Benioff zone. This nest of seismicity coincides with the northern termination of the Manila Trench subduction zone and the transition to full arc-continent collision. Focal mechanisms of two events just north of Luzon and four events south of Taiwan can be interpreted as east- or southeast-directed underthrusting and would thus corroborate active plate convergence along the Manila Trench. A belt of large and moderate-sized shallow earthquakes with strike-slip focal mechanisms extends from southern Taiwan northeast to the western termination of the Ryukyu Trench. This belt may represent an incipient dextral shear zone developing into a new segment of the northwestern boundary of the Philippine Sea plate.

1

Introduction

The Philippine archipelago, which comprises the western margin of the Philippine Sea between 5° and 20°N, separates the Philippine Sea basin from the South China, Sulu, and Celebes seas marginal basins. Like other convergent plate boundaries, the archipelago is characterized by high seismicity, Quaternary volcanism, and intense recent deformation. In detail, however, it proves to be one of the most complex regions of plate interaction of the entire circum-Pacific belt. The arc as a whole is complicated by zones of plate convergence of opposite polarity on its western and eastern margins. At its northern end the arc is colliding with the passive margin of Asia and terminates near a junction with an arc of opposite polarity at the Ryukyu Trench. Rocks exposed on the Philippine Islands are a jumble of arc magmatic units, subduction melange, volcaniclastic sediments, fragments of continental crust rifted from the Asian margin, and depositional basins of various ages. Thus far, regional studies have examined only the large-scale features of Philippine geology [Karig, 1973; Gervasio, 1973; Murphy, 1973; Hamilton, 1979; Balce et al., 1979; DeBoer et al., 1980] and of regional seismicity [Katsumata and Sykes, 1969; Fitch, 1970, 1972; Rowlett and Kelleher, 1976; Seno and Kurita, 1978; Wu, 1970, 1978; Acharya and Aggarwal, 1980; Cardwell et al., 1980]. The complexity of geological and geophysical observations in the Philippines has led to widely diverging opinions about the evolution and active tectonics of the archipelago.

Previous seismicity studies have included the northern Philippines in broader regional reviews of the entire Philippine-Indonesian system or of the northern Philippine-Taiwan-Ryukyu system. Because of the broad regional scope of these studies, they have neglected fine-scale seismicity patterns and focal mechanism solutions of smaller earthquakes. The rapidly accumulating seismological data from the worldwide network, now including readings from the regional UNESCO/UNDP stations, allows a detailed reexamination of this complex and enigmatic region of plate convergence. In this study we examine variations in the mode of plate convergence along the eastern and western margins of the Philippine arc from Luzon to southern Taiwan.

In this paper we analyze in detail the three-dimensional distribution of earthquake hypocenters by carefully selecting events located by the worldwide network. We then correlate the selected seismic data with newly obtained bathymetric and marine geophysical observations of the northern Philippine region [e.g., Lewis and Hayes, this volume]. In addition, we present 14 new focal mechanism solutions for the northern Philippine arc and review their relation to inferred Philippine Sea-Eurasia plate motions. Two critical areas of plate interaction in the northern Philippine arc are emphasized in this study: (1) the eastern margin of Luzon, where a nascent subduction zone has been inferred from seismicity and marine geophysical data, and (2) the region between Luzon and Taiwan (which we shall refer to as the Luzon Strait region), where a complex zone of ridge/trough topography, active volcanism, and diffuse shallow seismicity provide information on active deformation of an island arc during the early stages of an arc-continent collision.

Tectonic Setting

In the broadest sense the northern Philippine archipelago (Figure 1) is a zone of convergence between the Eurasian and Philippine Sea plates. The complexity of seismicity patterns and stress distribution in the Luzon-Taiwan region led Katsumata and Sykes [1969] to describe the Luzon-Taiwan region as a broad zone of complex internal deformation between these two plates. Extreme variations in morphology and geologic structure are observed along strike of the arc. However, the gross patterns of volcanism and seismicity can be interpreted in a relatively simple framework of plate interaction (Figure 2). The tectonics of the region are dominated by northwest-directed convergence between the Philippine Sea and Eurasian plates [Fitch, 1972; Karig, 1975; Seno, 1977]. Plate motion is accommodated by varying degrees of convergence along its eastern and western margins as well as by internal deformation within the arc.

An active, west-facing arc is represented by a continuous linear belt of Quaternary volcanism extending northward from Mindoro through western Luzon and the North Luzon Ridge. The association of this volcanic arc with a deep-sea trench (the Manila Trench), a sediment-filled forearc basin (the West Luzon and North Luzon troughs), and a trench-slope break [Ludwig et al., 1967; Ludwig, 1970] as well as an east-dipping Benioff zone [Cardwell et al., 1980], demonstrates active subduction of South China Sea lithosphere beneath western Luzon and the North Luzon Ridge. The North Luzon Ridge is a largely submarine volcanic ridge, produced by Neogene arc volcanism [Karig, 1973; Ho, 1979], that has continued into the Quaternary in the southern (Babuyan) islands and in northern Luzon [Philippine Bureau of Mines, 1963].

A second, east-facing arc is manifested by a linear belt of active volcanism in southeastern Luzon (the Bicol Peninsula). This volcanism is associated with a west-dipping Benioff zone and is related to westward subduction of West Philippine Basin lithosphere at the Philippine Trench [e.g., Divis, 1980; Cardwell et al., 1980].

A possible third site of active plate convergence is the East Luzon Trough, a bathymetric low east of central Luzon. A series of large earthquakes with shallow underthrust fault plane solutions have been cited as evidence for incipient westward subduction of West Philippine Basin lithosphere [Fitch, 1972]. No Quaternary magmatism is associated with westward subduction at the East Luzon Trough.

At the northern end of the arc the island of Taiwan lies

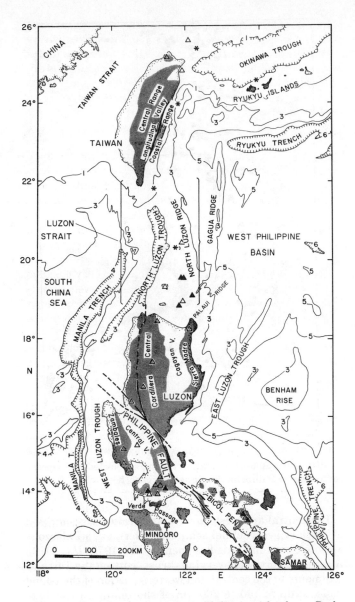

Fig. 1. Location map, northern Philippine island arc. Bathymetry in thousands of meters, taken from D. E. Hayes and S. D. Lewis (unpublished manuscript, 1982) for 118°–121°E, from *Lewis and Hayes* [this volume] for 121°–125°E and from *Chase and Menard* [1973] east of 125°E. Philippine Fault taken from *Philippine Bureau of Mines* [1963]. Historically active volcanos shown by closed triangles [from *Neumann van Padang*, 1953; *Simkin et al.*, 1981]. Other Quaternary volcanos shown by open triangles (from analysis of LANDSAT imagery and Philippine Bureau of Mines [1963]). Asterisks denote submarine eruptions. Areas of high relief shown by shading.

near the junction of the west-facing northern Philippine arc and the southeast-facing Ryukyu arc. The Neogene uplift and deformation of the Central and Coastal ranges of Taiwan are the result of a collision of the northern

Philippine island arc with the passive continental margin of Asia [*Chai*, 1972; *Karig*, 1973].

The islands of Luzon and Taiwan are cut by major sinistral strike-slip faults: the Philippine Fault in Luzon and the Longitudinal Valley of Taiwan [*Allen*, 1962]. These faults accommodate a significant portion of Philippine Sea-Eurasia plate convergence in those areas [*Fitch*, 1972; *Acharya and Aggarwal*, 1980].

Tectonic History

Luzon

Phase of westward subduction. A late Mesozoic-Early Cenozoic east-facing island arc in eastern Luzon is inferred from the presence of Cretaceous and Paleogene volcanics and pyroclastics in the southern Sierra Madre [*Gervasio*, 1973; *Karig*, 1973]. Eocene and Oligocene granitic plutons in the northern Sierra Madre may represent the intrusive component of this magmatic arc [*Wolfe*, 1981]. This magmatic belt extends through the northern and southern Sierra Madre to western Bicol, Samar, and farther south [*Balce et al.*, 1979]. The forearc may be represented by an ophiolite and melange zone in the northern Sierra Madre [*Balce et al.*, 1979]. A series of northeast-trending basement ridges in the northwestern portion of the West Philippine Basin were interpreted by *Mrozowski et al.* [1982] to be pre-Mio-

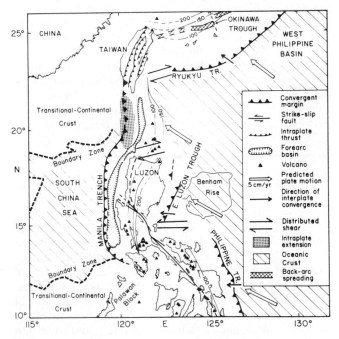

Fig. 2. Tectonic framework of the northern Philippine island arc. Plate motion taken from *Seno* [1977]; continent-ocean boundary zone from *Taylor* [1982]; Okinawa Trough from *Lee et al.* [1980]; Quaternary volcanos and Philippine Fault as in Figure 1. Depth contours (in km) are to the tops of Benioff zones, dashed where uncertain; see text for discussion.

cene compressional deformation structures related to this phase of plate convergence.

The eastward convexity, presence of east-directed thrusts, the presence of a relict subduction zone off northeastern Luzon [*Karig and Wageman*, 1975; *Lewis and Hayes*, this volume] and the absence of exposed melange along the western margin of the southern Sierra Madre (D. Karig, personal communication, 1981) all corroborate the east-facing polarity of the Sierra Madre arc.

Phase of eastward subduction. The progressive tilting and uplift of the Zambales ophiolite complex in western Luzon suggests that initiation of subduction at the Manila Trench occurred in the Middle Oligocene [*Schweller and Karig*, 1982].

Igneous activity in western Luzon appears to have been virtually continuous since the Late Oligocene [*Balce et al.*, 1980; *Wolfe*, 1981]. Although there may be overlap with the ages of igneous activity related to the early Tertiary east-facing arc, the spatial and temporal correlation of magmatism in western Luzon with initiation of subduction at the western margin of Luzon [*Balce et al.*, 1980] suggests that the magmatism is related to eastward subduction at the Manila Trench. This 'flip' from westward subduction in eastern Luzon to eastward subduction at the Manila Trench is accepted by most geologists [*Murphy*, 1973; *Karig*, 1973; *Bowin et al.*, 1978; *DeBoer et al.*, 1980; *Balce et al.*, 1980]. However, estimates of the timing of this polarity reversal vary from Early Oligocene to Pliocene.

A westward migration of the subduction zone is indicated by a change in forearc deposition from the Central Valley to the West Luzon Trough [*Balce et al.*, 1979]. The magmatic arc produced by this new subduction geometry is presently superimposed on the Zambales ophiolite, having migrated westward from central Luzon. The timing of this shift is not clear, but it appears to have taken place between Late Miocene and Pleistocene time [*Balce et al.*, 1979].

Taiwan

The eastern Coastal Range of Taiwan is dominated by Tertiary andesitic volcanic and pyroclastic rocks [*Ho*, 1979]. The continuity of these magmatic rocks and of a large free-air gravity high [*Bowin et al.*, 1978] indicate that the Coastal Range represents the northward extension of the North Luzon Ridge. Thus the Late Tertiary uplift and deformation of Taiwan is related to a collision of the northern end of the Philippine magmatic arc with the passive continental margin of Asia [*Chai*, 1972; *Karig*, 1973]. The ages of orogenic sedimentation in western Taiwan demonstrate a southward-propagating collision beginning in the north in Late Pliocene [*Page and Suppe*, 1981]. The Longitudinal Valley may represent the suture between these two tectonic elements [*Chai*, 1972; *Ho*, 1979; *Page and Suppe*, 1981]. The most recent movement along the Longitudinal Valley, however, is dominated by left-lateral strike-slip movement with a significant vertical component [*Allen*, 1962; *York*, 1976; *Hsu*, 1976].

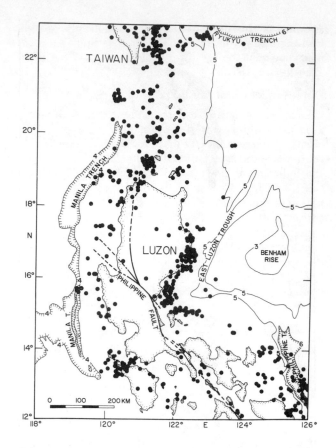

Fig. 3.   Shallow earthquakes ($h < 70$ km) 1961–1980, located by PDE using 20 or more station readings. Selected bathymetric contours and Philippine Fault as in Figure 1.

The central and western portion of Taiwan is composed of Tertiary geosynclinal sediments overlying a Paleozoic-Mesozoic metamorphic basement [*Chai*, 1972; *Ho*, 1979]. A series of northwest-verging thrusts cutting the Tertiary sediments in the central and western portion of the island indicate up to 150 to 200 km of shortening in the Late Pliocene to Pleistocene [*Page and Suppe*, 1981]. Geomorphological studies [*Bonilla*, 1977] suggest that this shortening has continued through the Quaternary.

Biq [1971] and *Murphy* [1973] proposed a Mesozoic east-facing subduction zone along the Asian continental margin, followed by a flip in subduction polarity similar to that described in Luzon. Considerable ambiguity in the dating of the basement rocks exposed in the eastern Central Range [*Chai*, 1972] makes this conclusion extremely speculative.

The unmetamorphosed Neogene rocks of the West Taiwan foothills represent folded and faulted Asian continental shelf sediments [*Ho*, 1979]. The deformation front of the orogenic belt (i.e., the frontal thrust of the West Taiwan fold belt, Figures 2 and 14) can be thought of as the on-land continuation of the deformation front of the Manila Trench (i.e., the base of the trench inner slope).

Arcward of this deformation front, the deeply subsiding Pintung Valley (just west of Taiwan's southern peninsula) may represent the northward continuation of the Manila Trench [*Biq*, 1972].

Data Sources and Methods

Earthquakes located by worldwide seismic stations were used to examine spatial and temporal patterns of seismicity of the northern Philippine region. The major data sources are the International Seismological Survey (ISS), the bulletins of the International Seismological Centre (ISC), and U.S. Geological Survey Preliminary Determination of Epicenters (PDE), as well as the catalogs of *Gutenberg and Richter* [1954] and *Rothé* [1969].

Regional patterns of seismicity are shown in Figures 3 (shallow earthquakes) and 4 (intermediate earthquakes). We present PDE locations of earthquakes during the period 1961–1980, based on 20 or more station readings. Exclusion of events located by fewer stations reduces the

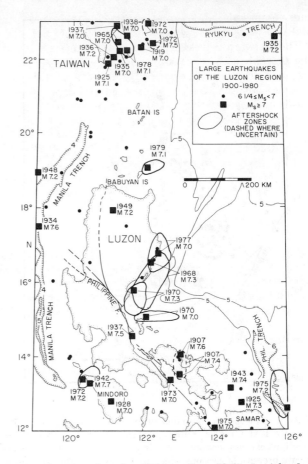

Fig. 5. Large earthquakes of the northern Philippine island arc, 1900–1980; see text for data sources.

scatter in spatial distribution of hypocenters resulting from mislocation errors. Locations of large shallow earthquakes (Figure 5) are taken from Gutenberg and Richter for the period 1900–1952, Rothé for 1952–1965, and PDE 1966–1980. We used *Rowlett and Kelleher*'s [1976] relocations of shallow earthquakes with $M_s \geq 6.9$ for the period 1915–1974. Aftershock zones were determined from PDE locations of events occurring within 1 month of the mainshock. The aftershock zone for the 1907 event is taken from Rowlett and Kelleher.

For the zones of special tectonic interest—the East Luzon region (14°–18°N, 121°–124°E) and the Luzon Strait region (18°–23°N, 118°–126°E)—we used the best quality teleseismic locations available from bulletins. We examined the ISS (1959–1963), ISC (1964–1978), and the U.S. Geological Survey Earthquake Data Report (EDR) (1978–1980) bulletins in addition to the hypocenter relocations of *Katsumata and Sykes* [1969] for the portion north of 20°N (1961–1967). The list of earthquakes was subdivided into four grades based on quality of azimuthal coverage, local station control, and depth control from surface reflection phases (*pP* and *sP*). The list of selected earthquakes located south of 20°N represents an updated ver-

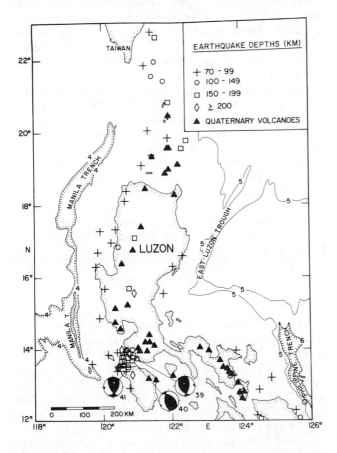

Fig. 4. Intermediate-depth earthquakes ($h \geq 70$ km), 1961–1980. Earthquake locations as in Figure 3. All Quaternary volcanos (as in Figure 1) shown as closed triangles. Focal mechanism solutions shown as lower hemisphere projections; compressional quadrants shown in black; *P* (pressure) and *T* (tension) axes shown as closed and open circles, respectively. Selected bathymetric contours as in Figure 1.

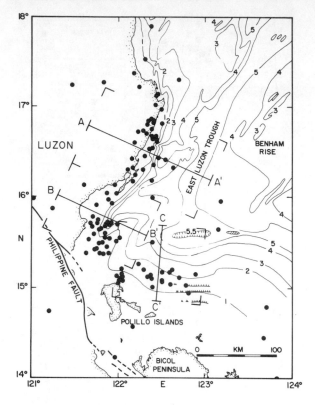

Fig. 6.   Selected earthquakes in the East Luzon region, 1959–1980; see text for data sources. Lines AA', BB', and CC' are projections of cross sections shown in Figures 7, 8, and 9. Brackets indicate breadth of sections. Hachured lines indicate locations of surface faulting [from *Lewis and Hayes,* this volume], possibly associated with transform faulting. Bathymetry and Philippine Fault as in Figure 1.

sion of the list compiled by *Cardwell et al.* [1980] through 1974. Detailed description of the method of selection can be found in *Cardwell et al.* [1980]. Those events whose epicentral locations are best constrained (grades A, B, and C) have been used to analyze the spatial distribution of earthquakes in map view (Figures 6, 11, and 12); those whose epicentral locations and focal depths are well constrained (grades A and B) are emphasized when examining the three-dimensional configurations of hypocenters in cross sections (Figures 7–9 and 13–14). Rather than presenting broad cross sections across the entire archipelago, we have attempted to isolate individual tectonic features to avoid obscuring subtle characteristics of the seismic zones.

Fourteen focal mechanism solutions of larger earthquakes have been determined by using *P* wave first motions and *S* wave polarization directions read from the long-period stations of the Worldwide Standard Seismograph Network (WWSSN). Newly obtained mechanisms, together with published mechanisms, provide the most up-to-date information on the tectonic regime of the northern Philippine arc. The mechanism parameters are presented

in Table 1 and in map view in Figures 3, 10, 13, 15, and 16. Individual mechanism diagrams for each event are shown in the appendix.

Plate Convergence Along the Eastern Margin of the Arc

Philippine Trench

Along the southeastern margin of Luzon, westward subduction of the Philippine Sea plate is indicated by the association of a deep trench with a linear belt of Quaternary arc volcanism [*Divis,* 1980] and a poorly defined Benioff zone extending to approximately 100-km depth [*Cardwell et al.,* 1980]. Focal mechanisms of four earthquakes located in the arc-trench gap beneath southeastern Luzon and Leyte show shallow underthrusting along west-dipping planes [*Cardwell et al.,* 1980]. The orientation of slip vectors for these underthrust-type events deviates substantially from that predicted by the direction of Philippine Sea/Eurasia plate convergence (see Figure 2).

Five large events in this century (two in 1907, 1925, 1943, and 1975) are associated with this subduction zone (Figure 5). The large 1975 event ($M_s = 7.2$) occurred seaward of the trench, and its mechanism [*Cardwell et al.,* 1980] indicates extensional bending stresses in the downgoing slab.

Although *Balce et al.* [1979] suggested continuous convergence at the northern portion of the Philippine Trench since the Eocene, seismic reflection profiles [*Karig and Sharman,* 1975] do not resolve a well-developed accretionary prism. Furthermore, the Benioff zone extending landward of the trench is less than 150 km in length [*Cardwell et al.,* 1980]. We note that the Benioff zone beneath southeastern Luzon does not clearly extend westward as far as the volcanic front; most of the well-constrained intermediate-depth events are farther south, beneath the islands of Samar, Leyte, and Mindanao [*Cardwell et al.,* 1980]. Thus, the Paleogene magmatic rocks exposed on the Bicol Peninsula [*Balce et al.,* 1979] may have been produced by an earlier phase of subduction.

The Role of the Philippine Fault

There has been considerable disagreement regarding the presence, the sense of motion, and the tectonic significance of the Philippine Fault (Figure 1). The fault that cuts Neogene volcanic rocks of the Bicol Peninsula is apparently a continuation of the fault system exposed along strike farther south in Masbate, Leyte, and possibly Mindanao [*Allen,* 1962; *Hamilton,* 1979]. The fault is presumed to continue into eastern Luzon, where it cuts the Cretaceous and Paleogene rocks of the Sierra Madre [*Rutland,* 1968] along a narrow linear valley. In central Luzon the fault forms the boundary between the mountainous Cordillera Central and the lowlands of the Central Valley basin. Farther west the fault appears to separate into minor splays [*Philippine Bureau of Mines,* 1963]. New seismic reflection data indicate significant faulting of the west

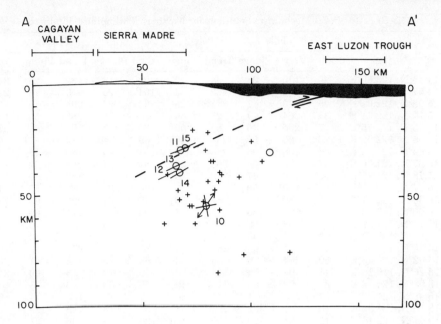

Fig. 7. Cross-section AA' (for location see Figure 6) with no vertical exaggeration. Earthquake locations as in Figure 6. Best-constrained hypocenters (grades A and B) shown by circles; fair quality (grade C) shown by crosses. Numbers refer to fault plane solutions shown in map view in Figure 10 and listed in Table 1. Back-hemisphere projections of inferred fault planes are shown by lines through hypocenters; diverging arrows show the back-hemisphere projection of axis of minimum compression. Bathymetric/topographic profile along axis of projection is shown at top of section. Location and sense of motion on inferred shallow thrust at East Luzon Trough are taken from *Lewis and Hayes* [this volume]. Dashed line indicates inferred upper surface of subducting plate.

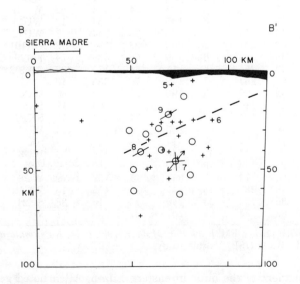

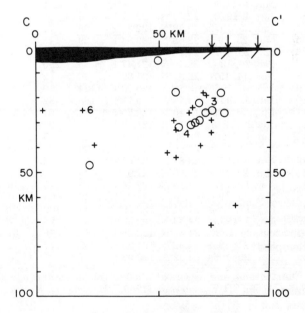

Fig. 8. Cross-section BB' (for location see Figure 6) with no vertical exaggeration. Earthquake locations as in Figure 6; symbols as in Figure 7. Upper surface of subducting plate is the same as that shown in Figure 7.

Fig. 9. Cross-section CC' (for location see Figure 6) with no vertical exaggeration. Earthquake locations as in Figure 6. Symbols as in Figure 7. Arrows and inclined slashes indicate location and approximate dip of surface faults [*Lewis and Hayes*, this volume].

TABLE 1. Focal Mechanism Solutions for Northern Philippine Island Arc

Event	Date	Position	Depth, km	Pole 1 Trend	Pole 1 Plunge	Pole 2 Trend	Pole 2 Plunge	P Axis Trend	P Axis Plunge	T Axis Trend	T Axis Plunge	B Axis Trend	B Axis Plunge	Reference*
1	Aug. 19, 1976	14.45°N, 123.68°E	26	136	30	233	10	91	13	188	29	340	59	***
2	Dec. 20, 1966	14.57°N, 122.17°E	32	56	34	323	2	104	20	5	26	226	60	F41
3	Apr. 12, 1970	15.08°N, 122.01°E	25	73	2	341	28	32	23	294	16	169	61	SK3
4	Apr. 15, 1970	15.11°N, 122.71°E	31	264	14	28	66	68	28	288	55	169	19	CIK12
5	Apr. 8, 1970	15.43°N, 121.75°E	7	236	40	64	50	60	5	198	84	329	4	CIK11
6	Aug. 28, 1968	15.55°N, 122.02°E	25	60	50	256	40	69	4	304	81	160	7	F53
7	Jul. 4, 1971	15.60°N, 121.85°E	46	108	85	288	5	108	40	288	50	18	0	CIK9
8	Feb. 13, 1976	15.67°N, 121.70°E	41	73	60	268	28	82	17	284	72	174	7	***
9	Apr. 7, 1970	15.78°N, 121.71°E	50	56	47	288	30	85	9	338	60	180	27	F51
10	Nov. 22, 1968	16.17°N, 122.17°E	60	100	80	280	10	100	35	280	55	10	0	F47
11	Aug. 1, 1968	16.30°N, 122.11°E	30	280	20	65	66	89	24	299	63	185	12	F44
12	May 22, 1972	16.60°N, 122.19°E	41	95	66	275	24	95	21	275	69	5	0	CIK5
13	Mar. 18, 1977	16.73°N, 122.29°E	37	68	62	272	25	84	18	296	70	178	10	***
14	Mar. 19, 1977	16.80°N, 122.34°E	40	70	59	268	30	82	15	291	73	173	8	***
15	Jul. 21, 1977	16.86°N, 122.39°E	29	74	66	264	24	81	21	272	69	161	6	***
16	Aug. 26, 1970	18.02°N, 120.48°E	50	244	16	85	73	68	29	235	61	336	6	CIK3
17	Sep. 3, 1974	18.26°N, 119.20°E	41	98	55	313	30	119	13	355	69	213	17	***
18	Feb. 3, 1974	18.93°N, 120.13°E	30	90	50	270	40	270	85	90	5	0	0	CIK2
19	Oct. 8, 1976	18.98°N, 121.27°E	57	5	4	181	86	5	49	185	41	95	0	***
20	Nov. 19, 1974	19.00°N, 121.39°E	40	127	32	292	58	301	13	148	26	32	8	***
21	Aug. 26, 1979	19.07°N, 122.10°E	15	84	20	350	11	129	6	36	21	235	67	***
22	Jun. 25, 1973	19.11°N, 121.19°E	36	101	32	265	56	275	13	130	75	6	8	***
23	Feb. 8, 1972	19.36°N, 122.06°E	54	280	33	79	55	92	11	314	75	184	10	CIK1
24	Jun. 6, 1963	19.87°N, 120.47°E	42	153	70	340	20	345	65	158	25	250	3	KS3
25	Jan. 8, 1972	20.95°N, 120.26°E	36	42	44	278	30	326	58	71	10	168	31	SK5
26	Apr. 26, 1965	21.00°N, 120.68°E	29	109	72	276	18	270	64	98	28	6	4	KS10
27	Jan. 7, 1977	21.20°N, 120.25°E	36	72	44	224	42	153	75	58	1	327	15	***
28	Mar. 4, 1967	21.40°N, 121.89°E	128	8	20	188	70	188	25	8	65	98	0	***
29	Oct. 20, 1971	21.94°N, 121.40°E	43	173	57	316	28	150	17	278	67	55	18	***
30	Feb. 14, 1973	22.27°N, 121.52°E	42	140	21	242	30	283	4	189	37	18	52	SK14
31	Sep. 22, 1972	22.37°N, 121.16°E	8	173	4	285	79	184	48	344	40	82	10	SK11
32	May 17, 1965	22.41°N, 121.26°E	80	14	4	106	32	155	20	55	26	278	56	KS11
33	Jan. 4, 1972	22.50°N, 122.07°E	6	36	10	307	4	82	4	351	10	194	79	SK4
34	Jan. 25, 1972	22.56°N, 122.37°E	29	71	0	161	0	116	0	26	0	---	90	WU14
35	Mar. 23, 1975	22.68°N, 122.84°E	29	148	7	237	3	103	5	193	5	322	83	WU20
36	May 23, 1975	22.70°N, 122.61°E	1	140	0	230	5	95	4	185	4	314	82	WU21
37	Feb. 26, 1968	22.76°N, 121.47°E	55	290	50	130	40	302	6	185	69	34	19	WU11
38	Nov. 14, 1970	22.82°N, 121.36°E	26	82	50	250	40	75	5	165	83	344	7	WU12
39	Jan. 10, 1966	13.80°N, 120.70°E	124	279	34	67	30	261	2	359	68	171	12	AA101
40	Jan. 5, 1967	13.78°N, 120.71°E	168	82	34	218	46	243	6	139	66	336	24	FM26
41	Aug. 30, 1966	13.40°N, 120.70°E	81	61	28	302	42	271	6	11	54	175	37	AA102
42	Jul. 23, 1978	22.19°N, 121.42°E	34	93	38	254	51	264	7	142	79	356	10	***
43	Jan. 18, 1964	23.09°N, 120.58°E	18	105	48	272	42	99	5	190	8	8	6	WU2
44	Apr. 24, 1972	23.60°N, 121.55°E	29	99	44	315	40	116	1	22	0	209	19	SK8
45	Nov. 9, 1972	23.87°N, 121.61°E	22	106	44	321	40	124	2	27	72	214	17	SK12
46	Mar. 23, 1966	23.86°N, 122.97°E	40	22	12	290	10	70	1	336	16	160	74	WU7
47	Apr. 17, 1972	24.10°N, 122.44°E	48	173	70	327	18	154	27	314	62	60	8	SK7
48	Mar. 12, 1966	24.24°N, 122.67°E	42	24	0	294	37	242	50	346	25	114	53	WU6
49	Feb. 13, 1963	24.41°N, 122.09°E	53	169	61	349	29	169	16	349	74	80	0	KS2
50	May 5, 1966	24.33°N, 122.50°E	53	28	7	298	8	72	1	343	11	159	79	WU8
51	Oct. 25, 1967	24.43°N, 122.25°E	73	76	31	282	46	267	15	47	21	175	15	WU10
52	Mar. 11, 1958	24.62°N, 124.29°E	65	277	12	173	50	126	23	240	43	16	38	SH21
53	Nov. 26, 1964	24.92°N, 122.02°E	17	27	8	207	82	27	53	207	37	298	1	WU3

AA, *Acharya and Aggarwal* [1980]; CIK, *Cardwell et al.* [1980]; F, *Fitch* [1972]; FM, *Fitch and Molnar* [1970]; KS, *Katsumata and Sykes* [1969]; SH, *Shiono et al.* [1980]; SK, *Seno and Kurita* [1978]; WU, *Wu* [1978]; ***, this study.

Luzon forearc along strike of fault splays observed at the coast (D. E. Hayes and S. D. Lewis, unpublished manuscript, 1982).

Although Rutland found little evidence for strike-slip offsets of recent deposits in a detailed study of a small segment of the fault in eastern Luzon, Allen noted geomorphic features in Luzon, Masbate, and Leyte that were indicative of consistent left-lateral movement along the fault. Considerable credibility was lent to these observations by surface rupture associated with the 1973 Ragay

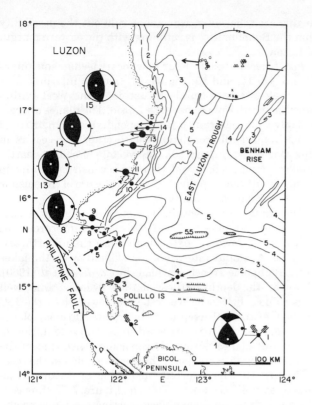

Fig. 10. Focal mechanism solutions for East Luzon region. New mechanisms shown as lower hemisphere projections as in Figure 4. Mechanism parameters are given in Table 1, and individual diagrams are shown in Appendix. Large circles denote earthquakes with $M_s \geq 6.9$. Slip vectors of shallow underthrust mechanisms shown as single arrow; horizontal projections of P axes of reverse-fault mechanisms shown as opposing arrows; single preferred plane or both planes of strike-slip solutions are shown with arrows indicating sense of movement; symbols dashed where interpretation is ambiguous. Inset shows poles to auxiliary plane (circles), poles to fault plane (triangles) and B or null axes (X) for shallow underthrust mechanisms. Heavy arrow indicates mean slip vector. Bathymetry and Philippine Fault as in Figure 1. Location of surface faulting as in Figure 6.

Gulf earthquake. Left-lateral surface offsets of up to 3.2 m were observed along a rupture zone over 75 km long [*Morante and Allen,* 1974]. Though seismicity along the fault north of about 15°N is quite low (Figure 3), considerable recent activity of the fault is implied by the sharpness of fault scarps, disrupted soil horizons, and streams offset in a left-lateral sense along the fault in the western portion of central Luzon (S. Lewis, personal communication, 1981).

Shallow earthquakes in southeastern Luzon (Figure 3) appear to correlate with the surface trace of the fault. The instrumental location (Figure 5) and focal mechanism [*Cardwell et al.,* 1980] of the 1973 Regay Gulf event ($M_s = 7.0$), as well as observation of surface rupture [*Morante and Allen,* 1974], are all consistent with left-lateral

strike-slip faulting along the Philippine Fault. The correlation of one additional fault plane solution [*Acharya and Aggarwal,* 1980] and several large earthquakes (1937, 1973, and 1975) with the mapped fault trace (Figure 5) corroborates the fault's activity from 12°N to at least 14°N. Event number 2 in Figure 10, which also suggests left-lateral strike-slip motion along a northwest-trending plane, is displaced significantly eastward of the mapped traces of the fault and may occur on a splay off the main fault.

Farther north there are few instrumentally located shallow events within Luzon (Figure 3). Seismicity is more clearly concentrated along the margins of the island. One moderate-sized earthquake ($M_s = 4.9$) appears to be associated with this northern portion of the fault, but a focal mechanism could not be obtained for the event. A very brief microearthquake survey reported no seismicity associated with the fault during a 16-day period of observations [*Acharya et al.,* 1979]. However, unpublished earthquake locations from the regional network of the Philippine Atmospheric, Geophysical and Astronomical Services Administration (R. L. Kintanar, personal communication, 1981) do show some shallow seismicity near fault splays in west central Luzon. The $M_s = 7.2$ event in 1949 (Figure 5) could be related to one of these fault splays. Despite the low seismicity of the fault north of 15°N, the geomorphic features, as well as rotation of slip vectors of interplate earthquakes (discussed below), lead us to suspect that the fault's seismic quiescence may only reflect the brevity of the period of observation. More geomorphological and geodetic constraints on the rate of slip and the role of aseismic creep in fault movement are essential to evaluate earthquake risk along the Philippine Fault.

Eastern Luzon

Fitch [1972] proposed an incipient subduction zone east of central Luzon and parallel to the coast on the basis of a belt of shallow seismicity (Figure 6) and underthrust-type focal mechanisms whose slip vectors trend approximately normal to the coastline. Aftershock zones of three large events (1968, 1970, and 1977; Figure 5) are elongated subparallel to the seismic zone and to the surface trace of the inferred fault planes. The belt of seismicity (Figure 6) occurs landward of a bathymetric low, averaging 5 km depth between the coastline and the Benham Rise.

A well-defined, negative, free-air gravity anomaly [*Bowin et al.,* 1978; *Lewis and Hayes,* this volume] coincides with the bathymetric low. The gravity low extends as far north as 20°N, which is farther north than the belt of seismicity. This gravity anomaly and possible normal faulting of the sediments north of 17°N [*Bowin et al.,* 1978] may reflect crustal downwarping preceding lithospheric rupture [*Bowin et al.,* 1978; *Lewis and Hayes,* this volume].

New seismic reflection profiles [*Lewis and Hayes,* this volume] show a strong subsurface reflector, interpreted as oceanic basement, dipping gradually (5°–10° apparent dip) westward beneath a wedge of deformed sediment. The degree of deformation of the sediment wedge appears to decrease rapidly north of 17°N, coincident with a similar decrease in seismic activity.

Lewis and Hayes proposed a westward jump in the locus of sediment deformation at 17.5°N, following the trend of the coastline rather than the bathymetric low (see Figure 2). The north rather than northeast trend of the zone of sediment deformation coincides with the trend of the gravity low [*Lewis and Hayes*, this volume]. The sparse seismicity data cannot resolve such a jump in the locus of subduction. Such a configuration would require tear-faulting in the subducting oceanic lithosphere. A simpler explanation may be that these loci of deformation represent discrete zones of back-arc thrusting similar to those observed in the Banda Arc [*Silver and McCaffrey*, 1981] and have not yet coalesced into a single throughgoing detachment fault. Nonetheless, the observations discussed above suggest that a zone parallel to the island margin, rather than the inferred relict trench off northeastern Luzon [*Karig and Wageman*, 1975], is being activated.

In cross-section (AA', Figure 7, and BB', Figure 8), the best-constrained hypocenters (shown as circles) do not form a well-defined west-dipping Benioff zone. Rather, they comprise a broad wedge, at least 40 km wide, with only a suggestion of a landward (eastward) dip. The zone extends to over 60-km, and possibly as much as 80-km, depth. The width of this seismic zone may be related to the poor depth control of teleseismically located shallow events. The detailed spatial configuration may only be resolved by extensive local coverage with land and ocean-bottom seismographs. Nonetheless, the thickness of the seismic zone is no greater, and may be slightly less, than the shallow portions of other well-developed Benioff zones [*Isacks and Barazangi*, 1977].

Although this deformation may be interpreted as back-arc thrusting associated with the west-facing Manila Trench, the size of the earthquakes in relation to those observed along the western margin (Figure 5), the linearity of the seismicity belt (Figure 6), and the indications that the underthrusting is being accommodated along a former convergent margin with a preexisting detachment fault all suggest that the East Luzon Trough is becoming a major zone of plate subduction. On the other hand, it is clear that the East Luzon Trough is not a simple plate boundary because the accretionary prism deformation [*Lewis and Hayes*, this volume] and seismicity (Figure 6) appear to vanish rather abruptly between 17°N and 18°N, without being connected to another plate boundary.

The Benham Rise (Figure 1) is a zone of thickened oceanic crust, probably produced by melting anomalies at a spreading ridge [*Karig*, 1975]. The striking similarity of its shape to the sharp bend in the Luzon coastline (Figures 1 and 2) may suggest a genetic relationship. Seismic reflection profiles [*Lewis and Hayes*, this volume] may suggest that there is a wedge of normal oceanic crust between the rise and the Luzon margin. Without detailed refraction studies

in this area, however, we cannot rule out the possibility that the Benham Rise is colliding with the eastern margin of Luzon.

Focal mechanisms. Four new focal mechanisms (numbers 8, 13, 14, and 15; Figure 10) may be interpreted as low-angle, westward underthrusting of the West Philippine Basin lithosphere beneath Luzon. The planes are reasonably well-constrained (see Appendix), and their trend shows remarkable consistency with other slip vectors in this zone (inset, Figure 10). Slip vectors of shallow earthquakes (numbers 8, 9, and 11–15) are consistent with slip along a fault zone dipping 20°–30° at the top of the seismic zone (see Figures 7 and 8).

The fault planes obtained from two slightly deeper mechanisms (numbers 7 and 10) may be interpreted as indicative of down-dip extension within the underthrusting slab. A similar interpretation of shallow earthquakes at the Philippine Trench was made by *Cardwell et al.* [1980]. However, the depths of these shallow events are not well constrained. If they prove to be shallower than the ISC depth, they could also represent interplate slip along planes with shallow (5°–10°) westward dips. The orientation of these fault planes is roughly consistent with that of the other interplate events, and they are included in the inset in Figure 10. The inferred subduction of oceanic lithosphere at the East Luzon Trough (Figure 7) is thus consistent with the shallow underthrusting mechanisms and geometry of the seismic zone.

Perhaps the most surprising result of this investigation is that seismicity and focal mechanisms associated with incipient subduction at the East Luzon Trough are not strikingly different from the shallow portion of well-developed subduction zones. If this incipient subduction zone were forming by fracture of previously unbroken lithosphere, considerable lateral complexity might be expected in the stress field near the tip of an actively propagating crack cutting through the lithosphere. No such complexity is observed. The presence of a preexisting zone of weakness—the Early Tertiary subduction zone—apparently allows the lithosphere to rerupture without the mechanical complexity expected from rupturing unfractured oceanic lithosphere.

Similar patterns of seismicity are observed at the Nankai Trough in Southwest Japan. Like the East Luzon Trough, the Nankai Trough has neither a well-developed trench and forearc, intermediate-depth seismicity, nor an associated volcanic arc. A wedge of seismicity extends to 70-km depth landward of a bathymetric low [*Kanamori and Tsumura*, 1971], and focal mechanisms of shallow earthquakes landward of the Nankai Trough [*Kanamori*, 1972; *Shiono*, 1977] indicate shallow underthrusting normal to the coastline. Unlike the East Luzon Trough, however, there is a dense cluster of seismicity at 40–60 km depth, interpreted by *Kanamori* [1972] as deformation at the leading edge of the oceanic lithosphere. *Shiono* [1977] explained focal mechanisms of events in this cluster by invoking thermally induced stresses in the continental lithosphere. The presence of subducted oceanic lithosphere is corroborated by studies of travel-time anomalies at stations in Southwest Japan [*Shiono*, 1974; *Kanamori and Tsumura*, 1971]. The overall similarity of the seismic zones in Southwest Japan and East Luzon leads us to a similar tectonic interpretation—reactivated subduction along a previously active convergent margin.

Implications for plate motions. Predicted Philippine Sea/ Eurasia slip vectors obtained from the poles of rotation calculated by *Fitch* [1972], *Minster and Jordan* [1979], and *Seno* [1977] show general agreement. Their poles predict northwest-southeast convergence ranging from 7.5 to 9.5 cm/yr for the northern Philippine region. *Karig* [1975] proposed low convergence rates in Southwest Japan that yield a pole much closer to Japan and convergence in the Philippine region of 6–6.5 cm/yr. These rotation vectors are all calculated without using the slip vectors of underthrust-type earthquakes for the Taiwan-Philippine region.

The trend of the shallow underthrust slip vectors at the East Luzon Trough averages 276° (inset, Figure 10). This orientation of plate convergence represents a 40°–50° counterclockwise rotation with respect to slip directions along the Southwest Japan and Ryukyu arcs. As shown in Figure 2, it deviates substantially from the northwest-directed plate convergence predicted by poles of rotation obtained by *Fitch* [1972], *Karig* [1975], *Seno* [1977], and *Minster and Jordan* [1979]. The deviation cannot be explained by nonrigid behavior of the Eurasian plate (i.e., relative motion between Southeast Asia and the Eurasian plate). The discrepancy between the predicted and observed convergence of Southeast Asia with the Australian plate at Java [*Cardwell et al.*, 1981] and the deformation of Eurasia caused by the India-Eurasia collision [*Molnar and Tapponnier*, 1975] both imply significant southeastward migration of Southeast Asia with respect to stable portions of Eurasia. This would require clockwise, rather than counterclockwise, rotation of the observed slip vectors.

Fitch [1972] suggested that the rotation of slip vectors in eastern Luzon might be accomplished by decoupling the eastern Philippines from Eurasia by strike-slip motion along the Philippine Fault. Such decoupling would create a long narrow microplate between the Philippine Fault and the Philippine Trench. The western margin was not thought to accommodate significant convergence. Thus Fitch contended that the northwest-directed plate convergence is separated into a normal component (westward underthrusting at the Philippine Trench and East Luzon Trough) and a parallel component (left-lateral strike-slip faulting at the Philippine Fault). The interplate slip observed along the East Luzon Trough (direction 276°) is close to that observed further south along the Philippine Trench (270°) [*Cardwell et al.*, 1980]. Thus the change in ratio of parallel to normal slip between the Philippine Trench and East Luzon Trough subduction zones invoked by *Fitch* [1972] would not be required by these data, although his decoupling hypothesis may still be correct.

On the other hand, *Seno and Kurita* [1978] proposed that the deviation between predicted and observed slip vectors might be explained by northwestward migration of the Philippine block as a whole, deemphasizing the tectonic significance of the Philippine Fault. They note the absence of shallow thrust mechanisms and the presence of three left-lateral strike-slip mechanisms along the western Philippine margin as support for this model. Underthrust mechanisms are, however, reported: one at the Manila Trench [*Acharya and Aggarwal*, 1980], one at the Cotabato Trench [*Stewart and Cohn*, 1977], and possibly several additional ones in the Luzon Strait region (discussed below). In addition, *Cardwell et al.* [1980] pointed out the diversity of orientations of nodal planes of the strike-slip events in western Luzon, which is not consistent with simple northward migration of the Philippine block. The lack of convincing evidence of northward movement at the Manila Trench suggests that motion along the Philippine Fault is largely responsible for the deviation in slip vectors.

Active subduction at the Manila Trench (discussed below) implies some decoupling of the arc from Eurasia. Because the direction and rate of convergence at the Manila Trench remain poorly constrained, and the rate of movement along the Philippine Fault is unknown, motion of the Philippine Sea and Eurasian plates, relative to the eastern and western Philippine blocks, cannot be quantitatively evaluated.

Transform faulting at 15°N. The aftershock zone of a large ($M_s = 7.0$) earthquake in 1970 (see Figure 5) forms a narrow east-trending belt at about 15°N (Figures 5 and 6). Its focal mechanism (number 3 in Figure 10) indicates left-lateral strike-slip faulting along a plane subparallel to the belt of seismicity. Seno and Kurita proposed that this zone is part of a transform fault linking subduction at the Philippine Trench and East Luzon Trough. This belt of seismicity is located significantly southward of the bathymetric low connecting the Philippine Trench and East Luzon Trough (Figure 6). The East Luzon Trough gravity low deviates from the bathymetric low in the same manner [*Lewis and Hayes*, this volume]. Lewis and Hayes showed seismic reflection profiles indicative of normal faulting (with an unknown strike-slip component) of sediments along several closely spaced zones near the belt of seismicity (hachured lines in Figures 6 and 10). In cross section (CC', Figure 9) the well-constrained hypocenters form a narrow (20–25 km wide) zone less than 35 km deep whose location is consistent with the northward dip of the faults observed at the surface. These pieces of evidence all point to active faulting along a narrow east-west zone at about 15°N. The bathymetric trend probably reflects progradation of the shelf (caused by rapid sedimentation from the Bicol Peninsula and eastern Luzon) rather than the actual location of the transform [*Lewis and Hayes*, this volume].

Several observations, however, indicate a pattern of deformation far more complex than a simple, linear trench-trench transform: (1) The large 1970 event (number 3 in

Figure 10) occurred west of the southern end of the East Luzon Trough, along the landward continuation of the proposed transform. There, a significant vertical component of interplate motion would be predicted. (2) Several fault plane solutions of smaller events near the proposed transform (numbers 4–6 in Figure 10) indicate reverse faulting on northwest-striking planes. (3) Strike-slip events just south of 15°N (numbers 1 and 2 in Figure 10) indicate a stress system rotated 45° from that required by a simple transform fault. These two events could be interpreted as deformation within the eastern Philippines block, accommodating motion similar to that documented on the Philippine Fault. The orientation of the P axis of events 4–6, and of the fault plane of the large 1970 event (number 3), with respect to the east-trending seismicity belt, suggest that this zone might better be modeled as a complex zone of distributed sinistral shear connecting the two convergent margins than as a simple trench-trench transform (Figure 2).

Large earthquakes. Contiguous aftershock zones of four large earthquakes along the east Luzon margin (Figure 5) suggest that the entire plate margin, from 17.3°N to 15°N and along the transform zone to 123°E, ruptured between 1968 and 1977. This may result in high accumulated stresses at the edges of this plate margin, i.e., along the East Luzon Trough north of 17.3°N and along the transform zone east of 123°E.

The northern boundary of this earthquake sequence coincides with a major structural and morphological boundary. The northern limit of this sequence occurs at the westward jump in the locus of subduction described by *Lewis and Hayes* [this volume]. In addition, this change in subduction coincides with a marked transition in morphology of the eastern Luzon coastline. There the north-trending structures of the Sierra Madre are sharply truncated, and the coastline makes a major S-shaped bend. A speculative east-west lineation of shallow epicenters (Figures 4 and 6) coincides with this morphological boundary. Thus this transition may mark a major transverse segmentation in both the subducting and overlying plates and may limit the possible rupture area of interplate events in eastern Luzon.

Plate Convergence Along the Western Margin of the Arc

The Manila Trench

A continuous west-facing subduction zone is associated with the Manila Trench (and its northward extension), stretching from southwestern Luzon to southern Taiwan. The trench can be traced as a bathymetric low to 20°N [*Ludwig et al.,* 1967], although sediment deformation associated with the trench can be traced to at least 21.5°N (D. E. Hayes and S. D. Lewis, unpublished manuscript, 1982). Subduction of the lithosphere of the South China Sea basin is indicated by an east-dipping Benioff zone extending to over 200-km depth [*Katsumata and Sykes,* 1969;

Cardwell et al., 1980]. Intermediate-depth earthquakes (Figure 4) form a continuous belt, coinciding with a continuous belt of Quaternary volcanism along the western margin of Luzon and the North Luzon Ridge to southern Taiwan (Figure 1). A marked, S-shaped bend in the trend of the intermediate depth events just north of Luzon coincides with a similar bend in the strike of the trench and forearc basin as well as an anomalously broad zone of arc volcanism.

The relative paucity of large shallow earthquakes (Figure 5) and the absence of shallow thrust mechanisms along the west Luzon margin have been cited as evidence for the slowing or cessation of convergence between Luzon and the South China Sea [*Rowlett and Kelleher,* 1976; *Seno and Kurita,* 1978]. Marine geophysical observations, however, do not corroborate this interpretation. No layer of undeformed sediment is observed blanketing the deformed sediments of the Manila Trench accretionary prism (D. E. Hayes and S. D. Lewis, unpublished manuscript, 1982). One thrust mechanism reported by *Acharya and Aggarwal* [1980], and several new mechanisms in the Luzon Strait and Taiwan regions (this study), may represent interplate thrusting at the Manila Trench. In addition, two large shallow earthquakes ($M_s = 7.2$ and $M_s = 7.6$) and several smaller ones ($M_s > 6.25$) are located along the Manila trench near northern Luzon (Figure 5). Though they are not well located, and focal mechanisms are unavailable for these events, they could represent interplate motion. Continuing convergence at the Manila Trench, though possibly slowed by the arc-continent collision at Taiwan [*Karig,* 1973], cannot be ruled out.

At both ends of the subduction zone—near southwestern Luzon and southern Taiwan—the dip of the Benioff zone steepens to near vertical. These zones are also characterized by intense shallow seismicity.

On the basis of marine geophysical observations, *Taylor* [1982] identified the boundaries of the oceanic crust of the South China Sea basin (shown by dashed lines in Figure 2). The intersection of these boundaries with the arc may mark major transitions in the mode of plate convergence at the Manila Trench. These transitions appear to be associated with zones of concentrated deformation of the overlying plate and contortion of the underthrusting plate. Three transitions in the mode of plate convergence along the Manila Trench may be responsible for the zones of intense seismicity along the western margin of the arc: (1) the transition from the Palawan arc-continent collision to subduction of South China Sea oceanic lithosphere at southwestern Luzon, (2) the speculative transition from subduction of South China Sea oceanic lithosphere to restricted subduction of Asian margin transitional lithosphere north of Luzon, and (3) the transition from this possibly restricted subduction to full arc-continent collision at Taiwan. These transitions may be responsible for the zones of intense seismicity along the western margin of the arc. The nature of each of these transition regions, and of the plate convergence between them, is discussed in the following sections.

Southwestern Luzon

At the southern end of the Manila Trench a dense cluster of large and moderate-sized shallow events (Figures 3 and 5) directly overlie a vertically dipping nest of intermediate-depth earthquakes (Figure 4). Although the intermediate-depth events are characterized by down-dip tension, typically observed in active subduction zones, shallow events in this area display strike-slip mechanisms with widely varying fault plane orientations [Cardwell et al., 1980].

This knot of seismicity is located adjacent to the boundary between South China Sea oceanic crust and transitional/continental crust of the Palawan (Calamian) continental block [Taylor and Hayes, 1980; Taylor, 1982]. Thus the concentration of seismicity coincides with the transition between subduction of oceanic lithosphere at the Manila Trench and collision of the continental block with a previously existing southward continuation of the Manila Trench. The collision has resulted in active southwest-verging thrusts in southern Mindoro (D. Karig, personal communication, 1981).

Western Luzon

Sparsely distributed, but well-constrained, intermediate-depth hypocenters define a Benioff zone dipping eastward from the Manila Trench to 220-km depth [Cardwell et al., 1980]. Few large earthquakes are associated with this portion of the Manila Trench (Figure 5). One focal mechanism obtained by Acharya and Aggarwal [1980] may represent east-northeast-directed underthrusting along a fault plane dipping eastward at 38°. Continuing deformation of accretionary prism sediments (D. E. Hayes and S. D. Lewis, unpublished manuscript, 1982) and Quaternary arc volcanism in western Luzon imply continuing convergence there.

Luzon Strait

Regional setting. A complex zone of ridge and trough topography, active volcanism, and diffuse shallow seismicity lies between the islands of Luzon and Taiwan. It is bounded to the west by the northward extension of the Manila Trench and to the east by the Palaui and Gagua ridges (Figures 1 and 11). The West Luzon Trough forearc basin can be traced northward as the North Luzon Trough [Ludwig, 1970] and is associated with a gravity minimum, even as the trough shallows approaching Taiwan [Bowin et al., 1978]. Seismic reflection profiles show that the trough is clearly fault bounded on its eastern flank, and there is suggestion of faulting along the west flank as well (S. Lewis, personal communication, 1981).

The Palaui Ridge and its northward extension, the Gagua Ridge (Figures 1 and 11), make up a north-south topographic high extending from northeast Luzon to the west-

ern Ryukyu arc. In the south this ridge appears to represent deformed accretionary prism sediments of the early Tertiary subduction system [Karig, 1973; Karig and Wageman, 1975]. North of 20°N, Bowin et al. [1978] assigned the Gagua Ridge an origin as a spreading ridge. The gabbros and amphibolites dredged there, however, show greater affinity to environments common to oceanic fracture zones than to those of oceanic ridges or accretionary prisms [Mrozowski et al., 1982].

The volcanic rocks of the North Luzon (or Lutao-Babuyan) Ridge are calc-alkaline volcanics and pyroclastics of Miocene and Pliocene age in the north. This arc volcanism was active through the Quaternary in the southern islands and in northeastern Luzon [Bowin et al., 1978]. The volcanic rocks of the Coastal Range of Taiwan are the northward continuation of the North Luzon Ridge. Volcanism there terminated in the Late Miocene [Chi et al., 1981], and this is consistent with cessation of subduction beginning in the north (Taiwan) and propagating southward [Page and Suppe, 1981].

Seismicity patterns. Shallow earthquakes in the Luzon Strait region (Figure 11) occur in a broad north-south belt extending from west of the Manila Trench to east of the North Luzon Ridge. Activity is concentrated east of southern Taiwan and at the southwestern end of the North Luzon Ridge. This belt of seismicity terminates abruptly along the north-northwest-trending escarpment that separates the rough topography of the North Luzon Ridge from a segment of deeper sea floor with subdued relief to the east.

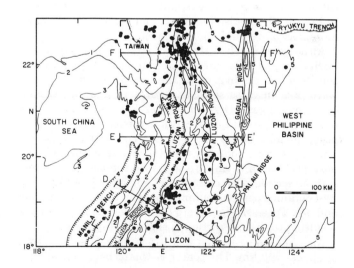

Fig. 11. Selected shallow earthquakes in the Luzon Strait region, 1959–1980. See text for data sources. All Quaternary volcanos (as in Figure 1) shown as open triangles. Bathymetry as in Figure 1. Lines DD′ and EE′ show projections of cross sections used to compile composite cross section in Figure 13. Line FF′ shows projection of cross section in Figure 14. Brackets indicate breadth of sections.

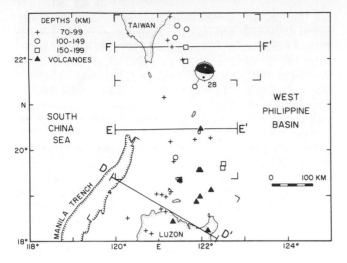

Fig. 12.   Selected intermediate-depth earthquakes in the Luzon Strait region, 1959–1980; see text for data sources. All Quaternary volcanos (as in Figure 1) shown as closed triangles. Selected bathymetric contours as in Figure 1. Lines DD′, EE′, and FF′ show projections of cross sections as in Figure 11. Focal mechanism shown as lower hemisphere projection as in Figure 4; mechanism parameters are given in Table 1.

A group of epicenters just west of the Paulaui Ridge occurs along strike of a major submarine canyon that cuts through the volcanic ridge. This canyon feeds a large sediment apron derived from the Cagayan River basin [*Karig and Wageman*, 1975]. The linearity of the canyon, the location of the large 1979 earthquake ($M_s = 7.1$; number 21 in Figure 15; see also Figure 5), and the orientation of its aftershock zone imply fault control.

Although sparsely distributed, intermediate-depth seismicity (Figure 12) extends continuously from Luzon to southern Taiwan. In general, the earthquakes deepen eastward, and in cross section (Figures 13 and 14), Benioff zones are suggested by dashed lines. Because of the paucity of seismicity in the Luzon Strait, a composite cross section perpendicular to the arc is presented for the region extending along-strike from northern Luzon through the northern Luzon Strait. To generate the composite cross section, sections DD′ and EE′ (see Figures 11 and 12) were first obtained perpendicular to the local trend of the trench and arc. These sections were then superimposed, and the resulting composite section is shown in Figure 13.

An east-dipping Benioff zone can be delineated to a depth of almost 200 km. The center of Quaternary volcanic activity of northern Luzon and the North Luzon Ridge occurs over a Benioff zone approximately 100 km deep, typical for other active arcs [*Isacks and Barazangi*, 1977]. In the southern portion of the Luzon Strait, however, the Benioff zone beneath the volcanic front appears to be anomalously shallow (approximately 60 km), and arc volcanism occurs over an anomalously broad zone (>125 km wide).

The absence of seismic stations in this region and the generally low magnitude of these events leaves us with only one intermediate-depth event graded A or B (most reliable) in these two sections. However, the continuity of this seismic zone with well-defined Benioff zones beneath western Luzon [*Cardwell et al.*, 1980] and southern Taiwan (section FF′, Figure 14) and its association with morphologic and volcanic features typical of active subduction zones support the interpretation that these events are associated with lithospheric subduction.

Focal mechanisms.   The shallow events in the Luzon Strait fall into three relatively distinct clusters. The westernmost cluster is located beneath and arcward of the Manila Trench (Figures 11 and 13). Focal mechanisms for several of these events are shown in map view in Figures 15 and 16 and in cross section in Figure 13. Events 18 and 24–27 indicate normal faulting. The scatter in trends of the T axes may be illusory, as the best constrained mechanisms (numbers 18, 25 and 26) have consistent east-trending T axes approximately perpendicular to the trend of the Manila Trench. The position of these events shown in Figure 13 suggests that they represent extensional deformation of the subducting slab. The observation that several of the events (numbers 25 and 27) occurred seaward (west) of the continuation of the trench supports this interpretation. This conclusion contradicts Seno and Kurita's model of incipient back-arc spreading at the North

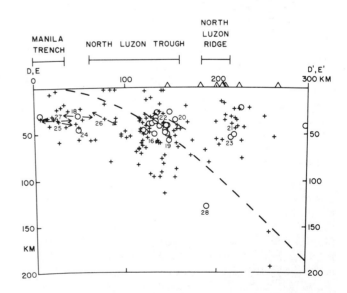

Fig. 13.   Composite cross section across the Luzon Strait obtained by superimposing cross-sections DD′ and EE′ perpendicular to the local trend of the trench and arc (for location see Figure 11), with no vertical exaggeration. Earthquake locations as in Figures 11 and 12; symbols as in Figure 7. Quaternary volcanos shown as triangles at top of section. Dashed line denotes upper surface of subducted plate. Numbers refer to fault plane solutions shown in Figure 15 and listed in Table 1.

Luzon Trough; it is inconsistent with the spatial configuration of the normal fault events.

Similar normal faulting earthquakes are observed near trenches at many convergent margins and are often ascribed to flexural bending stresses in the downgoing slab. Event number 17, which occurred beneath the Manila Trench just south of the cross section in Figure 13, has a reverse-faulting mechanism with a P axis oriented normal to the trench. Similar compressional mechanisms have also been observed at many convergent margins and may be related to flexural bending stresses in the lower portion of the subducting plate [Chapple and Forsyth, 1979].

The normal fault events appear to be restricted to a zone extending from the Asian margin to the northwestern tip of Luzon (Figure 2). The proximity of this zone to the continent-ocean boundary zone is significant. Oceanic lithosphere already subducted at the Manila Trench may exert considerable gravitational pull on the lithospheric slab at the surface. If transitional to continental crust has entered the Manila Trench north of this boundary zone (at about 19°N), gravitational stresses induced by the previously subducted oceanic lithosphere could produce an area of high tensional stress concentration near the boundary between oceanic and continental lithosphere. An analogous model has been suggested for the great normal-faulting earthquake near Sumba (Indonesia) at the boundary of continental (Australian Shelf) and oceanic (Indian Ocean) lithosphere subducting at the Banda arc [Cardwell et al., 1981]. The boundary zone between continental and oceanic crust, cut by normal faults inherited from an original rifting phase, is thus a zone easily susceptible to extensional reactivation.

A second cluster of events occurs beneath and slightly arcward of the North Luzon Trough (Figures 11 and 13). The best-located events occur in a tight cluster between 25- and 55-km depth, although locations of more poorly constrained hypocenters extend to about 110 km. Two reverse fault mechanisms (numbers 20 and 22) may represent interplate slip along shallow east-dipping planes. Their orientation agrees well with the inferred plate boundary (dashed line in Figure 13); their slip vectors indicate east-southeast-directed underthrusting of the Eurasian plate beneath the Philippine Sea plate. This direction, though poorly constrained, is close to the predicted plate convergence for this zone [Seno, 1977].

Two other mechanisms in this cluster are not so readily interpretable. Event number 16 shows reverse faulting along a very steep east-dipping plane or a shallow west-dipping plane. Event number 19, with a near-vertical fault plane, may indicate dip-slip movement on an east-striking fault (south wall down-dropped). Whether this event represents upper- or lower-plate deformation remains ambiguous because of the poor depth constraint on these shallow events. The strike of the vertical fault plane for event 19 correlates with an abrupt truncation of a south-trending limb of the North Luzon Trough as well as easterly trends

in the bathymetric contours (Figure 11). This event is also near the marked S bend in the trends of the trench and the intermediate-depth events (Figure 2). If the hypocentral depth proves to be greater than that shown here, it could represent deformation within the subducting slab near this contortion.

A third cluster of shallow events is located beneath the volcanic arc. Focal mechanisms for two of these events (numbers 21 and 23) both demonstrate compressional deformation of the upper plate in a west to northwest direction. Event 23 shows reverse faulting on a north-striking plane. The new mechanism for the large 1979 event (number 21) and the orientation of its aftershock zone (Figure 5) suggest right-lateral strike-slip faulting on an east-northeast-trending plane. The mode of deformation indicated by this large earthquake helps us assess the validity of several models of plate deformation in the Luzon Strait. Karig [1973] suggested an 'arc vs. buttress' model, pointing out left-lateral offsets of ridges and troughs along northeast-trending faults. This would reflect upper plate deformation of the Philippine arc following its collision with a more stable buttress of the continental margin of Asia at Taiwan. Seno and Kurita [1978] proposed a 'transform belt' model on the basis of strike-slip mechanisms southeast of Taiwan. This transform belt would connect westward subduction of the Philippine Sea plate (along the East Luzon Trough) with limited eastward subduction of the South China Sea (Eurasian) plate southwest of Taiwan. Bowin et al. [1978] minimized the importance of deformation of the North Luzon Ridge, implying that the present morphology of the ridge is close to its original constructional form. That the largest earthquake in the Luzon Strait indicates right-lateral strike-slip faulting on an east-northeast-trending plane is consistent with none of the above models. The bathymetric trends [Lewis and Hayes, this volume] and orientation of the aftershock zone of this earthquake (Figure 5) clearly favor Karig's observation of northeast-trending faults in this portion of the Luzon Strait, but the sense of motion on these faults appears to be opposite that proposed by Karig. Though the overall pattern of faulting in the northern Luzon Strait remains poorly known, the bathymetric trends and this well-constrained fault plane solution clearly imply dextral shearing on northeasterly faults, at least in the southern portion of the Luzon.

The concentration of seismicity in the Luzon Strait near its southern end is demonstrated by (1) small earthquakes recorded over the last 21 years (Figure 11); (2) the largest earthquakes during that period, indicated by the concentration of events for which focal mechanisms could be obtained (Figures 15 and 16); and (3) the largest event in this century (Figure 5). This concentration of seismicity coincides with the arcward projection of Taylor's [1982] continent-ocean boundary zone (Figure 2). This boundary zone is also adjacent to the marked S shaped bend in the trench, the forearc basin, and the zone of intermediate-

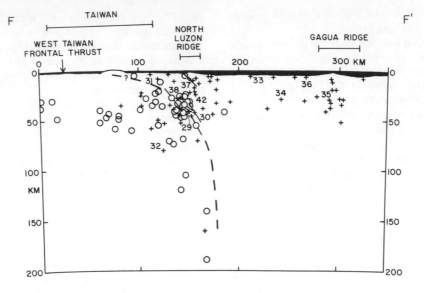

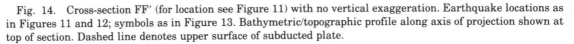

Fig. 14.   Cross-section FF′ (for location see Figure 11) with no vertical exaggeration. Earthquake locations as in Figures 11 and 12; symbols as in Figure 13. Bathymetric/topographic profile along axis of projection shown at top of section. Dashed line denotes upper surface of subducted plate.

depth seismicity. To the north of this boundary zone, subduction may be restricted because of the buoyancy of transitional or continental lithosphere, and continued plate convergence may be accommodated by compressive deformation of the overlying plate. Further south, in central Luzon, the subduction zone, though probably influenced by the collision in Taiwan, has not encountered buoyant transitional or continental lithosphere, and oceanic lithosphere of the South China Sea basin is still entering the subduction zone. Along-strike of the continent-ocean boundary zone, the convergence system may respond to differential subduction on either side of the boundary zone by concentrated shearing of the upper plate. This upper-plate shearing, accommodated by strike-slip, reverse, and normal faulting, occurs near this transition, in the southern portion of the Luzon Strait (Figure 16). A recent large earthquake (M_s = 6.5; Nov. 22, 1981, 18.75°N, 120.84°E), whose mechanism is reported in the *Monthly Listing, Preliminary Determination of Epicenters* (PDE), is consistent with this interpretation. The event is characterized by pure normal faulting, with a horizontal T axis oriented northeast-southwest. The orientation of the T axis is consistent with the more nearly east-west direction of right-lateral shear in the region. The existence of this shear zone may also provide an explanation for the anomalous breadth of the volcanic arc in this area.

Taylor [1982] inferred a sharp change in trend of the boundary zone as it approaches Luzon. Thus a thin wedge of unsubducted oceanic crust may remain between the boundary zone and the Manila Trench. Nonetheless, the coincidence of the strike of the boundary zone with the concentrated upper plate deformation implies that the

northward-thickening trench fill and the proximity to rigid continental crust may produce a similar resistance to subduction.

Southern Taiwan

Subducted lithosphere is continuous from southwestern Luzon to southern Taiwan (Figure 4). There is no evidence for well-located, intermediate-depth earthquakes north of 22.8°N (Figures 2 and 12) other than the earthquakes in northern Taiwan associated with the Ryukyu Trench subduction zone. Well-constrained intermediate-depth events define a steeply dipping Benioff zone beneath southern Taiwan (section FF′, Figure 14). The Neogene volcanos of the North Luzon Ridge are located above the vertical segment of the Benioff zone. The steepening of the Benioff zone is accompanied by a narrowing of the arc-trench gap, where the 'trench' is the northward extension of the Manila Trench to the frontal thrust of the West Taiwan fold belt. A single submarine eruption (Figure 1) was reported near southern Taiwan in 1854 [*Simkin et al.*, 1981]. Although its exact location may be unreliable, the presence of active volcanism near southern Taiwan helps substantiate the presence of subducted lithosphere there. Similar near-vertical dips of the subducted slabs are documented at both ends of the Manila Trench—southwestern Luzon and southern Taiwan—at the transition between subduction and arc-continent collision.

The earthquake distribution does not constrain the nature of the transition from the near-vertical slab at Taiwan to the shallower dip (approximately 45°) farther south. Although the Benioff zone contours in Figure 2 imply a smooth transition zone, a fault plane solution of an intermediate depth event near this transition (number 28, Figures 12 and 13) could be interpreted as tear faulting of the subducted lithosphere along a near-vertical east-trending plane. The sense of motion along this fault would be pure dip-slip, with the north side down-dropped. Alternatively, the north-dipping T axis of this event may reflect the lateral component of stresses in a contorted, continuous slab. Detailed study of microseismicity south of Taiwan may resolve the geometry of the subducted plate.

An extraordinarily dense cluster of large shallow earthquakes (Figure 5) occurs offshore southern Taiwan and coincides with the northern termination of the Manila Trench subduction zone. It marks the transition from limited subduction of transitional lithosphere of the Asian continental margin at the Luzon Strait to full arc-continent collision at Taiwan.

Four events near the northern end of the North Luzon Ridge (numbers 30, 37, 38, and 42) have in common a steeply eastward-dipping plane (see Figures 15 and 16). Event 37, whose depth is not well constrained, is placed well above the plate boundary, and event 30 indicates a significant strike-slip component of motion. Nonetheless, the dip of this east-dipping plane agrees well with that of the inferred plate boundary (Figure 14), and may document interplate slip near the northern end of the Manila Trench subduction zone. Although there is considerable scatter in the inferred slip directions, the mean is 111°, which is close to the east-southeast-directed convergence predicted by Philippine Sea/Eurasia plate motions. Without additional consistent slip directions from underthrust-

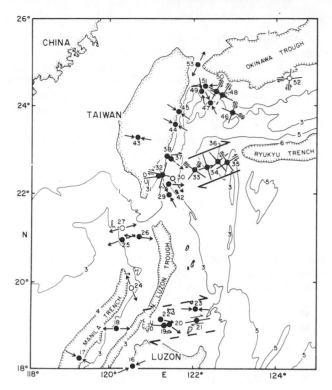

Fig. 16. Interpreted focal mechanism solutions for the Luzon Strait and Taiwan regions. Mechanism parameters are given in Table 1. Slip vectors of shallow underthrust mechanisms shown as single arrow. Opposing and diverging arrows represent horizontal projections of P and T axes of reverse-fault and normal-fault mechanisms, respectively. Single preferred plane or both planes of strike-slip and vertical dip-slip mechanisms are shown as lines with arrows for strike-slip and U (up) and D (down) for dip-slip mechanisms. Heavy arrows north of Luzon and southeast of Taiwan denote sense of motion on inferred zones of distributed shear. Symbols are dashed where interpretation is ambiguous. Selected bathymetric contours as in Figure 1.

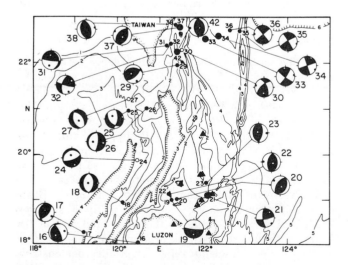

Fig. 15. Focal mechanism solutions for the Luzon Strait region. New and published mechanisms shown as lower-hemisphere projections as in Figure 4. Mechanism parameters are given in Table 1, and individual diagrams are shown in Appendix. Large circles denote earthquakes with $M_s \geq 6.9$; open circles denote poorly constrained mechanisms. Selected bathymetric contours and Quaternary volcanos as in Figure 1.

type mechanisms, we cannot confidently constrain the direction of plate convergence in this zone.

Event 29 in this cluster shows reverse faulting on east-northeast-trending planes. Its mechanism, while well-constrained, does not reflect this east-directed underthrusting. Two nearby mechanisms (numbers 31 and 32) have a similar near-vertical east-striking plane. One is characterized by nearly pure dip-slip movement (south flank uplifted), and one is dominated by a right-lateral strike-slip component. Though its depth is poorly constrained, event 32 apparently occurred at moderate depth (80 km); it would thus represent deformation of the subducting slab. If it proves to be shallower, the alternate plane of this event would imply sinistral shearing along a fault parallel to the Taiwan coastline. This is consistent with observed sense of movement further north along the Longitudinal Valley fault.

Southern Taiwan to the Ryukyu Trench

A striking ENE-trending belt of epicenters extends from southern Taiwan to the Ryukyu Trench (Figures 3 and 11). These large and moderate-sized events are characterized by strike-slip focal mechanisms (numbers 33–36). Their northeast-trending fault planes are subparallel to the trend of epicenters and suggest right-lateral strike-slip faulting along this belt. However, two observations are not consistent with this interpretation. First, microearthquakes recorded by the Taiwan network [*Wu,* 1978] show that the aftershock zones of two moderate-sized events within this zone (numbers 35 and 36) are elongated in the northwest direction, indicating faulting on the conjugate fault plane. Second, one of the two large strike-slip events (number 34; $M_s = 7.5$, 1972) was followed immediately by another large event $M_s = 7.0$) whose mechanism could not be obtained. The proximity of these events suggests a possible association with a northwest-striking feature (dashed 1972 aftershock zone in Figure 5). Neither of these two large events, however, has a well-defined aftershock zone, and we have no reliable constraint on the trend of their fault planes.

The striking linearity of this belt and the similarity of this orientation to the east-northeast plane of these mechanisms could thus be spurious. However, its coincidence with the termination of the Ryukyu Trench suggests a tectonic interpretation for this lineament. Following arc-continent collision at Taiwan, a shear zone with this orientation would function to divert Philippine Sea/Eurasia plate motion from the collision zone to a site of less-resistant convergence further south. That several of the earthquakes within this zone appear to have ruptured along conjugate planes does not necessarily contradict this hypothesis. Observations of incipient shear zones, both in laboratory experiments and in field examples [e.g., *Bornyakov,* 1980; *Tchalenko,* 1970] show that early stages of shear failure are marked by formation of a relatively broad zone of strike-slip faulting along en echelon 'Riedel shears' cut by transverse conjugate faults ('conjugate Riedel shears'). This sense of motion on such an incipient shear zone is shown in Figure 16 by the large force couple and individual shears by the individual earthquake fault planes within the zone.

Taiwan

At the island of Taiwan, widespread intense shortening and thickening of both the converging plates has taken place because of the collision of the west-facing arc with the Asian margin [e.g., *Page and Suppe,* 1981]. Much of the deformation now appears to have moved outboard (east) of the main collision zone.

Additional focal mechanisms to the north of our study area reveal a complex pattern of deformation in and around Taiwan. Mechanism numbers 43–53 presented in Figure 16 generally reflect subhorizontal compression. In addition

to the compressive deformation observed within Taiwan (reverse fault mechanism numbers 43–45), a dense cluster of seismicity occurs east of northern Taiwan [*Katsumata and Sykes,* 1969; *Wu,* 1978]. *Wu* [1978] suggested that this zone represents a small westward continuation of the Ryukyu Trench that is displaced to the north from the trench by 100 km. This segment would be connected to the Ryukyu Trench by a northeast-trending trench-trench transform. Events 46, 48, and 50 were interpreted by *Wu* [1978] as strike-slip faulting along this transform. Events 47 and 49 would represent interplate thrusting at this subduction zone. However, such a trench segment is not defined bathymetrically, and submarine volcanism (see Figure 1) and one of the 'interplate thrusts' (event 47) are located seaward of this 'trench segment.' It is perhaps more appropriate to consider this east-trending belt of seismicity to be a zone of compressional stress concentration, forming the northern boundary of a small, relatively stable block to the south of it. The stable block might be thought of as a microplate, squeezed northward relative to the Eurasian plate (along the Longitudinal Valley fault and Wu's 'trench-trench transform') and eastward relative to the Philippine Sea plate (along the incipient shear zone) because of the northwestward movement of the Philippine Sea plate. Intermediate-depth seismicity beneath northern Taiwan is part of a continuous Benioff zone associated with subduction at the Ryukyu Trench (Figure 2); a separate trench segment east of Taiwan need not be invoked. Event 53 indicates that active back-arc spreading at the Okinawa Trough [*Lee et al.,* 1980] extends into northeastern Taiwan [*Wu,* 1978]. Event 52 reflects northwest-directed deformation of the upper (Eurasian) plate associated with convergence at the Ryukyu Trench.

Conclusions

1.   The northwest-southeast convergence between the Philippine Sea and Eurasian plates is accommodated by plate consumption on the eastern margin (Philippine Trench/East Luzon Trough) and the western margin (Manila Trench) of the Philippine arc as well as by internal deformation within the arc. The Philippine Fault, whose activity is demonstrated by seismicity, earthquake ground rupture, and geomorphic observations, also plays a significant role in accommodating plate convergence.

2.   Westward subduction of West Philippine Basin oceanic lithosphere may be beginning at the East Luzon Trough. A series of large earthquakes landward of the trough with underthrust-type focal mechanisms, together with multichannel seismic reflection profiles, demonstrate convergence from 15°N to at least 17.5°N. The observed direction of slip along the East Luzon Trough deviates substantially from that predicted by Philippine Sea/Eurasia plate convergence. The eastern Philippine Islands are decoupled from Eurasia because of movement along both the Philippine Fault and the Manila Trench. The lack of evidence of northward movement along the Manila Trench

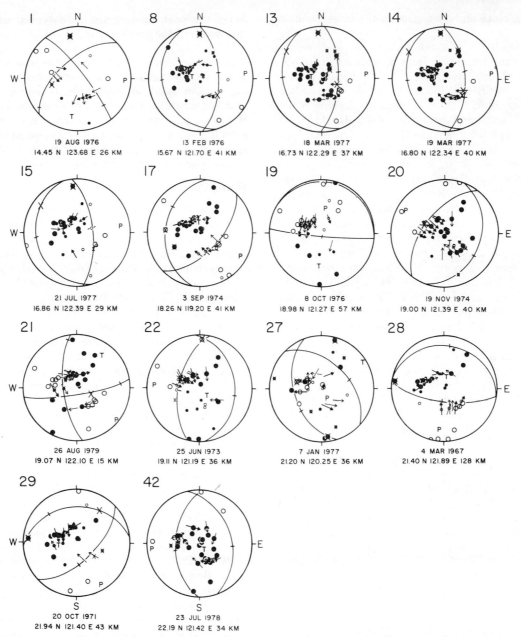

Fig. 17.   Focal mechanism solutions for the northern Philippine island arc; see text for explanation.

suggests that transcurrent motion along the Philippine Fault is largely responsible for this discrepancy.

The geometry and fault plane solutions in the East Luzon seismic zone provide no evidence of major flexural or other anomalous deformation that might be associated with lithospheric rupture in an incipient subduction zone. The presence of a relict subduction zone east of central Luzon may facilitate the formation of a new convergent margin.

3.   A zone of distributed sinistral shear connects the Philippine Trench and East Luzon Trough. The geometry of the seismicity belt and the focal mechanisms of several

events within it require a model more complex than a simple trench-trench transform fault.

4.   A striking finding of this study is that subduction associated with the Manila Trench (and its northward extension) can be traced continuously from southwestern Luzon into southern Taiwan. A Benioff zone extending to over 200-km depth, a belt of active volcanism, several new underthrust-type focal mechanisms, and sediment deformation at the trench all corroborate its activity. Major transitions in plate convergence along the Manila Trench appear to be responsible for zones of concentrated deformation of

the overlying plate and contortion of the underthrusting plate.

5. A dense cluster of earthquakes near southwestern Luzon marks the transition from oceanic subduction beneath western Luzon to collision of the Philippine arc with the Palawan continental block. A concentration of large shallow events coincides with a vertically dipping Benioff zone near this southern termination of the Manila Trench.

6. A more speculative transition to restricted subduction of buoyant Asian margin lithosphere appears to take place just north of Luzon. A cluster of seismicity occurs where the trench and subducted slab form an S-shaped bend along-strike of the boundary between oceanic and transitional (Asian margin) lithosphere. Strike-slip and dip-slip mechanisms may describe another zone of distributed shear located along the arcward projection of this boundary zone. A cluster of normal-faulting events near the Manila Trench just north of this boundary zone suggest that buoyant transitional lithosphere may be influencing the Manila Trench subduction zone. The gravitational pull of previously subducted oceanic lithosphere and the resistance of this thick buoyant lithosphere of the Asian margin provide an explanation for the observed concentration of extensional stresses in the subducting slab.

7. At the northern end of the Manila Trench subduction zone, as at its southern end, an extraordinary concentration of large shallow earthquakes coincides with a vertically dipping Benioff zone. With the termination of the subduction zone, plate convergence further north takes place by full arc-continent collision. There, much of the interplate convergence must be accommodated by upper plate deformation, resulting in intense seismicity offshore eastern Taiwan. A northeast-trending belt of large and moderate-sized strike-slip events forms an incipient shear zone, marking the northwestern boundary of the Philippine Sea plate. This shear zone may provide the convergence system with a simpler connection between the Manila Trench and Ryukyu Trench subduction zones, eventually shunting off the Taiwan collision zone from Philippine Sea/Eurasia plate motion.

Appendix

Table 1 lists the solution parameters for all the focal mechanism solutions in the northern Philippine island arc region. The 14 new solutions are illustrated by equal area projections of the lower hemisphere of the focal sphere (Figure 17). Large open and closed circles represent clear dilatational and compressional first motions, respectively. Uncertain first motions are shown as small open and closed circles. Large and small crosses indicate clear and uncertain compressional wave data, respectively, that are judged to be near a nodal plane. Heavy arrows with closed heads and lighter arrows with open heads indicate clear and uncertain S wave polarization directions, respectively. Tick marks on the nodal planes indicate the poles to the opposite plane. The P (pressure) and T (tension) axes are located at the base of the letters P and T. The information

below the focal sphere gives the date, latitude, longitude, and depth of the event.

Acknowledgments. The authors wish to thank Stephen Lewis for his generosity in sharing data and ideas with us; we also thank M. Barazangi, D. Karig, S. Bachman, and W. Schweller for helpful discussions. E. Farkas and P. Bulack provided tireless technical assistance. This research was part of the SEATAR Philippines-Marianas transect and was supported by NSF grant EAR-79-19516. This is Cornell contribution 735.

References

Acharya, H. K., and Y. P. Aggarwal, Seismicity and tectonics of the Philippine Islands, *J. Geophys. Res., 85,* 3239–3250. 1980.

Acharya, H. K., J. F. Ferguson, and V. Isaac, Microearthquake surveys in the central and northern Philippines, *Bull. Seismol. Soc. Am., 69,* 1889–1903, 1979.

Allen, C. R., Circum-Pacific faulting in the Philippines-Taiwan region, *J. Geophys. Res., 67,* 4795–4812, 1962.

Balce, G. R., A. L. Magpantay, and A. S. Zanoria, Tectonic scenarios of the Philippines and northern Indonesian region, paper presented at Ad Hoc Working Group Meeting on the Geology and Tectonics of Eastern Indonesia, ESCAP-CCOP-IOC-SEATAR, Bandung, Indonesia, July 9–14, 1979.

Balce, G. R., R. Y. Encina, A. Momongan, and E. Lara, Geology of the Baguio District and its implication on the tectonic development of the Luzon Central Cordillera, *Geol. Paleontol. Southeast Asia, 21,* 265–287, 1980.

Biq, C. C., A fossil subduction zone in Taiwan, *Proc. Geol. Soc. China, 14,* 146–154, 1971.

Biq, C. C., Dual-trench structure in the Taiwan-Luzon region, *Proc. Geol. Soc. China, 15,* 65–75, 1972.

Bonilla, M. G., Summary of Quaternary faulting and elevation changes in Taiwan, *Geol. Soc. China Mem., 2,* 43–55, 1977.

Bornyakov, S. A., The modeling of strike-slip zones in elastic-viscous materials (English translation), *Sov. Geol. Geophys., 21:11,* 64–71, 1980.

Bowin, C., R. S. Lu, C. S. Lee, and H. Schouten, Plate convergence and accretion in the Taiwan-Luzon region, *Am. Assoc. Petrol. Geol. Bull., 62,* 1645–1672, 1978.

Cardwell, R. K., B. L. Isacks, and D. E. Karig, The spatial distribution of earthquakes, focal mechanism solutions, and subducted lithosphere in the Philippine and northeastern Indonesian Islands, in *The Tectonic and Geological Evolution of Southeast Asian Seas and Islands, Geophys. Monogr. Ser. 23,* edited by D. E. Hayes, pp. 1–35, AGU, Washington, D. C., 1980.

Cardwell, R. K., E. S. Kappel, M. B. Lawrence, and B. L. Isacks, Plate convergence along the Indonesian arc, *Eos Trans. AGU, 62,* 404, 1981.

Chai, B. H. T., Structure and tectonic evolution of Taiwan, *Am. J. Sci., 272,* 389–422, 1972.

Chapple, W. M., and D. W. Forsyth, Earthquakes and bending of plates at trenches, *J. Geophys. Res., 84,* 6729–6749, 1979.

Chase, T. E., and H. W. Menard, Bathymetric Atlas of the North-Western Pacific Ocean, 164 charts, *Publ. 1301-2-3,* U.S. Nav. Oceanogr. Office, Washington, D. C., 1973.

Chi, W. R., J. Namson, and J. Suppe, Stratigraphic record of plate interactions in the Coastal Range of eastern Taiwan, *Geol. Soc. China Mem., 4,* 491–530, 1981.

DeBoer, J., L. A. Odom, P. C. Ragland, F. G. Snider, and N. R. Tilford, The Bataan orogene: Eastward subduction, tectonic rotations and volcanism in the western Pacific (Philippines), *Tectonophysics, 67,* 251–282, 1980.

Divis, A. F., The petrology and tectonics of recent volcanism in the central Philippine Islands, in *The Tectonic and Geological Evolution of Southeast Asian Seas and Islands, Geophys. Monogr. Ser. 23,* edited by D. E. Hayes, pp. 127–144, AGU, Washington, D. C., 1980.

Fitch, T., Earthquake mechanisms and island arc tectonics in the Indonesian-Philippine region, *Bull. Seismol. Soc. Am., 60,* 565–591, 1970.

Fitch, T., Plate convergence, transcurrent faults and internal deformation adjacent to Southeast Asia and the western Pacific, *J. Geophys. Res., 77,* 4432–4460, 1972.

Fitch, T. J., and P. Molnar, Focal mechanisms along inclined earthquake zones in the Indonesia-Philippine region, *J. Geophys. Res., 75,* 1431–1444, 1970.

Gervasio, F. C. Geotectonic development of the Philippines, in *The Western Pacific, Island Arcs, Marginal Seas, Geochemistry,* edited by P. J. Coleman, pp. 307–324, Crane Russak & Co., New York, 1973.

Gutenberg, B., and C. F. Richter, *Seismicity of the Earth and Associated Phenomena,* 310 pp., Princeton Univ. Press, Princeton, N.J., 1954.

Hamilton, W., Tectonics of the Indonesian Region, *Geol. Surv. Prof. Pap. (U.S.) 1078,* 345 pp., 1979.

Ho, C. S., Geologic and tectonic framework of Taiwan, *Geol. Soc. China. Mem., 3,* 57–72, 1979.

Hsu, T. L., Neotectonics of the Longitudinal Valley, eastern Taiwan, *Bull. Geol. Surv. Taiwan, 25,* 53–62, 1976.

International Seismological Centre, *International Seismological Summary,* Newbury, Berkshire, U.K., 1918–1963.

International Seismological Centre, *Bulletin of the International Seismological Centre,* Newbury, Berkshire, U.K., 1964–1978.

Isacks, B. L., and M. Barazangi, Geometry of Benioff zones: Lateral segmentation and downwards bending of the subducting lithosphere, in *Island Arcs, Deep Sea Trenches and Back-Arc Basins, Maurice Ewing Ser.,* vol. 1, edited by M. Talwani and W. C. Pitman III, pp. 99–114, AGU, Washington, D. C., 1977.

Kanamori, H., Tectonic implications of the 1944 Tonankai and the 1946 Nankaido earthquakes, *Phys. Earth Planet. Int., 5,* 129–139, 1972.

Kanamori, H., and K. Tsumura, Spatial distribution of earthquakes in the Kii peninsula, Japan, south of the

Median Tectonic Line, *Tectonophysics, 12,* 327–342, 1971.

Karig, D. E., Plate convergence between the Philippines and the Ryukyu Islands, *Mar. Geol., 14,* 153–168, 1973.

Karig, D. E., Basin genesis in the Philippine Sea, *Initial Rep. Deep Sea Drilling Proj., 31,* 857–879, 1975.

Karig, D. E., and G. F. Sharman III, Subduction and accretion in trenches, *Geol. Soc. Am. Bull., 86,* 377–389, 1975.

Karig, D. E., and J. M. Wageman, Structure and sediment distribution in the northwest corner of the West Philippine Basin, *Initial Rep. Deep-Sea Drilling Proj., 31,* 615–620, 1975.

Katsumata, M., and L. R. Sykes, Seismicity and tectonics of the western Pacific: Izu-Mariana-Caroline and Ryukyu-Taiwan regions, *J. Geophys. Res., 74,* 5923–5948, 1969.

Lee, C. S., G. G. Shor, Jr., L. D. Bibee, R. S. Lu and T. W. C. Hilde, Okinawa Trough: Origin of a back-arc basin. *Mar. Geol., 35,* 219–241, 1980.

Lewis, S. D., and D. E. Hayes, Northward-propagating subduction along eastern Luzon, Philippines, this volume.

Ludwig, W. J., The Manila Trench and West Luzon Trough, 3, Seismic refraction measurements, *Deep Sea Res., 17,* 553–571, 1970.

Ludwig, W. J., D. E. Hayes, and J. E. Ewing, The Manila Trench and West Luzon Trough, 1, Bathymetry and sediment distribution, *Deep Sea Res., 14,* 533–44, 1967.

Minster, J. B., and T. H. Jordan, Rotation vectors for the Philippine and Rivera plates, *Eos Trans. AGU, 60,* 958, 1979.

Molnar, P., and P. Tapponnier, Cenozoic tectonics of Asia: Effects of a continental collision, *Science, 189,* 419–426, 1975.

Morante, E. M., and C. R. Allen, Displacement on the Philippine Fault during the Ragay Gulf earthquake of March 17, 1973 (abstract), *Geol. Soc. Am. Abstr. with Programs, 5,* 744–745, 1974.

Mrozowski, C. L., S. D. Lewis, and D. E. Hayes, Complexities in the tectonic evolution of the West Philippine Basin, *Tectonophysics, 82,* 1–24, 1982.

Murphy, R. W., The Manila Trench-west Taiwan foldbelt: A flipped subduction zone?, *Geol. Soc. Malaysia Bull., 6,* 27–42, 1973.

Neumann van Pandang, M., *Catalog of the Active Volcanoes of the World, Including Solfatara Fields,* part 2, *Philippine Islands and Cochin China,* International Volcanological Association, Naples, Italy, 1953.

Page, B. M., and J. Suppe, The Pliocene Lichi Melange of Taiwan: Its plate tectonic and olistostromal origin, *Am. J. Sci., 281,* 193–227, 1981.

Philippine Bureau of Mines, Geological map of the Philippines, scale 1:1,000,000, 9 sheets, Manila, 1963.

Rothé, J. P., *The Seismicity of the Earth, 1953–1965,* 336 pp. UNESCO, Paris, France, 1969.

Rowlett, H., and J. Kelleher, Evolving seismic and tectonic patterns along the western margin of the Philippine Sea plate, *J. Geophys. Res., 81,* 3518–3524, 1976.

Rutland, R. W. R., A tectonic study of part of the Philippine Fault zone, *Q. J. Geol. Soc. London, 123,* 293–325, 1968.

Schweller, W. J., and D. E. Karig, Emplacement of the Zambales Ophiolite into the West Luzon margin, Hedberg Conference, *Amer. Association Petrol. Geol. Mem.,* in press, 1982.

Seno, T., The instantaneous rotation vector of the Philippine Sea plate relative to the Eurasian plate, *Tectonophysics, 42,* 209–226, 1977.

Seno, T., and K. Kurita, Focal mechanisms and tectonics of the Taiwan-Philippine region, *J. Phys. Earth, 26* (suppl.), S249–S263 1978.

Shiono, K., Travel-time analysis of relatively deep earthquakes in Southwest Japan with special reference to the underthrusting of the Philippine Sea Plate, *J. Geosci. Osaka City Univ., 18,* 37–59, 1974.

Shiono, K., Focal mechanisms of major earthquakes in Southwest Japan and their tectonic significance, *J. Phys. Earth, 25,* 1–26, 1977.

Shiono, K., T. Mikumo and Y. Ishikawa, Tectonics of the Kyushu-Ryukyu arc as evidenced from seismicity and focal mechanism of shallow to intermediate depth earthquakes. *J. Phys. Earth, 28,* 17–43, 1980.

Silver, E. A., and R. McCaffrey, Active back-arc thrusting in the eastern Sunda arc, Indonesia: An example of incipient arc-polarity reversal (abstract), *Geol. Soc. Am. Abstr. with Programs,* 554, 1981.

Simkin, T., L. Siebert, L. McClelland, D. Bridge, C. Newhall, and J. H. Latter, *Volcanos of the World,* 232 pp., Hutchinson Ross, Stroudsburg, Pa., 1981.

Stewart, G. S., and S. N. Cohn, The August 16, 1976, Mindinao, Philippine earthquake (M_s = 7.8)—Evidence for a subduction zone south of Mindinao (abstract), *Eos Trans. AGU, 58,* 1194, 1977.

Taylor, B., On the tectonic evolution of marginal basins in northern Melanesia and the South China Sea, Ph.D. thesis, Columbia Univ., New York, 1982.

Taylor, B., and D. E. Hayes, The tectonic evolution of the South China Basin, in *The Tectonic and Geologic Evolution of Southeast Asian Seas and Islands, Geophys. Monogr. Ser. 23,* edited by D. E. Hayes, pp. 89–104, AGU, Washington, D. C., 1980.

Tchalenko, J. S., Similarities between shear zones of different magnitudes. *Geol. Soc. Am. Bull., 81,* 1625–1640, 1970.

U.S. Geological Survey, *Earthquake Data Report,* Arlington, Va., 1978–1980.

U.S. Geological Survey, *Preliminary Determination of Epicenters,* Denver, Colo., 1961–1980.

Wolfe, J. A., Philippine geochronology, *J. Geol. Soc. Philipp., 35,* 1–30, 1981.

Wu, F. T., Focal mechanisms and tectonics in the vicinity of Taiwan, *Bull. Seismol. Soc. Am., 60,* 2045–2056, 1970.

Wu, F. T., Recent tectonics in Taiwan, *J. Phys. Earth, 26* (suppl.), S265–S299, 1978.

York, J. E., Quaternary faulting in eastern Taiwan, *Bull. Geol. Surv. Taiwan, 25,* 63–72, 1976.

Origin and History of the South China Sea Basin

BRIAN TAYLOR[1] AND DENNIS E. HAYES

Lamont-Doherty Geological Observatory of Columbia University
Palisades, New York 10964

Magnetic anomaly data acquired on R/V *Vema* cruises in the South China Sea in 1979 have allowed us to refine the previously identified pattern of seafloor spreading in the South China Basin. East trending magnetic lineations 5D to 11 identified in the eastern half of the basin date seafloor spreading as mid-Oligocene through early Miocene (32–17 m.y. B.P.). Heat flow data are consistent with these ages. Half spreading rates varied between 2.2 and 3.0 cm/yr. An east trending chain of seamounts occurs near the center of symmetry of the magnetic lineations. Basalts dredged from these and other seamounts in the basin have major element chemistries which are alkalic or transitional between tholeiitic and alkalic. Thick sediments (1–2.5 km) and relatively smooth oceanic basement characterize the older portions of the basin, whereas thinner sediments (300 m to 1 km) and a blocky basement fabric characterize the younger central part of the basin. A broad basement arch topped by fault-bounded tilted blocks occurs in the east-central area of the basin. The South China Basin is an 'Atlantic-type' marginal basin, bounded by passive continental margins to the north and south. Opening of the basin moved microcontinental blocks including northern Palawan and Reed Bank from their Paleogene position adjacent to the China mainland. Fracture zones in the eastern half of the basin and the inferred transform margin east of Vietnam indicate that the South China Basin opened in a north-south direction and require a distant pole to describe the opening. However, the spreading fabrics of the western and eastern halves of the basin are significantly different. Free air gravity and limited seismic reflection data delineate a southwest trending relict spreading center in the southwest of the South China Basin. Much less oceanic crust was generated at this relict spreading center than at the east trending spreading center farther east. We speculate that opening of the western half of the basin was dominantly by crustal stretching of the complex of microcontinental blocks inferred to occupy a large part of the area. The landward boundary of normal oceanic crust in the South China Basin is marked by a change in basement structure and magnetic anomaly signature, by a free air gravity anomaly low, and by a steep landward gradient in the isostatic gravity anomaly. Analysis of the gravity data suggests that the boundary is associated with a density contrast at crustal depths, which is compensated below the Moho. Comparison with other Atlantic-type continental margins suggests that a positive density contrast between oceanic and continental crust may be a common feature of continent-ocean boundary zones. The rifting of the proto-China margin which preceded seafloor spreading in the South China Basin probably began in the latest Cretaceous or Paleocene (~65 ± 10 m.y.). The rifting was localized along a former Andean-type arc terrain at which volcanism had ceased by approximately 85 m.y. B.P. North Palawan and Reed Bank are inferred to have been forearc areas in the Mesozoic. The margins of the South China Sea record a regional mid-Oligocene unconformity which we interpret as caused

---

[1]Now at Hawaii Institute of Geophysics, Honolulu, Hawaii 96822.

by the superposition of breakup and sea level effects. Seafloor spreading in the basin ended slightly before the late middle Miocene cessation of subduction at the Palawan subduction zone to the south.

Introduction

The South China Basin (Figure 1) is one of several marginal basins in the western Pacific and southeast Asia whose crust and sediments have not been sampled by deepsea drilling. Initial attempts to date the basin by correlating observed magnetic anomaly lineations with the geomagnetic time scale were unsuccessful [Ben-Avraham and Uyeda, 1973; Bowin et al., 1978]. Consequently, early scenarios for the tectonic evolution of the basin were based largely on the geological history of the surrounding land areas. For example, Karig [1973] suggested that the South China Basin opened as an interarc basin behind the Late Cretaceous–early Tertiary east facing Philippine arc. Ben-Avraham and Uyeda [1973], on the other hand, proposed that major WNW trending sutures in north Vietnam and southwest Borneo were formerly contiguous and that the South China Basin formed in the Late Jurassic–Early Cretaceous as Borneo moved south with respect to China.

More recently, Taylor and Hayes [1980] identified a symmetric pattern of east trending late Oligocene and early Miocene magnetic lineations in the eastern half of the South China Basin. They proposed that the South China Basin is a small 'Atlantic-type' marginal basin bounded by passive continental margins to the north and south and a transform margin to the west. To test Taylor and Hayes' identification of magnetic anomalies and to define better the magnetic lineation pattern, a marine geophysical survey of the eastern half of the basin was conducted in November and December 1979 using the R/V Vema (see Figure 1). This survey was part of a cooperative program between the Lamont-Doherty Geological Observatory of Columbia University and the Ministry of Geology of the People's Republic of China.

In the first half of this paper we discuss the geological evolution of the South China Basin as revealed by the new and previously existing marine geological and geophysical data. This information is then integrated with geological data from the surrounding continental shelves and land areas to develop an evolutionary scenario for the region.

Magnetic Anomaly Data and Seafloor Spreading History

Over 90% of the unclassified marine magnetics data in the South China Basin have been collected by Lamont-Doherty Geological Observatory and most of that in the last 5 years. Figure 2 shows this data plotted as magnetic anomalies along ship tracks. Also shown is the pattern of magnetic lineations and fracture zones we have identified. This refinement of Taylor and Hayes' [1980] magnetic anomaly pattern is one result of the large number of newly acquired NNW trending survey lines. Several details of

the anomaly correlations are shown in Figures 3a–3c. Magnetic anomalies along the solid track in Figure 2 are compared in Figure 4 to theoretical seafloor spreading anomaly profiles based on the geomagnetic reversal time scale of LaBrecque et al. [1977].

Following Taylor and Hayes [1980], the east trending magnetic lineations in the South China Basin have been correlated with the late Oligocene through early Miocene sequence of magnetic anomalies 11 to 5d (32–17 m.y.). In general, the magnetic lineations are symmetrically disposed about an east trending chain of seamounts between 15° and 16°N (Figure 2)—subsequently referred to as the Scarborough seamounts, after the shoal by that name. However, in detail the magnetic pattern is quite complicated as a result of fracture zones, ridge jumps, and seamounts. Several fracture zones have been recognized (Figure 2) both on the basis of offset magnetic anomalies and basement morphology. Most of the offsets are small, less than 20 km. We have not been able to detect the offsets of the 6b-6 sequence, observed on the northern side of the basin, in the south.

As seen in Figure 2, a 10° to 15° reorientation in spreading direction occurred between anomaly 8 and 7a time. The subsequent return to the original spreading direction between anomaly 7 and 6c time was accompanied by a ridge jump to the south which left both sides of the 7a-7 sequence on the northern plate. As a result, anomalies 6c and 8 are juxtaposed on the southern plate. Furthermore, the location of fracture zones observed in the northern half of the basin are different on either side of the 7a-7 sequence, the ridge-transform geometry having been reorganized when the ridge jumped to the south.

A magnetic quiet zone characterizes much of the northern and southern continental margins (Figure 2). An obvious exception is the high-frequency anomalies near 20.5°N, 117.5°E which are associated with volcanic intrusions of probable Pliocene age. In the north-central portion of the basin, magnetic anomaly 11 is identified just seaward of the magnetic quiet zone on profiles 1–3 and 6 (Figures 2 and 3a). Anomalies 10–8 are well lineated and trend 7° north of east. On the southern side of the basin, anomaly 11 is not observed, and anomaly 10 has low amplitude (Figure 3b). Even anomaly 9 is only well developed on one profile, so that the first correlation on the south flank is that of anomaly 8. Despite the further complication that both sides of the anomaly 7a-7 sequence are interpreted to occur in the northern half of the basin, we believe that this interpretation is the simplest one that agrees with the available data. The different trend of the 7a-7 sequence, and the duplication of at least anomaly 7, is clear (Figure 2). The 6c-6 sequence of magnetic anomalies is easily identified in the northern half of the basin (Figure

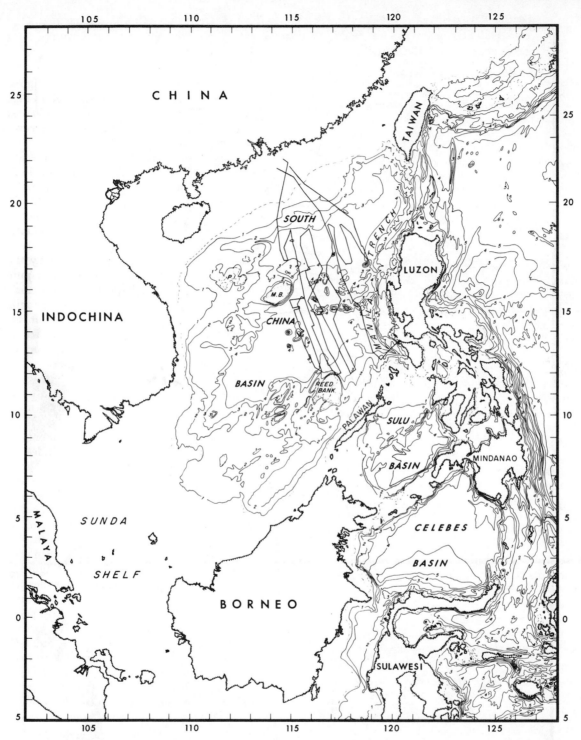

Fig. 1.   Bathymetry of the South China Sea and surrounding areas, modified after *Mammerickx et al.* [1976]. MB, Macclesfield Bank; PI, Paracel Islands. Also shown are the tracks of R/V *Vema* cruises V3604 and V3605.

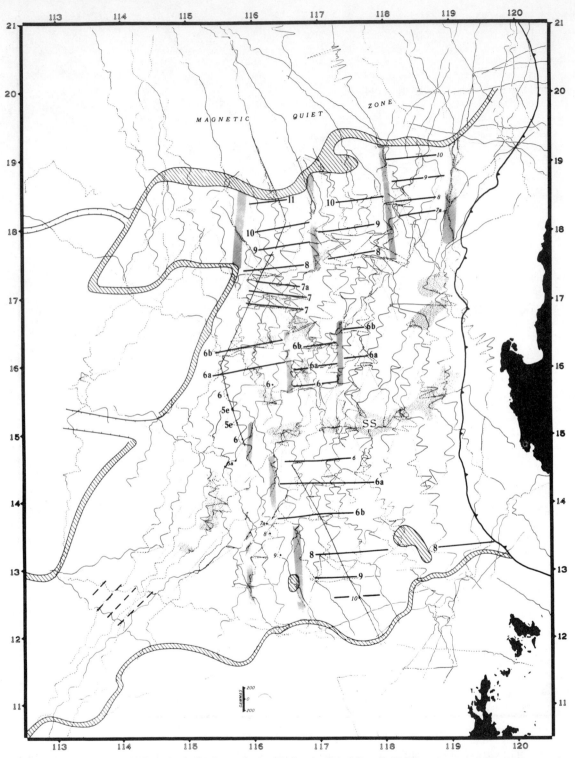

Fig. 2.   Magnetic anomalies along ship's tracks in the South China Sea. Bold line segments indicate magnetic correlations between tracks. The magnetic lineations have been identified according to the numbering scheme of *LaBrecque et al.* [1977]. The smaller numbers indicate less confident identifications. Fracture zones are represented by diagonal, finely hatchured lines. The oceanic crust of the South China Basin is bounded to the east by the Manila Trench (barbed line) and on all other sides by the narrow continent-ocean boundary zone (coarsely hatchured) which is defined in the text. The coarsely stippled areas delineate seamounts. SS, Scarborough shoals.

3c). The symmetric counterparts of these anomalies to the south are similarly well lineated, although they do not match the 6c-6 sequence as well (Figures 3b and 4).

The youngest magnetic anomalies in the basin are difficult to identify because generally they are overprinted by anomalies associated with the Scarborough seamount chain. These seamounts occur at the center of symmetry of the 6b-6 sequence (Figure 2) and were probably extruded along the relict spreading center at or after the time seafloor spreading ceased [Taylor and Hayes, 1980]. An age of approximately 17 m.y. (anomaly 5d) can be predicted for the youngest crust in the basin, assuming spreading rates remained nearly constant following anomaly 6. One magnetic profile along a track 50 km west of the seamount chain records a sequence of anomalies symmetric about anomaly 5e (~18 m.y.). The distance between the pair of anomaly 6's along this track is 45 km less than that observed to the east (Figure 2). This may be evidence of a slightly diachronous cessation of spreading.

Magnetic anomalies along this track segment form part of the projected anomaly profile that is compared in Figure 4 to model magnetic profiles based on the geomagnetic reversal time scale of LaBrecque et al. [1977]. At the top of the figure is a constant spreading rate model (half rate = 2.5 cm/yr). The lower preferred model incorporates a ridge jump between anomalies 6c and 7 and half spreading rates varying between 2.2 and 3.0 cm/yr. The correlation between the observed and modeled anomalies is quite good with the not surprising exception of the segments immediately adjacent to the proposed ridge jump. Anomaly 11 is not observed in the south, possibly because of a general decrease in anomaly amplitudes from anomaly 8 to 11. This decrease in anomaly amplitudes is best developed in the south but is also present in the north. The cause for the amplitude modulation of the observed magnetic profiles remains speculative.

The magnetic model profiles have been phase shifted −135° from symmetry to match the observed profiles. Given the present magnetic inclination and declination at 16°N, 117°E of 15° and 0°, respectively, a skewness parameter of −135° ± 10° and a magnetic lineation strike of 080° to 095° imply an effective remnant inclination of 30° ∓ 10° [Schouten and Cande, 1976]. In the absence of a significant rotation of the plate and assuming a dipole field, a remnant inclination of 30° ± 10° corresponds to a paleolatitude of 16° ± 6°N. In other words, the observed phase of the magnetic anomalies is consistent with no latitudinal motion of the South China Basin since its formation, though small shifts could not be resolved.

The accompanying bathymetric and free air gravity profiles as well as the single-channel seismic data (not shown) indicate the lack of significant basement structure along the profile shown in Figure 4. The gravity profile is very flat and near zero, varying between −5 and +12 mGal. The landward gradient in the free air gravity at both ends of the profile is associated with the transition from oceanic to continental crust (see later section).

There remain two basin areas where magnetic lineations have not been mapped. The first is an area north of Macclesfield Bank centered on 18°N, 115°E (Figure 2). The anomaly 8 to 11 sequence is not observed west of the fracture zone near 116°E. Although there are numerous tracks in this northwest subbasin, lineated magnetic anomalies are not apparent. Consequently, the exact age of this oceanic crust remains undetermined. Seismic profiles suggest that an east trending remnant rift axis may be present at 18°N. A failed rift clearly defined by bathymetric and seismic data extends west from the western edge of the northwest subbasin to at least 112°E (Figures 2, 5, and 10). Some of the faults bounding this system are still active, as seen by the offset of surficial sediments on the seismic profiles across the rift shown in Figure 5.

The second area occupies the southwest third of the basin (Figures 1 and 2). The available magnetics data suggest a northeast trending magnetic fabric [Taylor and Hayes, 1980]. On the basis of limited seismic profiler and gravity data (Figure 6), Taylor and Hayes [1980] inferred the presence of a relict spreading center. A central rift is bounded by basement highs not unlike the rift walls of slow spreading, accreting plate boundaries. Some lateral extent to the bounding highs is indicated by their ability to dam sediments outside the rift axis. This rift has an associated free air gravity anomaly of amplitude 50 mGal and wavelength approximately 60 km on profile D (Figure 6). Similar gravity anomalies with wavelengths and amplitudes varying between 30 and 60 km and 15 and 50 mGal, respectively, characterize slow spreading centers, both active and extinct [Watts, 1982]. A gravity and seismic profile over a relict Eocene spreading center in the Coral Sea Basin [Weissel and Watts, 1979] is provided for comparison (Figure 6).

Six east-west gravity profiles collected by USS Archerfish (Figure 7) extend the documented presence of this gravity anomaly, and presumably the associated relict spreading center, southwest to the edge of the Sunda Shelf. The profiles at 9.5°, 11°, and 13°N have wavelengths and amplitudes of 30–60 km and 25–30 mGal, respectively. The inferred rift axis is quite linear, with the exception of a small transform offset near 12°N (Figure 7). In the basin area shoaler than 4 km the gravity low is at the center of a 90-km-wide trough bounded by structural highs. Unfortunately, magnetic data were not collected from the submarine, so that the age of this southwest subbasin remains unknown. The role of this spreading center in the total opening history of the South China Basin will be considered in the last section.

In 1965 a magnitude 5.9 earthquake occurred at 12.5°N, 114.5°E (Figure 7). Two magnitude 5.3 earthquakes occurred within 0.1° of this location, one in 1966 and 1969. Wang et al. [1979] determined the focal mechanism of the 1965 event to be almost pure thrusting at a very shallow depth (about 5 km). They noted that the event did not occur on a bathymetric feature and interpreted its mechanism to represent the overall state of stress in the South

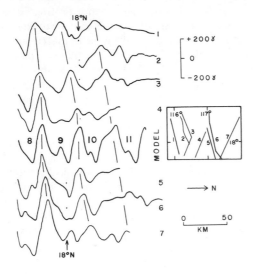

Fig. 3a

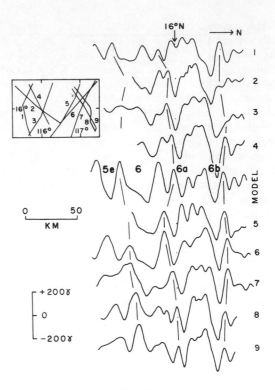

Fig. 3c

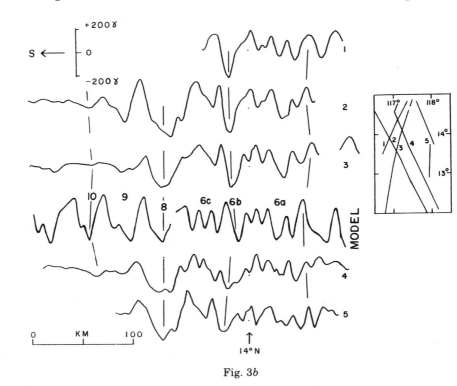

Fig. 3b

Fig. 3.   Detailed correlations of north-south projected magnetic anomalies observed in three areas of the South China Sea. The areas are located on Figure 3d. The model profiles in Figures 3a and 3c assume the same parameters as the model profiles in Figure 4 and a half spreading rate of 2.5 cm/yr. Geophysical data from the four tracks located on Figure 3d are shown in Figure 16.

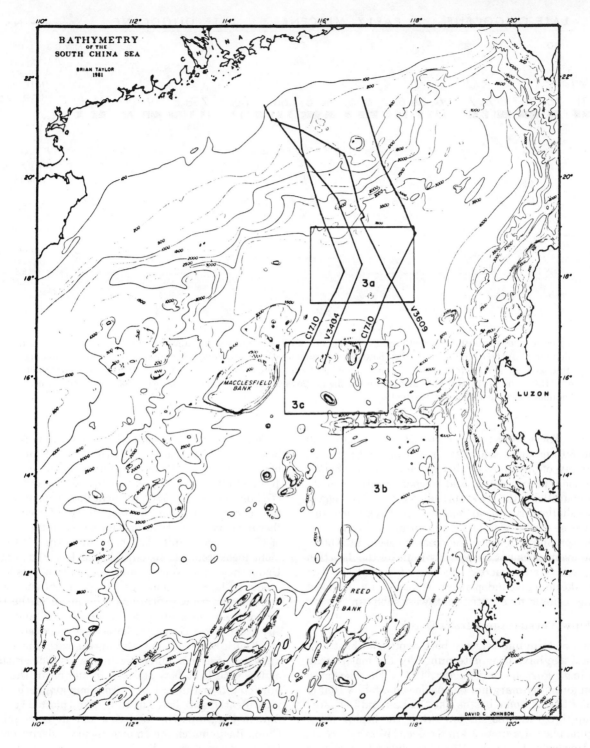

Fig. 3d

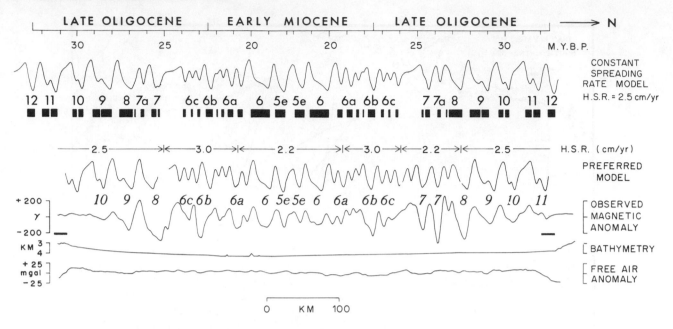

Fig. 4.   North-south projected profiles of magnetic anomaly, bathymetry, and free air gravity anomaly data along the solid track in Figure 2. Also shown are two computed magnetic anomaly profiles based on the geomagnetic time scale of *LaBrecque et al.* [1977], an assumed uniform magnetization of 0.007 emu, and a skewness parameter of −135°. The horizontal bars at each end of the geophysical profiles mark the positions of the continent-ocean boundary zone as defined in the text.

China Sea. These earthquakes occurred approximately 30 km and 60 km southeast of the rift walls and rift axis, respectively, of the inferred relict spreading center (Figure 7). The nodal planes of the focal mechanism strike northeast, subparallel to the relict spreading center. We suggest that this shallow crustal earthquake occurred along a fault plane whose strike was largely determined by the pre-existing spreading fabric of the oceanic basement. Therefore the event does not necessarily indicate the direction of compressive stress across the South China Basin. The earthquake does give independent evidence for a northeast trending tectonic zone in the southwest subbasin.

Heat Flow Versus Age Versus Depth

Empirical formulae relating both depth and heat flow to crustal age have been determined for the major ocean basins [e.g., *Parsons and Sclater*, 1977]. The age-depth relation for many marginal basins does not follow that of the major basins; some basins like the Lau and Bismarck being up to 0.5 km too shallow, while the basins of the Philippine plate are up to 1 km too deep [*Watanabe et al.*, 1977]. However, *Anderson* [1980] has shown that several marginal basins of the western Pacific follow the Pacific Basin heat flow versus age relation. Consequently, we may use the heat flow versus age relation to provide an independent test of our age determination for the South China Basin, though we may expect some deviation from the depth versus age relation.

Thirty five measurements of heat flow in the South China Sea are listed in Table 1. As the South China Basin is blanketed by over 1 km of sediment in the north and by 400 and 800 m of sediment in the center and southwest [*Taylor and Hayes,* 1980], the heat flow measurements are considered reliable and unlikely to be affected by hydrothermal circulation. Heat flow measured in the basin north of 17.5°N is very uniform. The mean of the 11 A environment measurements listed in Table 1 is 90 ± 8 mW m^{-2}. Using the empirical heat flow–age relation, this heat flow predicts a late Oligocene age, in agreement with our magnetic anomaly identifications. The mean of the three A environment measurements between 14° and 16°N is 107 ± 4 mW m^{-2}. The early Miocene age predicted for oceanic crust with this heat flow also agrees with our age determinations based on magnetic anomalies. The crustal ages at the nine heat flow stations in the basin that can be predicted to within ± 1 m.y. from our magnetic anomaly identifications are listed in Table 1 and plotted versus heat flow in Figure 8. The heat flow–age data for the South China Basin match the empirical curves derived from the Pacific Basin very well. The three stations 21 ± 1 m.y. average 107 ± 4 mW m^{-2} and the six stations 30.5 ± 1.2 m.y. average 87 ± 6 mW m^{-2}.

The scatter of the observed heat flow in the southwest portion of the basin, which is not dated by magnetics, is somewhat greater. The mean of the eight B environment measurements listed in Table 1 is 114 ± 20 mW m^{-2}. If the

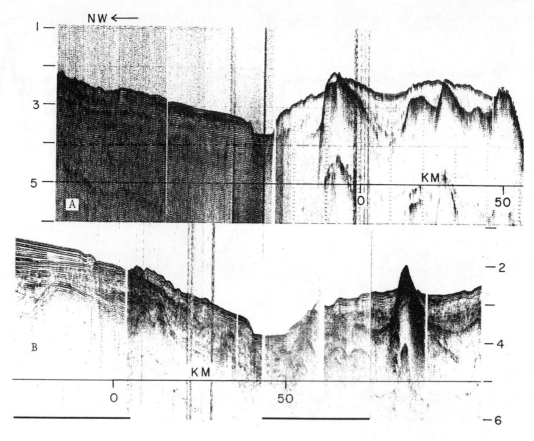

Fig. 5.   Single-channel seismic profiles across the east trending graben near 18°N (Figure 10). The horizontal bars locate sections of the lower seismic profile with horizontal scales expanded by a factor of 2 with respect to the rest of the profile. Vertical scales are in seconds of two-way travel time. The northwest half of the profile was recorded through a lower band-pass filter. We suggest that the graben represents a failed arm of the network of rifts along which the South China Basin opened in the mid-Oligocene. The offset of surficial sediments shows that some of the bounding faults are still active.

significantly higher value at station V28-192 is excluded, the mean is 108 ± 13 mW m^{-2}. These values are similar to those for the central portion of the basin and are consistent with an early Miocene age for this crust.

Water depths above oceanic crust in the South China Basin range from 4.3 to 3.7 km. Basement depths, corrected for sediment loading, range from 4.4 to 4.8 km. This is some 400 m deeper than the depths predicted (3940–4480 m) for oceanic crust of this age (17–32 m.y.) using the empirical relation of *Parsons and Sclater* [1977] ($d = 250 \pm 350t^{1/2}$ m). Like the basins of the Philippine plate, the South China Basin is deeper than the global age-depth curves would predict. This may result from cooling from a deeper ridge crest depth than the major ocean basins (see *Anderson* [1980] for a full discussion).

Seamounts

Numerous seamounts, 20 to 50 km wide at their base, rise 2 to 4 km from the abyssal floor of the South China Basin. Apparently, the location of some of these seamounts is controlled by the seafloor spreading fabric of the basin. As noted previously, the Scarborough seamounts were probably extruded at or very close to the relict spreading center. Three seamounts occur along the center of the southwest portion of the basin (Figures 2 and 10), and the seamount at 17.65°N, 117°E is near a fracture zone (Figure 2).

Prior to the 1979 *Vema* 36 cruise, none of these seamounts had been dredged successfully. We dredged three of the seamounts, seismic profiles across which are shown in Figure 9. The location of the dredges and chemical analyses and molecular norms of some of the rocks recovered are given in Table 2. The dredged rocks are generally manganese-coated pillow fragments, moderately weathered. They range from aphyric and highly vesicular (dredge 10) to phyric and sparsely vesicular. Basalts from dredge 9 have plagioclase phenocrysts (An_{70}) up to 5 mm across, while those from dredge 10 have very fine laths of plagio-

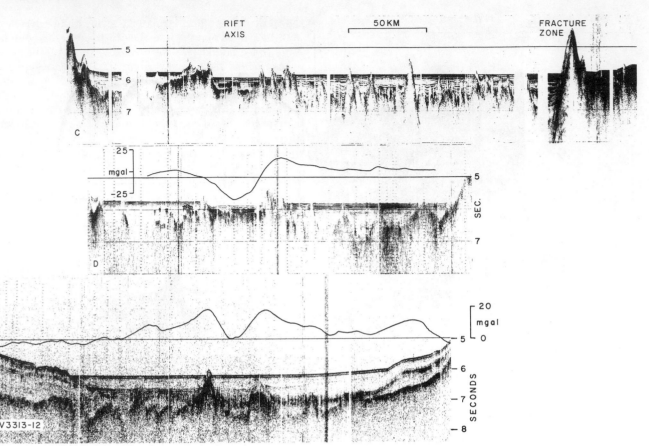

Fig. 6.   Free air gravity and seismic profiles over inferred relict spreading centers in the southwest South China Basin (profiles C and D, Figure 10) and the Coral Sea Basin (V3313-12; *Weissel ar d Watts* [1979]). The rift axes are marked by short-wavelength (50–60 km) gravity lows superimposed on broader gravity highs.

clase. A large pillow boulder recovered in dredge 8 has a glassy rind partially altered to orange brown secondary minerals.

The major element chemistries of these rocks (Table 2) identify them as alkalic basalt (dredge 10) and basalts transitional between tholeiitic and alkalic (dredges 8 and 9). Their petrochemistry is similar to that of other alkalic and transitional basalts dredged from seamounts formed at the spreading centers of major ocean basins [e.g., *Engel et al.*, 1965; *B tiza*, 1980]. Basalts similar to V36D8 have also been recovered from several back arc basins [e.g., *Hart et al.*, 1972; *Marsh et al.*, 1980]. However, basalts similar to V36D9 and V36D10 have rarely been reported from back arc basins; one of the few occurrences being an alkaline sill drilled at DSDP site 444 in the Shikoku Basin [*Marsh et al.*, 1980]. The Kinan seamounts were extruded at or near the spreading axis of the Shikoku Basin within a few million years of the cessation of spreading [*Kobayashi and Nakada*, 1979]. Their chemistry, however, is tholeiitic not alkalic [*Tokuyama and Fujioka*, 1976]. Likewise, rocks dredged from the many seamounts formed during

the complex spreading history of the Lau Basin are tholeiitic [*Hawkins*, 1976]. The apparent paucity of alkalic basalts in back arc basins may be the result of inadequate sampling, although the Lau Basin and Mariana Trough have been extensively sampled. Alternatively, it may reflect the slightly different petrogenesis of mid-ocean ridge and back arc basin basalts, the higher water contents of the latter [*Garcia et al.*, 1979; *Meunow et al.*, 1980] suppressing the formation of alkalic melts. The petrochemistry of the rocks dredged from the seamounts in the South China Sea is more similar to the petrochemistry of rocks recovered from seamounts extruded near the spreading centers of major ocean basins than those dredged from back arc basins.

Sediment Distribution and Basement Topography

Except where interrupted by seamounts, the South China Basin is floored by a topographically smooth abyssal rise and plain. To determine the basement topography beneath this relatively featureless bathymetry, we examined over 25,000 km of mainly single-channel seismic profiles col-

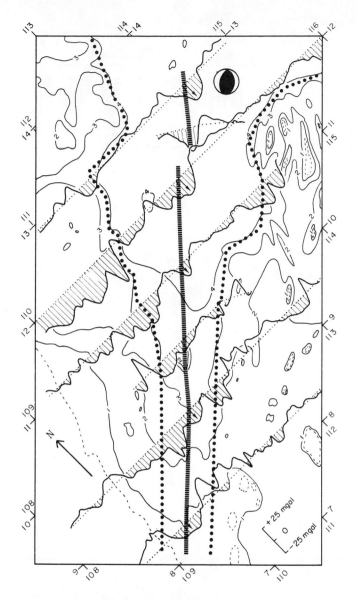

Fig. 7. Free air gravity anomalies along ships' tracks in the southwest South China Basin. The bathymetric base is contoured in kilometers, with the 200-m contour dashed. The inferred rift axis associated with the gravity low near the center of the basin is marked by the dashed line. The dotted line delineates the continent-ocean boundary zone defined in the text. Also shown is the focal mechanism determined by *Wang et al.* [1979] for a magnitude 5.9 earthquake which occurred at 12.5°N, 114.5°E in 1965.

the data density and structural style change across the Manila Trench and the inferred continent-ocean boundary zone, these boundaries were used to limit the basin isopachs. Where data permit, structural symbols depict the relative basement relief landward of these boundaries.

Numerous seismic profiles located on Figure 10 are reproduced in this paper (Figures 5, 6, 9, and 11–15) to illustrate the various basement structures in the South China Basin. Sections A and B (Figure 5) showing the faulted basement and failed rift near 18°N (Figure 10) have already been discussed. Several other graben, defined both bathymetrically and structurally, occur on the western and southern margins of the basin, some of which are depicted in Figure 10.

Sections C and D (Figure 6) illustrate the SW trending rift and the very rough basement in the southwest portion of the basin. The eastern end of profile C also crosses a major fracture zone near 116°E, which is the westernmost of a series of north trending fracture zones. The outline of these fracture zones, which bound the area where magnetic lineations are mapped (Figure 2), can be seen clearly in the isopach data of Figure 10. A section across another fracture system outlined by the isopach data is shown in profile H, Figure 11. This fracture system in the northeast part of the basin (Figures 2 and 10) comes within 50 km of the Manila Trench axis and is seen in cross section as two normal faults downdropping basement to the east. The original seismic records show that the throw on these faults has been increased following their reactivation to accomodate bending of the oceanic crust beneath the trench. Sediments are offset across the western fault nearly to the surface.

The volume of sediments delivered to the Manila Trench changes appreciably along strike; the thickness of the sediment infilling the trench axis varys from less than 0.2 s at 15.6°N to over 2.5 s at 18.5°N (Figure 10). The slope of the trench inner wall decreases to the north with the increase in width and volume of the underlying prism of accreted sediments (e.g., Figure 11). The West and North Luzon troughs are thickly sedimented forearc basins formed behind the accreted outer arc highs.

The thick sediments and relatively smooth basement in the South China Basin north of 17.5°N are in marked contrast to the thinner sediments and blocky basement character in the center of the basin. As seen in Figure 10 and on profiles H and I (Figure 11), the northern basin is filled with 1.2 to over 2.0 s of sediment; the central basin only 0.3 to 1.0 s. The zone of closely spaced reflectors at approximately 0.5-s subbottom on profile H can be mapped throughout the basin north and west of the 0.8-s contour (Figure 10). The time thickness of the sediments above this horizon varies from 0.7 s at the northern margin to 0.3 s farther south. The blocky basement fabric characteristic of the center of the basin generally has no preferred orientation. The highs are capped by acoustically transparent, presumably pelagic, sediments; all but the highest

lected in the area. Sediment thickness, measured in two-way travel time, was picked to 0.1 s every half hour (4 to 8 km) along track and was contoured at 0.4-s intervals to produce the sediment isopach map shown in Figure 10. The new and larger data set allowed significant improvement over the previous maps of *Mrozowski and Hayes* [1978], *Ludwig et al.* [1979], and *Taylor and Hayes* [1980]. As both

TABLE 1.   Heat Flow Measurements in the South China Sea

Station Number	Latitude N	Longitude E	Depth, m	Heat Flow, mW m^{-2}	Experiment Quality*	Environment*	Age,† m.y.
Northern Basin							
C14-37	17°58′	115°22′	3789	98	10	A	
C17-91	18°02′	114°40′	3631	79	10	A	
C17-87	17°42′	116°19′	3928	83	10	A	28.5
C17-88	18°53′	118°03′	3776	78	10	A	31.5
C17-92	18°15′	116°48′	3905	91	10	A	30.5
V36-07	18°03′	116°22′	3903	95	9	A	30
V36-08	18°14′	116°11′	3821	91	7	A	30.5
V36-10	18°36′	116°09′	3678	100	10	A	32
V28-189	18°33′	119°05′	4059	103	10	A	
C17-93	19°36′	119°42′	3680	90	10	A	
C17-96	20°19′	119°49′	3300		10	A	
Central Basin							
C17-89	16°59′	117°10′	3970	80	8	B	
C17-90	15°49′	116°37′	4168	107	10	A	20
C17-86	15°49′	115°21′	4196	111	10	A	21
TAS5-12	14°24′	116°00′	4182	103		A	22
SW Basin							
TAS5-13	13°10′	114°43′	4446	109		B	
TAS5-14	11°59′	113°29′	4329	104		B	
TAS5-15	11°05′	112°45′	4211	129		B	
TAS5-16	10°14′	111°51′	3918	118		B	
V28-192	12°53′	113°01′	4309	152	10	B	
V28-193	12°19′	115°49′	4390	98	8	B	
C17-84	11°03′	112°49′	4219	112	10	B	
C17-85	13°40′	114°26′	4322	88	10		
Margins							
C17-95	19°58′	117°59′	3070	76	10	A	
V28-190	20°10′	118°39′	2833	62	10	A	
V28-191	16°26′	113°33′	2403	119	8	C	
C12-180	7°33′	111°13′	1952	46	7	A	
C12-181	5°02′	113°35′	1225	52	7	B	
C17-83	6°21′	110°33′	1341	79	8	A	
Palawan Trough							
ANT13-176	6°35′	113°51′	2350	65		B	
ANT13-177	6°46′	114°15′	2863	56		A	
ANT13-178	6°58′	114°40′	2882	64		A	
ANT13-179	7°08′	115°07′	2844	58		B	
C12-182	6°00′	114°01′	2305	27	10	C	
C12-183	9°27′	114°33′	1476	14	8	C	

Data provided by M. G. Langseth, M. A. Hobart and R. N. Anderson (1980). The four Antipode 13 measurements were previously published by *Sclater et al.* [1976]. Most measurements were included in the compilation of heat flow, thermal conductivity, and thermal gradients in southeast Asia by *Anderson et al.* [1978].

*The classification of experiment quality and environment of the Lamont stations follows that of *Anderson et al.* [1977].

†The crustal age at the heat flow station is given when it is known to within ±1 m.y. based on our magnetic anomaly identifications.

are covered by the abyssal plain (e.g., profiles I and L, Figures 11 and 13). In the area 16°–17°N, 117°–119°E there is a broad basement arch with less than 0.4 s of sediment cover (Figure 10). Profiles M to Q (Figure 12) illustrate the fault-bounded tilted blocks which top this feature. The basement highs form east trending ridges (Figure 10), subparallel to the magnetic fabric, that have acted as a partial barrier to sediment transport from the north. The origin of this relatively high standing and faulted basement is not known. The magnetic anomalies and basement morphology do not support *Bowin et al.*'s [1978] speculation that this 'basement arch' is an extinct spreading ridge.

Figure 13 shows three V3605 single-channel seismic profiles across the southern portion of the basin. The large thickness of sediments adjacent to the Reed Bank margin was previously unrecognized owing to the shallow penetration of small-volume air gun sound sources (for example, see far right of profiles J and K). Most of these profiles were shot using two 466 in.^{3} air guns. The sediment thickness toward the southern ends of the profiles is greater than 2 s. The thickness increases to the east from ~2 s on profile J to over 2.4 s on profile L, probably as a result of sediment derived from the post middle Miocene uplift and tilting of the northern Palawan block. Typ-

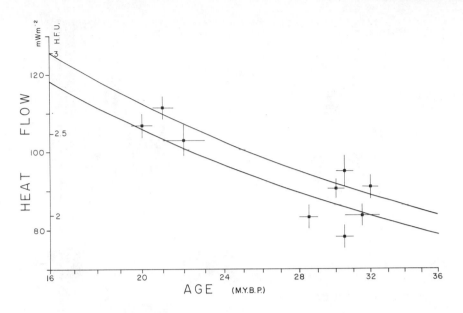

Fig. 8.   Plot of heat flow versus age for nine A environment heat flow stations in the central and northern South China Basin (Table 1). The solid curves show empirical heat flow versus age relations derived from data from the Pacific plate by *Parsons and Sclater* [1977], the lower curve, and *Davis and Lister* [1977], the upper curve.

ical oceanic basement is clearly visible only on the northern end of these profiles. To the south the oceanic basement defined by the refraction data is masked by a zone of closely spaced reflectors at 6.5 to 7.0 s below sea level (bsl) with velocities of 3.5 to 4.0 km/s. The sonobuoy data (Table 3) reveals the presence of over 2 km of sediments on the southern halves of profiles J and K (Figure 13).

The sediment fill in the frequency passband of the recordings (~15–60 Hz) is moderately stratified, the apparent stratification increasing to the west. The horizon at and just below 6 s on profile K can also be traced on profiles J and L. In the absence of drill hole data the stratigraphy of these and other horizons in the South China Basin remains unknown. Piston cores (≤12 m) raised from the basin consist of Quaternary radiolarian hemipelagic clays; the basin is presently below the carbonate compensation depth (CCD). The reflectivity contrast between the basal, high-velocity (3.5–3.9 km/s) sedimentary horizons seen on the profiles in Figure 13 and the overlying sections implies a corresponding lithological contrast. The highly reflective horizons just above oceanic basement may represent carbonate sediments deposited before the basin subsided through the CCD that were subsequently lithified through compaction.

Crustal Structure

It has been known for over a decade that the crust of the South China Basin is oceanic in structure [*Ludwig*, 1970; *Ludwig et al.*, 1979]. However, the spatial variation in crustal parameters (velocity, thickness) within the basin has remained largely undetermined. The new *Vema 36*

sonobuoy results, together with *Ludwig et al.*'s [1979] solutions (see Table 3), now provide quite reasonable control on the crustal structure of all but the basin center. Sonobuoy results from the center of the basin have been poor, generally because of the rough and blocky basement topography characteristic of the area.

The unreversed sonobuoy results from the South China Basin (Table 3) evidence crustal thicknesses varying between 4.7 and 6.1 km and depths to mantle ranging from 10.3 to 12.8 km. The upper ends of these ranges are typical of Pacific oceanic crust [e.g., *Woollard*, 1975]. Crustal thickness and velocity solutions may vary locally as much as they do over the whole basin (e.g., compare 8V36 and 9V36, Table 3). The abnormally large layer 2 and small layer 3 thicknesses determined by *Ludwig et al.* [1979] for sonobuoys 124C17 and 126C17 were not observed on our nearby sonobuoys 25V36 and 35V36, respectively. However, most of the crustal sections determined for the northwestern part of the South China Basin are thinner than those of the rest of the basin or of the Pacific. This does not appear to be an artifact of the reduction of the refraction data since multichannel seismic (MCS) reflection data also indicate a thinner crust in this area.

MCS line 2 (Figure 14), for example, shows oceanic basement at 6.9 to 7.1 s bsl and intermittent reflections from the Moho occur at 8.6 to 8.7 s bsl. The crustal structure determined from a digitally recorded sonobuoy (35V36) shot contemporaneously with this 40 km of MCS data is shown at the left of Figure 14. Both the time thickness of the oceanic crust observed on the profile (1.6–1.7 s) and the crustal thickness determined from the refraction data

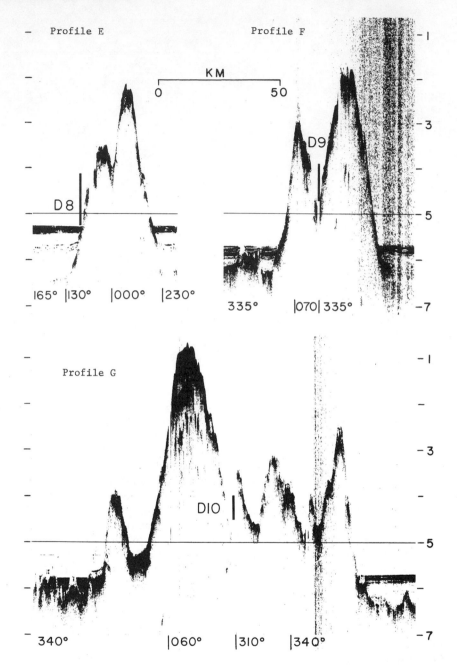

Fig. 9.   Single-channel seismic profiles E, F, and G (Figure 10) showing seamounts dredged during R/V *Vema* cruise 36-05. The vertical scales are in seconds of two-way travel time, and changes in the ship's course are labeled at the base of each profile. The vertical bars mark the maximum depth range of each dredge haul.

(5.3 km) are significantly less than crustal thicknesses (~2.0 s and 6–7 km) characteristic of the Pacific.

The Continent-Ocean Boundary Zone

The transition between the oceanic crust of the South China Basin and the continental crust of the Asian mainland occurs over a poorly defined zone, probably of varying width. On the China margin, as on other passive margins, there is an absence of adequate deep crustal structure information to characterize this transition zone. Nevertheless, there is considerable evidence to indicate that the now submerged margins surrounding the South China Basin to the north, west, and south were once continental areas contiguous with China and Indochina. The published stra-

TABLE 2. Major Element Composition of Basalts Dredged From South China Sea Seamounts

	V36D8	V36D9 Sample 1	V36D9 Sample 2	V36D10
Location	17°39'N, 117°00'E	14°59'N,	116°30'E	14°00'N 115°36'E
Depth range, fm	1640–2120	1560–1880		1580–1800
Chemical analyses*				
SiO_2	49.71	48.90	48.97	46.73
TiO_2	2.10	3.74	3.77	3.40
Al_2O_3	16.22	16.23	16.22	16.70
FeO^t	10.80	10.15	11.25	10.11
MnO	0.11	0.14	0.15	0.19
MgO	6.03	3.06	2.88	6.12
CaO	10.86	9.18	8.96	9.52
Na_2O	3.08	3.57	3.55	3.28
K_2O	0.51	2.11	2.12	2.19
P_2O_5	0.30	1.11	1.26	0.62
Cr_2O_3	.01	0.02	0.02	0.04
Total	99.73	98.21	99.15	98.90
Molecular norms†				
Orthoclase	3.02	12.48	12.53	12.95
Albite	26.06	30.21	30.04	18.07
Anorthite	28.94	22.03	22.07	24.38
Nepheline	—	—	—	5.25
Diopside	19.12	14.18	12.60	15.70
Hypersthene	8.45	1.89	4.03	—
Olivine	7.87	6.34	6.26	13.18
Magnetite	1.74	1.64	1.81	1.62
Ilmenite	3.99	7.10	7.16	6.46
Apatite	0.65	2.42	2.75	1.35
Chromite	0.01	0.03	0.03	0.06

*These whole rock microprobe analyses were done by E. Geary of Cornell University and represent the average of four analyses.
†Norms were calculated assuming 10% FeO^t expressed as Fe_2O_3.

tigraphy of the Sampaguita 1 well in the Reed Bank area (Figure 10) exhibits a continental provenance for the Paleogene and Cretaceous sections and provides strong evidence that this area is continental [*Taylor and Hayes*, 1980]. By analogy, the reefs and shoals southwest of Reed Bank, together with Macclesfield Bank and the shallow areas around the Paracel Islands (Figure 10), are inferred to be built on continental crust. Wells drilled by the Petroleum Company of the People's Republic of China have encountered Paleozoic and possibly pre-Cambrian basement beneath the Paracel Islands and Cretaceous granites and metavolcanics beneath the China shelf.

We use the term continent-ocean boundary (COB) zone to refer to and delineate the landward boundary of normal oceanic crust and not to refer to the total transition zone from continental to oceanic crust (see also Figures 2, 7, and 10). We have defined and mapped the COB zone on the basis of magnetic, gravity, and reflection seismic data. In areas where these data are not available, the COB zone has been interpolated solely on the basis of the position of the base of the continental slope. As noted previously, the northern and southern margins are characterized by magnetic quiet zones. An abrupt change from seafloor spread-

ing anomalies to longer wavelength, very low amplitude magnetic anomalies is especially evident along the northern margin of the basin (Figures 2 and 16). In contrast to some other Atlantic-type continental margins, this phenomenon can not plausibly be attributed to a geomagnetic period of no reversals or low field intensity preceding seafloor spreading in the basin. The change in magnetic anomaly character must be due to changes in the nature of the basement magnetization. We suggest that in the South China Sea these changes are related to the transition from oceanic to continental crust. Note, however, that this criterion is not used alone to locate the COB zone. As can be seen on Figures 2, 3, and 16, there are some oceanic areas of low-amplitude, long-wavelength magnetic anomalies, notably just seaward of the northern and southern COB zones between 117° and 118°E. While it is not surprising to observe a poorly organized spreading system in the initial stages of seafloor spreading, the exact processes by which contemporaneous magnetic lineations are well developed in one area and not in another are unclear.

The COB zone is also marked by a change in basement structure observed on seismic reflection records. Profile S in Figure 15 shows this change very well. The relatively

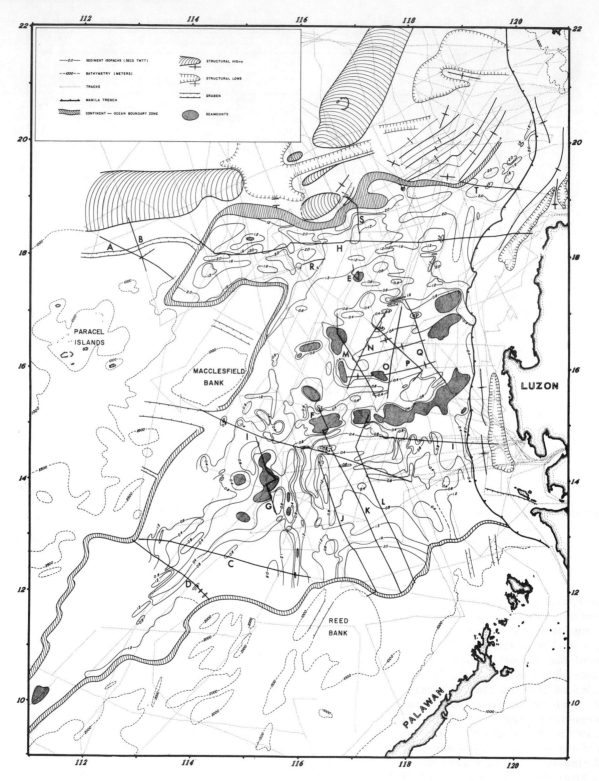

Fig. 10.   Sediment isopach map contoured at intervals of 0.4 s of two-way reflection time. Structural symbols depict the relative basement relief landward of the Manila Trench and the continent-ocean boundary zone. Seismic data along track segments labeled A to T are shown in Figures 5, 6, 9, and 11–15.

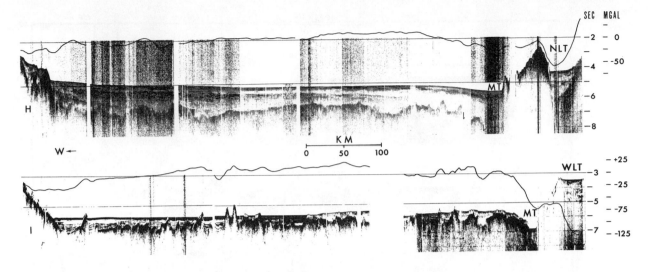

Fig. 11. East-west free air gravity and single-channel seismic profiles H and I across the northern and central South China Basin, respectively (Figure 10). MT, Manila Trench; WLT, West Luzon Trough; NLT, North Luzon Trough.

smooth basement near 6.5 s on the right of profile S is oceanic crust with associated seafloor spreading magnetic anomalies. The north dipping block faulted basement with a flat magnetic signature on the left of the profile is clearly different from the oceanic crust farther south. We place the COB zone between 1130 and 1230 on this profile. Note that the oceanic crust is down faulted to the north at 1230 and 1240. A graben structure associated with the COB zone is a common feature of the northern margin between 117° and 119°E. Another section across this feature, which is outlined by the 2.0-s contour in Figure 10, is shown on profile T (Figure 15b). Seismic penetration on profiles across the southern margin (Figure 13) is insufficient to determine whether similar structures are present. The basement scarp on the far right of profile L (Figure 13) may represent the COB zone.

The COB zone around the South China basin is also identified by a characteristic gravity anomaly. Stacked profiles of marine geophysical data from four crossings of the northern margin illustrate this feature (Figures 3d and 16). The profiles are aligned with respect to the COB zone inferred from magnetic and seismic reflection data alone. Near the transition from continental slope to abyssal plain, there is a minimum in the free air gravity anomaly profile [Taylor and Hayes, 1980]. A steep landward gradient in the free air gravity anomaly is observed just seaward of this minimum. For comparison with other 'Atlantic-type' continental margins, two-dimensional isostatic anomalies are also shown in Figure 16. The isostatic anomalies were calculated assuming local Airy compensation of the mass deficit associated with the deepening water layer solely by a corresponding shoaling of the crust/mantle interface [see Talwani and Eldholm, 1972]. The isostatic anomalies were computed using the method of

Karner and Watts [1982], assuming a water/crust density contrast of -1.67 g cm^{-3} and a distance between the mean surface and root topographies of 20 km. This is equivalent to a lower crust/upper mantle density contrast of 0.334 g cm^{-3} for a depth of compensation of 32 km and an oceanic Moho at 12 km.

The isostatic anomaly profiles are characterized by positive and negative anomalies over the oceanic and continental sides of the COB zone, respectively. The COB zone inferred on the basis of the magnetic and seismic reflection data is at the approximate midpoint of the steep landward gradient in the isostatic anomalies. If a regional compensation mechanism were used to calculate the isostatic anomalies, the anomalies would have even larger magnitudes and gradients. To minimize the effects of local differences in crustal structure and sediment distribution, the four bathymetric profiles and the four isostatic anomaly profiles were stacked and averaged using the COB zone as the horizontal reference point. The resulting mean bathymetry and isostatic anomaly profiles are shown at the top of Figure 16. The mean isostatic anomaly is approximately antisymmetric about the COB zone, has an amplitude of 30 mGal, a maximum gradient over the COB zone of 0.85 mGal km^{-1}, and a wavelength of about 400 km.

The isostatic anomalies indicate the presence of density contrasts across the COB zone other than those associated with the water/crust and the assumed lower crust/upper mantle boundaries. The persistence of the isostatic anomalies after stacking with respect to the COB zone suggests that these density contrasts are a characteristic of the COB zone independent of local differences in crustal structure and sediment distribution. The isostatic gradients are too steep to be caused solely by very deep struc-

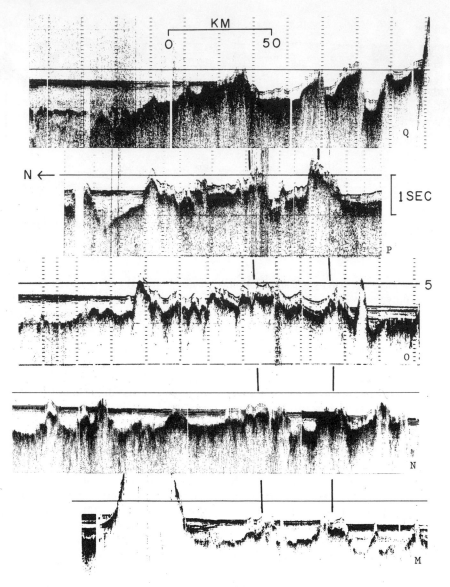

Fig. 12. Single-channel seismic profiles M through Q across the broad basement arch in the east-central South China Basin (Figure 10). The arch is capped by fault-bounded tilted blocks, some of which form east trending ridges and troughs.

ture, such as in the upper mantle. It seems reasonable to assume therefore that the mean isostatic anomaly is due, at least in part, to a density contrast across the COB zone at crustal depths. Furthermore, the mean isostatic anomaly very nearly sums to zero (Figure 16), implying that . the mass excess associated with this density contrast is regionally compensated at a scale of about 400 km. The magnitude of the isostatic anomaly requires that the crustal density contrast be of order $+0.1$ to 0.2 g cm^{-3} and that its compensation occur relatively deep in the mantle (compensation at Moho depths would necessitate unrealistically large density contrasts to preserve the amplitude of

the anomaly). Although the crustal structure and sediment distribution of the oceanic basin is relatively well known, that of the continental margin is not. Therefore we have not determined a particular density model to match the observed anomalies.

The inferred positive density contrast between oceanic and continental crust may plausibly be compensated by invoking a different thermal structure of oceanic versus continental lithosphere that gives rise to a corresponding negative density contrast at considerable depths. For example, 6 km of oceanic crust with an average positive density contrast of 0.1 g cm^{-3} could be balanced by 60 km

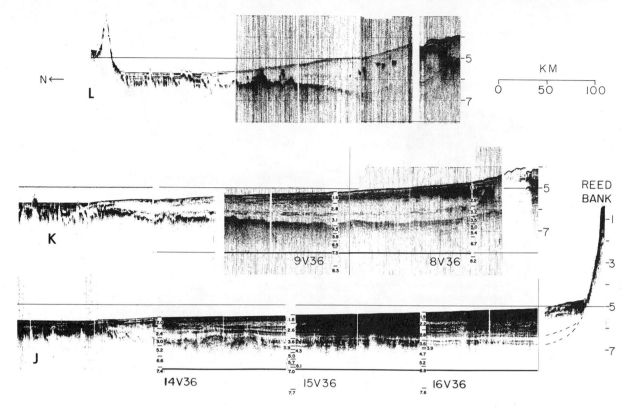

Fig. 13. Single-channel seismic profiles J, K, and L, showing the thick sedimentary section beneath the southern half of the South China Basin (Figure 10). Vertical scales are in seconds. Sonobuoy solutions (Table 3) are drawn at the beginning of the corresponding profile sections. Over the length of the sonobuoy experiments the horizontal scale of the seismic profiles is expanded by a factor of 2 with respect to the rest of the profiles.

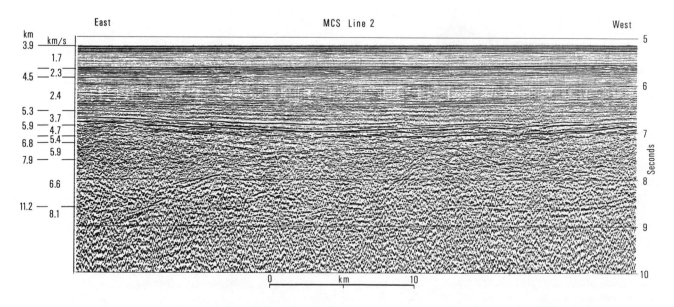

Fig. 14. Multichannel seismic line 2 (profile R, Figure 10) together with the velocity structure solution determined from the contemporaneously recorded sonobuoy 35V36 (Table 3). The seismic profiles show oceanic basement at 6.9 to 7.1 s and intermittent reflections from the Moho at 8.6 to 8.7 s.

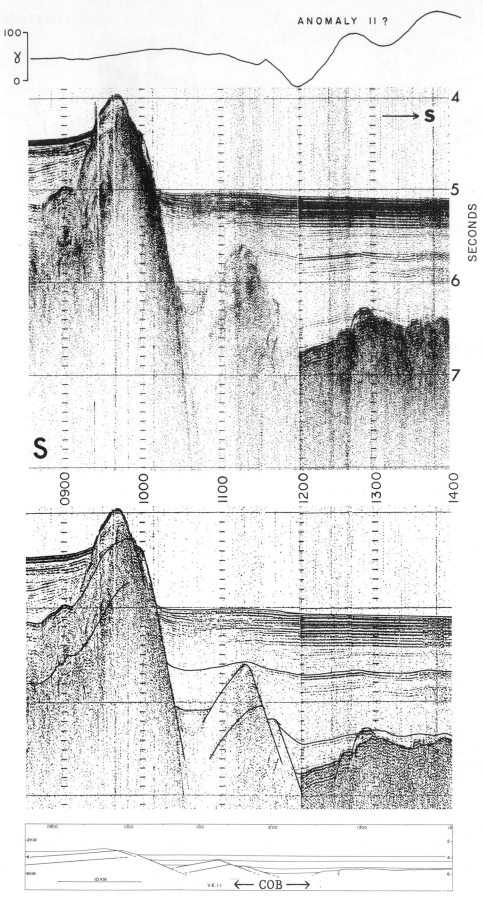

ANOMALY II ?

COB

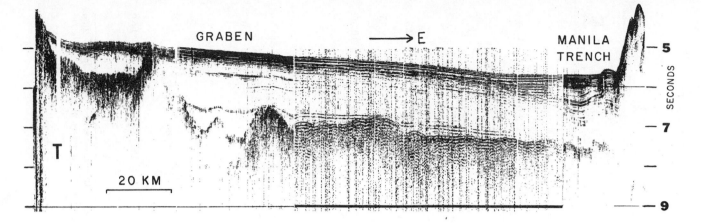

Fig. 15b. Seismic profile T illustrating the graben structure commonly associated with the continent-ocean boundary zone between 117° and 119°E (Figure 10). East of the graben, oceanic basement is observed dipping east from approximately 7 to 7.5 s. Folding of the turbidite wedge at the base of the landward wall of the Manila Trench gives evidence of crustal shortening across the trench. The horizontal scale along the section marked by the horizontal bar at its base is twice that of the rest of the profile.

of oceanic lithosphere with an average negative density contrast of 0.01 g cm^{-3}. Given an upper mantle density of 3.33 g cm^{-3} and a thermal expansion coefficient of 3×10^{-5} °C^{-1}, a density difference of 0.01 g cm^{-3} would correspond to an average temperature differential between the upper 60 km of oceanic versus continental lithosphere of only 100°C.

Many other Atlantic-type continental margins are characterized by a steep landward gradient in the isostatic anomaly, and this gradient has been inferred to mark the COB zone [see, e.g., *Talwani and Eldholm*, 1972; *Rabinowitz and LaBrecque*, 1977, 1979; *Kahle et al.*, 1981]. These margins are also characterized by basement highs or ridges on the oceanic side of the COB zone. *Rabinowitz and LaBrecque* [1979] proposed that the isostatic anomalies could be accounted for if the basement highs were uncompensated. The China COB zone has a similar isostatic anomaly to those observed at other passive margins, but there are no known basement ridges on the ocean side of the COB. We suggest that a positive density contrast between oceanic and continental crust may very likely prove a common characteristic of COB zones.

Fig. 15a. (Opposite) Magnetic anomaly and seismic profiles along track segment S (Figure 10) illustrating the change in magnetic signature and basement structure across the continent-ocean boundary (COB) zone. Our interpretation of the basement structure observed on the upper seismic profile is shown on the lower profile and the basal line drawing. See text for discussion.

Complexities in the Opening History of the South China Basin

As proposed by *Taylor and Hayes* [1980] and further supported by the stratigraphic compilations of *Fontaine and Mainguy* [1981] and *Holloway* [1981], the microcontinental blocks which separated from the proto-China margin by mid-Oligocene through early Miocene seafloor spreading in the South China Basin include northern Palawan, Reed Bank, and the other banks and shoals commonly known as Dangerous Ground. The southernmost blocks probably form the basement for the Miocene carbonate reefs of Luconia Shoals (Figure 17), which through prograding sedimentation have now been incorporated into the Sunda Shelf. Other large blocks such as the Parcel Islands and Macclesfield Bank were 'left behind' following their early but limited separation from the northern margin.

The precise reconstruction of northern Palawan, Reed Bank, and the other microcontinental blocks to their pre-rift configuration as part of the proto-China margin unfortunately is not tightly constrained by the known pattern of seafloor spreading in the South China Basin. The northwest and southeast boundaries of the oceanic crust can not be restored by any single, simple rotation (see Figure 17). Apparently substantial parts of the continental margins of the basin did not remain rigid during rifting and seafloor spreading but rather underwent considerable stretching and differential subsidence. Bathymetry and industry multichannel seismic data [e.g., *Mammerickx et al.*, 1976; *Holloway*, 1981] indicate that the continental margins were block faulted and distended in a complex manner.

The orientation of magnetic anomalies and the geometry of fracture zones in the South China Basin east of 115.5°E

TABLE 3. Crustal Structure Results From Sonobuoy Measurements in the South China Basin

Sonobuoy	Latitude °N	Longitude °W	Water	Sediments				Layer 2			Layer 3		Mantle	Σ Sediment	Σ L2	Σ L3	Moho
				S_1	S_2	S_3	S_4	$L2_1$	$L2_2$	$L2_3$	$L3_1$	$L3_2$					
45C14	17.89	115.24	3.75	0.46 / 2.02	0.83 / 2.63	0.71 / 2.65	0.27 / 3.10	1.37 / 5.55						2.3	1.4		
125C17	17.78	116.38	3.94	1.00 / 2.02	0.45 / 2.40			1.48 / 5.15	/ 5.9		/ 6.7			1.45			
126C17	18.13	116.51	3.92	0.36 / 1.73	0.31 / 2.07	0.48 / 2.44	0.67 / 2.49	1.44 / 5.0	1.54 / 6.2		(2.07) / 6.7		(8.1)	1.6	3.0	(2.1)	(10.6)
133C17	18.23	116.58	3.91	0.76 / 1.92	0.42 / 2.68	0.70 / 2.71		1.14 / 5.0	/ 5.55								
35V36	17.95	116.42	3.87	0.40 / 1.67	1.04 / 2.40	0.58 / 3.73		0.56 / 4.7	0.38 / 5.35	1.07 / 5.95	3.31 / 6.6		8.15	2.0	2.0	3.3	11.2
132C17	18.71	117.95	3.81	0.81 / 1.91	1.34 / 2.72	0.75 / 3.84		0.86 / 4.95	/ 5.75					2.9			
210V28	18.71	119.46	4.17	2.25 / 2.43				2.22 / 5.25			1.16 / 6.75	1.52 / 7.4	8.3	2.25	2.2	3.7	12.3
212V28	18.73	119.34	4.15	1.16 / 1.91	1.15 / 2.95	0.27 / 3.40		2.13 / 4.95			2.36 / 6.5	1.62 / 7.4	8.15	2.6	2.1	4.0	12.75
124C17	15.63	115.20	4.22	0.34 / 1.97	0.35 / 2.27	0.42 / 3.24		1.52 / 4.95	1.75 / 5.9		(1.73) / 7.1		(8.1)	1.1	3.3	(1.7)	(10.35)
24V36	15.90	115.54	4.20	0.99 / 2.04	0.19 / 3.0			1.84 / 4.6	/ 5.7					1.2	2.0		
25V36	16.19	115.48	4.15	1.07 / 2.2				0.72 / 4.6	1.24 / 5.8		/ 6.95			1.1	2.0		
8V36	12.34	117.73	3.51	0.42 / 1.5	0.83 / 2.9	0.75 / 3.2	0.59 / 3.46	0.59 / 5.01	0.98 / 5.42		3.1 / 6.7	/ 7.4	8.2	2.6	1.6	3.1	10.7
9V36	12.74	117.51	3.88	0.43 / 1.6	0.67 / 2.8	0.87 / 3.1		0.82 / 4.5	1.02 / 5.9		1.0 / 6.5	3.1 / 7.25	8.3	2.0	1.8	4.1	12.0
14V36	13.08	116.80	4.17	0.53 / 2.0	0.49 / 2.4	0.55 / 3.0		1.24 / 5.15			1.65 / 6.6	/ 7.4		1.6	1.2		
15V36	12.80	116.96	4.10	0.34 / 1.8	0.89 / 2.6	0.59 / 3.6	0.30 / 3.9	0.32 / 4.3	0.75 / 5.0	0.74 / 5.7	3.6 / 7.0	/ 7.7		2.1	2.1	>3.6	>11.9
16V36	12.55	117.13	4.00	0.31 / 1.85	0.49 / 2.2	0.69 / 2.8	0.66 / 3.7	1.12 / 4.65	0.84 / 5.15		3.45 / 6.85	/ 7.6		2.15	2.0	>3.5	>11.6
19V36	12.68	116.26	4.31	0.84 / 2.12	0.24 / 3.6			0.41 / 4.8	1.28 / 5.35	1.41 / 5.95	/ 7.2			1.1	3.1		

V36 sonobuoys are new solutions. The rest were published by *Ludwig et al.* [1979]. For each sonobuoy the upper line gives thicknesses in km, the lower line, velocities in km/s. Thickness values in parentheses were determined by *Ludwig et al.* [1979] from the range and reflection time to the onset of critical mantle reflections, assuming a mantle velocity of 8.1 km/s.

how the 700 km of north-south opening east of 116°E can be accommodated farther west where the basin narrows and trends southwest. The recognition of a southwest trending relict spreading center also raises questions regarding the relationship between the eastern and southwestern spreading systems.

The fracture zones near 116°E and the transform margin off Vietnam near 110°E inferred herein and also by *Ben-Avraham and Uyeda* [1973] (also see Figure 17) suggest that the overall extension in the western half of the South China Sea has also been in a north-south direction. Within this region the rift at 18°N trends east, other graben and fault blocks trend both WNW and NNE, and the SW subbasin and its central rift trend northeast. Little more can be inferred given the sparse marine geophysical data coverage. The age and rate and direction of seafloor spreading in the southwest subbasin are not known owing to the lack of magnetic anomaly data. Plausible spreading directions range between NW-SE (orthogonal to the inferred relict spreading center) and N-S (parallel to the spreading direction east of 116°E). Heat flow data provide an important constraint on the age of spreading. As discussed previously, the mean of seven heat flow measurements is 108 ± 13 mW m^{-2}, consistent with an early Miocene age for this crust. The heat flow data suggest that there is no large age difference between the eastern and southwestern spreading systems.

The width of the SW subbasin progressively narrows to the southwest (Figures 1 and 17). A possible explanation of this narrowing, given the linearity of the inferred relict spreading center (Figure 7), is that spreading propagated southwestward. Whatever the direction of opening, the southwestern 300 km of the SW subbasin, being less than 100 km wide (Figures 7 and 17), can account for only 10–20% of the north-south extension observed east of 116°E. To accommodate the rest of the extension implied by a distant pole of opening for the basin would require up to 100% net north-south extension of the western continental margins. Whether the network of rifts, graben, and faults on these margins can account for so much extension is not known. In the absence of clear data to the contrary, we suggest that the timing of such crustal extension, to whatever extent it did occur, was essentially synchronous with seafloor spreading in the east [cf. *Holloway*, 1981].

The Tertiary Evolution of the South China Basin

We now develop a hypothetical evolutionary scenario for the area shown in Figure 17 based on our interpretation of the geological history of the South China Basin and surrounding land areas. We seek to provide the simplest and most plausible model of the tectonic evolution of the area, fully aware that there may be alternative explanations and that several aspects are necessarily speculative. Many details of the geology of the land areas are not known or not published. Of the published literature, we have relied heavily on geological syntheses by *Allen and Stephens*

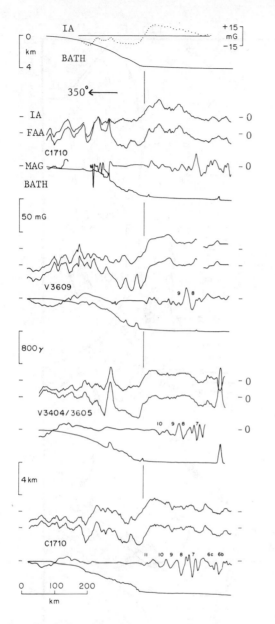

Fig. 16.   Stacked profiles of marine geophysical data from four crossings of the China margin located on Figure 3d. Projected bathymetry, magnetic anomaly, free air gravity anomaly, and isostatic gravity anomaly profiles are shown for each track. The four sets of profiles are aligned with respect to the continent-ocean boundary zone inferred from magnetic and seismic reflection data. The mean of the four bathymetry and the four isostatic anomaly profiles if plotted at the top of the figure.

(Figures 2 and 17) indicates that seafloor spreading directions in this part of the basin were essentially north-south and require a quite distant pole to describe the opening. *Taylor and Hayes* [1980] previously noted the implicit geometrical problems in reconstructing the prerift geometry of the proto-China margin. In particular, it is not clear

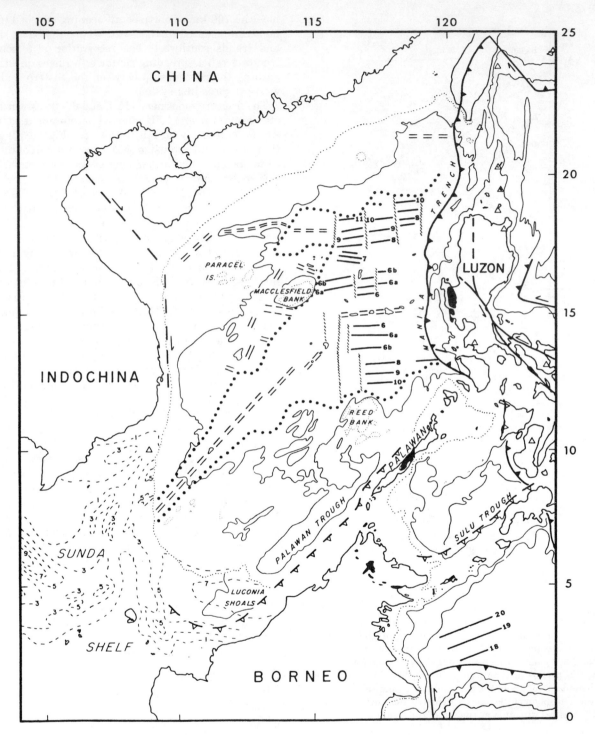

Fig. 17.   Summary of inferred tectonic elements in the South China Sea region on a bathymetric base showing the 200-m (dotted) and 2- and 4-km contours. Dashed and solid barbed lines represent inactive and active subduction zones, respectively. Dashed parallel lines locate relict rifts. Magnetic lineations identified in the South China Basin (this paper) and the Celebes Basin [*Weissel*, 1980] are numbered. Dashed contours depict generalized isopachs of sediment thickness (in kilometers) beneath the Sunda Shelf [*Hamilton*, 1979]. The line of large dots bounds normal oceanic crust of the South China Basin. Triangles locate historically active volcanoes, and ophiolite complexes are shown as shaded areas.

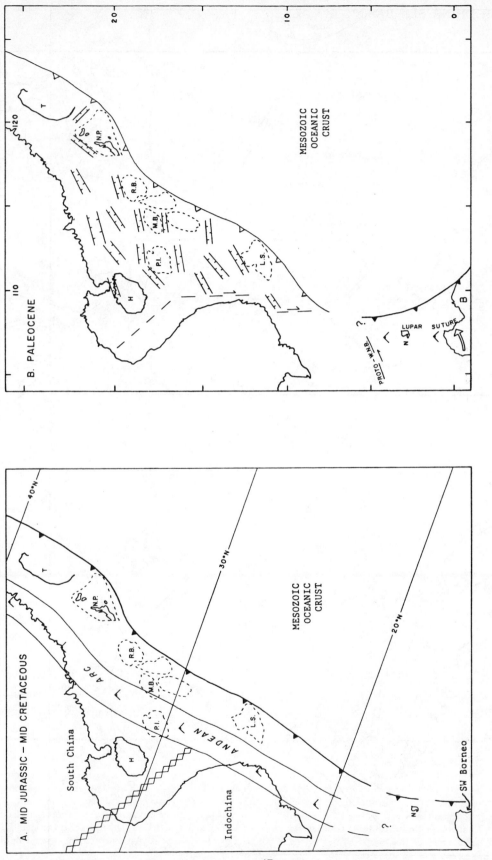

Fig. 18. Hypothetical reconstructions illustrating our evolutionary scenario for the development of the South China Sea region. The basis for each reconstruction is discussed in the text.

47

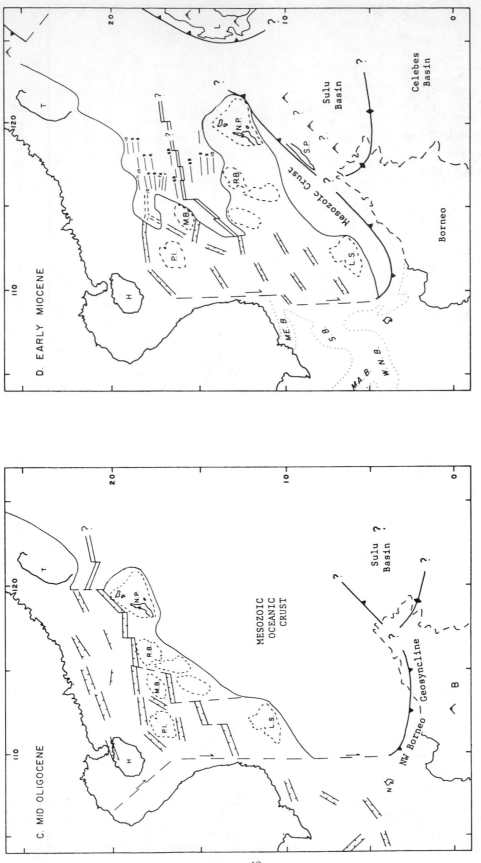

Fig. 18. (continued)

LEGEND

TECTONIC ELEMENTS

⋀	VOLCANO
⊢	SUBDUCTION ZONE
⊢•⊣	SUBDUCTION ZONE OF UNCERTAIN POLARITY
⊤	INACTIVE SUBDUCTION ZONE
◇◇◇◇	SUTURE ZONE
—∣—	STRIKE-SLIP FAULT
—⊥—	NORMAL FAULT
═══	SPREADING CENTER
= = =	RELICT SPREADING CENTER
··········	SEDIMENTARY BASIN OUTLINE
———	MODERN COASTLINE
———	CONTINENT-OCEAN BOUNDARY
- - - -	MODERN BANK OR SHELF
⟹	PLATE MOTION
—•—	MAGNETIC LINEATION
-30°N	PALEO-LATITUDE

PHYSIOGRAPHIC FEATURES

P.I.	PARACEL ISLANDS
M.B.	MACCLESFIELD BANK
R.B.	REED BANK
L.S	LUCONIA SHOALS
T	TAIWAN
H	HAINAN
L	LUZON
N.P.	NORTH PALAWAN
S.P.	SOUTH PALAWAN
B	BORNEO
N.	NATUNA
MA. B.	MALAY BASIN
W N B	WEST NATUNA BASIN
S.B.	SAIGON BASIN
ME. B.	MEKONG BASIN

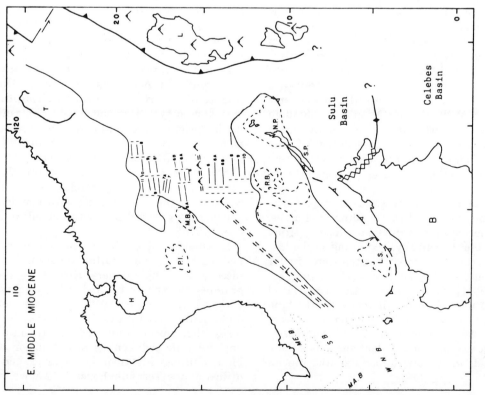

Fig. 18. (continued)

49

[1971], *Jahn et al.* [1976], *Hamilton* [1979], and *Holloway* [1981].

Unfortunately, only a qualitative reconstruction of the prerift configuration of the microcontinental blocks which separated from the proto-China margin can be made. Their relative longitudinal position is reasonably well constrained by the north-south direction of opening of the South China Basin. However, their latitudinal position relative to one another and to the China mainland is poorly constrained, being dependent on the unknown amount of crustal stretching involved in the breakup of the margin.

On the basis of regional correlations between the basement complexes of northern Palawan and southwest Mindoro, *Hamilton* [1979], *Taylor and Hayes* [1980], *Holloway* [1981], and others have suggested that southwest Mindoro is one of the microcontinental blocks which once formed part of the Asian mainland. The Mindoro basement was considered pre-Jurassic, since it is unconformably overlain by an ammonite-bearing clastic sequence dated as Callovian to Oxfordian [*Hashimoto and Sato,* 1968]. However, recent work has shown that many of the fossils in this sequence were reworked in the Eocene (D. E. Karig, personal communication, 1981). At present, there remains little basis for the regional correlations with northern Palawan, and hence we no longer infer a mainland origin for southwest Mindoro.

We will assume that sufficient crustal extension occurred in the western half of the South China Basin to match the opening by seafloor spreading observed in the east. The age of seafloor spreading based on our magnetic anomaly identifications in the eastern half of the basin is late Oligocene through early Miocene (32–17 m.y.). We will further assume that crustal extension in the western half of the basin occurred contemporaneously with seafloor spreading in the east and that the SW subbasin opened about a spreading center which propagated southwestward from the eastern basin to the Sunda Shelf in the early Miocene. Subduction at the Borneo, Manila, and Ryukyu trenches has removed nearly all evidence of the ocean basins formerly east and south of the present South China Basin, so that we can only speculate whether the latter's opening was related to other spreading systems, such as in the Sea of Japan.

Figures 18a–18e summarize our evolutionary scenario for the South China Basin and surrounding land areas. The reconstructions maintain China in a fixed position and are presented chronologically, beginning in the Mesozoic. Several aspects of our reconstructions are similar to those recently published by *Holloway* [1981]: first, because *Holloway* [1981] provides the only documentation of much of the stratigraphy of the margins of the South China Basin; second, because Holloway's reconstructions were based, in part, on data presented by *Taylor and Hayes* [1980]; and last, because our independent analysis of these and other data leads inevitably to many similar conclusions.

Mid-Jurassic–Middle Cretaceous (Figure 18a)

Following the collision of Sundaland (including Indochina) and south China along the Red River and Black River lines in the latest Triassic [*Hutchinson,* 1975], the areas west and north of what was to become the present-day South China Basin formed a continuous landmass, with sedimentation being almost exclusively nonmarine [*Fontaine and Workman,* 1978]. A hypothetical reconstruction of the southeast China margin in the Jurassic is shown in Figure 18a. The proto-Philippines lay southeast of the map area and do not enter into the picture until the mid-Tertiary. The reconstruction of the numerous pre-Tertiary complexes of East Borneo, Sabah (northeast Borneo), and the eastern Philippines is beyond both the scope of this paper and the existing data base. We speculate that these complexes probably represent former intraoceanic terrains, including island arcs.

Several independent reconnaissance paleomagnetic studies of Jurassic and Cretaceous red beds from China and Indochina consistently suggest that this region has moved 4° to 19° south and rotated about 20° clockwise since the late Mesozoic [*Lee et al.,* 1963; *Haile and Tarling,* 1975]. In contrast, Cretaceous paleomagnetic poles for the Malay Peninsula and southwest Borneo and Jurassic to Early Cretaceous and Paleogene poles for southwest Sulawesi all indicate a counterclockwise rotation of these areas by 45° to 50°, with Malaya moving south about 15° and southwest Borneo and Sulawesi remaining within 5° of their present latitudes [*McElhinny et al.,* 1974; *Haile et al.,* 1977; *Haile,* 1978; *Sasajima et al.,* 1980]. As the reconstructions shown in Figures 18a–18c maintain China in a fixed position, southwest Borneo has been rotated 70° clockwise in Figure 18a from its present position.

From mid-Jurassic through mid-Cretaceous times the proto-southeast Asian margin was an Andean-type arc (Figure 18a). Northwest subduction beneath the continent is evidenced by widespread rhyolitic volcanism and granitic intrusions along southeast China and southeast Vietnam [*Allen and Stephens,* 1971; *Jahn et al.,* 1976; *Hamilton,* 1979]. K-Ar and Rb-Sr dates on these rocks evidence two major thermal episodes within this Yenshanian orogenic belt: 140 to 170 m.y. (Middle and Late Jurassic) and 85–120 m.y. (mid-Cretaceous), with a predominance of the younger ages toward the southeast side of the belt [*Jahn et al.,* 1976].

It is questionable whether the Andean-type arc continued south of Vietnam during the Jurassic and Early Cretaceous. Two single-mineral K-Ar dates from SW Borneo granites give ages of 127 and 154 m.y. [*Haile et al.,* 1977]. However, *Hutchinson* [1973] comments that from Middle Jurassic to Early Cretaceous times Sundaland is characterized by igneous quiescence, continental molasse-type and fluviatile sedimentation, and major strike slip and block faulting. Metamorphosed lavas and amphibolites drilled at the Terubuk 1 and AB-1X wells northwest of

Natuna give radiometric ages of 169 and 171 m.y., but these are only minimum ages and the actual volcanic event is probably Triassic (Indosinian) [*Pupilli*, 1973]. The Bunguran beds and associated mafic rocks on Natuna indicate the presence of melange older than the 73-m.y. granites which intrude them [*Haile and Bignell*, 1971], but it is questionable whether the melange is older than Late Cretaceous. During the Late Jurassic and Early Cretaceous the eastern edge of Sundaland may have been predominantly a strike slip boundary linking the Indochina subduction zone with another convergent margin stretching from southeast Borneo (Meratus Mountains) to western Java [*Holloway*, 1981].

The inferred forearc position for North Palawan and Reed Bank (Figure 18a) is consistent with their known Mesozoic stratigraphy. Jurassic and Cretaceous rocks are not present onshore northern Palawan, though they have been drilled offshore to the east and west [*Holloway*, 1981]. On the east, ultrabasic rocks were encountered beneath poorly sorted quartzose sandstones interbedded with clays and silts of probably Late Cretaceous age, while on the west, Upper Jurassic to Lower Cretaceous shales and carbonates are present. On Reed Bank the published stratigraphy of the Sampaguita 1 well includes a thick sequence of marginal marine clastics of Lower Cretaceous and possibly Jurassic age [*Taylor and Hayes*, 1980].

The significance of the Mesozoic Andean arc in the origin of the South China Basin is that it appears to have marked the approximate locus for the subsequent rifting of the China margin. The second major thermal episode of the Yenshan orogeny continued from about 120 to 85 m.y. [*Jahn et al.*, 1976], with one of the latest dated intrusions being an 86-m.y. quartz diorite from Taiwan [*Yen and Rosenblum*, 1963]. The position of this arc may have been somewhat seaward of that shown in Figure 18a, that is, largely under the present China shelf. Unlike the uncertainty in the southern extent of the Jurassic arc, the mid-Cretaceous arc certainly continued south of Vietnam along the Natuna arch and SW Borneo. Fourteen granitic samples from SW Borneo gave K-Ar ages between 76 and 115 m.y. [*Haile et al.*, 1977], and numerous dates between 70 and 120 m.y. have been determined for granitic rocks, both outcropping and penetrated in wells, on the Natuna arch [e.g., *Haile and Bignell*, 1971; *Pupilli*, 1973].

Paleocene (Figure 18b)

Volcanism along the Natuna arch and SW Borneo continued through the Late Cretaceous and into the earliest Paleocene. *Gueniot et al.* [1976] quote Rb-Sr and K-Ar whole rock dates as young as 64 m.y. The southeast China volcanism had ceased 20 m.y. earlier. Continued subduction along the Lupar line of SW Borneo was presumably the result of Sundaland beginning its 60° to 70° counterclockwise rotation with respect to Indochina at approximately 75 to 80 m.y. This initiated the 'northwest Borneo geosyncline' and the accretion of thick abyssal clastics on

the subducting Mesozoic crust against the Lupar suture [*Haile*, 1969; *Hamilton*, 1979].

With the exception of the Borneo subduction complex and thin continental clastics in a few areas such as Hong Kong [*Allen and Stephens*, 1971], South China, Indochina, and Sundaland were characterized by a regional unconformity in the latest Cretaceous and early Paleocene. Presumably the Cretaceous arcs were being rapidly eroded, feeding voluminous turbidities into the NW Borneo trench. *Holloway* [1981] interprets this regional unconformity to represent the onset of rifting and the inception of tensional block faulting along the South China margin. We had previously inferred from the Sampaguita 1 well on Reed Bank that the rift onset unconformity was mid-Eocene [*Taylor and Hayes*, 1980]. However, the large amount of additional well and seismic data presented by *Holloway* [1981] for China, Taiwan, Mindoro, and, particularly, Palawan, indicates that the mid-Eocene unconformity associated with block faulting in the Reed Bank area was not regional and that the onset of rifting is most likely latest Cretaceous or Paleocene (65 ± 10 m.y.). This does not conflict with the stratigraphy of Sampaguita 1, which has the entire Upper Cretaceous section missing. Further support for this timing comes from the age of intrusive events on the China margin. Dolerite dykes intruded at Hong Kong gaves ages of 63 and 76 m.y. [*Allen and Stephens*, 1971], and porphyrites in the Penghu Islands of Taiwan Strait are dated as 56 m.y. [*Jahn et al.*, 1976].

The Paleocene reconstruction shown in Figure 18b depicts the South China and east Vietnam margins as a zone of widespread rifting bounded by the Mesozoic trench and oceanic crust to the southeast. The rotation of SW Borneo may have been facilitated in part by strike slip faulting in the area which subsequently became the West Natuna Basin. The Paus-Ranai ridge just east of Natuna is composed by phyllitic rocks [*Pupilli*, 1973] and probably formed the trench slope break (outer arc high). The Paleocene and subsequent reconstructions place China and Indochina in their present latitude and orientation, as suggested by the limited paleomagnetic data available [*Jarrard and Sasajima*, 1980].

Mid-Oligocene (Figure 18c)

Global plate reorganizations occurred in the middle to late Eocene, including the collision of greater India with Asia and the change in Pacific plate motion with respect to the Hawaiian-Emperor hot spot. Strike slip and block faulting of the Sunda Shelf at this time resulted in the development of the Thai, Malay, and West Natuna basins, into which continental clastics were deposited [*Du Bois*, 1981]. This may have been associated with the beginning of Himalayan events and/or with the continued counterclockwise rotation of Malay-Borneo-SW Sulawesi. Paleogene subduction of the Mesozoic oceanic crust beneath NW Borneo accreted the voluminous Crocker-Rajang subduction complexes [*Hamilton*, 1979]. This convergent terrain is unusual for its almost complete lack of associated Paleogene plutonic and volcanic rocks.

Rifting of the South China margin continued through the Eocene and early Oligocene. Sedimentation was nonmarine (graben and half-graben fill) with the exception of some carbonate platforms developed on outer margin areas such as north Palawan [*Holloway*, 1981]. The mid-Oligocene reconstruction (Figure 18c) shows the area at the time of breakup, just prior to the formation of magnetic anomaly 11. A quartz diorite intrusion in central Taiwan is dated as occurring at this time (33 m.y.; *Yen and Rosenblum* [1963]). Most of the margin was subaerial, as evidenced by unconformities encountered in wells on the China shelf, Taiwan, Reed Bank, and north Palawan and by a prominent unconformity recognizable on multichannel seismic lines across the south China and northwest Palawan shelves [*Taylor and Hayes*, 1980; *Holloway*, 1981]. We interpret this unconformity as due to the superposition of breakup and sealevel effects.

In the mid-Oligocene, Luzon was still southeast of our map area. Recent paleomagnetic work suggests that the Zambales ophiolite on west-central Luzon has moved 20° to 30° north and rotated 75° counterclockwise since its formation in the late Eocene [*Fuller et al.*, this volume]. Whether similar movements characterize all of Luzon and the rest of the Philippines remains unknown. Uplift of the western edge of the Zambales ophiolite in the middle to late Oligocene may date the initiation of southeastward subduction beneath Luzon at the Manila Trench [*Schweller and Karig*, 1982].

The northeast Borneo (Sabah) suture zone (Figures 18c–18d) resulted from a collision of oceanic crust and accreted terrains with Borneo in the mid-Oligocene to mid-Miocene [*Bell and Jessop*, 1974; *Hutchinson*, 1975; *Hamilton*, 1979]. The Sabah complex records a long history of subduction involving materials as old as Jurassic. The geology of this area is poorly understood. Northwest Borneo, southern Palawan, and Sabah all evidence similar Late Cretaceous-Paleogene melange and broken formations, but their relation to each other and the evolution of the inferred triple junction between them is still speculative. It has been suggested by several authors, on the basis of tectonic position and the empirical heat flow-depth-age relations, that the Sulu Basin northeast of Sabah is of a similar age to the South China Basin. However, there are conflicting interpretations regarding the timing and direction of subduction beneath the arcs surrounding the Sulu Basin [*Hutchinson*, 1975; *Hamilton*, 1979; *Weissel*, 1980; *Holloway*, 1981]. Consequently, the original extent of the Sulu Basin and its relation to other basins is unclear.

Early Miocene (Figure 18d)

By early Miocene times, seafloor spreading in the eastern South China Basin had created a basin over two thirds its final size. The reconstruction in Figure 18d shows the area at anomaly 6A time, approximately the end of the Aquitanian. Spreading north of Macclesfield Bank had now jumped to the south. The westernmost South China Sea was experiencing widespread extension that was bounded by the Vietnam transform to the west. Sedimentation on what is the present China shelf was of dominantly nonmarine clastics, whereas the relatively isolated blocks to the south such as Reed Bank and north Palawan developed carbonate platforms [*Taylor and Hayes*, 1980; *Holloway*, 1981]. Fault-controlled subsidence led to the accumulation of continental clastics in the Thai, Malay, West Natuna, Saigon, and Mekong basins on the Sunda Shelf [*White and Wing*, 1978; *Du Bois*, 1981]. Northeast trending subbasins within the eastern Saigon Basin may have developed as en echelon pull-apart basins forming part of the north-south dextral shear system on the edge of the Sunda Shelf.

The counterclockwise rotation of Borneo was probably complete by the early Miocene. Little remained of the Mesozoic oceanic crust. Subduction at the Borneo-Palawan Trench continued, to accommodate the opening of the South China Basin. However, little sediment was being fed into the trench, so that major accretion of Crocker-type sediments came to an end. There are no known associated volcanic or plutonic rocks of early Miocene age. It has been suggested that the Cagayan Ridge southeast of Palawan may have been a volcanic arc at this time [*Hamilton*, 1979]. The only volcanics which outcrop are Plio-Pleistocene (on Cagayan de Sulu Island; *Bell and Jessop* [1974]. Nevertheless, the existence of a lower Miocene island arc seems necessary in order to bound the sedimentary basins southeast of Palawan, which have been interpreted as outer arc basins [*Bell and Jessop*, 1974; *Beddoes*, 1976; *Hamilton*, 1979]. The ophiolites on southern Palawan were part of the outer arc high. The north Palawan block probably began to collide with the northeastern continuation of this arc-trench system in the early Miocene (Figure 18d). At the same time, Luzon was rotating counterclockwise and advancing westward, forming the proto-Manila Trench. Luzon now constituted a single entity, with sedimentary basins developing over the sutures between older component island arcs [*Mineral Fuels Division*, 1976]. The oldest dates of intrusions related to the current period of subduction along the Ryukyu arc northeast of Taiwan are 21 m.y. [*Bowin and Reynolds*, 1975]. Subduction may have begun in the earliest Miocene, with a transform bounding the trench on the southwest (Figure 18d).

Middle Miocene (Figure 18e)

Shortly after the oceanic rift is inferred to have propagated all the way to the southwest of the South China Basin (Figure 18c), seafloor spreading ceased. The youngest identifiable magnetic anomaly is anomaly 5E, though the youngest crust may correspond to anomaly 5D (see previous discussion). Thus by approximately 17 m.y. (latest early Miocene) spreading had stopped. At or after this time, seamounts were extruded along several sections of the relict spreading center, particularly in the east.

The collision of the Reed Bank–Palawan blocks with the arc-trench system to the south was virtually complete by the middle Miocene. Early Miocene carbonate deposition offshore Palawan was smothered by a clastic influx in the middle Miocene [*Hatley*, 1978; *Holloway*, 1981]. However, imbricate thrusting, as evidenced by thrice repeated lower and middle Miocene sections in the Phillips Albion head 1 well offshore SW Palawan, continued into the middle Miocene [*Holloway*, 1981]. The cessation of subduction along Palawan was associated with regional uplift, the unconformity being firmly dated as late middle Miocene (N13-N14; *Holloway* [1981]).

There was no major tectonic event associated with the cessation of subduction along the northwest Borneo trench from Luconia shoals to Sabah [*White and Wing*, 1978; *Hamilton*, 1979; *Holloway*, 1981]. Southwest of Palawan the continental fragments of the China margin never quite collided with the trench. There is still a bathymetric expression of the former trench off northwest Borneo (the Palawan Trough). *Haile* [1969] and others have observed that the volume of sediments increases, and the folding appears to become younger, northeastward along the NW Borneo geosyncline. Several authors [e.g., *Hamilton*, 1979; *Holloway*, 1981] have inferred that subduction progressively ceased from southwest to northeast along northwest Borneo. These observations could alternatively be explained by the rotation of Borneo, causing more sediments to be accreted and the folding to be tighter toward the northeast. As there is little Tertiary igneous activity and no collisional tectonics associated with this region, there is little data to support a progressive cessation of subduction.

In the middle Miocene clastics prograded out over the former outer arc high east of Natuna [*White and Wing*, 1978]. However, Luconia shoals remained a province of carbonate banks [*Doust*, 1977]. Following the propagation of the oceanic rift to the edge of the Sunda Shelf, the Natuna high was breached and a marine environment penetrated the West Natuna Basin for the first time [*White and Wing*, 1978].

Luzon continued its counterclockwise rotation, advancing westward over the Manila Trench. Paleomagnetic data from the Baguio and Zambales regions of western Luzon evidence an approximately 14° counterclockwise rotation of Luzon since the middle Miocene, although the rates of rotation may not have been constant [*Fuller et al.*, this volume]. The Manila Trench may have been connected by a transform to the opposite polarity Ryukyu Trench (Figure 18e) [*Karig*, 1973].

Late Miocene to Recent (*Figure 17*)

This period is marked by substantial subsidence of the South China Basin and its margins, presumably as a result of continued thermal contraction. Growth of some reefs such as Reed Bank and Macclesfield Bank kept up with the subsidence. Extensive deltas prograded out over the northern Sunda Shelf, burying the former trench and ending carbonate deposition on most of Luconia shoals [*Doust*, 1977].

Subduction of South China Basin crust beneath the Manila Trench brought the Luzon arc into collision with Taiwan and north Palawan in the late Miocene [*Bowin et al.*, 1978; *Suppe*, 1980; *Holloway*, 1981]. On Taiwan the thick pre-late Miocene passive continental margin sequence continues to be imbricated and thrust westward.

A major event of unknown relation to the evolution of the rest of the area was the development in middle to late Miocene time of an extensive continental volcanic province of alkalic basalts and tholeiites in Indochina [*Barr and Macdonald*, 1981]. Much of the volcanism is late Pliocene to Recent, the youngest eruption being of nepheline hawaiite at Ile des Cendres off the southeast Vietnam coast in 1923. Similar volcanism has occurred along the South China coast and shelf from Hainan and the Luichow Peninsula to the Penghu Islands in Taiwan Strait [*Chen*, 1973; *Barr and Macdonald*, 1981]

Summary

Our hypothetical tectonic evolution of the South China Basin and surrounding areas may be summarized as follows:

1. From mid-Jurassic through mid-Cretaceous times the proto-southeast Asian margin was an Andean-type arc. The north Palawan–Reed Bank–Luconia Shoals microcontinental blocks were forearc areas of this margin.

2. Arc volcanism along the proto-China margin ceased about 85 m.y. B.P. Subduction of inferred Mesozoic oceanic crust beneath west Borneo continued through the Late Cretaceous and early Tertiary, presumably as a result of the 60° to 70° counterclockwise rotation of Sundaland with respect to Indochina.

3. The onset of rifting of the proto-China margin was during the latest Cretaceous or Paleocene (65 ± 10 m.y. B.P.).

4. Strike slip and block faulting of the Sunda Shelf resulted in the development of the Thai, Malay, and West Natuna basins in the Eocene. Rifting of the proto-China margin continued through the Eocene and early Oligocene.

5. Seafloor spreading in the eastern half of the South China Basin occurred during the late Oligocene and early Miocene (32–17 m.y. B.P.). The western half of the basin is inferred to have opened by a combination of seafloor spreading and crustal extension, the amount of crustal extension increasing from east to west. Opening of the South China Basin separated the north Palawan–Reed Bank–Luconia Shoals blocks from the Asian mainland.

6. The collision of the north Palawan and Reed Bank blocks with the south Palawan convergent margin began

in the early Miocene but was not complete until the late middle Miocene.

7. Subduction of South China Basin crust beneath the Manila Trench brought the Philippines into collision with Taiwan and north Palawan in the late Miocene. The opening of the South China Basin was not related to back arc spreading behind a Philippine arc-trench system.

Acknowledgments. We would like to thank Manik Talwani, Jeff Weissel, Greg Moore, Dan Karig, and Richard Murphy for their critical review of the manuscript and for their many helpful suggestions. We also gratefully acknowledge the outstanding efforts of the officers, crew, and technical staff of the R/V *Vema,* and especially Captain H. C. Kohler. *Vema* cruises 34–05 and 34–06 initiated the first phase of our cooperative program with the Ministry of Geology (Marine Geology Division) of the People's Republic of China. The bulk of that cooperative work was focussed on the outer continental shelf and will be jointly reported elsewhere. Brian Taylor was the recipient of the Paul A. Gorman Fellowship, provided by International Paper, during much of the period of this study. This study was primarily funded by the Office of Naval Research under contract N0014-80-C. Partial support also derived from National Science Foundation grant OCE79-19069. Lamont-Doherty Geological Observatory Contribution 3370.

References

Allen, P. M., and E. A. Stephens, Report on the geological survey of Hong Kong, 107 pp., Hong Kong Government Printer, Hong Kong, 1971.

Anderson, R. N., 1980 update of heat flow in the east and southeast Asian seas, in *The Tectonic and Geologic Evolution of Southeast Asian Seas and Islands, Geophys. Monogr. Ser.,* vol. 23, edited by D. E. Hayes, pp 319–326, AGU, Washington, D. C., 1980.

Anderson, R. N., M. G. Langseth, and J. G. Sclater, The mechanisms of heat transfer through the floor of the Indian Ocean, *J. Geophys. Res., 82,* 3391–3409, 1977.

Anderson, R. N., M. G. Langseth, D. E. Hayes, T. Watanabe, and M. Yashui, Heat flow, thermal conductivity, thermal gradient, A Geophysical Atlas of East and Southeast Asian Seas, *Map Chart Ser. MC-25,* edited by D. E. Hayes, Geol. Soc. of Am., Boulder, Colo., 1978.

Barr, S. M., and A. S. MacDonald, Geochemistry and geochronology of late Cenozoic basalts of Southeast Asia, *Geol. Soc. Am. Bull., 92,* 1069–1142, 1981.

Batiza, R., Origin and petrology of young oceanic central volcanoes: Are most tholeiitic rather than alkalic?, *Geology, 8,* 477–482, 1980.

Beddoes, L. R., The Balabac subbasin, southwestern Sulu Sea, Philippines, SEAPEX Program, paper presented Offshore Southeast Asia Conference, Southeast Asian Pet. Explor. Soc., Singapore, 1976.

Bell, R. M., and R. G. C. Jessop, Exploration and geology of the West Sulu Basin, Philippines, *APEA. J., 14,* 21–28, 1974.

Ben-Avraham, Z., and S. Uyeda, The evolution of the China Basin and the Mesozoic paleogeography of Borneo, *Earth Planet. Sci. Lett., 18,* 365–376, 1973.

Bowin, C., and P. H. Reynolds, Radiometric ages from Ryukyu Arc region and an $^{40}Ar/^{39}Ar$ age from biotite dacite on Okinawa, *Earth Planet. Sci. Lett., 27,* 363–370, 1975.

Bowin, C., R. S. Lu, C.-S. Lee, and H. Schouten, Plate convergence and accretion in the Taiwan-Luzon region, *Am. Assoc. Pet. Geol. Bull., 62,* 1645–1672, 1978.

Chen, J.-C., Geochemistry of basalts from Penghu Islands, *Proc. Geol. Soc. China, 16,* 23–26, 1973.

Davis, E. E., and C. R. B. Lister, Heat flow measured over the Juan de Fuca Ridge: Evidence for widespread hydrothermal circulation in a highly heat transportive crust, *J. Geophys. Res., 82,* 4845–4860, 1977.

Doust, H., Geology and exploration history of offshore Central Sarawak, paper presented at the Asian Council on Petroleum, Conference, Jakarta, Indonesia, 1977.

Du Bois, E. P., Review of principal hydrocarbon-bearing basins of the South China Sea area, *Energy, 6(11),* 1113–1140, 1981.

Engel, A. E., C. G. Engel, and R. G. Havens, Chemical characteristics of oceanic basalts and the upper mantle, *Geol. Soc. Am. Bull., 76,* 719–734, 1965.

Fontaine, H., and M. Mainguy, Pre-Tertiary hydrocarbon potential of the South China Sea, *Energy, 6(11),* 1165–1177, 1981.

Fontaine, H., and D. R. Workman, Review of the geology and mineral resources of Kampuchea, Laos and Vietnam, in *Proceedings of the 3rd Regional Conference on Geology and Mineral Resources of Southeast Asia,* John Wiley, New York, 1978.

Fuller, M., I. S. Williams, R. McCabe, R. Y. Encina, J. Almasco, and J. A. Wolfe, Paleomagnetism of Luzon, this volume.

Garcia, M. O., N. W. K. Liu, and D. W. Muenow, Volatiles in submarine volcanic rocks from the Mariana island arc and trough, *Geochim. Cosmochim. Acta, 43,* 305–312, 1979.

Guenoit, J.-P., G. Hirlemann, and B. Penet, The pre-Tertiary structural evolution of Borneo, paper presented at the 25th International Geological Congress, Aust. Acad. of Sci., Geol. Soc. of Aust., Int. Union of Geol. Sci., Sydney, Australia, 1976.

Haile, N. S., Geosynclinal theory and the organisational pattern of the Northwest Borneo Geosyncline, *Q. J. Geol. Soc. London, 124,* 171–194, 1969.

Haile, N. S., Reconnaissance paleomagnetic results from Sulawesi, Indonesia, and their bearing on paleogeographic reconstruction, *Tectonophysics, 46,* 77–85, 1978.

Haile, N. S., and J. D. Bignell, Late Cretaceous age based on K/Ar dates of granitic rock from the Tambelan and Bunguran Islands, Sunda Shelf, Indonesia, *Geol. Mijnbouw, 50,* 687–690, 1971.

Haile, N. S., and D. H. Tarling, Note on reconnaissance paleomagnetic measurements on Jurassic redbeds from Thailand, *Pac. Geol., 10,* 101–103, 1975.

Haile, N. S., M. W. McElhinny, and I. McDougall, Paleomagnetic data and radiometric ages from the Cretaceous of West Kalamantan (Borneo), and their significance in interpreting regional structure, *J. Geol. Soc. London, 133,* 133–144, 1977.

Hamilton, W., Tectonics of the Indonesian region, *U.S. Geol. Surv. Prof. Pap., 1078,* 345 pp., 1979.

Hart, S. R., W. E. Glassley, and D. E. Karig, Basalt and sea-floor spreading behind the Mariana Island arc, *Earth Planet. Sci. Lett., 15,* 12–18, 1972.

Hashimoto, W., and T. Sato, Contribution to the geology of Mindoro and neighboring islands, the Philippines, in *Geology and Palaeontology of Southeast Asia,* vol. 5, Tokyo University Press, pp. 192–210, Tokyo, 1968.

Hatley, A. G., Palawan oil spurs Philippine action, *Oil Gas J., Feb. 27,* 112–118, 1978.

Hawkins, J. W., Petrology and geochemistry of basaltic rocks of the Lau Basin, *Earth Planet. Sci. Lett., 28,* 283–297, 1976.

Holloway, N. H., The stratigraphic and tectonic relationship of Reed Bank, north Palawan and Mindoro to the Asian mainland and its significance in the evolution of the South China Sea, *Bull. Geol. Soc. Malaysia,* in press, 1981.

Hutchinson, C. S., Tectonic evolution of Sundaland: A Phanerozoic synthesis, *Bull. Geol. Soc. Malaysia, 6,* 61–86, 1973.

Hutchinson, C. S., Ophiolite in southeast Asia, *Geol. Soc. Am. Bull., 86,* 797–806, 1975.

Jahn, B.-M., P. Y. Chen, and T. P. Yen, Rb-Sr ages of granitic rocks in southeastern China and their tectonic significance, *Geol. Soc. Am. Bull., 87,* 763–776, 1976.

Jarrard, R. D., and S. Sasajima, Paleomagnetic synthesis for southeast Asia: constraints on plate motions, in *The Tectonic and Geologic Evolution of Southeast Asian Seas and Islands, Geophys. Monogr. Ser.,* vol. 23, edited by D. E. Hayes, pp. 293–316, Washington, D. C., 1980.

Kahle, H.-G., B. R. Naini, M. Talwani, and O. Eldholm, Marine geophysical study of the Comorin Ridge, north central Indian Basin, *J. Geophys. Res., 86,* 3807–3814, 1981.

Karig, D. E., Plate convergence between the Philippines and the Ryuku Islands, *Mar. Geol., 14,* 153–168, 1973.

Karner, G. D., and A. B. Watts, On isostasy at Atlantic-type continental margins, *J. Geophys. Res., 87,* 2923–2948, 1982.

Kobayashi, K., and M. Nakada, Magnetic anomalies and tectonic evolution of the Shikoku inter-arc basin, *Adv. Earth Planet. Sci., 6,* 391–402, 1979.

LaBrecque, J. L., D. V. Kent, and S. C. Cande, Revised magnetic polarity time scale for Late Cretaceous and Cenozoic time, *Geology, 5,* 330–335, 1977.

Lee, C., H. Lee, H. Lio, G. Lio, and S. Yen, Preliminary study of paleomagnetism of some Mesozoic and Cenozoic redbeds of South China, *Acta Geol. Sin., Engl. Transl., 43,* 261–264, 1963.

Ludwig, W. J., The Manila Trench and West Luzon Trough, 3, Seismic refraction measurements, *Deep Sea Res., 17,* 553–571, 1970.

Ludwig, W. J., N. Kumar, and R. E. Houtz, Profiler-sonobuoy measurements in the South China Sea Basin, *J. Geophys. Res., 84,* 3505–3518, 1979.

Mammerickx, J., R. L. Fisher, F. J. Emmel, and S. M. Smith, Bathymetry of the east and southeast Asian seas, *Map Chart Ser. MC-17,* Geol. Soc. of Am., Boulder, Colo., 1976.

Marsh, N. G., A. D. Saunders, J. Tarney, and H. J. B. Dick, Geochemistry of basalts from the Shikoku and Daito Basins, Deep-Sea Drilling Project leg 58, *Initial Rep. Deep Sea Drill. Proj., 58,* 805–841, 1980.

McElhinny, M. W., N. S. Haile, and A. S. Crawford, Paleomagnetic evidence shows Malay Peninsula was not part of Gondwana, *Nature, 252,* 641–645, 1974.

Meunow, D. W., N. W. K. Liu, M. O. Garcia, and A. D. Saunders, Volatiles in submarine volcanic rocks from the spreading axis of the east Scotia sea back-arc basin, *Earth Planet. Sci. Lett., 47,* 272–278, 1980.

Mineral Fuels Division, Bureau of Mines, Manila, The Philippines, A review of oil exploration and stratigraphy of sedimentary basins of the Philippines, *Tech. Bull. U. N. Econ. Soc. Comm. Asia Pac., Comm. Co-ord. St. Prospect. Miner. Resourc. South Pac. Offshore Areas, 10,* 55–102, 1976.

Mrozowski, C. L., and D. E. Hayes, Sediment isopachs, A Geophysical Atlas of East and Southeast Asian Seas, *Map Chart Ser. MC-25,* edited by D. E. Hayes, Geol. Soc. of Am., Boulder, Colo., 1978.

Parsons, B., and J. G. Sclater, An analysis of the variation of ocean floor bathymetry and heat flow with age, *J. Geophys. Res., 82,* 803–827, 1977.

Pupilli, M., Geological evolution of South China Sea area: Tentative reconstruction from borderland geology and well data, *Proc. Annu. Conv. Indones. Pet. Assoc., 2,* 223–241, 1973.

Rabinowitz, P. D., and J. L. LaBrecque, The isostatic gravity anomaly: Key to the evolution of the ocean-continent boundary at passive continental margins, *Earth Planet. Sci. Lett., 35,* 145–150, 1977.

Rabinowitz, P. D., and J. L. LaBrecque, The Mesozoic South Atlantic ocean and evolution of its continental margins, *J. Geophys. Res., 84,* 5973–6002, 1979.

Sasajima, S., S. Nishimura, K. Hirooka, Y. Otofuji, T. V. Leeuven, and F. Hehuwat, Paleomagnetic studies combined with fission track datings on the western arc of Sulawesi, East Indonesia, *Tectonophysics, 64,* 163–172, 1980.

Schouten, H., and S. C. Cande, Palaeomagnetic poles from marine magnetic anomalies, *Geophys. J. R. Astron. Soc., 44,* 567–575, 1976.

Schweller, W. J., and D. E. Karig, Emplacement of the Zambales ophiolite into the west Luzon margin, *Mem. Am. Assoc. Pet. Geol.,* in press, 1982.

Sclater, J. G., D. E. Karig, L. A. Lawver, and K. Louden, Heat flow, depth, and crustal thickness of the marginal basins of the South Philippine Sea, *J. Geophys. Res., 81,* 309–318, 1976.

Suppe, J., A retrodeformable cross section of northern Taiwan, *Proc. Geol. Soc. China., 23,* 46–54, 1980.

Talwani, M., and O. Eldholm, Continental margin off Norway: A geophysical study, *Geol. Soc. Am. Bull., 83,* 3575–3606, 1972.

Taylor, B., and D. E. Hayes, The tectonic evolution of the South China Basin, in *The Tectonic and Geologic Evolution of Southeast Asian Seas and Islands, Geophys. Monogr. Ser.,* vol. 23, edited by D. E. Hayes, pp. 89–104, AGU, Washington, D. C., 1980.

Tokuyama, E., and K. Fujioka, The petrologic study on basalt from Kinan Seamount and DSDP site 54 (in Japanese with English abstract), *Mar. Sci. Mon., 8,* 184–191, 1976.

Wang, S.-C., R. J. Geller, S. Stein, and B. Taylor, An intraplate thrust earthquake in the South China Sea, *J. Geophys. Res., 84,* 5627–5631, 1979.

Watanabe, T., M. G. Langseth, and R. N. Anderson, Heat flow in Back-arc basins of the western Pacific, in *Island Arcs, Deep Sea Trenches and Back-Arc Basins, Maurice Ewing Ser.,* vol. 1, edited by M. Talwani and W. C. Pitman III, pp. 137–161, AGU, Washington, D. C., 1977.

Watts, A. B., Gravity anomalies over oceanic rifts, in *Continental and Oceanic Rifts, Geodyn. Ser.,* vol. 8, edited by G. Pálmason et al., pp. 99–105, AGU, Washington, D. C., 1982.

Weissel, J. K., Evidence for Eocene oceanic crust in the Celebes Basin, in *The Tectonic and Geologic Evolution of Southeast Asian Seas and Islands, Geophys. Monogr. Ser.,* vol. 23, edited by D. E. Hayes, pp. 37–47, AGU, Washington, D. C., 1980.

Weissel, J. K., and A. B. Watts, Tectonic evolution of the Coral Sea Basin, *J. Geophys. Res., 84,* 4572–4582, 1979.

White, J. M., and R. S. Wing, Structural development of the South China Sea with particular reference to Indonesia, *Proc. Annu. Conv. Indones. Pet. Assoc. 7,* 159–177, Jakarta, 1978.

Woollard, G. P., The interrelationships of crustal and upper mantle parameter values in the Pacific, *Rev. Geophys. Space Phys., 13,* 87–137, 1975.

Yen, T. P., and S. Rosenblum, Potassium-Argon ages of micas from the Tananao Schist Terrane of Taiwan—A preliminary report, *Proc. Geol. Soc. China, 7,* 80–81, 1963.

The Tectonics of Northward Propagating Subduction Along Eastern Luzon, Philippine Islands

STEPHEN D. LEWIS[1] AND DENNIS E. HAYES[1]

Lamont-Doherty Geological Observatory of Columbia University Palisades, New York 10964

Marine multichannel seismic reflection and other geophysical data acquired along the eastern margin of Luzon indicate that subduction of the Philippine Sea plate is occurring south of approximately 18°N. The thick sediments of the East Luzon Trough are deformed by folding and thrust faulting near the base of the continental slope, and the acoustic basement underlying the sediment section can be traced as it dips landward beneath the east Luzon margin. Earthquake activity in this region indicates that underthrusting is presently occurring. The intensity of observed deformation decreases to the north between 16°N and 18°N, and north of approximately 18°N no recent compressive deformation can be recognized. These observations indicate that the locus of subduction is propagating northward along the east Luzon margin. A narrow linear negative free air gravity anomaly is associated with the zone of recent deformation and subduction. A zone of closely spaced faults disrupts the entire sediment section near 15.25°N and is coincident with a linear east-west trending belt of shallow earthquake epicenters. This zone may represent the surface expression of an east trending transform fault that connects the Philippine Trench to the newly developing east Luzon subduction zone. From approximately 17°N to 19°N, the Sierra Madre Basin, which contains up to 4.5 km of sediment, lies between the East Luzon Trough and the Luzon coast. NNE striking basement ridges can be traced from beneath the Sierra Madre Basin into the deep ocean environment of the adjacent West Philippine Basin. The East Luzon Trough, the Sierra Madre Basin, and the intervening bathymetric ridge, the Isabella Ridge, may represent the trench, forearc basin, and subduction complex, respectively, of a relict subduction system which was active during Oligocene time along eastern Luzon. Arcuate NNE to NNW trending bathymetric troughs divide the North Luzon Ridge into a series of discrete blocks. Regional seismicity and focal mechanism data may indicate that these troughs represent the traces of presently active strike slip and thrust faults. This faulting within the North Luzon Ridge may absorb the relative convergence between Luzon and the Philippine Sea plate which, to the south, is taken up along the active Philippine Trench and eastern Luzon subduction zones.

Introduction

The Cenozoic tectonic history of the Philippine Islands is dominated by geological processes associated with convergent plate margins. The complex collage of active and inactive island arc terrains, belts of deformed strata, ophiolite assemblages, and deep sedimentary basins which make up the Philippines may be the result of several episodes of subduction and collision that occurred during the Cenozoic [*de Boer et al.*, 1980; *Hamilton*, 1979; *Karig*, 1973].

The regional geology of the northern Philippines, particularly that of Luzon Island, suggests that subduction has occurred along both the eastern and western sides of the archipelago [*Balce et al.*, 1980; *Karig*, 1973, *Murphy*, 1973].

Geological and geophysical data available from offshore western Luzon suggests that subduction is presently occurring along the Manila Trench (Figure 1) [*Cardwell et al.*, 1980; *Lewis and Hayes*, 1981]. Activity along the Manila Trench probably began in the early Miocene [*Schweller et al.*, this volume].

The Philippine Trench marks the site of subduction of the Philippine Sea plate beneath the eastern Philippines between the Bicol Peninsula of southeastern Luzon and

---

[1]Also at Department of Geological Sciences, Columbia University, New York, New York 10027.

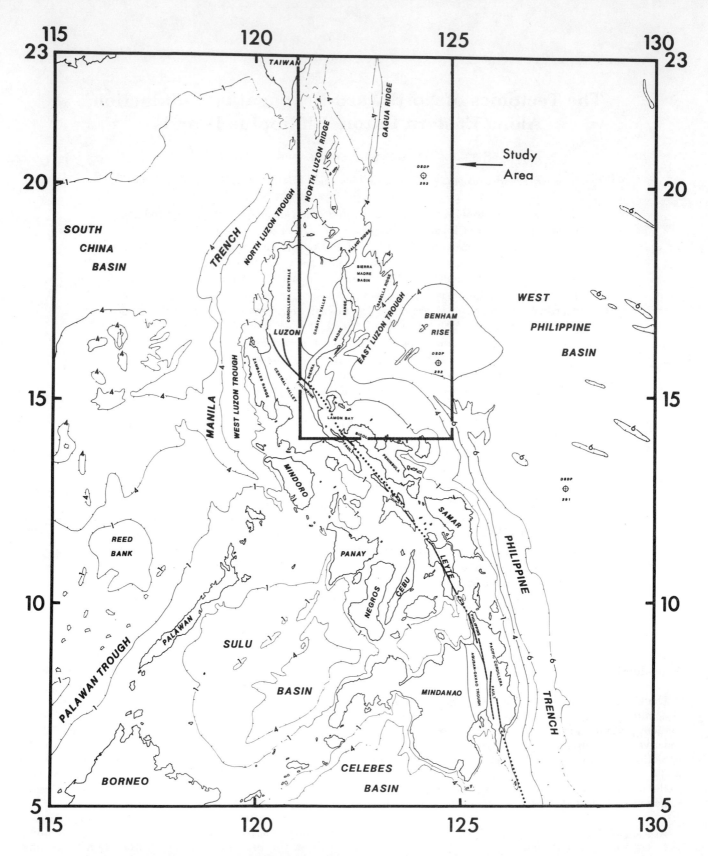

Fig. 1. Regional index map of the Philippines and the surrounding regions. Bathymetric contours in kilometers. Deep Sea Drilling Project (DSDP) holes are indicated by numbered circles. The region of detailed study is shown enclosed by a box. Philippine place names mentioned in the text are located on this figure.

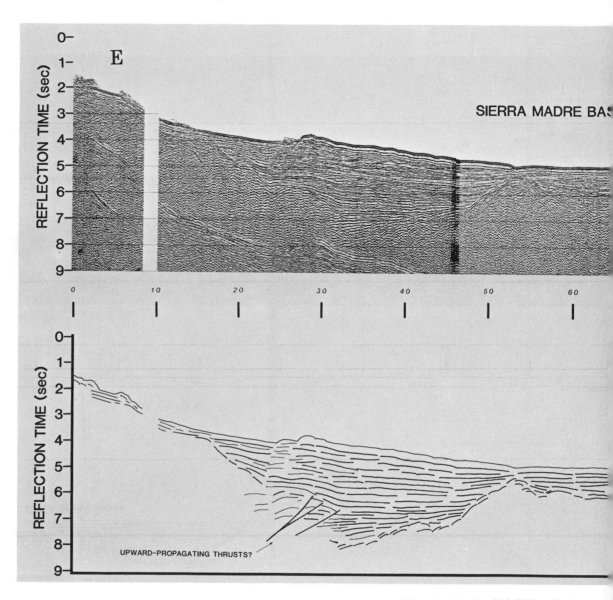

Plate 2. Twelve-fold CDP reflection pro
slope basin present near 5 km is steeply t
basement ridge between 50 and 60 km. Bet
reflector sequence is present beneath the u
the East Luzon Trough between 7- and 7
between 20 and 40 km is mildly disrupted a
along eastern Luzon.

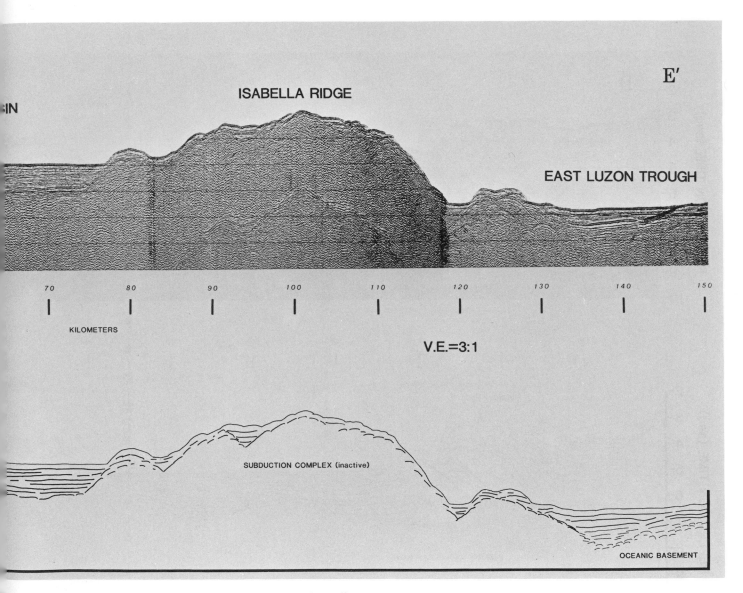

E'

ISABELLA RIDGE

EAST LUZON TROUGH

70   80   90   100   110   120   130   140   150

KILOMETERS

V.E.=3:1

SUBDUCTION COMPLEX (inactive)

OCEANIC BASEMENT

...le across the east Luzon margin. Track location in Figure 3. A small
...lted oceanward. The Sierra Madre Basin is divided by a prominent
...een 30 and 50 km and 6.5- and 8-s reflection time, a landward dipping
...conformity discussed in the text. The unconformity is also present in
...5-s reflection time. The western portion of the Sierra Madre Basin
...d may represent the beginning of folding and faulting at this latitude

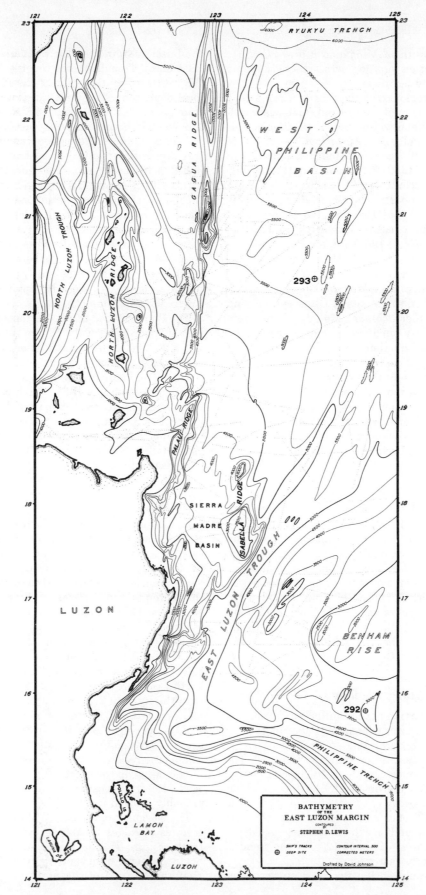

Fig. 2.   Bathymetric map of the eastern margin of Luzon, the Gagua Ridge, and the North Luzon Ridge. Contours are in meters. DSDP holes are indicated by numbered circles.

the Talaud Islands south of Mindanao [*Cardwell et al.*, 1980]. Seismic reflection profiles across the Philippine Trench indicate that no significant accretionary prism has formed in the forearc region [*Hamilton*, 1979; *Karig*, 1975]. This, coupled with a lack of significant Quaternary volcanism on eastern Mindanao and Samar [*Philippine Bureau of Mines*, 1963], and a Benioff zone which extends to less than 200-km depth beneath Mindanao and Talaud Island [*Cardwell et al.*, 1980] may mean that this subduction pulse has only recently begun along the southern portion of the Philippine Trench [*Cardwell et al.*, 1980; *Hamilton*, 1979; *Karig*, 1975]. However, the geology of the Sierra Madre Mountains of eastern Luzon [*Christian*, 1964; *Philippine Bureau of Mines;* 1963] and the Pacific Cordillera of eastern Mindanao [*Moore*, 1981] may indicate that at least one early mid-Tertiary subduction pulse occurred along the eastern Philippines. Mafic and ultramafic rocks and associated deformed sediments of Oligocene age and older exposed in the Sierra Madre Mountains may represent an early Tertiary subduction complex associated with an east facing subduction zone [*Karig*, 1973]. The Davao-Agusan Trough of central Mindanao is interpreted by *Moore* [1981] to represent the forearc basin of an early mid-Tertiary subduction zone which was also east facing. Thus present-day activity along the Philippine Trench represents a rejuvenation of subduction along the eastern Philippines. This subduction zone may be propagating to the south, east of Talaud Island and Halmahera Island, along the Philippine Trench [*Cardwell et al.*, 1980; *Murphy*, 1973].

This paper presents marine multichannel seismic reflection and other geophysical data from the margin off eastern Luzon. The level of seismicity [*Cardwell et al.*, 1980] and the degree of recent deformation [*Lewis and Hayes*, 1980] shows strong variation along strike, from areas of well-defined underthrusting and sediment accretion to areas of no recent deformation or shallow earthquake activity. Subduction of the Philippine Sea plate beneath eastern Luzon appears to be propagating northward toward Taiwan and the Ryukyus, perhaps representing an arc polarity reversal in the Luzon-Taiwan region [*McKenzie*, 1969; *Dewey and Bird*, 1970; *Chai*, 1972; *Fitch*, 1972; *Karig*, 1973].

Bathymetry and Offshore Geology

The northernmost portion of the Philippine Trench, the Benham Rise, the East Luzon Trough, the Palaui and Gagua ridges, and the North Luzon Ridge are prominent bathymetric features of the east Luzon margin (Figure 2). The Philippine Trench is a continuous deep ($\geq$6000 m) trending approximately N20°W from about 2°N latitude to about 14°N latitude [*Mammerickx et al.*, 1976]. Near 14°N the axis of the Philippine Trench curves to the west, where its trend becomes approximately northwest and its axial depth shallows to 5500 m. West of about 123.5°N the trend of the deep becomes nearly east-west.

The East Luzon Trough, a flat-floored deep, separates

the Luzon margin from the Benham Rise to the east. The East Luzon Trough strikes roughly N30°E and intersects the westward continuation of the Philippine Trench in a local depression. The East Luzon Trough is well defined from about 15.5°N to about 18.5°N, where it merges with the West Philippine Basin. The East Luzon Trough is floored up to about 2 km of sediment fill.

A broad, flat region lies east of Luzon between 17°N and 18°N (Figure 2), separated from the East Luzon Trough by a series of discontinuous ridges. Seismic reflection profiles show this region, herein named the Sierra Madre Basin, to be underlain by a thick (up to about 3.5-s reflection time, or about 4.5 km) accumulation of acoustically stratified sediment (profiles D, E, and F; locations in Figure 3). The basin floor generally deepens toward the north and, between 18°N and 19°N, merges into the turbidite apron of the West Philippine Basin described by *Karig and Wageman* [1975].

The Benham Rise is a broad topographic high centered east of Luzon at about 17°N. The crest of the Benham Rise is 2000–3000 m deep and reaches to within 38 m of the sea surface at Benham Bank (site 292 report; *Karig et al* [1975]). The Benham Rise is underlain by a series of basement ridges trending N-S or NNE-SSW (K. Hegarty, personal communication, 1980), roughly parallel to the trend of the East Luzon Trough.

The northeastern peninsula of Luzon extends offshore as the Palaui Ridge [*Karig*, 1973]. The Palaui Ridge may be structurally continuous with the Sicalao-Casiggayen High (Figure 4), an east-west trending antiformal structure consisting of early Tertiary metasediments [*Durkee and Pederson*, 1961] in the Cagayan Valley of northeastern Luzon, which may be of subduction-related origin [*Karig*, 1973]. At approximately 19.25°N the Palaui Ridge is truncated by a complex submarine canyon system which drains the Cagayan River of northern Luzon (Figure 2). This canyon system supplies material to the large sediment apron that extends into the West Philippine Basin [*Karig and Wageman*, 1975; *Karig*, 1973].

There is no morphologic expression of the Palaui Ridge north of the Cagayan submarine canyon system. However, the Gagua Ridge, a prominent steep-sided N-S trending ridge, continues the general trend of the Palaui Ridge northward from approximately 20.75°N to its intersection with the Ryukyu Trench at about 23°N (Figure 2) [*Mammerickx et al.*, 1976]. Reflection profiles across the Palaui and Gagua ridges exhibit many differences, compare Figure 9 of *Mrozowski et al.* [1982] with Figure 5. This, as well as the recovery of rocks of ocean crustal affinity during dredging on the Gagua Ridge discussed by *Mrozowski et al.* [1982], suggests that the Palaui and Gagua ridges probably evolved independently.

The basement of the Sierra Madre Basin is composed of a series of ridges which can be correlated from track to track (Figure 4). The ridges strike generally NNE and begin to emerge from beneath the sediment fill of the basin near the latitude of the north end of Luzon (Figure 4)

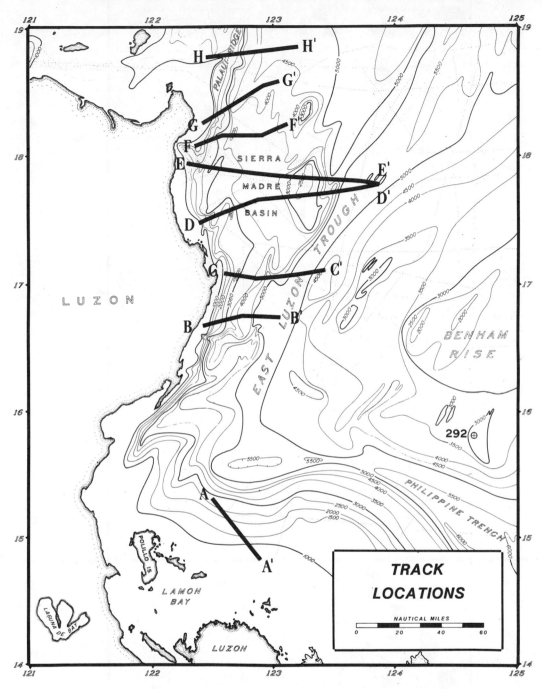

Fig. 3.   Location map of seismic reflection profiles illustrated in subsequent figures.

[*Lewis and Hayes*, 1980], where they were also mapped by *Karig and Wageman* [1975]. Some individual ridges appear to be continuous features for up to about 200 km along strike, the most prominent of which we have named the Isabella Ridge (Figure 2). The Isabella Ridge is a series of basement highs that stretches from about 17°N to about 18.5°N (Figure 2) and represents the eastern boundary of the Sierra Madre Basin. The sediment fill of the western

region of the Sierra Madre Basin is involved in ongoing compressive deformation between 17°N and 18°N. The nature of this deformation will be discussed in detail in a later section.

The North Luzon Ridge trends roughly N-S and extends from northern Luzon to east of Taiwan (Figure 2). The North Luzon Ridge rises above sea level to form the Babuyan and Batanes islands. The islands are composed of

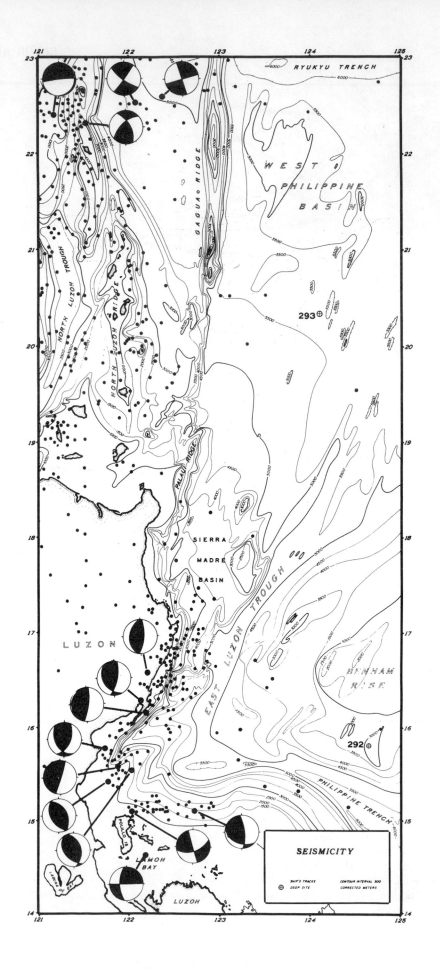

SEISMICITY

SHIP'S TRACKS   CONTOUR INTERVAL 500
⊕ DSDP SITE   CORRECTED METERS

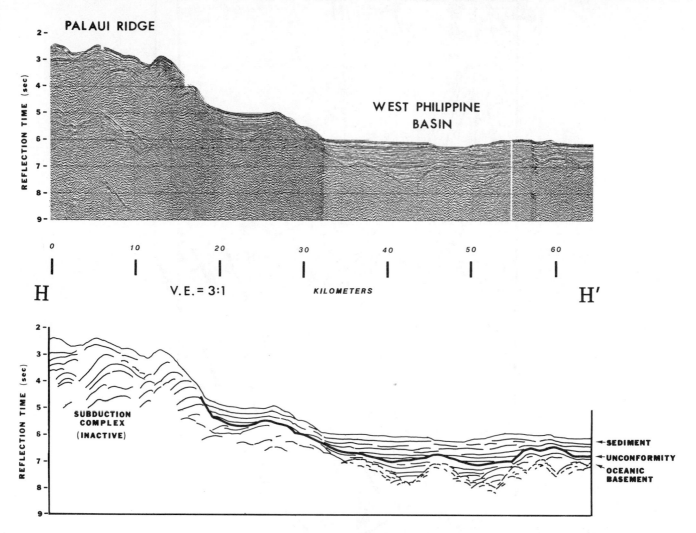

Fig. 5. Twenty-four fold CDP reflection profile and interpretation across the northern Palaui Ridge and the Luzon margin. Compare the internal structure of the Palaui Ridge to that of the zone of deformation in Figure 8. Track location in Figure 3. This and subsequent reflection profiles were shot using large-volume air guns. Digital processing consists of Burg deconvolution, bandpass filtering, water velocity moveout, predictive deconvolution, CDP stack, Burg deconvolution, and time-varying filter. Some lines are 12-fold CDP stacks, but source and processing parameters are similar.

recently active to Miocene basaltic and andesitic volcanic centers and uplifted Tertiary limestones [*Philippine Bureau of Mines*, 1963]. The active volcanism on the North Luzon Ridge is probably related to subduction of South China Basin lithosphere along the west facing Manila Trench [*Cardwell et al.*, 1980]. *Karig* [1973] states that the north and northeast trending arcuate bathymetric troughs of the North Luzon Ridge may represent the traces of left-

lateral strike slip faults which divide the North Luzon Ridge into a series of en echelon blocks. The diffuse pattern of shallow seismicity that typifies the region between Taiwan and Luzon [*Seno and Kurita*, 1978] probably indicates that faulting is still active.

Seismicity

Figure 6 shows all the available earthquake focal mechanism solutions for the east Luzon region, compiled from *Cardwell et al.* [1980], *Seno and Kurita* [1978], *Fitch* [1972], and *Katsumata and Sykes* [1969]. Seismic activity is concentrated along eastern Luzon south of approximately 17°N and in a diffuse zone some 200 km wide that extends be-

Fig. 4. (Opposite) Structural/tectonic summary map of the east Luzon margin. Ship tracks used in the structural compilation are shown as dotted lines. Aspects of the geology of eastern Luzon and the North Luzon Ridge are discussed in the text.

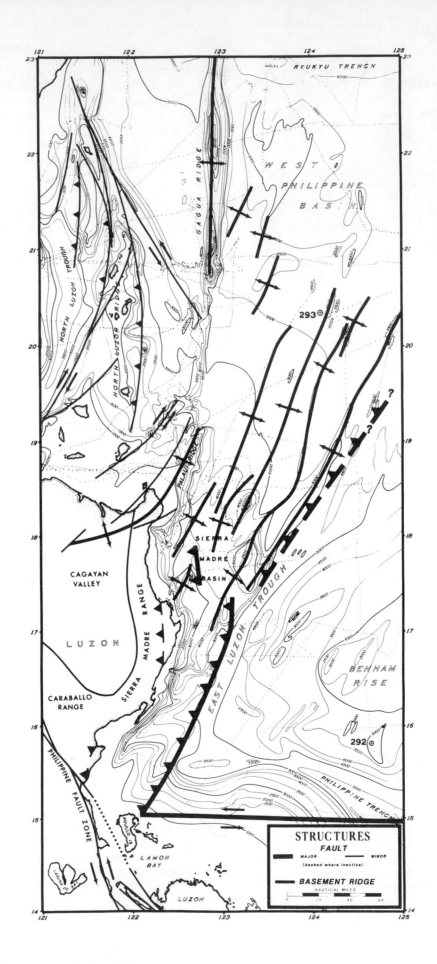

tween northern Luzon and Taiwan [*Cardwell et al.*, 1980; *Seno and Kurita*, 1978]. Hypocentral depths are all shallower than 200 km, with the majority of events occurring at depths less than 70 km. A region of relatively low seismic activity occurs between about 17°N and 18.5°N along eastern Luzon [*Cardwell et al.*, 1980; *Seno and Kurita*, 1978].

The seismicity between 15°N and 17°N is characterized by thrust fault-type focal mechanism solutions, with some mechanisms indicative of strike slip faulting. In contrast, the region east of Taiwan is dominated by strike slip focal mechanism solutions, implying a major change in the nature of tectonism along eastern Luzon and the North Luzon Ridge.

There is no well-defined west dipping Benioff zone along eastern Luzon and the East Luzon Trough [*Cardwell et al.*, 1980]. However, the thrust fault focal mechanism solutions between approximately 15°N and 17°N indicate that underthrusting of the Philippine Sea plate is presently occurring in this region. The absence of shallow depth thrust fault focal solutions north of about 17°N may indicate that subduction is not occurring or has only recently begun in this area of apparently low seismicity. Seismic reflection data discussed later support this conclusion.

Two earthquakes with strike slip focal mechanism solutions occur near Polillo Island (Figure 6). One event, with north and east striking nodal planes is probably associated with the Phillippine fault (Figure 4). In this region, strands within the Philippine fault zone trend roughly parallel to the north striking nodal plane of this focal mechanism solution [*Philippine Bureau of Mines*, 1963], and the left-lateral motion along this trend indicated by the focal mechanism solution is consistent with the apparent sense of offset along the Philippine fault determined from field observations [*Allen*, 1962; *Ranneft et al.*, 1960].

The other strike slip event, north of Polillo Island, also indicates faulting along either north or west striking planes. Aftershocks associated with this event define an east-west trending zone, indicating that the slip plane for this event was probably oriented east-west [*Seno and Kurita*, 1978]. This event may be related to left-lateral transform faulting between the Philippine Trench to the south and the East Luzon Trough to the north [*Seno and Kurita*, 1978; *Karig*, 1973].

The distribution of focal mechanism solutions and seismic activity along eastern Luzon helps define two regions. The high level of seismic activity and the thrust-type focal mechanism solutions observed south of about 17°10'N along eastern Luzon provide compelling evidence in support of the presence of an active convergent plate

boundary between Luzon and the Philippine Sea plate in this region. Left-lateral strike slip focal mechanism solutions associated with an east-west trending zone of shallow seismicity near 15°N may reflect the presence of a transform fault which connects the active Philippine Trench to the south with the east Luzon subduction zone to the north. The absence of a well-defined Benioff zone together with the absence of seismicity deeper than 200 km beneath eastern Luzon may indicate that this subduction episode is relatively young.

The pattern of seismicity north of about 17°10'N does not lend itself to any simple interpretation (Figure 6). Unlike the area south of 17°10'N, the locations of plate boundaries cannot be inferred on the basis of seismicity alone in this region.

Recent Deformation Along the East Luzon Margin

A series of single and multichannel seismic reflection profiles across the east Luzon margin reveals subduction-related deformation of sedimentary strata from approximately 15.5°N to 18°N (tracks in Figure 3). The intensity of folding and faulting along the margin increases from north to south. At about 18.5°N, the latitude of northern Luzon, there is no recognizable recent deformation of near-surface strata. In sharp contrast, south of about 17.5°N, much of the sedimentary section is involved in recent faulting and folding.

Very well-developed structures characteristic of active subduction zones are present between approximately 15.5°N and 17.5°N (Figures 7 and 8 and Plate 1). In the south, where the sediment fill in the East Luzon Trough is up to about 2 km thick, folds are developed at the landward edge of the East Luzon Trough with wavelengths of 3–5 km and amplitudes at the seafloor of 150–300 m. In profiles B, C, and D (Figures 7 and 8 and Plate 1) much of the entire sediment section overlying oceanic basement appears to be involved in folding and thrust faulting, above a possible décollement surface near the oceanic basement. Reflections from the top of the oceanic crust can be traced dipping landward at 5°–10° under the deformed sediment of the trench-slope on profiles south of 17.5°N.

The onset of deformation occurs very suddenly along a sharply defined 'deformation front' [*White*, 1977; *Montecchi*, 1974]. The fold farthest to seaward is generally of smaller amplitude than folds closer to the coast and presumably is the most recent fold to nucleate. Larger folds landward of the deformation front have roughly constant amplitude and wavelength (profiles B, C, and D) and collectively form the outer arc basement high of *Karig* [1970, 1971]. Recent sediments are ponded behind the deformed material, forming a series of small but probably discontinuous slope basins which coalesce to the north to form the Sierra Madre Basin. The small basins of east Luzon are often disturbed, either by seaward tilting (profile C) or by minor folding and/or faulting (profile D). Tilting of the slope basin strata is best displayed near 18°N, near

Fig. 6. (Opposite) Seismicity map of eastern Luzon compiled from various sources that are cited in text. All hypocenters are shallower than 200 km. Earthquakes for which focal mechanism solutions have been determined are shown by large dots.

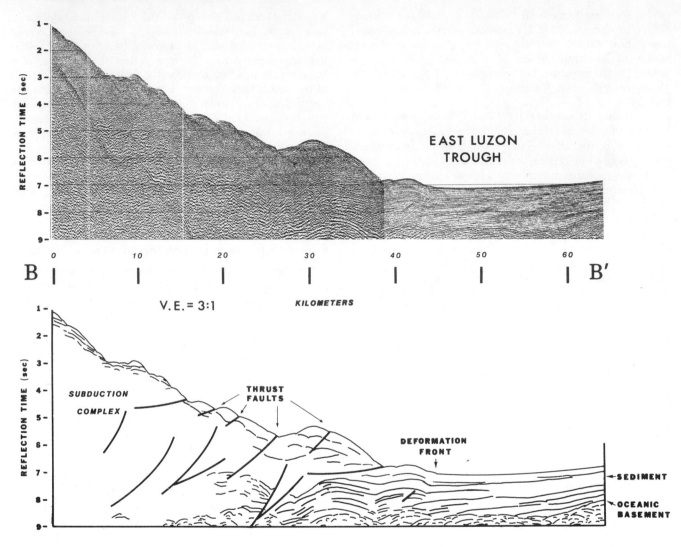

Fig. 7.   Twenty four fold CDP seismic reflection profile and interpretation from the east Luzon margin. Track location in Figure 3. Reflections from the top of the oceanic crust can be continuously traced at least to about the 27-km position at 8.8-s reflection time. A major shallow-dipping thrust fault is presented at 7-s reflection time between 30 and 40 km.

the western end of profile E (Plate 2) at the northern limit of recognizable recent deformation. Seaward tilting of the ponded sediments may be a result of regional downwarping prior to the commencement of underthrusting.

Seismic lines north of 18°N show no evidence of recent tectonism. Flat-lying reflectors fill local depressions in the 'basement' and remain essentially undisturbed as they bury the basement topography. In some places (profile H; Figure 5), laminated 'turbidite' horizons exhibit angular uncomfortable onlap relationships onto a sediment layer that appears conformable with oceanic basement. This horizon may represent pelagic or hemipelagic sediments draped over the rough topography of oceanic crust prior to turbidite-style deposition.

Thrust faulting of accreting sediments begins essentially contemporaneously with folding. On most lines that show recent deformation, the folds immediately landward of the deformation front are cut by landward dipping thrust faults (profiles B, C, and D). As deformation progresses, anticlines are tightened and uplifted, and slip along thrust faults cuts off the limbs of the anticlines and inhibits further formation of synclines during continued deformation. Track spacing is too large to determine the degree of along-strike continuity of individual folds or faults, but the zone of deformation appears to be continuous over about 200 km along strike. The zone of deformation displays a greater degree of along-strike continuity than the back-arc thrusting recognized behind the Banda and Sunda arcs [*Silver*,

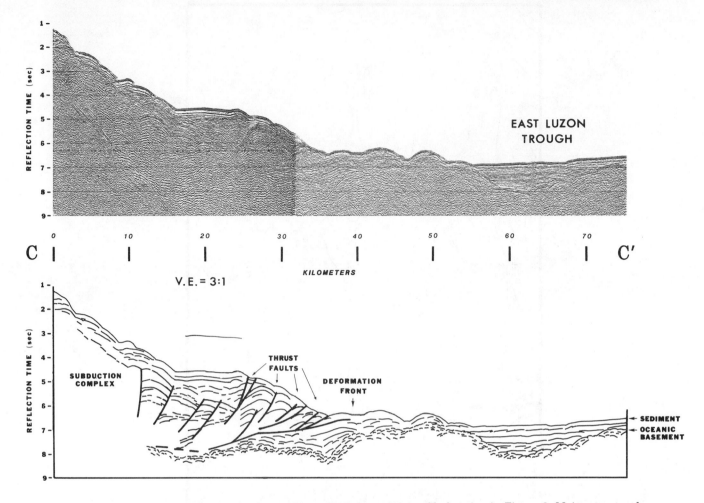

Fig. 8.   Twelve-fold CDP reflection profile and interpretation, profile location in Figure 3. Major structural disharmony is present across a décollement surface between 25 and 40 km at about 7-s reflection time. The décollement occurs within the sedimentary section, well above the oceanic crust. The sediments of the East Luzon Trough onlap the basement topography between 55 and 60 km, suggesting that the basement topography between 40 and 55 km is not the result of recent deformation. The seaward half of the sediment basin present between 15 and 25 km is tilted about 5° toward the East Luzon Trough.

1979]. On profiles where thrust faults are well defined, the dips of individual fault surfaces tend to increase toward land (profiles B, C, and D). The seawardmost faults within the accreted material have shallow dips, generally about 10°–15°, and the faults well within the subduction complex have dips of 20°–30°. Some thrust faults may offset the oceanic basement (profile B).

The surfaces of many small upper slope sediment ponds north of 17°N are tilted to seaward, with inclinations up to approximately 5°. In the upper slope basins, deep reflectors are upturned as they onlap the deformed material of the subduction complex (profile D). To the east, reflector units thin onto the subduction complex, suggesting that uplift and tilting occur contemporaneous with deposition of the basin sediments. Slope basins whose surfaces are

horizontal or landward dipping are generally associated with the larger and better developed segments of the subduction complex (profiles C and D). More steeply seaward dipping slope basins are associated with poorly developed portions of the subduction complexes that presumably have experienced less recent compressive deformation (profile E). Thus significant accretion of material to the overriding plate during subduction may be an important mechanism for producing uplift of the trench-slope and outer forearc region [Dickinson and Seely, 1979; Karig and Sharman, 1975].

The zone of subduction-related deformation is located along the landward margin of the East Luzon Trough south of 17°N (Figure 4). North of 17°N, however, the western boundary of the East Luzon Trough is formed by the Is-

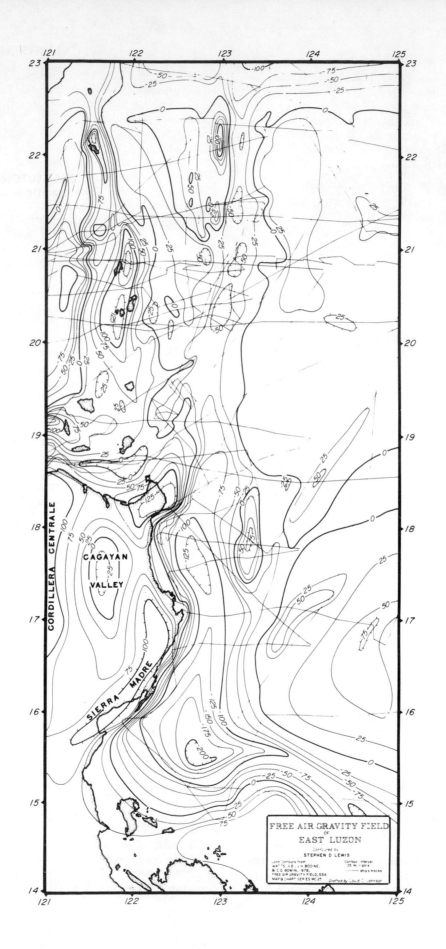

FREE AIR GRAVITY FIELD
OF
EAST LUZON

Contoured by
STEPHEN D. LEWIS

Land Contours from   Contour Interval
WATTS, A.B., J.H. BODINE,   25 milligals
& C.O. BOWIN, 1978.   —— ship's tracks
FREE AIR GRAVITY FIELD, GSA
MAP & CHART SERIES MC-25

Drafted by David L. Johnson

abella Ridge, and the East Luzon Trough is separated from the Luzon margin by the Sierra Madre Basin (Figure 2). The zone of recent subduction-related deformation extends northward to almost 18°N along the landward flank of the Sierra Madre Basin, following the northerly trend of the Luzon coastline rather than the northeasterly trend of the East Luzon Trough (Figure 4).

Gravity

A strongly negative north trending linear gravity anomaly is present in the offshore free air gravity field of eastern Luzon (Figure 9). The axis of the anomaly closely follows the zone of observed deformation and does not follow the trend of the East Luzon Trough. Near 16°N, however, the zone of deformation, the East Luzon Trough, and the negative gravity anomaly are coincident. The gravity anomaly extends much farther north than any observed recent deformation, up to about 19.5°N at the −50 mGal level, and to the Ryukyu Trench at the −25 mGal level. The largest gravity minimum, −200 mGal, occurs at the southern end of the East Luzon Trough, near 15.5°N. The observed gravity anomaly minimum may be related to and therefore serve to define the active tectonism along eastern Luzon. The mass distribution that gives rise to the gravity anomaly may be caused in part by local downwarping along the narrow zone characterized by the observed recent deformation. This may account for the seaward tilting of turbidite-filled basins along the margin. The observation that the gravity anomaly extends much farther to the north than any recognizable recent deformation leads us to speculate that crustal downwarping along a narrow linear belt may be a precursory phenomenon to actual plate rupture and near-surface deformation, at least along eastern Luzon.

The observation that the zone of active tectonism as defined by gravity and recent deformation does not follow the trend of the East Luzon Trough suggests that the East Luzon Trough does not presently constitute an important tectonically active element, despite its 'trenchlike' morphology. Between about 16°N and 17°N, the active tectonic zone does coincide with the western edge of the East Luzon Trough, but north of 17°N the active tectonic zone and the East Luzon Trough diverge (Figure 4). The East Luzon Trough may, however, mark the locus of past subduction episodes.

Transform Fault

Geometric constraints imposed by rigid plate theory suggest that the west striking bathymetric trough which connects the northern end of the Philippine Trench to the East Luzon Trough near 15.5°N (Figure 2) may be the location of a transform fault separating the Philippine Sea plate and Luzon. Westward directed underthrusting of the Philippine Sea plate along the Philippine Trench and east Luzon imply a component of left-lateral strike slip motion along this zone, a prediction that is supported by earthquake focal mechanism solutions (Figure 6).

Two multichannel seismic tracks that cross the transform zone contain regions of near-surface faulting that occur within the west trending zone of seismicity of *Cardwell et al.* [1980] located near 15°N. Profile A (Figure 10) shows a series of northward dipping faults that disrupt and offset the thick sediment section of the shelf and slope (track location in Figure 3). Apparent fault spacing is roughly 2–5 km. Most faults show relatively small vertical offset, generally a few tens of meters, but the cumulative vertical surface offset across the observed faults is at least 1000 m. No estimate of horizontal offset can be made directly from the seismic reflection data, but it may be at least several tens of km. Most faults exhibit surface displacement, which, together with the observed seismic activity, indicates that this fault zone is presently active. The trend of the zone of active faulting is constrained by the seismicity and seismic reflection data to be roughly east-west, but it may trend more to the south than is shown in Figure 4.

The bathymetric deep which connects the Philippine Trench to the East Luzon Trough is not coincident with the active transform fault zone as located by seismicity and recent faulting, but lies 50–70 km farther to the north (Figure 4). Thus the seismic and structural expressions of the transform zone are roughly coincident but fall to the south of the bathymetric trough. The northward displacement of the bathymetric trough relative to the structural and seismicity expressions of active transform faulting may reflect a northward dip of the fault zone at depth, as suggested by the observed dip of some of the near-surface faults. Also, progradation of the Lamon Bay shelf northward may have buried the active transform zone under a thick sequence of sediment, which would result in a northward displacement of the bathymetric trough from the transform fault zone.

Thus several lines of evidence indicate that an active trench-trench transform fault takes up the left-lateral strike slip motion that results from plate convergence at the subduction zones of the Philippine Trench to the south and the east Luzon subduction zone to the north (Figure 4). This relatively simple model of present-day plate geometry does not require a 'leaky' transform fault which connects the north end of the Philippine Trench and the south end of the Manila Trench, as inferred by *Divis* [1980], nor does it require a throughgoing 'fracture zone,' as hypothesized by *de Boer et al.* [1980] and *Wolfe* [1981]. However, the apparent lack of active underthrusting north of about 17.75°N, coupled with ongoing subduction to the south, implies some degree of nonrigid behavior, probably of the overriding plate, on a regional scale.

Fig. 9. (Opposite) Free air gravity map of the east Luzon margin. Contour interval is 25 mGal. The major tectonic segments of the region exhibit characteristic gravity signatures. See text for discussion.

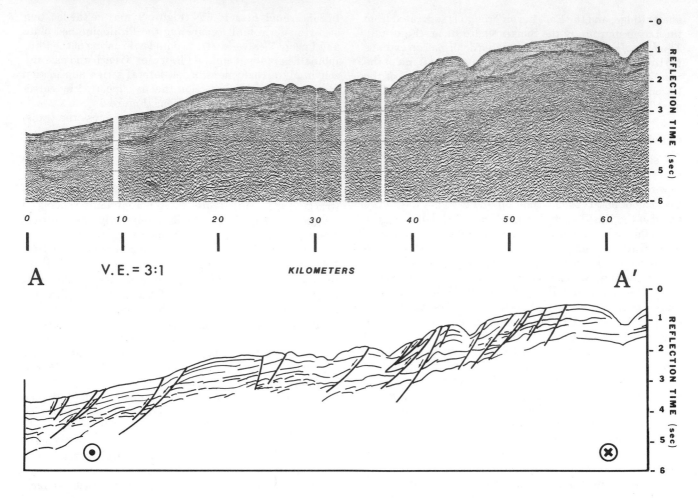

Fig. 10.   Twenty-four fold CDP reflection profile and interpretation across the east Luzon transform zone. The track is located in Figure 3. Many fault traces show surface offset, suggesting recent activity. The normal component of motion on these faults is apparent, but the fault zone is dominantly one of strike slip offset.

Previous Subduction Episodes

The seismic reflection and seismicity data discussed above provide evidence for a presently active subduction zone along eastern Luzon. The geology of the Sierra Madre Mountains [*Christian*, 1964; *Philippine Bureau of Mines*, 1963] suggests that subduction also took place along eastern Luzon during the early Tertiary [*Karig*, 1973]. Radiometric dating of granitic plutons in the Sierra Madre Range [*Wolfe*, 1981] identifies two pulses of igneous activity, one during the middle Eocene (49–43 m.y. B.P.) and one during the late Oligocene (33–27 m.y. B.P.). These periods of igneous activity may be related to subduction pulses along eastern Luzon. Structures that may be associated with Tertiary subduction along eastern Luzon are delineated by the offshore seismic reflection data as well. Seismic lines which cross the Palaui Ridge (profile H) of northeastern Luzon reveal its internal structure to consist of folded and faulted material similar in style to the defor-

mation observed in the trench-slope regions associated with active subduction to the south and along other active trenches (for example, profile D). The Palaui Ridge may represent an early Tertiary subduction complex [*Karig*, 1973]. The deformed material is overlain by a substantial thickness (up to 1.0 s locally) of sediments that are only mildly deformed. A prominent angular unconformity can be identified on segments of some seismic reflection profiles that separates the older deformed material from the overlying strata (Figures 5, 9, 11, and 12). The unconformity can be traced to the south along eastern Luzon as far as about 16.75°N within the undeformed sediment fill along the margin and to about 17.5°N within the deformed region of the slope. The unconformity cannot be identified on the slope south of about 17.5°N, where it becomes lost within the zone of intense recent deformation. The unconformity is also present within the sediment section of the East Luzon Trough (profiles D and E).

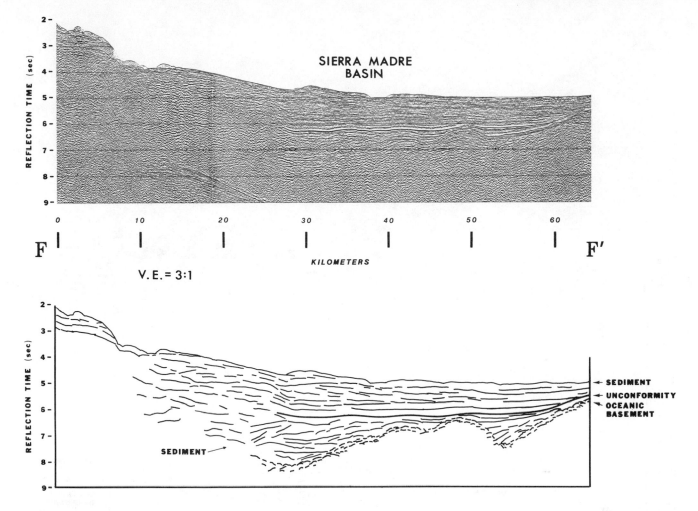

SIERRA MADRE
BASIN

F   F'

KILOMETERS

V. E. = 3:1

SEDIMENT
UNCONFORMITY
OCEANIC
BASEMENT

SEDIMENT

Fig. 11.   Twelve-fold CDP reflection profile and interpretation across the Sierra Madre Basin. Location in Figure 3. The regional unconformity is present as a strong reflector on this profile. A thick sediment sequence is apparent near 30 km beneath the unconformity. There is no evidence for recent compressive deformation within the Sierra Madre Basin at this latitude.

Within the Sierra Madre Basin, the unconformity separates a landward dipping planar reflector sequence from the overlying horizontally bedded sequence (profile E). The overlying sequence exhibits an onlap relationship to the east and west over the surface of unconformity. The sediment section above the unconformity is essentially horizontal. Below the unconformity, the planar reflectors diverge, or 'fan' as well as dip toward Luzon.

The reflector units observed below the regional unconformity are (1) a zone of incoherent or intensely folded reflectors beneath the upper slope represented by the Palaui Ridge, and (2) a landward dipping divergent planar reflector sequence whose dips increase with depth and which directly overlies acoustic basement beneath the Sierra Madre Basin. The unconformity cannot be traced with confidence across the Isabella Ridge.

An interpretation of the observed reflector sequences is suggested by the regional relationships between the East Luzon Trough, the Sierra Madre Basin, and the Isabella Ridge (profiles D and E). The unconformity beneath the East Luzon Trough is disrupted by a series of normal faults whose western sides are downthrown (profiles D and E). Overlying strata are not displaced by the faults that offset the horizons below the unconformity. Thus motion on these faults predates deposition onto the unconformity surface. Weak reflections can be resolved beneath the unconformity. They define a landward dipping and landward thickening planar sequence that is offset by the faulting which deforms the unconformity surface.

The planar sequence can be traced westward within the East Luzon Trough to where it abuts the incoherent internal reflector pattern of the Isabella Ridge. Structures

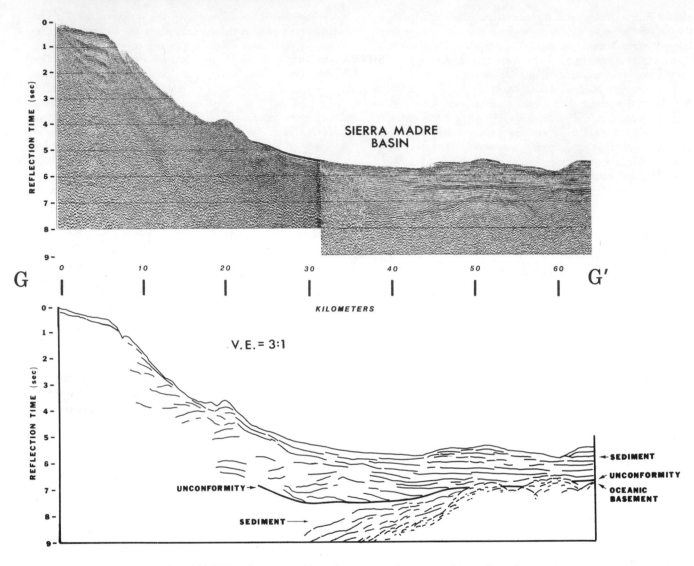

Fig. 12.   Twenty-four fold CDP reflection profile and interpretation across the northern Sierra Madre Basin. The deepest resolvable reflectors are present near 40 km. Track location in Figure 3.

cannot be resolved with confidence within the Isabella Ridge. However, a thin unit (<0.2 s thick) overlies acoustic basement on many portions of the ridge. Sediments within the Sierra Madre Basin are upturned and thinned as they onlap the 'basement' of the ridge (profiles D and E). The stratigraphy of this basin was discussed earlier.

This assemblage of stratigraphic and structural elements is very similar to a trench-subduction complex-forearc basin system. The regional unconformity may represent the seafloor of a subduction zone active during the early Tertiary along eastern Luzon. The trench and forearc basin of this subduction zone are now covered by sediments deposited after subduction ceased. The subduction complex, the Isabella Ridge, is not yet completely buried by postsubduction sediments. The inactive normal faults observed within the East Luzon Trough are identical in character to the normal faults observed in oceanic crust seaward of active trenches (profiles D and E). The chaotic nature of the internal structure of the Isabella Ridge and its relatively low p wave velocity (about 4.0 km/s) are similar to those observed within the subduction complexes of active subduction systems [Lewis and Hayes, 1981], and the stratigraphy and geometry of the Sierra Madre Basin that is now undergoing deformation are similar to that of the forearc basins of other subduction zones.

This tectonic interpretation identifies the East Luzon Trough as the active trench during early Tertiary subduction along eastern Luzon, and the Sierra Madre Basin represents the forearc basin associated with that subduction pulse. The Isabella Ridge represents the subduction complex formed over the subducting oceanic crust. The amplitude of the gravity anomaly associated with the Isabella Ridge, its chaotic internal structure, and its relatively low *p* wave velocity suggest that it is composed largely of deformed sedimentary material. The trends of these structures (Figure 4) indicate that the trend of the subduction zone paralleled the East Luzon Trough.

If the Sierra Madre Mountains and the Palaui Ridge represent the along-strike landward continuation of any of the offshore tectonic elements, then the Sierra Madre range and the Palaui Ridge may be composed in part of uplifted and deformed forearc basin sediments and arc-related volcanic and plutonic rocks rather than a subduction complex of accreted materials, as suggested by *Karig* [1973].

Alternatively, the deformed rocks of the Sierra Madre Mountains and the Palaui Ridge may represent a subduction complex composed of off-scraped and accreted material. As discussed earlier, the same regional unconformity directly overlies the deformed rocks of the Palaui Ridge as well as the structures of the Sierra Madre Basin and the East Luzon Trough which are related to the early subduction pulse. This observation indicates that the Palaui Ridge is probably not significantly older (or younger) than the structures beneath the unconformity farther to the east that were also formed during early Tertiary subduction along eastern Luzon. Hence, if the Palaui Ridge and portions of the Sierra Madre Range actually represent a subduction complex, then the subduction complex was formed during the early phases of the same subduction pulse that was responsible for the offshore structures, the Sierra Madre Basin, the Isabella Ridge, and the East Luzon Trough (profiles D and E).

Some aspects of the geology of the Sierra Madre Mountains may provide constraints on the timing of the cessation of early Tertiary subduction, and hence the age of the unconformity that separates the synorogenic and post-orogenic rock units identified in the seismic data. A north-south trending 40-km-wide band of granite, quartz diorite, and tonalite batholiths, part of the Coastal Batholith of the Sierra Madre Mountains (Figure 4) [*Philippine Bureau of Mines*, 1963], has been dated by K-Ar methods as middle Eocene to late Oligocene [*Metal Mining Agency of Japan*, 1977; *Wolfe*, 1981]. The batholith zone falls approximately 100–120 km landward of the hypothesized fossil trench, which is within the range of spacings between trenches and volcanic arcs observed by *Dickinson* [1973]. Late Eocene to early Oligocene basaltic and andesitic lavas, tuffs, and volcanic breccias of the Caraballo and Mamparang formations form the basement of the southeastern Cagayan Valley [*Metal Mining Agency of Japan*, 1977]. The Eocene-Oligocene granitic batholiths together with the Eocene-Oligocene volcanic rocks may collectively represent the volcanic arc which was associated with the fossil subduction zone identified offshore along east Luzon. A pure white limestone lacking a substantial component of volcanic detritus, the Columbus Formation, directly overlies the volcanogenic Mamparang Formation. The unit is paleontologically dated as upper Oligocene on the basis of benthonic

foraminifera [*Metal Mining Agency of Japan*, 1977]. The apparent absence of volcanic debris in the Columbus limestone suggests that volcanic activity effectively ceased or slowed in the Sierra Madre during the late Oligocene. The youngest radiometric age date obtained from granitic plutons within the Sierra Madre is 27 m.y. B.P. [*Wolfe*, 1981], which also implies a late Oligocene diminution of plutonic activity. We infer that subduction ceased or slowed along eastern Luzon during the late Oligocene. The accumulation rate of volcanic ash to the east of the Sierra Madre Range at DSDP site 292 on the Benham Rise (Figure 1) determined by *Donnelly* [1975] also shows a steep decline near the Oligocene-Miocene boundary. Hence, the age of the offshore regional unconformity separating the syn-subduction and postsubduction strata observed on the seismic profiles may be late Oligocene. If the cessation of subduction occurred along east Luzon in latest Oligocene time, it may have been roughly comtemporaneous with the beginning of tectonism on western Luzon along the Manila Trench [*Schweller et al.*, this volume]. We cannot presently suggest a cause for this 'flip' of subduction polarity.

Several studies have proposed that the northwest portion of the West Philippine Basin has experienced a deformational event since its formation [*Mrozowski et al.*, 1982; *Karig and Wagemen*, 1975]. Perhaps the most compelling evidence in support of this hypothesis is the strongly lineated north-northeast trending basement fabric [*Mrozowski et al.*, 1982; *Karig*, 1975] mapped in this region, the absence of lineated seafloor spreading-type magnetic anomalies [*Mrozowski et al.*, 1982], and the recovery of a Miocene metamorphosed fault breccia at DSDP site 293 [*Karig et al.*, 1975]. In contrast, farther to the east, the 'main basin' of *Mrozowski et al.* [1982] exhibits a northwest trending basement fabric and well-developed seafloor-spreading magnetic anomalies roughly parallel to the basement fabric. Drilling results provide no evidence for a Miocene tectonic event in the central West Philippine Basin [*Karig et al.*, 1975]. *Mrozowski et al.* [1982] suggest that the origin of the northeast trending basement fabric of the West Philippine Basin near Luzon may be related to regional compression or shear. *Lewis and Hayes* [1980] correlated the basement ridges mapped by *Karig and Wageman* [1975] in the deep basin with basement ridges on the east Luzon margin (Figure 4). They hypothesized that the ridges were formed by regional compression related to the initiation of subduction along east Luzon, perhaps analogous to the deformation observed in the Indian Ocean by *Weissel et al.* [1980].

The late Oligocene (?) unconformity is uplifted and gently folded over many basement ridges (Figure 5), which suggests that at least some of the deformation that gave rise to the northeasterly basement fabric in the West Philippine Basin may have occurred after late Oligocene time. The drilling results from DSDP hole 293 indicates that deformation occurred before middle Miocene time and before most of the turbidite apron of east Luzon was deposited

[*Karig and Wagemen,* 1975]. These age constraints help date the deformation event which generated the northeast trending basement fabric of the West Philippine Basin near Luzon as early Miocene. We recognize no features in the seismic reflection data typically associated with recent major transcurrent faulting. Furthermore, seismicity studies give no indication of any presently active strike slip faulting offshore of northeast Luzon.

North Luzon Ridge

The diffuse pattern of shallow seismicity which characterizes the North Luzon Ridge (Figure 6) is evidence that deformation is presently occurring in this region. *Seno and Kurita* [1978] propose a model for the North Luzon Ridge based on seismicity data which consists of a series of northwest striking strike slip faults which together comprise a 'transform belt' [*Ishibashi,* 1978]. The transform belt takes up the relative motion between Taiwan and the Philippines. *Seno and Kurita* [1978] choose the northwesterly trending nodal planes of the strike slip focal mechanisms between 22°N and 23°N in Figure 6 to represent the active fault planes.

In contrast, *Karig* [1973] argues that the arcuate north and northeasterly trending bathymetric deeps which incise the North Luzon Ridge (Figure 2) represent the traces of active faults (Figure 4). On the basis of a simple mechanical model [*Karig,* 1973, Figure 2] Karig states that left-lateral motion occurs across the faults of the North Luzon Ridge. The focal mechanism solutions presented by *Seno and Kurita* [1978] argue against left-lateral motion along northeasterly trending faults as proposed by *Karig* [1973]; yet the bathymetric data (Figure 2) provide little evidence of northwest trending faults within the North Luzon Ridge north of about 21.5°N, as *Seno and Kurita* [1978] hypothesize.

We interpret the bathymetric data (Figure 2) and the seismicity data (Figure 6) together to indicate that the North Luzon Ridge region between Luzon and Taiwan is presently undergoing faulting along arcuate curvilinear fault traces whose sense of displacement is right lateral along northeasterly trends and left lateral along northwesterly trends. Also, portions of the North Luzon Ridge may be bounded by northwest trending left-lateral faults (Figure 4). Many north trending faults also probably exhibit a component of thrust motion as well [*Hamburger et al.,* this volume]. The direction of relative motion between northeastern Luzon and the Philippine Sea plate determined from the seismic slip vectors along east Luzon [*Cardwell et al.,* 1980] and the strike of the transform fault discussed earlier is roughly east-west. Most relative plate motion models suggest a more northwesterly direction of convergence between northeastern Luzon and the Philippine Sea plate [e.g., *Minster and Jordan,* 1979; *Chase,* 1978; *Seno,* 1977], although most workers concede large uncertainties in determining relative motion directions

in this region. This apparent discrepancy may indicate that the left-lateral Philippine fault system accommodates the transcurrent component of Philippine Sea plate–Luzon relative motion [*Fitch,* 1972], and roughly orthogonal convergence takes place along east Luzon, as the seismicity data and transform fault trend indicate. Right-lateral motion along the northeast striking bathymetric troughs of the North Luzon Ridge implies a more westerly direction of relative plate motion in this region than suggested by *Karig* [1973], but his 'arc versus butress' mechanical model of Eurasia-Taiwan-Luzon interaction may still be applicable. The low level of seismicity along the Gagua Ridge (Figure 6) suggests that it probably does not now constitute an active tectonic element.

The collision of Taiwan with the Eurasian continental margin began in the late Pliocene [*Page and Suppe,* 1981]. From at least the mid-Miocene through much of the Pliocene Taiwan, the North Luzon Ridge, and northern Luzon probably moved together as part of the Philippine Sea plate since there is no evidence to suggest that a well-defined plate boundary was active in the region during that time other than the Manila Trench. Faulting within the North Luzon Ridge may have begun during the Pliocene as Luzon continued to move westward or northwestward relative to Eurasia over the still active Manila Trench, while Taiwan slowed in its motion relative to Eurasia as a consequence of the onset of collision and suturing. The recent active convergence along east Luzon may also have begun at about the same time (latest Pliocene).

Conclusions

The mode of plate interaction between the Philippine Islands and the Philippine Sea plate is highly variable along strike. In the south the Philippine Trench represents a young but apparently well-defined subduction system east of Mindanao and the central Philippines [*Cardwell et al.,* 1980]. The Bicol Peninsula of southeast Luzon is in contact with the Philippine Sea plate along a left-lateral trench-trench transform fault [*McKenzie and Morgan,* 1969]. The plate boundary along eastern Luzon is represented by a very young zone of active underthrusting and compressive deformation that decreases in intensity to the north until earthquake and seismic reflection data indicate that relative motion evidently ceases. Plate interaction north of Luzon along the North Luzon Ridge–Gagua Ridge system is probably represented by a broad zone (some 200 km wide) of diffuse faulting and deformation rather than by a relatively simple plate boundary. The northward decrease in observed recent compressive seismic activity indicates that the 'simple' convergent plate boundary represented by the Philippine Trench in the south may be propagating northward along eastern Luzon and the North Luzon Ridge system. A linear zone of subsidence manifested by seaward tilting of sediment ponds may be a precursor of actual plate rupture and underthrusting.

The East Luzon Trough and the Sierra Madre Basin probably represent the morphologic/structural expressions of a trench and forearc basin, respectively, along east Luzon which were parts of a subduction system active in the early Tertiary. Stratigraphic relationships in the eastern Cagayan Valley and radiometric dating of plutonic rocks in the Sierra Madre Mountains suggest that subduction probably stopped in late Oligocene time, roughly contemporaneous with early tectonism within the Manila Trench system of west Luzon.

Slip vectors calculated from earthquake focal mechanisms [*Cardwell et al.*, 1980] and the trend of the transform fault which connects the Philippine Trench to the east Luzon subduction system suggest a roughly east-west relative motion direction between east Luzon and the Philippine Sea plate. Global plate relative motion models predict a more northwesterly convergence direction between east Luzon and the Philippine Sea plate. Left-lateral motion along the Philippine fault may absorb much of the northward component of relative motion [*Fitch*, 1972], allowing nearly westerly directed convergence along the east Luzon margin. However, there may be larger uncertainties in relative plate motion determinations in this region than in many other regions [*Minster and Jordan*, 1979].

There is no evidence to date of active plate convergence between the Philippine Sea plate and the North Luzon Ridge–Gagua Ridge complex. However, if the relative motion that occurs to the south along east Luzon extends to the north, it may be taken up along numerous strike slip and thrust faults that cross the North Luzon Ridge along arcuate northeasterly to northwesterly trends. Focal mechanism solutions calculated from events east of Taiwan suggest that the sense of motion along northeasterly trending faults is right lateral, and the sense of motion along northwesterly trending faults is left lateral. The North Luzon Ridge may represent a region of complex tectonism flanked to the north and south by relatively simple plate boundaries, the Ryukyu Trench system to the north and the Philippine Trench–east Luzon subduction system to the south. If the east Luzon system continues to propagate to the north, it may eventually join with the Ryukyu Trench, forming a continuous convergent plate boundary between Taiwan, Luzon, and the Philippine Sea plate. Renewed subduction along East Luzon may represent the beginning of the second subduction polarity reversal to occur along the Luzon margins during the Cenozoic.

Acknowledgments. We thank Mike Hamburger, Kerry Hegarty, Cary Mrozowski and Will Schweller for many fruitful discussions. Critical reviews by S. Bachman, J. Ladd, M. Langseth, J. C. Moore, R. Murphy, E. Silver and J. Weissel helped improve the manuscript. Technical advice and assistance were provided by E. Free, M. Garland, D. Johnson, and A. Lewis. This work was supported by NSF grant OCE 79-19069 as part of the IDOE/SEATAR project. Lamont-Doherty Geological Observatory contribution 3371.

References

Allen, C. R., Circum-Pacific faulting in the Philippines-Taiwan region, *J. Geophys. Res., 67,* 4795–4812, 1962.

Balce, G. R., R. Y. Encina, A. Momongan, and E. Lara, Geology of the Baguio District and its implications on the tectonic development of the Luzon Central Cordillera, *Geol. Paleont. Southeast Asia, 21,* 265–287, 1980.

Cardwell, R. K., B. L. Isacks, and D. E. Karig, The spatial distribution of earthquakes, focal mechanism solutions, and subducted lithosphere in the Philippine and northeastern Indonesian Islands, in *The Tectonic and Geologic Evolution of Southeast Asian Seas and Islands, Geophys. Monogr. Ser.,* vol. 23, edited by D. E. Hayes, pp. 1–36, AGU, Washington, D.C., 1980.

Chai, B. H. T., Structure and tectonic evolution of Taiwan, *Am. J. Sci., 272,* 389, 1972.

Chase, C. G., Plate kinematics: The Americas, east Africa, and the rest of the world, *Earth Planet. Sci. Lett., 37,* 355–368, 1978.

Christian, L. B., Post-Oligocene tectonic history of the Cagayan Basin, Philippines, *Philipp. Geol., 18,* 114–147, 1964.

de Boer, J., L. A. Odom, P. C. Ragland, F. G. Snider, and N. R. Tilford, The Bataan Orogene; eastward subduction, tectonic rotations, and volcanism in the western Pacific (Philippines), *Tectonophysics, 67,* 251–282, 1980.

Dewey, J. F., and J. M. Bird, Mountain belts and the new global tectonics, *J. Geophys. Res., 75,* 2625, 1970.

Dickinson, W. R., Widths of modern arc-trench gaps proportional to past duration of igneous activity in associated magmatic arcs, *J. Geophys. Res., 78,* 3376–3389, 1973.

Dickinson, W. R., and D. R. Seely, Structure and stratigraphy of forearc regions, *Am. Assoc. Pet. Geol. Bull., 63,* 2–31, 1979.

Divis, A. F., The petrology and tectonics of recent volcanism in the central Philippine Islands, in *The Tectonic and Geologic Evolution of Southeast Asian Seas and Islands, Geophys. Monogr. Ser.,* vol. 23, edited by D. E. Hayes, pp. 127–144, AGU, Washington, D.C., 1980.

Donnelly, T. W., Neogene explosive volcanic activity of the western Pacific: Sites 292 and 296, DSDP leg 31, *Initial Rep. Deep Sea Drill. Proj., 31,* 577–598, 1975.

Durkee, E. F., and S. L. Pederson, Geology of northern Luzon, Philippines, *Am. Assoc. Pet. Geol. Bull., 45,* 137–168, 1961.

Fitch, T. J., Plate convergence, transcurrent faults, and internal deformation adjacent to Southeast Asia and the western Pacific, *J. Geophys. Res., 77,* 4432–4460, 1972.

Hamburger, M. W., R. K. Cardwell, and B. L. Isacks, Seismotetonics of the northern Philippine island arc, this volume.

Hamilton, W., Tectonics of the Philippine Region, *U.S. Geol. Surv. Prof. Pap., 1078,* 338 pp., 1979.

Ishibashi, K., Plate convergence around the Izu collision zone, central Japan: Development of a new subduction boundary with a temporary transform belt, paper presented at the International Geodynamics Conference, Tokyo, 1978.

Karig, D. E., Ridges and basins of the Tonga-Kermadec island arc system, *J. Geophys. Res., 75,* 239, 1970.

Karig, D. E., Structural history of the Mariana island arc system, *Geol. Soc. Am. Bull., 82,* 323, 1971.

Karig, D. E., Plate convergence between the Philippines and the Ryukyu Islands, *Mar. Geol., 14,* 153–168, 1973.

Karig, D. E., Basin genesis in the Philippine Sea, *Initial Rep. Deep Sea Drill. Proj., 31,* 857–879, 1975.

Karig, D. E., and G. F. Sharman III, Subduction and accretion in trenches, *Geol. Soc. Am. Bull., 86,* 377–389, 1975.

Karig, D. E., and J. M. Wageman, Structure and sediment distribution in the northwest corner of the West Philippine Basin, *Initial Rep. Deep Sea Drill. Proj., 31,* 615–620, 1975.

Karig, D. E., J. C. Ingle, Jr., A. H. Bouma, C. H. Ellis, N. Haile, I. Koizumi, H. Y. Ling, I. MacGregor, J. C. Moore, H. Ujiie, T. Watanabe, S. M. White, and M. Yasui, *Initial Rep. Deep Sea Drill. Proj., 31,* 927 pp., 1975.

Katsumata, M., and L. R. Sykes, Seismicity and tectonics of the western Pacific: Izu-Mariana-Caroline and Ryukyu-Taiwan regions, *J. Geophys. Res., 74,* 5923–5948, 1969.

Lewis, S. D., and D. E. Hayes, Northward-propagating subduction along eastern Luzon, Philippine Islands (abstract), *Eos Trans. AGU, 61,* 1105, 1980.

Lewis, S. D., and D. E. Hayes, Structure of the Manila Trench system, offshore northern Luzon (abstract), *Eos Trans. AGU, 62,* 1042, 1981.

Mammerickx, J., R. L. Fisher, F. J. Emmel, and S. M. Smith, Bathymetry of the east and southeast Asian seas, *Map Chart Ser. MC-17,* Geol. Soc. of Am., Boulder, Colo., 1976.

McKenzie, D. P., Speculations on the consequences and causes of plate motion, *Geophys. J. R. Astron. Soc., 18,* 1, 1969.

McKenzie, D. P., and W. J. Morgan, Evolution of triple junctions, *Nature, 224,* 125–133, 1969.

Metal Mining Agency of Japan, Report on the Geological Survey of Northeastern Luzon, consolidated report, 106 pp., Jpn. Int. Cooperation Agency, Tokyo, 1977.

Minster, J. B., and T. H. Jordan, Rotation vectors for the Philippine and Rivera plates (abstract), *Eos Trans. AGU, 60,* 958, 1979.

Montecchi, F., Shallow tectonic consequences of possible phenomenon of 'subduction' and its meaning to hydrocarbon explorationist, paper presented at the Circum-Pacific Energy and Mineral Resources Conference, AAPG/CCOP/SOPAC, Honolulu, 1974.

Moore, G. F., Geology of southern Mindanao, Philippines (abstract), *Eos Trans. AGU, 62,* 1086, 1981.

Mrozowski, C. L., S. D. Lewis, and D. E. Hayes, Complexities in the tectonic evolution of the West Philippine Basin, *Tectonophysics, 82,* 1–24, 1982.

Murphy, R. W., The Manila Trench-West Taiwan foldbelt: A flipped subduction zone, *Bull. Geol. Soc. Malays., 6,* 27–42, 1973.

Page, B. M., and J. Suppe, The Pliocene Lichi melange of Taiwan: Its plate tectonic and olistostromal origin, *Am. J. Sci., 281,* 193–227, 1981.

Philippine Bureau of Mines, Geological map of the Philippines, scale 1:1,000,000, 9 sheets, Manila, 1963.

Ranneft, T. S. M., R. M. Hopkins, A. J. Froelich, and J. W. Gwinn, Reconaissance geology and oil possibilities of Mindanao, *Am. Assoc. Pet. Geol. Bull., 44,* 529–568, 1960.

Schweller, W. J., D. E. Karig and S. B. Bachman, Original setting and emplacement of the Zambales ophiolite, Luzon, Philippines, from stratigraphic evidence, this volume.

Seno, T., The instantaneous rotation vector of the Philippine Sea plate relative to the Eurasian plate, *Tectonophysics, 42,* 209–226, 1977.

Seno, T., and K. Kurita, Focal mechanisms and tectonics in the Taiwan-Philippine region, *J. Phys. Earth, 26,* S249–S263, 1978.

Silver, E. A., Back-arc thrusting (abstract), *Eos Trans. AGU, 60,* 958, 1979.

Weissel, J. K., R. N. Anderson, and C. A. Geller, Deformation of the Indo-Australian plate, *Nature, 287,* 284–291, 1980.

White, R. S., Recent fold development in the Gulf of Oman, *Earth Planet. Sci. Lett., 36,* 85, 1977.

Wolfe, J. A., Philippine Geochronology, *J. Geol. Soc. Philipp., 35,* 1–30, 1981.

Paleomagnetism of Luzon

M. Fuller, R. McCabe, and I. S. Williams

Department of Geological Sciences, University of California, Santa Barbara, California 93106

J. Almasco, R. Y. Encina, and A. S. Zanoria

Philippine Bureau of Mines, Manila, Philippines

J. A. Wolfe

Pan-Asian Technical Services, Manila, Philippines

Paleomagnetic studies of 50 Cenozoic sites in the northern Philippines are reported and used to constrain tectonic models of the evolution of the region. Plio-Pleistocene results demonstrate that paleomagnetically distinguishable rotations of the regions have not taken place in the past 5 m.y. Early and mid-Miocene results reveal counterclockwise rotation of approximately 20°, but give no evidence of detectable N/S translation. Oligocene results from the Cordillera of Luzon near Baguio give similar rotations to those found in the early and mid-Miocene formations. Eocene sites from the Zambales ophiolite and overlying sediments indicate stronger rotations and appear to have been formed at equatorial latitudes. A tectonic model for the past 17 m.y. is proposed which is constrained by the counterclockwise rotation of early- and mid-Miocene sites. The model invokes the combined action of the northwestward motion of the Philippine Sea Plate and the pinning of the northern Philippines at its southern margin by collision with the Calamian continental fragment. The continued advance of the leading edge of the Philippine Sea Plate causes counterclockwise rotation of Luzon and the Manila trench, until the northern end of the trench becomes pinned by collision with Taiwan.

1. Introduction

The Philippine Islands lie off the Asian landmass between the South China and Philippine Seas, part of the string of islands from the Kuriles in the north to Indonesia in the south. At the latitude of the Philippines, more than two thousand kilometers separate the Pacific plate from the Asian landmass. This tectonically active region is the site of formation of new ocean crust, of subduction, and of major transform faulting. In addition to these fundamental plate interactions, new continental crust is being formed. The processes forming these new landmasses may well be similar to those that operated, in more remote geological time, to produce the continents we now see. An understanding of the geology of the Philippines and of the tectonically active region in which they lie may therefore be of some general interest in explaining the growth of continents.

The geology of the Philippines records the complicated history of a mobile region caught between two major plates. The first paper on the geology of the Philippines was written at the very end of the last century [Becker, 1899]. Since that time, many workers have attempted regional descriptions [e.g., Corby, 1951]. During the 1960's, the Philippine Bureau of Mines compiled much of this work and published the *Geologic Map of the Philippines* [Philippine Bureau of Mines, 1964], at a scale of 1 : 1,000,000. In the late 1960's, *Gervasio* [1967] published a tectonic synthesis of the Philippines. With the advent of plate tectonics and in particular the development of ideas on the origin of marginal basins [*Karig,* 1971], the geology of the area in which the Philippines lie became a matter of considerable general interest; papers appeared interpreting the Philippines in terms of plate tectonic concepts [e.g., *Karig,* 1973] in such a complicated mobile region, an essential aid to our understanding is a paleomagnetic framework against which tectonic models can be tested.

Hsu [1972] carried out the first paleomagnetic study in Luzon, analyzing a large collection of samples. His conclusions were that Luzon had suffered continuous counterclockwise rotation throughout the Tertiary and had moved from south to north at a rate of 10 cm per year.

This study may well have established major elements of the paleomagnetic framework of Luzon. However, Hsu did encounter serious experimental difficulties. Moreover, the study was carried out before the advent of cryogenic magnetometers and before recent improvements in analysis of paleomagnetic data. It is therefore important that the study be refined and extended. More recently, *De Boer et al.* [1980] reported a study of Plio-Pleistocene volcanics in central Luzon, suggesting that the region had suffered counterclockwise rotation of about 35° in the past 4 million years. This rotation rate is remarkable, even for so tectonically active a region as the Philippines, and is not supported by the earlier study of *Hsu* [1972]. An excellent review of paleomagnetic studies in Southeast Asia, including an analysis of the tectonic implications of the seafloor magnetic anomalies in the area, has been given by *Jarrard and Sasajima* [1980].

The paleomagnetic study described in this paper is designed to extend and refine the earlier studies and, in particular, to answer the following two questions. (1) What were the tectonic rotations suffered by Luzon during the Tertiary? (2) Has Luzon acted as a single tectonic unit during this time, or can separate histories be determined for different units within the island?

Clearly, the answers to these questions have important bearing on the geological history of the Philippines themselves and of the surrounding regions.

2. Paleomagnetic Techniques

Standard paleomagnetic techniques were used throughout this study. The samples were mostly collected with a portable drill, although, at some sites, it was more convenient to collect hand samples, which were later drilled at the laboratory. The measurements of remanent magnetization were made for the most part with an SCT horizontal access magnetometer. Particularly strongly magnetized samples were measured with a Schonstedt spinner. The magnetometers have dedicated North Star minicomputers, which carry out the data reduction. Six measurements are made of each sample with the SCT magnetometer and Fisher statistics [*Fisher,* 1953] used to assess the measurement of the individual sample.

Alternating field and thermal demagnetization were used to isolate the primary natural remanent magnetization (NRM) from secondary magnetizations. Static AF demagnetization was used with a coil capable of giving a 1000 Oe (79.58 × 10^3 At/m) H field. The coil was placed in a three-stage μ-metal shield and driven at resonance by a Bellman tunable AC power supply. No single level of AF demagnetization was chosen on the basis of pilot studies, but, rather, each sample was carried through a sequence of demagnetizations. Preferred directions were chosen by using Zijderveld diagrams [*Zijderveld,* 1967]. Nonconvergence to the origin on Zijderveld diagrams and vector differences, which remained distinct from the demagnetized directions, were used as criteria to exclude results.

The North Star minicomputer drives a Houston plotter, so that the Zijderveld diagrams and equal area plots of unit vectors may be produced and studied during demagnetization.

Thermal demagnetization was carried out with two systems. One, which relied upon field cancelation by a coil array, gave cancelation to a part in 10^3 of the field. The other system, which had a three-stage μ-metal shield, gave cancelation to a part in 10^4. The results of thermal demagnetization were analyzed in a similar manner to that used with the AF demagnetization.

The analysis of these samples from the Philippines was unusually difficult. Part of this difficulty arose from weathering of some of the samples, which had generated stable secondary magnetizations not readily separated from the primary NRM. About 20% of the sites had to be rejected because the NRM after demagnetization was still erratic. Those sites which are reported generally had a α_{95} of less than 10.

Tectonic corrections were applied in the standard manner, but in no case did we have sufficient geological field data to apply a correction for plunging folds. The tectonic corrections applied to dikes was that of the bedding of the oldest sediments not intruded by the dike. The choice of the tectonic correction in the Zambales ophiolite presented special problems and is discussed below.

Fisher [1953] statistics were used throughout. The final analysis is given in terms of directions of magnetization.

3. Geological Setting of Sampling Sites

The principal collections were made in the vicinity of Baguio in Benguet Province and in Zambales and Pangasinan. Subsidiary collections were made in the various provinces shown in the map of the northern Philippines (Figure 1), such as Rizal, Albay, and Marinduque. The samples range in age from Eocene to Pleistocene. Although they do not give total coverage across Luzon, they do represent different major structural regions of the island. For example, the two major collections permit comparison of the tectonic history of the Central Cordillera represented by Baguio and the Zambales ophiolite and associated formations.

The geology of the Baguio region has recently been reviewed by *Balce et al.* [1980]. As they note, the tectonic development of the Luzon Central Cordillera is related to the north-south subduction zones lying offshore on either side of the Philippines. To the west lies the east dipping Manila subduction zone. To the east there is the west dipping subduction zone of the Philippine trench. Baguio itself is located at the southern end of the Central Cordillera, where the Cordillera is terminated by the Philippine fault. The area is of great economic interest, providing major contributions to the gold and copper production of the Philippines. As a consequence, its geology has received considerable attention.

The Baguio region affords access to important Tertiary

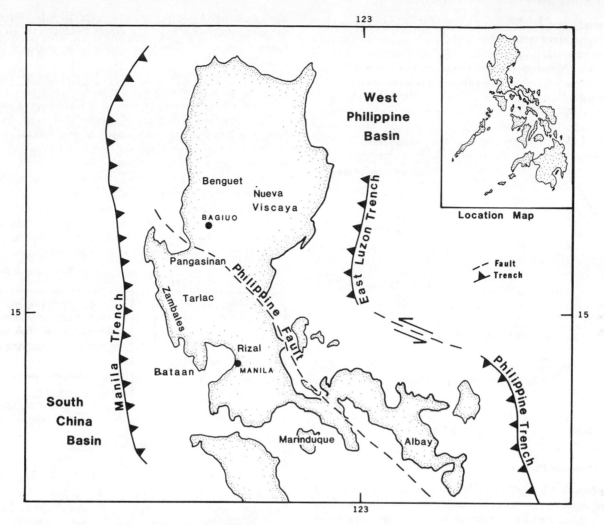

Fig. 1.   Sketch map of the northern Philippines showing the major structural features and the provinces from which paleomagnetic samples were taken.

sedimentary sections, as well as to flows and intrusions of the same age. Exposures are good due to the rapid uplift and associated high relief in the region. Figure 2 gives a simplified stratigraphic column and Figure 3 illustrates the various sites. Immediately to the south of Baguio, the Kennon Road follows the Bued River through a region of rugged relief in which Oligocene to Pliocene sediments are exposed. There are also minor intrusions and flows. This formed one of the main collecting sections. To the west of Baguio, the Asin, Naguillian Roads, and the new Marcos Highway give somewhat comparable sections through upper Miocene and Pliocene sediments. Some 30 sites are reported from the region.

Zambales lies to the southwest of Baguio and is separated from the Central Cordillera by the Central Basin. The region is dominated by the Zambales range composed of the ultrabasics and gabbros of the Zambales ophiolite

complex [e.g., *Hawkins and Evans,* this volume]. Our collection sites are illustrated in Figure 4 and the stratigraphic column is given in Figure 5. The main collections were from the gabbros at the Acoje and Coto mines. In addition, we sampled sills, gabbros, and a quartz diorite to the north near Sual. At Barlo, we collected Oligocene sediments. On the east side of the range, sills in the Camelling River, and the Eocene Aksitero formation, which lies conformably above the pillow basalts of the ophiolite complex [*Schweller and Karig,* 1982]. Above that, the Miocene Moriones formation was also sampled.

4.   Paleomagnetic Results

The paleomagnetic results are presented in Tables 1–5 and Figures 6–10. They are discussed in order of increasing age from Plio-Pleistocene to Eocene. In each of the figures, the small stereonets show the directions of mag-

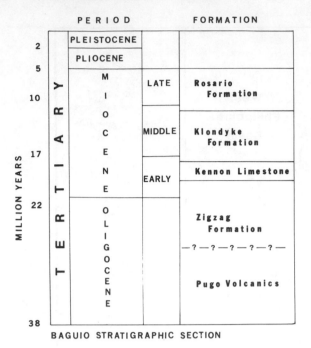

Fig. 2.   Stratigraphic section for the Baguio area (Benguet province); after *Balce et al.* [1980].

netization of the individual samples from each site. The large stereonets give the site mean directions with their associated α_{95} ovals.

4.1. *Pliocene-Pleistocene Results*

There are 14 sites in Pliocene-Pleistocene age formations from five provinces of Luzon. The results from Benguet, Tarlac, and Rizal have declinations indistinguishable from zero; however, their inclinations are consistently shallower than that predicted by the geocentric axial dipole. The direction from the Albay site is marginally distinguishable at the 95% confidence level. However, these results come from volcanic units that cool rapidly, giving spot readings of the field; they do not, therefore, average secular variation. The Fisher mean of all the Bataan sites yields a declination of 354.7, indistinguishable from zero at the 95% confidence level.

These results suggest that the mean declination for Pliocene-Pleistocene formations on Luzon is not displaced from zero. The mean inclination is 17.1 and is statistically distinguishable from that of the axial geocentric dipole. The mean virtual geomagnetic pole (VGP) estimate from all the Luzon sites is located at N83.8 W26.9 with an α_{95} of 3.6.

4.2 *Late Miocene*

The late Miocene results come from Benguet, Marinduque, and Albay provinces. The sites on Marinduque are indistinguishable from the geocentric axial dipole direc-

tion. Although two sites near Legaspi, in Albay Province, are distinguishable from the dipole field, the mean of the three sites is not. Thus, the departures could be due to secular variation.

The four sites from the Baguio City region of Benguet Province give inconsistent results. One site has a declination of 009. The two reversed sites have equivalent normal declinations of 018 and 033. Site 78-10 is an andesite flow and has a single component of magnetization. Site 79-26 is a sediment from the Rosario formation a few hundred meters stratigraphically beneath the flow. Site 81-29 is a flow in the Santo Tomas area, which, like the other flow, is paleomagnetically well behaved. Site 79-25, which has a declination of 348, is stratigraphically lower in the Rosario section and has a direction close to the present field before tectonic correction. It therefore appears that the most reliable sites from the Baguio region

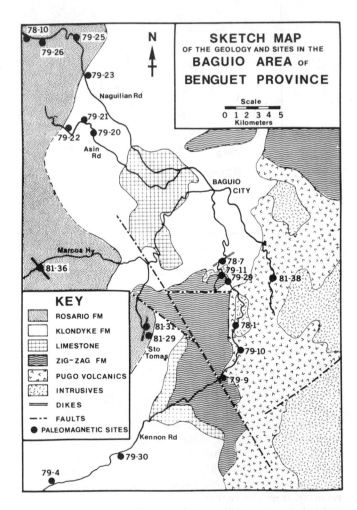

Fig. 3.   Sketch map of the geology of the area around Baguio city in Benguet province. The geology is simplified and is based upon that shown by *Balce et al.* [1980], *Pena* [1970], and the *Philippine Bureau of Mines* [1964].

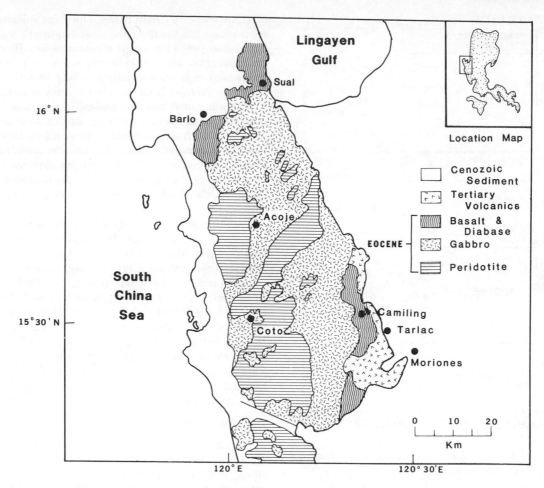

Fig. 4.   Sketch map of the geology of the Zambales Ophiolite, in Zambales and Pangasinan provinces, based upon that by *Hawkins and Evans* [this volume].

of late Miocene age all have easterly declinations. However, the mean direction of these three sites is not statistically distinguishable from the geocentric axial dipole field direction.

The late Miocene results from the Baguio region present something of a puzzle, which is not likely to be resolved until more extensive collections are made. Taken at face value, they suggest that the region may have recorded a different declination from those found in Marinduque and Albay. However, until the result is demonstrated by more extensive collections, one should be cautious in its use.

4.3.   Early and Mid-Miocene Results

The early and mid-Miocene results give consistently westerly declinations. All of the sites from Benguet, Marinduque, and Tarlac show this westerly declination, although the determinations from Tarlac and Marinduque do not provide as high quality data as those from Benguet.

The results from Baguio in Benguet come from sediments, a dyke, and a stock. Before tectonic correction, the

Klondyke sites have a direction close to, but distinguishable from, the present field. The reversed sites from the Zig-Zag formation and from the dyke are, however, antipodal to the tectonically corrected Klondyke results. Since the reversed sites are clearly not remagnetized in the present field, their antipodal relation to the Klondyke sites gives increased confidence in the Klondyke results. The mean direction for the Baguio results is D. 344.2 I. + 30 with an α_{95} of 5.2.

The Miocene results present one of the clearest patterns we have found. Over a considerable region a variety of rock types define a westerly declination. The results from Baguio are particularly convincing. The mean inclination of all sites is indistinguishable from the geocentric axial dipole field.

4.4.   Oligocene Results

As Figure 9 and Table 4 reveal, the Oligocene results are much sparser than those for other periods. The results come from Benguet, Nueva Vizcaya, and Pangasinan. The

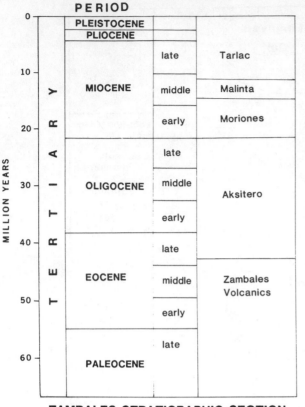

ZAMBALES STRATIGRAPHIC SECTION

Fig. 5. Stratigraphic section for the Zambales area (Zambales and Pangasinan provinces); after *Schweller and Karig* [1982].

latter province is in the northern part of the region dominated by the Zambales ophiolite.

The results from Benguet and Nueva Vizcaya (immediately to the east) reveal declinations of between 320° and 340°, which are indistinguishable from the early and mid-

Miocene westerly declinations. The Pugo volcanics series, from which the results from near Baguio City come, has been subjected to low grade metamorphism. However, the magnetization does not seem to have been affected; a single component is seen, and it is unlikely that the magnetization would have been completely reset by such low temperature regional metamorphism. In any case, the metamorphism must have occurred prior to the early Miocene; the Zig-Zag formation, which lies unconformably upon the Pugo, is unmetamorphosed. It therefore appears that the Pugo is carrying westerly declinations acquired during the Oligocene. The single result from an unmetamorphosed flow from Caraballo in Nueva Vizcaya gave a similar result.

The Oligocene sediments from Pangasinan lie on the northern part of the Zambales ophiolite and give puzzling results. They will have to be re-sampled. The sediments consist of buff and very light-colored siltstones. The light-colored facies gave a good grouping at 300 Oe as shown in Figure 9. The darker siltstones had negative inclinations and were distributed in declination from about 030 to 080 after AF demagnetization. Thermal demagnetization did not bring about a better grouping. Taken at face value, the appropriate recorded direction is that of the light-colored siltstones. However, the inconsistency between the two sediment types suggests that additional analysis is required.

4.5. *Eocene Results*

The Eocene results come exclusively from the Zambales ophiolite and associated formations in Zambales, Pangasinan, and Tarlac provinces. Most of the igneous units sampled gave paleomagnetically well-behaved directions, but there is considerable difficulty in assigning appropriate tectonic corrections. The sediments, which are not such good paleomagnetic recorders, do provide some control for tectonic corrections.

On the beach at Sual, there are tabular intrusives that

TABLE 1.   Pliocene-Pleistocene Paleomagnetic Results

Province	Site	Formation	Location		N	Demagnetization*	Inc.	Dec.	α_{95}	K	VGP Latitude	VGP Longitude
Benguet	81-31	dike	N16.3	E120.5	7	400/450	14.0	359.5	8.8	48	N80.8	W56.4
	81-36	dike	N16.3	E120.5	4	—/450	−18.3	170.2	10.1	24	S78.2	W175.8
Tarlac	81-51	Bamban	N15.3	E120.6	7	200/450	26.1	1.0	7.4	68	N88.2	W91.7
	81-52	Bamban	N15.3	E120.6	9	200/550	−16.3	176.3	9.8	28	S82.1	W148.4
Rizal	81-61	Antipolo	N14.5	E121.1	6	900/400	−27.9	175.4	5.1	175	S85.5	W144.2
	81-99	Guadalupe	N14.6	E121.0	9	100/400	−12.9	182.1	7.7	45	S81.8	E106.6
	81-100	Guadalupe	N14.6	E121.2	8	150/450	−23.8	183.4	10.7	24	S86.1	E63.3
	81-101	Guadalupe	N14.6	E121.2	11	—/450	−18.2	183.8	8.1	36	S83.6	E85.6
Albay	81-203	flow	N13.5	E123.6	9	300/550	19.1	352.6	4.8	117	N81.9	E7.5
Bataan	81-110	Mariveles†	N14.4	E120.4	6	300/500	3.3	4.2	6.7	190	N76.6	W78.0
	81-111	Mariveles	N14.4	E120.4	11	300/550	−22.3	171.1	1.9	518	S80.9	W166.5
	81-112	Mariveles	N14.4	E120.4	9	300/550	−19.7	177.5	3.2	260	S85.1	E150.6
	81-113	Mariveles	N14.4	E120.4	5	300/—	−12.3	175.6	4.1	346	S80.7	E147.7
	81-114	Mariveles	N14.4	E120.4	6	300/—	−4.6	165.2	7.4	83	S71.0	E172.1

N is the number of individually oriented paleomagnetic samples; Inc., inclination; Dec., declination; VGP, virtual geomagnetic pole.
*Demagnetization values are listed for first A.F. demagnetization in Oersteds, then thermal demagnetization in degrees Celsius.
†K-Ar age of 3.26 ± 0.6 m.y. [*Wolfe*, 1981].

TABLE 2. Late Miocene Paleomagnetic Results

Province	Site	Formation	Location		N	Demagnetization*	Inc.	Dec.	α_{95}	K	VGP	
											Latitude	Longitude
Benguet	79-25	Rosario	N16.5	E120.5	5	100/—	22.2	348.2	7.3	110	N77.6	E9.4
	79-26	Rosario	N16.5	E120.5	6	300/380	−9.4	213.1	9.8	62	S55.5	E46.0
	78-10	Rosario†	N16.5	E120.5	13	400/—	−40.0	198.0	4.5	86	S71.9	E7.3
	81-29	flow	N16.3	E120.5	4	—/425	12.3	9.3	11.9	60	N76.4	W102.5
Albay	81-202	flow	N13.5	E123.6	8	300/550	13.4	352.5	3.4	265	N80.1	W7.4
	81-204	flow	N13.5	E123.8	6	300/550	15.3	357.6	5.8	132	N84.1	W32.5
	81-205	flow	N13.5	E123.8	8	300/550	24.3	339.6	6.8	67	N70.1	E34.6
Marinduque	81-303	Gasan	N13.5	E122.0	9	150/—	21.6	0.2	9.1	38	N87.7	W62.9
	81-304	dike	N13.5	E122.0	5	200/—	16.1	6.6	5.1	212	N81.6	W109.5

N is the number of individually oriented paleomagnetic samples; Inc., inclination; Dec., declination; VGP, virtual geomagnetic pole.
*Demagnetization values are listed for first A.F. demagnetization in Oersteds, then thermal demagnetization in degrees Celsius.
†K-Ar age of 9.6 ± 0.5 m.y.

TABLE 3. Early Mid-Miocene Paleomagnetic Results

Province	Site	Formation	Location		N	Demagnetization*	Inc.	Dec.	α_{95}	K	VGP	
											Latitude	Longitude
Benguet	78-1	intrusion†	N16.4	E120.6	9	200/—	23.7	353.3	8.9	34	N82.4	E0.2
	79-4	Klondyke	N16.2	E120.5	6	150/—	27.4	350.9	7.6	78	N81.1	E20.3
	79-20	Klondyke	N16.4	E120.5	7	150/—	30.3	351.0	9.1	45	N81.2	E31.7
	79-21	Klondyke	N16.4	E120.5	6	150/—	36.8	346.6	8.9	57	N76.6	E50.9
	79-22	Klondyke	N16.4	E120.5	9	150/—	35.9	344.0	6.7	61	N74.4	E46.2
	79-30	Klondyke	N16.2	E120.5	5	100/—	26.6	336.6	13.1	35	N67.2	E28.0
	78-7	dike‡	N16.3	E120.6	8	200/500	−35.0	158.0	6.7	70	S68.9	W137.9
	79-9	zigzag	N16.3	E120.6	7	150/—	39.0	342.0	7.0	74	N72.1	E52.1
	79-11	zigzag	N16.3	E120.6	4	200/300	16.0	339.0	11.1	69	N67.9	E11.4
	79-28	zigzag	N16.3	E120.6	10	200/—	−28.0	160.7	5.3	85	S71.4	W151.2
Tarlac	79-14	Moriones	N15.5	E120.7	6	200/300	20.7	333.4	20.1	11	N63.6	E23.1
Marinduque	81-307	Mahinhin§	N13.3	E122.0	5	150/400	−28.6	141.2	17.7	20	S53.3	W140.4
	81-301	dike	N13.5	E122.0	5	—/300	−16.3	126.5	21.2	14	S37.3	W147.8

N is the number of individually oriented paleomagnetic samples; Inc., inclination; Dec., declination; VGP, virtual geomagnetic pole.
*Demagnetization values are listed for first A.F. demagnetization in Oersteds, then thermal demagnetization in degrees Celsius.
†K-Ar age of 12.4 ± 0.6 m.y.
‡K-Ar age of 15.0 ± 1.6 m.y.
§K-Ar ages of 17.5 and 15.0 m.y. [Wolfe, 1981].

TABLE 4. Oligocene Paleomagnetic Results

Province	Site	Formation	Location		N	Demagnetization*	Inc.	Dec.	α_{95}	K	VGP	
											Latitude	Longitude
Benguet	79-10	Pugo	N16.5	E120.5	4	200/—	26.8	339.2	5.5	284	N69.5	E27.3
	81-38	Pugo	N16.5	E120.5	6	300/—	−9.2	146.8	6.1	121	S53.0	W163.9
Vizcaya	81-40	flow	N16.5	E121.2	12	200/500	2.5	326.7	3.6	148	N15.8	E9.7
Zambales	79-7	sediments	N16.1	E120.1	4	300/550	9.0	81.5	5.0	334	N9.0	W151.9

N is the number of individually oriented paleomagnetic samples; Inc., inclination; Dec., declination; VGP, virtual geomagnetic pole.
*Demagnetization values are listed for first A.F. demagnetization in Oersteds, then thermal demagnetization in degrees Celsius.

TABLE 5. Eocene Paleomagnetic Results

Province	Site	Formation	Location		N	Demagnetization*	Inc.	Dec.	α_{95}	K	VGB	
											Latitude	Longitude
Tarlac	79-12	Aksitero	N15.5	E120.4	8	400/520	−0.9	288.2	10.9	27	N17.4	E26.7
	81-50	sills	N15.5	E120.3	10	200/550	31.3	108.1	7.1	47	S12.0	W171.3
Zambales	79-15	Ophiolite	N15.7	E120.1	10	400/—	17.2	98.8	7.2	46	S6.0	W160.8
	79-16	Ophiolite	N15.7	E120.1	6	400/—	−28.7	289.3	9.3	52	N13.6	E9.6
	79-18	Ophiolite	N15.7	E120.1	8	400/—	−12.3	115.6	6.5	73	S26.3	W150.7
	79-40	Ophiolite	N15.6	E120.0	12	400/—	1.3	120.0	6.3	49	S28.6	W159.6
	79-41	Ophiolite	N15.6	E120.0	5	300/—	−7.2	110.6	9.9	61	S20.8	W152.1
Pangasinan	79-42	stock	N16.1	E120.1	5	300/—	7.2	129.1	5.3	206	S35.9	W166.8
	81-43	sills	N16.1	E120.1	10	200/500	17.4	126.9	7.4	44	S31.9	W171.6
	81-45	gabbro	N16.1	E120.1	4	300/—	−11.8	125.5	12.8	53	S35.7	W154.2

N is the number of individually oriented paleomagnetic samples; Inc., inclination; Dec., declination; VGP, virtual geomagnetic pole.
*Demagnetization values are listed for first A.F. demagnetization in Oersteds, then thermal demagnetization in degrees Celsius.

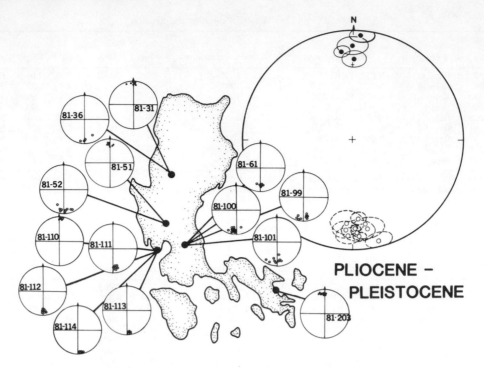

Fig. 6.   Paleomagnetic results from sites in Luzon of Plio-Pleistocene age. All paleomagnetic directions are field directions (after tectonic correction), displayed on Lambert equal-area stereographic projections. The small stereo nets represent the paleomagnetic directions, after either thermal or alternating field demagnetization, for each sample from the indicated site. The solid circles show the approximate location of the sites. The large stereo net shows the site Fisher mean directions and their associated ovals of 95% confidence.

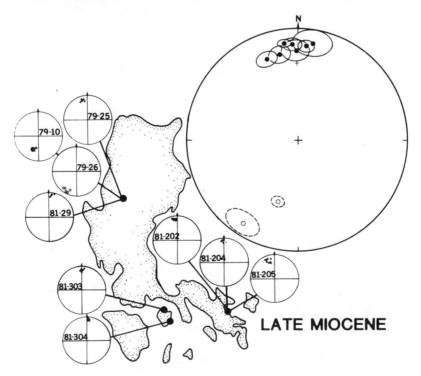

Figure 7.   Paleomagnetic results for sites in Luzon and Marinduque of late Miocene age. (See legend for Figure 4.)

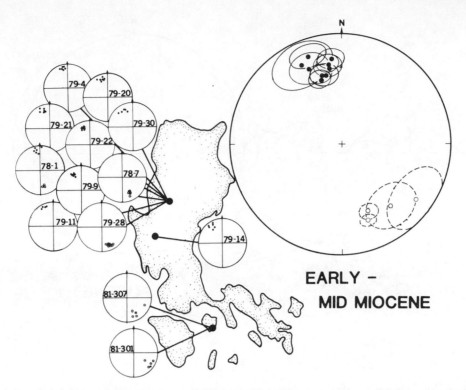

Figure 8. Paleomagnetic results from sites in Luzon and Marinduque of early to mid-Miocene age. (See legend for Figure 4.)

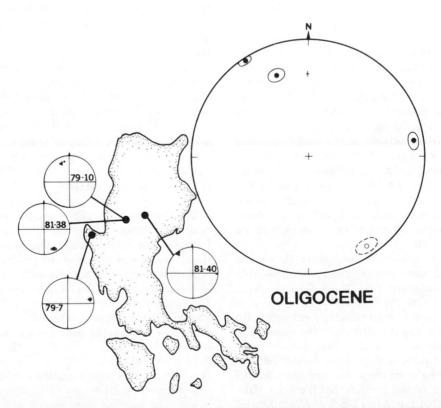

Figure 9. Paleomagnetic results from sites in Luzon of Oligocene age. (See legend for Figure 4.)

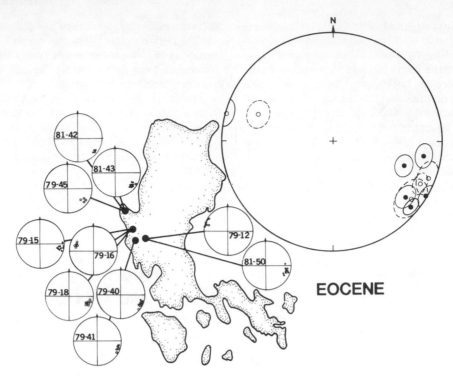

Figure 10.   Paleomagnetic results from sites in Luzon of Eocene age. (See legend for Figure 4.)

are interpreted as sills [e.g., *Schweller and Karig,* 1982; *Hawkins and Evans,* this volume]. They (81-43) and the gabbros (81-45), into which they are intruded, are paleo-magnetically good recorders, giving a southeasterly dec-lination and shallow inclinations. A small stock (79-42), thought to be part of the ophiolite (J. W. Hawkins, personal communication, 1981) gives a similar direction before tec-tonic correction. A tectonic correction can be made by using the dip of the Oligocene sediments collected at Barlo; they lie conformably upon the pillows of the ophiolite there. The mean direction for the sills, the gabbro, and the diorite remains to the southeast as shown in Figure 10. A second possibility is to take the dip of the sills as an indicator of the horizontal at the time of acquisition of magnetization of these units. This gives a similar declination, but rotates the magnetization on to the upper hemisphere with an inclination of about $-20°$.

At the Acoje mine, samples from the gabbros, the olivine cumulate dunite, and from minor dykes gave self-consist-ent results before tectonic correction. The declinations were roughly 120 and 300, and the inclinations shallow with normal and reversed results approximately antipodal. In-itially, we attempted to use the attitude of the layering in the gabbro as a tectonic correction. This, however, de-stroyed the antipodal relationship of the normal and re-versed results. We then tried the same correction as had been used in the north, on the grounds that the Acoje Mine sites are part of northern or Acoje block of *Hawkins and*

Evans [this volume]. This brought the results into agree-ment with those from the north, except the declinations were a little nearer to east-west. The mean inclination for these sites is 13°, using this tectonic correction. If, instead, we argue that the Aksitero sediments are closer to the Acoje Mine and use that correction, the declination is not changed much, but the inclination is made somewhat shal-lower, ~9°. The magnetization of these igneous units is carried by magnetite. The olivine cumulate is somewhat serpentized, so the origin of some of the magnetite in it may be associated with the serpentinization, but, in the gabbros it appears to be primary. The failure of the lay-ering as a tectonic correction is probably because it is not a good indicator of horizontal. It may well be a reflection of the inclined walls of the magma chamber at the time of origin [*Irvine,* 1980; *Pallister and Hopson,* 1981]. We interpret this magnetization as a record of the field when the ophiolite was a part of the seafloor.

The southern, or Coto, block of the ophiolite was sampled on the eastern side of the Zambales range, where the upper reaches of the body can be seen with sediments on top. The Aksitero sediments, which lie on top, are particularly important because the sedimentary environments they re-cord constrain tectonic models [*Garrison et al.,* 1979; *Schweller and Karig,* 1982]. Unfortunately, the Aksitero is not a very good paleomagnetic recorder; only the pelagic limestones at the base gave satisfactory results. Upsection, the results become inconsistent and confusing. We are making a detailed investigation of the magnetic carriers throughout the section to see if we can interpret a reliable direction from upsection. For the moment, we prefer to use only the results from the Eocene pelagic limestones in the bottom 5 m of the section. They gave a declination of 288 and essentially zero inclination. The intrusives in the Camelling River are interpreted as sills. Using the tectonic correction from the Aksitero, we determine a declination of 108 and an inclination of 31. If we use the sills as an indication of horizontal, the declination is rotated to the southeast, but the inclination is not strongly changed.

In interpreting the paleomagnetic results from units that were magnetized at low latitudes and subsequently subjected to large tectonic rotations, we encounter the all too familiar problem of polarity ambiguity. If we assume polarity, we know the sense of rotation. Alternatively, if we know the sense of rotation, we know the polarity. However, if we cannot safely demonstrate polarity, we do not know the sense of rotation. For Zambales, we can argue that the normal vector is to the west because the Miocene Moriones formation on top of the Aksitero has a declination of about 330° and a positive inclination. However, this interpretation could be contested on the grounds that the Moriones determination is not very good and because there is considerable time between the Miocene and the Eocene, in which large rotations could invalidate our interpretation.

In assessing the Eocene results, greater uncertainties have become apparent in the analysis than we encountered with the younger material. The results from the northern block give northwest-southeast declinations with shallow inclinations. At the Acoje Mine, the directions are more nearly east-west with shallow inclinations. The few results from the Coto block on the east side of the range are very similar to those from the Acoje Mine. Assuming that the Aksitero is a westerly, normal declination because of the Moriones result, all sites give a westerly normal declination.

4.6. *Summary and Discussion of Paleomagnetic Results*

The paleomagnetic results described above come from some 50 sites from the Cenozoic of the northern Philippines, so that although it is far from an exhaustive collection, it should give some indication of the regional paleomagnetic framework. The results are principally from Luzon.

The Plio-Pleistocene results give strong evidence of declination, over the past 5 million years, consistent with a geocentric axial dipole, although the inclinations are systematically shallow for reasons that we do not understand. It may of course be indicative of a strong g_2^0 component in the field, as has been suggested by others [e.g., *Merrill and McElhinny*, 1977]. In giving zero declination, our results differ from those of *De Boer et al.* [1980] who observed westerly declinations of as much as 35° in the past 4 m.y. These were interpreted by those authors as evidence of

tectonic rotations. Although we did see individual flows in Bataan that had westerly normal declinations of as much as 15°, the mean result from Bataan was indistinguishable from zero. In other regions, there was no indication of nonzero declination.

The early and mid-Miocene, results give clear and unmistakable evidence of westerly declinations. The samples come from igneous and sedimentary units and include normal and reversed directions, and there are no particular difficulties in interpreting the tectonic corrections. We are therefore very confident that these early and mid-Miocene directions are reflecting westerly declinations of the geomagnetic field. The inclination is indistinguishable from the geocentric axial dipole field direction for the latitude of the Philippines.

In the light of the well-established early and mid-Miocene result, the late Miocene result is puzzling. The easterly declinations exhibited by the three sites from the Baguio region of Benguet are paleomagnetically reliable and should not be disregarded. They necessarily imply a very rapid change from the easterly declinations to the observed Plio-Pleistocene directions. The possibility of such easterly declinations introduced other complications into the analysis; secondary magnetizations of older rocks may be acquired not only with westerly declination, but also if they were acquired in the late Miocene could they have easterly declinations.

The Oligocene results from Baguio and Caraballc give westerly declinations comparable to those of the early mid-Miocene results. The results from the Barlo sediments are puzzling. Taken at face value, they suggest an eastern declination. However, the result depends upon the resolution of the polarity ambiguity, and the site is not totally reliable.

Given the resolution of the polarity ambiguity based upon the magnetization of the Miocene Moriones formation, the Eocene results all have westerly declinations and shallow inclinations. It must, however, be emphasized that these results are from units magnetized in low latitudes and subsequently strongly rotated. Hence, they are extremely difficult to interpret because of the polarity ambiguity. We prefer to interpret a westerly declination, but, until a lot more work is done, the result should be used with caution.

5.   Tectonic Implications of the Paleomagnetic Results

The paleomagnetic results do not yet point clearly to any single model of the evolution of the Philippines. Indeed, in such a complex region, in which plate boundaries are changing radically on the time scale of the observations, it may not be possible to establish a unique history. Nevertheless, the paleomagnetic results do suggest certain shortcomings of some of the earlier tectonic models.

In this last section of the paper we note the constraints the paleomagnetic data can place on the tectonic models.

Next, we see where discrepancies exist between the models and those constraints. Finally, we present, as a working hypothesis, the outline of a model of evolution consistent with the paleomagnetic data.

5.1. *Paleomagnetic Constraints on Tectonic Models of the Northern Philippines*

In making use of paleomagnetic data for tectonic reconstructions, a definition of discordance of a particular paleomagnetic result is necessary so that one can establish which tectonic units have been displaced. *Beck* [1980] has suggested a definition of discordance in terms of rotation and flattening. These quantities measure the departure of a particular paleomagnetic result from the direction proscribed by the appropriate reference pole. As Beck points out, this has the advantage of giving the discordance in terms of two parameters, which are readily related to tectonic rotations and N/S displacements. Unfortunately, in a region such as the Philippines, it is not at all clear what the appropriate reference pole should be. However, we assume that for at least the past 17 m.y. the Asian landmass to the north has behaved as a single tectonic unit. The paleomagnetic data for China for this period are not distinguishable from the geocentric axial dipole field (J. Lin, personal communication, 1981). For the past 17 m.y., we therefore use departures from this dipole as the measure of tectonic rotation and N/S displacements. For older periods, this would clearly not be justified and Lower Cretaceous rocks from China show declinations of about 20° east.

We suggest, from the paleomagnetic results the following tectonic constraints: (1) counterclockwise rotation of the region, predominantly since mid-Miocene time; (2) northward motion of the region, which ceased by early-Miocene time; (3) the northward motion of the Zambales complex from equatorial latitudes accompanied by counterclockwise rotation, since Eocene time; (4) the absence of paleomagnetically detectable rotations in the Plio-Pleistocene.

The value of such constraints clearly depends upon the reliability of the relevant paleomagnetic result. Thus, constraint 1 is the best established result, so any model should involve counterclockwise rotation since mid-Miocene. The northward motion of Benguet in the Oligocene is less well established. However, the results from the whole region from early-Miocene rocks strongly suggest that there was no important northward motion since that time. The northward motion of the Zambales complex and its counterclockwise rotation depend upon the resolution of the polarity ambiguity, but it is included as a weaker constraint. During the Miocene it became a part of Luzon and so subject to the constraint of no further northward motion. The Plio-Pleistocene results justify constraint 4. The possible clockwise rotation of the region north of the Philippine fault in northern Luzon has not been included as a

constraint, because the paleomagnetic dates do not at present justify it.

In all use of paleomagnetic data for tectonic analysis, it should be remembered that determination of declination detects relatively small tectonic rotations, perhaps to the limit of geological interest. However, the resolution in the paleomagnetic determination of latitude, from inclination, is no better than ±500 km, even with very good data.

5.2. *Tectonic Models of the Philippines*

We now look at some of the earlier models in light of the paleomagnetic constraints established. It is then possible to see where discrepancies arise between models and the new data.

It is hard to reconcile with the paleomagnetic data models (such as that of *Murphy* [1973]) in which there is no plate boundary between Asia and the Philippines in the Cenozoic until middle Pliocene times. Our first constraint requires that sometime after mid-Miocene and prior to the Pliocene the region was rotating counterclockwise. Paleomagnetic results from the Asian landmass [e.g., *McElhinny,* 1973] suggest that it was not rotating in this sense. A plate boundary, or at least a collision zone, is then required between Asia and Luzon at this time; certainly, it must have been there before Pliocene times.

Karig [1973] proposed a model that invoked commencement of subduction, to the west of Luzon, in late Miocene time. This model is therefore more consistent with paleomagnetic data. It does not, of course, include the rotations that were reported after the model was proposed. The effect is to rotate the northern Philippines and the eastern subduction zone in Karig's model into a NE/SW orientation during the Early Tertiary.

Models in which the Philippines are rifted and rotated away from the Asian landmass, from China, in the Early Tertiary [e.g., *De Boer,* 1980] satisfy the counterclockwise rotation revealed by the Paleomagnetic work, but not the northward motion. They also encounter a fundamental difficulty in that spreading of the present South China Sea is younger than, for example, the Zambales ophiolite.

The documentation of the opening of the South China Sea by *Taylor and Hayes* [1980] has permitted more realistic models of the evolution of the region. They have shown that the South China Sea opened between anomaly 11 (32 m.y. or mid Oligocene) and anomaly 5D (17 or mid-Miocene). In a number of discussions that have followed, an essential role has been played by the southward motion of a continental block rifted from China [e.g., *Taylor and Hayes,* 1980; *Holloway,* 1982; *McCabe et al.,* 1982]. This continental block consisted of what is now the northern part of Palawan, Mindoro, Panay, and the Sulu Sea. Its importance is that it eventually collides with the subduction zone to the south of the South China Sea and stops subduction in this area. In the discussion of *McCabe et al.* [1982], the idea is taken further and used to account for

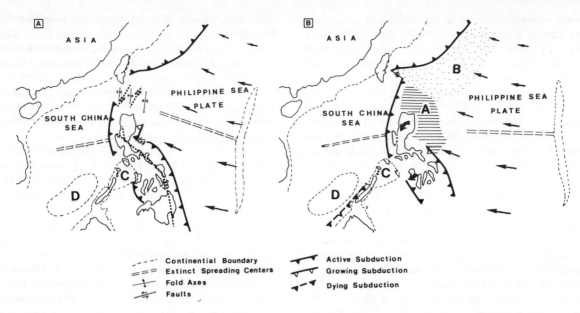

Figure 11.   Tectonic maps for the South China Sea, Philippine Islands, and part of the Philippine Sea plate (a) at present and (b) proposed for approximately 17 m.y. ago (end of the mid-Miocene). The motion of the Philippine Sea plate is shown by a set of arrows, whose length approximates the velocity with respect to the Asian continent. Region A is Philippine Sea plate convergence during the time interval between the two maps. Region B is Philippine Sea plate that will be subducted by the Ryukyu trench during this same time interval. Region C is the Calamian continental fragment. Region D is the Reed bank continental fragment.

the difference in paleomagnetic directions found in Panay and the northern Philippines by invoking a collision between this block and the Philippines themselves. The collision of the block with the subduction zone is coincident with the cessation of spreading in the South China Sea. It appears that this collision indeed terminated spreading of the South China Sea.

5.3. *Paleomagnetically Based Tectonic Model of the Philippines*

The essence of a successful paleomagnetically based tectonic model is to define the motions of different tectonic units within the area of interest. This requires the identification of so-called terranes, whose boundaries are established. We cannot do this. However, we can give some indication of the motion of the region and begin to isolate these tectonic units. As a point of departure, we use the schematic tectonic map in Figure 11a based upon *Karig* [1982].

Our interpretation is that the Plio-Pleistocene results do not reveal any significant departures from the present configuration during the past 5 million years. Clockwise rotation of the late Miocene sites north of the Philippine fault in Luzon can be readily accommodated by rotating a unit whose southern boundary is the Philippine fault, but it is not required by the data.

The principal observation to be explained is the change

between the situations shown in Figures 11a and 11b; the latter gives the tectonic configuration at the beginning of the middle Miocene (17 m.y.). It is immediately clear that we cannot explain this by invoking Philippine Sea plate motion directly, as it is in the opposite sense from that required. However, in an indirect manner this plate motion may cause the observed counterclockwise rotations.

At the cessation of spreading in the South China Sea, the Calamian continental fragment (labeled C in Figure 11b) collides with the trench at the southern boundary of the South China Sea and stops subduction. Therefore, the Philippine Sea plate cannot continue to move westward at this point. At this time there is evidence of volcanic activity on Samar [*Balce et al.*, 1981], which indicates the beginning of the development of the Philippine trench and subduction of the Philippine Sea plate under the central Philippines. This subduction takes up the relative motion between the Philippine Sea plate and Asia. Similarly, to the north, along the Ryukyu trench, the Philippine Sea plate is subducted (the portion of the plate labeled B in Figure 11b).

Between the growing Philippine trench and the Ryukyu trench, the leading edge of the Philippine Sea plate overrides the oceanic crust of the South China Sea. The Manila trench is effectively pinned by the collision of Calamian in the south, so that it is only free to migrate westward, toward Taiwan in the north. The extent of this migration of the Manila trench is not sufficient to account for the amount of convergence of the Philippine Sea plate (labeled A in Figure 11b). Therefore, in conjunction with the subduction along the Manila trench, there is crustal shortening in the region north of Luzon, as indicated in Figure 11a. The westward migration of the northern section of the Manila trench, while the southern end is pinned by the Calamian continental block, results in a counterclockwise rotation of northern Luzon. Eventually, the northern end of the Manila trench collides and is pinned by Taiwan, so that counterclockwise rotation is terminated.

Prior to 17 m.y., the constraints from paleomagnetism are inadequate to justify an attempt at a model of tectonics comparable with that proposed for subsequent time. It appears that Zambales was formed at equatorial latitudes. It seems possible that the northward motion and associated counterclockwise rotation of Zambales may be related to the counterclockwise rotation of the north Borneo-Palawan spur of the Asian landmass [Haile et al., 1977]. We have preliminary indications that the Cordillera of Luzon does not have a similar tectonic history to that of Zambales. Again, we have preliminary results from the Angat ophiolite, which suggest it was strongly clockwise rotated. It is now crucial to get paleomagnetic data to outline the various tectonic units, or terranes, and to see their rotations and displacements for this older time interval.

This study affords an example of how paleomagnetism can help in the interpretation of the tectonics of a complicated region. The paleomagnetic data focus on certain aspects of the geologic history, which must be explained by successful models of the evolution of the region. In this instance, the paleomagnetic data focus attention on the required counterclockwise rotation of the northern Philippines in the Miocene. To explain this, the combined action of the Philippine Sea plate motion and the pinning of the region in the south was recognized. It therefore appears that the paleomagnetic results have helped to establish the manner by which the Philippines is being accreted to the Asian landmass. The results therefore help to document the growth of Asia by the addition of products of island arcs.

Appendix: Sample Locality Descriptions

Pliocene-Pleistocene

81-31.   Basaltic dike, which cuts through a hornblende andesite flow on Mt. Santo Tomas. Dike is not corrected for bedding, the intruded volcanics attitude is 305,18,S.

81-36.   Basaltic dike that cuts vertically through the lower section of the Rosario Formation. Dike is located on Marcos Highway just west of city limit of Baguio 1 : 50,000 sheet. No bedding correction was used.

81-51; 81-52.   Ash flow tuff. Two sites located just west of the McArthur Highway at kilometer 89 road post. Sites taken from two flow units. The site with the normal inclination is stratigraphically below the reversed site. These flows have small dip of less than 5 that is depositional.

81-61.   Porphyritic plagioclase basalt flow with holocrystalline texture. Flow located 4 km from above Antipolo. Flow is horizontal.

81-99.   Fine-grained volcanic ash fall deposit collected from water pipe trenches recently dug along the side of the geology building of the University of the Philippines campus. Ash deposits are horizontal.

81-100; 81-101.   Fine-grained volcanic ash deposit (81-100) and ash fall deposit (81-101). Location is in a rock quarry located 15 km east of Cubao, Quezon City, on the northern side of the Marcos Highway. Layering of the ash is horizontal.

81-3.   Hornblende andesite flow located along the east coast at a small fishing Barrio just east of Mt. Malingao. Dip is primary, not structural.

81-110; 81-111; 81-112; 81-113, 81-114.   These five sites are located on the east side of the Mariveles Bay. Site 81-110 is located at the eastern edge of the harbor along the eastern boundary of the ship building pier. This unit makes up the basement on which the other four sites from this area sit. Bedding assumed horizontal. The other sites are from plagioclase hornblende dacite flows, which are exposed in three different quarries and along the sea cliffs. The bedding of these sites is horizontal, and the initial relationship of these samples with the erupting vents is easily seen in the field.

Early to Mid Miocene

78-1.   Diorite pluton located on the east side of the Kennon Road at kilometer 238. Used homocline bedding of 320,20,W for tilt correction.

79-4.   Well-bedded sandstone located on the east side of the Kennon Road 1 km of Camp 1. Attitude of bedding is 131,19,S.

79-20.   Pyroclastic ash deposits located on the north side of the Asin Road, west of timber creek at a small north-south running stream. Attitude of layering deposits is 319,22,SW.

79-21.   Fine-grained ash deposit located 1 km west of the Asin Road from site 79-20, along the Irisan Creek. Attitude of bedding is 318,29,SW.

79-22.   Fine-grained tuffaceous sandstone located 7 km outer north side of the Asin Road near the Rosario-Klondyke content; bedding is 318,29,S.

79-30.   Fine-grained sandstone located at kilometer 219 on the Kennon Road. Attitude of bedding is 147,30,SW.

78-7.   Basaltic dike located at kilometer 241. Dike has been carved into large lion head. Used general Baguio homocline (320,20,W) to correct for tilt.

79-9.   Fine-grained sandstone in Bued River along side of the Kennon Road, 1 km south of Camp 4. Attitude is 340,40,W.

79-11.   Well-bedded green sandstone located in the river north of the houses located on the Kennon Road just below the dike carved in the form of a lion head. Attitude of beds is 0,25,W.

79-14. Calcareous sandstone located in the Moriones River near the town of Moriones, Tarlac. Attitude of bedding is 340,20,E.

79-28. Well-bedded red sandstone in the river 200 m south of site 79-11. Attitude is 336,63,W.

81-307. Hornblende quartz diorite. Samples taken from two mining pits, one belonging to Mar-Copper at drainage portal for south pit. The second site is located at Consolidated Mining, Inc., located around 30 km northwest of Mar-Copper. Samples from both locations were taken from barrier area of mines. Sediments above Mahinhin intrusions are horizontal, and, therefore, no bedding correction is used.

81-301. Basaltic dikes that intrude the Mar-Copper pits. Samples were collected from five dikes that proved to be nonmineralized. No bedding correction was applied because they intrude flat lying sediments about the Marinduque intrusions.

Late Miocene

79-25. Light-brown shale, which was collected around Barrio Calot on the Naguilian Road. Attitude of the bedding is 200,34,W.

79-26. Medium-to-coarse grained sandstone located on the Naguilian Road west of 3 km/site 79-25. Attitude of the bedding is 202,45,W.

78-10. Hornblende andesite flow in the Rosario Formation. Located 200 m west of 79-26 site on the Naguilian Road. Bedding correction for 79-26 is used.

81-29. Andesite flow on the road 200 m north of the road end at Mt. Santo Tomas. Attitude of interbedded ash on top of volcanic is 305,18,S.

81-202. Hornblende andesite flow located on the highway along the eastern slope of Mt. Malingao in the folded section of older volcanic basement of Mt. Isarog.

81-204. Hornblende andesite located southeast of Legaspi City, 10 km along the highway between Legaspi and Manila City. Attitude is horizontal, based on relationship with overlying ash.

81-205. Hornblende andesite, 3 km along the highway toward Manila from 81-204. Attitude of bedding is 355,42,W.

81-303. Fine-grained calcareous siltstone. Located at Barrio Balogo, across from the gate to the Mar-Copper Mining Company. Bedding is horizontal.

81-304. Dacitic dike that cuts vertically through Torrijos volcanics at Cagpoc Point, southeast of Marinduque. Sample not corrected for bedding of host volcanics. Attitude of Torrijos volcanics is 330,24,E.

Oligocene

B 79-10. Basalt flow located on the east side of the Kennon Road near kilometer 235 post. Attitude of upper surface is 153,32,W.

81-39. Andesite flow located around 10 km west of the airport on the west side of the road toward Philex Mines. Attitude at site is 270,55,N.

81-43. Red andesite flow on the east side of the road from Manila to Isabela at kilometer 212. Attitude dip of overlying ash bed is 72,28,N.

79-7. Tan colored tuffaceous siltstone situated above pillow section on the road to Barlo Mine. Attitude of siltstone is 284,13,N.

Eocene

79-12. Bottom 5 m of Aksitero section in type section of Aksitero river; bedding attitude is 320,20,E.

81-50. Sills in Camelling River to the west of Tarlac as shown on the map; bedding attitude of Aksitero used as tectonic correction.

79-15. Layered gabbro 700 m east of the contact with the Olivine cumulate at the Acoje Mine. Tectonic correction of following sites in ophiolite are discussed in text.

79-16. Olivine cumulate 1400 m west of the contact in the Acoje Mine.

79-18. Layered gabbro in stream section 15 km east of the Acoje Mine on the continuation of the Acoje Mine Road.

79-40. Layered gabbro on the Coto Access Road 30 km from Coto Mine.

79-41. Dyke that intrudes layered gabbro of site 79-40.

79-42. Stock near Sual on the coast road opposite Sauscito restaurant close to kilometer 349 post.

81-43. Sills on beach near Sual.

81-45. Gabbros intruded by sills of site 81-43.

Acknowledgments. The opportunity to carry out this study arose as part of the program in East Asian Tectonics and Resources of the United Nations Committee for Coordination of Joint Prospecting for Mineral Resources in Asian Offshore Areas. To this group and to the National Science Foundation we wish to express gratitude for financial support. Those of us from Santa Barbara wish to thank our many friends in the Philippines Bureau of Mines whose expertise was invaluable in the fieldwork, which was itself made possible by the logistic support of the bureau. We also thank K. Burton of Philippine Billiton Inc. and H. Pagauitan of Electro-Copper for many discussions and assistance with the field part of the studies. R. McCabe thanks S. Sasajima and members of his group for assistance in measurements.

References

Balce, G. R., R. Y. Encina, A. Momongan, and E. Lara, Geology of the Baguio district and its implications on the tectonic development of the Luzon central cordillera, *Geol. Paleo. S.E. Asia, 21,* 265–287, 1980.

Balce, G. R., O. A. Crispin, C. M. Samaniego, and C. R. Miranda, Metallogenesis in the Philippines: Explanatory text for the CGMW metallogenic map of the Philippines, *Geol. Surv. Jpn, 261,* 125–148, 1981.

Beck, M. E., Paleomagnetic record of plate-margin tectonic processes along the western edge of North America, *J. Geophys. Res., 85,* 7115–7131, 1980.

Becker, F. G., Brief memorandum on the geology of the Philippine Islands, 20th Annual Report, U.S. Geol. Surv., 1899.

Chase, C. G., Plate kinematics: The Americas, East Africa, and the rest of the world, *Earth Planet. Sci. Lett., 37,* 355–368, 1978.

Corby, C. W., Geology and oil possibilities of the Philippines, *Philippine Dept. Agr. Nat. Res. Tech. Bull., 21,* 1–363, 1951.

DeBoer, J., L. A. Odom, P. C. Ragland, F. G. Snider, and N. R. Tolford, The Bataan orogene: Eastward subduction, tectonic rotations, and volcanism in the Western Pacific (Philippines), *Tectonophysics, 67,* 251–282, 1980.

Fisher, B. A., Dispersion on a sphere, *Proc. R. Soc. London, Ser. A, 217,* 295–305, 1953.

Garrison, R. E., E. Espiritu, L. J. Horan, and L. E. Mack, Petrology, sedimentology and diagenesis of hemipelagic limestones and tuffaceous turbidites in the Aksitero Formation (Eocene-Oligocene), Central Luzon, Philippines, *U.S. Geol. Surv. Prof. Pap., 1112,* 1–16, 1979.

Gervasio, F. C., Age and nature of orogenesis of the Philippines, *Tectonophysics, 4,* 379–402, 1967.

Haile, N. S., M. W. McElhinny, and I. McDougall, Paleomagnetic data and radiometric ages from the Cretaceous of West Kalimantao (Borneo) and their significance in interpreting regional structure, *Geol. Soc. London Q. J., 133,* 133–144, 1977.

Hawkins, J. W., and C. A. Evans, Geology of the Zambales Range, Luzon, Philippine Islands: Ophiolite derived from an island arc–back arc basin pair, this volume.

Holloway, N. H., The stratigraphic and tectonic relationship of Reed bank, north Palawan and Mindoro to the Asian mainland and its significance in the evolution of the China Sea, *Geol. Soc. Malaysia Bull.,* in press, 1982.

Hsu, I., Magnetic properties of igneous rocks in the northern Philippines, Ph.D. Thesis, Washington Univ., St. Louis, Mo., 1972.

Irvine, T. N., Magmatic density currents and cumulus processes, *Am. J. Sci., 280-A,* 1–58, 1980.

Jarrard, R. D., and S. Sasajima, Paleomagnetic synthesis for southeast Asia: Constraints on plate motions, in *The Tectonic and Geologic Evolution of Southeast Asian Seas and Islands, Geophys. Monogr.,* vol. 23, edited by D. E. Hayes, pp. 293–316, AGU, Washington, D. C., 1980.

Karig, D. E. Origin and development of marginal basins in the western Pacific, *J. Geophys. Res., 76,* 2542–2561, 1971.

Karig, D. E., Plate convergence between the Philippines and the Ryukyu islands, *Mar. Geol., 14,* 153–168, 1973.

Karig, D. E., Accreted terranes in the northern part of the Philippine archipelago, in *Proceedings of the SEATAR Workshop on Luzon-Marianas Transert,* in press, 1982.

McCabe, R., J. Almasco, and W. Diegor, Geologic and paleomagnetic evidence for a possible Miocene collision in western Panay, central Philippines, *Geology, 10,* 325–329, 1982.

McElhinny, M. W., *Paleomagnetism and Plate Tectonics,* Cambridge University Press, New York, 1973.

Merrill, R. T., and M. W. McElhinny, Anomalies in the time-averaged paleomagnetic field and their implications for the lower mantle, *Rev. Geophys. Space Phys., 15,* 309–323, 1977.

Murphy, R. W., The Manila trench-West Taiwan foldbelt: A flipped subduction zone, *Geol. Soc. Malaysia Bull., 6,* 27–42, 1973.

Pallister, J. S., and C. A. Hopson, Samail ophiolite plutonic suite: Field relations, phase variations, cryptic variation and layering and a model of a spreading ridge magma chamber, *J. Geophys. Res., 86,* 2593–2644, 1981.

Pena, R., Brief geology of a portion of the Baguio mineral district, *J. Geol. Soc. Phil., 24,* 41–43, 1970.

Philippine Bureau of Mines, Geologic map of the Philippines, 1964.

Schweller, W. J., and D. E. Karig, Emplacement of the Zambales Ophiolite into the West Luzon margin, *Am. Assoc. Petrol. Geol. Mem.,* in press, 1982.

Taylor, B., and D. E. Hayes, The tectonic evolution of the South China Sea Basin, in *The Tectonic and Geologic Evolution of Southeast Asia Seas and Islands, Geophys. Monogr.,* vol. 23, edited by D. E. Hayes, pp. 89–104, AGU, Washington, D. C., 1980.

Wolfe, J. A., Philippine geochronology, *J. Geol. Soc. Phil., 35,* 1–30, 1981.

Zijderveld, J. D. A., A. C. demagnetization of rocks: Analysis of results, in *Methods in Paleomagnetism,* edited by D. W. Collinson, K. M. Creer, and S. K. Runcorn, pp. 254–286, Elsevier, New York, 1967.

Geology of the Zambales Range, Luzon, Philippine Islands: Ophiolite Derived from an Island Arc–Back Arc Basin Pair

James W. Hawkins and Cynthia A. Evans

Geological Research Division, Scripps Institution of Oceanography, La Jolla, California 92093

The Zambales Range Ophiolite comprises peridotite, gabbro-norite, diabase dikes and sills, and a range of basaltic rock types including pillows, massive and brecciated rocks. The mafic and ultramafic section of the ophiolite is typical of other ophiolites and of rocks presumed to form oceanic crust and upper mantle. The peridotite is mainly saxonite, but extensive areas are underlain by dunite; it has been serpentinized to varied extent throughout all of the range. Chromite is an abundant constituent of the peridotite, and in many areas it forms very large concentrations, some of which are currently being mined. Clinopyroxene is a minor constituent in most of the peridotite mass, but locally there are occurrences of lherzolitic rocks. The entire peridotite mass shows effects of deformation at high temperature (e.g., >600°C) as well as widespread evidence for intense shearing and recrystallization at low temperature (e.g., 200–500°C) with the development of serpentine, talc, and amphibole on fault zones. We recognize two major blocks of crust and upper mantle, Acoje block and Coto block, which are distinguished on the basis of their crustal section thicknesses and on the geochemistry of the crust and mantle rocks. A thick zone (0.7–1.3 km) of cumulate textured ultramafic rocks with varied amounts of olivine, clinopyroxene, and orthopyroxene structurally overlies the peridotite of the Acoje block. This layered series also contains chromite and sulfide minerals; the layered series grades upward into gabbro-norite and plagioclase is a minor component of layered series rocks near the petrologic contact. The plutonic mafic rocks which overlie the layered ultramafic rocks of the Acoje block are mainly cumulate textured norite. Both massive and layered rocks are present; the layered units of the norite series include anorthosite, pyroxenite, olivine norite, and a wide range of norite varieties having variable proportions of the major minerals. Calcic amphibole is a minor late magmatic mineral. The upper levels of the Acoje block norite have quartz-rich norite and hypersthene quartz diorite layers and intrusive masses. An upper norite phase merges with massive and tabular (sill-dike) diabase and with a tonalite-trondhjemite series. The Coto block lacks the layered ultramafic series, and the mafic units are mainly gabbro. The Coto-layered gabbros include gabbroic anorthosite, pyroxenite, olivine gabbro, troctolite, allivalite, and some layers which are nearly anorthosite in compositions. Cumulate textures and both rhythmic and cryptic layering are present. Hypersthene is a very rare component of this series except as a reaction rim on olivine. Diabase and basaltic rocks structurally overlie both the norite and gabbro series. In many areas, such as on the Camiling and Moriones Rivers of the Coto block and on the Sual coast of the Acoje block, there are structural-textural gradations from plutonic to hypabyssal rocks. A similar gradation from the diabase sills-dikes to pillow basalts is well shown on the Camiling River, the North Balincaguin River (Acoje block) on the Bucao River (Coto block), and on the coast near Olongapo. Tonalite to trondhjemite rocks form fine grained dikes, medium grained plutonic intrusive masses and intrusion breccias of mafic blocks mixed with fine grained silicic material. These rocks are exposed mainly near Barlo and in the Iba-Botolan area. The field relations indicate that they formed at the same time as the mafic units. Andesitic to rhyodacitic volcanic rocks formed a belt on the east side of the Zambales Range in late Tertiary and Quaternary time, but they are petrologically different as well as being exposed in features which obviously postdate the uplift of the range. Although the Zambales silicic rock series is exposed only in a few areas at present, it probably was originally of much greater extent and perhaps formed a carapace over much of the Acoje block. The chemistry and petrology of Acoje block rock units resemble arc-tholeiite series rocks from intraoceanic island arcs. Pillow basalts and dikes with boninite series characteristics form part of this assemblage. The Coto block is typical of a back-arc basin rock series. The sedimentary rocks which overlie the pillow basalts indicate a late Eocene age of formation of the crustal rocks of the ophiolite and presumably dates a time during arc and back arc generation. Miocene conglomerates containing clasts derived from the ophiolite indicate that it had been uplifted and was being eroded not long after its generation. We interpret the range as consisting of two major blocks which have been uplifted, tilted down to the east and juxtaposed by left lateral strike slip motion. The range has been thrust westward, probably only a small distance, over crust containing both mafic and sialic rocks. The Zambales ophiolite predates the opening of the South China Sea and cannot be considered an obducted fragment derived from it. The short life span of the Zambales arc–back arc system must indicate rapidly changing plate geometry in Early Tertiary time.

Introduction

Ophiolite assemblages are believed to represent remnants of oceanic crust and upper mantle emplaced in orogenic belts by lithosphere shortening and imbrication (eg., by obduction). Nearly all of the oceanic crust in the present oceanic basins has been generated at mid-ocean spreading ridges, and it has been postulated that the ophiolites of orogenic belts preserve material originally formed at mid-ocean ridges (*Coleman* [1977] and references therein). However, oceanic crust is also formed in back arc basins, at intraplate magma leaks (linear volcanic chains and isolated seamounts), on leaky fracture zones, and as oceanic plateaus. Intraoceanic island arcs are, in our opinion, sites of oceanic crust generation as well. The rock series developed in young arcs are chemically similar to other oceanic material, and the total crustal thicknesses, while greater than 'normal' ocean crust (e.g., 9–20 km, *Murauchi et al.* [1968] and *Gill* [1981]) are similar to that of some linear volcanic chains and plateaus. Thus it is important to be able to discriminate between the various sites of origin for an ophiolite as it places important constraints on making paleotectonic and paleo-oceanographic reconstructions. Methods for using various petrologic and geochemical criteria have been proposed for recognizing the site of origin of ophiolites, and many appear to offer useful insights [*Pearce and Cann*, 1973; *Saunders et al.*, 1980; *Hawkins*, 1980; *Cameron et al.*, 1980]. In this paper we will make use of some of these criteria and field relations to discuss the origin of mafic and ultramafic rocks of the Zambales Range of northwestern Luzon. We recognize rocks typical of an ophiolite assemblage which we believe to be remnants of oceanic crust and depleted upper mantle.

The Zambales Range includes uplifted and tilted blocks of ultramafic and mafic rocks (the ophiolite series) and a chain of basaltic to rhyodacitic volcanos which extends southward to the Bataan Peninsula. The ultramafic rocks have great concentrations of both high-Cr and high-Al chromite [*Stoll*, 1958; *Fernandez*, 1960; *Rossman*, 1964; *Bacuta*, 1978; *Evans and Hawkins*, 1980, 1981]. Cu-Fe sulfides are mined from blanket-like deposits interbedded with pillow basalt and basaltic breccia of the ophiolite near Barlo at the north end of the range. Cu-sulfides are also mined in porphyry copper deposits which are in late Tertiary plutons near the north end of the volcanic belt. Miocene and younger shallow water marine sediments surround the ophiolite along the coast of the South China Sea and Lingayen Gulf. The central valley of Luzon, which delineates the eastern edge of the range, is filled with a thick accumulation of Tertiary and Quaternary clastic sediments. The base of this sedimentary series is the upper Eocene Aksitero formation, which is primarily pelagic carbonate sediments but grades upward into silicic volcanic ash and tuffaceous sediments [*Amato*, 1964; *Garrison et al.*, 1979; *Schweller and Karig*, 1979; *Schweller et al.*, 1981]. The basal part of the Aksitero fm. is in depositional contact with brecciated basalt which is considered to be part of the Zambales ophiolite series. Thus an upper Eocene age for at least part of the ophiolite is inferred.

The ophiolite part of the Zambales Range consists of a north trending belt of rocks which may be separated into two major units, or blocks, which have been tilted down to the east. At the north end of the range there is evidence that, at least locally, part of the range has been tilted down to the northeast. Because the present relief of the range is about 1600 m, the effects of erosion on the tilted block have provided excellent exposures and a good three-dimensional view of the ophiolite. The western edge of the ophiolite shows extensive evidence for shearing and serpentinization of peridotite. Although in several areas there

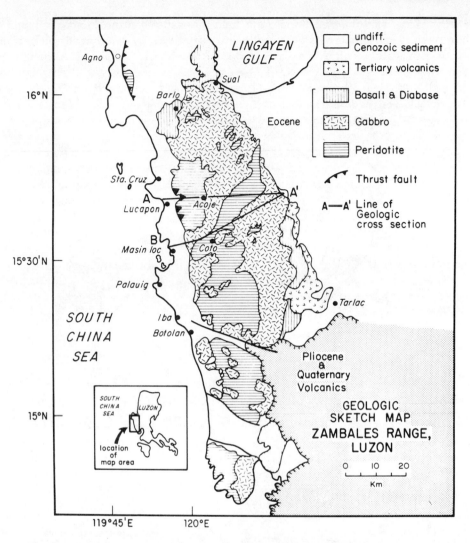

Fig. 1.   Generalized geologic map of the Zambales Range based on the Geological Map of the Philippines, mapping by the Philippine Bureau of Mines, and our work. Lines A-A′ and B-A′ are locations of cross sections shown in Figure 2. Note the small sliver of peridotite and thrust fault zone near Agno. This area includes fragments of chert, pillow basalt, gabbro, and serpentinized peridotite which resemble the Zambales ophiolite series.

is field evidence for both east dipping low-angle thrusts and high-angle reverse faults, it is by no means clear that there has been extensive westward overthrusting. At least some of the field relations are compatible with high angle strike slip motion which has been modified by minor west directed thrusting. About 5 km east of Lucapon, on the Cabaluan River (Figures 1 and 2), there is a good exposure which shows that the serpentinized peridotite has been thrust westward onto overturned beds of Miocene conglomerate. The conglomerate consists largely of rounded cobbles of serpentinized peridotite, very minor amounts of gabbro, and fragments of coralline reef material. The planar structure of the serpentinized rocks parallels the contact with the conglomerate.

Even though much of the area is covered with fairly

dense jungle and, locally, thick laterite, there are excellent exposures in river valleys and on many ridge crests. We have made a complete traverse of the range on foot at about 15°42′N and have made our way up the major rivers well into the interior of the range. We also have had access to the underground areas at the Acoje and Coto (BCI) chromite mines and have made detailed sampling traverses at various levels in the mines. Our interpretations (Figures 1–4) are based on our own mapping and sampling, but we have drawn extensively on the excellent regional mapping of *Rossman* [1964] and the continuing mapping and field studies of the Philippine Bureau of Mines [*Bacuta,* 1978]. In addition, we have used the detailed maps of mine areas prepared by *Rossman* [1964] and by the geologists at Acoje, Barlo, and Coto mines.

Zambales Range - Cross section A-A' Through Acoje Mine Area

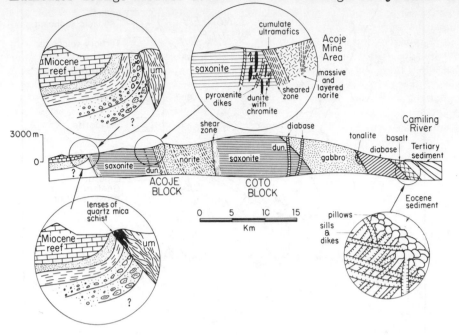

Zambales Range - Cross section B-A' Through Coto Mine Area

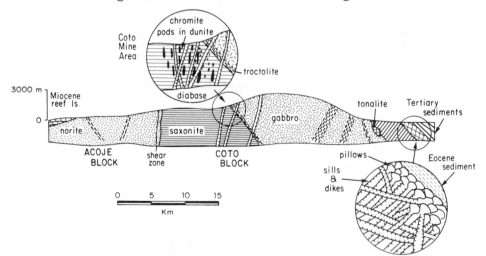

Fig. 2.   Geologic cross sections on lines shown in Figure 1. Sections traverse both Acoje block and Coto blocks. On A-A' the western (Acoje) block is in fault contact with Miocene conglomerate, filled with peridotite clasts, which is overturned to west. Lenses of quartz-rich schists are interlayered with sheared serpentinite. Inset shows schematically the podiform chromite bodies and pyroxenites of Acoje peridotite. The gabbro-norite unit includes a basal wehrlite-dunite cumulate unit. Eastern (Coto) block has peridotite, cumulate gabbro, diabase, pillow basalts, and Eocene hemipelagic limestone. Miocene reefs are built on the conglomerate unit on western side. B-A' section crosses through Coto mine area. Western gabbro-norite unit is part of the Acoje block. Inset shows schematically the diabase dikes cutting the peridotite—they rarely cut the overlying gabbro. Black lenses in inset are the podiform chromite ore bodies. Diabase sills and dike swarms and pillow basalts are overlain by Eocene hemipelagic limestone and Miocene conglomerate with ophiolite debris.

Our impression of the internal structure of the range is that the ophiolite series appear to be relatively intact and shows little or no evidence for internal disruption by low-angle faulting. There are many minor high-angle faults within the range but none cause repetition of the overall 'stratigraphy.' The Zambales Range is not part of a melange although there is a zone of polymict tectonic breccias (melange, in our view) at the west edge of the Zambales Range near Agno on the Balincaguin River. A major fault appears to separate a southeastern block (here informally termed the Coto block) from a northwestern block (here informally termed the Acoje block). These two blocks (Figures 1–4) are delineated primarily on the basis of their petrologic properties as described below, but there is a large NNE-SSW high-angle shear zone on the presumed boundary between blocks which supports our interpreta-

tion. We will show that each of these blocks has petrologic-geochemical features which serve to differentiate them and which we will use to relate them to the probable petrologic-tectonic setting at their time or origin.

We propose that the Zambales Range ophiolite consists of two major crust/mantle units (lithosphere blocks), which were derived from an early Tertiary island arc and back arc basin system. The arc was active for a very short time (a few m.y.?). This arc–back arc basin pair has been rotated nearly 90° in an anticlockwise direction since it formed in late Eocene time [*Fuller et al.,* this volume]. The original orientation of the trench and arc must have been east-west with the arc (Acoje block) lying north of the Coto back arc basin. A north facing trench is implied which presumably was the site of subduction of Philippine Sea lithosphere. *Fuller et al.* [this volume] suggest that the arc system

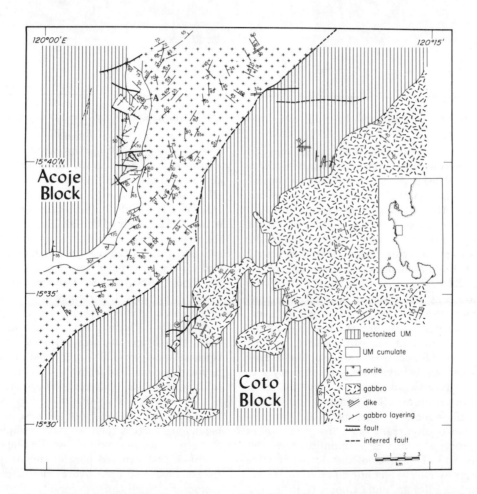

Fig. 3.   Geologic map of the Zambales Range between Acoje Mine (A) and Coto Mine (C) showing distribution of main rock types and major structures. The northeast trending fault marks the boundary between the Acoje block on the west and Coto block on the east. Both blocks are tilted down to the east. The northeast fault shows evidence for intense deformation; the most recent movement seems to have been lateral slip as shown by slickensides in the fault zone. Faults shown with hachures are near vertical; all faults have dips of greater than 40°. Main displacement on the northeast trending fault may have been left slip, but some significant vertical slip is implied as well.

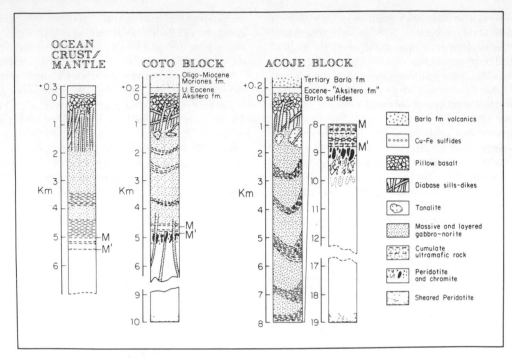

Fig. 4.   Columnar sections comparing oceanic crust with thicknesses of Acoje and Coto Blocks.

formed near the equator; it has moved northward and has been rotated since Eocene time. Our interpretation differs from *Schweller et al.* [1981], who postulate that the central valley of Luzon and the Zambales are parts of a forearc. This rotation and northward movement has undoubtedly caused some dislocation within the ophiolite, but, except for the lateral slip and tilting down to the east, the blocks of the ophiolite appear to retain their internal structure remarkably intact. Field evidence on the west side of the Zambales Range shows that there has been minor west directed thrust faulting, but there is no indication that this caused extensive lateral transport. In a separate section we discuss this deformation and the possible significance of fragments of continental crustal rocks in the fault zone. We see no evidence for obduction of the ophiolite over large areas of crust and believe that the main style of deformation involved eastward tilting of large blocks of oceanic crust and upper mantle. The apparent linearity of the boundaries of these blocks suggests that they may be located along former transcurrent fault directions. *Jakes and Gill* [1970] proposed that many ophiolites have not moved from their site of formation and do not involve obduction or similar extensive tectonic dislocation. We believe that the Zambales Range is an example of an 'in situ' ophiolite in the sense that it is not deep ocean basin crust obducted over a continental margin or island arc.

Petrology of the Acoje and Coto Blocks

We distinguish two major blocks of oceanic crust and upper mantle (Figures 1–4); each is composed of peridotite,

gabbro-norite, basaltic units, and sediment in an upward stratigraphic sequence which includes a transition from depleted upper mantle rocks to oceanic crustal material [*Hawkins and Batiza,* 1977; *Hawkins,* 1980; *Evans and Hawkins,* 1980, 1981; *Hawkins and Evans,* 1980; *Hawkins et al.,* 1981]. Each of the blocks has been tilted down to the east thereby exposing a cross section of peridotite and crustal rocks. In the section which follows we will discuss and compare the petrology of major rock types in each of the blocks. A summary of the major differences for the blocks is in Table 1. A comparison of the two blocks with typical ocean basin lithosphere is in Figure 4.

Peridotite.   The western, structurally lowest, parts of each of the blocks are serpentinized peridotite. The actual thicknesses of peridotite preserved in each block can only be roughly estimated from the dip of the contact with crustal layers and the width of the peridotite belts. The Acoje block peridotite is about 10–12 km thick and the Coto block unit is 8–10 km thick. If these estimates, discussed below, are correct then we have samples of mantle from as deep as about 10 km below the Moho.

The lowest exposed part of the Acoje block has been thrust westward onto overturned beds of Miocene conglomerate which consists largely of cobbles of serpentinized dunite and saxonite (Figure 2). This thrust contact, although clearly indicative of some westward thrusting of the periodotite, may be a minor feature developed on a fault zone which was primarily a north trending transcurrent fault. The western edge of the Coto block is well marked in some places by a northeasterly trending zone of intense shearing in which huge lenses of serpentinized peridotite are interspersed with contorted masses of serpentinite. The fault zone is nearly vertical, and the lineation of peridotite lenses and slickenside orientation indicate that lateral slip was the most recent type of dislocation. We have no way to measure the magnitude of this slip but the nature of rocks in the fault zone suggest that it has experienced very extensive movement and tens of kilometers of slip may have occurred. The probable sense of motion was left lateral slip. The peridotite of each of the blocks is more or less similar. The Acoje block has saxonite with from 5–25% enstatite (equivalent to harzburgite—but we retain the local terminology), dunite, lherzolite, websterite-clinopyroxenite, and massive chromite. The abundance of each of these units is not well known, but probably the relative proportions are about 50–55:40–45:<5:<5:<5. Clinopyroxene is scarce in most of the peridotite except for the irregular 'lherzolitic' masses near the western (basal) part of the complex. The irregular masses (ranging from tens of centimeters to a few meters maximum dimension) appear to be solidified pockets of melt which either were in the process of aggregation or were not able to escape (Table 2). Clinopyroxenite bodies are not at all common in the Coto peridotite but CPX is an important primary constituent in many samples. We did find some CPX-rich 'melt aggregates' forming small irregular masses similar to those of Acoje. Based on our sampling and mapping, the approximate proportions of saxonite, dunite, chromite, and lherzolite are 55–60:30–35:5:≪5. In general, both the Acoje and Coto blocks show an increase in the OPX content towards the west (increase in depth), and dunite is more common near the top of the mantle section. Chromite concentrations also increase towards the east, and the major ore bodies at Acoje and Coto are essentially at the paleo–crust/mantle boundary.

There are substantial differences in the bulk chemistry and mineralogy of the two peridotite masses (Table 2) which must indicate either different initial composition or different melting histories or both. We interpret the peridotite masses as fragments of depleted upper mantle ma-

TABLE 1.   Summary of Distinctive Features of Acoje/Barlo and Coto/East Side Blocks

	Acoje/Barlo			Coto/East Side	
Silicic Rocks	Zr:	average (eight samples) range	51 ppm 39–75 ppm	Zr:	average (four samples) range	128 ppm 62–190
	TiO$_2$:	average range	0.26 wt % 0.12–0.42 wt %	TiO$_2$:	average range	0.19 wt % 0.09–0.40
	FeO*:	average range	3.81 wt % 2.25–6.77	FeO*:	average range	0.83 wt % 0.42–1.31
	CaO:	average range	4.29 wt % 2.35–6.12	CaO:	average range	1.21 wt % 0.45–2.06
Basalt/Diabase	Zr:	average (eight samples) range	20 ppm 14–28	Zr:	average (seven samples) range	61 ppm 24–120 ppm
	TiO$_2$:	average range	0.35 wt % 0.22–0.52	TiO$_2$:	average range	1.05 wt % 0.21–2.61
	FeO*/MgO:	average range	0.90 0.62–1.40	FeO*/MgO:	average range	1.57 (0.99–3.10)
	Ni:	average range	105 ppm 43–251 ppm	Ni:	average range	52 ppm 7–102 ppm

Gabbro thick (0.7–1.3 km) basal cumulate section; dominated by OL and CPX (dunite-wehrlite-pyroxenite)

thick (6–7 km) section of gabbro-norite PL-CPX-OPX-(OL) assemblage

abundance of OPX

calcic PL (AN 95)

Peridotite abundance of dunite OPX-poor saxonite (generally <10% OPX, although up to 25% may be found)

section locally cut by pyroxenite-dikes

more refractory mineral compositions:

OL (average of 10) = Fo 91.6
range = Fo 89.9–93.3

CHR (average of 32) Cr/Cr + Al = 0.73
range = 0.60–0.81

thin (<300 m) basal cumulate section (dunite-troctolite; CPX-bearing units are absent)

thinner (~3 km) gabbroic section

PL-CPX-(OL) assemblage

OPX rare (except as reaction rim on OL)

less calcic PL (An 85)

less dunite, minor lherzolite

OPX-rich saxonite (generally 15–20% OPX, although a range of 5–25% is observed)

section pervasively dissected by diabase dikes

less refractory mineral compositions:

OL (average of 5) = Fo 89.9
range = Fo 87.6 – 91.7

CHR (average of 20) Cr/Cr + Al = 0.47
range = 0.39–0.62

TABLE 2. Peridotite, Pyroxenite, Chromite

	1	2	3	4	5	6	7	8	9	10	11	12	13	14	15	16
SiO_2	36.87	42.47	41.04	41.21	42.11	0.00	0.01	53.56	42.13	42.95	42.36	41.59	...	...	54.20	56.62
TiO_2	0.02	0.02	trace	0.04	0.05	0.25	0.27	0.04	trace	0.02	0.07	0.06	...	0.21	0.04	0.01
Al_2O_3	0.49	0.56	0.24	1.57	1.97	12.86	13.33	1.40	0.27	0.61	0.56	0.57	20.35	33.77	2.14	1.52
FeO^*	7.76	8.94	8.28	8.94	8.79	20.21	16.28	2.43	7.91	8.45	11.46	11.73	20.98	12.96	2.29	5.59
MnO	0.14	0.16	0.13	0.15	0.15	0.31	0.20	0.15	0.18	0.17	0.19	0.19	0.32	0.16	0.11	0.15
MgO	40.65	46.82	48.45	45.35	44.49	11.56	14.34	17.86	48.63	46.75	44.28	45.00	10.39	17.98	17.74	34.44
CaO	0.03	0.03	0.12	1.16	1.17	0.03	0.00	23.19	0.57	0.94	0.75	0.53	...	...	21.76	0.98
Na_2O	0.00	0.00	trace	0.02	0.02	0.00	0.00	0.16	0.01	0.01	...	...	...	...	0.31	...
K_2O	0.00	0.00	0.00	trace	trace	0.00	0.00	0.00	trace	trace	trace	trace	...	...	...	...
NiO	0.27	0.31	0.33	0.36	0.34	0.11	0.03	0.08	0.40	0.30	0.27	0.28	0.09	0.31	0.08	0.10
Cr_2O_3	0.61	0.70	0.47	0.32	0.39	54.56	55.82	0.89	0.40	0.45	0.46	0.56	48.65	35.35	1.38	0.67
Total	86.82	100.01	99.06	99.12	99.48	99.89	100.28	99.76	100.50	100.65	100.40	100.51	100.78	100.74	100.05	100.08
Ni	2090	2440	2600	2800	2310	860	240	630	3150	2300	2100	2200	710	2400	9440	4580
Cr	4140	4790	3200	2200	2700	3.7×10^5	3.8×10^5	6100	2700	3100	3100	3800	3.3×10^5	2.4×10^5	630	790

Analyses 3, 4, 5, 9, 10, 11, and 12 are based on modal analyses and microprobe data. Values are in weight percent. Column 1. Serpentinized dunite with disseminated chromite, upper level of peridotite at Acoje Mine, LUZ-180-C XRF analysis. Column 2. LUZ-180-C, recalculated anydrous. Column 3. Saxonite, upper level at Acoje Mine; OL 92.1% (Fo 91.6), OPX7.3% (En 89.2 Fs 8.1 Wo 2.7), CHR 0.5% (Cr number 0.642). Column 4. Saxonite, ~1 km east of base of peridotite sheet, Acoje Mine access road, Cabaluan River; OL 82.8% (Fo 90.2), OPX 11.3% (En 89.4 Fs 9.5 Wo 1.1) CPX 4.9% (En 48.1 Fs 4.2 Wo 47.7), CHR 1.1% (Cr number 0.140), LUZ-292-A. Column 5. Saxonite base of peridotite near guard station on Acoje Mine access road, Cabaluan River; OL 76.9% (Fo 90.2), OPX 17.0% (En 89.4 Fs 9.5 Wo 1.1), CPX 4.7% (En 48.2 Fs 4.2 Wo 47.7), CHR 1.3% (Cr number 0.140), LUZ-290. Column 6. Disseminated chromite, 1260 level Acoje Mine, LUZ-181-7, Cr/(Cr + Al) = 0.740, coexists with Fo 92.8 microprobe analysis. Column 7. Massive chromite, 1260 level Acoje Mine, LUZ-181-R, Cr/(Cr + Al) = 0.738; microprobe analysis. Column 8. Clinopyroxene (Wo 46.4 En 49.8 Fs 3.8) from pyroxenite vein, Acoje Mine, LUZ-184-A, coexists with Fo 89.2 and Wo 1.5 En 88.4 Fs 10.1; microprobe analysis. Column 9. Saxonite, Coto Mine; OL 91.2% (Fo 91.9) OPX 6.1% (En 90.0 Fs 8.2 Wo 1.8), CPX 2.2% (En 51.2 Fs 3.7 Wo 45.1), CHR 0.4% (Cr number 0.300), LUZ-214-B. Column 10. Saxonite, Coto Block, interior of Zambales Range; OL 81.2%, (Fo 90.7), OPX 14.6% (En 89.9 Fs 9.1 Wo 1.0), CPX 3.7% (En 48.6 Fs 3.3 Wo 48.1), CHR 0.6% (Cr number 0.446), LUZ-394. Column 11. Saxonite, lower exposed level of Coto Block, interior of Zambales Range; OL 81.9% (Fo 87.5) OPX 14.8% (En 86.4 Fs 11.9 Wo 1.7), CPX 2.6% (En 47.9 Fs 4.7 Wo 47.4) CHR 0.7% (Cr number 0.470) LUZ-193. Column 12. Saxonite, lowest exposed level of Coto Block, interior of Zambales Range; OL 86.3% (Fo 87.5), OPX 11.1% (En 86.4 Fs 11.9 Wo 1.7), CPX 1.7% (En 47.9 Fs 4.7 Wo 47.4), CHR 0.9% (Cr number 0.470) LUZ-195-A. Column 13. Disseminated chromite, Coto Mine, coexists with Fo 91.9, LUZ-214-B, Cr/(Cr + Al) = 0.616. Column 14. Massive chromite, Coto Mine, LUZ-214-L, Cr/(Cr + Al) = 0.422. Column 15. Clinopyroxene in saxonite, Coto Mine, LUZ-214-B, En 51.2 Fs 5.2 Wo 41.9. Column 16. Orthopyroxene in saxonite, Coto Mine, LUZ-214-B, En 90.0 Fs 8.2 Wo 1.8. Estimates of accuracy and precision for analyses in this table and others are in Table 8.

terial which are genetically related to the crustal rocks which overlie them. For simplicity, we will refer to these mantle derived saxonite-dunite units as MSD.

The presumed upper mantle peridotite unit (MSD) of the Acoje block consists mainly of partly serpentinized saxonite and, locally, partly serpentinized dunite. The saxonite in the western part of the block, presumed to be the deeper part of the series, has up to 20–25% enstatite (En_{89-91}) but there appears to be a regional trend toward the east (upward) to 5–10% enstatite. Some saxonite samples in the eastern part of the belt are interlayered with lenses and layers of dunite. The olivine varies through a small compositional range (Fo_{93-90}, Ni 3600–2700 ppm) with the more magnesian olivine common near the top of the series and in the dunite. Clinopyroxene (e.g., $Wo_{48.0}En_{48.1}Fs_{3.9}$) is present only as a minor constituent (<10%, usually <5%) in some samples from the western part of the unit; these rocks are lherzolites (OL + OPX + CPX). Chromite is an important accessory mineral throughout the unit; it forms ore bodies in the Acoje mine area, and at several other localities in the block there are chromite claims and inactive chromite mine sites [*Rossman*, 1964; *Bacuta*, 1978]. The chromite has high Cr/(Cr + Al), low Fe'''/Fe'', and Mg/(Mg + Fe'') of 0.5–0.6 (Table 2). Our data for the minerals, as well as the bulk rock composition, support the interpretation that the unit is derived from a part of the mantle which has been through one or more previous episodes of fractional melting and both a high-T ductile deformation and low-T brittle deformation. The high Mg and Ni content of both olivine (e.g., 2400–3600 ppm Ni) and orthopyroxene (e.g., 800–700 ppm Ni), low Ti, Al, and Ca of orthopyroxene, the lack of Ca- and Al-bearing minerals except in the presumed deeper samples, and the high Cr content of the chromite (Cr/(Cr + Al) = 0.65–0.78) all support this interpretation. This is consistent with experimental work done on partial melting of peridotites [*Dickey et al.*, 1971; *Mysen and Kushiro*, 1977]. Studies of the Josephine peridotite [*Dick*, 1974, 1975a, b, 1977] and the Ronda peridotite [*Dickey*, 1975] led to similar conclusions that podiform and massive chromite are refractory residue left behind in mantle material depleted of its basaltic constituents. The MSD peridotite has been serpentinized to varying extent, and we have attempted to reconstruct the original chemistry from microprobe analyses of the constituent minerals and modal analyses. The rocks analyzed in this way are essentially undeformed, and it was not difficult to recognize the pseudomorphed form of the original minerals. Mineral data and our estimates of the rock composition are in Table 2. We note that there is a progression from relative Ca and Fe enrichment in the western (lower) part of the mass to strong depletion toward the east (top). The western part is characterized by minor amounts of clinopyroxene as well as a less depleted bulk rock chemistry. Some of these lherzolitic rocks may represent trapped pockets of 'partial melt' rather than unmelted residue. Pyroxenite masses and dikes found in various parts of the

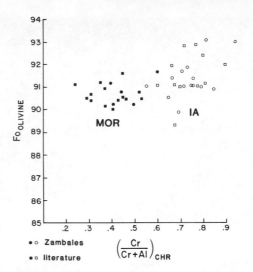

Fig. 5.   Plot of olivine composition versus Cr/(Cr + Al) of coexisting chromite. Circles are data for Zambales range: solid circles are from Coto block; open circles are from Acoje block. Squares are data from descriptions in literature of seafloor (MOR) and island arc (IA) olivine-chromite pairs. Boundary between the two petrologic settings is proposed at Cr ratio of 0.6 [*Dick and Bullen*, 1982].

MSD may be derived from liquids trapped in transit to the crustal levels after or during melting. The pyroxene is low Ti-Al diopside (e.g., $En_{47.7w}Wo_{46.8}Fs_{5.6}$).

Peridotite of the Coto block differs from that of the Acoje block in that both orthopyroxene and clinopyroxene of Coto have slightly lower Mg and Ni content than those of the Acoje block. For example, olivine ranges from Fo_{88} to $Fo_{91.5}$, and Ni ranges from 2900 to 3200 ppm in saxonite. Chromites are much more aluminous, with Cr/(Cr + Al) ranging from 0.39 to 0.62. The geochemical differences between the two peridotite units is best displayed by the Fo versus Cr number$_{CHR}$ plot (Figure 5). The peridotites occupy distinct fields. These fields are separated by a Cr number of about 0.60. This is consistent with compilations of data from various oceanic environments [*Dick and Bullen*, 1982] which show different mineral chemistries from different geologic settings. Orthopyroxene is more abundant in the Coto saxonite (Table 2) than at Acoje. It ranges in composition from $En_{45.1-47.9}Wo_{47.4-51.2}Fs_{3.7-4.7}$, Cr_2O_3 0.70–1.38% and Al_2O_3 2.14–2.30%. Numerous dikes of diabase textured amphibolite cut the Coto peridotite—the Acoje block lacks these dikes. The dikes, discussed in more detail in a separate section, are chemically similar to crustal rocks of the Acoje block. They do not resemble the basal dikes and pillows of the east side of the Coto block. The Coto dikes obviously were not derived from the depleted peridotite which they intrude, and a deeper mantle source is required.

The MSD unit of both blocks has from 20 to 60% serpentine in most samples. The serpentine is largely lizar-

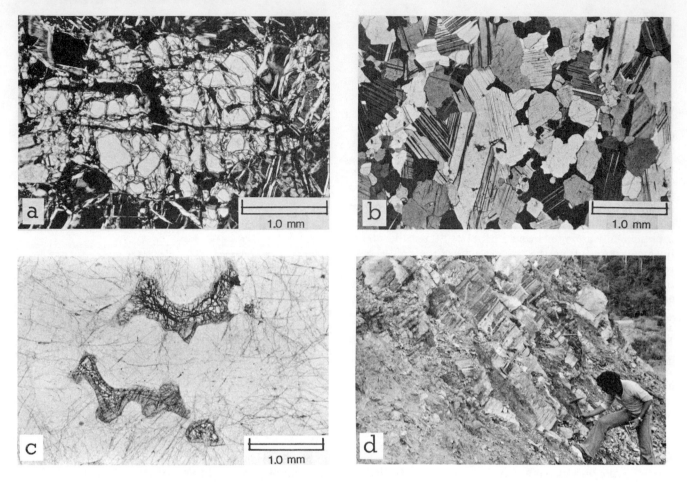

Fig. 6. Photomicrographs of rock textures and outcrop of layered rocks. (a) Partly serpentinized olivine in dunite, $Fo_{92.9}$, LUZ 181-T, plane light. (b) Cumulus texture in norite of Acoje block, LUZ-199, crossed polarizers; $An_{82.5}$, $En_{60.0}Fs_{37.7}Wo_{2.4}$, and $En_{40.2}Fs_{17.0}Wo_{42.9}$. (c) Cumulus texture in troctolite of Coto block, LUZ 213 F, intercumulus olivine is $Fo_{84.5}$, plagioclase is $An_{86.1}$, plane light. (d) Layered wehrlite-dunite unit, Acoje Mine, near portal to 1250 level, LUZ-171 (geologist 1.7 m tall).

dite and is accompanied by magnetite. Both talc and Mg-chlorite are also present as replacement products of orthopyroxene. Clinopyroxene and chromite are apparently unaltered, although some chromite grains have films of Fe oxide; serpentine fills cracks in some chromite grains as well as rimming them. The olivine has been replaced by anastomising networks of serpentine, which give the grains an 'exploded' appearance (Figure 6), but the grain shape has been retained. There is little or no evidence for volume change and no evidence for extensive penetrative deformation. The serpentinized planar structures in the MSD unit trend northerly more or less parallel to the contact with the layered ultramafic cumulate section and to the overall trend of the sheared zones on the western margin of the Acoje block. The strike of the serpentinized zones may reflect the strike of slip planes formed during emplacement of the ophiolite. We emphasize that these serpentinized zones are relatively minor features in terms

of size and distribution, and we suspect that there was no extensive east-west shortening associated with the uplift and emplacement of the range.

There is a marked difference in structure between the essentially undeformed cumulate layered series of ultramafic rocks and the uppermost 500 m or so of the MSD. The upper part of the MSD shows evidence for strong deformation of chromite films and layers, many of which are isoclinally folded. Saxonite layers also show evidence for intense folding. Microscopic features, including kink bands and twinning of olivine and partly deformed and recrystallized aggregates of OL and OPX, indicate that this deformation took place at elevated (probably near solidus) temperatures. The overall effect is the development of a plastic type of deformation discussed above. This gives rise to a primary mantle tectonite fabric which seems to be concentrated near the upper level of the unit and in the more strongly depleted rocks. This fabric may be the result of ductile flow of the unmelted residual mantle during the extraction of the partial melts which formed the crustal rocks.

We interpret the differences in bulk chemistry of the two blocks and the mineralogical differences, especially the chromite, as being due to differences in their melting history [*Evans and Hawkins*, 1980, 1981]. The Acoje block is much more extensively depleted in magmaphilic elements than the Coto block. There are fundamental petrologic differences in the crustal units which overlie each of the peridotite units which support this interpretation. The Acoje peridotite may have been through more than one episode of fractional melting, but there is no evidence for the rocks formed by the previous episode(s). It seems more likely to us that it was extensively melted during generation of the Acoje block crust. This accounts for the thick crustal section present and for the extreme depletion of the peridotite. The Coto block, on the other hand, seems to have been less extensively melted, and there is a less thick crustal unit with it. We are convinced that the two peridotite units are petrologically distinct and are not the same unit split and imbricated by faulting. They must represent different parts of the mantle each of which has had a different petrologic evolution.

Cumulate ultramafic rocks.   A layered series of cumulate textured olivine-pyroxene-chromite-nickel sulfide rocks forms a belt about 2 km wide which trends approximately north-south through the Acoje mine area (Figure 3). This layered series consists mainly of wehrlite and dunite with lesser amounts of saxonite, clinopyroxenite, lherzolite. A few thin layers of norite-gabbro are found near the middle of the section. We informally call it the layered wehrlite-dunite (LWD) to distinguish it from the ultramafic rocks of the mantle saxonite-dunite (MSD) unit. Although the unit we call LWD has been mapped as a unit with an extreme N-S elongation ('black dunite' unit on some maps of the area), we wish to emphasize that the well-developed compositional layering strikes NE with NW dips and that the apparent N-S trend of the map pattern is due to repetition by a series of high-angle, NW trending en echelon faults. Some of these faults have been mapped underground in the mines where they offset ore bodies in the MSD. The locations of other faults are inferred from offsets in the map patterns which adjoin NW trending gullies. Thick jungle vegetation and alluvium mask critical exposures in these gullies, but the faults must be steeply dipping and probably had considerable dip slip movement. The best exposures of the LWD are in the 1000 and 1250 levels at Acoje Mine, in outcrops near the portal of the 1250 level, and on roads leading south to Levante Creek. We measured approximately 1.3 km of layered cumulate ultramafic rock on the 1000 and 1250 level drifts. The LWD may vary in thickness from about 0.7 to 1.3 km in this area as determined from the map pattern of 'black dunite' and the measured dip of layering.

The LWD series is characterized by rocks having compositional layering (rhythmic layering) with layers ranging from about 1 cm to 1 m in thickness (Figure 6). Individual layers appear to be continuous for distances of at least 20–30 m in outcrops and probably extend for hundreds of meters or more. The minerals forming the layers have cumulus textures. Euhedral to subhedral grains of olivine, clinopyroxene, and chromite range in size from 1 to 4 mm in maximum dimension: anhedral intercumulus grains typically are 1–2 mm in size. Many olivine-rich layers have been serpentinized, and they have bands and zones of lizardite plus magnetite. Relict textures in the serpentinized rocks as well as the relatively unaltered layers preserve both orthocumulus and mesocumulus grain relations.

The sequence of appearance of the cumulus phase minerals from the base upwards is olivine (OL), chromite (CHR), clinopyroxene (CPX), orthopyroxene (OPX), plagioclase (PL). Nickel-sulfide forms irregular drop-like masses (0.5–1 mm) in the middle and upper part of the series. In general, OL predominates at the base of the series and the base is compositionally gradational into the underlying depleted mantle dunite (Figure 7). Minor amounts of chromite coprecipitated with the olivine. CPX becomes a more abundant mineral upsection, although periodic fluctuations (new injections of melt) in mineral proportions yield dunite horizons. The main rock type is wehrlite. The reappearance of some early minerals (e.g., OL) and accompanying disappearance of later minerals (e.g., OPX and PL) leads to interlayering of dunite and wehrlite with pyroxenite, lherzolite, and gabbro. In general, the upward trend of rock types in the layered series is (1) OL or (2) OL + CHR; (3) OL + CPX (+ CHR); (4) OL + CPX + OPX (+ CHR); (5) OL + CPX + OPX + PL. Nickel sulfide may be present in assemblages 3, 4, and rarely in 5; CHR is common in 3 and 4 but usually is less abundant than in 2. We present microprobe data for the individual minerals of these assemblages in Table 3. In general there is a decrease in Mg, Cr, Ni upward and a relative increase in Fe, Ca, Al which gives rise to cryptic layering (Figure 7). There are minor fluctuations in mineral chemistry—especially in OL and CHR. This presumably reflects new injections of mafic melts with corresponding shifts in liquidus phases and oxygen fugacities. Because of the fine scale of the apparent layering we did not sample each of the hundreds of recognizable layers, and we can only show the general compositional trend rather than the intricate compositional layering which must be present.

Although the mineral species present in the LWD are the same as those in the MSD the entire section contrasts with the underlying peridotite in terms of structure, textures, and chemistry. The LWD is essentially an undeformed stratiform sequence. Cumulus textures are well developed in all of the rock types. There are distinctive differences in mineral chemistry which clearly separate the two units in a petrologic sense. In general, the rocks are enriched in Ca, Al, Ti, Fe relative to the MSD unit.

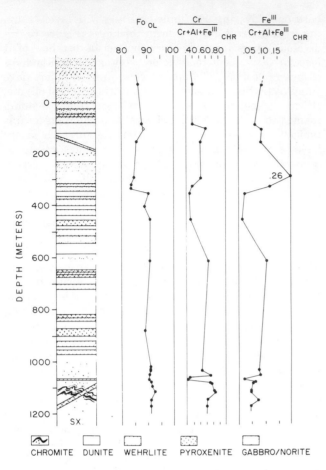

Fig. 7.   Layered wehrlite-dunite unit; geologic section based on mapping of nearly continuous exposures in underground workings of Acoje Mine on 1000 and 1250 levels. Olivine and chromite compositions, based on microprobe data, show cryptic variation superposed on the rhythmic layering. Cryptic variation probably reflects fractional crystallization, changes in fO₂ and new injection of Mg-rich tholeiite. Saxonite and dunite (SX) begins at −1100 m, but the crust mantle transition may be gradational over several hundred meters. The cumulate layered norite extends upwards for more than 6000 m.

Minerals are zoned, and their chemistry varies significantly between layers (Figure 7). Data in Tables 2 and 3 show that the LWD is more enriched in Fe and magmaphilic elements than the MSD. Our interpretation of the chemical and textural data is that the LWD formed as a result of fractional crystallization and crystal accumulation of mafic melts. The most likely source of these melts, which probably were a Mg-rich tholeiite, was the (now depleted) residual mantle material of the MSD. The layered and massive gabbro-norite unit which stratigraphically overlies the LWD is the probable complement to the basal ultramafic cumulate section.

The Coto block lacks a well developed LWD unit, although OL-PX rich layers are present in the basal gabbros at Coto mine. D. L. Rossman (personal communication, 1981) described a small area of rocks similar to the Acoje LWD in an area east of Coto mine. We have no data for these rocks. The basal cumulate material at Coto which we have studied is a thin (~300 m) dunite-troctolite series. The abundance of plagioclase and lack of pyroxene is one of the major differences between this unit and the Acoje LWD.

Gabbro-norite. The most abundant rock types in the crustal series of each block are medium grained (2–4 mm) plagioclase-rich gabbro or norite. Rhythmically layered cumulus-textured rocks (Figure 6), with individual layers ranging from a few millimeters up to 5 m or more thick, are most common but there are also massive homogeneous units. The layering is due to planar orientation of plagioclase and pyroxene and also to variations in mineral abundances. This gives rise to both a textural and compositional layering.

The layering of the Acoje block has steep dips, and, even when the effect of tilting of the range is removed, primary dips of up to 40° are indicated for some of the layers. The Acoje norite unit may be as much as 6–7 km thick. Although this seems unusually thick when compared to the gabbro section of many ophiolites, *Hopson et al.* [1981] report up to 5 km of cumulate gabbro in parts of the Semail ophiolite, and *Juteau and Whitechurch* [1980] estimate a 5–6 km thick cumulate section in the Antalya ophiolite. We estimate that the Coto gabbro series is of the order of 2–3 km thick on the basis of the dip of layering, the outcrop pattern, and the assumption that the upper contact with the diabase-basalt series initially was nearly horizontal. Much of the internal compositional layering of the Coto gabbro series appears to be nearly parallel to the contact between the pillow basalt and the overlying sediments; thus, by inference, this internal layering was also originally nearly horizontal. Locally, both the Acoje and Coto units show sharp angular discordance in the lithologic layering, which suggests that initially there were many layers or beds of cumulus minerals which were steeply inclined. These steep dips could have resulted from in situ nucleation and crystallization on the walls of the magma chamber as suggested by *Casey and Karson* [1981]. Our field observations suggest that these steep initial dips could also be due to effects of crystal cumulate slumping, magma scour and fill process, or primary deposition at high angles of repose (e.g., foreset beds in magma chamber 'deltas'). Probably all of these processes were active. The widespread occurrence of compositional layering, and development of textures and structures typical of clastic sedimentary rocks, are an important aspect of the gabbro series; we interpret these features as an indication that the gabbro series formed by fractional crystallization and crystal accumulation in convecting magma chambers. The rela-

TABLE 3. Mineralogy of Layered Wehrlite-Dunite Unit

Sample Rock Type	Depth m	Fo Ni, ppm	Chromite Cr/(Cr + Al)	Clinopyroxene			Orthopyroxene			Plagioclase, An
				En	Wo	Fs	En	Wo	Fs	
LUZ 171A W	1	85.7 700		47.3	47.5	5.3				
LUZ 171D₁ W	2		0.42	47.8	47.3	4.9				
LUZ 171D₂ L	2	86.6 1500	0.44	48.2	47.0	4.8	86.2	1.4	12.4	
LUZ 171G₁ W	3	87.4 1100		48.8	47.4	4.8	86.6	1.3	12.1	
LUZ 171G₂ W	3	86.5 1100		48.4	46.7	4.9	85.8	1.6	12.7	
LUZ 176B* W	90	87.9 1450	0.40	48.4	47.4	4.3				
LUZ 176D S	120	89	0.66							
LUZ 176F W	130	84.4 1350		45.9	48.6	5.5				
LUZ 176H L	150	85.7 2000	0.58	48.8	46.3	4.9	84.3	2.4	13.3	
LUZ 177S N	295		0.69							85.1
LUZ 177V* W	320	85.0 1000		47.8	46.4	5.7	84.9	1.1	14.1	
LUZ 177Wa L	330	83.9 1250		46.1	47.6	6.4	83.3	1.9	15.9	94.1
LUZ 177Wb P				46.0	47.1	6.9	82.2	1.9	15.9	93.7
LUZ 177Wc W		84.1 1400	0.44	47.5	46.3	6.2				
LUZ 177Wd N				47.6	45.5	6.9	70.4	15.1	14.1	
LUZ 177We	335			47.0	46.2	6.8	82.4	1.5	16.1	
LUZ 177 Xb W	340	83.8 1200		47.4	46.1	6.5	83.0	1.2	15.8	
LUZ 276L W	355	90.6 3300	0.36	49.4	46.2	4.4	87.8	3.3	8.9	
LUZ 177Zg*	400	88.4 400		48.0	47.0	5.0	87.1	1.5	11.4	
LUZ 276N L	435	91.00 3200	0.37	49.3	47.0	3.7	89.6	1.4	9.0	
LUZ 276S D	600	91.0 2400	0.71							
LUZ 276Zb* W	870	89.0 1700		49.7	46.3	4.0				
LUZ 176Zb SD	970	89–90	0.72							
LUZ 276Zh D	1090	91.3 2900								
LUZ 276Zo W	1125	90.8 3000	0.33	48.4	48.3	3.3	88.4	3.1	8.5	
LUZ 181Za D	1135	91.7 2350	0.70							
LUZ 181T D-CHR	1140	92.9 3200	0.74							

TABLE 3. (continued)

Sample Rock Type	Depth m	Fo Ni, ppm	Chromite Cr/(Cr + Al)	Clinopyroxene			Orthopyroxene			Plagioclase, An
				En	Wo	Fs	En	Wo	Fs	
LUZ 181M4 SD-CHR	1150	93	0.74	49.6	48.9	0.7				
LUZ 181A SX	1200	91.4 2900	0.64				89.3	2.7	8.1	

Datum plane for depth is the base of the cumulate norite series. Key for rock types: N = cumulus norite; P = pyroxenite; W = wehrlite; D = dunite; SD = serpentinized dunite; S = serpentinite; CHR = chromite; * = Fe-Ni sulfides and Cu-Fe sulfides present.

tively fine grain size suggests either that small batches of magma were involved or that these were magma chambers close to the surface or both. The inferred bulk chemistry of both series indicates a picritic or Mg-rich tholeiitic parental magma with trace element chemistry resembling N-type MORB [Sun et al., 1979]. The cyclic nature of mineralogic 'stratigraphy,' as well as the presence of both rhythmic and cryptic layering, indicates that there were repeated injections of this magma which subsequently experienced fractional crystallization.

The sequence of appearance of minerals during crystallization of the Coto series was OL-PL-CPX; this sequence is typical of that inferred for MORB and back arc basin basalts. Crystal settling and accumulation has given rise to OL-PL rocks (allivalite and troctolite), OL-PL-CPX (olivine gabbro), and PL-CPX-minor OL (gabbro). There are

minor amounts of wehrlite (OL-CPX) and anorthosite. Orthopyroxene is extremely uncommon and, when present, only appears as enstatite-bronzite rims on OL; amphibole is not a primary mineral except for small amounts which form the intercumulus phase in the uppermost gabbros of the series. Oxides of Fe, Fe-Ti, Cr, and Fe sulfides are rarely found as primary minerals in the gabbros and then only in trace amounts, but magnetite does form aggregates with amphibole, talc, chlorite, and serpentine as a replacement of olivine and pyroxene. There are traces of chromite, pyrrhotite, pyrite, and chalcopyrite in some of the wehrlite layers. There is a general upward variation in the composition of both bulk rocks and minerals which goes from higher Mg, Ca, Cr, Ni assemblages to those with lesser concentrations (Table 4). However, there are several compositional cycles in which layers having high Mg, Cr, etc.,

TABLE 4. Gabbros and Norites of Coto and Acoje Blocks—Chemical Analyses

	1	2	3	4	5	6	7	8	9	10	11	12	13	14
SiO_2	47.60	46.05	47.25	47.51	48.34	48.60	48.39	48.91	48.59	49.54	51.41	54.11	45.44	48.82
TiO_2	0.10	0.01	0.12	0.12	0.17	0.30	0.10	0.12	0.21	0.07	0.15	0.21	0.38	0.11
Al_2O_3	18.30	29.96	18.46	18.43	18.03	17.76	18.42	17.22	16.08	20.15	5.95	2.15	22.22	17.73
FeO^*	4.06	2.06	5.29	6.28	4.62	6.83	5.00	5.03	9.17	3.66	6.03	7.72	7.17	5.24
MnO	0.05	0.06	0.10	0.11	0.11	0.19	0.13	0.10	0.22	0.09	0.16	0.18	0.16	0.14
MgO	11.58	5.26	11.44	12.48	10.92	11.92	11.57	12.14	11.28	10.75	17.40	20.45	9.85	11.56
CaO	17.82	14.61	13.76	12.78	15.33	13.68	15.97	16.39	14.67	15.06	17.57	14.77	13.52	15.62
Na_2O	0.77	1.40	2.24	1.51	1.50	1.41	0.43	0.37	0.52	0.47	0.20	0.45	0.91	0.43
K_2O	0.05	trace	trace	trace	trace	trace	0.02	0.03	0.14	0.02	0.01	0.01	0.02	0.04
P_2O_5	0.01	0.79	0.73	0.67	0.82	0.59	0.01	0.01	0.02	0.05	0.01	0.02	0.01	0.02
Total	100.41	100.14	99.68	100.01	99.93	101.28	100.01	100.32	100.90	99.85	99.03	100.07	99.38	99.71

Trace Elements, ppm

Cr	450	15	...	470	350	145	530	520	105	870	1700	280	70	430
Ni	390	140	240	220	128	250	185	150	90	110	275	85	65	145
V	90	10	390	70	110	110	130	140	250	100	220	210	565	225
Ba	35	30	35	30	25	25	30	30	30	35	40	35	15	30
Rb	<1	...	...	...	...	...	1.2	...	0.8	1	...	...	<1	2
Sr	320	185	106	134	126	158	90	75	130	80	30	80	135	90
Zr	8	9	12	12	9	17	8	7	0	5	8	8	4	7

Analyses by XRF, recalculated anhydrous with all Fe as FeO. Values are in weight percent. Column 1. Basal gabbro, Coto Mine, LUZ-213-B. Column 2. Basal OL-anorthosite, Coto Mine, LUZ-213-G. Column 3. Basal gabbro, Coto Mine, LUZ-217-M. Column 4. Middle level gabbro, center of Zambales Range, LUZ-410-A. Column 5. Middle level gabbro, center of Zambales Range, LUZ-414-D. Column 6. Middle level gabbro, center of Zambales Range, LUZ-416-A. Column 7. Basal norite within layered wehrlite-dunite, Acoje Mine, LUZ-176P. Column 8. Basal norite, pumphouse, Acoje Mine LUZ-206-A. Column 9. Lower level norite, 4 km east of Acoje Mine, LUZ-205-A. Column 10. Lower level norite, 5 km east of Acoje Mine, LUZ-204-E. Column 11. Lower level wehrlite layer, 5 km east of Acoje Mine, LUZ-204-C. Column 12. Middle level wehrlite layer, 6 km east of Acoje Mine, LUZ-202-B. Column 13. Middle level norite, 6 km east of Acoje Mine, LUZ-201. Column 14. Estimate of average composition of the norite series, Acoje block.

reappear, and we interpret this as an indication of periodic replenishment of batches of less differentiated parental magma as the gabbro series was being formed.

Plagioclase is the major mineral component of the Coto series and it constitutes from 50 to 70% of the mode of most samples. Plagioclase concentration does not show any correlation with stratigraphic position in the gabbro unit. It varies in composition from An_{60} to An_{65} in some of the higher level gabbros to An_{90-97} in the basal gabbros and several of the layers which appear throughout the series. An approximate average plagioclase composition for the whole series probably is around An_{75} to An_{80}. The plagioclase is almost always nearly euhedral or has minor irregularities which probably are due to adcumulus growth. Most grains are nearly homogeneous and commonly show, at most, 1 or 2 mol % differences in An between core and rim; the rare zoned grains show normal zoning. Plagioclase textures dominate the rocks and delineate the rhythmic lithologic layering by subparallel alignment of grains which typically are 2–5 mm by 0.5–1.5 mm in size.

Olivine also formed as a cumulus phase and is interspersed between plagioclase grains (Figure 6). Most of the olivine is anhedral to subhedral in form. Lenticular grains, 2–4 mm by 1–2 mm in size, are common, but in some samples the olivine has a highly irregular amoeboid form which partly engulfs plagioclase. Olivine composition ranges from Fo_{92} to Fo_{80} with Ni varying from 2000 to 1000 ppm, respectively. Basal gabbros have the most Mg-rich OL, but there is no good correlation between olivine composition and stratigraphic level. This suggests that these were repeated injections of relatively unfractionated magma during formation of the gabbro series.

Clinopyroxene is present in nearly all samples; it ranges from <5 to 40 modal % in the gabbros and up to 30% in the wehrlite. It typically forms 1–7 mm subhedral grains, but, when present only as a trace component, it is restricted to a thin (<0.2 mm) film around OL and PL. In some of the layers it forms poikilitic irregular patches up to 8 × 12 mm in size which enclose both PL and OL. The clinopyroxene appears to have been the last phase to form in most samples, and the intercumulus liquid probably was essentially CPX. However, some samples have subhedral poikilitic CPX grains with inclusion of OL and PL which are smaller than the adjacent cumulus OL and PL. The PL and OL inclusions have higher An and Fo content than the adjacent cumulus grains. These data suggest that CPX also was a cumulus phase in some samples as well as being an important intercumulus material. Compositional data for the CPX are in Figure 8a. They show that the CPX is mainly a low TiO_2 (~0.5%), low Al_2O_3 (~2–3%) diopside-salite. Cr content is about 5000 ppm.

The total thickness of the Coto gabbro is of the order of 2–3 km. This seems compatible with oceanic layer 3 thicknesses and with gabbro thicknesses in many other ophiolites such as Marum, Papua New Guinea [*Davies*, 1971; *Jacques*, 1981] and Bay of Islands [*Williams and Stevens*,

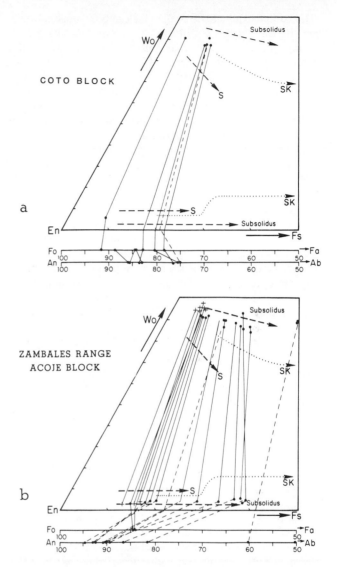

Fig. 8. Mineral data for CPX, OPX, PL, and OL of gabbro and norite units. Pyroxene data are plotted on part of the Wo-En-Fs ternary diagram. Solid tie lines connect coexisting Ca-rich and Ca-poor pyroxene. Trends for Skaergaard pyroxenes is shown, and solidus (s) and subsolidus paths are shown as heavy dashed lines with arrows [*Jacques*, 1981]. Dashed tie lines connect plagioclase and Ca-pyroxene. All data are microprobe analyses. Figure 8a is for Coto block, Figure 8b is for Acoje block.

1974]. The Acoje block has a much greater thickness of gabbro-norite, and it may be as thick as 6–7 km; this, plus the 0.7–1.3 km of cumulate ultramafic rocks, makes it much thicker than layer 3 of back arc basin or deep sea floor crust and more like areas such as the Palau-Kyashu Ridge which has up to 9 km of crust [*Murauchi et al.*, 1968]. *Hopson et al.* [1981] report up to 8 km of ophiolite crustal rocks, with up to 6 km of gabbro, in the Ibra section of the Semail ophiolite.

The composition of the Acoje block gabbro-norite differs from the Coto series mainly in the common occurrence of orthopyroxene as a cumulus mineral (Figure 6), in the relatively minor importance of olivine, and in the appearance of CPX before PL in the crystallization series. We will refer to the Acoje PL-CPX-OPX-(OL) rocks as norite or olivine norite for simplicity and to emphasize their silica saturated character, but technically they are mainly orthopyroxene gabbros according to the classification of *Streckeisen* [1973]. In addition to the abundant calcic plagioclase-bearing noritic rocks, there are also repeated layers of CPX-OPX, CPX-OPX-OL, and, less commonly, quartz norite and hornblende quartz diorite. Cumulus textures with evidence of adcumulus overgrowth are well developed (Figure 6). The repeated appearance of OL-bearing layers and the cyclic variations in mineral chemistry are considered good evidence that crystal fractionation took place in an open system and that there were periodic reinjections of new magma batches. The general coherence in mineral chemistry for a given layer (i.e., covariance of An, En) suggests that the magma chambers were relatively small and that mixing of crystals from different episodes of melt injection was minimal. The amount of intercumulus melt trapped in the layers was generally quite small as indicated by the low Zr content of layers (Table 4). It is likely that during adcumulus growth most of the intercumulus liquid was expelled.

The sequence of crystallization appears to have been OL and CHR-CPX-PL-OPX giving rise to rocks with OL + CHR;   OL + CPX;   (OL) + PL + CPX;   PL + CPX + OPX. The common tendency for CPX to appear with OL in the basal LWD unit and for CPX to appear before PL in much of the layered gabbro-norite is distinctly different from the Coto series. This mineral sequence is common in the arc series rocks dredged from the forearc region of the Mariana Trench [*Hawkins et al.*, 1979; *Bloomer and Hawkins*, this volume; J. Natland, manuscript in preparation, 1982]. The relative sequence of appearance of PL-CPX-OPX seems to have changed periodically during development of the layered series—e.g., CPX-PL-OPX near the base, PL-CPX-OPX higher in the series, and PL-CPX-OPX or PL-OPX-CPX near the upper part. These changes probably reflect differences caused by injection of new magma batches during fractional crystallization in addition to possible flotation of PL as the melt evolved. The mineral chemistry (Figure 8b) shows progressive evolution of the magma series, but the stratigraphy indicates that the melts were injected periodically and did not evolve as a single magma batch. Ca-hornblende is a rare mineral as a primary phase and is always intercumulus. Both ilmenite and magnetite are important accessory minerals; pyrite and chalcopyrite are present in a few of the OPX-rich samples. The plagioclase is more calcic than that of the Coto block, and in the norite varieties it varies from $An_{93}Ab_7$ to $An_{85}Ab_{15}$. Most of the plagioclase is homogeneous, but some has reversed zoning of a few percent An.

The norite series is notably lacking in Na, K, and Rb; there is no primary mica even in the hornblende quartz diorite, no K-feldspar, and the plagioclase typically had 0–0.1% Or component. There is a poorly developed trend from more calcic plagioclase and En-rich pyroxenes near the base to more sodic plagioclase and Fs-rich pyroxene near the central part of the layered series and a return to basal compositions in the upper levels. This general trend is complicated by numerous reversals in both relative proportions of PL:PX and their composition. There is as much compositional variation within some 10-m-thick segments of the layer series as there is over the entire norite section. The spectrum of mineral composition in the layered series seems to conform to predictable variations caused by fractional crystallization, but it is important to note that the cryptic variation developed in the series is not progressive in a stratigraphic sense. This must be due to the repeated injection of new magma into an evolving open system 'magma chamber.'

Upper level Gabbro-Norite.   Massive gabbro and norite, lacking obvious layering or cumulus textures, structurally overlie the cumulus gabbro and norite of both blocks. They form a transition zone between the well-developed cumulate rocks and the diabase series. The thickness of the upper level unit is of the order of 500–1000 m thick but it is mixed with both the diabase and the tonalite-trondhjemite series so the true thickness and volume are difficult to estimate.

The upper level gabbro-norite is more fractionated (Table 5; e.g., enriched in Ti, Zr, Fe) than the cumulate gabbro-norite but has the same mineralogy as the cumulate series, i.e., PL-CPX, with amphibole as an additional primary mineral; OPX is present in the norite but commonly shows reaction to amphibole. Talc-tremolite pseudomorphs of OPX are in many 'noritic' samples. Plagioclase is very calcic (An_{75-95}) and varies in abundance from 40 to 60%. CPX varies from 8 to 35%, amphibole (primary plus secondary) varies from 15 to 25%. Opaque minerals rarely exceed 5%. The gabbro data are plotted in Figure 9 to show how TiO_2 varies with fractionation using FeO*/MgO as an index. There is little distinction in the upper level gabbros from the two crustal blocks, although Coto block samples show a greater TiO_2 enrichment with fractionation than the Acoje block samples. Neither set of data follow a mid-ocean ridge basalt (MORB) trend at the more fractionated end of the series and appear to be more typical of island arc trends. The depletion in Ti at any given stage in fractionation must be a reflection of the general depletion of Ti in the whole magma series and in the source. One of the main differences between the two upper level gabbro series is that those of the Acoje block show complex intrusive contact relations with the tonalite-trondhjemite series plutons and chemically equivalent dikes. The Coto block upper level gabbros commonly are intruded by diabase dikes or form intrusion breccias with veins of tonalite.

TABLE 5.   Upper Level Gabbro, Tonalite, Trondhjemite Series—Chemical Analyses

	1	2	3	4	5	6	7	8
SiO_2	51.80	48.21	62.28	73.32	75.55	68.22	71.84	66.4
TiO_2	0.25	0.41	0.54	0.24	0.18	0.42	0.49	0.40
Al_2O_3	16.40	13.13	15.95	14.68	12.66	15.13	13.27	14.8
FeO^*	8.58	8.45	9.10	4.29	2.40	2.42	4.94	4.6
MnO	0.15	0.13	0.08	0.09	0.04	0.05	...	0.06
MgO	9.80	13.82	2.85	0.57	0.68	1.14	1.62	3.1
CaO	11.40	13.50	6.68	3.11	5.13	8.18	3.45	5.2
Na_2O	0.84	0.76	2.07	3.21	1.50	2.75	4.05	4.3
K_2O	0.03	0.20	0.09	0.16	0.12	0.16	0.28	1.1
P_2O_5	0.03	0.76	0.36	0.06	0.04	0.46	0.05	0.11
Total	99.28	99.31	100.01	99.74	98.33	98.77	99.99	100.07
FeO^*/MgO	0.88	0.61	3.19	7.53	3.53	2.12	3.05	1.48

Trace Elements, ppm

	1	2	3	4	5	6	7	8
Ni	50	290		5	2		2	3
V	265	165		290	10		5	70
Sr	60	95		115	95		120	120
Zr	14	24		30	53		45	80
Y	7	...		20	...		35	40
Ba	30	20		20	60		20	20
Rb	<1	...		2	3		3	...

Values are in weight percent. Column 1. Upper level gabbro, Acoje block, Balincaguin River LUZ-142-A. Microprobe analysis of fused glass, trace elements by XRF. Column 2. Upper level hornblende gabbro, Coto block, Camiling River LUZ-430-A. XRF analysis, recalculated anhydrous. Column 3. Tonalite, Acoje block Balincaguin River LUZ-137. XRF analysis, recalculated anhydrous. Column 4. Trondhjemite, Acoje Block, Balincaguin River LUZ-138. Probe analysis of fused glass. Column 5. Dacite dike, Acoje Block, Balincaguin River, LUZ-136-B. Column 6. Leucotonalite, Acoje Block, Alaminos River, LUZ-388-A. Column 7. Ave plagiogranite, Troodos, Cyprus [*Coleman and Peterman*, 1975]. Column 8. Tonalite 1576, Papua New Guinea [*Davies, 1971*].

Silicic plutonic and volcanic rocks. The upper level gabbros are one end-member of a compositional spectrum which includes diorite, tonalite, quartz-rich tonalite, and leuco-trondhjemite plutonic rocks. (For our discussion we use diorite for quartz normative rocks with 52–56% SiO_2, normative PL less calcic than An_{75}, mafic minerals 30–40%; tonalite = 56–62% SiO_2, PL < An_{75}, mafic minerals 20–30%; quartz tonalite = 62–68% SiO_2, PL < An_{75}, mafic minerals <20%; trondhjemite = >68% SiO_2, PL < An_{60}, mafic minerals ≪20%. Granodiorite and quartz monzonite are not used because the series lacks K-feldspar and has virtually no K. The series is characterized by rocks with high CaO and normative An.) We believe that this continuum is an important characteristic of the series and offers insights to its origin. In addition to the plutonic and hypabyssal rocks, there are 1–3 m wide dikes and irregular small masses of fine grained and aphanitic rocks which are compositionally equivalent. For simplicity, we will refer to the rock suite as the tonalite series. Modal and normative data are plotted in Figure 10, and Table 5 lists some representative compositional data. It is important to note that all of the rocks have high CaO content, and plagioclase typically is more calcic than An_{50}.

The tonalite series plutons and dikes have intruded upper-level gabbros and diabase, and they appear to be restricted to crustal levels near the diabase gabbro contact. There are good exposures of both plutons and dikes on the Balincaguin River east of Barlo, on the Alaminos River near Sioasio, on the Camiling and Moriones Rivers, and east of Iba on the Bucao River (Figure 11 and 12). The tonalite series is present in both of the crustal blocks, but the larger plutons are at the north end of the Acoje block. This may only be due to relative preservation, rather than a petrologic characteristic of the blocks, but we believe that it actually is a fundamental geochemical difference between the crustal units.

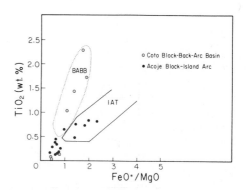

Fig. 9. TiO_2 versus FeO^*/MgO for upper level gabbros. MORB-BABB field is for mid-ocean ridge and back arc basin basalts. IAT is for island arc tholeiite series.

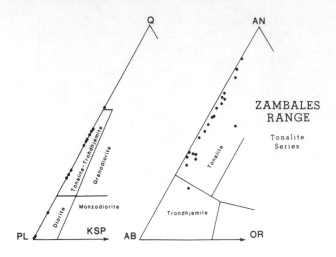

Fig. 10.   Modal and normative data for plutonic and volcanic rocks of the tonalite-trondhjemite series and upper level gabbros.

There are also many small plutons (hundreds of square meters in outcrop) and irregular masses, including intrusion breccias with anastomosing veins of tonalite, exposed on the east side of the range in the Camiling and Moriones River drainage system. Although they are widespread, the tonalite series probably constitutes only about 3–5% of the crustal rocks. Since much of the upper crustal levels has been removed by erosion, there has been a significant amount of quartz-bearing detritus deposited in the sediment traps flanking the Zambales Range.

In most areas it is apparent that the tonalitic series was the latest to form as the plutons have intruded the gabbro and diabase. Fine grained and aphanitic dikes (0.5–3 m wide) cut both the mafic and silicic rocks. However, in a few areas there are dike swarms with both silicic and basaltic dikes, and at least some of the basalts are younger than the silicic dikes.

Although the tonalite series plutons have compositions which could be derived by fractional crystallization of the gabbro series, the two rock types show mutually intrusive contact relations. Gabbro xenoliths are present on the mar-

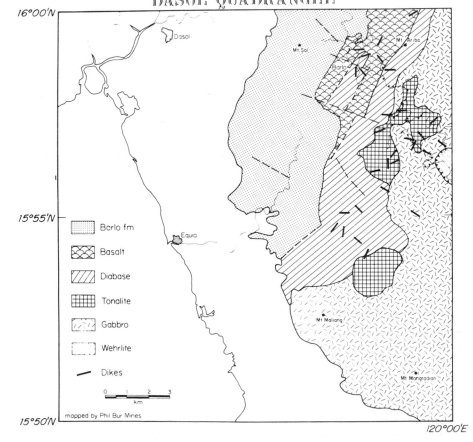

Fig. 11.   Geologic map of Barlo area, Dasol quadrangle, based on mapping of Philippine Bureau of Mines and Geosciences and our data. Area along coast without pattern is Tertiary and Quaternary sediments.

gins of some tonalite plutons, and irregular offshoots of tonalite invade the gabbro, but at some gabbro-tonalite contacts the gabbro has a border zone of finer grained rock which suggests a chilled contact against the tonalite. These complex contacts suggest that the two magma types were being formed and emplaced essentially simultaneously and thus probably are genetically related. An example of the

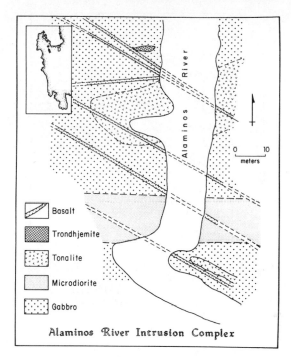

Fig. 13. Geologic map of area of mixed gabbro, tonalite, trondhjemite, and younger dikes near Sioasio on Alaminos River. Inset shows location of map area (arrow) and Dagupan City (D).

complex intrusion relations is shown in Figure 13, which is a map of a small area on the Alaminos River at the north end of the Acoje block. We suggest that intrusion of successive batches of magma—some essentially unfractionated, others derived from a compositionally stratified magma chamber—could explain the field relations.

The tonalite series rocks are medium grained (1–4 mm) quartz-plagioclase rocks with varied amounts of Ca-hornblende (<5–20%) and rare clinopyroxene. Plagioclase generally is more calcic than An_{50}; the trondhjemites have up to 4–5% CaO, and most should be termed calcic trondhjemite [Barker, 1979]. In a few samples remnants of hypersthene are rimmed by amphibole. They lack K-feldspar or biotite, and primary Fe-Ti oxides are only a minor constituent (1–5%). Quartz is abundant in the tonalite-trondhjemite part of the series. It forms large irregular grains, patches enclosing plagioclase and amphibole, granophyric textured haloes around plagioclase, and it fills veins a few millimeters wide. Some of the quartz may be the result of silicification during hydrothermal alteration, but most of it appears to be primary.

The plutonic rocks have been altered by hydrothermal activity which has formed turbid plagioclase, spots of epidote in plagioclase and clusters of epidote, chlorite, magnetite between plagioclase grains and as a replacement of hornblende. Much of the hornblende has been replaced by chlorite, Fe oxides, and fibrous actinolitic hornblende. The secondary mineral assemblage is indicative of albite-epi-

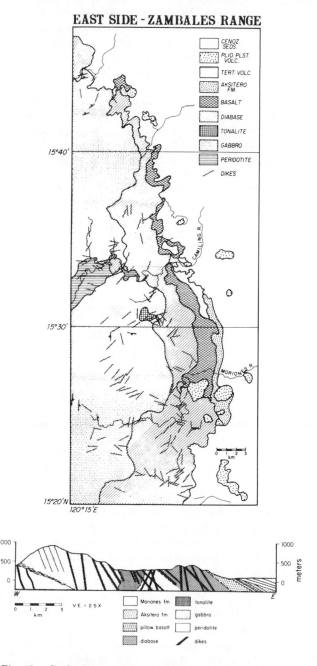

Fig. 12. Geologic map and cross section of eastern side of Zambales Range based on mapping of Philippine Bureau of Mines and Geosciences and our data. Section is along 15°30′N.

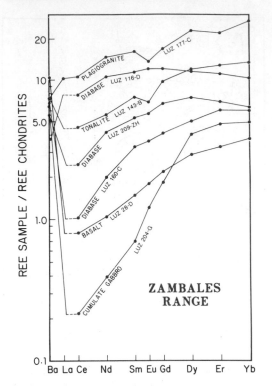

Fig. 14.   Chondrite normalized REE data for selected rocks of Zambales Range. Diabase 116D is BABB, diabase 209ZH and 160C are island arc tholeiite, basalt 28-D is a depleted arc tholeiite series rock with chemical characteristics of boninite series. Tonalite is part of arc tonalite series, plagiogranite is from dike cutting Acoje peridotite, cumulate gabbro is from Acoje gabbronorite series.

dote hornfels metamorphic conditions (e.g., 1–2 kbar, 300°–350°C). The rocks retain their igneous textures; therefore the metamorphic overprint must have developed under hydrous static conditions, presumably while hydrothermal circulation was active during cooling of the plutons.

Major element, trace element, and normative data for the tonalite series are in Table 5, and REE are in Figure 14. The modal and normative data of Figure 10 illustrate their extreme depletion in K. The series has high Na/K and Fe/Mg, especially in the diorite and tonalite. All of the samples have relatively high CaO and normative An. Normative Hy is present in all samples, in leuco-trondhjemites as well as diorites.

The petrologic characteristics of the tonalite series, especially the trondhjemites, are similar to 'plagiogranites' [*Coleman and Peterman*, 1975], and their association with an ophiolite makes this a useful descriptive name. We agree that 'oceanic plagiogranites' of many ophiolites probably formed by differentiation of subalkaline magmas at slowly spreading ridges where there is differentiation of subcrustal gabbroic magma chambers [*Coleman and Peterman*, 1975]. The differentiation process probably is en-

hanced by stoping of blocks of hydrothermally altered mafic crust. We believe that similar processes operated during generation of the Zambales tonalite series but choose to use the more traditional rock names (tonalite, etc.) to avoid the implication that the series formed at a slowly spreading midocean ridge.

Tonalite-trondhjemite plutonic rocks are minor but important components of many ophiolites (e.g., Papua New Guinea *Davies*, [1971], Bay of Islands, *Malpas* [1979], and Oman, *Hopson et al.* [1981]). In Oman it seems likely that they are intrinsic to the fractionation history of the upper level gabbro-diabase series and are 'oceanic plagiogranites' in the sense implied by the name. The Camiling and Moriones River areas have good examples of intrusion breccias in which fine to medium grained (0.5–2 mm) trondhjemitic material is mixed with angular to subrounded blocks of ampibole-rich diabase. Many contacts between trondhjemite and diabase blocks are blurred, and it appears that the diabase may have been partly melted by the hydrous trondhjemite magma. These may be 'oceanic plagiogranite' in the strict sense. The association of island arc series plutonic rocks with ophiolites is also well known, e.g., Papua, New Guinea [*Davies*, 1971]; Twilingate Granite [*Williams and Payne*, 1975; *Payne and Strong*, 1979]; and Little Port Complex [*Malpas*, 1979]. This spatial association is not unusual if one considers the geometry of arc–back arc basin systems and the likelihood that at least some ophiolites preserve both arc and back arc basin rocks [*Hawkins*, 1980].

We will show that the Zambales tonalite series of the Acoje block bears strong similarities to rocks of intra-oceanic island arcs and offers further support for our interpretation of the Acoje block as the remnant of island arc crust and depleted upper mantle. In considering the chemical characteristics of the tonalite series, we have also considered the fact that there is a nearly continuous range in rock types from gabbro to trondhjemite (Figures 10, 15) and that the tonalites are 'stratigraphically' above the layered cumulate gabbro-norite units.

The mineralogy and composition support the likelihood that rocks of the tonalite series are related to the gabbro by fractional crystallization processes. We have used Zr as an index of fractionation (Figure 15) to show the covariance with TiO_2. A similar plot is shown in Figure 16, where TiO_2 is plotted against FeO*/MgO. In both plots it is apparent that the most-fractionated magmas follow a calc-alkaline trend rather than a MORB trend. The TiO_2-Zr plot shows that the tonalite series resembles the trend for the West Mariana Ridge [*Mattey et al.*, 1980] and for the Antilles island arc [*Brown et al.*, 1977], but the source for the Zambales series must have been relatively depleted in Ti as well as K, Rb, Zr, Y, and Ba. The TiO_2 versus FeO*/MgO plot includes data for mafic dikes (which follow an arc tholeiite trend) and tonalites from the Agno and Cordillera batholith of northern Luzon. The main point of the diagram is to demonstrate that the tonalite series of the

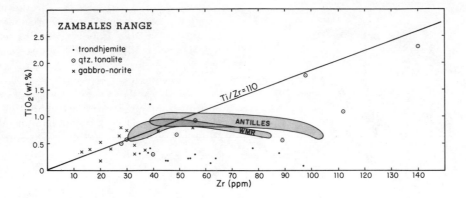

Fig. 15. Plot of Zr versus TiO_2 for rocks of the Acoje Block tonalite series. Chondritic Ti/Zr ratio (110) is followed by MORB and BABB and by some of the gabbro-norite and tonalite. More fractionated Zambales rocks follow the trends for the West Mariana Ridge (WMR) and Antilles arc but have lower TiO_2 concentrations for equivalent Zr concentrations.

Zambales Range closely resembles island arc series rocks and that they may be genetically related to the Eocene-Oligocene plutonic rocks of Northern Luzon.

In addition to the Ti, Zr, FeO*/MgO trends which resemble arc series rocks, the limited data for REE (Figure 12) indicate an arc tholeiite pattern for mafic dikes-sills associated with the tonalite series. Other arc characteristics include normative Quartz and Hypersthene in all samples, an Fe enrichment trend, and high Ca and normative An. The tonalite series resembles low-Al_2O_3 trondhjemites [Barker, 1979] even though they have up to 3.7% normative corundum. The transition from Di normative to C normative rocks is at about 62% SiO_2; this transition seems to be a characteristic of many calc-alkaline series trondhjemites. Barker [1979] suggests that fractionation of low-K 'andesitic' magma could give rise to low-Al tonalites and trondhjemites and leave cumulate residues of plagioclase, hypersthene, and augite.

We cannot rule out the possibility that stoping of altered roof rocks has played a role in forming the tonalite, but we suggest that the major factor was fractional crystallization of the low-K, silica-saturated magmas which were parental to the layered cumulate norite series. The norite and tonalite formed in an intra–oceanic island arc setting, and they represent the root zone of a (late?) Eocene island arc which may have been a part of the more extensive arc volcanic-plutonic complex of the Agno-Central Cordillera batholith. The limited extent of the tonalite series leads us to speculate that the 'Acoje block island arc' was a very short-lived feature, and its short life span must have implications for the paleo-tectonic configuration of oceanic plates in early Tertiary time.

Pillows-dikes-sills-breccias. The crustal units of both the Acoje and Coto blocks have pillow basalts, massive basalt, and basaltic breccias as well as dikes or sills which appear to be feeders to them (Figure 17). The best exposure of pillow basalts and their associated dikes and breccias

is at Barlo and on the North Balincaguin River in the Acoje block (Figure 11). The Barlo area pillows are interbedded with Fe and Cu-Fe sulfides which form fault-bounded, blanket-like deposits in narrow grabens. They are considered to be a Cyprus-type sulfide deposit. Pillows are also well exposed on the eastern side of the Coto block along the Camiling and Moriones Rivers (Figure 12), east of Botolan on the Bucao River and near Olangapo. The pillow complex east of Botolan is heavily altered to greenstone assemblages and has been permeated by secondary pyrite and chalcopyrite, but apparently it lacks the blanket type sulfide deposits which are seen at Barlo. All of the pillow complexes have 0.5–3 m diameter pillows with 1–2 cm thick chilled rinds, radial joints, interpillow hyaloclastite, and crosscutting dikes or sills. They are submarine extrusive rocks. Basalt breccias, interbedded with

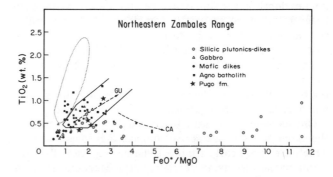

Fig. 16. Plot of TiO_2 versus FeO*/MgO for Acoje Block tonalite series, similar rocks from the Agno batholith (Luzon), and cobbles from the Oligocene-lower Miocene Pugo fm. (Luzon). The dotted field outlines the field for mid-ocean ridge and back arc basin basalts, the solid line outlines the field for island arc series rocks from the Mariana arc. The calc-alkaline trend is shown by the dashed line CA and island arc series rocks from the Umatac fm., Guam, by the line GU.

Fig. 17.   Diabase dikes 3–4 m wide cutting saxonite at Coto mine. Power shovels and trucks on benches give scale. These are island arc series dikes cutting depleted mantle material of the (back arc basin) Coto block.

the pillows, probably indicate talus accumulated at the foot of small scarps near a spreading center. Massive basaltic layers may represent either thick sills or lava lake accumulations. Some of the pillows have vesicular interiors, and some breccia clasts are also vesicular. These textures need not indicate emplacement in shallow water, as we do not know the initial volatile content of the magmas. For example, we have collected vesicular basalts from 2.8 km depth in the Mariana Trough back arc basin. Elongated pillows are especially common near Barlo, and in that area they have a preferred orientation to the northeast with a general plunge of about 50°. Elongated pillows must have accumulated with an initial inclination, but this trend suggests that there has been some regional tilting to the northeast. Perhaps as much as 30° tilting to the northeast has been involved.

Our geochemical data discussed below indicate that the Barlo area basalt series comprises pillows and dikes which are chemically similar to island arc tholeiites from primitive oceanic island arcs (e.g., the Mariana Arc). The Barlo area also has basalts which have the chemistry of boninite series rocks. These appear to be diagnostic of the early stages of evolution of island arcs. The Barlo area basalts, dike-sill complexes near the Lingayen Gulf, and the dikes cutting the peridotite at Coto and elsewhere in the Coto block all appear to be related to arc series rocks, while the pillows and dikes of the east side of the range probably are MORB or back arc basin basalts.

On the east side of the range there are numerous dikes in the transition zone between the gabbro and the pillow basalts, and many of these dikes also intrude the pillow complex. Locally, there are dike swarms in which 0.5–3 m wide dikes share a common orientation, and a few of them show evidence for dike on dike intrusion which formed sheeted split dikes. The general pattern is, however, for a dike swarm to have numerous full dikes with a strong preferred orientation but to also have many dikes with apparently random trends. Figure 18 shows data for 129 dikes from a small area near the Moriones River. These dikes are primarily individual bodies (i.e., not swarms of split dikes) which have only a weak tendency to trend NE. If the effect of Cenozoic tilting of the range is removed from the dips, then most of these dikes appear to have been originally subhorizontal. The weak NE maximum in present trends may be the general trend of the axis of tilting. Maximum stress release must have been in a vertical sense rather than horizontal. The zone of diabase and basaltic dikes probably is about 0.5–1.0 km thick as determined from the map pattern (Figure 11) and the regional structure.

The peridotite and norite of the Acoje block are notably lacking in basaltic dikes, and the few dikes seen are mainly pyroxenite or, even more rarely, PL-rich gabbro which cut the peridotite. In contrast, the Coto block peridotite is cut by numerous dikes of diabase which range from 1 to 6 m wide (Figure 17). They make sharp contacts with serpentinized peridotite, but they almost always have a border zone of chlorite, serpentine, and magnetite a few centimeters wide. These border zones often show slickensided surfaces, many of the dikes have been truncated by faults, and some have been broken up into large wedge-shaped or lenticular masses. Joint planes, perpendicular to the dike margin, are common. Many of the dikes are closely spaced—e.g., 10–20 m apart—and tend to be nearly parallel in a given area. The exposures in the mine pits show this well and give a cross section view more than 100 m high. However, the common orientation does not prevail throughout the block or even over areas of a few hundred square meters. None of the dikes formed a sheeted dike complex, none appear as split dikes in multiple dike injection units. The original orientation of the Coto dikes is not easy to determine because we have no direct indication of the nature of postdike deformation of the peridotite. If we assume that the peridotite has experienced the same northeasterly tilting that this overlying pillow basalt and other crustal units have experienced, then many of the near vertical dikes originally had dips of from 30° to 40° to the east or southeast, some dikes originally were near vertical, and some dipped westerly. The most important fact about dike orientations is that they do not indicate simple lateral stress release but also indicate that major stress release was in a vertical direction. This does not seem to be indicative of the stress field inferred for active crust/mantle dilation at seafloor spreading centers.

In addition to the mafic dikes described above, there are local concentrations of silicic dikes which range in composition from andesitic to Si-rich rhyolite. These are closely associated with the quartz diorite–trondhjemite–leucogranodiorite plutons and are discussed with them.

The chemistry of the mafic dike-pillow units offers the best evidence for the petrologic character of the crustal units. In order to eliminate the obscuring effects of low-grade metamorphism and alteration, we have relied most

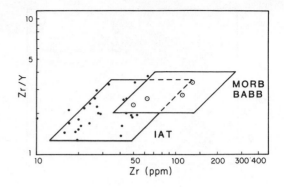

Fig. 19. Plot of Zr/y versus Zr for mafic dikes and pillows. Circled dots are presumed to be back arc basin basalts, solid dots are presumed island arc samples. Fields shown were defined by *Pearce and Norry* [1979]. MORB-BABB field is for mid-ocean ridge and back arc basin basalts, IAT field is for island arc tholeiite series.

heavily on abundances and ratios of elements such as the high field strength elements (HFS) in making an estimate of the tectonic setting of the crustal units when they formed. For purposes of comparison, we have used our own data and others [e.g., *Jakes and Gill*, 1970; *Ewart and Bryan*, 1972; *Hawkins*, 1974, 1976, 1977; *Miyashiro*, 1974; *Stern*, 1979; *Dixon and Batiza*, 1979; *Mattey et el.*, 1981; *Gill*, 1981] to distinguish between basalts derived from back arc basins and island arc environments. In addition to the HFS elements, the Fe/Mg and Ca/Al ratios, and transition metal abundances, all serve as useful discriminants [e.g., *Pearce and Cann*, 1973; *Pearce and Norry*, 1979; *Pearce*, 1980].

The basaltic dikes and pillows exposed along the Camiling and Moriones River (Figure 12) have the chemical signature of back arc basin (BABB) or mid-ocean ridge (MORB) basalts. For example, the chondrite normalized REE pattern for a sill from the Camiling River (Figure 13) Y and Zr (Figure 19) and TiO_2-FeO^*/MgO (Figure 20) all show the mafic dikes to be similar to MORB or to BABB from the Mariana Trough and Lau Basin. We have used the ratio $(Mg \times 100/(Fe'' + Mg))$ = Mg number as an index of fractionation to distinguish relatively 'primitive' basalts from more fractionated samples. The most primitive Zambales basalts have Mg number = 66–69. We accept the suggested values of Mg number = 63–73 [*Green*, 1971] to indicate 'primitive' samples; these magmas would have been in equilibrium with Fo_{87-92}. The residual dunites and saxonites of Acoje and Coto have olivine of Fo_{88-93}, and the basal layered wehrlite-dunite at Acoje has Fo_{85-92}. Thus the postulated mantle source for the basalt series appears to have residual olivine compositions which would have been in equilibrium with the least fractionated basalts. For least fractionated samples the TiO_2, Zr, and Y will vary from 1.5%, 80 ppm, 30 ppm, respectively, for relatively small amounts of melting (e.g., <15%) to 0.7% TiO_2, 40

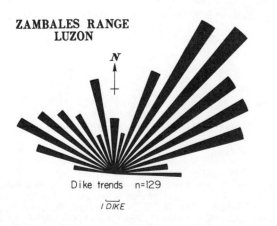

ZAMBALES RANGE LUZON

Dike trends   n=129

Fig. 18. Diagram showing strike of 129 'dikes' on east side of Zambales range near Moriones River. Dike azimuths were plotted in 5° groupings. Dips varied from vertical to 40°.

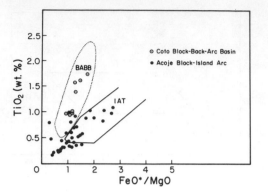

Fig. 20.   Plot of TiO_2 versus FeO^*/MgO for mafic pillows and dikes. MORB-BABB is the field for mid-ocean ridge and back arc basin basalts, respectively; the island arc (IAT) field shown is based on western Pacific are volcanics.

ppm Zr, 15 ppm Y for greater amounts of melting (e.g., ~25%). The ratios CaO/TiO_2 and Al_2O_3/TiO_2 show good correlation with TiO_2 and are useful indicators of source melting in unfractionated rocks [*Sun et al.*, 1979]. The basalts we consider to be MORB-BABB type from the Coto

block have Mg number 0.66–0.69, TiO_2 0.6–1.0%, 50–60 ppm Zr 20–25 ppm Y and a MORB-like REE pattern. They could represent about 20% melting of a source like the source of N-type MORB or basalts of many back arc basins. Additional chemical data for Coto block basalts are in Table 6. The basalts are all HY-OL normative and plot close to the PL-OL cotectic [*Shido et al.*, 1971] as do most BABB and MORB. The petrologic data do not permit us to distinguish between a mid-ocean ridge or back arc basin source, but the geologic association with the island arc series strongly suggests the latter. The random pattern to dike orientations and the numerous occurrences of sills suggest that it was a very slowly spreading basin.

The Acoje block basalt pillows and dikes and the diabase dikes cutting the peridotite at Coto (Figures 3 and 17) differ from the basalts described above mainly in their HFS element abundances. Acoje basalts show a strong depletion in Ti, Zr, REE (Figure 14) as well as Ni and Cr. These are characteristics of island arc tholeiite series [*Jakes and Gill*, 1970; *Pearce and Norry*, 1979; *Mattey et al.*, 1981]. The plot of Y vs Zr (Figure 19) shows that all of the Acoje samples plus the Coto mine area diabase dikes lie in the island arc field, and only the Coto dikes feeding the pillow complex lie in the MORB-BABB field. We also have used the ratio TiO_2 versus FeO^*/MgO and find that the island

TABLE 6.   Basalt-Diabase, Pillows, Dikes, and Sills

	1	2	3	4	5	6	7	8	9	10	11	12	13	14	15	16
SiO_2	50.42	53.91	53.49	56.07	51.28	54.14	52.81	55.13	49.86	51.89	49.67	50.87	51.39	52.90	50.15	49.99
TiO_2	0.25	0.26	0.22	0.28	0.51	0.56	0.24	0.34	0.46	0.49	0.97	0.91	0.59	1.47	0.98	1.01
Al_2O_3	14.06	13.74	13.67	16.31	16.33	18.02	13.94	16.72	15.57	16.09	17.74	15.93	16.71	15.97	16.47	15.49
FeO^*	9.29	8.32	8.49	8.17	8.92	8.81	8.74	8.04	7.31	7.30	8.92	8.94	8.27	10.87	8.58	8.75
MnO	0.33	0.12	0.23	0.16	0.17	0.17	0.22	0.27	0.14	0.16	0.15	0.21	0.22	0.36	0.19	0.17
MgO	15.88	9.94	13.08	8.79	7.64	6.31	14.06	7.77	10.64	8.61	7.73	8.15	8.56	5.48	9.12	8.07
CaO	8.72	10.22	7.76	4.06	11.52	9.26	7.76	7.84	13.80	12.96	11.23	11.91	9.54	9.41	12.19	11.26
Na_2O	0.65	1.16	2.01	5.67	2.95	2.26	1.44	2.45	1.68	2.28	2.50	1.83	1.64	3.09	1.62	3.09
K_2O	0.38	2.17	0.70	0.03	trace	0.26	0.37	0.58	0.05	0.08	0.17	0.01	0.97	0.05	0.22	0.16
P_2O_5	0.02	trace	0.02	0.10	0.62	0.03	0.05	0.02	0.11	trace	0.23	trace	0.17	0.17	0.09	0.21
Total	100.31	99.84	99.83	100.07	99.94	99.82	99.89	99.14	99.62	99.86	99.31	98.75	98.06	99.77	99.61	99.20
Mg number	75.3	68.1	73.3	65.7	60.4	56.1	74.1	63.3	72.2	67.8	60.7	61.9	64.9	62.3	65.5	62.2
Trace Elements, ppm																
Cr	1200	800	495	216	56	45	800	14	380	230	...	70	90	...	...	243
Ni	365	278	250	105	40	25	320	45	160	90	90	60	60	20	155	85
V	220	235	240	270	270	245	230	235	230	220	275	240	220	340	220	275
Zr	15	10	15	20	30	30	20	20	25	30	50	60	50	130	60	30
Y	10	10	11	13	16	14	8	13	<10	13	21	24	21	...	...	27
Sr	28	35	50	40	75	105	40	80	200	110	125	130	90	105	120	90
Rb	4	3	5	1	...	2	3	4	3	<1	<1	<1	≈10	...	...	3
Ba	40	910	30	25	20	30	40	50	20	15	15	15	30	<10	15	10

Values are in weight percent. Column 1. Altered pillow basalt (boninitic) Barlo Mine, Acoje Block LUZ-28C. Column 2. Altered pillow basalt (boninitic) Barlo Mine, Acoje Block, LUZ-28D. Column 3. Diabase dike (boninitic) Balincaguin River, Acoje Block, LUZ-159. Column 4. Pillow rind (boninitic, spilitized) Balincaguin River, Acoje Block, LUZ-160-C. Column 5. Diabase dike, Uyong River, Acoje Block, LUZ-129-B. Column 6. Diabase dike, Sual coast, Acoje Block, LUZ-130-A. Column 7. Diabase dike, Balincaguin River, LUZ-160-A. Column 8. Diabase dike (or sill) Balincaguin River, Acoje Block, LUZ-152. Column 9. Diabase dike, Coto Mine, LUZ-209-ZH. Column 10. Diabase dike, Coto Mine, LUZ-4A. Samples 1–10 presumed island arc series. Column 11. Basalt dike, Maguisguis, Coto Block, LUZ-258-C2. Column 12. Diabase dike, Camiling River, Coto Block, LUZ-116-0. Column 13. Diabase sill, Moriones River, Coto Block, LUZ-112. Column 14. Diabase sill, Camiling River, Coto Block, LUZ-118C. Column 15. Diabase sill, Camiling River, Coto Block, LUZ-116-F. Column 16. Basalt dike, Maguisguis, Coto Block, LUZ-258-B. Samples 11–16 presumed back-arc basin series.

TABLE 7. Averaged Analyses of Basaltic and Boninitic Rocks

	1	s.d.	2	3	s.d	4	s.d	5	s.d	6	7
SiO_2	50.80	0.96	49.7	50.1	1.91	53.6	1.09	54.17	2.08	53.0	55.06
TiO_2	1.20	0.37	1.5	0.70	0.22	0.65	0.27	0.29	0.07	0.85	0.25
Al_2O_3	16.4	1.42	16.6	16.8	1.56	16.4	1.17	14.66	1.55	15.9	12.91
FeO^*	9.4	1.32	9.8	9.0	0.60	9.4	1.45	8.45	0.32	10.6	7.51
MnO	0.2	0.07	0.2	0.2	0.07	0.2	0.07	0.20	0.06	0.2	0.14
MgO	7.5	1.05	6.8	7.7	1.60	6.4	1.27	11.01	2.49	5.9	11.02
CaO	10.4	2.01	12.6	13.3	2.59	9.5	1.27	7.63	2.65	10.5	8.05
Na_2O	2.4	0.59	2.2	1.5	0.69	2.8	1.23	2.52	1.12	2.5	1.55
K_2O	0.22	0.27	0.24	0.07	0.03	0.20	0.15	0.41	0.24	0.56	0.79
P_2O_5	0.2	0.06	0.1	0.4	0.24	0.3	0.23	0.05	0.06	0.1	0.05
Total	98.72		99.74	99.77		99.45		99.39		100.11	97.33
Mg number	58.7		55.3	60.4		54.8		69.9		49.8	72.3

Trace Elements, ppm

	1		2	3		4	s.d	5	s.d	6	7
Cr	145	65	300	100	75	60	65	560	260	20	629
Ni	70	35	90	70	45	40	20	210	115	25	186
V	270	65	330	310	70	315	75	245	20	330	164
Zr	80	35	70	35	10	40	15	20	5	55	54
Y	30	10	15	10							
Sr	120	35	150	125	40	105	35	50	15	145	88
Rb	5		3	1	0.5	2	2	3	1.5	2	14
Ba	14	7	30	20	2	30	10	35	10	50	20

s.d. = standard deviation of sample population from mean. Column 1. Coto Block dikes, sills and pillow basalts. $SiO_2 < 53\%$ average of 13 samples. Back arc basin series. Column 2. Lau Basin (back arc basin) basalt, least and moderately altered, fractionated basalts, average of nine samples [Hawkins, 1977]. Column 3. Acoje Block dikes, sills and pillow basalts, diabase dikes cutting Coto peridotite, least fractionated basalts, average of eight samples, arc-tholeiite series. Column 4. Acoje Block dikes, sills and pillow basalts, diabase dikes cutting Coto peridotite, fractionated basalts, average of 14 samples, arc-tholeiite series. Column 5. Acoje Block boninite series dikes and pillows, $SiO_2 > 52\%$, average of seven samples (least fractionated samples in Table 6). Column 6. Average island arc tholeiite, Mariana Arc (J. W. Hawkins, work in progress, 1982). Column 7. Boninite, MARA 50-23, from Mariana Trench (2.9% volatiles (S. Bloomer and J. W. Hawkins, work in progress, 1982).

arc trend for the Acoje basalts follows the trend for Mariana Arc basalts (Figure 19). Additional data are listed in Table 6. Note that some of the pillow basalts from the Barlo area (Figure 11 and Table 6) have unusual compositions. They have low Ti, Zr, Y, and REE, especially LREE, and have high Mg, Cr, Ni. They resemble boninite series from the Mariana arc [Hawkins et al., 1979; Bloomer et al., 1979; Bloomer, 1982]; however, rocks with the distinctive boninite texture have not been identified. Boninite series rocks appear to be distinctive in the early stages of evolution of an island arc. There is no indication that they are part of 'normal' oceanic crust, and they have not been found in back arc basins.

The chemistry and petrology of the Acoje basaltic units strongly support the evidence from the gabbro and tonalite series rocks that they all formed in an island arc setting. The small volume of tonalitic rocks requires that magma generation ended before the arc evolved beyond the initial stages of formation of the basaltic units and the generation of a number of small fractionated tonalite-trondhjemite plutons. The lack of evidence for volcanic-clastic rocks which could have been derived from this arc may be an indication of the limited extent to which it evolved. It may never have risen above sea level. An alternate explanation may be that these clastic rocks lie buried in the central basin to the east or form part of the sedimentary prism off the west coast of Luzon. The general lack of intermediate composition volcanic rocks of early Tertiary age in the Zambales region suggests that the first explanation is more likely.

TABLE 8. Estimates of Analytical Accuracy and Precision

Oxide or Element	Accuracy	Precision
SiO_2	0.61%	1.09%
TiO_2	0.03%	0.01%
Al_2O_3	0.20%	0.23%
FeO^*	0.14%	0.10%
MnO	0.01%	0.01%
MgO	0.12%	0.33%
CaO	0.10%	0.09%
K_2O	0.05%	0.05%
P_2O_5	0.01%	0.01%
Rb	1.2 ppm	3.4 ppm
Sr	5 ppm	5 ppm
V	5 ppm	5 ppm
Zr	6 ppm	2 ppm
Ni	4 ppm	2.5 ppm
Ba	4 ppm	4 ppm
Y	1.4 ppm	2.5 ppm

Above elements determined by XRF, using SIO Philips AXS unit on fused glass discs (for major elements) and pressed powder pellets (trace elements). Cr precision: 6 ppm for <200 ppm; 33 ppm for 200–400 ppm; 55 ppm for >400 ppm. Cr determined by atomic absorption with Perkin-Elmer 430 unit.

Basal metamorphic rocks. The Western margin of the ophiolite is clearly fault-bounded, although the prevailing sense of motion and timing are still equivocal. There probably has been both westward thrusting and N-S strike slip movement.

The base of the ophiolite is exposed in only a few places (Figure 2). Generally, it is overlain by Oligocene-Miocene fluvial and shallow marine clastics and by Miocene reefs. Dense vegetation covers most of the area. Topographically, these limestone reefs are quite obvious. They form prominent flat-topped N-S linear ridges along the edge of the ophiolite.

There is an exposure of what may be the sole of the ophiolite on the eastern flank of a limestone capped ridge extending south from the Cabaluan River (The hill with these exposures is about 5 km east of Lucapon near the guard station (in 1981) on the Acoje Mine access road.) (Figure 2). This ridge has intensely sheared lenticular blocks of ultramafic rocks, quartzite, and quartz-feldspar-mica schist which form a tectonic breccia consisting of the blocks enveloped in a matrix of platy serpentine, chlorite and minor talc. The main rock type is sheared serpentinized saxonite. The general trend of the schistosity is between NNW and NNE, but it wanders due to folding and wrapping around the elongated blocks and lenses which typically range in size from 1 to tens of meters in long dimension. Some lenticular masses are up to 150 m long. The schistosity dips at steep angles to the NW or SE, and in some places it is vertical. The overall structure is controlled by imbrication, folding, and refolding of schistose layers which have been deformed by intense dislocation.

The large-scale structural complexities are also apparent in thin section. Most of the peridotite has experienced several episodes of serpentinization. Schistosities are much less obvious, being obscured by seemingly random orientation of serpentine fibers. Some relict OPX pseudomorphs may be seen in some samples, and patches of chlorite (replacing pyroxene?) are in others. Relict primary opaques (chromite), secondary opaques (magnetite), and relict OPX are aligned in many samples to give foliation. When OPX are aligned, they appear to have been deformed and rotated. Serpentine veins and kinks within serpentine veins also give rise to foliations. These veins generally do not have the same orientation as the aligned opaque minerals and OPX.

Most of the rocks on the ridge appear to be ultramafic schist and breccia, but there is a small exposure (about 150×25 m) which is largely quartzite and quartz-rich schist. The rocks have a platy appearance and a poorly defined foliation due to layers of different grain size and preferred orientation of mica-rich layers, lenticular quartz, and quartz-feldspar aggregates. A few areas have intensely folded layers with axial planes aligned on a new schistosity. In thin section the rocks have a very fine grained (<0.1 mm) granoblastic texture with irregular and deformed clasts of feldspar (1–3 mm) recrystallized lenses and layers of quartz (1–5 mm) and planar alignment of

mica and accessory minerals. Texturally, they are blastomylonites. The dominant mineral is quartz (50–90%); plagioclase ($An_{99}Ab_{0.5}Or_{0.5}$) forms about 20–30% and K-feldspar ($Or_{98-99}Ab_{1-2}$) forms about 10–15% of the rock. Because of the very fine grain size, the modal abundances are hard to estimate and the values given are based on grain scans with the microprobe. In addition to the lenticular quartz aggregates, which appear to be stretched and recrystallized pebbles, there are subidioblastic 0.3×0.6 mm grains of grid-twinned microcline ($Or_{97.9-99.1}Ab_{0.9-2.1}An_0$) and albite ($Ab_{96.9-99.5}An_{0.1-1.6}Or_{0.4-1.6}$). Fe-rich colorless mica (probably phengite) is interspersed through the fabric of the rocks, and mica-rich layers help to delineate the schistosity. Both clinozoisite and piedmontite are present in minor amounts as mineral segregations and tend to be aligned in the schistosity. Hematite (0.01–0.1 mm) outlines folded quartz layers in some samples. Very minor amounts of idioblastic 10–50 micron diameter spessartine garnets (SP 67, ALM 17, GRO 15, AND 0.2, PYR 0.8) are aligned in filmy streaks in some quartz schists. This mineral assemblage is indicative of middle to higher greenschist facies conditions. Veins of quartz with radiating bundles of idioblastic clinopyroxene cut the schistosity at a high angle. The pyroxene is Mn-rich ferro salite (Wo 48–50 En 11–15 Fs 37–41) with 3–3.8% MnO, 1.3–1.5% Al_2O_3, and 0.8–1.2% Na_2O. The veins and radiating pyroxene clusters show no signs of deformation and indicate static metamorphism in amphibolite facies which must postdate the synkinematic greenschist recrystallization. Another set of veins, filled with orange brown phyllosilicate having the elemental proportions of stilpnomelane, records a lower temperature metamorphic event.

Two very important problems are posed by the presence of the quartzose schists: What was their protolith? When and where were they metamorphosed? The quartz-rich schists are unlike any rocks in the Zambales ophiolite or any other rocks in western Luzon. They could not have been derived from cherts because albite and K-feldspar are so abundant. They are very low in Ca and Mg, and, except for some hematite bearing quartzschists, Fe is also very low, therefore a 'graywacke' protolith is not likely. The bulk composition of the samples requires a 'granitic' composition parent. Because some samples are as much as 90–95% quartz, and because some compositionally layered samples include mica and feldspar-rich layers, it seems likely that the protolith was a quartzo-feldspathic sediment derived from a granitic provenance. The Mn minerals are volumetrically very minor and could be due to MnO coating on the parent rocks. The fragments of continental rock in the shear zone could have been dragged up from deeper levels during westward thrusting of the Zambales Range and may be representative of deep crustal rocks under western Luzon. Alternatively, they could be fragments translated from the south along major lateral slip faults which were subsequently caught in the westward thrusting of the Zambales ophiolite.

The metamorphism of the schists could not have taken

place during the shearing episode which developed the tectonic breccia and the serpentinization of the peridotite. The intense shearing and recrystallization of the quartz schists and the ferro-salite vein filling record a style and intensity of metamorphism different from that of the serpentine breccia. We believe that the quartz schists were dragged up along the shear zone and that their initial metamorphism was at a deeper level and perhaps at a different time and place. We propose that the quartz schists are a fragment of continental material (Asian block?) which may be equivalent to the continental rocks known on Mindanao, Panay, and Palawan. The Zambales occurrence may be a single exotic fragment, or it may be a fragment of a much more extensive terrane, including schists and quartz-rich rocks, which extends south along the South China Sea coast towards Mindanao and Palawan. We can only speculate about the extent of this material, but it seems plausible that fragments of Asia, rifted off during opening of the South China Sea, could have been stranded along the western edge of Luzon and incorporated in the tectonic breccia at the base of the Zambales ophiolite.

Conclusions

We have used rock and mineral chemistry to distinguish between several possible tectonic settings for the generation of oceanic crust. The data best fit an origin in a back arc basin for the Coto block and in a nascent island arc for the Acoje block. We have made an estimate of layer thicknesses of the crust/mantle columns for the two blocks (Figure 4) which shows that the Coto block is similar to the presumed structure of ocean basin lithosphere. Coto block crust could be either from the deep ocean basin or from a back arc basin. The Acoje block crust is much thicker and could best be explained as island arc crust. The best chronologic evidence at hand shows that the Coto pillow basalts have Eocene pelagic limestone in depositional contact (Aksitero fm., *Garrison et al.* [1979]). The Coto mine area dikes give radiometric ages of ($\sim$36 $\pm$ 5 m.y., *Wolfe* [1981]). These ages may be too low because of alteration effects but may be reasonably close to true ages because the dikes cut the depleted mantle section on which the basalts and Aksitero fm. were accumulated. The Acoje arc may have been partly imprinted on part of the Coto back arc basin. The thick layered wehrlite-dunite and cumulate norite section of the Acoje block requires extensive melting of a peridotitic source. The subjacent mantle section at Acoje shows effects of strong depletion which has left an olivine-rich peridotite with great quantities of Cr-rich chromite. The Coto block also has a layered cumulate gabbro section, but it is closer to ocean crust thicknesses. The accompanying mantle is less depleted and the chromite deposits are Al-rich rather than Cr-rich. The main cause for the chemical differences between the chromite deposits of Acoje and Coto is the extent of depletion by melting of the host peridotite. The Acoje and Coto ore bodies are not the result of crystal setting from Cr-rich magmas (as in

the Stillwater complex) but are unmelted mantle residue accreted into large masses by plastic, high-T deformation of the refractory peridotite as basaltic melt was extracted. We see a genetic relationship between the depleted mantle material and the crustal rocks. The structural and 'stratigraphic' evidence indicates that the ophiolite is essentially intact, although the two lithosphere blocks are separated by a major fault which probably had left lateral slip. Paleomagnetic data [*Fuller et al.*, this volume] show that both blocks have been rotated nearly 90° anticlockwise since they formed in late Eocene time. A position close to the equator is suggested. This eliminates any possible relation between the Zambales and the adjacent South China Sea as a source for obducted oceanic lithosphere—the present South China Sea did not exist when the Zambales was formed [*Taylor and Hayes*, 1980, this volume].

The Zambales Range began to emerge in late Oligocene or early Miocene time as evidenced by the abundance of conglomerates formed largely of serpentinite clasts. There has been tilting down to the east of each block and perhaps only minor westward movement which has formed overturned beds of Miocene conglomerate. The presence of quartz-muscovite schist lenses along the basal contact zone of the peridotite in the thrust zone offers the intriguing possibility that fragments of continental crust underlie part of the offshore area west of the Zambales.

Reconstruction of the paleotectonic setting is at best highly speculative, but we suggest that the Zambales Range may have formed as part of an east-west trending arc–back arc basin complex which may have been directly associated with the Eocene seafloor spreading which generated the West Philippine Sea. This postulated Eocene east-west trending arc system would have included parts of the central Cordillera of Luzon, southwestern Luzon, parts of Samar, and eastern Mindanao and Palau. All of these areas include arc volcanic plutonic material of Eocene, or older, age. Tertiary rotation of the West Philippine Sea has eliminated the Eocene subduction zones which formed these arcs.

Acknowledgments. Our studies in the Philippines would not have been possible without the extensive assistance of the Philippine Bureau of Mines and Geosciences and the staffs of the mines of Acoje, Barlo, and Coto. We are especially grateful to Oscar Crispin, Chief Geologist, Bureau of Mines and Geosciences, and Ferdinand Ignacio of Acoje Mines. We thank our field colleagues G. Bacuta, T. Apostol, R. Villones, J. Gappe, and C. Danig for their help and many useful discussions on the outcrops. D. L. Rossman, USGS, has given as many valuable ideas and insights based on his years of work in the Zambales. R. G. Coleman reviewed the manuscript, and we thank him for many helpful comments. Our special thanks go to Marilyn Orona, who patiently and expertly typed several drafts of the manuscript. The field and lab work was supported by NSF grants OCE 75-19148, OCE 79-20483, and OCE 80-24894 as part of the IDOE/SEATAR project.

References

Amato, F. L., Stratigraphic paleontology in the Philippines, *Philipp. Geol., 19,* 1–24, 1964.

Bacuta, G. C., Geology of some alpine-type chromite deposits in the Philippines, report, Bureau of Mines, Manila, Philippine Islands, 1978.

Barker, F., Trondhjemite: Definition, environment and hypotheses of origin, in *Trondhjemites, Dacites and Related Rocks,* edited by F. Barker, pp. 1–12, Elsevier, New York, 1979.

Bloomer, S., Mariana Trench petrologic and geochemical studies; implications to the structure and evolution of the inner slope, Ph.D. dissertation, Univ. of Calif., San Diego, 1982.

Bloomer, S., and J. W. Hawkins, Gabbroic and ultramafic rocks from the Mariana Trench: An island arc ophiolite, this volume.

Bloomer, S., J. Melchior, R. Poreda, and J. Hawkins, Mariana arc trench system: Petrology of boninites and evidence for a boninite series, *Eos Trans. AGU, 60,* 1968, 1979.

Brown, G. M., J. G. Holland, H. Sigurdsson, J. F. Tomblin, R. J. Arculus, Geochemistry of the Lesser Antilles volcanic island arc, *Geochim. Cosmochim. Acta, 41,* 787–801, 1977.

Cameron, W. E., E. G. Nisbet, and V. J. Dietrich, Petrographic dissimilarities between ophiolitic and ocean floor basalts, in *Ophiolites, Proceedings International Ophiolite Symposium, Cyprus, 1979,* edited by A. Panayitou, p. 192, Geological Survey Department, Nicosia, Cyprus, 1980.

Casey, J. F., and J. A. Karson, Magma chamber profiles from the Bay of Islands ophiolite complex, *Nature, 292,* 295–301, 1981.

Coleman, R. G., *Ophiolites,* p. 229, Springer-Verlag, New York, 1977.

Coleman, R. G., and Z. E. Peterman, Oceanic plagiogranite, *J. Geophys. Res., 80,* 1099–1108, 1975.

Davies, H. L., Peridotite-gabbro-basalt complex in eastern Papua: An overthrust plate of oceanic mantle and crust, *Bull. Bur. Miner. Res. Geol. Geophys. Aust., 128,* 48, 1971.

Dick, H., The Josephine peridotite, a refractory residue of andesite, *Eos Trans. AGU, 56,* 464, 1974.

Dick, H. J., The Josephine peridotite, the refractory residue of the generation of andesite, *Eos Trans. AGU, 56,* 464, 1975a.

Dick, H. J., Alpine peridotites and ocean lithosphere a comparison, *Eos Trans. AGU, 56,* 1077, 1975b.

Dick, H., Partial melting in the Josephine peridotite: The effect on mineral composition and its consequence for geobarometry and geothermometry, *Am. J. Sci., 277,* 801–832, 1977.

Dick, H., and T. Bullen, Chromian spinel as a petrogenetic indicator in oceanic environments, *Bull. Geol. Soc. Am.,* in press, 1982.

Dickey, J. S., A hypothesis of origin for podiform chromite deposits, *Geochim. Cosmochim. Acta, 39,* 1061–1074, 1975.

Dickey, J. S., H. Yoder, and J. F. Schairer, Chromium in silicate-oxide systems, *Year Book Carnegie Inst. Washington 1970–1971,* 118–120, 1971.

Dixon, T., and R. Batiza, Petrology and chemistry of recent lavas in the northern Marianas: Implications for the origin of island arc basalts, *Contrib. Mineral. Petrol., 70,* 167–181, 1979.

Evans, C., and J. W. Hawkins, Petrology of Zambales Range Ophiolite, Luzon, P.I.: Ultramafics and chromite, *Eos Trans. AGU, 61,* 1154, 1980.

Evans, C., and J. W. Hawkins, Island arc and back arc basin sections within the Zambales Ophiolite, Philippines: Differences in the upper mantle peridotites, *Eos Trans. AGU, 62,* 1086, 1981.

Ewart, A., and W. F. Bryan, The petrology and geochemistry of the igneous rocks from Eua, Tongan Islands, *Bull. Geol. Soc. Am., 83,* 3281–3298, 1972.

Fernandez, N. S., Notes on the geology and chromite deposits of the Zambales Range, *Philipp. Geol., 14,* 1–8, 1960.

Fuller, M., I. S. Williams, R. McCabe, R. Y. Encina, J. Almasco, and J. A. Wolfe, Paleomagnetism of Luzon, this volume.

Garrison, R. E., E. Espiritu, L. J. Horan, and L. E. Mack, Petrology, sedimentology and diagenesis of hemipelagic limestone and tuffaceous turbidites in the Aksitero Formation, Central Luzon, Philippines, *U.S. Geol. Surv. Prof. Pap. 1112,* 16, 1979.

Gill, J. B., *Orogenic Andesites and Plate Tectonics,* p. 390, Springer-Verlag, New York, 1981.

Green, D. H., Composition of basaltic magmas as indicators of conditions of origin: Application to oceanic volcanism, *Philos. Trans. R. Soc. London Ser. A, 268,* 707–722, 1971.

Hawkins, J. W., Geology of the Lau Basin, a marginal sea behind the Tonga Arc, in *Geology of Continental Margins,* edited by C. Burk and C. Drake, pp. 505–520, Springer-Verlag, New York, 1974.

Hawkins, J. W., Petrology and geochemistry of basaltic rocks of the Lau Basin, *Earth Planet. Sci. Lett., 28,* 283–298, 1976.

Hawkins, J. W., Petrologic and geochemical characteristics of marginal basin basalts, in *Island Arcs, Deep Sea Trenches, and Back-Arc Basins, Maurice Ewing Ser.,* vol. 1, edited by M. Talwani and W. C. Pitman III, pp. 355–365, AGU, Washington, D. C., 1977.

Hawkins, J. W., Geology of marginal basins and their significance to the origin of ophiolites, in Ophiolites, Proc. of Intern. Symposium on Ophiolites, Cyprus, 1979, edited by A. Panayitou, pp. 244–254, Geological Survey Department, Nicosia, Cyprus, 1980.

Hawkins, J. W., and R. Batiza, Petrology and geochemistry of an ophiolite complex, Zambales Range Luzon, Republic of Philippines, *Eos Trans. AGU, 58,* 1244, 1977.

Hawkins, J. W., and C. Evans, Petrology of Zambales Range Ophiolite, Luzon, P. I.: Gabbro, diabase and basalt, *Eos Trans. AGU, 61,* 1154, 1980.

Hawkins, J., S. Bloomer, C. Evans, and J. Melchior, Mariana arc-trench system: Petrology of the inner trench wall, *Eos Trans. AGU, 60,* 968, 1979.

Hawkins, J., C. Evans, and S. Bloomer, Ophiolite series rocks from an island arc complex: Zambales Range, Luzon, Philippine Islands, *Eos Trans. AGU, 62,* 409, 1981.

Hopson, C., R. G. Coleman, R. T. Gregory, J. S. Pallister, and E. H. Bailey, Geologic section through the Samail ophiolite and associated rocks along a Muscat-Ibra transect, southeastern Oman Mountains, *J. Geophys. Res., 86,* 2527–2544, 1981.

Jacques, A. L., Petrology and petrogenesis of cumulate peridotites and gabbros from the Marum ophiolite complex, northern Papua-New Guinea, *J. Petrol., 22,* 1–40, 1981.

Jakes, P., and J. Gill, Rare earth elements and the island arc series, *Earth Planet. Sci. Lett., 9,* 17–28, 1970.

Juteau, T., and H. Whitechurch, The magmatic cumulates of Antalya (Turkey): Evidence of multiple intrusions in an ophiolitic magma chamber, in *Ophiolites, Proceedings International Ophiolite Symposium, Cyprus, 1979,* edited by A. Panayitou, Geological Survey Department, Nicosia, Cyprus, 1980.

Malpas, J., Two contrasting trondhjemite associations from transported ophiolites in western Newfoundland: Initial report, in *Trondhjemites, Dacites and Related Rocks,* edited by F. Barker, pp. 465–487, Elsevier, New York, 1979.

Mattey, D. P., N. G. Marsh, and J. Tarney, The geochemistry, mineralogy, and petrology of basalts from the West Philippine and Parece Vela Basins and from Palau-Kyushu and West Mariana Ridges, DSDP leg 59, *Initial Rep. Deep Sea Drill. Proj., 59,* 753–800, 1981.

Miyashiro, A., Volcanic rock series in island arcs and active continental margins, *Am. J. Sci., 274,* 321–355, 1974.

Murauchi, S., et al., Crustal structure of the Philippine Sea, *J. Geophys. Res., 73,* 3143–3171, 1968.

Mysen, B. O., and I. Kushiro, Compositional variations of coexisting phases with degree of melting of peridotite in the upper mantle, *Am. Mineral., 62,* 843–856, 1977.

Payne, J. G., and D. F. Strong, Origin of the Twilingate Trondhjemite, north central Newfoundland: Partial melting in the roots of an island arc, in *Trondhjemites, Dacites and Related Rocks,* edited by F. Barker, pp. 489–516, Elsevier, New York, 1979.

Pearce, J. A., Geochemical evidence for the genesis and eruptive setting of lavas from Tethyan ophiolites, in *Ophiolites, Proceedings International Ophiolite Symposium, Cyprus, 1979,* edited by A. Panayitou, pp. 261–272, Geological Survey Department, Nicosia, Cyprus, 1980.

Pearce, J. A., and J. R. Cann, Tectonic setting of basic volcanic rocks determined using trace element analyses, *Earth Planet. Sci. Lett., 19,* 296–300, 1973.

Pearce, J., and M. J. Norry, Petrogenetic implications of Ti, Zr, Y and Nb variations in volcanic rocks, *Contrib. Mineral. Petrol., 69,* 33–47, 1979.

Rossman, D. L., Chromite deposits of the north-central Zambales Range, Luzon, Philippines, *U.S. Geol. Surv. Open File Rep. 65,* 1964.

Saunders, A., J. Tarney, N. G. Marsh, and D. A. Wood, Ophiolites as ocean crust or marginal basin crust: A geological approach, in *Ophiolites, Proceedings International Ophiolite Symposium, Cyprus, 1979,* edited by A. Panayitou, pp. 193–204, Geological Survey Department, Nicosia, Cyprus, 1980.

Schweller, W. J., and D. E. Karig, Constraints on the origin and emplacement of the Zambales ophiolite, Luzon, Philippines, *Geol. Soc. Am. Abstr. Programs, 11,* 512–513, 1979.

Schweller, W. J., S. B. Bachman, and D. E. Karig, Evolution of a fore-arc basin in Luzon from stratigraphy and detrital mineralogy, *Geol. Soc. Am. Abstr. Programs, 13,* 550, 1981.

Shido, F., A. Miyashiro, and M. Ewing, Crystallization of abyssal tholeiites, *Contrib. Mineral. Petrol., 31,* 251–266, 1971.

Stern, R. J., On the origin of andesite in the Northern Mariana Island Arc: Implications from Agrigan, *Contrib. Mineral. Petrol., 68,* 207–219, 1979.

Stoll, W. C., Geology and petrology of the Masinloc chromite deposit, Zambales, Luzon, Philippine Islands, *Bull. Geol. Soc. Am., 69,* 419–448, 1958.

Streckeisen, A., Classification and nomenclature of plutonic rocks: IUGS subcommission on systematics of igneous rocks, *Geotimes, 18,* 26–30, 1973.

Sun, S. S., R. W. Nesbitt, and A. Y. Sharaskin, Geochemical characteristics of mid-ocean ridge basalts, *Earth. Planet. Sci. Lett., 44,* 119–138, 1979.

Taylor, B., and D. E. Hayes, The tectonic evolution of the South China Basin, in *The Tectonic and Geologic Evolution of Southeast Asian Seas and Islands, Geophys. Monogr. Ser.,* vol. 23, edited by D. E. Hayes, pp. 89–104, AGU, Washington, D. C., 1980.

Taylor, B., and D. E. Hayes, Origin and history of the South China Basin, this volume.

Williams, H., and J. G. Payne, Twilingate granite and nearby volcanic groups, *Can. J. Earth Sci., 12,* 982–995, 1975.

Williams, H., and R. K. Stevens, The ancient continental margin of eastern North America, in *The Geology of Continental Margins,* edited by C. Burk and C. L. Drake, pp. 781–796, Springer-Verlag, New York, 1974.

Wolfe, J. A., Philippine geochronology, *J. Geol. Soc. Philipp., 35,* 1–30, 1981.

Original Setting and Emplacement History of the Zambales Ophiolite, Luzon, Phillipines, From Stratigraphic Evidence

W. J. Schweller[1], D. E. Karig, and S. B. Bachman

Department of Geological Sciences, Cornell University, Ithaca, New York 14853

A detailed study of sedimentary rocks associated with the Zambales ophiolite outlines its original setting and uplift history. The oldest sedimentary unit is upper Eocene pelagic limestone with thin ash layers that depositionally overlies the volcanic complex of the ophiolite. This limestone, the Aksitero Formation, was deposited at 1–4 km depth on newly formed ocean floor and has remained isolated from coarse volcanic and continental detritus for several million years. A comparison of this limestone to numerous DSDP sites from the western Pacific shows that the Aksitero Formation resembles pelagic sediments from marginal basins, but not those of island arcs. Sedimentation rates increase from 3–5 m/m.y. in the late Eocene to about 10 m/m.y. in the middle to late Oligocene, as volcaniclastic turbidites begin to dilute the pelagic limestone. Thick lower Miocene sandstones change from volcaniclastic to ophiolitic composition over a few million years, indicating rapid uplift and erosion of the ophiolite during this time. Seismic reflection profiles in the Central Valley, just east of the Zambales Mountains, show as much as 2 km of pre-Miocene strata onlapping the buried eastern flank of the ophiolite. This onlap apparently represents original bathymetric relief of the oceanic crust. The seismic reflection profiles also demonstrate that eastward tilting of the ophiolite began in the early Miocene and continued through the late Miocene. The uplift and eastward tilting is believed to be related to subduction of the Manila Trench, although the cause of the initial detachment of the ophiolite may be tied to transform faulting along the present western edge of the Zambales Mountains.

Introduction

The genesis of a large ophiolite is a poorly understood geologic process involving the uplift of a fragment of oceanic crust to a structurally elevated position where it eventually becomes exposed. Many of the world's better-known ophiolites are in settings so far removed in time and/or space from their original settings that cogent reconstructions of their primary tectonic setting and their emplacement history are difficult, if not impossible, to formulate. The Zambales ophiolite is a rare example of a large, complete, and relatively young ophiolite that is, at least at first glance, in a clear tectonic setting that is probably related to its emplacement. As such, it offers excellent prospects for the study of ophiolite genesis in an island arc setting.

Well-preserved sedimentary sections in stratigraphic contact with ophiolites can contain valuable information about the original setting and emplacement history of the ophiolite. Ideally, the sediments would reflect both the original paleoenvironment of the oceanic crust and the changes in depositional setting, detrital provenance, and paleogeography as the ophiolite was emplaced. Unfortunately, the tectonic events that cause the emplacement of an ophiolite also usually remove or obliterate most or all of the associated sediments.

Reconnaissance surveys of the Zambales Mountains indicated the presence of several sedimentary formations that appeared to be related to either the early history or the emplacement of the ophiolite complex. This project was designed to study these strata and evaluate the paleogeographic and paleotectonic history of the Zambales ophiolite through a combination of methods, including sedimentary facies and paleoenvironmental analysis, detailed biostratigraphy and paleodepth studies, detrital mineralogy for provenance, and structural history from seismic reflection profiles with well control. This information is then used to evaluate the existing tectonic models of Luzon, which are based mainly on marine geophysical data, and the models for the origin of the ophiolite, which are based primarily on geochemical data.

[1]Now at Gulf Science and Technology Co., Pittsburgh, Pennsylvania 15230.

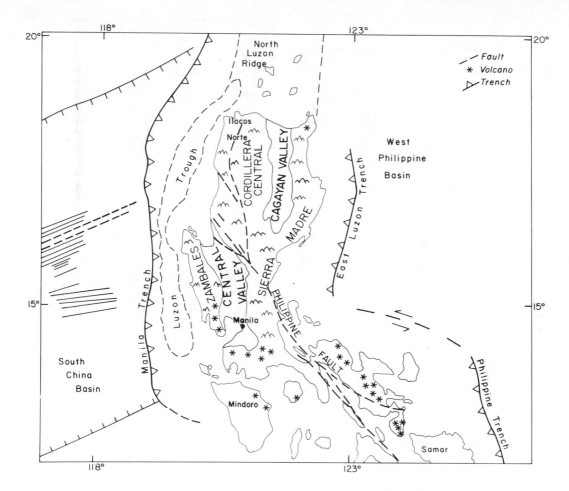

Fig. 1.   Physiographic and tectonic features of the northern Philippine region.

Plate Tectonics and Regional Geology

Luzon forms part of the western edge of the Philippine plate, where the complex interactions of trenches, island arcs, marginal basins, and continental fragments create a rapidly changing tectonic configuration [*Hamilton*, 1979; *Cardwell et al.*, 1980]. At present, two trenches border Luzon: the Manila Trench to the west and the northern end of the Philippine Trench along the southwest peninsula (Figure 1). The seismic zone of the Manila Trench extends to a depth of about 220 km, while the seismic zone of the northern Philippine Trench reaches to 200 km, with an isolated cluster of events below 550 km [*Cardwell et al.*, 1980]. The seismic zone of the Manila Trench bends sharply downward to nearly vertical but has no thrust-type focal mechanisms to indicate active eastward subduction, whereas the East Luzon Trough has abundant shallow seismicity with several low-angle westward thrust focal mechanisms [*Fitch*, 1972; *Cardwell et al.*, 1980]. These thrust mechanisms have been cited as evidence that the East Luzon Trough is an incipient trench [*Karig*, 1973].

Marginal basins of various ages flank the Philippines.

Oceanic crust in the South China Basin formed at an east-west spreading ridge during the middle Oligocene to early Miocene [32–17 m.y. b.p., *Taylor and Hayes*, 1980]. *Weissel* [1980] mapped marine magnetic anomalies 18–20 in the Celebes Basin southwest of Mindanao that corresponded to a middle to late Eocene crustal age and suggested an Oligocene age for the Sulu Basin. The West Philippine Basin has several different genetic interpretations [*Hilde et al.*, 1977; *Bowin et al.*, 1978; *Mrozowski et al.*, 1982], the most recent of which concludes that spreading from near the Central Basin Fault was active during most of the Eocene [*Mrozowski et al.*, 1982]. The Benham Rise, an oceanic plateau just east of Luzon (Figure 1), was dated at 38 m.y. by paleontology of the oldest sediment at Deep Sea Drilling Project (DSDP) site 292 [*Karig et al.*, 1975]. Thus, most of the marginal basins around Luzon were formed during the Eocene and Oligocene, perhaps indicating one or more active subduction zones near the Philippines during this period.

Many authors have proposed reversals in the polarity of subduction zones around the Philippines, but as Figure 2 illustrates, there is little agreement as to the timing of

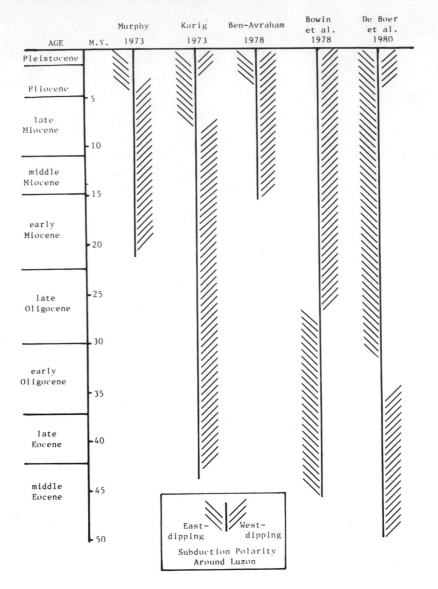

Fig. 2. Conflicting models for the history of subduction polarity around Luzon as proposed by various authors.

these changes. The diversity of these models reflects the lack of reliable geologic constraints on subduction from either land or marine studies.

Three major north-south trending mountain belts dominate the structure of Luzon: the Sierra Madre, Cordillera Central, and Zambales ranges. These ranges partially enclose two large basins: the Cagayan Valley and the Central Valley (Figure 3). The Zambales Mountains consist of a large igneous massif in the north and a line of recent volcanos overlying the massif in the south. The Luzon Central Valley contains up to 10 km or more of Tertiary sediments [Bachman et al., 1982] and separates the Zambales range from the rest of Luzon.

The geology of the Sierra Madre and Cordillera Central

ranges is poorly known at present. The geologic map of Luzon [Philippine Bureau of Mines, 1963] shows that both ranges consist predominantly of Tertiary intrusions with associated sedimentary rocks that have been deformed and metamorphosed to various degrees. The Cordillera Central contains numerous Miocene quartz diorite to granodiorite batholiths intruded into calc-alkaline volcanics, diorite, and volcanogenic sedimentary rocks [Divis, 1980]. Balce et al. [1980] identified a major orogenic period of uplift and batholith intrusion near the end of the Oligocene in the southern Cordillera Central, with two smaller pulses of magmatic activity in the middle Miocene and the late Miocene to Pleistocene. They interpret the Cordillera Central as a fully developed magmatic arc during the late

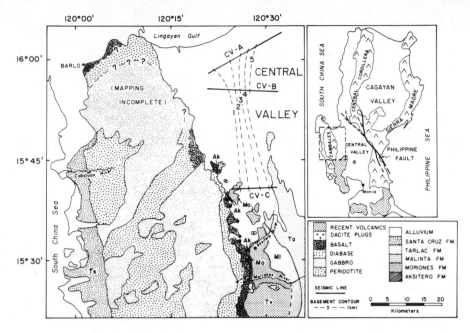

Fig. 3.   Generalized geology of the northern Zambales Range, Luzon. Seismic lines are shown on Figures 7 and 8.

Oligocene and early Miocene, probably related to eastward subduction along the western margin of Luzon, while the Sierra Madre was probably related to subduction along the eastern margin of Luzon in Eocene to Oligocene time.

Upper Cretaceous strata in the southern Sierra Madre [*Reyes and Ordonez,* 1979; *Hashimoto,* 1980] represent an older core of central Luzon that predates the Eocene and Oligocene magmatic activity. Field work by D. Karig in 1981 outlined a large Cretaceous ophiolite along the west flank of the southern Sierra Madre, probably formed as marginal basin crust behind an east-facing arc.

Recent volcanism in the Philippines follows two north-south belts that bracket the archipelago from central Luzon to north of Mindanao [*Divis,* 1980]. The Zambales volcanos are generally related to subduction at the Manila Trench [*Divis,* 1980; *Cardwell et al.,* 1980], with radiometric ages predominantly Pliocene and younger [*de Boer et al.,* 1980].

Previous Studies of the Zambales Ophiolite

The earliest geologic overviews of the Zambales Mountains recognized it as a complex of basic and ultrabasic intrusions of unknown age or origin [*Irving,* 1950; *Corby,* 1951]. Most later reports dealt primarily with details of the chromite deposits associated with the ultramafic complex [*Stoll,* 1958; *Fernandez,* 1960] and the copper deposits near Barlo at the northern end of the range [*Buangan,* 1962; *Bryner,* 1967]. Despite its great size, the Zambales escaped notice as a possible large ophiolite until quite recently. *Coleman*'s [1977] compendium on ophiolites men-

tions ophiolite occurrences in the Philippines but contains no reference to the Zambales ophiolite. The present study is part of a multifaceted investigation of the ophiolite by several investigators at Cornell University and other institutions under the Southeast Asia Tectonics and Resources (SEATAR) project. Previously published reports from this project are cited within the text where appropriate.

Stratigraphy of Sedimentary Rocks Associated With the Zambales Ophiolite

Aksitero Formation

The Zambales ophiolite contains an unusually complete and well-preserved sedimentary section, especially along the eastern flank of the range. This section begins with a thin pelagic limestone that depositionally overlies the volcanic complex of the ophiolite. This limestone was initially considered as the basal number of the clastic strata of the Miocene Moriones Formation [*Corby,* 1951; *Divino-Santiago,* 1963; *Bandy,* 1963] until *Amato* [1965] revised its age to late Eocene through Oligocene. Amato renamed this unit the Aksitero Formation and identified it as a deep marine deposit, probably deposited in at least 1000 m of water. *Garrison et al.* [1979] concurred with Amato's age and paleoenvironment analysis and studied the lithology of the Aksitero limestone in detail, although they did not address the significance of these sediments for the history of the ophiolite. *Schweller et al.* [1982] presented a comprehensive evaluation of the sedimentology of the Aksitero

Formation, including detailed biostratigraphy, using calcareous nannofossils.

The Aksitero Formation directly overlies volcanic breccia and pillow basalt of the Zambales ophiolite along the eastern edge of the Zambales Mountains in a narrow swath about 35 km long but only a few hundred meters wide (Figure 3). This formation contains two members distinguished by an abrupt change in lithology: the Bigbiga Limestone, a sequence of pelagic limestone with thin tuff interbeds, totaling about 35 m in thickness, and an upper member of interbedded limestone and sand volcaniclastic turbidites approximately 75 m thick (Figure 4).

The Bigbiga Limestone rests directly on massive to brecciated altered basalt with isolated zones of pillow basalts and sills. The pillows and sills become more predominant downward in the volcanic complex and are transitional into a sheeted sill complex that in turn overlies the gabbroic complex of the Zambales massif [*Geary and Kay*, this volume; *Hawkins et al.*, 1981]. Previous studies [*Amato*, 1965; *Roque et al.*, 1972; *Ingle*, 1975; *Garrison et al.*, 1979] reported that the Aksitero Formation unconformably overlies the Zambales volcanic complex. However, careful study of outcrops along the Aksitero and Moriones rivers revealed no evidence of an hiatus between the eruption of the volcanic rocks and the onset of limestone deposition. Typical outcrops show limestone interfingering down the angular volcanic breccia and suggest deposition of limestone immediately following volcanic activity [*Schweller and Karig*, 1979]. At two localities, up to 2 m of red mudstone overlies basalt and is in turn conformably overlain by limestone at a gradational contact. This succession of red mudstone and pelagic limestone over basalt is analogous to the sediment sequence on parts of the Semail (Oman) ophiolite [*Tippit et al.*, 1981].

The Bigbiga Limestone is fine-grained, white to cream or light buff limestone with thin interbeds of light gray to white lithified fine-grained ash. The limestone contains abundant foraminifera tests, many of which have been partially crushed by burial compaction. Garrison et al. described this limestone as a hemipelagic biomicrite because of a relatively high amount of noncarbonate content (10%–30%), including plagioclase crystals, grains of pumice and tuffaceous mudstone, volcanic glass shards, and dispersed clay. Because most of the noncarbonate material appears to be of volcanic origin (perhaps deposited at large distances from the volcanic source), we prefer to term this lithology a tuffaceous pelagic limestone.

Numerous thin lithified ash layers make up 5% to 10% of the Bigbiga Limestone member. These ash layers are normally graded from silty tuff at the base to very fine-grained porcelanite at the top. About half of the layers contain a thin layer of sand-sized grains, mainly plagioclase crystals and pumice fragments, near the base of the layer. The ash layers range from 2 to 20 cm thick, although most are less than 8 cm thick. Bottoms of the layers are sharp and flat, while tops are diffuse and bioturbated, with

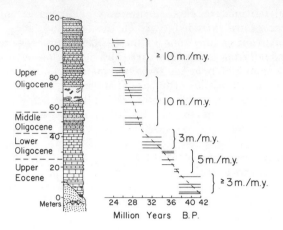

Fig. 4.   Stratigraphy and nannofossil age zonation of the Aksitero Formation type section with calculated sedimentation rates.

burrows penetrating several centimeters down into the thicker layers. The ash layers are distributed approximately uniformly through the lower 20 m of the Bigbiga Limestone, averaging one to three layers per meter of section, then become sparse for the next 10 m before reappearing in the upper 5 m of the section.

The proportion of ash in the lower Aksitero Formation increases northward along the east flank of the Zambales range. Limestone along the Moriones River, about 10 km south of the type section, contains no ash layers, while outcrops north of the type section become progressively more tuffaceous, reaching 75% or more insoluble residue in the northernmost localities near Suaco. Garrison et al. attributed the ash to extrusive or shallow intrusive dacitic or trachyandesitic sources on the basis of the coarser-grained fragments. The absence of continental detritus, particularly of quartzose sand, and the lack of both pumic fragments coarser than sand size and any basaltic or andesitic fragments suggest that the ash beds were not deposited near the flanks of a large island arc or a continental margin.

An abrupt change in lithology, from white limestone to interbedded brown sandy turbidites and greenish blue marly limestone, marks the contact between the lower and upper members of the Aksitero Formation. The lithologic transition occurs within a zone less than 1 m thick, but there is no apparent hiatus in the section (Figure 4). The limestone layers between the turbidites resemble the underlying Bigbiga Limestone, except for the color change (probably a diagenetic effect) and a slight increase in the silt content. The turbidites range from 10 cm to over 50 cm thick, with limestone interbeds 3 to 20 cm thick. Individual turbidites grade upward from a medium- or coarse-grained, volcanic, lithic-rich sandstone at the base, through lighter brown siltstone, and finally into silty limestone. The bases of the turbidites are sharp and planar, while the transition from siltstone to limestone interbeds is gradual over several centimeters. Flute casts and other paleocurrent indicators are extremely rare on the turbidite bases. The proportion of limestone in the upper member decreases gradually upsection from 20%–30% in the lower 10 m to less than 10% near the top of the member.

About 20 m above the base of the upper member, a 2–3 m thick zone with contorted bedding and recumbent folding interrupts the well-bedded and undisturbed (other than regional tilting) Aksitero strata. This folded section is overlain by about 5 m of unbedded sandy mudstone, which is in turn overlain by well-bedded, undisturbed turbidites above a sharp, planar contact. This disturbed zone is restricted to a single stratigraphic horizon and probably represents a submarine slump of unconsolidated sediments during the middle Oligocene.

Sedimentation rates increase from 3–5 m/m.y. in the Bigbiga Limestone to 10 m/m.y. or more in the upper member, reflecting the addition of the clastic component to the pelagic limestone. The 10 m/m.y. rates for the upper member are surprisingly low for a turbidite sequence and indicate very long periods between turbidite events.

Miocene and Younger Formations

The lower Miocene Moriones Formation, which overlies the Aksitero Formation along most of the western Central Valley, consists of several hundred meters or more of marine mudstone and sandy siltstone with occasional conglomerate lenses in the middle and upper parts of the formation (Figure 5). The strata at the base of this formation are similar in lithology to the upper member of the Aksitero Formation but lack the blue limestone in-

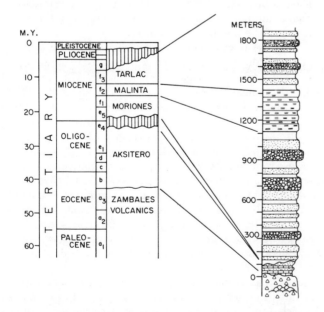

Fig. 5. Generalized stratigraphy of the western Luzon Central Valley, after Roque et al. (1972) and Schweller et al. (submitted).

terbeds of the latter unit. Amato suggested that a significant structural and stratigraphic break separates the two formations. However, our field work and recent paleontological studies [*Schweller et al.*, 1982; *Hashimoto*, 1980] indicate no evidence for a major hiatus at this boundary. There is no change in the eastward tilting of beds of the two at the type section and, at most, only a brief (1–2 m.y.) gap in biostratigraphic zones [*Hashimoto*, 1980].

The lowest part of the Moriones Formation consists of well-bedded mudstone and sandstone turbidites up to 1 m thick. About 200-m upsection, channelized conglomerates appear together with thick zones of unbedded mudstone within the turbidite units. The unbedded mudstone zones contain well-rounded cobbles and pebbles of serpentinized peridotite and gabbro suspended in a mudstone matrix with abundant foraminifera tests. Fragments of carbonized wood and coralline limestone give evidence for subaerial and shallow marine sources for these sediments. Paleocurrent indicators (mainly flute casts and migrating ripple forms) suggest sediment transport from the west and southwest.

The first conglomerates in the Moriones Formation contain cobble-sized mudstone clasts in a sandy matrix. Later conglomerates contain cobble- to boulder-sized clasts of basalt, diabase, gabbro, and serpentinized peridotite and clearly represent erosion of an ophiolite suite during the early Miocene. Although the paleontological age control for the Moriones Formation is not detailed enough to resolve a more precise data for the first appearance of the ophiolite suite conglomerates, they occur in the lower part of the formation and probably postdate the end of the Aksitero Formation by a few million years or less.

Several kilometers of Miocene and younger strata overlie the Moriones Formation eastward into the Central Valley. The middle Miocene Malinta Formation consists of 300 to 600 m of tuffaceous sandstone and occasional conglomerates with the formation thickness decreasing to the north [*Roque et al.*, 1972]. Conformably over these units lies the Tarlac Formation, a late Miocene to early Pliocene sequence of interbedded shale, sandstone, and conglomerate over 1200 m thick [*Roque et al.*, 1972]. *Ingle* [1975] indicates that paleodepths in the Central Valley remained deep until the late Miocene, although fragments of coral limestone in the Moriones Formation show that some source areas around the basin were near sea level early in the basin's history. Recent alluvium directly abuts the ophiolite along the northeast flank.

West Flank Stratigraphy and Structures

The stratigraphy of the west flank of the Zambales range is much less complete than that of the east side and apparently records only the later stages of uplift and erosion of the ophiolite. There is no evidence of the diabase or volcanic (pillow basalt) upper layers of the ophiolite, nor of any pelagic sediments, equivalent to the Aksitero Formation either in outcrop or as detrital remnants within the sediments along the west flank. However, there are anomalous rock types along the western edge of the ophiolite near the Cabaluan River (Figure 3). A diverse assemblage of radiolarian chert, quartz-sericite schist, tectonized ultramafic rocks, and sheared serpentinite appears to follow a north-trending shear zone roughly coincident with the present western fault boundary of the Zambales range. Banded chert is not found in any stratigraphic sequence around the ophiolite, but similar associations of chert, schist, and serpentinite crop out in scattered localities northwest of the Zambales range and along the west coast of northernmost Luzon [*Irving and Quema*, 1948; G. Haeck, personal communication, 1981]. These rock types are exotic to the Zambales range and may have been juxtaposed against the ophiolite as part of a large zone of transform motion along western Luzon (D. E. Karig, unpublished manuscript, 1982).

The oldest sediments along the western flank of the Zambales range occur within a narrow zone along the truncated western edge of the ophiolite massif, east of a low coastal platform of younger Tertiary shallow marine deposits (Figure 3). These older strata are upturned against the edge of the ophiolite to steep westward dips, and at the Cabaluan River are overturned to a 55° east dip. These strata consist of a basal conglomerate of ultramafic cobbles and boulders overlain by pebbly sandstone and mudstone of predominantly serpentine composition. Abundant large gastropods, mollusks, and coral fragments indicate a shallow marine paleoenvironment for most of this formation, although thin coal layers suggest that parts of the sequence were littoral or fluvial. The detrital minerals of the sandstone and the lithology of the conglomerate clasts indicate an entirely ultramafic provenance for these sediments, similar to the composition of modern sediments in rivers that drain the western Zambales range. No sedimentary record of earlier stages of erosion of the ophiolite have been found along the western flank.

Northern Zambales Stratigraphy

Fine-grained, white tuffaceous siltstone depositionally overlies parts of the volcanic complex near Barlo. Although they appear to overlie steeply tilted pillow basalt masses, these sediments dip only about 20° northeast. Fossils in these sediments are extremely rare and poorly preserved. Roth [*Schweller et al.*, 1982] tentatively identified a few species of calcareous nannofossils that suggested a middle Oligocene age, while A. Sanfillipo (personal communication, 1982) identified possibly late Eocene radiolaria in other samples from the same locality. The tuffaceous siltstones near Barlo may be correlative with the Aksitero Formation, but the ages are so poorly constrained that further dating is imperative before reaching any conclusions concerning correlations or more complex structural history in the area.

Uplift History From Sandstone Compositions

The detrital mineralogy of sandstones as determined by point counting of thin sections is a potent and widely used means of evaluating the geologic character of the detrital source (provenance) in a variety of tectonic settings [*Dickinson and Suczek*, 1979; *Dickinson and Valloni*, 1980]. However, to our knowledge this technique has not been applied to the emplacement of ophiolites. Several problems with ophiolite-derived sediments contribute to this lack of studies: Most of the existing methods of analyzing sandstone compositions are based on quartz-rich mineral suites, whereas ophiolites are very quartz-poor. The mafic minerals that predominate in ophiolites (olivine, pyroxene) are easily altered and do not survive extensive transport. Detrital mineralogy techniques cannot easily distinguish between volcaniclastic sediment derived from the upper parts of an ophiolite complex and sediment from island arc volcanic sources. Finally, ophiolites constitute only a small fraction of the source terrane for most basins, so that the ophiolite detritus is severely diluted by other detrital sources.

Most of these problems are minimized or can be circumvented in the Zambales sediments. The ancient sandstones are well preserved, and modern sands from streams that drain various parts of the ophiolite can be used as standards for the detrital products of the different igneous layers. The mafic-rich mineral suite required the development of a variation of the standard QFL (quartz-feldspar-lithic) methods of *Dickinson* [1970]. The complete results of this analysis of ophiolite-derived sandstones are given elsewhere (W. J. Schweller and S. B. Bachman, unpublished manuscript, 1982). Here we briefly summarize the crucial findings and apply this information to the uplift history of the Zambales ophiolite.

The mineralogy of sandstones within the upper Aksitero and lower Moriones formations can be classified into three compositional categories: lithic volcanics (Lv; grains of aphanitic volcanic rocks), plagioclase (P), and mafic grains (Mu; olivine, pyroxene, serpentine, and chromite), defined in *Schweller and Karig* [1982]. Sands derived from the basalt and diabase zones of the upper part of the ophiolite should contain predominantly Lv grains, with some plagioclase from phenocrysts. The gabbro complex should yield sands rich in plagioclase and pyroxene with other minor mafic minerals. Sands from the ultramafic complex should be rich in olivine (usually altered to serpentine), pyroxene, and chromite. Modern sands from the Zambales range generally follow these trends, although alteration of some units adds metamorphic components to the suite (W. J. Schweller and S. B. Bachman, op. cit.).

Ideally, as an ophiolite is uplifted, exposed, and eroded, the detrital products should follow a gradual change from Lv-rich sands to Mu-rich sands that corresponds to the exposure of progressively deeper igneous layers of a typical ophiolite-layered igneous complex [*Coleman*, 1977]. The

Aksitero and Moriones formations show these changes quite well, as illustrated in Figure 6. The changing mineral suites represent an ophiolite suite eroded progressively deeper into the layered igneous complex, beginning with volcaniclastic sandstone in the upper Aksitero and lowermost Moriones formations and changing rapidly through a plagioclase-rich interval into a suite dominated by serpentine, chromite, and pyroxene.

The volcaniclastic sandstones in the upper Aksitero Formation may be derived from erosion of the top of the ophiolite complex or may represent detritus from a nonophiolite volcanic source area. The rapidity of the transition in the Moriones Formation from volcaniclastic sandstone to sandstone dominated by ultramafic detritus is surprising in that it appears to document an extremely fast erosional exposure of the ultramafic through at least 5 km of gabbro, diabase, and volcanic rocks of the upper part of the ophiolite. Erosion of several kilometers of igneous rocks in, at most, a few million years seems unlikely, even in a humid tropical climate. An alternative is that some part of the ultramafic complex was structurally elevated and exposed to erosion before the bulk of the Zambales was exposed, and this is what contributed heavily to the sediments along the present east flank. This hypothesis would help to explain the dearth of mixed plagioclase and mafic sandstone expected from erosion of a gabbro complex.

The detrital mineralogy documents the exposure of some part of the deeper igneous layers of the ophiolite during the early Miocene. The timing of the initial uplift and erosion of the top of the ophiolite is less well constrained but was probably within the earliest Miocene or the late Oligocene. This represents an interval of only 15 to 20 million years between the origin of the Zambales crust on the ocean floor and its uplift and subaerial exposure as a detrital source.

Tilting History From Seismic Reflection Profiles

A number of multichannel seismic reflection profiles were collected across the Luzon Central Valley in conjunction with an oil exploration effort during the 1960's and 1970's, and they resulted in the drilling of several wells in the valley. A comprehensive study of the evolution of the Central Valley Basin, using these seismic profiles and well-log data, is presented in another paper [Bachman et al., 1982]. The present analysis uses three of these profiles together with stratigraphic control and velocity information from Bachman et al. to examine the tilting history of the east flank of the Zambales ophiolite.

Three seismic reflection profiles near the exposed edge of the Zambales ophiolite (Figure 3) show acoustic basement dipping eastward at a moderate angle beneath a thick wedge of less steeply dipping reflectors (Figure 7). Extrapolation of the acoustic basement westward to the surface approximately matches the exposed edge of the volcanic complex of the ophiolite and strongly suggests that the acoustic basement represents the eastward subsurface continuation of the Zambales ophiolite. These seismic profiles were converted from time sections (vertical scale in seconds of two-way travel time) to depth sections (vertical scale in meters), using a time-depth function derived from velocity surveys in wells and stacking velocities used in processing the multichannel profiles [Bachman et al., 1982]. Simplified drawings of the depth sections (Figure 8) were plotted at no vertical exaggeration, and prominent reflection horizons were assigned approximate ages by following the horizons laterally to wells with biostratigraphic control [Bachman et al., 1982]. A sharp anticline along the southern profile breaks the continuity of reflectors and precludes assigning of ages to this profile; the horizons shown for profile C are prominent reflectors but do not necessarily correspond in age to the horizons of profiles A and B (Figure 8).

These dated horizons can be used to 'untilt' the sections to various times during the Tertiary if it is assumed that reflection horizons roughly represent time-synchronous sediment surfaces that were subhorizontal at the time of deposition. Outcrops of Miocene sediments along the edges of this basin contain mainly turbidites, which should sat-

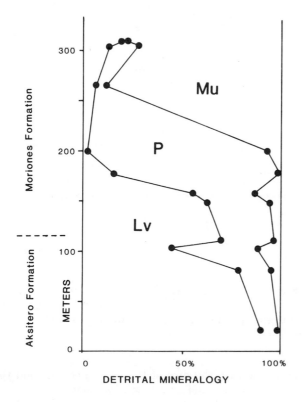

Fig. 6.   Graph of changes in sandstone composition through the Aksitero River stratigraphic section. Compositions are normalized to Lv + P + Mu = 100%. Stratigraphic columns and ages are given in Figure 4 for the Aksitero Formation and in Figure 5 for the Moriones Formation.

CV-A

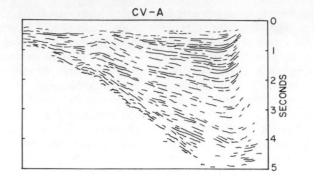

CV-B

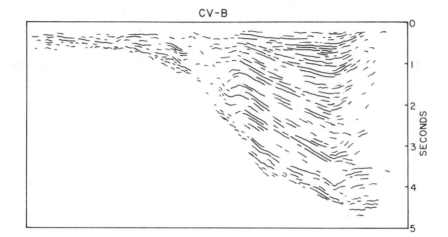

CV-C

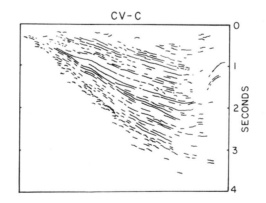

Fig. 7.   Line-drawing interpretations of multichannel seismic reflection profiles across the northwestern Central Valley. Locations shown in Figure 3.

isfy this requirement. Layers were sequentially removed from the simplified depth sections, and the underlying horizons were adjusted upward by the amount of the removed section. The depth sections are uncorrected for compaction, which would thin the more deeply buried strata relative to their less deeply buried equivalents. Adding a compaction correction to the profiles in Figure 8 would slightly increase the dips of the tilted horizons and the amount of onlap relief on acoustic basement.

The reconstructed profile sequences (Figure 8) illustrate several important points about the evolution of the east flank of the ophiolite. There has been progressive eastward tilting of the western Central Valley during the Neogene; the total post-Oligocene tilting along profiles A and B is

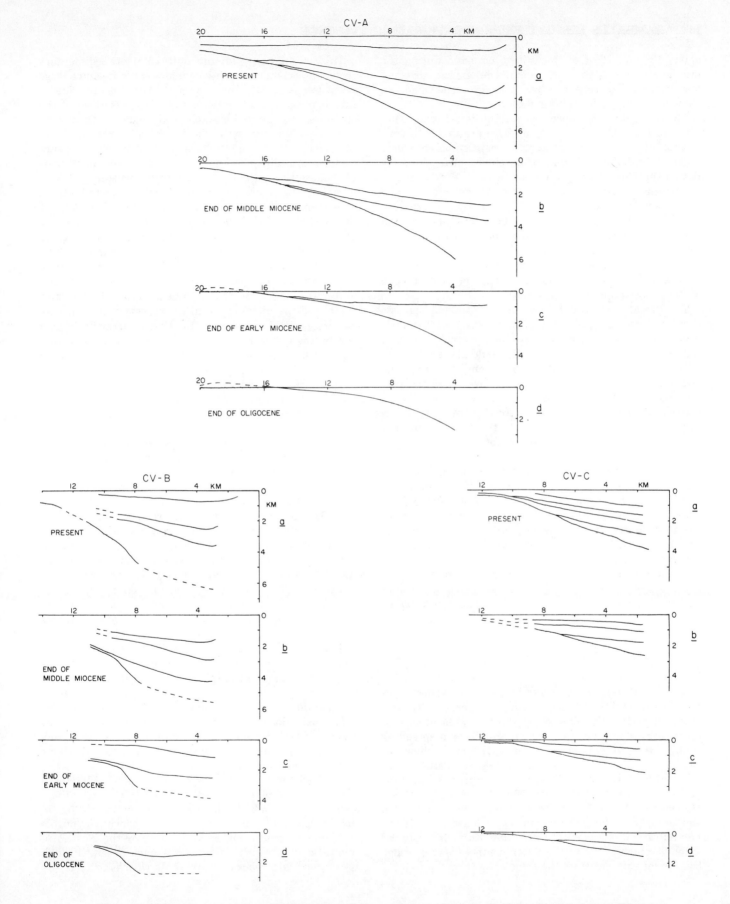

Fig. 8. Tilting history of the western Central Valley and the eastern flank of the Zambales Ophiolite as reconstructed from seismic reflection profiles (Figure 7). Locations of profiles CV-A, CV-B, and CV-C are shown in Figure 3. Methods and implications are discussed in the text.

about 12°–15°. Most of this tilting occurred during the middle Miocene, with lesser amounts during the early and late Miocene. The reflectors below the top of the Oligocene show no appreciable additional tilting.

The amount of pre-Miocene strata thickens rapidly eastward into the Central Valley on both profiles with time control (Figure 8). The Oligocene reconstructions show over 2000 m of relief on the acoustic basement surface relative to paleohorizontal and have paleoslopes of 6° to over 20°, steepening eastward into the basin. There is a dramatic increase in the thickness of pre-Miocene strata concurrent with this paleoslope, with over 2000 m of Oligocene and older strata that onlap the acoustic basement ramp. This thickness of Paleogene strata far exceeds the 120 m of upper Eocene through Oligocene sedimentary rocks at the Aksitero Formation type locality along the eastern edge of the exposed ophiolite (Figure 4).

The pre-Miocene relief on acoustic basement suggests that the east flank of the Zambales was a bathymetric high during its early history. A sharp lateral change in sediment facies related to this relief could also explain the discrepancy between the 120 m of Eocene and Oligocene strata at the Aksitero type section and over 2000 m of pre-Miocene section less than 20 km to the east. By this hypothesis, the pre-Miocene strata in the Central Valley Basin would be rapidly deposited basinal turbidites, while the upper member of the Aksitero Formation would be topographically isolated from most of the turbidites until the basin filled to that level. The rapid increase in sedimentation rate at the beginning of the Moriones Formation may represent the progradation of the basin turbidite facies over the Aksitero Formation type section (Figure 5).

An alternative hypothesis is that the crust beneath the thicker Central Valley sediments is older than the apparent age of the igneous basement at the Aksitero section (late Eocene). Given the order-of-magnitude difference in thickness of the sections, a lateral facies change would still probably be needed to account for part of the eastward increase in thickness of the section.

Comparison of Ophiolite Sediments With Philippine Sea DSDP Sites

A current major controversy in the study of ophiolites revolves around the tectonic setting of the original oceanic crust; in particular, whether ophiolites represent oceanic crust formed at midoceanic or marginal basin spreading centers or fragments of island arcs. The latter view has been advocated by *Miyashiro* [1973, 1975] and other workers mainly on the basis of geochemical affinites. Hawkins and his coworkers [*Hawkins*, 1980; *Hawkins et al.*, 1980; *Hawkins et al.*, 1981] have presented geochemical data in support of an island arc origin for at least parts of the Zambales ophiolite, while *Geary and Kay* [this volume] found geochemical patterns that mimic those of ocean ridge basalts for samples from the same ophiolite.

The combination of an unusually complete sedimentary section associated with the Zambales ophiolite and a large set of Deep Sea Drilling Project (DSDP) sites provides a rare opportunity to evaluate the tectonic setting of the ophiolite through sedimentological evidence, then compare these findings with the geochemical results. More than 20 DSDP sites in the Philippine Sea provide reference samples of sediments and igneous basement in various tectonic settings (Figure 9), many with crustal ages comparable to the Zambales (Eocene-Oligocene). Data for these sites is drawn from initial site reports for DSDP legs 31 [*Karig et al.*, 1975], 58 [*Klein et al.*, 1980], and 59 [*Kroenke et al.*, 1980], plus sedimentological summaries of legs 58 [*White et al.*, 1980] and 59 [*Scott et al.*, 1980; *Rodolfo*, 1981].

Several factors interact to control the types and rates of sedimentation in the Philippine Sea and other marginal basins; the most important of these are probably the water depth and the distance to an active volcanic arc [*Karig and Moore*, 1975]. Most of the DSDP sites in the Philippine Sea basins penetrated little or no carbonate, indicating that the sites were near or below the calcite compensation depth (CCD) for most of their history (Figure 9, sites 290, 291, 293, 294, 444, 450). Some of these sites (291, 293, 294) contain pelagic clay directly overlying basaltic basement, while other sites (290, 444, 447, 450) nearer the arcs began with relatively coarse-grained volcaniclastic breccia and ash and became progressively less volcaniclastic as arc volcanism waned. The carbonate midway through a section of pelagic clay in site 291 may reflect the passage of the site beneath the equatorial zone of a depressed CCD caused by high surface productivity.

DSDP sites near or on island arcs are dominated by volcaniclastic breccia and tuff during the active volcanic period, followed by carbonate deposition when arc volcanism abates. The coarse-grained volcaniclastic material accumulates rapidly, with rates of 50 to over 300 m/m.y. The thickness of the volcaniclastic sections on arcs generally exceeds several hundred meters and decreases out onto the flanking marginal basin crust. Carbonate sediments cap the volcaniclastic section on most arc sites (Figure 9, sites 296, 445, 448, 451) and reach thicknesses of up to 700 m on older arcs (site 445). Carbonate ooze accumulates less rapidly than tuff and breccia, typically at rates of 5 to 20 m/m.y.

Transitional zones between volcanic arc ridges and marginal basin floors, here referred to as arc flanks, show aspects of both island arc and basin floor sedimentation with variable amounts of volcaniclastic breccia and pelagic clay (sites 447 and 290, Figure 9).

The sedimentary sequence on the Benham Plateau, a broad aseismic platform east of Luzon (Figure 1), differs significantly from both island arcs and marginal basins in the Philippine Sea. Site 292, on the southern part of this plateau and about 2 km above the level of the West Philippine Basin floor, contains 368 m of upper Eocene through Pleistocene calcareous ooze and chalk with ash-rich layers

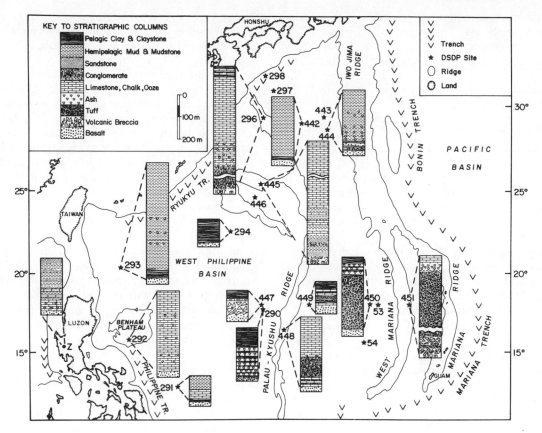

Fig. 9.   Deep-Sea Drilling Project sites in the Philippine Sea, with the type section of the Aksitero Formation (Z) added for comparison. See text for discussion.

at several levels (Figure 9). The ash layers represent explosive volcanism from sources in Luzon several hundred kilometers to the west [*Donnelly,* 1975]. Although site 292 is presently near 3 km water depth, *Karig et al.* [1975] suggest that, on the basis of basalt vesicularity, the site may have formed at depths of less than 1 km.

Several important points can be inferred about the original depositional and tectonic settings of the east flank of the Zambales ophiolite by comparing these DSDP sites to the Aksitero Formation strata described above. First, the pelagic limestone interbedded with thin ash layers (the Bigbiga Limestone member) immediately overlying the ophiolite is not typical of island arc sediment and is somewhat anomalous, compared to most Philippine Sea basin floors. Most of the DSDP sites in the West Philippine Basin entirely lack calcareous sediments, while a few sites (290, 291, 447, 449) contain thin sections of calcareous ooze or slightly calcareous brown clay over volcanic basement. Other marginal basins in the western Pacific, such as the Fiji and Lau basins, are much shallower and receive more carbonate sediment than do most Philippine Basin sites [*Griffin et al.,* 1972]. The Bigbiga Limestone suggests an open ocean site that remained above the CCD for at least

10 m.y., and thus was shallower than most of the West Philippine Basin.

The best DSDP analog for the lower Aksitero Formation from the Philippine Basin sites is site 292 on the Benham Plateau. This site includes about 200 m of late Eocene to late Oligocene nannofossil chalk with thin ash layers similar in age and sediment type to the Aksitero Formation type section. The sedimentation rate of 8.5 to 12 m/m.y. is comparable to the sedimentation rate of the Bigbiga Limestone (3–5 m/m.y.) when both are converted to accumulation rates to compensate for the greater compaction of the Aksitero section. Accumulation rates range from 1000 to 1700 g/cm^2/m.y. for the 100- to 300-m levels of site 292, compared to 1050 to 1250 g/cm^2/m.y. for the lower 40 m of the Aksitero Formation. (Rates for site 292 calculated from porosity, wet bulk density, and sedimentation rate data given in Karig et al.; rates for Aksitero from samples and sedimentation rates in Schweller et al.).

Another possible setting for the eastern flank of the Zambales ophiolite, based on DSDP site comparisons, is an elevated flank of an island arc with relatively weak volcanic activity. The stratigraphy of such a site might be intermediate between sites 448 and 449 (Figure 9), but with slightly more carbonate at the base of the section than either of these sites. Weak arc volcanism would account for a low proportion of ash at such a site, while topographic relief could isolate it from volcaniclastic turbidites. A third alternative is a marginal basin with a shallower original depth than the basins sampled by DSDP sites in the Philippine Sea region.

Discussion: Paleogeographic Implications of Stratigraphy

Two major questions dominate the study of the paleogeography of ophiolites: What was the original setting of the igneous complex, and what sequence of events led to the emplacement of the ophiolite into a convergent continental margin or island arc setting? The sediments overlying the Zambales Ophiolite give insights into both of these questions.

The pelagic limestone and fine-grained ash deposited on the Zambales volcanic complex indicate that the initial setting of the Zambales ocean crust was in a deep marine basin, deeper than 1 km, but above the calcite compensation depth (CCD). The CCD in the Pacific Ocean was between 3 and 4 km depth in the late Eocene, then deepened abruptly to between 4 and 5 km in the Oligocene [van Andel, 1975]. Since the Bigbiga Limestone shows no change in sediment character or sedimentation rate through the late Eocene and early Oligocene (Figure 4), it must have been above the CCD throughout this period. Nannofossil assemblages also indicate open ocean conditions for the entire Aksitero depositional period. The Bigbiga Limestone's low sedimentation rate, absence of detritus from land sources, and lack of shallow water fauna all support an original setting of the Zambales crust far from a large island system and more distant from the Asian mainland than Luzon's present position.

The thin layers of volcanic ash interbedded with the Bigbiga Limestone apparently represent the windborne ejecta of violent eruptions of silicic volcanos at some distance from the Zambales' original setting. Studies of similar tephra layers in the Mediterranean [Keller et al., 1978], the Lesser Antilles [Sigurdsson et al., 1980], and offshore Sumatra [Ninkovich et al., 1978] demonstrate that ash layers several centimeters thick can spread out as far as 2500 km from their volcanic sources, but the patterns of distribution are greatly influenced by prevailing winds. The spacing of the ash layers in the Bigbiga Limestone, together with the low sedimentation rate of the limestone, indicate average recurrence intervals of 50,000 to 200,000 years for these events. Most island arcs produce large eruptions at much shorter intervals: for example, ash layers in the Mediterranean average one to three per 10,000 years [Keller et al., 1978]. It is possible that the infrequent events in the Aksitero Formation represent very large explosive eruptions, perhaps comparable to the Toba eruption [Ninkovich et al., 1978], at distances of at least several hundred kilometers. The preponderance of very fine-grained ash and the complete lack of any tephra particles larger than sand size also suggest a distant source for the ash. Although there is a northward increase in the ash content of the Aksitero Formation, possible tectonic rotations of the Zambales range and of Luzon must be taken into account before this trend can be applied to a regional tectonic reconstruction; counterclockwise rotations have been suggested by the initial work of Fuller et al. [this volume].

Volcaniclastic turbidites gradually overwhelmed limestone deposition along the east flank of the Zambales during the middle and late Oligocene periods, as recorded in the upper Aksitero Formation. These turbidites may represent debris from an island arc or erosion of a distant, uplifted part of the volcanic cap of the ophiolite suite. If derived from a volcanic arc, the debris could represent either the first volcaniclastic products of a new arc or the encroachment of a volcaniclastic apron across a back arc basin, as modeled by Karig and Moore [1975]. An encroaching volcaniclastic apron or fan usually spreads large amounts of silt and clay in advance of the sandy units, whereas in the Aksitero Formation, sandy turbidites first appear interbedded with relatively pure pelagic limestone. Large amounts of silt and clay do not appear in the Aksitero Formation until several million years later (Figure 4). Only a few paleocurrent indicators were found in the upper Aksitero Formation volcaniclastic turbidites; these suggest eastward transport. If more measurements prove this direction valid, it would strongly favor an eroding ophiolite as the source for the turbidites, since the only nonophiolite volcanics in the Zambales range are the much later, Plio-Pleistocene Bataan volcanos [de Boer et al., 1980].

The Zambales ophiolite was the dominant source of detritus along the present western side of the Central Valley in the early Miocene. Conglomerates in the Moriones Formation contain abundant ultramafic cobbles, as well as diabase, gabbro, basalt, and fine-grained limestone. Detrital minerals in the lower Morines Formation sandstones show a progression from dominantly volcaniclastic grains to an ophiolite provenance that includes abundant serpentine, some pyroxene, and significant amounts of chromite (Figure 6) [Schweller and Karig, 1982; W. J. Schweller and S. B. Bachman, unpublished manuscript, 1982]. Paleoenvironments become shallower through the Miocene as the Central Valley basin filled with clastic sediments [Ingle, 1975; Roque et al., 1972]. Seismic reflection profiles show a progressive eastward tilting of the western half of the Central Valley, beginning in the early Miocene and continuing to the Pliocene.

These combined lines of evidence can be integrated and used as constraints to test the various tectonic histories proposed for Luzon (Figure 2). The Zambales crust formed in a large marginal basin or at an oceanic spreading center during the later Eocene. This age eliminates an origin as part of the South China Basin, which did not begin to form oceanic crust until 8 to 10 million years later, during the late Oligocene. During the first 10 million years of its history, the Zambales basin was isolated from coarse clastic material and remained above the CCD, but in a deep (1–4 km) open marine paleoenvironment.

Compared to DSDP sites around the Philippine Sea (Figure 9), this stratigraphy suggests a shallow marginal basin or a bathymetric high not in the immediate vicinity of an active island arc. Both the Celebes Basin and the Philippine Sea were opening during late Eocene, but it is not clear whether the Zambales crust was a part of either of these marginal basins.

Eastward tilting and erosion of the ophiolite began during the early Miocene and were probably related to the initiation of eastward subduction along the ancestral Manila Trench during the late Oligocene. This timing approximately fits the subduction history proposed by *de Boer et al.* [1980] but conflicts with the other tectonic history models for the Philippines (Figure 2). Uplift and tilting continued through the late Miocene but appear to have slowed or ceased in the Plio-Pleistocene.

Acknowledgments. We thank the Philippine Bureau of Mines' Zambales mapping group for their able assistance with the fieldwork and Darwin Rossman of the U.S. Geological Survey and James Hawkins and Cindy Evans of the Scripps Institution of Oceanography for discussion on the igneous massif. This work was supported by National Science Foundation grant OCE-79-19164 to D. K. and grant OCE-80-25282 to S. B.

References

Amato, F. L., Stratigraphic paleontology in the Philippines, *Philipp. Geol. 20,* 121–140, 1965.

Bachman, S. B., W. J. Schweller, and S. D. Lewis, Evolution of a forearc basin, Luzon Central Valley, Philippines, *Am. Assoc. Pet. Geol. Bull.,* in press, 1982.

Balce, G. R., R. Y. Encina, A. Momongan, and E. Lara, Geology of the Baguio district and its implication on the tectonic development of the Luzon central cordillera, *Geol. Paleontol. Southeast Asia, 21,* 265–287, 1980.

Bandy, O. L., Cenozoic planktonic foraminiferal zonation and basinal development in the Philippines, *Am. Assoc. Pet. Geol. Bull., 47,* 1733–1745, 1963.

Ben-Avraham, Z., The evolution of marginal basins and adjacent shelves in east and southeast Asia, *Tectonophysics, 45,* 269–288, 1978.

Bowin, C., R. S. Lu, C. S. Lee, and H. Schouten, Plate convergence and accretion in Taiwan-Luzon region, *Am. Assoc. Pet. Geol. Bull., 62,* 1645–1672, 1978.

Bryner, L., Geology of the Barlo Mine and vicinity, Dasol, Pangasinan Province, Luzon, Philippines, *Rep. Invest. 60,* Phillip. Bur. Mines, Manila, 1967.

Buangan, A. S., Barlo Mine Area, *Rep. Invest. 41,* Philipp. Bur. Mines, Manila, 1962.

Cardwell, R. K., B. L. Isacks, and D. E. Karig, The spatial distribution of earthquakes, focal mechanism solutions, and subducted lithosphere in the Philippine and northeastern Indonesian islands, *The Tectonic and Geologic*

Evolution of Southeast Asian Seas and Islands, Geophys. Monogr. Ser., vol. 23, edited by Dennis E. Hayes, pp. 1–36, AGU, Washington, D. C., 1980.

Coleman, R. G., *Ophiolites—Ancient Oceanic Lithosphere?,* 229 pp. Springer-Verlag, New York, 1977.

Corby, W. G., et al., Geology and oil possibilities of the Philippines, *Tech. Bull. 21,* 363 pp., Philipp. Bur. Mines, Manila, 1951.

de Boer, J., L. A. Odom, P. C. Ragland, F. G. Snider, and N. R. Tilford, The Bataan orogene: Eastward subduction, tectonic rotations, and volcanism in the western Pacific (Philippines), *Tectonophysics, 67,* 251–282, 1980.

de Vries Klein, G., et al., *Init. Rep. Deep Sea Drilling Proj., 58,* 1980.

Dickinson, W. R., Interpreting detrital modes of graywacke and arkose: *J. Sediment. Petrol., 40,* 695–707, 1970.

Dickinson, W. R., and C. A. Suczek, Plate tectonics and sandstone compositions, *Am. Assoc. Petrol. Geol. Bull., 63,* 2164–2182, 1979.

Dickinson, W. R., and R. Valloni, Plate settings and provenance of sands in modern ocean basins, *Geology, 8,* 82–86, 1980.

Divino-Santiago, P., Planktonic foraminiferal species from west side of Tarlac Province, Luzon Central Valley, *Philipp. Geol., 17,* 69–99, 1963.

Divis, A. F., The petrology and tectonics of recent volcanism in the central Philippine Islands, in *The Tectonic and Geologic Evolution of Southeast Asian Seas and Islands, Geophys. Monogr. Ser.,* vol. 23, edited by D. E. Hayes, pp. 105–125, AGU, Washington, D. C., 1980.

Donnelly, T. W., Neogene explosive volcanic activity of the western Pacific: Sites 292 and 296, DSDP Leg 31, *Init. Rep. Deep Sea Drilling Proj. 31,* 577–598, 1975.

Fernandez, N. A., Notes on the geology and chromite deposits of the Zambales Range, *Philipp. Geol., 14,* 1–14, 1970.

Fitch, T. J., Plate convergence, transcurrent faults and internal deformation adjacent to Southeast Asia and the western Pacific, *J. Geophys. Res., 77,* 4432–4460, 1972.

Fuller, M., I. S. Williams, R. McCabe, R. Y. Encina, J. Almasco, and J. A. Wolfe, Paleomagnetism of Luzon, this volume.

Garrison, R. E., E. Espiritu, L. J. Horan, and L. E. Mack, Petrology, sedimentology and diagenesis of hemipelagic limestone and tuffaceous turbidites in the Aksitero Formation, Central Luzon, Philippines, *Geol. Surv. Prof. Pap. (U.S.) 1112,* 16 pp., 1979.

Geary, E. E., R. W. Kay, Petrological and geochemical documentation of ocean floor metamorphism in the Zambales ophiolite, Philippines, this volume.

Griffin, J. J., M. Koide, A. Hohndorf, J. W. Hawkins, and E. D. Goldberg, Sediments of the Lau Basin—Rapidly accumulating volcanic deposits, *Deep-Sea Res., 19,* 139–148, 1972.

Hamilton, W., Tectonics of the Indonesian region, *Geol. Surv. Prof. Pap. (U.S.) 1078,* 345 pp., 1979.

Hashimoto, W., Geologic development of the Philippines, *Contributions to the Geology and Paleontology of Southeast Asia,* vol. 217, pp. 83–190, University of Tokyo, Tokyo, Japan, 1980.

Hawkins, J. W., Petrology of back-arc basins and island arcs: Their possible role in the origin of ophiolites, in *Ophiolites—Proceedings, International Ophiolites Symposium, Cyprus, 1979,* edited by A. Panayiotou, pp. 244–254, Geological Survey Department of Cyprus, Nicosia, 1980.

Hawkins, J. W., C. A. Evans, and G. Bacuta, Petrology of Zambales range ophiolite, Luzon, Philippine Islands: Gabbro, Diabase, and Basalt (abstract), *Eos Trans. AGU, 61,* 1154, 1980.

Hawkins, J. W., C. A. Evans, and S. Bloomer, Ophiolite series rocks from an island arc complex: Zambales Range, Luzon, Philippine Islands (abstract), *Eos Trans. AGU, 62,* 409, 1981.

Hilde, T. W., S. Uyeda, and L. Kroenke, Evolution of the western Pacific and its margin, *Tectonophysics, 38,* 145–165, 1977.

Ingle, J. C., Jr., Summary of late Paleogene-Neogene insular stratigraphy, paleobathymetry and correlations, Philippine Sea and Sea of Japan region, *Init. Rep. Deep Sea Drilling Proj., 31,* 837–855, 1975.

Irving, E. M., Review of Philippine basement geology and its problems: *Philip. J. Sci., 79,* 267–307, 1950.

Irving, E. M., and J. C. Quema, Reconnaissance geology of the Burgos-Pasuguin area, Ilocos Norte: A demonstration of modern reconnaissance methods: *Philip. Geol., 2,* 1–17, 1948.

Karig, D. E., Plate convergence between the Philippines and the Ryukyu Islands, *Marine Geol., 14,* 153–168, 1973.

Karig, D. E., and G. F. Moore, Tectonically controlled sedimentation in marginal basins, *Earth Planet. Sci. Lett., 26,* 233–238, 1975.

Karig, D. E., et al., *Init. Rep. Deep Sea Drilling Proj., 31,* 927 pp., 1975.

Keller, J., W. B. F. Ryan, D. Ninkovich, and R. Altheir, Explosive volcanic activity in the Mediterranean over the past 200,000 yr. as recorded in deep-sea sediments, *Geol. Soc. Am. Bull., 89,* 315–336, 1978.

Kroenke, L., et al., *Initial Rep. Deep Sea Drilling Proj, 59,* 820 pp., 1981.

Miyashiro, A., The Troodos ophiolite complex was probably formed in an island arc, *Earth Planet Sci. Lett., 19,* 218–224, 1973.

Miyashiro, A., Classification, characteristics, and origin of ophiolites, *J. Geol., 83,* 249–281, 1975.

Mrozowski, C. L., S. D. Lewis, and D. E. Hayes, Complexities in the tectonic evolution of the West Philippine Basin, *Tectonophysics,* in press, 1982.

Murphy, R. W., The Manila Trench-West Taiwan fold belt: A flipped subduction zone, *Bull. Geol. Soc. Malays.,* 27–42, 1973.

Ninkovich, D., R. S. J. Sparks, and M. T. Ledbetter, The exceptional magnitude and intensity of the Toba eruption, Sumatra: An example of the use of deep-sea tephra layers as a geological tool, *Bull. Volcanol., 41,* 285–298, 1978.

Philippine Bureau of Mines, Geological map of the Philippines, sheet ND 51: Manila, 1963.

Reyes, M. V., and E. P. Ordonez, Philippine Cretaceous smaller foraminifera: *J. Geol. Soc. Philip. 24,* 1–67, 1979.

Rodolfo, K. S., Sedimentologic summary: Clues to arc volcanism, arc sundering, and back-arc spreading in the sedimentary sequences of Deep Sea Drilling Project Leg 59, *Init. Rep. Deep Sea Drilling Proj., 59,* 621–624, 1981.

Roque, V. P., Jr., B. P. Reyes, and B. A. Gonzales, Report on the comparative stratigraphy of the east and west sides of the mid-Luzon Central Valley, Philippines, *Mineral Eng. Mag.,* 11–15, Sept. 1972.

Rossman, D. L., Chromite deposits of the north-central Zambales Range, Luzon, Philippines, *U.S. Geol. Surv. Open File Report,* 65 pp., 1964.

Schweller, W. J., and D. E. Karig, Constraints on the origin and emplacement of the Zambales Ophiolite, Luzon, Philippines, *Geol. Soc. Am. Abstr. with Programs, 11,* 512–513, 1979.

Schweller, W. J., and D. E. Karig, Emplacement of the Zambales Ophiolite into the West Luzon margin: *Mem. Am. Assoc. Pet. Geol.,* in press, 1982.

Schweller, W. J., P. H. Roth, D. E. Karig, and S. B. Bachman, Sedimentation history and biostratigraphy of ophiolite-related Tertiary sediments, Luzon, Philippines, submitted to *Geol. Soc. Am. Bull.,* 1982.

Sigurdsson, H., R. S. J. Sparks, S. N. Carey, and T. C. Huang, Volcanogenic sedimentation in the Lesser Antilles Arc, *J. Geol., 88,* 523–540, 1980.

Stoll, W. C., Geology and petrology of the Masinloc chromite deposits, Zambales, Luzon, Philippine Islands, *Geol. Soc. Am. Bull., 69,* 419–440, 1958.

Taylor, B., and D. E. Hayes, The tectonic evolution of the south China Basin, in *The Tectonic and Geologic Evolution of Southeast Asian Seas and Islands,* edited by D. E. Hayes, *Geophys. Monogr. Ser.,* vol. 23, pp. 89–104, 1980.

Tippit, P. R., E. A. Pessagno, Jr., and J. D. Smewing, The biostratigraphy of sediments in the volcanic unit of the Semail ophiolite, *J. Geophys. Res., 86,* 2756–2762, 1981.

van Andel, Tj. H., Mesozoic/Cenozoic calcite compensation depth and the global distribution of calcareous sediments, *Earth Planet. Sci. Lett., 26,* 187–194, 1975.

Weissel, J. K., Evidence for Eocene oceanic crust in the Celebes Basin, in *The Tectonic and Geologic Evolution of Southeast Asian Seas and Islands,* edited by D. E. Hayes, *Geophys. Monogr. Ser.,* vol. 23, pp. 37–47, AGU, Washington, D. C., 1980.

White, S. M., H. Chamley, D. Curtis, G. deVries Klein, and A. Mizuno, Sediment synthesis: Deep Sea Drilling Project Leg 58, Philippine Sea, *Init. Rep. Deep Sea Drilling Proj., 58,* 1000, 1980.

Petrological and Geochemical Documentation of Ocean Floor Metamorphism in the Zambales Ophiolite, Philippines

E. E. GEARY AND R. W. KAY

Department of Geological Sciences, Cornell University, Ithaca, New York 14853

Metamorphism in the Zambales ophiolite is characterized by (1) a general increase in metamorphic grade from zeolite facies in the basaltic breccias to upper actinolite facies in the upper gabbros, (2) a steep metamorphic gradient (200°–400°C/km), (3) lateral variations in the extent of metamorphism within igneous pseudostratigraphic layers, and (4) the presence of coexisting mineral assemblages which represent disequilibrium chemical reactions and indicate varying oxygen fugacities and water-rock ratios during metamorphism. These features are typical of processes associated with ocean floor hydrothermal metamorphism and indicate the importance of hot circulating fluids as agents of heat and element transfer within the upper portion of the ophiolite. In addition, mineralogical transformations resulting from these metamorphic processes yield important insights into element mobility at different pseudostratigraphic levels within oceanic crustal layers. Petrological investigations and geochemical analyses of mafic rocks (gabbro-diabase-basalt) within the upper part of the Zambales ophiolite document this important crustal alteration process. Together, these data may be used to better constrain mass balance models of element recycling at convergent plate margins.

Introduction

Ocean floor metamorphism has been documented in oceanic dredge rocks from ridges, rises, fracture zones, trenches, and Deep Sea Drilling Project (DSDP) holes [*Spooner and Fyfe*, 1973; *Wolery and Sleep*, 1976; *Humphris and Thompson*, 1978; *Staudigel et al.*, 1980, 1981; *Mevel*, 1981; *le Roex and Dick*, 1981]. It has also recently been described in ophiolite terraines from southern Chile [*Stern et al.*, 1976; *Elthon and Stern*, 1978; *Stern and Elthon*, 1979] and Taiwan [*Liou*, 1979; *Liou and Ernst*, 1979]. In addition, several of the petrological and geochemical descriptions of other ophiolite complexes [*Gass and Smewing*, 1973; *Colemen*, 1977; *Spooner et al.*, 1977; *Suen et al.*, 1979; *Furnes et al.*, 1975; *Capedri et al.*, 1980] have metamorphic characteristics that are probably attributable to ocean floor metamorphic processes.

Some of the more characteristic features of this type of hydrothermal metamorphism include (1) a general increase in metamorphic grade with depth, ranging from low temperature weathering and zeolite facies in the uppermost basalts and breccias to lower and upper actinolite (lower amphibolite) facies in the upper gabbros and diabases; (2) steep metamorphic gradients, commonly in the range of 200–500°C/km, [*Spooner and Fyfe*, 1973]; (3) a general decrease in the extent of metamorphism within igneous pseudostratigraphic layers below depths of 2–3

km; and (4) the presence of coexisting disequilibrium mineral assemblages. The widespread occurrence of features commonly associated with ocean floor metamorphism, seen both in oceanic dredge rocks and in many ophiolitic complexes, implies that this type of subseafloor hydrothermal alteration plays an important role in the transfer of both elements and heat between the upper 3–4 km of oceanic crust and the overlying seawater-sediment interface.

The Zambales Range, located on Western Luzon, is an ultramafic-mafic rock complex of late Eocene age (40 m.y.) which exhibits structural, stratigraphic and igneous features similar to ophiolite complexes found in other parts of the world. It also preserves, in its upper 2–3 km of gabbro, diabase, and basalt, the mineralogical and geochemical changes that occurred during ocean floor metamorphism. In this paper we first use petrological and geochemical studies to document the effects and style of ocean floor metamorphism in the Zambales ophiolite. We then combine this data with mineralogical evidence to trace the mobility and mass transfer of different elements within the oceanic crust during ocean floor metamorphism. Finally, we develop a model for the expected distribution of major elements in oceanic crust subsequent to metamorphism and note that these data help to better constrain the sediment, seawater, and oceanic crustal components of subduction, arc magma generation models.

139

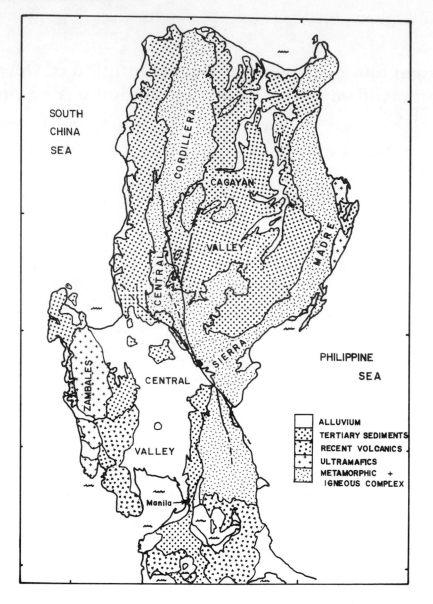

Fig. 1. Geological reference map of Luzon.

Regional Setting

The Zambales ophiolite forms a 130 km long, 30–40 km wide, north-south trending mountain range on western Luzon (see Figure 1). To the east, the ophiolite is bordered by a thick (>10 km) sequence of Tertiary sediments which fill the Luzon Central Valley and by a linear chain of dacite plugs. To the southeast, the plugs give way to an arcuate chain of andesite volcanoes which form the Bataan Peninsula. These Plio-Pleistocene to recent calc-alkaline volcanic rocks comprise the northernmost portion of the Zambales-Zamboanga arc defined by *Divis* [1980] and appear to be related to the eastward subduction of South China Sea basin crust beneath Luzon.

To the south, the ophiolite appears to plunge offshore toward the island of Ambil, where it is abruptly terminated against an east-west trending tectonic lineament (D. Karig, personal communication, 1982), while to the north it plunges gently beneath the South China Sea where it can be followed along trend for several tens of kilometers (S. Lewis and D. Hayes, personal communication, 1982). To the west, the ophiolite is tectonically juxtaposed against upturned sediments along a high angle fault which may have a significant strike-slip component. [*Schweller et al.*, this volume]. A more detailed discussion of the regional setting, geological history, and tectonic emplacement of the ophiolite is given by *Schweller et al.* [this volume].

Sampling and Analytical Procedures

Samples used in this study were collected along the eastern flank of the Zambales in the vicinity of Tarlac (see Figure 2). The outcrop exposures in this area are generally poor owing to the lush tropical vegetation, rugged topography and limited access. However, with the aid of Philippine Bureau of Mines (PBM) personnel and unpublished map data we were able to make several traverses into the upper igneous units of the ophiolite, systematically sampling the overlying sediments, the basaltic flows and breccias, the sill and dike complex, and the upper portion of the massive gabbros and diabases.

Unlike many geochemical and petrological studies which concentrate on locating and collecting the least altered rocks, our efforts focused on attaining a representative sampling of all the aforementioned rock types, from the least altered to the most extensively metamorphosed. As such, metamorphosed samples (with surficial weathering features removed) constitute the bulk of our petrological descriptions and geochemical analyses. It should also be noted that vein material in these samples is included as

part of the bulk rock compositions. Although this simplifies our analysis of whole rock alterations and element mobilities, it is recognized that attempts to quantitatively evaluate the net transport of some elements requires a closer examination of the distributions and compositions of vein material than is undertaken in this paper.

Whole rock chemical compositions of the samples are reported in Tables 1, 2, and 3. Major element concentrations (Si, Ti, Al, Fe, Mn, Mg, Ca, Na, K, P, Cr) were ascertained on fused (1100°C) glass beads by using a JEOL 733 electron microprobe and a Bence-Albee matrix correction program. Mineralogical compositions given in Tables 4, 5, 6, 7, 8, and 9 were determined by microprobe on polished thin sections. All sample analyses were compared to known standards (A-99, Juan de Fuca, etc.) run along with the samples. Precision and accuracy are generally better than ±5%, except for element concentrations below 0.10 wt %. A more detailed discussion of this technique, accuracy, and analytical errors can be found in *Frey et al.* [1974].

Trace element concentrations (Rb, Sr, Y, Zr) were measured on pressed powder discs by using X ray fluorescence. Statistical analyses of known standards (BHVO-1, AGV-1, G-2, and RGM-1) run concurrently with the samples indicate the precision and accuracy of the technique to be

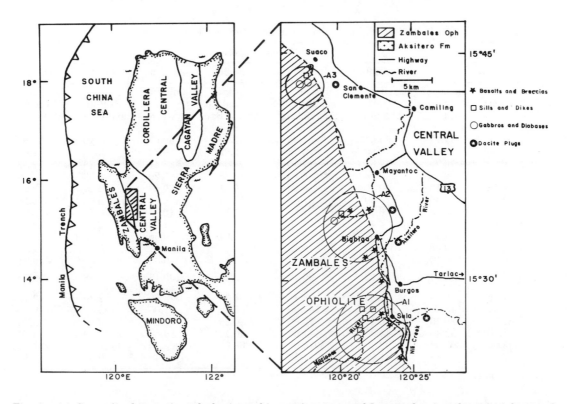

Fig. 2. (a) Generalized tectonic and physiographic provinces map of Luzon, showing the principle sample location area. (b) Enlargement of the principal sample location area. Almost all of the samples mentioned in this study were collected in areas A1, A2, and A3, but only a few representative samples are shown in this figure.

TABLE 1.   Upper Gabbros and Diabases: Major and Trace Elements

				Sample Number			
	EG12	EG20	EG59	EG60	EG61	EG63	EG23
Rock Type	diabase	gabbro	(Ol) gabbro	Z-Pyx diabase	(Ol) gabbro	(Ol) gabbro	(Hbl) gabbro
Metamorphic Minerals	Act-Feox	Act-Feox	Act Serp	Act-Feox	Act-Feox Serp	Act-Py Serp	Act-Fe oxides
Element							
SiO_2	47.26	51.95	51.52	50.03	50.66	48.30	56.33
TiO_2	3.24	0.93	0.59	1.50	0.70	0.45	2.26
Al_2O_3	14.82	16.17	15.63	15.23	15.61	17.95	15.36
FeO^*	13.31	9.32	8.09	10.75	8.45	7.99	10.57
MnO	0.18	0.13	0.16	0.14	0.12	0.14	0.09
MgO	8.83	7.68	8.54	7.11	9.06	8.64	3.64
CaO	10.42	10.86	13.62	11.58	12.17	15.20	6.58
Na_2O	1.70	1.90	2.35	2.93	2.39	1.55	3.14
K_2O	0.19	0.19	0.05	0.05	0.02	0.06	0.38
P_2O_5	...	...	0.09	...	...	0.08	0.29
Cr_2O_3	...	0.03	0.06	...	0.06	...	0.03
Total	99.95	99.16	100.70	99.32	99.89	100.36	98.67
Rb	2.00	2.30	1.80	2.60	2.50	1.90	3.60
Sr	148.00	139.00	118.00	110.00	115.00	156.00	175.00
Y	17.80	23.40	17.30	24.70	23.00	12.30	47.60
Zr	27.30	48.50	30.50	41.60	38.90	10.70	97.00
K/Rb	790.00	687.00	231.00	160.00	66.00	262.00	875.00
FeO^*/MgO	1.51	1.21	0.95	1.51	0.93	0.92	2.90

FeO^* = total iron as FeO.

TABLE 2.   Sill and Dike Complex: Major and Trace Elements

					Sample Number				
	EG26	EG27	EG29	EG34	EG36	EG35	EG62	EG64	EG40
Sample type	basaltic	diabasic	basaltic	diabasic	basaltic	diabasic	diabasic	basaltic	
Metamorphic Minerals	Chl-(Act) Ab	Act-Sph Feox	Chl and Ep	Act and Chl Sph and Ab	Chl and Ep Ab and Sph	Act-Feox	Act-Sph and Feox	Chl and Ep (Act)	Ep-Ab Chl
Element									
SiO_2	52.12	51.77	53.01	52.94	41.33	50.62	51.40	51.94	55.08
TiO_2	1.35	1.09	0.90	0.72	0.92	0.73	0.95	1.06	1.35
Al_2O_3	16.25	16.28	17.65	17.50	21.08	18.41	16.29	17.38	16.61
FeO^*	9.75	8.33	9.68	8.78	18.86	8.29	8.93	11.06	10.50
MnO	0.13	0.13	0.22	0.15	0.21	0.16	0.17	0.24	0.27
MgO	7.87	9.45	6.50	7.83	13.14	9.50	9.05	6.16	10.41
CaO	7.72	10.73	6.76	7.18	1.18	10.25	10.51	8.95	2.23
Na_2O	3.63	2.37	4.48	3.70	3.10	1.78	2.47	2.87	3.51
K_2O	0.56	0.31	0.02	0.54	0.04	0.36	0.42	0.06	0.01
P_2O_5	0.11	0.08	...	0.13	...	...	0.13	...	...
Cr_2O_3	0.03	0.10	...	trace	...	0.03	0.06	0.01	...
Total	99.52	100.64	99.12	99.47	99.86	100.13	100.38	99.73	99.94
Rb	4.50	3.20	1.80	4.40	4.90	3.10	2.30	2.10	1.80
Sr	145.00	143.00	140.00	141.00	28.00	131.00	129.00	138.00	36.00
Y	32.80	24.30	26.40	20.80	20.20	19.30	22.10	22.80	29.10
Zr	66.80	52.30	43.90	47.60	40.10	34.80	50.50	53.00	40.80
K/Rb	1033.00	803.00	92.00	1018.00	68.00	965.00	1517.00	237.00	46.00
FeO^*/MgO	1.24	0.88	1.49	1.12	0.86	1.49	0.99	1.80	1.65

FeO^* = total iron as FeO.

TABLE 3.   Metabasalts and Basaltic Breccias Whole Rock Analyses: Major and Trace Elements

	Sample Number							
	EG31	EG44	EG49	EG50	EG30	EG06	EG68	RK16A
Sample type	metabasalt	breccia	metabasalt	breccia	clast form	pillow Chl/ Smec	metabasalt	flow
Metamorphic Minerals	Chl Smec Hem-Qtz Ep	Cc-Vein Pg	Act-Chl/Smec Cc-Amyg	Pg-Cc Smec/ Chl	Chl-Sph (Ep) Qtz, Py	Pg- Hem	Chl-Sph (Ep)-pump	Cc (vesicles)
Element								
SiO_2	60.59	46.80	50.95	41.84	54.15	51.67	55.85	54.45
TiO_2	1.21	0.15	0.56	0.36	1.22	0.97	0.64	1.46
Al_2O_3	16.63	21.93	16.26	15.35	16.92	16.07	16.55	19.14
FeO*	9.11	3.03	9.09	6.65	14.79	11.64	8.16	8.32
MnO	0.22	2.34	0.24	0.20	0.17	0.23	0.21	0.20
MgO	8.05	1.15	9.62	3.97	7.82	5.88	8.02	1.19
CaO	1.00	16.92	8.85	25.31	2.01	6.91	4.91	8.00
Na_2O	4.08	7.58	3.75	5.13	1.89	3.88	5.39	4.50
K_2O	0.01	1.00	0.22	1.55	0.38	1.83	0.06	1.51
P_2O_5	...	0.13	0.11	0.14	...	0.35	...	0.30
Cr_2O_3	...	...	trace	trace	...	trace	trace	...
Total	100.90	101.03	99.65	100.50	99.35	99.43	99.79	99.07
Rb	3.00	14.90	4.40	29.60	6.10	31.10	1.50	27.80
Sr	34.00	121.00	213.00	78.00	46.00	429.00	116.00	183.00
Y	15.60	16.80	11.60	18.10	24.50	26.20	18.50	56.40
Zr	47.40	13.30	34.50	24.80	25.80	42.10	24.00	71.30
K/Rb	28.00	557.00	416.00	436.00	516.00	489.00	332.00	450.00
FeO/MgO	1.07	2.63	0.94	1.67	1.89	1.98	1.02	6.99

FeO* = total iron as FeO.

generally ±3% for Rb and Sr, ±5% for Y, and ±7–8% for Zr.

Igneous Stratigraphy

Overview

Recent field studies by a number of different workers (Schweller et al. [this volume]; De Boer et al. [1980]; Hawkins [1980]; the Philippine Bureau of Mines) have shown the Zambales Range, on western Luzon, to be a relatively complete and stratigraphically coherent ultramafic and mafic complex analogous to ophiolite terrains from other parts of the world. The igneous pseudostratigraphy (see Figure 3) of the deeper levels consists of a sequence of tectonized and serpentinized peridotites overlain by a cumulate section of gabbros. These cumulate gabbros grade upsection into massive, noncumulate gabbros and diabases, occasional leucocratic rocks (leucogabbros and gabbro breccias), a basaltic and diabasic complex of sills and dikes, and an overlying sequence of highly brecciated pil-

TABLE 4.   Actinolite Analyses: Sill and Dike Complex

	Sample Number							
	EG26	EG26	EG34	EG34	EG34	EG35	EG35	EG35
Element								
SiO_2	48.09	51.14	50.09	51.03	50.66	50.61	51.21	52.88
TiO_2	0.30	0.43	0.33	0.33	0.51	0.37	0.55	0.17
Al_2O_3	6.11	3.08	5.45	3.80	3.98	6.19	4.91	2.69
FeO*	16.65	14.92	15.56	16.66	15.23	12.42	15.23	13.98
MnO	0.21	0.27	0.31	0.25	0.21	0.18	0.27	0.20
MgO	11.27	14.01	13.27	12.45	13.10	15.87	13.71	14.79
CaO	11.68	11.22	10.93	11.67	11.63	10.97	11.55	11.83
Na_2O	0.54	0.27	0.54	0.37	0.45	0.49	0.46	0.24
K_2O	0.09	0.07	0.08	0.07	0.12	0.06	0.07	0.04
Cr_2O_3	0.98	0.30	0.55	0.33	0.76	N.D.	0.10	0.25
Total	97.64	95.71	97.11	96.95	96.65	97.17	98.06	96.82
Ca-Mg-Fe	$Ca_{29}Mg_{39}Fe_{32}$	$Ca_{27}Mg_{46}Fe_{28}$	$Ca_{26}Mg_{44}Fe_{29}$	$Ca_{28}Mg_{41}Fe_{31}$	$Ca_{28}Mg_{44}Fe_{28}$	$Ca_{26}Mg_{52}Fe_{23}$	$Ca_{27}Mg_{45}Fe_{28}$	$Ca_{27}Mg_{48}Fe_{25}$

FeO* = total iron as FeO.

TABLE 5. Chlorite Analyses: Sill and Dike Complex

	Sample Number							
	EG26	EG26	EG35	EG36	EG36	EG36	EG40	EG40
Element								
SiO_2	26.59	26.29	29.14	28.34	26.57	26.46	30.40	31.02
TiO_2	trace	trace	0.25	trace	trace	trace	trace	...
Al_2O_3	20.19	19.44	19.56	18.18	19.55	19.14	17.89	17.25
FeO*	20.36	20.92	19.02	20.96	26.77	28.36	18.46	18.36
MnO	0.15	0.16	0.21	0.21	0.19	0.43	0.52	0.53
MgO	18.13	18.38	20.32	18.57	14.95	13.67	20.23	20.61
CaO	trace	trace	0.19	trace	trace	trace	0.17	0.22
Na_2O	trace	...	trace	trace	trace	trace	...	...
K_2O	trace	trace	trace	trace	trace	trace	trace	trace
Total	85.42	85.19	88.69	86.98	88.03	88.06	87.68	87.99
Comments	pheno	pheno	groundmass	groundmass	vein	vein	Amygdule	Amygdule

FeO* = total iron as FeO.

TABLE 6. Epidote Analyses: Sill and Dike Complex

	Sample Number							
	EG26	EG26	EG36	EG36	EG37	EG37	EG40	EG40
Element								
SiO_2	37.42	37.44	37.40	37.93	38.19	37.90	38.63	38.08
TiO_2	0.24	0.43	0.20	trace	trace	trace	trace	0.11
Al_2O_3	24.25	23.45	23.94	24.81	25.04	24.56	27.11	24.57
FeO*	10.76	11.06	11.47	10.05	9.61	10.60	8.18	11.01
MnO	trace	trace	0.15	0.12	0.14	0.14	0.21	0.21
MgO	trace	trace	trace	trace	trace	trace	trace	trace
CaO	22.95	22.89	22.82	23.22	23.42	23.10	23.79	23.50
Na_2O	...	...	...	trace	trace	trace	trace	trace
K_2O	trace	trace	trace	trace	trace	trace	trace	trace
Total	95.62	95.27	95.98	96.13	96.40	96.30	97.92	97.48
Comments	vein	vein	vein	vein	part of glomerocryst	groundmass	part of glomerocryst	part of glomerocryst

FeO* = total iron as FeO.

TABLE 7. Plagioclase Analyses: Sill and Dike Complex

	Sample Number									
	EG26	EG26	EG34	EG34	EG36	EG37	EG40	EG40	EG35	EG35
Element										
SiO_2	65.38	57.50	55.68	66.11	68.62	67.58	68.77	68.50	46.19	50.70
TiO_2	...	...	...	...	...	...	...	...	trace	trace
Al_2O_3	21.20	26.41	27.97	20.83	19.91	19.49	20.06	20.18	33.97	31.10
FeO	0.17	0.18	...	...	0.13	0.12	0.12	0.14	0.39	0.68
MnO	trace	trace	trace	trace	trace	...	trace	trace	trace	trace
MgO	...	...	...	...	...	...	...	...	0.17	0.14
CaO	2.02	8.48	9.85	1.66	0.28	0.20	0.08	0.18	17.29	14.35
Na_2O	10.56	6.16	6.05	9.83	11.62	11.75	10.77	10.80	1.54	3.37
K_2O	trace	0.70	trace	1.76	trace	trace	trace	trace	trace	trace
Total	99.33	99.43	99.55	100.19	100.56	99.14	99.80	99.80	99.55	100.34
An content	An_{11}	An_{43}	An_{49}	An_8	An_2	An_1	An_1	An_2	An_{87}	An_{72}
Comments	ghost pheno	ghost pheno	relict pheno	relict pheno	part of glomerocryst	relict pheno	part of glomerocryst	relict pheno	diabase core	lath Rim

TABLE 8. Clinopyroxene Core-Rim Analyses: Upper Gabbros and Diabase

				Sample Number				
	EG60	EG60	EG60	EG60	EG61	EG61	EG61	EG61
Mineral	CPX core	CPX rim	CPX core	CPX rim	CPX core	CPX/ACT rim	CPX core	CPX/ACT rim
Element								
SiO_2	52.34	51.14	52.23	51.73	52.80	49.16	51.82	47.56
TiO_2	0.45	0.36	0.47	0.51	0.03	0.18	0.41	0.55
Al_2O_3	2.84	1.62	2.83	1.50	0.12	3.74	2.63	8.22
FeO*	5.69	13.72	5.70	11.51	8.86	19.57	5.62	12.30
MnO	0.11	0.19	0.12	0.22	0.21	0.17	0.13	0.17
MgO	17.03	15.19	16.78	14.09	13.17	11.06	16.04	15.06
CaO	21.06	16.33	21.25	20.40	24.72	11.38	21.83	10.85
Na_2O	0.26	0.25	0.28	0.33	0.05	0.68	0.30	1.64
K_2O	0.01	0.02	0.02	0.01	0.02	0.26	0.03	0.10
P_2O_5	...	...	...	...	...	...	...	...
Cr_2O_3	0.48	0.44	0.45	0.04	...	0.06	0.39	0.03
Total	100.27	99.26	100.13	100.34	99.98	96.26	99.20	96.48
Ca-Mg-Fe	$Ca_{43}Mg_{48}Fe_9$	$Ca_{34}Mg_{44}Fe_{22}$	$Ca_{43}Mg_{48}Fe_9$	$Ca_{42}Mg_{40}Fe_{18}$	$Ca_{49}Mg_{37}Fe_{14}$	$Ca_{27}Mg_{37}Fe_{36}$	$Ca_{45}Mg_{46}Fe_9$	$Ca_{26}Mg_{51}Fe_{23}$

FeO* = total iron as FeO.

low lavas and massive basalts. The igneous series is commonly capped by conformable pelagic limestones of late Eocene age [*Amato*, 1965; *Garrison et al.*, 1979], but locally thin (<2 m thick) red beds of what appear to be exhalative, hydrothermally derived mudstones are found between the upper ophiolite volcanics and the overlying sediments [*Schweller et al.*, this volume]. A younger volcanic sequence, consisting of tuffs, basalts, and more acidic material, also occurs in the study area and overlies the ophilitic basalts in several places (see Figure 2). The exact relationship of this younger, extrusive series to the ophiolite is not clear, but it may be related to a subsequent phase of off-axis volcanism such as those noted in other ophiolites [*Gass and Smewing*, 1973; *Liou*, 1979] or to the formation of the ophiolite in a back arc basin (BAB)–island arc tectonic environment. Samples from the calc-alkaline volcanic chain to the east and south of the ophiolite were also collected but will not be discussed in this report.

The Upper Gabbros and Diabases

The gabbros and diabases described herein are located stratigraphically 2–3 km below the sediment-basaltic breccia contact. They represent the uppermost (0.5 km) part of an extensive gabbro section mapped by Darwin Rossman in the 1960's and by the PBM and other workers [*Hawkins*, 1980] in the 1970's. Structurally, the upper gabbros and diabases appear to be intact, grading upward into an intrusive complex of sills and dikes. However, the total

TABLE 9. Plagioclase Core-Rim Analyses: Upper Gabbros and Diabases

				Sample Number				
	EG60	EG60	EG60	EG60	EG61	EG61	EG61	EG61
Mineral	plag core	plag rim	plag core	plag rim	plag core	plag rim	plag core	plag rim
Element								
SiO_2	51.23	53.47	49.69	53.85	50.83	54.43	52.18	55.37
TiO_2	0.01	0.02	...	0.01	0.04	0.09	0.04	0.09
Al_2O_3	31.36	29.49	32.10	29.84	30.71	28.60	30.65	28.21
FeO*	0.31	0.29	0.28	0.28	0.41	0.81	0.44	0.40
MnO	0.02	0.03	0.02	0.01	0.02	0.03	0.04	0.04
MgO	0.01	0.01	0.01	0.01	0.04	0.39	0.03	0.03
CaO	13.96	11.81	15.02	12.00	13.72	10.81	12.89	10.30
Na_2O	3.26	4.41	2.79	4.33	3.43	4.63	3.84	5.20
K_2O	0.04	0.07	0.04	0.07	0.06	0.25	0.06	0.08
P_2O_5	...	...	...	...	...	...	...	...
Cr_2O_3	...	...	...	...	...	...	...	...
Total	100.20	99.60	99.95	100.40	99.26	100.04	100.17	99.72
An content	An_{72}	An_{61}	An_{76}	An_{62}	An_{70}	An_{59}	An_{67}	An_{54}

FeO* = total iron as FeO.

LITHOLOGY

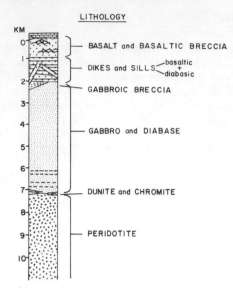

Fig. 3. Stratigraphic column of the Zambales ophiolite (igneous units). Only the upper 2–3 km have been affected by ocean floor metamorphism. Thicknesses of the different units are only approximate (adapted from *Schweller et al.* [this volume].

thickness of the massive and cumulate gabbro section is controversial. It is estimated to be 5–8 km by *Hawkins* [1980] but D. Rossman (personal communication, 1981) indicates that this figure may be too high and some tectonic thickening may have occurred.

At the contact with the overlying sill and dike complex, more leucocratic rocks consisting predominantly of hornblende-bearing gabbro breccias (see Table 1, EG-23) were occasionally observed. These rocks are volumetrically insignificant, at least in the vicinity of Tarlac. Farther north, however, near the Barlo Mine, *Hawkins* [1980] has noted a greater abundance of leucocratic-breccia swarms.

The upper gabbros and diabases are characterized, relative to the overlying sills, dikes, and basaltic rocks, by their pristine petrological appearance. They consist predominantly of holocrystalline subhedral to euhedral interlocking grains of augitic clinopyroxene (40–50% modally) and calcic plagioclase ($>An_{50}$, 50–60% modally) (see Tables 8 and 9). Grain boundaries tend to be distinct, and ophitic and subophitic textures are common. No signs of stress or tectonic deformation have been found, but the presence of fibrous green actinolite-tremolite amphibole and occasional blocky brown hornblende crystals indicates that medium-grade thermal metamorphism has affected some of these rocks.

In addition to clinopyroxene and calcic plagioclase, the gabbros and diabases commonly contain 1–2% of iron oxides (Fe-ox), usually with intergrowths of titanomagnetite (Mt) and/or ilmenite (Il). Olivine and/or orthopyroxene are sometimes present but not in sufficient abundance to be major rock-forming phases. It should be noted that where olivine is present, it is usually anhedral and is apparently in disequilibrium with the clinopyroxene and plagioclase. A typical olivine to Cpx-Plag reaction series consists of olivine patches surrounded by iddingsite-magnetite-serpentine intergrowths, which in turn are surrounded by thin rims of chlorite. Orthopyroxene is also usually anhedral, but does not exhibit obvious reaction textures with clinopyroxene or plagioclase. It may, however, show signs of alteration to actinolite-tremolite amphibole.

The Sill and Dike Complex

A complex of basaltic and diabasic sills and subhorizontal dikes lies stratigraphically between the upper gabbro-diabase unit described above and a series of overlying massive basalts and basaltic pillow breccias. The thickness of the sill and dike unit (estimated from map outcrop patterns and assuming an average dip of 25°E for the upper igneous units) varies from 0.5 to 1.5 km. Individual sills and dikes are generally 0.5–5 m wide and often exhibit fine-grained, basaltic chill margins immediately adjacent to medium-grained, diabasic interiors.

The predominance of sills and subhorizontal dikes is one of the unusual structural differences between the Zambales ophiolite and other ophiolites. However, despite the fact that several ophiolites (Troodos, Bay of Islands, Samail, Sarmiento, etc.) display mafic rock swarms consisting of 100% near-vertical or steeply dipping sheeted dike complexes, it has been noted by *Coleman* [1977] that subhorizontal dikes and sills are not uncommon structural features in ophiolites.

The sill and dike complex, unlike the underlying gabbros, is extensively metamorphosed, and primary igneous textures and mineralogies are frequently destroyed or overprinted. Medium-grained diabasic sill and dike interiors tend to exhibit the least alteration. Typically, they are composed of holocrystalline ophitic and subophitic intergrowths of altered clinopyroxene and calcic plagioclase similar in texture and composition to the underlying gabbros and diabases. Clinopyroxenes in these sill and dike interiors exhibit a moderate to strong actinolite-tremolite overprint, rims being generally more altered than cores. However, plagioclase analyses of some diabase sills (see Table 7, EG-35) show anorthite contents (An_{72-87}) which are equivalent to or even higher than most of the gabbro plagioclases (An_{67-72}) (see Table 9).

The fine-grained, basaltic sill and dike margins exhibit more alteration and greater textural diversity. They have hypocrystalline, xenomorphic granular, cryptocrystalline, felted, and resorbed textures. Primary phenocrysts are rare, but relict glomerocrysts and occasional relict or 'ghost' plagioclase cyrstals are sometimes present. Veins are also common but not ubiquitous.

The mineralogy of sill and dike rocks is typically dominated by fine-grained sub- to anhedral microlites of intermediate plagioclase (An_{15-50}) plus a host of metamorphic minerals, including chlorite, actinolite-tremolite amphibole, epidote, and sphene, with or without subordinate magnetite and ilmenite and occasional sulfides (usually pyrite). As discussed in greater detail in a later section, these metamorphic minerals may occur in any combination and amount, although assemblages dominated by chlorite and intermediate plagioclase are the most common. Albitic plagioclase ($<An_{15}$) also commonly occurs, replacing intermediate plagioclase as the dominant feldspar phase.

Metabasalts

Upsection, sills and subhorizontal dikes become less frequent as they become interspersed with screens of altered pillow basalt and pillow breccias. Massive basalts, basaltic breccias, and occasional altered pillows make up the upper portion of this unit, the total thickness of which is approximately 0.5–1.0 km. The contact with the overlying manganiferous pelagic limestones of the Aksitero formation has been described as conformable by *Schweller et al.* [this volume]. This contention is supported by geochemical analyses of the brecciated matrix material immediately underlying the limestones, which show local, dramatic increases in manganese concentrations (see Table 3, EG-44).

Texturally, the metabasalts may be similar to the felted and cryptocrystalline basaltic sills and dikes described above. However, some basalts tend to be finer grained and often display microcrystalline and hyalopilitic textures. Vesicles, amygdules, and veins are more common in these rocks than in the underlying sills and dikes. Metamorphism of the metabasalts is moderate to extensive. However, the preservation of relict anhedral, but generally well-preserved, clinopyroxene crystals indicates that metamorphism was not as extensive or as high grade as in the underlying sill and dike unit.

Mineralogically, the metabasalts are composed of microlites of albitic plagioclase ($<An_{15}$), palagonitized glass, and chlorite-clay intergrowths. Other minerals, such as quartz, calcite, epidote, and pumpellyite are often found in vein and amygdule assemblages, and quartz, calcite, hematite, and sulfides may also appear as minor groundmass phases. Zeolites are not common in the ophiolite metabasalts, which may reflect the dominance of palagonite and Smec-Chl mineral formation during low temperature weathering on the ocean floor.

Metamorphism

As witnessed from the preceding discussion, the Zambales ophiolite shows signs of a progressive increase in metamorphic grade with depth. More specifically, and in conventional metamorphic terminology, this progression from low to moderate grade is roughly equivalent to the following equilibrium assemblages:

1. Zeolite facies (Z): Various zeolite minerals, palagonitized glass (Pg) $\pm$ smectites (Smec) $\pm$ calcite (Cc) $\pm$ quartz (Qtz) $\pm$ sulfides $\pm$ sphene (Sph) $\pm$ albite (Ab).

2. Greenschist facies (G): Chlorite (Chl), epidote (Ep), Ab, Sph $\pm$ biotite (Bio) $\pm$ Cc.

3. Lower Actinolite facies (LA): Green fibrous actinolite-tremolite amphiboles (Al_2O_3 = 2.5–5.0 wt %) (Act), calcic plagioclase ($>An_{50}$) (Ca-plag), Sph $\pm$ Bio $\pm$ Cc.

4. Upper Actinolite facies (UA): Brown, blocky amphibole (Hbl), Act (Al_2O_3 = 5.0–8.0 wt %), Ca-plag, titanomagnetite (Mt) $\pm$ ilmenite (Il) $\pm$ Bio (see Figure 4).

Metamorphism in the upper gabbro through basalt units does not precisely follow the above scheme. This is probably due to variable thermal and water flow conditions in the different igneous pesudostratigraphic layers during successive stages of metamorphism. Such conditions are

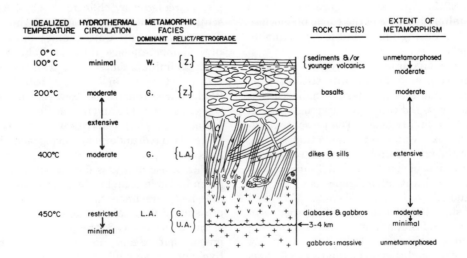

Fig. 4.   Idealized cross section of the ocean crust showing the general distribution of metamorphic facies and features.

documented by metamorphic nonequilibrium mineral associations such as Act + Chl + Ep + Inter − plag + Sph + Qtz + (Ab) and by skeletal intergrowths of Mt − Il and Sph.

Other authors [e.g., *Stern and Elthon,* 1979] have noted that such disequilibrium features often result from the incomplete transformation of higher temperature assemblages to lower temperature assemblages during subsequent retrograde events. These observations are supported by our petrological studies, which often show actinolite being replaced by chlorite, intermediate plagioclase by Ab + Ep, and chlorite by clays. It is, however, possible that the assemblage Act − Chl − Ep − Ab − Sph − Qtz is a stable transitional greenschist facies assemblage, since *Liou et al.* [1974] have shown the stability field of Act − Chl − Plag − Qtz to cover a wide range of T, PH_2O, and fO_2 conditions.

As noted previously, the upper gabbros and diabases show the least extensive alteration. Table 8 shows representative core-rim analyses for several clinopyroxene grains. These data indicate that primary Cpx's were normally zoned during igneous fractional crystallization processes, with Fe increasing in the margin with respect to Mg. These data also show that the subsequent actinolitization of some Cpx margins resulted in increases in Fe, Al, and volatiles (note the low 96–97% totals, probably reflecting the addition of a large H_2O component), and corresponding decreases in Mg and Ca. Al_2O_3 concentrations in the actinolitic margins range from 1.60–8.2 wt % and indicate lower and upper actinolite facies conditions, respectively [*Liou et al.,* 1974].

The unaltered nature of the calcic plagioclases is evident from both optical and microprobe studies. Grain boundaries are sharp, and there are no signs of sausseritization in either cores or rims. Table 9 shows core-rim analyses of several plagioclase laths indicating primary compositional zoning from calcic (An_{76}) cores to more sodium-rich (An_{54}) rims. The unaltered nature of the calcic plagioclase attests to its stability during upper and lower actinolite facies metamorphism.

The presence of occasional brown, blocky pleochroic amphiboles (Hbl) occurring as anhedral interstitial grains between Cpx and Ca-plag and the 'fresh' (unaltered) nature of the Fe-oxides also indicate metamorphism within the upper and lower actinolite facies. The presence of altered and unaltered Cpx crystals in the same thin section, and similar disequilibrium features on outcrop scale, suggest that the circulation of hydrothermal fluids at this pseudostratigraphic level was inhomogeneous.

Experimental studies by *Mottl and Holland* [1978] and *Mottl et al.* [1979] show the primary oxidizing species in seawater to be sulfate and that at temperatures exceeding 300°C and water/rock ratios of 10 or less, it is readily reduced to sulfide by the addition and oxidation of ferrous iron from the mafic country rocks. However, at water/rock ratios exceeding 10, this reaction is inhibited, and slightly

oxidizing solutions may predominate. If oxidizing conditions were present at this level (2–3 km) in the ophiolite, there is no evidence for it. The Fe-oxides appear fresh (no hematite is present) and epidote, which has been noted to form in oxidizing ocean crust environments [*Stern and Elthon,* 1979], is absent. It should be noted, however, that even under slightly oxidizing conditions epidote might not have been a stable phase. *Liou et al.* [1974] have shown that at 2 kbar, where P fluid = P total and fO_2 is defined by the QFM buffer, the upper stability limit for the Ab + Ep + Chl + Act (+ Qtz) assemblage is between 475°C and 550°C. Above this temperature, oligoclase (Inter-plag) and aluminous amphibole are the stable Ca-bearing phases, and chlorite and epidote break down. *Liou and Ernst* [1979] also point out that in the presence of plagioclase (>An_{15}) and at temperatures above 350°C, where P fluid = $\frac{1}{3} P$ total, epidote is not likely to form. Thus, based on our petrological data and the above experimental results, we suggest that the water/rock ratio in the upper gabbro-diabase unit did not exceed 10:1, the hydrothermal solutions at this level were predominantly reducing, and temperatures were at least >350°C. More probably, temperatures were even higher (>500°C), since there is not extensive fracturing within this unit and P fluid probably approached P total.

A more complex metamorphic history is observed in the sill and dike complex. In these rocks, the general metamorphic succession is from upper (relict) and lower actinolite facies assemblages near the contact with the underlying gabbros and diabases to greenschist facies assemblages in the upper portion of the unit. These assemblages are not, however, restricted to specific portions of the sill and dike complex and vary in abundance in both lateral and vertical dimensions. In the sill and dike complex, clinopyroxenes have been either completely uralitized, altered to mixtures of actinolite and patches of chlorite, or completely replaced by chlorite. Table 4 shows eight representative actinolite analyses from the sill and dike unit. These analyses are quite similar to the actinolites found in the gabbro-diabase unit (see Table 8).

The aluminum contents of sill and dike actinolites (Al_2O_3 2.7–6.2 wt %) are indicative of both lower and upper actinolite facies conditions. However, the presence of chlorite replacing actinolite and intermediate to albitic plagioclase replacing Ca-plag clearly show that upper actinolite facies minerals have undergone retrograde, disequilibrium effects.

Plagioclase analyses for the sill and dike complex are given in Table 7, and reflect the stability of more sodium-rich feldspars under greenschist facies conditions. The presence of very calcic plagioclase laths (An_{87-72}) in some of the rocks (see EG-35) helps to characterize their original igneous nature and suggests that they came from a relatively primitive mafic source.

Chlorite is ubiquitous in the sill and dike complex. As noted earlier, it may occur in association with actinolite (after clinopyroxene) or as a groundmass phase. It is also present in amygdules, veins, and occasionally as euhedral porphyroblasts. Table 5 gives a representative sampling of chlorite analyses. The observed spread in the SiO_2 (26.3–31.0 wt %), Al_2O_3 (17.2–20.2 wt %), FeO (18.4–28.4 wt %), and MgO (13.7–20.6 wt %) values provides additional evidence that the chemical composition of the circulating hydrothermal fluids (particularly the Fe and Mg components) was highly variable. However, these data probably also reflect some minor heterogeneity in the original whole rock composition.

Epidote is a common mineral in some sill and dike rocks, but is completely absent in others. It sometimes occurs along with albite replacing intermediate plagioclase (An_{15-50}), but more commonly it is found in massive intergrowths with quartz and albite. It is also found scattered throughout the groundmass and in veins. The narrow range in chemical analyses seen in Table 6 is typical of epidote analyses from other hydrothermally altered oceanic rocks [*Humphris and Thompson*, 1978; *Liou and Ernst*, 1979]. It should also be noted that, even where abundant, epidote is never present in greater amounts than chlorite.

Fe-oxides at this level are present in amounts similar to those found in the gabbros and diabases (1–2%) but generally appear as smaller, more anhedral crystals scattered throughout the groundmass. Sphene is the stable Ti-bearing phase under greenschist facies conditions and is often seen replacing ilmenite. It also occurs as very fine, light brown patchy aggregates associated with disseminated Fe-oxides. Microprobe analyses of Fe-oxides and sphene yield values similar to those reported in the literature [*Deer et al.*, 1966].

At higher levels in the sill and dike complex, Qtz, calcite, and sulfides (usually pyrite) are common accessory minerals. Qtz and calcite usually occur in amygdules and in veins, while sulfides tend to be found primarily in the groundmass.

The above observations indicate that although greenschist facies conditions predominated in the sill and dike complex, chemical equilibration during successive stages of metamorphism was not complete. This is in accord with observations of greenschist assemblages from other ophiolites [*Gass and Smewing*, 1973; *Liou*, 1979; *Elthon and Stern*, 1978].

Water/rock ratios in the sills and dikes are difficult to estimate owing to the highly variable nature of the metamorphic conditions and the lack of isotopic data. However, the presence of epidote, the replacement of ilmenite by sphene, and the widespread chloritization of these rocks are suggestive of relatively high (>10:1) values. In a similar assemblage of rocks from the Troodos ophiolite, *Spooner et al.* [1977] noted water/rock ratios of at least 15:1 based on Sr isotope data. Temperatures in the sill and dike unit undoubtedly varied dramatically, but the general lack of zeolites and clay minerals and the apparent stability of Ab + Chl + Qtz assemblages roughly confine

the range to between 300°C and 500°C.

Metamorphism of the overlying basalts and breccias, although generally less extensive than in the sill and dike unit, is complicated by subsequent low-temperature metamorphic effects. The dominant metamorphic assemblage consists of Smec − Chl + Pg + Ab. However, secondary Qtz + Cc often dominate whole rock chemical compositions (see Table 3).

The term smectite is used here to denote all clay mineralogies, but it should be noted that it is usually defined to include only montmorillonite, beidellite, nonienite, and other similar 2:1 clay species.

Although no smectite analyses are included in this report, microprobe studies show that nearly all of them contain a significant chlorite component and many of them are potassium rich (K_2O = 2–3 wt %). In addition, petrological observations often show cryptocrystalline, yellow-brown aggregates of low birefringence smectite replacing radiating patches of light green chlorite in the groundmass.

Albite, when present, exists entirely as microlites interstitial to basaltic and palagonitized glass. The glass itself exhibits various degrees of alteration ranging from relatively clear to muddy yellow-brown textures. *Staudigel et al.* [1981] have described similar features in oceanic rocks from DSDP holes and suggest that such features represent the development of submicroscopic protosmectite/protoceladonite material. The breakdown of basaltic glass to form palagonite is also a potentially important source of material for the formation of secondary minerals.

As noted previously, chlorite, quartz, and/or calcite are frequently found in vesicles, amygdules, veins, and groundmass phases, and petrological data indicate that most if not all of the silica and carbonate in these rocks is secondary. These data, when coupled with the geochemical evidence in Table 3, indicates that under low temperature (zeolite facies) conditions, elements such as Mg, Si, and Ca may be easily mobilized and redeposited.

The presence of hematite in some rocks and pyrite in other rocks reflects oxidizing and reducing conditions, respectively. This apparent contradiction is discussed by *Mottl et al.* [1979], who have shown in experimental studies that both hematite and pyrite may form by the introduction of ferrous iron from the basaltic material into solution. As the ferrous iron enters solution, it is oxidized to ferric iron in association with the reduction of seawater sulfate to sulfide. Under these initially oxidizing conditions (before much sulfide has formed), hematite may precipitate out. As further reduction of seawater sulfate to sulfide continues, the solution becomes more reducing (FO_2 decreases and fS_2 increases) until the pyrite stability field is reached. Thus when both pyrite and hematite are present, hematite may be a metastable phase.

Pumpellyite has been identified thus far in only one metabasalt. In this rock (e.g., Table 3, EG-68), it occurs as a vein mineral and is shown by microprobe analysis to be Fe-rich (11–12 wt % FeO* (total iron as FeO)). The apparent scarcity of pumpellyite and the lack of prehnite and zeolites in the ophiolite metabasalts is not readily understood but may reflect low temperature retrograde metamorphism and chemical exchanges controlled by the formation of smectite-chlorite and palagonite.

Studies of metamorphic mineral stability [*Liou*, 1979; *Mevel*, 1981] indicate that pumpellyite may form at temperatures of between 190°C and 250°C and low (less than 1 kbar) pressure. In addition, experimental studies by *Seyfried* [1977] have shown that at low temperatures potassium may be removed from seawater into K-rich clays and palagonite. These findings and the predominance of potassium-rich smectite-chlorite and palagonite assemblages support the contention that the metabasaltic rocks from the Zambales ophiolite were metamorphosed under low temperature (approximately 0°–300°C) ocean floor and zeolite facies conditions.

In summary, the above data indicate that (1) the Zambales ophiolite was metamorphosed under conditions ranging from zeolite facies to upper actinolite facies (0° to greater than 500°C and low (less than 1 kbar) pressures) and (2) higher grades of metamorphism can be generally correlated to greater depths and higher temperatures, but considerable lateral variations in metamorphic assemblages exist.

The predominate mineral transformations observed at successively higher pseudostratigraphic levels include (1) gabbros and diabases:

$$Cpx \rightarrow Act$$

(2) sills and dikes:

$$Cpx \rightarrow Act + Chl$$
$$Ca - plag \rightarrow Inter - plag \rightarrow Ab ; (Ep)$$
$$Mt - Il \rightarrow Mt - Sph$$

(3) metabasalts:

$$Chl \rightarrow Smec - Chl$$
$$basaltic\ glass \rightarrow Pg$$

In addition, secondary deposition of quartz, calcite, and sulfides is more widespread in the upper levels.

All of these data serve to document ocean floor metamorphism of the ophiolite and support the earlier contention that the upper 2–3 km of the Zambales ophiolite was metamorphosed near an active oceanic spreading center with a high geothermal gradient (at least 200°C–400°C/km) and fluctuating hydrothermal flow conditions.

Element Mobility and Mass Transfer

From the above discussion and documentation of metamorphism in the upper portions of the Zambales ophiolite, it seems logical to assume that the mineralogical transformations which occur during successive stages of metamorphism play an important role in the control of element mobility in the oceanic crust. Experimental studies are useful in helping to predict the relative movement of elements into and out of basaltic water/rock systems under different temperature, pressure, and fO_2 conditions but by themselves are insufficient to explain the observed element distributions in altered oceanic crust and ophiolites.

We have developed a model which uses both our metamorphic data from the Zambales (and other ophiolites) plus some of these experimental results to trace the transfer and general mobility of elements between and within the upper 2–3 km of oceanic crust and the overlying sediment/water column. Following is an element by element analysis of the progressive changes in whole rock and metamorphic mineral chemistries observed in the Zambales ophiolite from the upper gabbros and diabases through the metabasalts and basaltic breccias.

Magnesium

Average magnesium concentrations throughout the upper ophiolite units are quite uniform except for the basaltic breccias (see Tables 1, 2, and 3). In the gabbro/diabase unit, magnesium stability is governed primarily by clinopyroxene. Orthopyroxene and olivine play only a minor role owing to their minor abundances. Alteration of clinopyroxene to actinolite removes Mg from the rocks and adds Fe and Al. In the sill and dike unit (under greenschist facies conditions), chlorite is generally the stable Mg-bearing phase. As such, it provides a ready sink for any Mg removed from the underlying gabbros and diabases and from circulating seawater. The formation of chlorite also appears to buffer the removal of Mg from these rocks during the alteration of Cpx and Cpx-Act. These empirical observations are supported by experimental studies [*Hajash and Archer*, 1980; *Mottl and Holland*, 1978; *Seyfried and Bischoff*, 1981] as well as by observations of element trends in oceanic rocks [*Hart*, 1973; *Humphris and Thompson*, 1978]. These investigators show that for water/basaltic rock reactions occurring at temperatures >200°C Mg is preferentially enriched in the basaltic rock fraction.

In the metabasalts, under greenschist facies conditions, chlorite is also the major Mg-bearing phase. Since basaltic glass is the dominant component of the original rocks, the formation of chlorite indicates a large metamorphic segregation of Mg. Some of the Mg to form the chlorite is probably made available during the breakdown of basaltic glass to palagonite, while some Mg undoubtedly is extracted from seawater. However, under low temperature weathering conditions where chlorite is absent or is altering to smectite, Mg may be removed from the basaltic material [*Humphris and Thompson*, 1978] (see Figure 5 and Table 3, basaltic breccias).

Calcium and Sodium

The distributions and mobilities of these two elements are controlled almost exclusively by the stability of different feldspar phases. In the Zambales ophiolite, calcium

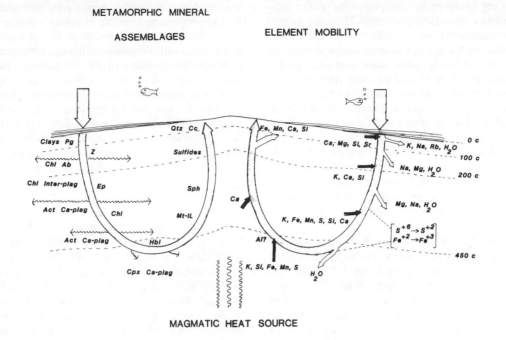

METAMORPHIC MINERAL
ASSEMBLAGES

ELEMENT MOBILITY

MAGMATIC HEAT SOURCE

Fig. 5. Schematic model of hydrothermal circulation near an active oceanic spreading center showing the element mobilities and mineral assemblages which control the distribution of elements in the upper 2–3 km of ocean crust. Dark arrows represent element transfer into the circulating seawater solution. Light arrows represent the transfer of elements into the surrounding rocks.

generally tends to decrease in the rocks from the upper gabbros and diabases through the sills and dikes while sodium shows a progressive increase. This behavior is most easily explained by the substitution of Na^+ for Ca^{++} during the alteration of Ca − plag → Inter − plag → Ab under greenschist facies conditions. Although the removal of calcium from the sills and dikes is somewhat inhibited by the formation of epidote and sphene, some of the calcium may make it into solution and be redeposited upward in the igneous section as secondary minerals in veins, amygdules, and groundmass phases (see Figure 5). In the metabasalts and basaltic breccias under zeolite and low-temperature weathering conditions, sodium continues to increase, probably reflecting more extensive albitization of the plagioclase and removal of Na from the seawater. However, calcium concentrations are variable and unpredictable owing to the great mobility of Ca under these conditions and to the increased likelihood of sediment (carbonate) contamination.

Potassium and Rubidium

Both potassium and rubidium are present in successively higher amounts in the gabbros, diabases, sills, dikes, and metabasalts and especially in the basaltic breccias (see Table 3). The consistent Rb values (2–5 ppm) throughout the lower units are probably in the range of primary concentrations. These concentrations are somewhat higher (0.5–2 ppm) than values reported for normal MORB [Sun

et al., 1979; Hart, 1973] but correspond well to Rb values for back arc basin basalt (BABB) from the East Scotia Sea [Saunders and Tarney, 1979], and the Mariana Trough [Reynolds, 1982]. The low and variable potassium and K/Rb values, on the other hand, suggest removal of K^+ from the rocks during upper actinolite through greenschist facies metamorphism. Enrichment of both potassium and rubidium in the basaltic breccias appears to occur under low temperature weathering conditions and is probably encouraged by the formation of illite, K-rich smectite clays, and palagonite in the breccia matrices. These interpretations are supported by experimental studies by Seyfried [1977] and Mottl and Holland [1978] which show the behavior of potassium to alter markedly between 70 and 150°C. Below this range, potassium is taken up in the basaltic material, but at temperatures exceeding 150°C it is leached from the basaltic material.

Iron and Manganese

Fe and Mn concentrations show no consistent trends from the gabbros and diabases through the metabasalts. Cpx, Act, and Chl are the Fe-bearing phases, while Mn is present in minor amounts in Cpx, Chl, and Ep. These phases probably act to buffer Fe and Mn removal from the rocks, even though experimental work by Mottl et al. [1979], Bischoff and Dickson [1975], and Seyfried and Bischoff [1981] indicate that at temperatures >200°C, Fe^{++} and Mn are readily leached from basaltic rocks. Below 200°C these elements are buffered by the formation of clay minerals. The positive correlation of Fe and Ti in the gabbro and diabase unit suggests that Fe-oxides (Mt and Il) may also exhibit some control on Fe mobility and distribution.

The introduction of ferrous iron into solution, as mentioned previously, should lead to the reduction of seawater sulfate to sulfide and the oxidation of soluble Fe^{++} to Fe^{+++} (relatively insoluble). Thus early forming metamorphic minerals should contain substantially more Fe^{+++} than later forming metamorphic minerals. Unfortunately, we do not yet have the data on the Fe^{+++} components of our metamorphic minerals to test this hypothesis. The presence of sulfides and hematite in both dike and sill rocks and metabasalts does, however, provide some support for the above model of Fe mobility and reduction of seawater sulfate.

Anomolously high Mn concentrations in some of the basaltic breccias (2–3 wt%) may be explained by either sediment contamination from the overlying manganiferous limestone unit or by chemical precipitation from hydrothermal solution. Mn analyses of smectite-chlorite intergrowths indicate, furthermore, that the Mn concentrations in the rock material may increase or be buffered under low temperature conditions.

Aluminum

Aluminum concentrations, like Fe, show no consistent trend from the gabbros through the metabasalts, and also like Fe, Al is distributed throughout a number of major rock-forming minerals. The predominant control of Al in the gabbros and diabases is in the Ca-plag, but Cpx also contains a small Al component. Alteration of Ca-plag to Inter-plag tends to release Al, but the formation of Act from Cpx acts to inhibit removal of Al from the rocks. One trend our data do show is that Al_2O_3 concentrations are somewhat higher (1–2%) in the sill and dike complex than in the gabbros and diabases. This probably reflects the abundance of major Al-bearing phases (Chl, Act, Ep, and Inter-plag) and might indicate limited transfer of Al from the gabbro-diabase unit into the sills and dikes, but it is also possible that these differences are related to primary igneous effects. In general, though, it appears that Al behaves conservatively during ocean floor metamorphism.

Silica

Silica concentrations tend to increase slightly from the gabbros through the metabasalts but are significantly depleted in the basaltic breccias. Examination of Tables 4 and 8 shows that alteration of Cpx to Act should not significantly alter the silica content of the whole rock. Therefore at most only a slight amount of SiO_2 should be added to solution under actinolite facies conditions. Transformations of Ca-plag to Inter-plag (SiO_2 in) and Act to Chl (SiO_2 out) in the sill and dike unit during greenschist facies metamorphism tend to be counteracting processes, but experimental data by *Seyfried and Bischoff* [1981] indicate that under greenschist facies conditions, SiO_2 should be leached from the basaltic material. While leaching does not appear to have taken place in the sill and dike unit, the presence of secondary silica in veins, amygudules, and groundmass phases in both the sill and dike and metabasalt units and the depletion of SiO_2 in the basaltic breccias provide evidence of extensive SiO_2 mobility under greenschist and zeolite facies conditions. It is therefore suggested that the increase in SiO_2 concentrations in the sills and dikes and the metabasalts is due primarily to the introduction of secondary silica into these rocks. The main source of the SiO_2 is probably from leached basaltic breccia material.

Titanium

Titanium concentrations are variable within and between different igneous pseudostratigraphic units, and their distributions appear to be controlled almost exclusively by the abundance of Fe-oxide phases (Mt + Il) and sphene. Although Ti is relatively mobile in a small scale, mineralogical sense (i.e., sphene is a common alteration phase of Fe-Ti bearing oxides), it is relatively immobile as far as large-scale, bulk rock translocations are concerned. Low Ti values in the basaltic breccias (see Table 3) indicate that at least some of the Ti is mobilized during low-temperature alteration of basalt.

Strontium

Strontium concentrations are generally consistent throughout the gabbro-diabase and sill and dike units ($\approx$ 135 ppm) and are considered to represent primary igneous values. Significant deviations from the above value which are common in the metabasalts, basaltic breccias, and some of the sills are considered to be due in part to Sr mobilization and redeposition in conjunction with calcium. However, it is also likely that much of the basaltic Sr has been supplemented or interacted with seawater Sr [*Spooner et al.*, 1977; *McCulloch et al.*, 1980].

Summary

A summary of the different element mobilities and mass transfers delineated above is as follows:

1.   K and minor amounts of Si, (Al?), Fe, Mg, and Mn are removed from the gabbro-diabase and sill and dike rocks at depths of 2–3 km and temperatures exceeding 300°C.

2.   During greenschist facies metamorphism Ca may be slightly leached from the rocks as Ca-Plag alters to Inter-Plag which in turn alters to Ab, while Na, Mg, and H_2O are added to the rocks to make Ab and Chl.

3.   During low-temperature greenschist and zeolite facies metamorphism, Si and Ca are readily mobilized causing depletion of these elements in some rocks and additions in others. Na, Mg, and H_2O continue to be removed from seawater to form Ab and Chl, while K continues to be added to solution.

4. Low-temperature zeolite facies and weathering conditions lead to the removal of K from seawater to form clays and palagonite glass. Fe and Mn may also be removed from solution to form clay and sulfide phases.

5. Sr and Rb are mobile under greenschist and zeolite facies conditions and are frequently incorporated into secondary carbonates, clays, and palagonite in the uppermost igneous units.

6. The formation of Act buffers the removal of Si, Al, and Fe under upper and lower actinolite facies conditions.

7. Transfer of ferrous iron into solution aids in the reduction of seawater sulfate to sulfides and leads to more reducing seawater solutions.

The above conclusions are pictorally represented in Figure 5 along with the mineral assemblages which appear to influence their behavior.

Tectonic Implications

In developing and evaluating element recycling models at convergent plate margins [e.g., *Kay*, 1980] the selection of appropriate element input values for the various subducted components, seawater, sediment, (altered) oceanic crust, and depleted mantle material is critical for the determination of the relative contributions each of these components makes to the subsequently generated melt. More specifically, in examining the contribution of the ocean crust component to arc magma generation two questions are of primary importance. First, what was the original geochemistry (and tectonic setting) of the oceanic rocks? And second, were the rocks altered by ocean floor metamorphic processes before subduction and partial melting led to the generation of arc magmas?

In the past, most of the element input values for mass

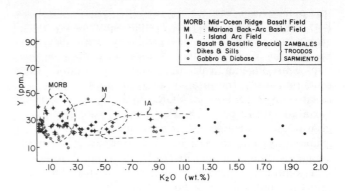

Fig. 7. Plot of Y (ppm) versus K_2O (wt %) for selected ophiolites, MOR rocks, Mariana Trough samples, and island arcs.

balance models have been based on 'typical' mid-ocean ridge concentrations ascertained from the freshest rock samples available. The growth of the geochemical data set for oceanic rocks has in recent years caused revisions in the average chosen input values to include the effects of seafloor alteration, but little has been done to evaluate and incorporate geochemical analyses of hydrothermally altered oceanic rocks from different tectonic settings.

A comparison of the metamorphic features in the Zambales and other ophiolites (see Figure 6) indicates that ocean floor metamorphism is a fairly common process and despite the fact that the degree of metamorphism varies with both depth and lateral extent, analogous element distributions and translocations might be expected in much of the upper oceanic crust. This observation is supported by other studies of ocean floor rocks [*Hart*, 1973; *Humphris and Thompson*, 1978; *Spooner et al.*, 1977; *Sleep and Wolery*, 1982; *Staudigel et al.*, 1981] and suggests that much of the crust formed at mid-ocean ridge and back arc basin spreading centers has been altered by subsequent metamorphic processes resulting in increases in Na, Mg, Rb, Sr, and H_2O into the rocks, and upward translocations of Si, (Al?), Ca, Fe, Na, and K. As a result of these observations we have recently compiled data from the literature, from personal communications with other workers, and from our own work which allow us to place additional constraints on some of the element input parameters to mass balance–element flux models at convergent plate margins.

One example of these constraints is seen by examining the potassium input components for *Kay*'s [1980] and *Karig and Kay*'s [1981] element recycling, mass balance model for the Mariana arc-trench system. In their model an input value of 0.20 wt % of K_2O was chosen for the ocean crust component.

In Figure 7 we have made a plot of K_2O (wt %) versus Y (ppm) values for a cross section of rocks from ocean ridges, the Mariana Trough, selected island arcs, and three ophiolites, the Zambales, Sarmiento, and Troodos. From

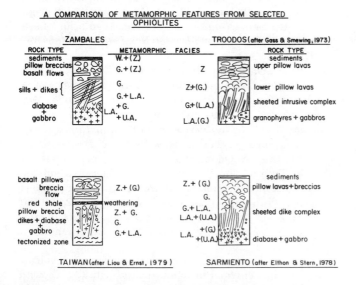

Fig. 6. Comparison of metamorphic features and facies distributions within selected ophiolites.

this plot several features are readily apparent. First, the K_2O values for the Mariana Trough samples are considerably higher than those for typical mid-ocean ridge rocks. Second, the Marianas samples are intermediate between mid-ocean ridge and volcanic-island arc rocks in K_2O content, and third, the altered ophiolitic rocks, while comparable in Y content to MORB and BABB, are highly variable with respect to K. Furthermore, the greatest degree of K_2O variability exists in the uppermost (stratigraphically) basalts and basaltic breccias. These rocks generally tend to be enriched in K when compared to both mid-ocean ridge rocks and the stratigraphically underlying dike and sill and gabbro and diabase units. These observations correlate well with the degree of alteration of the rocks.

Such data indicate that the input value of 0.20% K_2O chosen for the altered ocean crust component in Mariana arc magmas is probably much too low if any or most of these rocks have been altered by ocean floor metamorphic processes prior to subduction and partial melting. On the basis of both the ophiolite data shown in the K_2O versus Y plot and consideration of the distribution of metamorphic assemblages in altered ocean rocks, a more appropriate K_2O input value might be 0.40% K_2O. (This value represents a 50-50 mix of seafloor and hydrothermally altered MOR rocks estimated to have 0.20 and 0.60 wt. % K_2O, respectively.)

Using this new value in *Karig and Kay*'s [1981] model, the effect of the change in K_2O input is to reduce by half the estimated amount of altered oceanic crust that must be melted to satisfy the Mariana mass balance calculations. It may of course be argued that a value of 0.40% K_2O is unrealistic, since our data set is not extremely large and also because most of the subducted oceanic material may be more geochemically characteristic of unaltered MOR rocks (i.e., 'fresh' basalts have been collected from DSDP holes). However, while this is indeed a possibility, we feel that our value being based on both altered and unaltered samples is better constrained and more representative of the true nature of oceanic material being consumed at convergent plate margins. Resolution of this uncertainty and further refinement of K_2O input values should be possible as the geochemical data set for both altered and unaltered oceanic and ophiolitic rocks increases.

Returning to the K_2O versus $\bar{Y}$ plot, another point that is revealed is that element flux, mass balance modeling of arc trench systems, and selection of element input values is dependent not only on the extent and grade of metamorphism but also on the original geochemistry (and tectonic setting) of the rocks. If, for example, the Marianas Trough samples (which are more likely to represent the type of oceanic crust involved in subduction processes in back arc regions) plotted in Figure 7 were subducted and recycled, the average K_2O value of the ocean crust input would be at least 0.33%, and that is for the freshest rocks available. Again, consideration of metamorphic processes

would undoubtedly increase that value. We are currently analyzing hydrothermally altered Mariana Trough basalts to determine if this supposition is correct. Although the above analysis has only examined one element, potassium, similar observations and syntheses appear to be useful for reevaluating the ocean crust contribution to arc magmas of other mobile elements (e.g., Rb, Sr, Ba, etc.).

From the above discussion it is clear that an understanding of the geochemical and petrological changes which accompany ocean floor metamorphism is critical in the determination of appropriate element input values for element recycling–mass balance models at convergent plate margins. As such, more detailed studies of the metamorphic geochemistry of the Zambales, other ophiolites, and altered oceanic dredge material are crucial if we are to quantify the net mobilization and redistribution of elements within the upper 2–3 km of oceanic crust. Additional work is now in progress along those lines and should lead to further constraints on element flux, mass balance models.

Acknowledgments. We thank D. Karig, H. Staudigel, and C. Stern for their helpful comments on the manuscript, the Philippine Bureau of Mines for their invaluable field and logistical support on Luzon, and the other SEATAR and IDOE participants for their contributions to our increased knowledge of the Philippines. Finally, we thank the National Science Foundation for supporting our work on grants OCE-761878, OCE-7816757, OCE-7919165, and OCE-8026484.

References

Amato, F. L., Stratigraphic paleontology in the Philippines, *Philipp. Geol. 19,* 121–140, 1965.

Bischoff, J. L., and F. W. Dickson, Seawater-basalt interactions at 200°C and 500 bars: Implications for origin of sea-floor heavy-metal deposits and regulation of seawater chemistry, *Earth Planet. Sci. Lett., 425,* 385–397, 1975.

Capedri, S., G. Venturelli, G. Bocchi, J. Dostal, G. Garut, and A. Rossi, The geochemistry and petrogenesis of an ophiolite sequence from Pindos, Greece, *Contrib. Mineral. Petrol., 74,* 189–200, 1980.

Coleman, R. G., *Ophiolites: Ancient Oceanic Lithosphere?,* 222 pp., Springer-Verlag, New York, 1977.

De Boer, J., L. A. Odom, P. C. Ragland, F. G. Snider, and N. R. Tilford, The Bataan orogene: Eastward subduction, tectonic rotations, and volcanism in the western Pacific (Philippines), *Tectonophysics, 67,* 251–282, 1980.

Deer, W. A., R. A. Howie, and J. Zussman, *An Introduction to the Rock-Forming Minerals,* 528 pp., Longman Group Ltd., London, 1966.

Divis, A. F., The petrology and tectonics of recent volcanism in the central Philippine Islands, in *The Tectonic and Geologic Evolution of Southeast Asian Seas and Islands, Geophys. Monogr. Ser.,* vol. 23, edited by D. E.

Hayes, pp. 127–144, AGU, Washington, D.C., 1980.

Elthon, D., and C. Stern, Metamorphic petrology of the Sarmiento ophiolite complex, Chile, *Geology, 6,* 464–468, 1978.

Frey, F. A., W. B. Bryan, and G. Thompson, Atlantic Ocean floor: Geochemistry and petrology of basalts from legs 2 and 3 of the Deep Sea Drilling Project, *J. Geophys. Res., 79,* (B5), 5507–5527, 1974.

Furnes, H., B. A. Stuart, and W. I. Griffin, Trace element geochemistry of metabasalts from the Karmby ophiolite, southwest Norwegian Caledonides, *Earth Planet. Sci. Lett., 25,* 213–238, 1975.

Garrison, R. E., E. Espirito, L. J. Horan, and L. E. Mack, Petrology, sedimentology and diagenesis of hemipelagic limestone and tuffaceous turbidites in the Aksitero formation, central Luzon, Philippines, *U.S. Geol. Surv. Prof. Pap. 1112,* 16 pp., 1979.

Gass, I. G., and J. D. Smewing, Intrusion, extrusion and metamorphism at constructive margins: Evidence from the Troodos massif, Cyprus, *Nature, 242,* 26–29, 1973.

Hajash, A., and P. Archer, Experimental seawater/basalt interactions: Effects of cooling, *Contrib. Mineral. Petrol., 75,* 1–13, 1980.

Hart, R. A., A model for chemical exchange in the basalt-seawater system of oceanic layer II, *Can. J. Earth Sci., 10*(6), 799–816, 1973.

Hawkins, J. W., Jr., Petrology of back-arc basins and island arcs and their possible role in the origin of ophiolites, in *Proceedings of the International Symposium on Ophiolites, Cyprus,* edited by A. Parayioutou, pp. 244–254, Geological Survey Department, Nicosia, Cyprus, 1980.

Humphris, S. E., and G. Thompson, Hydrothermal alteration of oceanic basalt by seawater, *Geochim. Cosmochim. Acta, 42,* 107–125, 1978.

Karig, D. E., and R. W. Kay, Fate of sediments on the descending plate at convergent margins, *Philos. Trans. R. Soc. London Ser. A, 301,* 233–251, 1981.

Kay, R. W., Volcanic arc magmas, implications of a melting-mixing model for element recycling in the crust–upper mantle system, *J. Geol., 88*(5), 497–552, 1980.

le Roex, A. P., and H. Dick, Petrography and geochemistry of basaltic rocks from the Conrad fracture zone on the America-Antarctica Ridge, *Earth Planet. Sci. Lett., 54,* 117–138, 1981.

Liou, J. G., Zeolite facies metamorphism of basaltic rocks from the East Taiwan ophiolite, *Am. Mineral., 64*(1 and 2), 1–14, 1979.

Liou, J., and G. Ernst, Ocean ridge metamorphism of the East Taiwan ophiolite, *Contrib. Mineral. Petrol., 68,* 335–348, 1979.

Liou, J. G., S. Kuniyoshi, and K. Ito, Experimental studies of the phase relations between greenschist and amphibolite in a basaltic system, *Am. J. Sci., 274,* 613–632, 1974.

McCulloch, M. T., R. T. Gregory, G. J. Wasserburg, and H. P. Taylor, Jr., A neodymium, strontium, and oxygen isotopic study of the Cretaceous Samail ophiolite and implications for the petrogenesis and seawater—Hydrothermal alteration of oceanic crust, *Earth Planet. Sci. Lett., 96,* 201–211, 1980.

Mevel, C., Occurrence of pumpellyite in hydrothermally altered basalts from the Vema fracture zone (Mid-Atlantic Ridge), *Contrib. Mineral. Petrol., 76,* 386–393, 1981.

Mottl, M. J., and H. D. Holland, Chemical exchange during hydrothermal alteration of basalt by seawater, I, Experimental results for major and minor components of seawater, *Geochim. Cosmochim. Acta, 42,* 1103–1115, 1978.

Mottl, M. J., H. D. Holland, and R. F. Carr, Experimental basalt-seawater interactions—Hydrothermal alterations, *Geochim. Cosmochim. Acta, 43*(6), 869–884, 1979.

Reynolds, J. C., The petrology and geochemistry of basalts from the Mariana trough, Master's thesis, 79 pp., Cornell Univ., Ithaca, N.Y., 1982.

Saunders, A. D., and J. Tarney, The geochemistry of basalts from a back-arc spreading center in the East Scotia Sea, *Geochim. Cosmochim. Acta, 43,* 555–572, 1979.

Schweller, W. J., D. E. Karig, and S. B. Bachman, Original setting and emplacement history of the Zambales ophiolite, Luzon, Philippines, from stratigraphic evidence, this volume.

Seyfried, W. E., Seawater-basalt interaction from 25°–300°C and 1–500 bars, Ph.D. dissertation, 242 pp., Univ. of Calif., Los Angeles, 1977.

Seyfried, W. E., and J. L. Bischoff, Experimental seawater-basalt interactions at 300°C, 500 bars: Chemical exchange, secondary mineral formation and implications for the transport of heavy metals, *Geochim. Cosmochim. Acta, 45*(2), 135–148, 1981.

Sleep, N. H., and T. J. Wolery, Egress of hot water from mid-ocean ridge hydrothermal systems: Some thermal constraints, *J. Geophys. Res.,* in press, 1982.

Spooner, E. T. C., and W. S. Fyfe, Sub-seafloor metamorphism, heat and mass transfer, *Contrib. Mineral Petrol., 42,* 287–304, 1973.

Spooner, E. T. C., H. J. Chapman, and S. D. Smewing, Strontium isotopic contamination and oxidation during ocean floor hydrothermal metamorphism of the ophiolite rocks of the Troodos massif, Cyprus, *Geochim. Cosmochim. Acta, 41,* 873–890, 1977.

Staudigel, H., F. A. Frey, and S. R. Hart, Incompatible trace-element geochemistry and 87/86 Sr in basalts and corresponding glasses and palagonites, *Initial Rep. Deep Sea Drill. Proj., 51,* 53, 1980.

Staudigel, H., S. R. Hart, and S. H. Richardson, Alteration of the oceanic crust, processes and timing, *Earth Planet Sci. Lett., 52,* 311–327, 1981.

Stern, C., and D. Elthon, Vertical variations in the effects of hydrothermal metamorphism in Chilean ophiolites: Their implications for ocean floor metamorphism, *Tectonophysics, 55,* 179–213, 1979.

Stern, C., M. J. DeWitt, and J. R. Lawrence, Igneous and metamorphic processes associated with the formation of Chilean ophiolites and their implication for ocean floor metamorphism, seismic layering and magnetism, *J. Geophys. Res., 81*(23), 4370–4380, 1976.

Suen, C. J., F. A. Frey, and J. Malpas, Bay of Islands ophiolite suite, Newfoundland: Petrologic and geochemical characteristics with emphasis on rare earth element geochemistry, *Earth Planet. Sci. Lett., 45,* 337–348, 1979.

Sun, S., R. Nesbitt, and A. Sharaskin, Geochemical characteristics of mid-ocean ridge basalts, *Earth Planet. Sci. Lett., 44,* 119–138, 1979.

Wolery, T. J., and N. H. Sleep, Hydrothermal circulation and geochemical flux at mid-ocean ridges, *J. Geol., 84*(3), 249–275, 1976.

Structural Lineaments and Neogene Volcanism in Southwestern Luzon

JOHN A. WOLFE

Taysan Copper, Inc., Makati, Metro Manila, Philippines

STEPHEN SELF

Department of Geology, Arizona State University, Tempe, Arizona 85287

The Philippine Islands have at least 15 active composite volcanoes and as many more that are fumarolic or dormant. About 20 calderas of Pleistocene age are known so far. Southwestern Luzon, one of the major volcanic districts of the country, contains three young composite volcanoes, four in a fumarolic stage, and over 200 vents of Pliocene-Pleistocene age within 150 km of Manila. There are three large calderas in this zone with a fourth a short distance south on Mindoro Island, plus four summit calderas. One of the most striking features is the Bataan Lineament, a chain of 27 volcanic vents, only one at present active, which marks the western side of the district. The main segment extends from Naujan caldera in the south (on Mindoro Island) on a strike of N31°W through Batangas Bay caldera, Mataas Na Gulod (a summit caldera), Corregidor Island (a small caldera), to Mount Mariveles and Mount Natib on the Bataan peninsula. With a bend of 30° at Mount Natib, the lineament continues northward for another 100 km, giving a total length of 320 km. Here it includes Mount Pinatubo, which is active, and several other vents. The Bataan Lineament is a volcanic arc, with perhaps some extensional element, above the subduction zone of the Manila Trench, dipping eastward under Luzon. Another major volcanic element is the Verde Island transform, which forms a zone across southwest Luzon, including 10 or more volcanoes. Activity extended from the lower Miocene with periodic eruptions until the late Pleistocene. Two volcanoes may be in a waning (fumarolic) stage and have thermal areas. Near the western end of this lineament, recent rifting may have occurred, and presently it is a zone of intense seismic activity. In the zone between the Bataan and Verde Island lineaments, several major volcanoes have developed including Laguna de Bay and Taal volcano-tectonic depressions. Large volume ignimbrite-forming eruptions may have taken place from Laguna de Bay caldera approximately 1.0 m.y. ago. In the area between Mount Banahao and the south shore of Laguna de Bay there is a large maar and scoria cone field. Major Pleistocene eruptions of andesitic ignimbrite (ash flow tuff) from Taal and Laguna de Bay calderas have 'cemented' together several Tertiary island volcanoes to form the present land mass of southwestern Luzon.

Introduction

The Philippine Archipelago has formed by a composite of construction and accretion by arc collision [*Hamilton*, 1979]. Construction has been largely due to voluminous-subaerial and shallow submarine volcanism and accompanying plutonism. Volcanic activity was extensive in northern Luzon in the late Oligocene, apparently related to a west dipping subduction zone, with the trench located east of the island [*Philippine Bureau of Mines*, 1963; *Wolfe*, 1981]. A second stage of volcanism commenced in the middle Miocene. Much of the activity lay west of the Cordillera Central of Luzon (Figure 1), apparently related to an east dipping subduction zone and the Manila Trench which 'flipped' to the western side of Luzon at the end of the Oligocene [*Karig*, 1973]. Volcanism waned in northern Luzon in the Pleistocene, but commencing in the late Pliocene, many new volcanoes formed in the southern Luzon

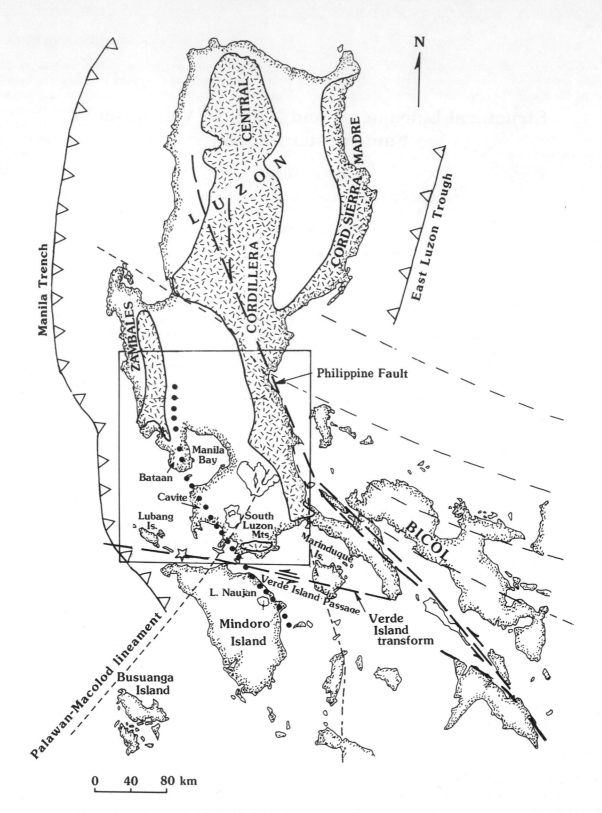

Fig. 1. Generalized map of the northern Philippines. Mountain ranges in light stipple. Manila Trench and Philippine fault of Pliocene-Pleistocene age after *Cardwell et al.* [1980] and *Hamilton* [1979]. Verde Island transform from *Divis* [1980]. Bold dotted line is the proposed Zambales-Zamboanga volcanic chain of *Divis* [1980], which approximates to the calc-alkaline arc of *de Boer et al.* [1980]. Star marks intensive seismic zone discussed in text. Box shows area of Figure 2.

area, and the region has continued to be very active. The volcanoes are largely above the east dipping subduction zone along the Manila Trench [*Cardwell et al.*, 19890; *Divis*, 1980].

A second active volcanic field is the Bicol peninsula of southeastern Luzon (Figure 1). It appears to be related to the west dipping Philippine Trench subduction zone. Older portions of the peninsula appear to be recently accreted from the Talaud arc [*Karig*, 1975]. The petrology and geochemistry of rocks from this zone have been described by *Divis* [1980].

The purpose of this paper is to document the recently active and 'live' Pliocene and Quaternary volcanoes of southwestern Luzon. (An active volcano is used here to describe one which has had recent historic activity or has obvious signs of recent eruptions, for example, young lava flows, steaming fissures, fumaroles. A 'live' volcano can be defined as one that has not been proven to be extinct by geologic relationships [*Walker*, 1974].) The area under discussion (Figure 2) extends northward to the Central Valley of Luzon and to the south is bounded by the Verde Island transform fault [*de Boer et al.*, 1980; *Divis*, 1980], located between Luzon and Mindoro islands. *Divis* [1980] describes the volcanoes as belonging to the northern part of the Zambales-Zamboanga chain, a major volcanic arc paralleling the Manila Trench.

The phreatomagmatic eruption of Volcano Island, Taal, in 1965 [*Moore et al.*, 1966] attracted wide attention to Luzon Island. This event was followed by the eruption of small pyroclastic flows (nuées ardentes) from the andesitic composite cone of Mount Mayon on the Bicol peninsula [*Moore and Melson*, 1969]. Recently, *Newhall* [1979] described compositional variations in the lavas of Mayon, and *Divis* [1980] discussed the petrology of volcanoes in the Central Philippines. In general, the magnitude of volcanism in the country has not been fully recognized. *Neumann van Padang* [1953] listed only four volcanoes in southwestern Luzon. There are many more than four 'live' volcanoes and two very young monogenetic cone and maar (explosion crater) fields in the region.

Some general papers which have included information about volcanism in this area are *Alcaraz* [1947], *Gervasio* [1966, 1968, 1971], *de Boer et al.* [1980], and *Divis* [1980]. *Wolfe et al.* [1978] described the South Luzon Mountain Range, and *Louden* [1977] described the Miocene activity on Marinduque Island (Figure 1). *Lo* [1981] recently investigated the chemistry of some tholeiitic basalts and low-Si andesites from Luzon.

Regional Structures Influencing Volcano Distribution

The Bataan Lineament

One of the most striking features of the South Luzon volcanic zone is the chain of 27 vents and volcanoes which mark the western side of the district (Figure 2). The main segment extends from Naujan caldera on Mindoro Island, at the southeastern end, through Batangas Bay (a possible caldera) to Mataas na Gulod (a composite volcano with a summit caldera) and Corregidor Island at the mouth of Manila Bay (a small caldera) to Mount Mariveles and Mount Natib on the Bataan peninsula (both composite volcanoes, the latter with a summit caldera). The length of this segment of the lineament is 220 km with a strike of N31°W. This corresponds roughly to structure 'H' of *Alcaraz* [1968, Plate 1].

With a change of strike of 30° at Mount Natib, the lineament continues northward for another 100 km, giving a total length of 320 km. Here it includes the volcanoes Mount Cuadrado, Mount Negron, Mount Pinatubo, several other vents, and two fields of diatremes. This group of volcanoes fills a gap in the Zambales Range (an ophiolite complex) where it narrows to half of its width to the north. The Bataan Lineament makes up a part of the calc-alkaline volcanic arc of the east dipping Manila Trench [*Gervasio*, 1971; *Karig*, 1973] and a portion of it *de Boer et al.* [1980] named the Bataan orogen.

The Verde Island Transform-Rift

A small volcano in the channel between Luzon and Mindoro, called Verde Island, has given its name to the passage between the islands. This channel is the locus of a pronounced structural feature which extends from the Manila Trench on the west to the Philippine fault zone and possibly (with offset) to the Philippine Trench.

At the western end of the Verde Island transform, Lubang, Ambil, and Golo islands appear to have recently rifted away from Mindoro. Further, the rotation of Bataan and Cavite noted by *de Boer et al.* [1980] may be associated with the rifting of this sector away from Lubang. There is intense seismicity of the western end of the Verde Island transform. At the southeastern corner of Golo is a seismically active zone (Figure 2) that led *de Boer et al.* [1980, p. 26] to conclude, 'The seismic swarm below Lubang (depth 50 km) may represent activity associated with rising magmas.' *Divis* [1980] has suggested that the volcanism in south Luzon may be related to a 'leaky transform.' The South Luzon Mountains are an older volcanic chain north of and parallel to the Verde Island transform extending from the west arm of Batangas Bay eastward for 50 km (Figure 2). If extended to include the activity on the Lubang group and the fumarolic volcano on Marinduque, the length of the active zone related to the rift is 190 km. Volcanism commenced here in middle Miocene and continued into the Pleistocene, concurrent with activity to the north.

Within the western end of the Verde Island Passage (the extensional zone) there are small volcanoes of Pleistocene age on Maricaban and Verde islands. The most recent activity on Verde Island, a phreatomagmatic eruption, was probably in the late Pleistocene, based on young terraces

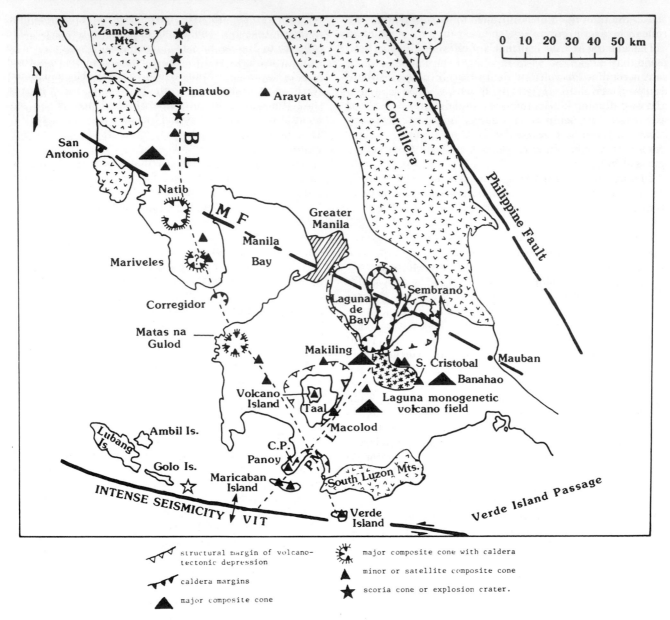

Fig. 2. Neogene volcanoes of southwestern Luzon. Centers discussed in text are given by name; there are many other late Tertiary structures that are not shown. Open star marks zone of intense seismicity discussed in text. Arrows at 90° to Verde Island transform indicate possible extension at western end of this trend. VIT, Verde Island transform; MF, Manila fault; PML, Palawan-Macolod Lineament; CP, Calumpan peninsula. Ornament with stars shows Laguna magnetic volcano field (Figure 4).

which transect accretionary lapilli-bearing tuffs and the active thermal area on the island.

Palawan-Macolod Lineament

The northwest face of Mount Macolod is a scarp marking one of the faults along which caldera collapse of the Taal volcanic structure occurred. The fault continues to the southwest and follows the western margin of Batangas Bay [*Philippine Bureau of Mines*, 1963]. Extending this line northeasterly, it passes through Mount Makiling and into Mount Sembrano within the Laguna de Bay structure (Figure 2). This line, including a volcano on Maricaban Island (Figure 2), is 100 km long, suggesting a significant crustal element. Also, there is a pronounced fault valley on northwestern Mindoro on this same trend, suggesting that the structure may continue to the southwest. If so, this segment of the fault must have formed after Mindoro rotated into its present position.

The lineament may continue into a major crustal element that marks the western edge of the Palawan shelf southwest of Mindoro Island, with the northwestern limb dropping into deep water. This has been called a trench and east-dipping subduction zone by *Hamilton* [1979]. It was active for a short time in the Miocene but probably became inactive before the Pliocene. This would imply that the Macolod segment extended northeast in the Pliocene and may have controlled the location of volcanic activity at several sites in southwestern Luzon. The total length of this element is over 800 km; it corresponds to structure 'E-E' of *Alcaraz* [1968, Plate 1].

The Manila Fault (?)

Wolfe [1981] suggested that there is evidence for an apparently inactive fault in the Manila area trending N50°W. It extends from the east coast near Mauban through the major structure of Laguna de Bay, to Mount Natib and the head of Subic Bay and out to sea northwest of San Antonio (Figure 2). Near the west coast it clearly cuts the Zambales ophiolite, forming the south limb of a graben.

While the true nature of the Manila fault is still debatable, it can be tentatively traced for some 260 km (Figure 2). Except for the northward extension of the Bataan Lineament, this structure is the northern limit of the volcanic zone of southwestern Luzon. It corresponds to the San Antonio fracture zone of *de Boer et al.* [1980] at the western end. However, the San Antonio graben is only a small part of the overall structure.

Minor Structures of South Luzon

De Boer et al. [1980] have defined the Iba fracture zone located in the middle of the Zambales coast (Figure 2). This structure is a small graben trending N7°W which intersects the Bataan Lineament and at that point Mount Pinatubo grew to become the highest volcano on the Bataan Lineament. This graben is not apparent through the Central Valley, but near the eastern margin of the basin *Sonido* [1978] has defined a graben of the same strike and width and approximately the same trend.

We see no evidence for the Corregidor fracture zone of *de Boer et al.* [1980]. The northeast trending fault on Volcano Island in the Taal structure seems to be confined to the caldera and does not extend beyond its margins as implied by *de Boer et al.* [1980]. It is one of a conjugate pair intersecting on the island and reflecting structures within the subsided caldera block. The main center occurs at their intersection.

The South Luzon Volcanic Zone

Southwestern Luzon contains three active volcanoes. Two are subaerial composite volcanoes, Mount Banahao (or Banahaw) and Mount Pinatubo; the other, Volcano Island

in Taal Lake (Figure 2), is a very young, largely subaquatic composite volcano that produces phreatomagmatic eruptions. Seven or more volcanoes may be classed as 'live' (they are in a dormant or fumarolic stage), and over 200 monogenetic vents of Pliocene-Pleistocene age occur within 150 km of Manila, one significant grouping being the Laguna volcanic field, south of Laguna de Bay (Figure 2). There are two volcano-tectonic depressions (encompassing large calderas) within the zone, Taal and Laguna de Bay. Another large collapse structure occurs to the south on Mindoro Island (Lake Naujan) (Figure 1). Three composite volcanoes of the region have summit calderas: Natib, Mariveles, and Mataas na Gulod.

As shown in Figure 2, the southwestern zone has been extended to include the volcanoes along the east flank of the Zambales ophiolite and also Mount Arayat in the Central Valley of Luzon. The easternmost volcano is Mount Banahao. In general, the southern limit is the Verde Island Passage.

Major Volcanic Centers of South Luzon

Mount Pinatubo

The highest volcano on the Bataan Lineament is a composite structure of calc-alkaline affinity rising to 1745 m. It is located at the point where the Iba fault zone, a graben, intersects the Bataan Lineament. There have been no historic eruptions of Mount Pinatubo, but there are reportedly two roaring vents on the summit and sulfur is actively depositing. Dates of 635, 2330, and 8050 years B.P. have been obtained by the ^{14}C method [*Wolfe*, 1981] from material within mudflow (lahar) deposits. Since there has been no thorough study of this volcano, there may have been younger eruptions. One of the most recent eruptions resulted in a voluminous plagioclase-rich crystal tuff (ignimbrite), which filled stream valleys southwest toward San Marcelino, northwest into the Iba graben and northeasterly toward Tarlac and Angeles.

Most samples reported from this volcano are high-silica hornblende andesites and dacites [*Philippine Atomic Energy Commission*, 1977]. Ignimbrite and glassy lavas appear to be common products. Table 1 lists an analysis of a typical andesite from Mount Pinatubo.

Mount Mariveles and Mount Natib

These massive composite volcanoes make up the Bataan peninsula which lies partially across the mouth of Manila Bay (Figure 2). The area of the summit caldera on Mount Natib (1253 m) is approximately 35 km². Mount Mariveles has a smaller summit crater about 10 km² and a maximum elevation of 1388 m. These volcanoes were recently studied [*Philippine Atomic Energy Commission*, 1977; *de Boer et al.*, 1980]: 31 K-Ar dates were obtained on Mount Natib volcanics, the range being 0.54 to 3.9 m.y. [*Wolfe*, 1981]. A 0.98 m.y. age may date the eruption which formed the summit caldera. There are 20 dates on Mount Mariveles

TABLE 1.   Analyses and Ages of Lavas and Volcaniclastic Rocks From Three Volcanic Centers in South Luzon (Data From *Philippine Atomic Energy Commission* [1977])

	Mount Natib				Mount Mariveles			Mount Pinatubo
	Sample 976-7GF	Sample 976-3GF	Sample 976-7FGM	Sample 976-3JN	Sample 976-2FLU	Sample 276-2F	Sample 176-7F	Sample 476-310
SiO_2, %	53.76	55.64	56.84	58.00	60.32	56.46	57.93	61.90
TiO_2, %	0.95	0.67	0.72	0.69	0.60	0.55	0.71	0.84
Al_2O_3, %	17.88	18.62	19.96	17.18	17.74	18.01	17.26	16.19
FC_2O_3, %	9.26	7.79	7.15	8.24	6.10	8.83	8.62	5.66
MgO, %	4.85	3.27	2.23	3.96	3.65	3.53	4.13	2.75
CaO, %	9.35	8.29	8.24	7.09	7.19	8.08	8.10	5.66
Na_2O, %	3.16	3.52	3.56	3.69	3.68	3.13	3.39	4.48
K_2O, %	0.98	0.92	1.28	0.83	1.45	1.15	1.01	1.66
Total	100.19	98.72	99.98	99.68	100.64	99.74	101.15	98.84
Age, m.y.	2.17 ± 0.35	2.06 ± 0.20	2.14 ± 0.08	0.76 ± 0.06	0.58 ± 0.06	0.47 ± 0.08	0.97 ± 0.08	<0.10

976-7GF lava flow, basaltic andesite; 976-3GF lava flow, andesite; 976-7FGM lava flow, andesite; 976-3JN lava flow, andesite; 976-2FLU pyroclastic (unspecified), andesite; 276-2F lava flow, andesite; 176-7F lava flow, andesite; 476-310 pyroclastic ?(lahar boulder), andesite-dacite.

ranging between 0.19 and 4.1 m.y.

Both volcanoes are considered to be extinct [*Philippine Atomic Energy Commission,* 1977]. The volcanoes are built largely of calc-alkaline andesites, with some high-Al basalts and dacites. The lavas are plagioclase rich and also contain augite [*Philippine Atomic Energy Commission,* 1977]. Selected analyses covering the range of lavas and pyroclastics are given in Table 1.

Corregidor Caldera

In the mouth of Manila Bay on the Bataan Lineament is Corregidor Island, an annular-shaped island enclosing the northern half of a small caldera which erupted in the marine environment. A postcaldera dacitic dome occurs near the central part. Caraballo Island is on the southeastern margin of the caldera rim, and reefs and islets nearly close the ring. The lagoon is only about 30 m deep, and the area of the caldera is 20 km². The andesites and dacites on the island fit the calc-alkaline trend of the Bataan Lineament. There is no date on the Corregidor caldera-forming eruption.

Mataas na Gulod Caldera

Lying on the Bataan Lineament, with its slopes constituting the south entrance to Manila Bay, is a sizeable composite volcano with an eroded summit caldera of nearly 30 km². On its western rim is Mount Palaypalay, 648 m high, and on the southeastern side is Mount Mataas na Gulod, 622 m high. The caldera is about 6 km in diameter and has an extrusive dome within it. Pumice-rich pyroclastic flows and basaltic lava flows extend 7 km north to Manila Bay where there are dates on the flows of 2.91, 2.92, and 2.95 m.y. [*Wolfe*, 1981]. To the west near the coast there is an exposed dioritic intrusion dated at 4.9 m.y. Volcanic debris flow deposits (lahars) to the west on the coastal area may be related to the ancestral volcano or may have come from a satellite cone. The highly dissected slopes expose composite features, suggesting this

was a stratovolcano which may have been 1200 m or more in height. Since the volcano is more dissected than other composite cones nearby, it is thought that activity terminated before 2 m.y. ago. The basalts and andesites of this caldera appear to be calc-alkaline, as are all of the volcanoes on the Bataan Lineament.

Batangas Bay

The scalloped coastline at several points in the Philippines suggests that calderas may have developed in the marine environment. One such location is Batangas Bay, 15 km in diameter and partly enclosed by small volcanoes such as Mount Casapao on Maricaban Island (Figure 2). Mount Panay on the Calumpan peninsula, the western arm of the Bay, may also be a postcaldera volcano. On the southeast side the South Luzon Mountains extend to the shore with Mount Pinamucan being a Pleistocene volcano.

Batangas Bay lies at the intersection of the Bataan Lineament with the Palawan-Macolod Lineament; the latter is probably one of the structures along which the caldera collapsed. The north limb of the Verde Island transform also passes through the zone. Thus structural conditions were favorable for a volcanic center to develop. The 200-m isobath passes close to the western shore, indicating a minimum amount of collapse on the Palawan-Macolod fault. The deepest portion, at the mouth of the bay, is 400 m. Ignimbrite sheets from the Taal center extend to the north shore of the bay and possibly accreted the Calumpan peninsula to mainland Luzon.

The volcanoes on the western side of the bay are composed of andesites. On the eastern side they are andesitic to dacitic. There are few volcaniclastic rocks in the area which can be related to a caldera in Batangas Bay, so the classification of this site as a flooded caldera is tentative. The volcanic activity is classed as Pleistocene [*Philippine Bureau of Mines*, 1963], and fumaroles are still active on the Calumpan peninsula.

Laguna de Bay

The largest lake in the Philippines, Laguna de Bay, lies directly east of Manila and is only 1 m above sea level (Figure 2). It may be, in part, a caldera-type structure with associated grabens. Young volcanic rocks surround Laguna de Bay, but as yet, detailed mapping has not been carried out. There are two $^{40}K/^{40}Ar$ dates on volcanic rocks, the older a basalt flow dated at 1.7 m.y. and the younger an ignimbrite dated at 1.0 m.y. [*Wolfe,* 1981], both from the northern edge of Laguna de Bay. The national geologic map [*Philippine Bureau of Mines,* 1963] shows an area of more than 1400 km² covered by tuffs extending to the east coast. The pyroclastics, at least 400 m thick at the east shore of the lake, are thought to be mainly from the eruptions of this caldera. Much greater volumes may have flowed into the Philippine Sea and Manila Bay.

Laguna de Bay has the appearance of a caldera of 25 × 12 km (300 km²) nested in a larger collapse structure of uncertain origin (Figure 3). This structure is 35 × 45 km, or 1500 km², which would make it one of the largest volcano-tectonic structures yet described, compared with, for example, the Taupo volcanic center, 1100 km² [*Cole and Nairn,* 1975]. The ignimbrite eruption dated at 1 m.y. was probably from the Middle Bay area. No available record classes this structure as a caldera.

The Laguna de Bay structure may be termed a volcano-tectonic depression [*Williams,* 1941], with bounding faults striking generally to the north (Figure 3). These faults extend into the Cordillera Central to the north. The locus of magmatic activity was apparently controlled by the intersection of the Palawan-Macolod Lineament and the Manila fault, which meet at approximately right angles within the region (Figure 2). There was apparently little in the way of a constructional, emergent volcanic edifice before collapse occurred.

Middle Bay (Figure 3) appears to be a collapsed caldera, the western side being Talim Island, a narrow sliver of volcanic rocks rising to 420 m. On the east side of Talim there is an echelon system of faults. A fault on the west side of Talim indicates collapse of more than 100 m into the West Bay of Laguna Lake. There is a major fault marking the east shore of the Morong peninsula north of Talim Island, the fault approximating the collapse scarp.

On the east side of Middle Bay, the caldera rim is the Jala Jala peninsula. Here Mount Sembrano is a volcanic peak rising to 743 m. Just east of Quisao, on the west side of the peninsula, three step faults account for collapse of over 300 m. These faults swing from a trend of about N20°E around the head of the bay (Figure 3).

There is also evidence of collapse into the East Bay of Laguna with a fault on the east side of the peninsula having an apparent down throw of about 400 m. Yet another major fault occurs on the eastern margin of East Bay with about 400 m of collapse on the downthrown (west) side into the caldera. This fault can be traced over 10 km to the south where it intersects Mount Banahao (Figure

2) and to the north for over 100 km.

The chemistry of some volcanic rocks from Laguna caldera is indicated in Table 2. *De Boer et al.* [1980] have classed this volcano as shoshonitic. The analyses in Table 2, though few in number, indicate a moderately high alkali content but not the high K_2O/Na_2O rations (>1) typical of shoshonites. The lavas and pyroclastics have low phenocryst contents; petrography is not available at the time of writing. Juvenile clasts from the ignimbrites examined are moderately mafic in composition (SiO_2 = 53–60 wt %; samples TA34, TA36, Table 2) and are dominantly black pumice with some mixed-magma clasts. Some of the ignimbrites show evidence of deposition in shallow water. As much as 50 to 60 km north of Laguna de Bay, extensive welded ignimbrites such as the Pleistocene Diliman tuff floor the Central Valley of Luzon [*Gonzales et al.,* 1971]. Gonzales et al. report thicknesses in excess of 150 m of ignimbrite indicating that these must be voluminous deposits. The area around Manila Bay and the Central Valley underlain by the Diliman tuff exceeds 5000 km².

Mount Banahao (or Banahaw)

Mount Banahao, the tallest mountain in southern Luzon (2165 m), lies 80 km southeast of Manila (Figure 2). It is an active volcano. The last two eruptions in 1730 and 1749 [*Andal and Yambao,* 1953] resulted in large mud flows (lahars) which swept down a valley from the breached crater and partially inundated the town of Sariaya 20 km to the south. Two lava flows on Mount San Cristobal, a major satellite cone (1470 m) on the west flank of Mount Banahao, were dated by $^{40}K/^{40}Ar$, yielding 1.6 and 1.31 m.y. [*Wolfe,* 1981].

The structural position of Mount Banahao is somewhat obscure. There is a major fault trending northerly from the volcano, possibly connecting with the eastern bounding fault of the Laguna de Bay depression and extending northward about 110 km into the Sierra Madre of eastern Luzon. *Divis* [1980] has suggested there may be an east-west fracture through Taal and Mount Banahao, complementary to the Verde Island transform, but this is only an inference.

De Boer et al. [1980] have classed this volcano and others around it as shoshonitic. However, the chemistry of lavas is similar to that of the Laguna de Bay center which borders it to the north and to Mount Makiling to the west. High-alumina basalts and alkaline pyroxene andesites are reported by the *Philippine Atomic Energy Commission* [1977].

Mount Makiling

This is a composite volcano (1090 m) to the south of Laguna de Bay. The Palawan-Macolod Lineament passes through Mount Makiling and there intersects a north-south fault that served as one of the lines of collapse for the Laguna de Bay depression to the north (Figure 2). Dates obtained on rocks from Mount Makiling are 0.51, 0.2, 0.18, and 0.1 m.y. [*Wolfe,* 1981]. La Mesa tuff cone on

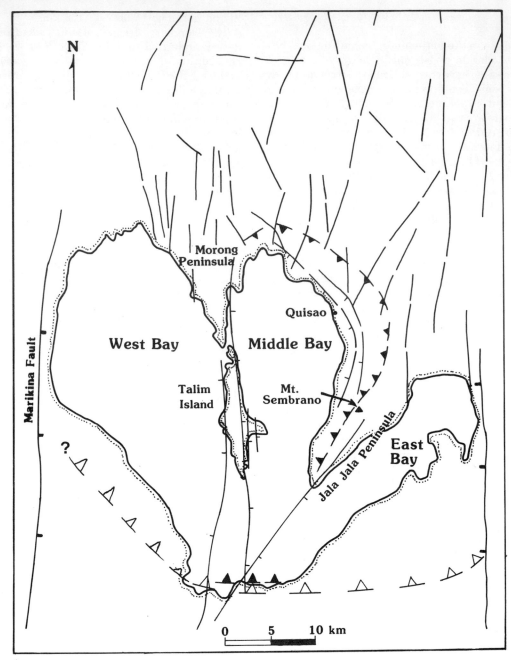

Fig. 3.   Tectonic map of the Laguna de Bay area showing major regional faults (down thrown side has tick),
possible caldera collapse faults (solid triangles), outline of volcano-tectonic depression (open triangles), and local
faults. The Manila fault (see Figure 2) is not shown.

the northwest flank is reported to have reversed magnetic
polarity [*Fernandez and Estupigan,* 1970], suggesting an
age of greater than 0.7 m.y. The volcano is classed as
fumarolic [*Neumann van Padang,* 1953] on the grounds of
hot springs on the northern flanks. The rocks are highly
potassic, porphyritic, glassy andesites and include, possibly, latites [*Philippine Atomic Energy Commission,* 1977].
Some crystal tuffs (ignimbrites) are also reported. A geo-

thermal project is obtaining 220 MW from a hydrothermal
reservoir on the south side of the volcano.

The Laguna Monogenetic Scoria Cone and Maar Field

Between Mount Banahao and Mount Makiling is a field
of monogenetic volcanoes consisting of 42 scoria cones and
36 maars (Figure 4; see Figure 2 for location). The rocks
exposed in the area are pyroclastic deposits including scoria

TABLE 2. Analyses and CIPW Norms of Volcanic Rocks From Taal and Laguna de Bay Volcanic Centers, South Luzon

	Taal				COMVOL A	COMVOL B	S_D	Laguna de Bay			
	TA1	TA2	TA21	TA20A				TA34	TA26	TA32	TA33
SiO_2, %	64.29	59.80	58.14	57.52	64.37	54.06	2.9	54.93	53.11	56.59	59.43
TiO_2, %	0.73	0.85	1.15	1.06	0.89	0.61	0.3	0.91	1.23	1.20	1.04
AlO_3, %	15.53	15.35	17.27	15.17	16.66	17.89	1.9	15.67	15.44	16.98	17.44
FC_2O_3, %	7.13	8.19	7.84	10.92	7.74	10.70	2.0	10.14	10.33	9.67	8.42
MgO, %	1.15	3.59	3.26	3.36	1.03	2.96	2.0	5.63	5.43	2.62	1.97
MnO, %	0.19	0.20	0.10	0.18	0.30	0.45	0.1	0.19	0.17	0.23	0.16
CaO, %	3.28	5.34	4.28	6.86	4.24	8.49	2.2	7.84	6.64	4.67	3.62
Na_2O, %	3.98	4.86	4.87	2.74	3.72	3.18	0.9	3.85	4.06	5.09	3.25
K_2O, %	3.02	2.14	3.13	1.83	1.52	1.22	0.4	1.50	2.67	2.52	2.99
P_2O_5, %	0.21	0.27	0.31	0.31	0.32	0.44	0.1	0.46	0.55	0.37	0.40
Total	99.51	100.59	100.35	99.95	100.79	100.00		100.12	99.63	99.94	98.72
Q	18.06	5.84	2.02	12.43				1.29	$\cdots$	1.10	15.72
or	17.85	12.65	18.50	10.81				8.86	15.78	14.89	17.67
ab	33.68	41.12	41.21	23.18				82.58	34.35	43.07	27.50
an	14.91	13.75	16.02	23.69				21.05	16.02	16.04	15.37
ne	$\cdots$	$\cdots$	$\cdots$	$\cdots$				$\cdots$	$\cdots$	$\cdots$	$\cdots$
ol	$\cdots$	$\cdots$	$\cdots$	$\cdots$				$\cdots$	8.93		$\cdots$
hy	10.26	12.89	14.17	16.10				18.62	6.17	14.16	13.23
di	$\cdots$	9.08	2.65	6.97				12.19	10.96	3.97	$\cdots$
ap	0.49	0.63	0.72	0.72				1.07	1.28	0.86	0.93
ie	1.39	1.61	2.18	2.01				1.73	2.34	2.28	1.98
c	0.23	$\cdots$	$\cdots$	$\cdots$				$\cdots$	$\cdots$	$\cdots$	3.22
mt	2.07	2.37	2.27	3.17				2.94	3.00	2.80	2.44
K/Na, wt %	0.75	0.40	0.64	0.66	0.41	0.38		0.39	0.66	0.49	0.92

TA1: dacitic pumice from Taal caldera wall (see Figure 7).
TA2: andesitic scoria from Taal caldera wall (see Figure 7).
TA21: black glass from primary pumice clast in ignimbrite, Malabon Bridge, Carmona—alkaline andesite.
TA20A: black glass from primary pumice clast in ignimbrite below TA21, Ilang Bridge—andesite.
COMVOL A: dacite xenolith from lava on Volcano Island.
COMVOL B: average of 17 analyses on Volcano Island basaltic andesites from historic eruptions. S_D is one standard deviation on B.
TA34: black juvenile pumice clast from ignimbrite quarry at Pililla—andesite.
TA26: black (juvenile?) clast from ignimbrite at Binagonan Quarry, Jala Jala Peninsula—basaltic andesite (alkaline).
TA32: black aphyric minor intrusion (dome) on Laguna ring fault near Pililla (postcaldera)—alkaline andesite.
TA33: similar to TA32, 3 km west of 32 on Pililla-Mabitac highway—andesite.

fall, base surge and fine air fall tuffs with abundant accretionary lapilli and lava flows. Compositions are similar to the Laguna de Bay–Banahao–Makiling centers.

There appear to be three generations of maars: the oldest are sediment-filled, the next youngest are nearly filled and swampy, and the youngest have deep lakes (Figure 4). The youngest maar is Sampaloc Lake (1.2-km diameter) in San Pablo City. There is a local myth about its origin that suggests the eruption was nearly historic, perhaps some 500–700 years ago. There are $^{40}K/^{40}Ar$ dates of 1.05 and 1.14 m.y. on older scoria cones within this field [Wolfe, 1981].

About 20 maars are aligned along a probable dike, here called the Mapula dike (Figure 4). This field is undoubtedly still 'live' and future monogenetic eruptions may depend on local conditions of groundwater access to magma-apophyses, resulting in phreatomagmatic eruptions and producing maars.

Taal

Taal is a volcano-tectonic depression some 26×25 km (650 km²), with scalloped margins typical of caldera collapse (Figure 5). It contains a lake 267 km² in area and only 3 m above sea level. In the middle of the caldera a young composite volcano, Volcano Island, is growing by the coalescence of tuff rings, scoria cones, and lava flows.

The most recent eruptive activity on Volcano Island commenced with a fairly strong phreatomagmatic eruption in September 1965, during which about 190 people were killed. This was described by Moore et al. [1966], where they pointed out that the phenomenon of 'base surge' occurred during the eruption and had in fact accounted for many of the deaths. Phreatomagmatic eruptions continued in 1967, but in 1968 and 1969, when lake water was excluded from the vents, they became strombolian with flows of aa lava which extended to the shore of Volcano Island. These are the only historic lava flows in southwestern Luzon. Activity was then mild until 1976, when phreatomagmatic eruptions occurred along the south side of the explosion crater of 1965 at four centers, and two of these centers exploded again in 1977 [Wolfe, 1980]. Although tectonic earthquakes frightened local residents in September 1981, it is probable that the eruptive phase of 1965–1977 is over. It was the longest period of activity in the documented his-

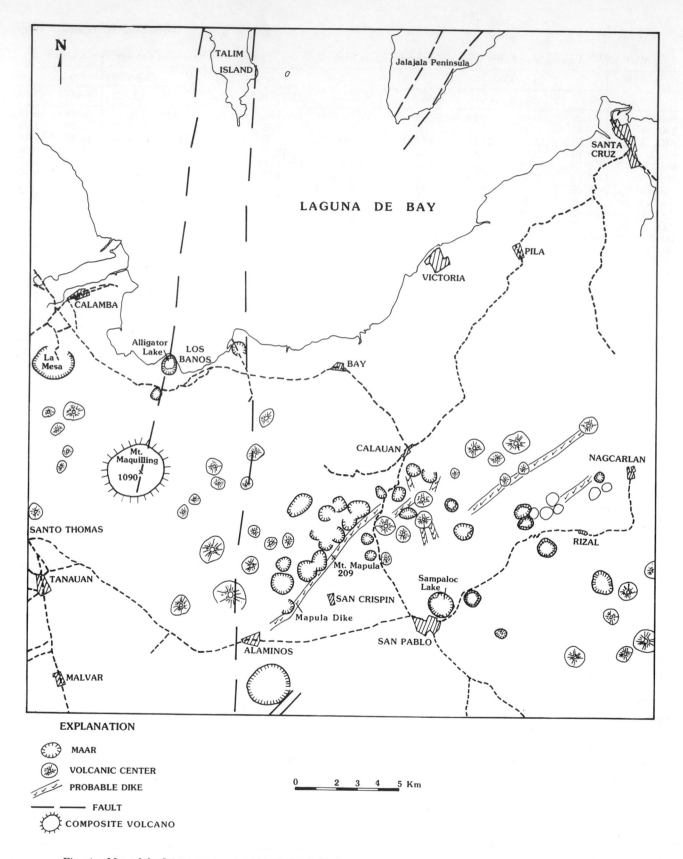

EXPLANATION

- MAAR
- VOLCANIC CENTER
- PROBABLE DIKE
- FAULT
- COMPOSITE VOLCANO

0   2   3   4   5 Km

Fig. 4.   Map of the Laguna monogenetic volcano field showing scoria and tuff cones and maars. Local lineaments
of vents suggest subsurface dikes, including the Mapula 'dike.' Roads are shown as dashed lines.

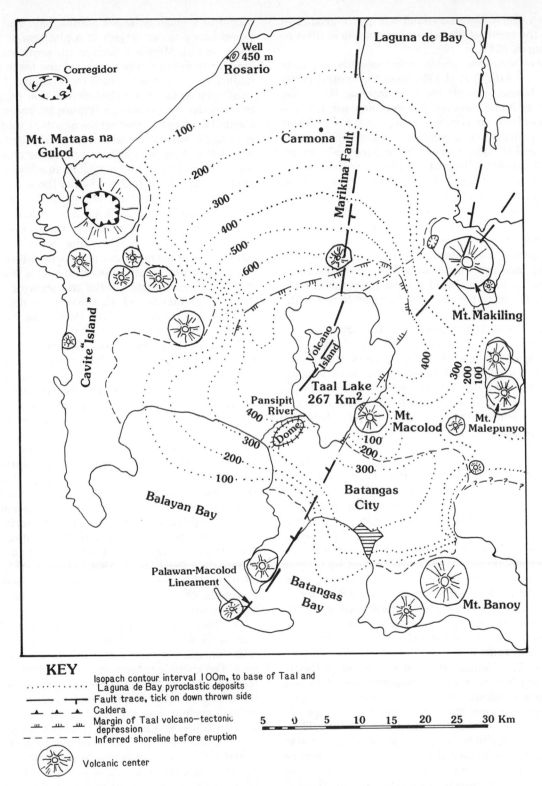

Fig. 5. The Taal area showing isopachs of estimated thickness to the base of the Taal pyroclastic apron (mainly consisting of ignimbrite). Thickness based on local bore holes and construction of geological sections. In the north, Taal ignimbrites must lap onto older Laguna de Bay ignimbrites and related debris flow deposits. Note the fanlike form of the Taal ignimbrite sheet and the former local coastline (dashed lines).

tory of the volcano. Volcano Island has grown about 25% in area as the result of historic eruptions, the earliest on record being in 1572.

There have been some particularly violent eruptions in historic times, with that of 1754 from the central crater of Volcano Island probably the most intense. Heavy ash fell in Manila and a reported 1–3 m of new ash fell tens of kilometers from the volcano during the 8-day eruption. It was at this time that a dome, 2 × 6 km, rose between Taal Lake and the sea (presumably pushed up by an intrusion). The doming diverted the Pansipit River, which serves as the outlet of Taal Lake (Figure 5). The abandoned river channel rose about 70 m.

A pair of conjugate faults pass through Volcano Island [Wolfe, 1980, Figure 1], and all eruptions since 1702 have been located on these faults or their intersection. The northeasterly striking fault appears to extend to the isthmus between the lake and Verde Island Passage. Apparently the eruption of 1754 included a large phreatomagmatic explosion adjacent to the shore, quite close to where the dome arose.

The south rim of Taal caldera is a spectacular drop of 700 m to the lake. This, the northwest face of Mount Macolod (Figure 5), is the scarp of the Palawan-Macolod Lineament upon which the southeast side of the caldera collapsed. At this point the strike of the fault is about N30°E. The trend of the north rim is about N60°E and is also fault controlled. It is scalloped by landslide scars and slide blocks, and lobes are piled to about 500-m elevation at several points along the scarp.

The major prehistoric eruptions of Taal greatly changed the topography of southwestern Luzon. Large volumes of andesitic ignimbrite were erupted and form an extensive plain sloping gently at about 1° toward the north and southwest coasts and abutting against older volcanic centers to the west (Figure 5). To the north and southeast, lobelike fans of ignimbrite reach up to 35 km from the volcano. Sections through these deposits show up to three ignimbrite sheets, some welded, with paleosols between them. Capping the ignimbrites are the widespread andesitic pyroclastic deposits from prehistoric eruptions of Taal. They are found up to tens of kilometers from Taal and include base surge deposits of extremely wide dispersal and fine air fall tuffs.

Before the Taal ignimbrites erupted, western Cavite (Figures 5 and 6) was probably a separate island, as was the Calumpan peninsula on the western side of Batangas Bay. The South Luzon Mountains had been a separate island in the Miocene and Pliocene [Gervasio, 1966; Wolfe et al., 1978], and there is subsurface evidence that 200 m of pyroclastics were deposited south of the Antipolo mine at Taysan, 35 km from the volcano. The surface elevation is about 150 m today; thus the connection of this island to Luzon may have also been initiated by an ignimbrite-forming eruption from Taal.

Until the Pleistocene, Laguna de Bay was probably connected to a much enlarged Manila Bay. The area in between now consists largely of a platform of Taal ignimbrites. In fact, Manila is built on the northern edge of this ignimbrite delta. To the northeast and north the Taal ignimbrites may lap onto those from Laguna de Bay. Thus small islands in the southwestern Luzon region were interconnected by ignimbrite (Figure 6). The mountains of Cavite Island to the west served as a dam, channelling the pyroclastic flows northward. At Rosario (Figure 5), a water well 450 m deep was still in tuffaceous material at the bottom [Oca, 1968]. It is impossible to say whether all this sequence came from Taal or whether the well went through Taal ignimbrites into those of Laguna de Bay, but it gives an idea of the thickness of ignimbrite deposition. Figure 5 gives approximate contours to the base of Taal pyroclastic deposits, showing the fanlike form. This pile of pyroclastic deposits forms a large ignimbrite delta which flooded into the sea and created the south side of Manila Bay. A simplified reconstruction of the area showing the new land added by ignimbrite and other pyroclastic deposits, as well as deltaic sediments from the Pampanga River basin, is shown in Figure 6.

The north rim of the caldera appears to have been somewhat built up by deposition from the later eruptions of the volcano. In the lower section there are chaotic units containing large boulders with interbedded units of conglomerates and sands representing a sequence of debris flow to torrent-type deposits. The middle section of the cliff is covered by landslide debris and is poorly exposed. The top 200 m is a mixture of coarse and fine pyroclastic deposits, mainly air fall material, with several reddish layers, minor soil horizons, that represent gaps in time between eruptions. A simplified stratigraphic column through the Taal caldera wall is given in Figure 7 to demonstrate the types of deposits exposed. No thick ignimbrite is exposed in the north wall, indicating that deposition from pyroclastic flows began several kilometers outside the caldera boundaries [Sparks et al., 1978]. Only one thin fall unit of dacitic pumice was seen in the caldera wall sections (Table 2, sample TA 1), the rest of the deposits being andesitic in composition.

The Taal ignimbrites are well exposed on the highway west of Carmona, and samples TA 21 and TA 20A (Table 2) are analyses of the black glassy pumice from a welded zone. The juvenile magma of the ignimbrite is alkali-rich basaltic andesite, the dominant magma type of the Taal center. COMVOL B is an average analysis of historically erupted magma from Volcano Island and is a basaltic andesite; COMVOL A is an average of two samples from rare dacite xenoliths from Volcano Island. Both of the Volcano Island samples are noticeably less alkaline and more typically calc-alkaline than the older Taal rocks. The Volcano Island basaltic andesite lavas are mainly fine grained and poorly porphyritic, with small plagioclase phenocrysts. The more alkaline pyroclastics of the main caldera eruptions are almost aphyric.

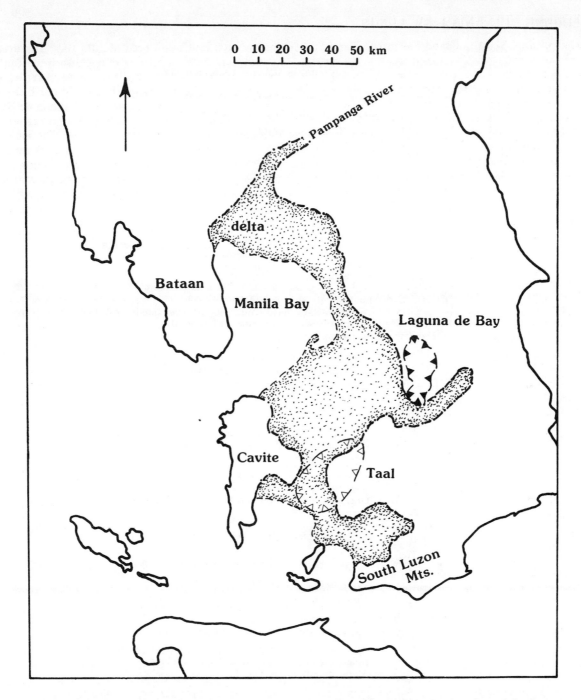

Fig. 6. Paleogeographic reconstruction of the south Luzon region prior to the main eruptions of the Laguna de Bay and Taal volcanoes (about 1 m.y. B.P.). New land (stippled), largely added by deposition of ignimbrite, has accreted the islands together to form the present land mass. The Taal ignimbrites may, in part, postdate the eustatic change and have substantially added to the land area.

The recent caldera-forming eruptions of Taal have not been dated, datable charcoal being apparently absent from the deposits so far examined.

The South Luzon Mountain Range

A small but rather rugged mountain range of volcanic and intrusive rocks forms the southern limit of Luzon.

Volcanism apparently commenced about 10 m.y. ago. There is a date of 9.2 m.y. on hornblende diorite intruding the base of Mount Banoy (986 m) [Wolfe, 1981]. There are several younger but eroded diatremes along the northern edge of this range. The eruptions of Mount Parayang Bilao (798 m) and Mount Lobo (946 m), a largely dacitic volcano, may have continued well into the Pleistocene. The volcanic

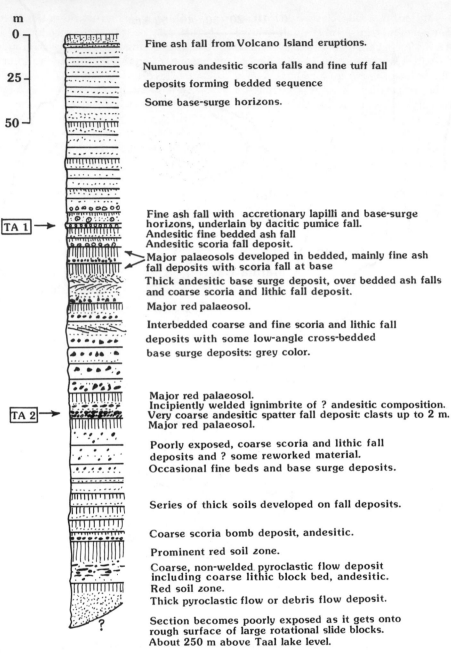

m

0 — Fine ash fall from Volcano Island eruptions.

Numerous andesitic scoria falls and fine tuff fall

25 — deposits forming bedded sequence

Some base-surge horizons.

50 —

TA 1 → Fine ash fall with accretionary lapilli and base-surge horizons, underlain by dacitic pumice fall.
Andesitic fine bedded ash fall
Andesitic scoria fall deposit.
Major palaeosols developed in bedded, mainly fine ash fall deposits with scoria fall at base
Thick andesitic base surge deposit, over bedded ash falls and coarse scoria and lithic fall deposit.
Major red palaeosol.

Interbedded coarse and fine scoria and lithic fall deposits with some low-angle cross-bedded base surge deposits: grey color.

Major red palaeosol.
Incipiently welded ignimbrite of ? andesitic composition.
TA 2 → Very coarse andesitic spatter fall deposit: clasts up to 2 m.
Major red palaeosol.

Poorly exposed, coarse scoria and lithic fall deposits and ? some reworked material.
Occasional fine beds and base surge deposits.

Series of thick soils developed on fall deposits.

Coarse scoria bomb deposit, andesitic.

Prominent red soil zone.

Coarse, non-welded pyroclastic flow deposit including coarse lithic block bed, andesitic.
Red soil zone.
Thick pyroclastic flow or debris flow deposit.

Section becomes poorly exposed as it gets onto rough surface of large rotational slide blocks.
About 250 m above Taal lake level.

TA 1-2 are analysed samples.

Fig. 7. Cross section through the upper part of the northwest wall of Taal caldera. This section reveals the major units and types of pyroclastic deposits exposed. Note that proximal ignimbrites are absent. Coarse air fall pyroclastic deposits and debris flow material are common.

and intrusive rocks are calc-alkaline, similar to younger volcanics of the Verde Island transform zone.

Tectonics

In discussing the tectonic framework of Luzon Island it must be kept in mind that there are many discrete seg-

ments which may respond sometimes as a unit, but part of the time they may interact independently.

At some point, probably near the end of the Miocene, collision of Palawan-Mindoro with Luzon resulted in the rupture of Mindoro from Palawan and this probably choked the southern portion of the Manila Trench. Southward motion of Mindoro resulted in counterclockwise rotation of the island. The Verde Island transform fault was active at that time and Marinduque, caught between two limbs of the fault, rotated counterclockwise about 60° [Hsu, 1972]. About 3.5 m.y. B.P. the Bataan Lineament may have opened as a rifting element, permitting primitive basaltic magma to erupt in Cavite and Bataan. The magma chambers differentiated and later, normal arc andesites and, finally, dacites erupted.

Meanwhile, more alkaline magma chambers developed further east above deeper portions of the Manila subduction zone and the older centers, for example, Mount Malepuno (Figure 5) and Laguna de Bay, erupted. By late Pliocene the only part of the Manila Trench still active was the sector west of Mindoro to northern Zambales. This continued through mid-Pleistocene when it may have ceased to be effective as a subduction zone. Karig [1973] stated that the Manila Trench was becoming inactive. A trench is again developing by arc polarity reversal on the eastern side of the island. Magma has continued to be generated in the Laguna de Bay and Taal chambers, but these volcanoes may now be in a late stage of activity. There is little evidence in the deposits from these volcanoes to suggest that the magma bodies are evolving to more silicic compositions. However, further study is needed to examine such a possibility.

Discussion and Summary

Southwestern Luzon is one of the major volcanic districts of the Philippines. In this region the volcanoes can be related to major structural elements, loosely defined here as fault or rift zones, that can often be traced for more than 100 km. Most volcanoes are situated at the intersection of two such elements. It appears that structures of this type are necessary to provide a conduit for magma to rise into the crustal zone. If one of the features has the characteristics of a rift, it provides ready access from the magma source. Alkaline volcanics present in several of the volcanic centers may reflect a tensional environment and tapping of deep magma sources.

There are 27 volcanic structures and vents on the Bataan Lineament, an arc that can be traced for 320 km. Mount Pinatubo is on the intersection with the Iba graben and Mount Natib is on the intersection with the Manila fault. All these volcanoes, with the exception of a group of explosion centers at the northern end, are calc-alkaline. Several authors [e.g., de Boer et al., 1980; Cardwell et al., 1980; Divis, 1980] relate this group of volcanoes to the east dipping Benioff zone associated with the subduction zone of the Manila Trench.

A second major feature is the Palawan-Macolod Lineament which is traceable for 800 km. On Luzon it is a fault passing through Batangas Bay (a possible caldera), Taal, and Laguna de Bay volcano-tectonic depressions. It apparently ends at the intersection with the Manila Fault north of Laguna de Bay.

A third major structure is the Verde Island transform which can be traced for 380 km or, if the fault through the Bicol peninsula is an extension cut off by the left lateral movement on the Philippine fault, for 660 km to the Philippine Trench. The calc-alkaline volcanic and plutonic activity is found on the western 200 km of this zone, most of it in the 50 km of southwestern Luzon. It is rather difficult to relate this to the subduction from the Manila Trench, and an association with a leaky transform [Divis, 1980] seems more plausible. However, this does not provide an adequate explanation for the source of the magma.

Between the two calc-alkaline trends is a group of volcanoes which have produced deposits with a distinctly alkaline affinity. Two of these, Taal and Laguna de Bay, are large volcano-tectonic depressions, encompassing calderas. The magmas which formed ignimbrites from these centers are andesitic. The others, Mount Malepuno, Mount Makiling, and Mount Banahao, are composite volcanoes. De Boer et al. [1980] relate this group to a deep magma source from the Manila Trench.

Alkali-andesitic rocks and closely related calc-alkaline volcanics, similar to those of southwestern Luzon, have been reported by Keller [1974] from the Aeolian arc, in the southern Tyrrhenian Sea. Barberi et al. [1978] further suggested that the alkalic association implies that the Aeolian arc is in a senile stage of evolution. The alkalic rocks typically form later in time, although both types of activity are still concurrent. Ninkovich and Hays [1972] also discussed the origin of the high-K magmas of the Calabrian arc, which have produced low-Si ignimbrites such as the Campanian tuff ($SiO_2 = 60 \pm 1$ wt %) [Barberi et al., 1978]. Sparks [1975] also presents analyses of Central Italian ignimbrites which are similar to those of Taal and Laguna de Bay, but rather more potassic.

Great depths to the Benioff zone under volcanoes producing high-K magmas, such as 200–300 km in the Italian region, are also plausible for southwestern Luzon, within the present ideas of the configuration of the subducted Manila Trench slab [Cardwell et al., 1980]. Further, Karig's [1973] contention that subduction at the Manila Trench is ceasing seems to indicate the same conditions for the production of alkalic magmas that are implied for the Aeolian arc.

Acknowledgments. We thank A. F. Divis and an unnamed reviewer for constructive criticism of this manuscript. Partial funding for reconnaissance field work in the Philippines by S. S. was made possible from NASA grant NSG5145.

References

Alcaraz, A. P., The major structural lines of the Philippines, *Philipp. Geol., 1,* 13–17, 1947.

Alcaraz, A. P., Crustal unrest in the Philippines. *Philipp. Geol., 22,* 163–170, 1968.

Andal, G. A., and B. Yambao, The Philippine volcanoes

and solfataric areas, *Philipp. Comm. Volcanol.*, 35, 1953.

Barberi, F., F. Innocenti, L. Lirer, R. Munno, T. Pescatore, and R. Santacroce, The Campanian ignimbrite: A major prehistoric eruption in the Neapolitan area, *Bull. Volcanol.*, *42*, 1–22, 1978.

Cardwell, R. K., B. L. Isacks, and D. E. Karig, The spatial distribution of earthquakes, focal mechanism solutions and subducted lithosphere in the Philippines and northeastern Indonesian islands, in *The Tectonic and Geologic Evolution of Southeast Asian Seas and Islands, Geophys. Monogr. Ser.*, vol. 23, edited by D. E. Hayes, pp. 1–35, AGU, Washington, D. C., 1980.

Cole, J. W., and I. A. Nairn, *Catalogue of the Active Volcanoes of the World*, part 22, New Zealand, p. 156, International Association of Volcanism and Chemistry of the Earth's Interior, Rome, 1975.

De Boer, J., L. A. Odom, P. C. Ragland, F. G. Snider, and N. R. Tilford, The Bataan orogene: Eastward subduction, tectonic rotation, and volcanism in the Western Pacific (Philippines), *Tectonophysics, 67,* 251–282, 1980.

Divis, A. F., The petrology and tectonics of recent volcanism in the central Philippine Islands, in *The Tectonic and Geologic Evolution of Southeast Asian Seas and Islands, Geophys. Monogr. Ser.*, vol. 34, edited by D. E. Hayes, pp. 127–144, AGU, Washington, D. C., 1980.

Fernandez, J. C., and P. Estupigan, Notes on magnetometer survey in the Cavite-Batangas-Laguna area, *J. Geol. Soc. Philipp., 24*(3), 30–35, 1970.

Gervasio, F. G., A study of the tectonics of the Philippine Archipelago, *Philipp. Geol., 20,* 51–75, 1966.

Gervasio, F. G., The geology, structures and landscape of Manila suburbs, *Philipp. Geol., 22,* 128–192, 1968.

Gervasio, F. G., Geotectonic development of the Philippines, *J. Geol. Soc. Philipp.*, 25(1), 18–38, 1971.

Gonzales, B. A., V. P. Ocampo, and E. A. Espiritu, Geology of the eastern Nueva Ecija and eastern Bulacan provinces, Luzon Central Valley, *J. Geol. Soc. Philipp., 25,* 1–14, 1971.

Hamilton, W., Tectonics of the Indonesian region, *U.S. Geol. Surv. Prof. Pap., 1078,* 345 pp., 1979.

Hsu, I., Magnetic properties of igneous rocks in the northern Philippines, Ph.D. thesis, 164 pp., Washington Univ., St. Louis, Mo., 1972.

Karig, D. E., Plate convergence between the Philippines and Ryukyu Islands, *Mar. Geol., 14,* 153–168, 1973.

Karig, D. E., Basin genesis in the Philippine Sea, *Initial Rep. Deep Sea Drill. Proj.*, 31 pp., 1975.

Keller, J., Petrology of some volcanic rock series of the aeolian arc, southern Tyrrhenian Sea: Calc-alkaline and shoshonitic associations, *Contrib. Mineral. Petrol., 46,* 29–47, 1974.

Lo, H. H., Geochemical investigation of some volcanic rocks from the Philippines, *Chem. Geol., 34,* 243–257, 1981.

Louden, A. G., Marcopper porphyry copper deposit, Philippines, *Econ. Geol., 71,* 721–732, 1977.

Moore, J. G., and W. G. Melson, Nuées ardentes of the 1968 eruption of Mayon volcano, Philippines, *Bull. Volcanol., 33,* 600–620, 1969.

Moore, J. G., G. Nakamura, and A. Alcaraz, The 1965 eruption of Taal volcano, *Science, 151,* 955–960, 1966.

Murphy, R. W., The Manila Trench–West Taiwan foldbelt, a flipped subduction zone, *Bull. Geol. Soc. Malays., 6,* 27–42, 1973.

Neumann van Padang, M., *Catalogue of Active Volcanoes of the World,* part II, Philippine Island and Cochin Sea, 49 pp., International Association of Volcanism and Chemistry of the Earth's Interior, Naples, 1953.

Newhall, C. G., Temporal variations in the lavas of Mayon volcano, Philippines, *J. Volcanol. Geotherm. Res., 6,* 61–83, 1979.

Ninkovich, D., and J. D. Hays, Mediterranean island arcs and origin of high potash volcanoes, *Earth Planet. Sci. Lett., 16,* 331–345, 1972.

Oca, G. R., The geology of greater Manila and its bearing to the catastrophic earthquake of August 2, 1968, *Philipp. Geol., 22,* 171–177, 1968.

Philippine Atomic Energy Commission, Basic geologic and seismic information, in *Preliminary Safety Analysis Report to Philippine National Power Company,* open file report, Manila, 1977.

Philippine Bureau of Mines, Geologic map of the Philippines, scale 1:1,000,000, Manila, 1963.

Sonido, E. P., A review of the gravity data on the Central Valley basin in the island of Luzon, Republic of the Philippines, in *Fifth Symposium on Mineral Resources Development,* p. 12, 1978.

Sparks, R. S. J., Stratigraphy and geology of the ignimbrites of Vulsini volcano, central Italy, *Geol. Rundsch., 64,* 497–523, 1975.

Sparks, R. S. J., L. Wilson, and G. Hulme, Theoretical modelling of the generation, movement, and emplacement of pyroclastic flows by column collapse, *J. Geophys. Res., 83,* 1729–1739, 1978.

Walker, G. P. L., Volcanic hazards and the prediction of volcanic eruptions, *Geol. Soc. Misc. Pap. London, 3,* 23–41, 1974.

Williams, H., Calderas and their origin, *Univ. Calif. Publ. Dep. Geol. Sci.,* 239–346, 1941.

Wolfe, J. A., Eruptions of Taal Volcano, 1976–1977, *Eos Trans. AGU,* 56–58, 1980.

Wolfe, J. A., Philippine geochronology, *J. Geol. Soc. Philipp.,* 35(1), 1–30, 1981.

Wolfe, J. A. M. S. Manuzon, and A. F. Divis, Taysan porphyry copper deposit, southern Luzon Island, Philippines, *Econ. Geol., 73,* 608–617, 1978.

The Geology and Geochemistry of Philippine Porphyry Copper Deposits

ALLAN F. DIVIS

Environmental Science Associates, Golden, Colorado 80403

Porphyry copper deposits and their relationship to felsic intrusions are of considerable economic as well as academic interest. They generally occur near continental margins and in island arcs. Although some conjecture exists as to the ultimate source of transition metals associated with the deposits, a growing body of evidence suggests the metals are principally derived from a magmatic source: the porphyry intrusions. These intrusions may also give rise to adjacent vein and precious metal deposits. Over 40 potential deposits have been reported in the Philippine island arc and approximately nine are or have been in production. The known and inferred reserves from these deposits exceed three billion metric tons of ore—approximately 10 to 20 million tons of metallic copper. Several deposits may ultimately have more than a billion tons of ore reserves. There appears to be a remarkable correlation between the timing of porphyry intrusions in the Philippines and that in other areas, particularly New Guinea and the Solomon Islands. These intrusions also show a close relationship to the timing of oceanic plate tectonic processes. Periods of increased subduction rate and accompanying dilation and/or shearing within the arc may be conducive to the generation of porphyry magmas on an episodic rather than continuous basis. Ultimately, the development of an economically significant ore body requires the presence of a hydrous magma and may be associated with other primary compositional characteristics. However, anomalous high primary magmatic concentrations of copper do not appear to be required for the formation of the porphyry deposits. The Philippine porphyry intrusions differ somewhat from 'continental' porphyries and are chemically similar to mineralized island arc intrusions of the Carribean and the southwest Pacific. The Philippine intrusions generally range from diorites and quartz diorites to low potassium granodiorites. Initial 87-Sr/86-Sr ratios are low, ranging from 0.7032 to 0.7040. Molybdenum concentrations are lower, and gold concentrations are generally higher than those reported for continental deposits. In the Baguio district, a series of at least three distinctly different magma pulses have occurred with an apparent hiatus separating the intrusions. The last of these is associated with the formation of porphyry and vein ore deposits. Geological evidence from other districts suggests similar complex intrusive histories, but geochronologic evidence is insufficient to confirm this suggestion. The pattern of alteration mineralogy surrounding the porphyry deposits is generally less well developed than in continental deposits, but much of this variation can be explained by examination of the structural and chemical characteristics of the country rocks and intrusions.

Introduction

The economic importance of porphyry copper deposits has made them the subject of extensive investigation relating to the latter stages of their evolution and ore formation. The occurrence of these deposits in island arcs and continental margins infers an association with plate convergence induced magmatism.

Speculation as to the primary tectonic-petrologic processes involved in the formation of porphyry magmas has been extensive [*Mitchell and Garson,* 1972; *Sillitoe,* 1972, 1973; *Oyarzun and Frutos,* 1974; *Keith,* 1979]. Several distinctly different hypothesis have been presented to account for the evolution of porphyry deposits. Formation of the deposits has been attributed to anomalous concentrations of metals and/or volatiles in oceanic sediments, volatile stripping of metals from the lithospheric slab overlying the subduction zone, 'normal' petrogenic processes associated with hydrous magma, hydrothermal concentration of metals from the wall rock or pluton, or combinations of these mechanisms.

In attempting to formulate models of porphyry evolution, it is desirable to examine the chemical, petrologic, and geochemical characteristics of unmineralized porphyry related intrusive rocks, since this approach may permit a separation of 'primary' magma characteristics from superimposed hydrothermal phenomena. This approach has been employed by *Kesler et al.* [1975] in the Carribean area and *Mason and McDonald* [1978] in the New Guinea-Solomon Island region. This study is intended to evaluate the various hypotheses with reference to the origin of Philippine porphyry deposits and to place constraints on the probable magmatic processes involved with their formation.

Approximately forty potential porphyry copper deposits have been discovered in six major groups in the Philippines (Figure 1). The deposits can be divided into the groups based on structural and limited geochronologic evidence (Figure 2):

1. The Baguio District-Cordillera area—located in the Cordillera of Central Luzon.

2. The Batangas-Marinduque area—located in the Batangas peninsula of Luzon and Marinduque Island.

3. The Negros-Occidental Batholith—located in the southern Visayan Islands.

4. The Atlas-Toledo deposit—on Cebu in the southern Visayan Islands.

5. The Eastern Mindanao area—the Surigao and Davao regions in Mindanao.

6. The Polillo-Paracale area—along the coast of the Bicol Peninsula, Luzon.

These various areas show a strong geologic similarity, and limited geochronologic data suggests that deposits within these clusters are closely associated in time, with ages differing between clusters. There has been a tendency to refer to a Philippine 'porphyry belt,' implying a common association in time that is not justified by the latter evidence. Existing data does suggest porphyry magmatism occurred in the Baguio district ca. 3–7 m.y. B.P. and possibly ca. 15–17 m.y. B.P.; in the Batangas-Marinduque area ca. 15–20 m.y. B.P., and in the Negros area ca. 30 m.y. B.P. and at Cebu-Atlas 95–110 m.y. [*Wolfe,* 1972; *PBM-JICA,* 1976; *Agawin et al.,* 1980; *Divis and Balce,* 1977; *Wolfe et al.,* 1978; *de Boer et al.,* 1980]. This chronology shows a notable correspondence to the history of plate convergence in the southwest Pacific area as docu-

mented by *Shih* [1980], *Lewis and Hayes* [1980], and *Taylor and Hayes* [1980]. An analogous correlation has been noted by *Titley* [1975] for porphyries in the New Guinea-Solomon Islands area (Figure 3).

The economic geology of selected Philippine porphyry deposits has been previously discussed by *Bryner* [1969], *Madamba* [1972], *Apostol* [1974], *Almogela* [1974], *Loudon* [1976], *Divis and Clarke* [1979], *Wolfe et al.* [1978], and *Balce* [1979]. General regional summaries have also been presented by *Kinkle* [1956], *Portacio* [1975], and *Saegart and Lewis* [1977]. This paper presents previously unpublished data on the geological characteristics and petrology of the Philippine porphyries and associated intrusions while summarizing critical features of the ore deposits. The discussion considers the evolution of porphyry copper deposits in the context of their relationship to closely associated (in time and space) barren intrusions. A prior paper [*Divis,* 1980] presented petrographic and geochemical studies of recent volcanism in the Philippines, particularly with reference to the distribution of ore forming trace elements.

It is suggested that there are strong similarities between the evolution of the porphyry magmas and hydrous volcanic series. There appears to be some geographic correlation between subduction zone depth and metal abundance and a fairly well developed association of porphyry mineralization with periods of higher rate of plate convergence. Although an association exists between porphyry mineralization and plate convergence rate, the correlation should not be overdrawn. The data presented herein supports the contention of *Titley and Beane* [1981]: '. . . the character of a porphyry deposit may be more strongly affected by its immediate environment than by its parentage.' The derivation of a porphyry magma probably involves a minimum of four stages:

1. Derivation of the protomagma by partial melting of a source terrain under the influence of a variable pressure and temperature regime.

2. Magma evolution during ascent by fractional crystallization and assimilation of country rock.

3. Fractional crystallization and re-equilibration in a shallow magma chamber, with the possibility of magma mixing and varying volatile activity.

4. Intrusion and hydrothermal interaction with country rock. Economic mineralization dependent upon structure and pressure-temperature environment of intrusion.

In attempts to characterize the formation of porphyry deposits, constraints are imposed by the pervasive overprint of hydrothermal alteration. This effect is somewhat diminished in comagmatic barren intrusive series. When such a series is sampled, it can yield significant data relative to the early evolution of porphry magmas.

Three principal lines of evidence are presented in an effort to address the question of porphyry mineralization. The descriptive geology and mineralogy of the deposits illustrates a common thread relating the intrusion of a hydrous porphyry magma into variable structural and

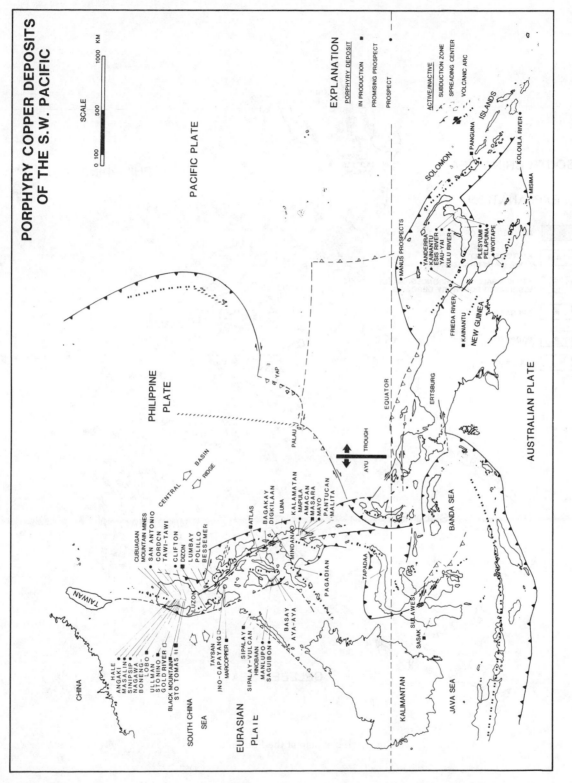

Fig. 1. Porphyry copper deposits of the southwest Pacific (modified after *Almogela* [1974], *Hamilton* [1974], *Hilde et al.* [1977], *Lowder and Dow* [1977], and *Titley* [1975]).

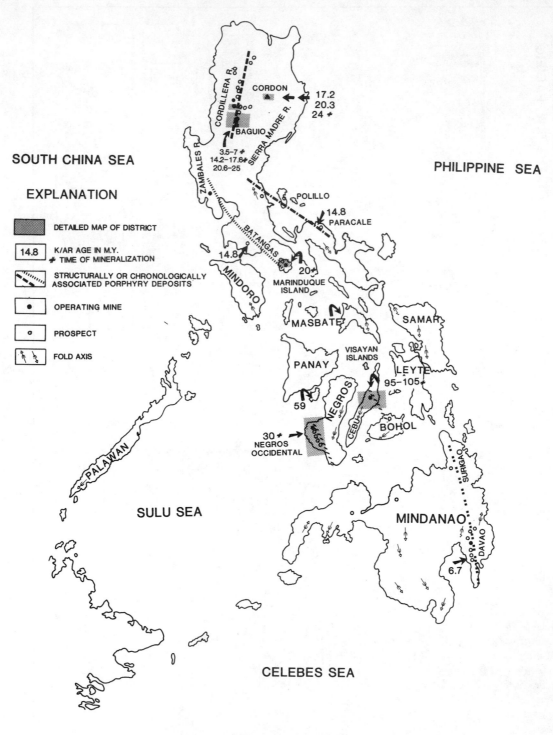

Fig. 2.   Index map of the Philippines.

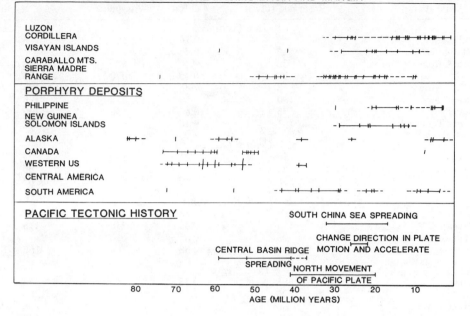

Fig. 3. Chronology of mineralization and tectonics in the Pacific area (based on data from *Titley* [1978], *Hollister* [1978], *Titley and Hicks* [1966], *Taylor and Hayes* [1980], *Shih* [1980], *Jackson et al.* [1972]).

lithologic environment to a set of conditions conducive to the concentration of iron, copper, gold, and other elements within the limits of hydrothermal circulation 'cell.' Petrologic and geochemical analysis of intrusive rocks associated with the porphyry deposits may give insight into crystalization and assimilation processes related to intermediate stages of magma evolution. Unfortunately, the early stages of magma evolution are obscured by a succession of potentially complex fractionation processes. Examination of recent volcanic series from the Philippine Arc has been used to draw tenuous analogies between the structure and history of lithologic plate convergence and the eruption of magmas very similar to those temporally and spacially associated with porphyry mineralization.

General Characteristics of Porphyry Deposits

Porphyry deposits are generally large (>100 million tons), low-grade ore deposits genetically related to felsic igneous intrusives. Deposits that have copper, molybdenum, gold, and tin as their primary mineral value have been reported. Copper and molybdenum deposits represent the most common and extensively mined porphyry occurrences. It is not unusual to find several of these elements concentrated in a single deposit. J. A. Wolfe (manuscript in preparation, 1982) emphasizes that the importance of disseminated deposits is closely related to the present dollar value of contained metals. Thus, base metal 'porphyries' may contain significant resources, but they do not represent 'ore' until exploitation becomes economically feasible. Gold fre-

quently represents substantial values in island arc deposits and may make a marginal deposit profitable. An extensive literature describing porphyry deposits exists and excellent summaries are provided by *Titley and Hicks* [1966], *Sutherland-Brown* [1976], *Hollister* [1978], *Titley and Beane* [1981], and *Beane and Titley* [1981].

Porphyry deposits occur both on and near the margins of continents and within island arcs. Several distinctive differences between continental and island arc deposits have been noted by *Kesler* [1973], *Kesler et al.* [1975], and *Titley* [1975, 1978], as outlined:

1. Continental porphyries tend to be dominantly granodiorite to quartz monzonite, while island arc intrusions are generally diorite to quartz diorite.

2. Continental intrusions tend to have higher initial strontium isotope ratios than island arc intrusions.

3. The relative abundances of copper, gold, and molybdenum differ between the two environments. Generally, gold is higher and molybdenum is lower with respect to copper in the island arc deposits.

4. Hydrothermal alteration and sulfide deposition patterns are more regular and consistently interrelated in continental in comparison with island arc deposits.

These first three contrasts suggest that the island arc porphyries may be more reflective of a 'parent' magma than are the continental intrusions that may have interacted with a thick sialic crust during ascent. *Titley* [1978] has pointed out that the copper-gold-molybdenum ratios of island arcs closely resemble those of oceanic basalts,

TABLE 1. Ore Quality and Reserves

Deposit	Reserves, 10^6 mt	Cu, %	Au, ppb	Cu/Au, $\times 100$
Baguio-Cordillera				
Cubuagan	50	0.70	400	175
Mt. Mines	12	0.45	· · ·	· · ·
San Antonio	32	0.40	390	103
W. Minolco	92	0.40	370	108
Santo Nino	53	0.38	310	123
Tawi Tawi	190	0.39	510	76
Black Mt.	26	0.45	150	300
Santo Tomas II	161	0.45	880	51
Central Luzon				
Lumbay	70	0.41	200	205
San Fabian	24	0.44	510	86
Marian	50	0.50	600	83
Batangas-Marinduque				
Dizon	85	0.43	750	57
Taysan	110	0.43	240	179
Ino/Iso-Pili	98	0.55	· · ·	· · ·
Marcopper	93	0.58	350	166
Cebu-Negros				
Atlas	898	0.46	250	184
Sipalay	626	0.50	40	1250
Hinobaan	89	0.50	240	208
Basay	200	0.41	<300	137
Aya-Aya	20	<0.50	· · ·	· · ·
Surigao-Davao				
Luna	50	0.30	· · ·	· · ·
Mapula	36	0.40	310	129
Amacan	129	0.42	500	84

(after *Saegart and Lewis* [1977])

while those of continental deposits do not. The characteristics of Philippine porphyries confirm these observations and also suggest a close tie between plate convergence and the generation of the copper deposits.

One of the most obvious generalizations that can be drawn from published literature and the work detailed herein is that the history of island arc porphyry copper deposits frequently involves a protracted (in time) series of intrusions and volcanism [*Grant and Nielsen*, 1975; *Titley*, 1975; *Loudon*, 1976; *Lowder and Dow*, 1977; *Titley et al.*, 1978; *Divis and Balce*, 1978; *Wolfe, et al.*, 1978]. This is most strikingly documented in the Baguio district [*Wolfe*, 1972; *Divis*, 1977]. At least four distinct magmatic episodes are present. Other districts in the Philippines may be less complicated and are not as well known. However, it is apparent that they also show a protracted history of episodic magmatism. *Hamilton*'s [1979] comments regarding the complexity of plate interactions in southeast Asia is very appropriate to this problem. One question that must be considered is whether this area represents an anomaly in geologic history or is typical of mature island arcs as they evolve into continental areas.

The proximity of events in the Philippines may allow us to perceive a much 'finer' chronologic structure (of the last 30 m.y.) than is possible in other areas where a greater time interval has obscured the geologic evolution. It ap-

pears that, in the Philippines, during a period of approximately 100 m.y. at least five and possibly six separate groups of porphyry intrusions have been assembled by a combination of tectonic processes (Figure 2, Table 1). Superficial examination could suggest that a single event formed the 'Philippine copper belt,' yet this conclusion is in error as demonstrated by the complex, yet internally consistent geologic history of the separate areas. Geochronologic data from the Baguio district suggests that the porphyry intrusions represent the last major phase of magmatism in the area, with the exception of recent volcanism.

During the initial stages of evolution the porphyry magma evolves by processes that are typical of a silicate melt. There appear to be some primary chemical characteristics associated with ore forming magmas, but these differences are not so profound that very obvious changes in the petrologic evolution occur.

The latter phases of evolution involve large scale hydrothermal processes that are responsible for the concentration of ore metals by a factor of 50 to 100 over average crustal abundances. Although it would seem that the magmas forming porphyry copper deposits should have comparatively higher copper concentrations [*Feiss*, 1978; *Mason and Feiss*, 1979] this may not be necessary [*Gustafson*, 1978; *Burnham*, 1979]. The activity of the hydrothermal

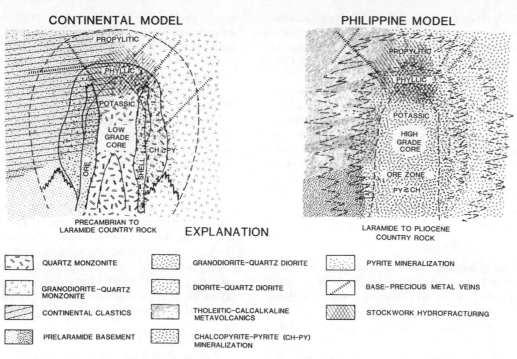

Fig. 4. Alteration-mineralization model; based in part on *Lowell and Guilbert* [1970].

system may be sufficient to form the deposit.

The hydrothermal system generally produces a relatively consistent pattern of silicate and sulfide mineral distribution in the area of the intrusion. *Lowell and Guilbert* [1970] and *Guilbert and Lowell* [1974] have presented a generalized portrayal of this pattern (Figure 3). Geologic structure, host rock lithology, and the chronology of multiple intrusions can significantly complicate the structure.

The model involves a stock-like intrusion with a series of concentric zones of alteration and sulfide mineralization. This hydrothermal silicate mineralogy consists of an inner potassic zone consisting of quartz, potassium feldspar, and biotite with accessory sericite and anhydrite. A phyllic zone surrounds the inner potassic zone. The phyllic zone mineralogy generally consists of secondary quartz and sericite. A pyritic shell generally corresponds closely to the phyllic zone of alteration. The highest grade pyrite-chalcopyrite shell generally occurs in the zone between the potassic and phyllic alteration. The outer most propylitic zone consists of chlorite, epidote, adularia, and albite as alteration minerals. This peripheral zone may show low-grade mineralization of chalcopyrite, galena, and sphalerite veinlets. Gold and silver vein mineralization may also be found in this outer zone.

Gustafson and Hunt [1975], *Sutherland-Brown* [1976], *Gustafson* [1978], and *Beane and Titley* [1981] have elaborated on the complexity and evolutionary character of the alteration-mineralization pattern found in the deposits of North and South America and the southwest Pacific. Although the departures from the *Lowell and Guilbert* [1970] 'generic' model may be significant, it is useful to employ

this model as a basis of reference. As *Gustafson and Hunt* [1975] suggested, these departures are 'variations on a common theme.'

The contrast between continental and island arc porphyry deposits has been previously discussed by *Titley* [1975], *Saegart and Lewis* [1977], and *Gustafson* [1978]. Figure 4 illustrates the contrast between these 'generic' models. In general, the source pluton of an island arc deposit tends to be a diorite or quartz diorite rather than the granodiorite to granite typical of continental deposits. The ore body in island arc deposits tends to be stock-like, tabular, or irregular in contrast to continental deposits that are generally cylindrical. The zonation of alteration assemblages tends to overlap and be rather indistinct in island arcs in comparison with continental deposits. The potassic zone of island arc deposits tends to be less prominently developed and dominated by biotite and chlorite with albite substituting for potassium feldspar. Near surface oxidation and leaching is generally very shallow and supergene enrichment is minor in the island arc deposits. As was previously mentioned, gold values tend to be higher and molybdenum values lower in comparison with continental deposits. Magnetite tends to be more common in the arc deposits.

Regional Geology

The Philippine island group represents a remarkable sampling of the history of a continental margin through nearly 500 m.y. of geologic time. By contrast to North America and South America, where Phanerozoic magmatism occurs adjacent to margin of Precambrian cratonic crust, the Philippines preserve their integrity as an evolving island arc structure with a relatively discrete 'stable area' of older crust separated from the main Asian land mass by an (apparently) oceanic basin (the South China Sea).

The imposition of widespread amphibolite facies metamorphism on preexisting basement during the Late Mesozoic or early Tertiary has obscured many earlier geologic events in the Philippines. However, the Tertiary magmatic and structural history is relatively well preserved and amenable to examination by geologic and geochemical techniques. A considerable accumulation of valuable geologic data exists in unpublished literature available from open file reports of the Philippine Bureau of Mines and local mining companies.

Pre-Tertiary Geologic Evolution

The existence of a stable tectonic block made up of older metamorphic rocks was recognized by *Corby et al.* [1951]. This area, including Palawan, Western Mindoro and Panay, and the Zamboanga Peninsula of Mindanao, may be related to the Upper Paleozoic rocks found in Western Borneo (Sabah). This older basement terrain consists principally of mafic amphibolite schists intruded by syn- and post-tectonic tonalite with rare quartz monzonites. Sparse paleontologic evidence and unpublished K-Ar data suggest a Paleozoic age for this area. On Mindoro, a 1500-m carbonaceous shale-greywacke-conglomerate sequence of the Mansalay Formation unconformably overlies the basement and contains Jurassic ammonites [*Teves,* 1953].

Compositionally banded amphibolite schists also occur as the oldest basement rock on Luzon, Mindanao, and the Visayas. Ages of analogous metamorphic rocks from Sabah range from 100 to 150 m.y. B.P. and *Walther et al.* [1981] have reported a Rb-Sr age of 108 m.y. B.P. for the porphyry intrusion at Atlas, Cebu. It is clear that an extensive premetamorphic history exists (although it is poorly understood) in the stable block and that the subsequent Late Mesozoic to Paleocene metamorphic and magmatic episode has been superimposed on the stable block and much of the rest of the Philippine basement. It is probable that the Paleocene K-Ar ages reported by *Wolfe* [1972] for Visaya quartz diorites and the schists of Rapu-Rapu Island reflect magmatism during the latter part of this event.

The geology of the Philippine arc suggests a closely spaced sequence of metamorphism (amphibolite), andesitic volcanism, tonalite intrusion, and greenschist metamorphism. In many areas a foliated amphibolitic basement predates the formation of tholeiitic and andesitic pillow lavas, radiolarian cherts, and greywackes that were folded and metamorphosed prior to the early Miocene.

This Late Cretaceous/early Tertiary volcanic-sedimentary series that post-dates the 'regional' amphibolite facies metamorphism may represent the first large scale, recognizable evolution of the Philippine Arc as a discrete, definable structure and is present as a relatively widespread geologic unit. Basal clastic sediments in the series frequently show lithologies reflecting the presence of adjacent amphibolite-diorite source terrain. *De Boer et al.* [1980] and J. A. Wolfe (manuscript in preparation, 1982) has suggested that many of these segments of older basement terrane may have been derived by rifting from the Asian mainland.

Gervasio [1966] assigned the tectonic emplacement of ophiolite and ultramafic masses to the earliest Tertiary as part of a 'geotectonic cycle.' There is evidence from field studies that these 'intrusions' are not readily correlatable as a single event and that their emplacement has occurred as a result of distinctly different tectonic processes (i.e., obduction versus emplacement as a tectonic horse in a fault zone). The first appearance of ultramafic derived detritus is in the Eocene sediments and clastic debris from the Luzon Zambales complex did not appear in the Central Valley sediments until the late Miocene. Detailed discussion of the petrology and tectonic history of the Zambales ophiolite is presented by *Villones* [1980], *Hawkins and Evans* [this volume], and *Schweller et al.* [this volume].

Early Tertiary Evolution

Paleocene and Eocene rocks suggest a hiatus in magmatic activity in the Philippine Arc. The sediments are dominantly marine shelf facies, volcanic wackes and shales intercalated with reef limestones, with thicknesses generally in excess of 1000 m, suggesting relative tectonic conditions during the erosion of pre-existing volcanic units. Andesitic flows and tuffs appear only in the uppermost horizons of these sediments.

Miocene Volcanism and Plutonism

Certainly one of the most distinctive features of the magmatic-tectonic history of the Philippines is the onset of voluminous andesitic volcanism in the early Miocene. This event probably is associated with a series of tectonic transitions in the Philippine Arc and the Pacific Basin. Contacts between Miocene and underlying sediments are usually unconformable, although local conformable sequences exist (i.e., Baguio district). Miocene sedimentation resulted in the accumulation of 5000 to 10,000 m of volcanic debris, clastic sediments, and intercalated coralline limestones in a series of NW-trending basins, reflecting a depositional/paleogeographic model very similar to that seen today.

In the Cagayan Valley, the best known of the sedimentary basins, deposition was controlled by a NW-trending basin axis during the early Miocene, which shifted to a N-S trend during the late Miocene with a distinct asymmetry resulting from greater subsidence in the west. In Luzon, contemporaneous volcanism seems to have been concentrated along the present trend of the Cordillera. Thick sequences of andesitic to dacitic pyroclastics and flows were intruded by tonalite plutons of batholithic dimensions during the middle to upper Miocene. The intrusion of batholithic plutons and later linear arrays of dacite porphyry stocks associated with porphyry copper mineralization appears to have been strongly influenced by preexisting north to NW-striking fault zones.

It is evident that continued activity along the Philippine fault system has exerted considerable control over localization of mineral deposits through extensive shattering of host rocks, particularly in the Baguio district. The Philippine fault system is represented by a NNW-trending series of lineaments extending from southern Mindanao to northern Luzon. Erosional valleys, sag ponds, fault scarps, and fault line scarps are certainly as impressive as those occurring along the San Andreas system. NW-SE-trending fold axes, concordant with fault offsets, occur in Miocene to recent sediments adjacent to the fault zone. Heavy vegetation and lack of extensive cultural features crossing the fault have made documentation of historic fault activity difficult. The best known example of recent activity along the system involves the 1973 Gulf of Ragay earthquake [*Morante and Allen,* 1974], with over 90 km of ground breakage and up to 3.5 m of left lateral offset. The southern boundary of the Philippine Cordillera in Luzon (about 90 km NNW of Manila) is an impressive scarp related to 'horsetailing' of the Philippine fault. Uplift in the area has been extremely rapid, over 2000 m in the Pliocene. The combination of rapid uplift and erosion associated with high rainfall has given rise to a 'badland' topography with hillslopes ranging from 30° to 40°. The rapid uplift has also resulted in chaotic sliding of poorly consolidated sediments off the topographic high of the Cordillera into the Central and Cagayan Valleys to form marginal folds and thrusts. The sedimentary records in these basins suggest that the present topographic expression of the Cordillera, and probably the Sierra Madre to the east, originated during the Plio-Pleistocene.

At least four distinct ages of mineralization are associated with porphyry copper deposits in the Philippines: (Figure 2, Table 1) at Atlas-Cebu 95–110 m.y., 30 m.y. Negros-Sipalay and Hinobaan, 15–20 m.y. at Marinduque Island-Marcopper, Isao-Pili, Ino-Capayang, and Taysan-Luzon, and 3.4–7 m.y. in the Baguio district (approximately ten mines and prospects). In eastern Mindanao (approximately seven mines or prospects) the porphyry intrusions appear to be of approximately the same age as the Baguio mineralization. The intrusions in the Polillo-Paracale area may be similar in age to those on Marinduque Island. It is interesting to note the close correspondence in time between these metallogenic epochs and those in Canada (50 m.y. and 8 m.y.) [*Christopher and Carter,* 1976], southwest United States (50–60 m.y.), Thailand ((tin) 50–60 m.y.), Andean porphyries (7–10 m.y., 15–16 m.y.) [*Sillitoe,* 1977], and the New Guinea-New Britain-Bismark Archipelago-Solomon Islands deposits of the South Pacific (1–2 m.y., 4–7 m.y., 13–16 m.y.) [*Page and McDougall,* 1972]. It is remarkable that these widely distributed deposits show such well developed synchroniza-

tion of mineralization, despite the fact that they are apparently associated with subduction zone processes at several different plate boundaries. This correlation suggests that mineralization is more closely related to the tectonic history of plate boundaries than to the composition of specific lithospheric plates.

Geology of the Deposits

Cordillera-Baguio District

Basement complex. The oldest rocks occurring in the Baguio area are a series of regionally metamorphosed igneous and sedimentary units inferred to be of Late Mesozoic to early Tertiary age (Ks-Figure 4). The low abundance of radiogenic elements and absence of paleontologic samples precludes an accurate estimate of an age of this complex. The rocks are generally of greenschist or upper epidote amphibolite facies with a metamorphic mineral assemblage composed of tremolite-actinolite, epidote, albite, chlorite, and biotite. Superimposed contact metamorphism and hydrothermal activity associated with Miocene intrusions frequently overprints the regional metamorphic mineralogy. Primary igneous and sedimentary structures are recognizable in spite of this multistage alteration. Local evidence of primary clastic mineralogy, cross and graded bedding persists in the meta-sedimentary units. *Fernandez and Damasco* [1979] have previously reported a variety of lithologic units and primary sedimentary textures. Textural pseudomorphs of pyroxene and plagioclase phenocrysts have been observed by the author in outcrops, even though such textures were not obvious in thin section. Hornblende pseudomorphs of pyroxene and amphibole phenocrysts are common along with relic tuff and agglomerate textures.

Shannon [1979] reported the existence of a metamorphosed trondhjemite stock near the confluence of the Ambalanga and Liang Rivers. This contributes to the overall picture of a tholeiitic magmatic sequence developing parallel to and westward of the early to middle Tertiary calc-alkaline and syenitic intrusions of the Caraballo and Sierra Madre ranges (Table 1). It would appear that this magmatism was associated with subduction related to the Manila Trench. However, reconstructions involving tectonic assemblage of the Western Cordillera, Caraballo, and/or Sierra Madre ranges are possible [*De Boer et al.* [1980] and J. A. Wolfe, manuscript in preparation, 1982].

Miocene Volcanic-Sedimentary Unit

The structural relationship between the basement complex and the Miocene Pugo volcanic sedimentary unit is obscured by the discordant intrusions of the composite Agno Batholith. Poorly defined stratigraphic relations and discontinuous exposures of this extensive unit have led to ambiguous nomenclature for the preintrusive sequence. Previous authors have referred to the succession as the Antamok series [*Leith,* 1938], Pugo metavolcanics and metasediments [*Schafer*, 1956; *Worley*, 1967; *Diegor*, 1980] and keratophyre series [*Fernandez and Damasco*, 1979]. It is probable that much of the ambiguity results from the fact that the sequence evolved synchronous with the early intrusions of the Agno Batholith.

The sequence is gently folded and shows local metamorphism in the zeolite facies with a substantial hydrothermal alteration overprint, particularly adjacent to porphyry copper intrusions. A pervasive pyrite-sericite mineralization zone with an elliptical outline exists in the immediate area of the Baguio district (Figure 4) and is suggested to be a large scale alteration halo associated with a poorly exposed subjacent granodiorite pluton. This halo is very apparent as a spectral anomaly in LANDSAT imagery.

Three distinct members of this unit have been defined by *Fernandez and Damasco* [1979] and *Shannon* [1979]. A lower volcanic unit consist of picritic and porphyritic andesite-basalt flows and minor tuffs, with minor intercalated sandstone lenses. The picritic basalts contain olivine, augite, and hornblende phenocrysts. Hypersthene occurs only rarely in rocks of the Baguio district. The andesitic basalts of the lower unit are generally hornblende-augite-plagioclase porphyries.

A middle sedimentary member overlies the lower volcanic member and may represent a local hiatus in volcanic activity. It consists of red, green, and gray graywackes and pebble conglomerates with some siltstone and minor shale and limestone. It is highly variable in thickness and appears to pinch out entirely in some areas.

Andesitic flows and agglomerates make up a discontinuous and poorly exposed upper member. The rocks are generally plagioclase and hornblende plagioclase porphyries that locally intertongue with the lower units of the Zig-Zag sediments.

Zig-Zag Sediment Sequence

The structural and stratigraphic relationships between the Pugo volcanic-sedimentary sequence and the younger Zig-Zag sediments are complex and generally poorly exposed. Paleontologic evidence from the lower Zig-Zag suggests an age of early Miocene [*Pena and Reyes*, 1970]. Much of the complexity probably results from the intrusion of the pluton of the Agno Batholith.

Agno Batholith

In the Baguio district the Agno Batholith is composed of three major intrusive sequences with distinctive petrogenic affinity [*Divis*, 1977; *Shannon*, 1979]. The oldest sequence is a tholeiitic magma series represented by a sheared and folded trondhjemitic gneiss in the metamorphic complex. This may be related to a magmatic and metamorphic episode in the Sierra Madre range of Luzon having potassium/argon dates in the range of 44 to 49 m.y. BP [*PBM-JICA*, 1976].

A younger series of intrusions, ranging in composition from gabbro to tonalite in the Sierra Madre and Caraballo mountains, have reported ages of 27 to 33 m.y. BP [*PBM-JICA*, 1976]. This latter intrusive sequence is probably represented by an episode of trondhjemitic to dioritic magmatism in the Baguio district. This is a magma series with tholeiitic affinities that shows moderate iron enrichment and relatively low potassium abundance (Tables 2 and 3, Figures 5, 6, and 7) and will subsequently be referred to as the Twin Rivers sequence following the nomenclature of *Shannon* [1979].

A third, younger series of intrusions, has calc-alkaline affinities and ranges in composition from gabbro to granodiorite. The latest intrusions of this series are inferred to be the source magma for the numerous porphyry copper intrusions in the Baguio area (Table 3). This series will be referred to as the Virac Sequence following the nomenclature of *Schafer* [1956] and *Shannon* [1979]. Potassium/argon dates in the Baguio district [*Wolfe*, 1972; *Divis*, 1977; *Wolfe*, 1981] suggest that a hiatus in magmatism separates the Virac series of intrusions from the Twin Rivers sequence.

Two distinct populations of ages may exist in this latter sequence, but data are insufficient to demonstrate this conclusively. Ages summarized by Wolfe [1981] tend to cluster in the ranges 9 to 15 m.y. BP and 3.6 to 7 m.y. BP. Obviously, the 2 m.y. 'gap' may be an artifact of sampling. Fission track ages reported by *Shannon* [1979] fall in this latter range and may reflect thermal annealing of the tracks. It does appear that there is a significant hiatus in magmatic activity in the Southern Cordillera between 15 and 24 m.y. BP. Geological evidence suggests that although the 7 to 9 m.y. gap may not be as profound, there is a petrologic gap between the plutons of the Virac sequence and younger discordant dikes and stocks associated with the porphyry intrusions. Geochronologic data indicates a period of porphyry mineralization in the Batangas-Marinduque and Paracale areas at approximately 15–20 m.y. BP that may overlap structurally and chronologically with mineralization in the Cordillera.

The Twin Rivers sequence is composed of intrusive units ranging in composition from gabbro to trondhjemite. The intrusions form large north-south elongate plutons with moderate flow foliation and contacts ranging from concordant to discordant. The majority of the southern Agno batholith is composed of diorite and tonalite related to the Twin Rivers sequence. Fission track age determinations by *Shannon* [1979] suggest that magmatic activity progressed westward with time from a north-south trending belt of the Twin Rivers sequence. This progression is concordant with a general westerly trend in intrusive activity indicated by potassium-argon dating from the Sierra Madre and Caraballo Ranges as previously indicated.

Unlike the Twin Rivers sequence, the intrusions of the Virac series tend to form discordant stocks and dikes, which intrude the Pugo and Zig-Zag sediments, the basement complex, and the rocks of the Twin Rivers sequence. It

TABLE 2. Baguio District

	BAL37	BAL24	BAL41	BAL51	KRL-BMS	KRL83	KRL82	LUCG*	KLOD†
SiO_2	71.80	62.93	58.50	58.95	65.02	59.60	68.47	50.63	53.70
TiO_2	0.40	0.48	0.81	0.80	0.55	0.28	0.11	0.93	0.85
Al_2O_3	15.85	18.25	18.17	17.89	17.10	17.57	17.13	17.53	18.57
Fe_2O_3	2.70	4.36	4.51	6.15	5.16	8.75	4.14	9.85	7.51
MnO	...	...	...	...	...	...	...	0.35	0.17
MgO	0.87	1.92	2.94	2.85	1.90	2.08	1.28	5.86	3.53
CaO	2.96	5.94	6.46	5.66	2.35	3.01	1.66	8.80	7.68
Na_2O	3.82	3.83	3.90	3.50	3.51	3.91	3.03	3.32	3.39
K_2O	0.25	0.70	1.90	2.63	2.21	3.65	3.15	0.49	1.34
P_2O_5	0.15	0.20	0.17	0.19	0.21	0.18	0.18	...	...
H_2O+	0.75	1.01	1.14	1.23	1.95	1.01	0.85	...	...
Sum	99.55	100.12	98.50	99.85	99.96	100.04	100.02	97.76	96.74
Ag, ppb	165	80	50	<70	155	76	80	...	...
Au, ppb	0.7	2.0	0.5	0.5	1.8	0.5	0.8	...	...
Ba, ppm	...	...	...	...	440	380	520	...	...
Cr, ppm	...	...	...	...	30	5	6	...	...
Cu, ppm	5	12	65	32	94	8	16	98	39
Li, ppm	...	...	...	...	9	14	11	...	...
Mn, ppm	920	755	975	1400	668	1027	943	...	...
Mo, ppm	...	...	...	...	...	0.05	0.44	...	...
Ni, ppm	50	15	40	25	19	13	12	...	...
Rb, ppm	15	26	35	55	42	90	77	25	32
Sn, ppb	130	1000	1100	850	1800	470	1900	...	...
Sr, ppm	210	380	820	465	463	323	335	470	489
Se, ppb	15	8	30	...	61	17	2	...	...
V, ppm	170	88	193	220	96	163	110	...	...
Zn, ppm	20	40	135	90	26	96	37	195	79

Normative Minerals

	BAL37	BAL24	BAL41	BAL51	KRL-BMS	KRL83	KRL82	LUCG*	KLOD†
Q	40.6	21.7	10.7	11.3	27.4	10.9	32.4	1.72	6.55
C	4.27	0.790	11.3	15.5	5.16	2.14	6.15	2.90	7.92
Or	1.48	4.14	33.0	29.6	13.1	21.6	18.6	28.1	28.7
Ab	32.3	32.4	26.5	25.3	29.7	33.1	25.6	31.5	31.5
An	13.7	28.2	3.59	1.22	10.3	13.8	7.06	9.81	5.35
Hy	3.81	7.55	8.00	10.3	7.98	11.7	6.41	15.9	10.6
Mt	1.57	2.52	2.61	3.57	2.99	5.07	2.41	5.71	4.35
Il	0.760	0.912	1.54	1.52	1.04	0.532	0.209	1.77	1.61
Ap	0.355	0.474	0.403	0.450	0.497	0.426	0.426	...	...
Sum	98.9	98.6	97.5	98.8	98.1	99.2	99.3	97.4	96.6

*Average of 3 Lucbuban Gabbro analyses by *Shannon* [1979].

†Average of 4 dikes in the Klondyke formation by *Shannon* [1979].

BAL37: Hornblende trondhjemite—Twin Rivers sequence.

BAL24: Biotite-hornblende quartz diorite—Ambaclao Reservoir.

BAL41: Biotite-hornblende diorite—Virac series—Santo Tomas II area.

BAL51: Biotite-augite diorite—Kennon Road area.

BMS30: Biotite-hornblende quartz diorite—Black Mountain stock.

KRL83: Biotite-hornblende diorite porphyry—Minor stock—Kennon Road.

KRL82: Biotite quartz diorite, interstitial myrmekite—Kennon Road.

includes the gabbro, diabase, and the multiphase Virac plutons. The porphyry copper stocks at Santo Nino, Santo Tomas (Philex), Black Mountain, Western Minolco, and Tawi-Tawi represent the latest phase of this activity, which may be associated with a subjacent pluton beneath the zone of alteration east of Baguio and a later series of dikes in the Kennon Road Emerald Complex (Figure 5).

Two relatively distinct series of dikes occur in the area. It is suggested that the older type I series (Figure 8) is related to the main sequence of Virac intrusions, with a younger type II series directly associated with the copper porphyry intrusions. The type II dikes show an interesting fractionation pattern of increasing Ag and Cu abundance with higher Au concentration than that found in the type I dikes. The type II dikes are frequently corundum normative and cross cut type I dikes and older Plutons. One of the most persistent and distinctive features of the older type I series is a group of hornblende megacryst and xenolith bearing porphyries. These dikes contain gabbroic hornblende-plagioclase xenoliths with well-developed comb layering. In some cases the xenoliths may make up as much as 50% of the rock mass and may be as large as 1 m in diameter. It is suggested that these xenoliths were formed as a result of fractional crystallization of a hydrous

TABLE 3. Kennon Road-Emerald Dikes

	Type I Dikes					Type II Dikes						
	KRL28	KRL26	KRL31	KRL18*	KRL16-3*	KRL15*	KRL17*	KRL19*	KRL23	KRL24	KRL25	KRL29
SiO_2	63.28	62.50	61.70	49.51	46.54	63.40	55.58	45.70	53.62	63.07	64.78	64.04
TiO_2	0.52	0.80	0.44	1.21	1.12	0.59	0.47	1.18	0.35	0.47	0.51	0.58
Al_2O_3	17.83	16.80	17.80	19.78	19.65	19.88	19.95	19.32	19.20	17.36	17.61	17.95
Fe_2O_3	6.47	5.31	7.91	9.53	9.93	2.46	6.46	9.57	9.65	7.45	6.43	6.70
MgO	2.95	3.02	1.93	5.30	7.15	2.02	3.87	7.90	5.30	2.22	2.06	2.38
CaO	2.85	3.54	3.42	9.22	10.95	3.73	8.15	11.97	2.84	2.75	2.69	2.64
Na_2O	2.46	3.62	3.65	3.62	2.28	5.37	3.25	2.31	6.72	3.60	3.59	2.50
K_2O	1.20	2.44	1.60	0.23	0.71	0.82	0.75	0.66	0.58	1.57	1.61	1.71
P_2O_5	0.22	0.25	0.16	1.14	0.10	0.16	0.18	0.12	0.19	0.23	0.21	0.18
H_2O+	1.88	1.80	1.36	1.66	1.15	1.40	1.05	1.29	1.55	1.33	1.45	1.25
Sum	100.16	100.08	99.97	101.20	99.50	99.83	100.21	100.02	100.00	100.05	100.94	99.93
Ag, ppb	538	181	169	...	...	144	70	48	487	114	171	50
Au, ppb	<0.5	0.6	3.9	...	...	2.3	2.2	0.5	1.9	1.3	27.4	5
Ba, ppm	365	415	440	...	...	...	...	...	375	310	280	470
Cr, ppm	46	8	5	...	...	...	...	...	...	...	14	33
Cu, ppm	95	79	18	132	19	98	33	92	160	15	14	442
Li, ppm	16	13	23	...	...	...	...	...	...	45	26	25
Mn, ppm	1340	1491	1525	1450	1700	490	1440	1365	1430	1220	1339	1091
Mo, ppm	...	...	0.15	...	...	0.14	<0.01	<0.01	0.87	1.15	0.20	0.22
Ni, ppm	40	25	13	27	30	5	23	60	...	34	12	17
Rb, ppm	31	55	33	8	12	15	12	16	...	43	30	33
Sn, ppb	890	1000	580	...	...	950	190	250	670	910	1100	870
Sr, ppm	430	630	540	805	540	800	565	420	...	395	720	520
Se, ppb	60	72	18	...	...	84	36	12	60	22	293	31
Te, ppb	2	1.6	<1	...	...	58.9	3.0	<1	...	<1	761	<1
V, ppm	178	193	102	315	460	60	190	380	225	135	154	124
Zn, ppm	58	62	122	100	88	37	64	68	82	62	46	46
Normative Minerals												
Q	32.5	19.3	22.0	.800	...	18.0	9.35	...	...	25.2	27.5	32.2
C	7.83	2.36	4.23	...	...	3.76	...	...	...	5.28	5.57	7.62
Or	7.09	14.4	9.46	1.36	4.31	4.85	4.43	3.90	2.81	9.28	9.51	10.1
Ab	20.8	30.6	30.9	30.6	19.3	45.4	27.5	17.8	3.43	30.5	30.4	21.2
An	12.7	15.9	15.9	37.0	41.2	17.4	37.6	40.4	56.9	12.1	12.0	11.9
Ne	...	...	...	...	...	...	...	0.958	12.8	...	...	...
Di	...	...	...	1.01	9.98	...	1.31	14.5	...	...	...	...
Ol	...	...	...	...	...	...	...	13.3	...	...	...	...
Hy	11.8	10.7	10.6	18.4	6.76	6.07	13.6	...	4.27	10.8	9.57	10.4
Mt	3.76	3.07	4.58	5.54	5.75	1.42	3.76	5.56	5.60	4.23	3.73	3.89
Il	0.988	1.52	0.836	2.30	2.13	1.12	0.893	2.24	0.665	0.893	0.969	1.10
Ap	0.521	0.592	0.379	2.70	0.237	0.379	0.426	0.284	0.450	0.545	0.497	0.426
Sum	98.0	98.5	98.8	99.8	89.7	98.5	98.9	98.9	86.9	98.9	99.7	98.8

*Rocks analyzed for rare earth elements. KRL15: Plagioclase andesite porphyry. KRL17: Hornblende-plagioclase porphyry. KRL18: Aphanitic andesite dike. KRL19: Hornblende-augite-plagioclase porphyry. KRL16-3: Hornblende-plagioclase andesite porphyry. KRL23: Plagioclase andesite porphyry. KRL24: Plagioclase andesite porphyry. KRL25: Hornblende dacite porphyry. KRL26: Hornblende-biotite dacite porphyry. KRL29: Quartz dacite porphyry. KRL28: Hornblende dacite porphyry. KRL31: Plagioclase andesite porphyry.

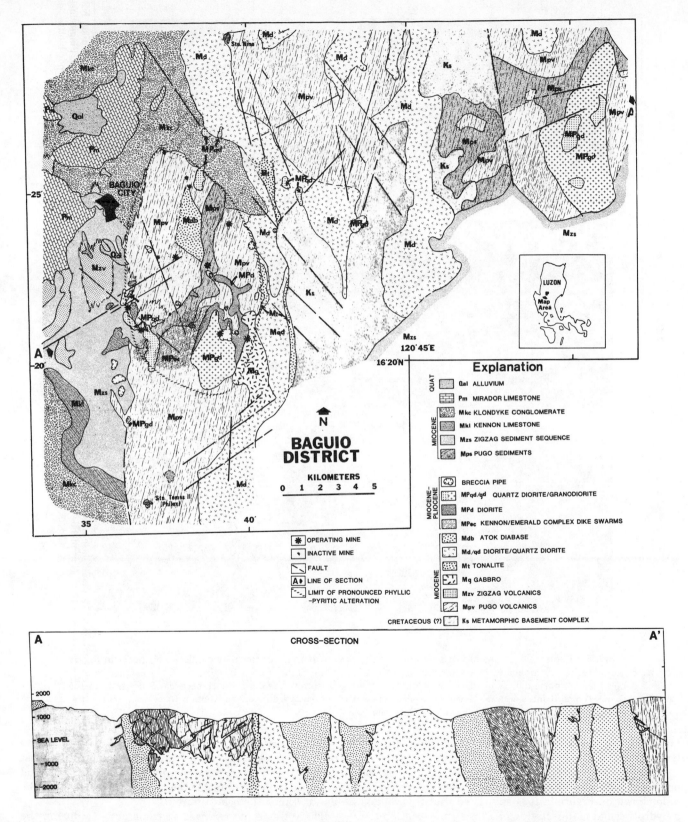

Fig. 5.   Geologic map of the Baguio district, based in part on *Shannon* [1979], *Fernandez* [1979] and J. A. Wolfe (unpublished manuscript, 1981).

185

Fig. 6.   Photomicrographs of intrusive rocks of the Baguio district. (*a*) Hornblende diorite of the Twin Rivers sequence. (*b*) Hornblende quartz diorite from the area of the Santo Tomas II mine. (*c*) Fine grained seriate hornblende quartz diorite of the Black Mountain stock. (*d*) Hornblende diorite with secondary biotite from the Santo Tomas II intrusion. (*e*) Hornblende-plagioclase porphyry-type I dike of the Kennon Road area. (*f*) Biotite-Hornblende porphyry from the Tawi-Tawi intrusion.

magma in an upper crustal chamber from which the intrusions of the Virac sequence were derived.

A series of strongly altered breccia systems appear to be associated with the elliptical pyrite-sericite alteration halo shown on Figure 4. It is suggested that this zone is a hydrothermal manifestation of a large subjacent pluton of granodiorite to quartz monzonite composition related to the type II dikes. Dacite porphyry is frequently the host rock for porphyry copper mineralization in the district and is associated with multiple coaxial intrusive phases at the Santo Tomas II stock (Figure 6).

There was some renewed volcanic activity during the

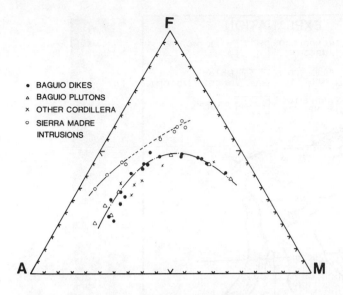

Fig. 7. AFM diagram for the intrusions of the Cordillera and Sierra Madre range. Data for the Sierra Madre range and 'other Cordillera' from *PBM-JICA* [1976]. $A = K_2O + Na_2O$, $F =$ Fe as FeO, and $M =$ MgO in molecular ratios.

Plio-Pleistocene accompanying the intrusion of the Bala-toc plug southeast of Baguio, a feature that has generated geothermal steam in the lower working levels of mines. Another episode of gold mineralization probably took place at this time.

Porphyry Deposits

The history of the Baguio mineral district dates back to the 17th century during the early years of the Spanish occupation. Gold was recovered by placer and shallow underground mining from this period to the present. The district represents one of the geologically best known areas in the Philippines owing to the numerous mining operations and relatively extensive exposures in the area. The porphyry copper mineralization found in the Baguio district extends north over 100 km along the Central Cordillera. Within the immediate area of Baguio, at least four major porphyry deposits are known and are or were actively mined. These deposits are Santo Nino, Santo Tomas II, Black Mountain, and Lobo-Boneng (Figure 5). Cubu-agan and Mountain Mines deposits also have developed reserves and are located approximately 150 km north of Baguio (Figure 1). A series of less well known prospects join these two areas along the axis of the Cordillera (Figure 2). The Santo Tomas II operation of Philex mining (Figure 9) represents the closest approach to an 'ideal model' of a porphyry intrusion shown in Figure 4.

The porphyry intrusion is a relatively small complex stock-like intrusion approximately 350 m in diameter. It intrudes metavolcanic rocks of the Pugo volcanic series. *Bryner* [1969] included a description of the Santo Tomas deposit in his summary 'Ore deposits of the Philippines-an introduction to their geology.' The stock appears to be a series of co-axial intrusions in plane and cross section. The youngest intrusion, of porphyritic andesite to dacite

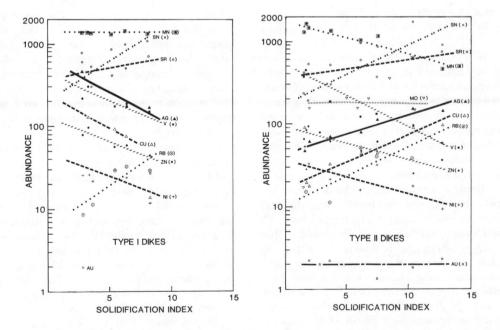

Fig. 8. Trace element distribution versus 'solidification index' $= 10(K_2O + Na_2O)/(FeO + CaO + MgO)$ in molecular percent. Ag, Au, and Sn abundances are in ppb; all others are in ppm.

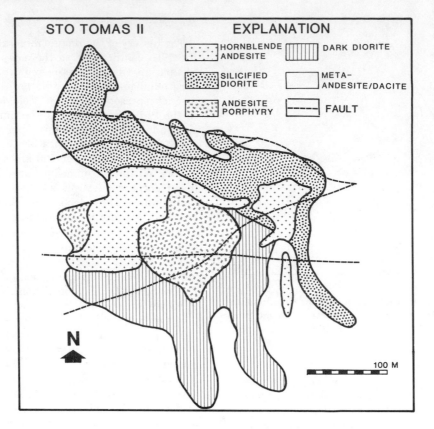

Fig. 9. Geologic map of the Santo Tomas II Intrusion, Baguio district.

composition forms the central area of the stock and is surrounded by older intrusions of dioritic and andesitic composition.

The ore zone is located in both the diorite and metavolcanics, with the majority of ore occurring as quartz-sulfide veinlets and stockworks in the metavolcanic country rock. A series of west to west-northwest striking faults occur in the mine area. A system of veins and stockwork mineralization generally parallels the orientation of these faults. Disseminated mineralization is relatively rare. Chalcopyrite is the major copper bearing sulfide and tends to dominate over pyrite in the ore zone. The ore tends to be within the potassic alteration zone. The dominant alteration minerals in this zone are biotite, sericite, and quartz, with chlorite replacing primary hornblende. Alteration zonation is comparatively well developed with a gradation from the potassic zone to a zone of sericite and chlorite alteration and then outward to a propylitic outer zone of chlorite-epidote-calcite. This outer zone merges with a very similar assemblage occurring in the Pugo volcanics. The outer zone is more distinctive owing to the presence of pyrite and magnetite than as a result of the silicate mineral assemblage.

The Santo Nino deposit is located along the contact zone between Miocene diorites and the Pugo volcanic series

(Figure 10). Two major ore bodies occur within the mine area. The southwest body occurs at the contact between a hornblende granodiorite complex and a dacite porphyry intruding Pugo volcanoclastics. The eastern, or Ulman, ore body is located adjacent to the contact between a plagioclase dacite porphyry intrusion and metavolcanic rocks. The youngest intrusions in the area are a series of northeasterly and east-northeasterly trending andesite porphyry dikes.

The area is strongly faulted with four major orientations apparent. Steeply dipping faults strike north, northwest, and west. A series of low angle faults strike predominantly northeast and dip to the southeast at an angle of approximately 30°. Slickenslides, gouge, and cataclastic textures are very common. *Bryner* [1970] indicates on steeply dipping faults these slickenslides show a dominant strike slip motion. The younger volcaniclastic unit shows over 100 m of downthrow relative to the older porphyries. Fault induced fracturing, particularly in the area of the southwest body, appears to have localized copper mineralization in relatively narrow, steep zones.

Mineralization within the two ore bodies tends to be concentrated in quartz-sulfide veinlets with some disseminated sulfide minerals present. Silicate alteration mineralization in the ore zone is dominantly potassic and is

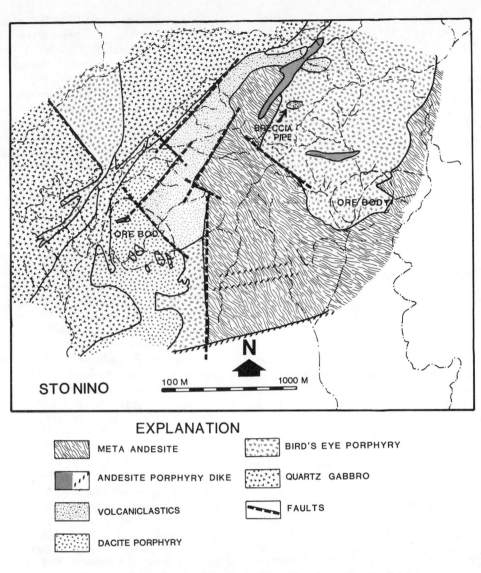

EXPLANATION

META ANDESITE		BIRD'S EYE PORPHYRY	
ANDESITE PORPHYRY DIKE		QUARTZ GABBRO	
VOLCANICLASTICS		FAULTS	
DACITE PORPHYRY			

Fig. 10. Geologic map of the Santo Nino Mine Area, Baguio district.

characterized by the presence of secondary biotite, quartz, and biotite vein filling. Surface oxidation and enrichment extends downward from the surface 50–100 m. As in the Santo Tomas II deposit, a propylitic halo of epidote-chlorite-calcite alteration grades irregularly outwards into the unaltered country rock.

Batangas-Marinduque

The deposits making up the Batangas-Marinduque group include Taysan, Ino-Capayang, Isao-Pili, Marcopper, and Dizon. The deposits appear to form a northwest striking belt extending from Marinduque Island northwest to the Batangas Peninsula and the southern Zambales Range. With the exception of Dizon, these deposits are located within 100 km of one another. Taysan is located on the

Batangas Peninsula and has been described by *Wolfe et al.* [1978]. Marcopper, Ino-Capayan, and Isao-Pili are located on Marinduque Island southeast of Taysan. The Marcopper deposit has been described by *Loudon* [1976]. Geological evidence and radiometric dating suggest that all the deposits were formed in the late Miocene approximately 15–20 m.y. BP. A number of minor diorite associated copper deposits have also been described along the projection of this trend along the eastern slope of the Zambales Range [*Kinkel,* 1956]. The deposits are generally quite similar in size, averaging approximately 100 million tons of reserves each (Table 1).

The Dizon deposit is located on the eastern edge of the Zambales Range and is included in this group due to the similarity in structural setting and geological evidence

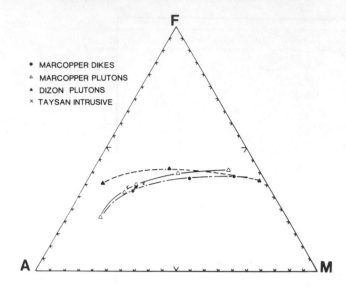

- MARCOPPER DIKES
△ MARCOPPER PLUTONS
▲ DIZON PLUTONS
× TAYSAN INTRUSIVE

Fig. 11. AFM diagram of Batangas-Marinduque area intrusions. $A = K_2O + Na_2O$, F = Fe as FeO, and M = MgO in molecular ratios.

suggesting a late Miocene age. The principal pluton associated with the Dizon deposit is a quartz diorite porphyry that intrudes altered andesitic volcanics and volcaniclastic material (Figure 11). The intrusion appears to be localized at the intersection of major northeast and northwest trending fracture sets and is overlain by andesitic flows and pyroclastic rocks derived from younger volcanic structures to the east. The mineralization is predominantly located in volcanics surrounding a tonalite intrusion and consists of quartz-chalcopyrite and bornite veinlets and chalcopyrite-chlorite veinlets [Saegart and Lewis, 1977]. Table 4 presents typical chemical analyses of unmineralized quartz diorite intrusions associated with the deposit. Copper concentrations are relatively low (10–24 ppm) in comparison with other porphyry associated intrusions. The older volcanic sedimentary basement includes fossil bearing units reported to be of Oligocene age (J. A. Wolfe, personal communication, 1982).

The Taysan deposit is located on the Batangas peninsula of southern Luzon. An older, basement complex similar to that found in the Baguio area makes up the oldest rock unit in the Taysan area. It consists of feldspathic gneiss, schist, and some marble. This unit is unconformably underlain by metabasalts and agglomerates. The metavolcanics are pillowed and appear to be metamorphosed marine basalts and diabases. This unit has been intruded and metamorphosed by rocks of the Tolos Batholith. The Tolos Batholith shows a northwest southeast elongation and biotite quartz diorite, hornblende quartz diorite, and hornblende diorite intrusive phases [Wolfe et al., 1978]. The Taysan deposit occurs in association with a series of younger discordant quartz dacite porphyry and hornblende dacite

porphyry intrusions located in the contact zone between the Tolos batholith and the older metavolcanics. A single whole rock K-Ar age of 14.8 ± 0.9 m.y. is similar to dates for the mineralization of the Marinduque Marcopper deposit of 20.8 m.y. [Walther et al., 1981].

Several major faults occur in the area of the deposit, the most notable being the Blay fault that strikes north-north-east and shows approximately 50 m of offset of a porphyry dike area [Wolfe et al., 1978]. Younger sediments in volcanic rock, including the Taal tuff overlie much of the area obscuring the geology of the deposit. Mineralization is associated with sulfide bearing quartz veinlets and fracture fillings consisting of three major sets. The principal set is steeply dipping and strikes northwest.

Wolfe et al. [1978] have noted the development of pebble dikes, intrusive breccias, eruptive breccias, and a collapsed breccia pipe suggesting near surface volcanic phenomena. One of the most notable of these structures is an inferred volcanic neck located northeast of the ore body. The breccia associated with this feature contains copper mineralization. Post-ore dikes originate from the area of the neck and appear to remobilize sulfide minerals.

The delineation of alteration zones has been impeded by poor exposures. Potassic alteration occurs as biotite replacements or as polycrystalline biotite aggregates forming psuedomorphs after hornblende. K-feldspar, when it occurs, appears as poorly developed potassic rims on plagioclase. The phyllic zone appears as strong sericitization of plagioclase and chloritization of hornblende. The phyllic zone irregularly intergrades with the potassic zone by the seritization and chloritization of biotite replacing hornblende and as a vein and fracture filling (Figure 16c).

The poor exposures and possibly biased view derived from drill hole data suggests that a multistage intrusive history has occurred in association with the Taysan deposit. The majority of these intrusions appear to be in the compositional range of tonalite and dacite porphyry. Evidence of more basic intrusions has not been encountered in the drilling program.

The Ino-Capayang, Isao-Pili, and Marcopper deposits all occur on Marinduque Island approximately 200 km south of Manila. Loudon [1976] detailed the economic geology of the Marcopper deposit, which is similar to that of the Ino-Capayang and Isao-Pili to the northwest (Figure 12). The oldest rocks on the island are spilitic basalts and andesite flows with minor intercalated siltstones and graywackes of the Marinduque basement complex. The basement rocks are very similar to the metavolcanic rocks located in the Taysan area.

The Eocene Tumicob Group overlies this basement and consists of a basal clastic unit overlain by andesitic to dacitic flows and pyroclastics. The Oligocene San Antonio Formation, composed of andesitic flows, agglomerates, and breccias with intercalated sediments and volcaniclastics discordantly overlies the Tunicob Group. A period of fold-

TABLE 4. Batangas-Marinduque Group

	DZL02	DZL01	TML34	TML-20	MCM20	MCM21	MCM06
SiO_2	54.55	43.59	65.50	65.15	62.48	60.55	52.17
TiO_2	0.79	0.95	0.40	0.35	0.25	0.29	0.60
Al_2O_3	18.51	18.20	18.15	18.20	19.20	18.75	18.70
Fe_2O_3	6.77	13.44	4.20	4.20	4.51	4.46	7.67
MgO	3.96	9.25	2.19	1.70	1.96	2.30	5.48
CaO	7.45	11.25	3.55	3.75	4.08	5.39	10.30
Na_2O	3.96	0.30	3.50	4.33	4.17	3.67	3.32
K_2O	0.70	0.02	1.23	1.30	2.35	2.32	1.83
P_2O_5	0.18	0.12	0.23	0.26	0.15	0.17	0.25
H_2O+	1.35	0.75	1.77	1.86	0.85	1.75	0.67
Sum	100.61	100.79	100.87	101.10	100.00	99.65	101.09
Ag, ppb	90	9	85	110	· · ·	· · ·	59
Au, ppb	1.5	<0.5	1.0	1.3	· · ·	· · ·	1.0
Cu, ppm	24	10	805	72	122	22	135
Mn, ppm	1135	820	222	253	690	665	1205
Ni, ppm	15	55	2	2	7	10	25
Rb, ppm	7	2	20	23	25	25	21
Sn, ppb	970	180	1850	2200	· · ·	· · ·	540
Sr, ppm	460	155	485	415	650	645	565
Se, ppb	9	17	32	35	· · ·	· · ·	19
Te, ppb	4.1	<1	· · ·	2.2	· · ·	· · ·	1.4
V, ppm	333	230	140	430	170	180	237
Zn, ppm	70	29	20	21	42	39	51
			Normative Minerals				
Q	8.31	1.58	29.0	23.9	16.5	14.4	· · ·
C	· · ·	· · ·	5.16	3.47	2.74	0.808	· · ·
Or	4.34	0.118	7.37	7.68	13.9	13.7	10.8
Ab	33.5	2.54	29.6	36.6	35.3	31.0	28.1
An	30.7	48.2	16.1	16.9	19.3	25.6	30.7
Di	1.01	5.45	· · ·	· · ·	· · ·	· · ·	15.1
Hy	8.59	29.4	8.09	6.95	8.13	8.87	4.32
Mt	5.89	7.80	2.44	2.44	2.61	2.58	4.45
Il	1.50	1.80	0.760	0.665	0.475	0.551	1.14
Ap	0.426	0.284	0.545	0.616	0.355	0.403	0.592
Sum	97.0	97.2	99.0	99.3	99.2	98.0	95.2

DZL02: Hornblende quartz diorite—Dizon Mine, Zambales.
DZL01: Augite-hypersthene gabbro—Dizon Mine, Zambales.
TML34: Biotite quartz diorite—Taysan, Batangas.
TML20: Hornblende-quartz dacite porphyry—Taysan.
MCM20: Hornblende quartz diorite—Marcopper, Marinduque.
MCM21: Hornblende-biotite-quartz diorite—Marcopper.
MCM06: Augite diorite—Marcopper mine, Marinduque Island.

ing and faulting apparently affected both the San Antonio and Tumicob units prior to the deposition of lower Miocene clastic sediments, volcanic rocks, and limestones. These latter units unconformably overlie the Oligocene-Eocene volcanics and sediments.

The four units are intruded by younger andesitic to dacitic sills and dikes that appear to be generically related to Miocene diorite and quartz diorite stocks (Figure 13). These intrusions tend to be elongated in a northwest/southeast trend similar to the Tolos batholith. The Marcopper ore body lies on the margin of one of these intrusions—the Mahinhin quartz diorite stock (Table 4).

The Mahinhin stock bears a striking resemblance to the Tolos intrusion. It is approximately 15 km long and 5 km wide with the axis of elongation trending in a northwest-

erly direction. Exposures in the mine vicinity range from diorite to quartz diorite with minor granodioritic phases. *Walther et al.* [1981] have reported a potassium-argon age of 20.8 m.y. B.P. for biotite from the Marcopper (Mount Tapian) deposit. Intra- and post-mineralization dike swarms have intruded the Mahinhin stock and surrounding metavolcanic units. The dominant direction of strike of the swarms is northwest/southeast suggesting a northeast/southwest dilation. Upper Miocene volcaniclastics, conglomerates, sandstones, and shales overlie the intrusions and older rocks.

The Marcopper ore zone has a mushroom-like shape similar to the Taysan deposit, with a high-grade 'cap' and a lower-grade 'stem' extending downward. Although the history of the hypabyssal intrusives associated with the Ma-

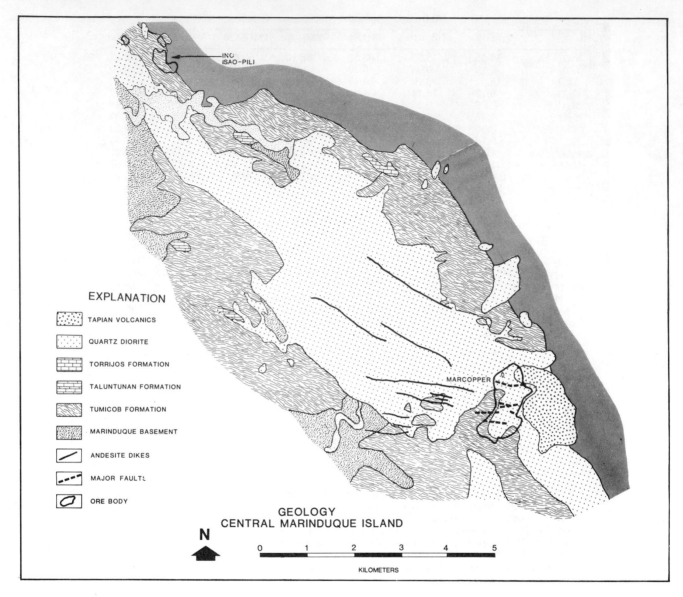

EXPLANATION

- TAPIAN VOLCANICS
- QUARTZ DIORITE
- TORRIJOS FORMATION
- TALUNTUNAN FORMATION
- TUMICOB FORMATION
- MARINDUQUE BASEMENT
- ANDESITE DIKES
- MAJOR FAULTS
- ORE BODY

GEOLOGY
CENTRAL MARINDUQUE ISLAND

N

0   1   2   3   4   5

KILOMETERS

Fig. 12.   Geology of central Marinduque Island, showing Marcopper and Isao-Pili mining areas (after Philippine Bureau of Mines Regional Map).

hinhin stock is complex, it appears that the majority of the dikes that cut the intrusion are significantly younger, showing chilled margins. Some may be very young and associated with the volcanic activity of Mount Malindig, a recent volcano on the southern shore of Marinduque Island. The Mount Tapian volcanics occur locally in the area of Mount Tapian and form a layer, up to 40 m thick, following the topography. The volcanics consist of volcanic breccia, tuff, and altered andesitic boulders [*Loudon,* 1976].

As in the Taysan deposit, the central core of alteration and mineralization zoning is associated with the potassic alteration and the highest grade of copper mineralization.

Biotite represents the dominant potassic mineral phase, with K-feldspar occurring in veinlets and as peripheral overgrowths on plagioclase. The sulfide mineralization occurs as disseminations and sulfide-quartz veinlets. The veinlets form a quartz network that is locally intensified to complete silica flooding. Chalcopyrite and pyrite are the dominant sulfides with minor amounts of bornite, covellite, and molybdenite. A phyllic alteration zone surrounds the inner potassic zone and tends to form an irregular outline constrained in part by the Eocene/Oligocene metasedimentary and metavolcanic rock. The propylitic alteration zone grades outward into a similar metamorphic

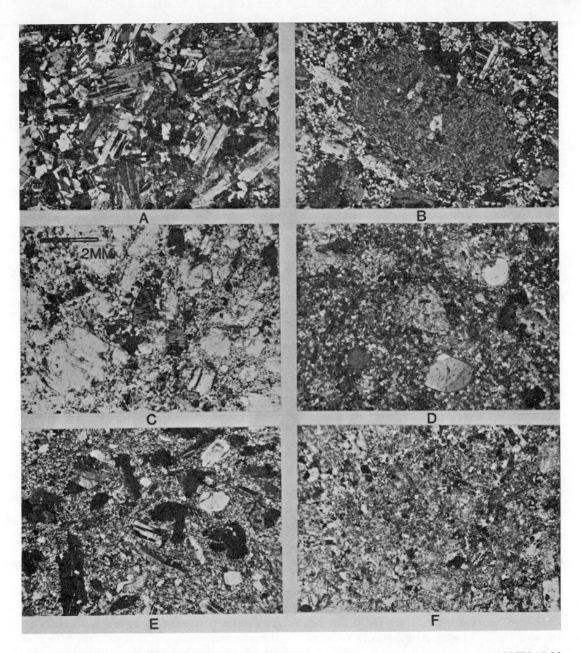

Fig. 13. Photomicrographs of Marcopper intrusive rocks and typical alteration mineralogy. (*a*) MCM-07 Mahinhin quartz diorite (crossed polarizer). (*b*) Typical potassic zone alteration of biotite replacing hornblende and sericitiztion of plagioclase. (*c*) MCM-07 mahinhin quartz diorite (plane light) showing secondary biotite after hornblende. (*d*) Phyllic zone alteration, partially resorbed quartz phenocrysts in a strongly sericitized matrix, showing relic porphyry texture. (*e*) Post-mineralization hornblende-plagioclase porphyry. (*f*) Strongly altered diorite porphyry—propyllitic zone—chlorite-sericite-calcite-magnetite replacement of original mineralogy.

mineral assemblage in the Eocene and Oligocene country rock. Irregular zones of argillic alteration occur between the propylitic and potassic zones. Oxidation and secondary enrichment are relatively limited, occurring within 20–100 m of the surface.

Marian Alkalic Deposit

The Marian prospect at Cordon, Isabela Province of Luzon is probably one of the more unusual porphyry copper deposits in the Philippines. Virtually all other deposits are

TABLE 5. Marian-Cordon

	M35-437	QU-4	QU-2	QU-5	B339*	K576*
SiO_2	51.60	57.10	54.15	58.50	57.95	48.32
TiO_2	0.40	0.48	0.98	0.38	0.42	0.88
Al_2O_3	19.78	19.91	19.40	19.80	19.95	15.78
Fe_2O_3	6.38	4.80	4.64	3.92	3.73	8.29
MnO	· · ·	· · ·	· · ·	· · ·	0.06	0.13
MgO	4.30	2.72	3.25	3.15	2.55	6.98
CaO	6.32	1.66	7.12	2.34	1.17	9.05
Na_2O	3.48	3.89	3.95	2.26	2.78	3.59
K_2O	5.76	5.58	3.80	7.23	8.55	3.99
P_2O_5	0.30	0.32	0.05	0.07	0.05	0.12
H_2O+	1.37	1.76	1.05	1.10	1.95	2.71
Sum	99.75	98.22	98.39	98.75	99.16	99.84
Ag, ppb	21	7	10	33		
Au, ppb	0.6	5.8	0.5	6.9		
Ba, ppm	890	740	850	1345		
Cu, ppm	160	265	160	950		
Mn, ppm	1000	390	1160	404		
Mo, ppm	0.92	0.23	0.32	3.58		
Ni, ppm	17	8	8	9		
Rb, ppm	115	62	60	123		
Sn, ppb	1000	1800	2600	2300		
Sr, ppm	1380	1180	3500	1850		
Se, ppb	10	65	2000	490		
Te, ppb	24	<1	3.8	8.9		
V, ppm	270	289	210	130		
Zn, ppm	40	63	850	40		
Normative Minerals						
Q	· · ·	5.03	· · ·	7.03	1.89	· · ·
C	· · ·	5.22	· · ·	4.17	4.11	· · ·
Hy	· · ·	9.78	· · ·	10.3	8.54	· · ·
Or	34.0	33.0	22.5	42.7	50.5	23.6
Ab	14.7	32.9	32.9	19.1	23.5	8.36
An	21.3	6.15	24.0	11.2	5.48	15.2
Ne	7.98	· · ·	0.271	· · ·	· · ·	11.9
Di	6.49	· · ·	8.79	· · ·	· · ·	23.1
Ol	8.73	· · ·	4.39	· · ·	· · ·	8.09
Mt	3.70	2.78	2.70	2.28	2.16	4.81
Il	0.760	0.912	1.86	0.722	0.800	1.67
Ap	0.711	0.758	0.118	0.166	0.118	0.284
Sum	98.5	96.5	97.5	97.7	97.2	97.0

*Data are from the Philippine Bureau Mines—Japan International Cooperation Agency report. Samples: Orthoclase-hornblende monzonites.

diorite to tonalite intrusions associated with composite batholiths of similar composition. These intrusions typically occur in the contact zone between the batholiths and older metavolcanic rocks of basaltic andesite to andesitic composition.

The Marian deposit is associated with syenitic intrusions associated with the alkalic Palali Batholith (Table 5, Figure 14). The occurrence of porphyry copper mineralization with an intrusion of alkalic composition may be unusual for the Philippines, but it is not a geological oddity. *Sutherland-Brown* [1978] has described several similar occurrences in the Coastal ranges of Western Canada.

Baquiran [1975] has outlined the geologic background of the Marian prospect. The country rocks are dominately medium to fine grained sandstones, conglomerates, and siltstones. In terms of composition and stratigraphy the sandstones and siltstones are volcanic graywackes. Rhythmic bedding defined by alternating fine and coarse grained units is common, as is graded bedding with crenulations and convolute laminations. The granular components are lithic fragments of andesitic affinity, mafic mineral fragments, and minor quartz granules. This sedimentary country rock is intruded by alkalic plutons probably related to the Palali batholith. General aspects of the Palali batholith have been described by *PBM-JICA* [1976], with potassium-argon ages ranging from 17.2 to 20.3. The majority of the intrusions in the Cordon area are alkali syenites and trachy-andesites. Some of these rocks are nepheline normative. Only minor mafic phases associated with stock-like intrusions have been noted (Figure 15).

The porphyries generally show flow banding as a result of subparallel orientation of the feldspar phenocrysts (Fig-

ERRATA

Geophysical Monograph 27
The Tectonic and Geologic Evolution of Southeast Asian Seas and Islands: Part 2
American Geophysical Union

The Table of Contents has reversed the order of the articles appearing at
pages 95 and 124 and at pages 326 and 349. The correct order is as follows:

ure 16e). The typical monzonite intrusion consists of subhedral potassium feldspar laths set in a medium grained matrix of plagioclase, orthoclase, and biotite. Trace minerals include apatite, sphene, and magnetite.

Mineralization is apparently associated with the intrusion of trachy-andesite porphyry stocks. U. Knittel (manuscript in preparation, 1981) reports a strontium isotope isochron age of 24.8 ± 0.6 m.y. B.P. with an initial 87-Sr/86-Sr ratio of 0.70374 ± 0.00014. As would be expected, the alteration-mineralization zones reflect the primary chemistry of the host rocks. Unlike most Philippine porphyry deposits, the potassic zone is well developed and represented by secondary potassium feldspar and biotite mineralization. Unaltered examples of the trachy-andesite porphyry generally carry phenocrysts of plagioclase, augite, and biotite in a matrix of potassium feldspar with rare quartz.

The ore body has a steeply dipping tabular core and is composed principally of chalcopyrite and bornite. Pyrite is rare, and the majority of the ore zone coincides with the zone of potassic alteration. Copper sulfides occur within the ore body as fracture coatings, disseminations, and microveinlets.

Atlas Deposit—Cebu Island

The Atlas mining operation at Toledo, Cebu, represents the largest developed porphyry ore body in the Philippines. The deposit lies approximately 30 km west of Cebu City on a 2-mile wide horst bounded by two parallel east-northeast trending structures (Figure 17). The discovery of the deposit was influenced by this horst structure, since erosion had removed overlying Tertiary units in the horst area. Exploration in the surrounding area is impeded by a thick overlying sedimentary sequence. The geology of the Atlas deposit has been described by Madamba [1972] and anonymous reports of the Atlas Consolidated Mining Company.

The oldest rocks known in the Atlas area are regionally metamorphosed pillow lavas and intercalated volcaniclastic rocks of the Mesozoic Pandan Formation. Flow banding, pillow structures, and amygdular structures are frequently observed. Sedimentary intercalations are typically lithic volcanic wackes and vitric volcaniclastic units. Walther et al. [1981] have reported five 'conventional' potassium-argon ages averaging 105 m.y. B.P. with an argon isochron age of 95 ± 5 m.y. for biotite samples from the Biga and Frank Pits. A biotite-whole rock rubidium strontium age of 107.6 ± 2.9 m.y. correspondes well with the potassium argon data. The single potassium-argon age reported by Wolfe [1972] of 59.7 m.y. may reflect a superimposed thermal event.

Diorite and tonalite intrusions occurring in a northeast/southwest trending belt intrude the metavolcanic unit and are associated with the porphyry copper mineralization (Table 6, Figure 18). The ore deposition is almost entirely

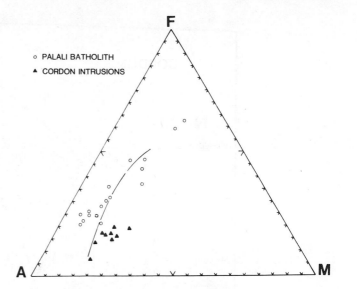

Fig. 14. AFM diagram showing intrusions of the Palali batholith [PBM-JICA, 1976] and Cordon area (this study). A = $K_2O + Na_2O$, F = Fe as FeO, M = MgO, molecular ratios.

confined to the dioritic intrusions and tends to correspond to zones of pervasive fracturing paralleling contacts between the intrusions and metavolcanic rocks.

Two major ore bodies occur at the Atlas mining operation, the Lutopan and Biga-Barot structures. They consist of four mineralized substructures mined as separate pits and an underground operation. Three subparallel lineaments designated as the Lutopan-Lantoy, Biga-Barot, and Kanapnapan-Luay appear to localize the intrusive bodies along a north-northeasterly orientation.

Sipalay

The Sipalay deposit is located along the contact zone between an outlying intrusion of a large batholithic complex (Negros Occidental Batholith) and mafic to andesitic metavolcanics (the Tabon-Tabon and Cansibit Formations) (Figure 19). The intrusion (the Lanipga granodiorite) ranges in composition from quartz diorite to a relatively low potassium granodiorite. The Cansibit dacite is a porphyritic aphanitic intrusion that occurs around the periphery of the Lanipga granodiorite (Figure 20). Partially resorbed quartz phenocrysts are set in a microcrystalline groundmass that has been subject to extensive hydrothermal alteration (Figure 13).

The Sipalay deposit is the second largest deposit in the Philippines and it is quite probable that further exploration in the peripheral zone to the Lanipga intrusion will discover additional reserves. The ore zones are steeply dipping tabular bodies associated with the Cansibit dacite porphyry. The majority of the ore occurs within the porphyry intrusion as stockworks of quartz-sulfide veinlets

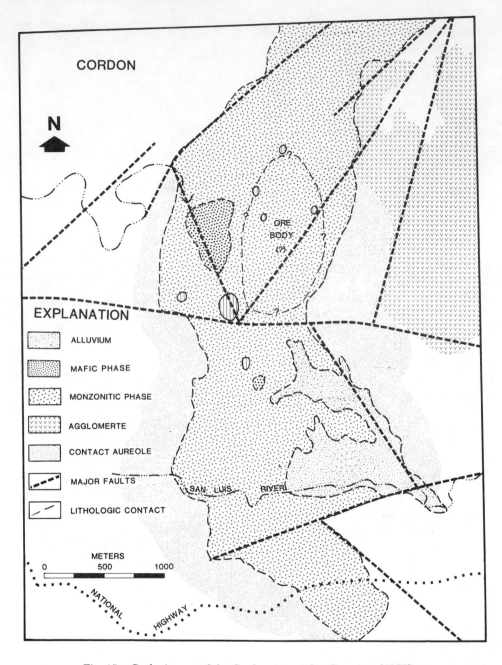

Fig. 15.   Geologic map of the Cordon area, (after *Baquiran* [1975]).

and minor disseminations. Bornite and chalcopyrite are the principal ore minerals and are found as sub-economic disseminations significant distances from the ore bodies in only slightly altered country rocks.

The Sipalay deposit is somewhat anomalous when compared with other porphyries of the Philippines in that it has an unusually low gold abundance and relatively high molybdenum. Biotite and sericite are the principal potassic zone minerals with a poorly defined gradation outwards

into sericite and chlorite. Chalcocite occurs as a blanketing supergene enrichment beneath a surface oxidized zone consisting of chrysocolla, chalcocite, malachite, azurite, and cuprite [*Saegart and Lewis,* 1977]. Ore mineralization was probably enhanced by extensive faulting and fracturing along the western margin of the Lanipga intrusion.

A northwest to northerly trending anticlinal structure extends from the older metavolcanics through the Lanipga intrusion. Secondary fold structures associated with the

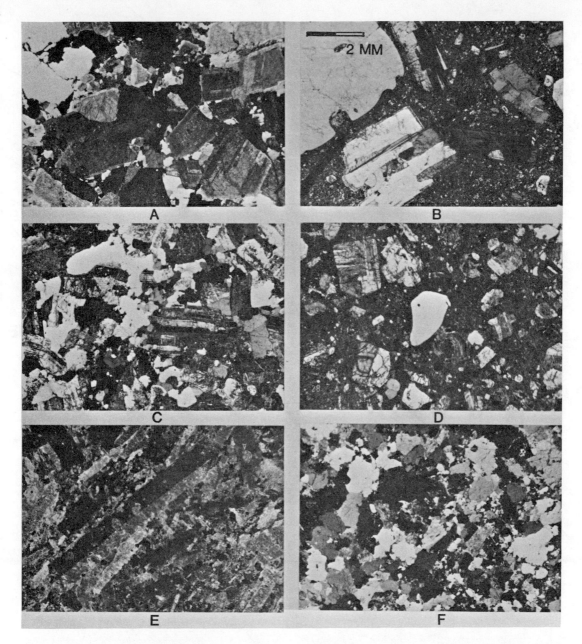

Fig. 16.   Intrusive rocks associated with porphyry copper intrusions. (a) Quartz diorite phase of the Lanipga intrusion, Sipalay Mine, Negros. (b) Partially resorbed quartz pheoncryst from dacite porphyry intrusion Sipalay mine. (c) Hornblende quartz diorite Taysan Mine, Batangas, and Luzon. (d) Quartz dacite porphyry Taysan intrusion. (e) Orthoclase phenocrysts Marian Prospect, Cordon, and Luzon. (f) Seriate quartz diorite, Samar Island.

intrusion of the Cansibit dacite strike north-northeast to northeast. Dacite and quartz diorite sills and dikes intrude the older metavolcanics and the Lanipga granodiorite.

Hinobaan

Aspects of the geology and exploration of the Hinobaan deposit have been previously discussed by *Malicdem* [1974].

The geology of Hinobaan is similar to that of the Sipalay. The mineralization occurs along the eastern contact of the composite Negros Occidental batholith. The ore body is approximately 1.3 km long and contains reserves of approximately 100 million metric tons. The mineralization occurs principally in basaltic to andesitic clastics equivalent to the Cansibit andesite tuff and Tabon-Tabon For-

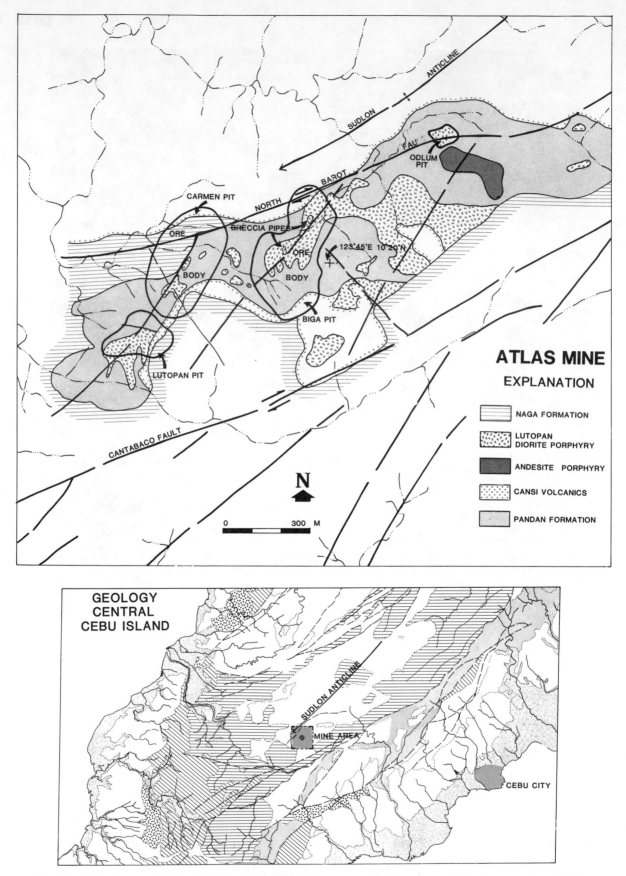

Fig. 17.   Geologic map of central Cebu and the Atlas Mine area (after Reyes, unpublished manuscript, 1977).

TABLE 6. Atlas Intrusions

	8-J100	2A3-980	10-2-208	6-P264	2-PB26	S-BR38	9PAN
SiO_2	68.04	65.88	63.73	67.21	66.68	64.78	61.02
TiO_2	0.45	0.50	0.53	0.37	0.40	0.51	0.52
Al_2O_3	18.47	18.55	17.55	18.62	17.74	16.15	18.57
Fe_2O_3	4.67	7.40	8.06	5.54	5.84	6.37	5.19
MgO	1.97	2.38	2.51	1.09	2.10	2.84	3.00
CaO	2.36	1.37	2.25	2.98	1.74	1.13	1.53
Na_2O	3.41	2.72	3.41	2.09	2.99	7.39	7.20
K_2O	0.59	1.16	1.91	1.84	2.39	0.70	2.96
P_2O_5	0.27	0.23	0.20	0.31	0.22	0.23	0.19
Ag, ppb	44	43	· · ·	38	97	163	43
Au, ppb	0.5	25.6	· · ·	2.7	1.7	39.4	0.5
Ba, ppm	590	440	550	385	580	470	420
Cr, ppm	3	35	30	14	18	· · ·	· · ·
Cu, ppm	12	475	523	25	105	480	22
Li, ppm	6	14	10	9	8	· · ·	· · ·
Mn, ppm	557	588	1005	635	790	1360	785
Mo, ppm	· · ·	3.10	· · ·	1.16	· · ·	2.22	· · ·
Ni, ppm	10	10	15	15	13	5	7
Rb, ppm	9	25	28	22	34	· · ·	· · ·
Sn, ppb	840	380	590	520	470	10800	410
Sr, ppm	760	460	378	610	820	· · ·	· · ·
Se, ppb	177	400	36	30	30	516	16
Te, ppb	21	15	3.3	8.2	6.3	50	1.6
V, ppm	80	92	132	113	96	85	112
Zn, ppm	23	34	53	51	50	140	42
			Normative Minerals				
Q	36.4	36.5	25.0	37.1	31.1	10.1	· · ·
C	8.58	10.9	6.26	7.51	7.60	1.73	1.19
Or	3.50	6.86	11.3	10.9	14.1	4.14	17.5
Ab	28.8	21.0	28.8	17.7	25.3	62.5	60.9
An	9.94	5.30	9.86	12.8	7.20	4.10	6.35
Hy	7.89	11.0	11.8	6.52	9.25	11.5	3.87
Mt	2.71	4.29	4.67	3.22	3.39	3.70	3.02
Il	0.855	0.950	1.01	0.703	0.760	0.969	0.988
Ap	0.639	0.545	0.474	0.734	0.521	0.545	0.450
Sum	99.3	99.3	99.3	97.1	99.2	99.3	94.3

8-J100: Hornblende quartz diorite.
2A3-980: Hornblende quartz diorite.
10-2-208: Hornblende diorite.
6-P26: Hornblende quartz diorite porphyry.
2-PB26: Hornblende quartz diorite porphyry.
5-BR38: Hornblende quartz diorite porphyry.
9-PAN: Hornblende andesite porphyry-post-mineralization.

mation found at Sipalay. In the area of Hinobaan, the Negros Occidental batholith is composed of diorite, tonalite, and hornblende tonalite porphyry (Figure 21). Thin dike-like bodies of granodiorite to quartz monzonitic composition intrude the other units.

Intrusive relations indicate that the hornblende tonalite porphyry is one of the youngest units in the sequence. Hornblende, partially resorbed quartz, and andesine constitute the phenocryst phases that are set in a fine grained hornblende, plagioclase, and quartz ground mass. A strongly altered quartz porphyry located near the northern portion of the ore body resembles the Cansibit dacite porphyry of the Sipalay deposit. It, like the Cansibit dacite porphyry, is characterized by the presence of large (2–10 mm) partially resorbed, bipyramidal quartz phenocrysts (Figure 15).

The Hinobaan deposit is located on an extension of the same broad northwest trending anticlinal structure in metavolcanic rocks as the Sipalay deposit. This structure was apparently formed as a result of the intrusion of the Negros Occidental batholith. The regional fault system is subparallel to the strike of the anticlinal structure. A majority of these faults strike north-northwest and the Hinobaan ore body is located in an approximately 300-m wide fault zone that intersects an easterly dipping contact between the metavolcanic rocks and the batholith. Post-mineralization andesitic sills and dikes occur in the area of Hinobaan and may be equivalent to a relatively thick column of basaltic andesite flows and breccias located to the east. Coralline limestone with interbedded shale and sandstone of the Canturay (Pliocene) Formation forms a cap on the erosion surface developed on older rocks.

The porphyry copper mineralization at Hinobaan occurs as quartz-sulfide veinlets and disseminations largely in

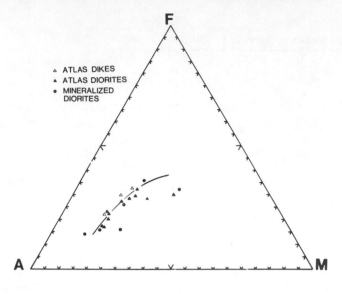

Fig. 18.   AFM diagram of Atlas intrusions. $A = K_2O + Na_2O$, F = Fe as FeO, M = MgO in molecular ratios.

the older pyroclastics rather than in the dioritic and dacitic intrusions. The orebody has been described as having the shape of a horizontal canoe by *Saegart and Lewis* [1977]. No well-defined potassic zone occurs in the deposit. The ore zone is typified by strong quartz 'flooding' and sericite alteration. The ore zone is essentially phyllic and as much as 60% of the host rock in the mineralized zone is occupied by a stockwork of quartz veinlets [*Malicdem*, 1974]. Outer alteration zones are poorly defined and grade into low grade metamorphic assemblages of epidote, chlorite, and sericite in the metamorphosed volcaniclastics.

The principal ore minerals are chalcopyrite, and bornite. As at Sipalay, supergene chalcocite and cuprite are common from the surface to a depth of approximately 60 m. This occurrence contrasts with younger Philippine deposits that rarely show extensive supergene mineral development.

Discussion

Tectonic Synthesis

It is difficult to draw any absolute conclusions regarding cause and effect relationships between subduction processes and the formation of porphyry copper deposits. There does appear to be a remarkable coincidence of porphyry intrusions with certain types of plate boundary interaction. *Titley* [1975] and *Titley and Beane* [1981] pointed out this correlation for the New Guinea-Solomon Island area and suggested that porphyry copper intrusions might be related to increased convergence rates and uplift and tension in the volcanic arc. At least three distinct periods of porphyry mineralization have occurred in the Philippines

at ca. 3–7 m.y. B.P. in the Baguio District of the Cordillera, with other major intrusive events at ca. 15–17 m.y. B.P. and 22–27 m.y. B.P.; the Marian deposit in the alkalic Palali batholith ca. 25 m.y. B.P. in the Batangas-Marinduque and possibly eastern Zambales ca. 14–20 m.y. B.P., in the Negros area ca. 30 m.y. B.P. and Cebu ca. 100 m.y. The time of mineralization in eastern Mindanao is not very well known. A K/Ar age of 6.7 m.y. B.P. for the Mati diorite in Davao and stratigraphic evidence suggests a tentative Late Miocene age for this group.

Several major tectonic events in the Pacific basin are closely associated in time with periods of porphyry mineralizations and barren intrusion (Figure 3):

1. Cebu (95–110 m.y. B.P.). Accelerated seafloor spreading and subduction, suggested by the magnetic quiet zone from 110 to 85 m.y. B.P. [*Larson and Pitman*, 1972]. Extensive batholithic intrusions in the Sierra Nevada. Porphyry mineralization occurred at Ely and Yerington, Nevada (109 m.y. B.P.) and El Arco, Baja California (107 m.y. B.P.) [*Hollister*, 1978]. This period may represent the time of initial rifting of the 'stable block' from the Asian Craton [*de Boer et al.*, 1980].

2. Barren calc-alkaline magmatism in the Sierra Madre of Luzon and the Visayan Islands (ca. 50 m.y. B.P.). Onset of spreading on the Central Basin Ridge (fault) with a half rate of 5 cm/year between anomalies 25 and 17, ca. 59 m.y. to 41 m.y. [*Shih*, 1980] or 52 m.y. to 37 m.y. B.P. [*Taylor and Hayes*, 1980]. In addition, an east-west to northwest/southeast orientation of the ridge axis is compatible with the northwest strike of folding concordant with the intrusions. In this same time period metallogenic episodes occurred in Canada (ca. 50 m.y. B.P.) [*Christopher and Carter*, 1976]; the southwestern United States (copper) and Thailand (tin) (ca. 50–60 m.y. B.P.) [*Sillitoe*, 1976].

3. Baguio magmatism and mineralization at the Marian deposit (22–27 m.y. B.P.) Marcopper (20 m.y.), and the deposits of the Negros Occidental Batholith (30 m.y.). During this time period changes occurred in direction of plate motion at 24.6 + 2.5 m.y. [*Jackson et al.*, 1972] and cessation of spreading on the Central Basin Ridge at ca. 26 m.y. B.P. [*Shih*, 1980]. Intrusions occurred at Plesyumi, ca. 22–24 m.y. B.P. and at New Britain, ca. 22–29 m.y. B.P. [*Page and Ryburn*, 1977]. Approximately 25 m.y. B.P. volcanism terminated on the Palau-Kyushu Ridge [*Karig*, 1975] the Shikoku Basin started opening [*Shih*, 1980] and subduction of the Pacific Plate accelerated ca. 24 m.y. B.P. [*Hilde et al.*, 1977]. *Taylor and Hayes* [1980] indicate that an east-west trending spreading center was active in the South China Sea during this period, ca. 17–32 m.y. B.P. Geochronologic data suggests that intrusion in the Sierra Madre-Caraballo Mountain-Cordillera became progressively younger east to west [*PBM-JICA*, 1977; *Wolfe*, 1981]. Alkali syenites and monzonites were intruded at approximately the same time as early calc-alkaline diorites in the Baguio district. This data suggests that the paleo-Manila trench system was active since the early Miocene.

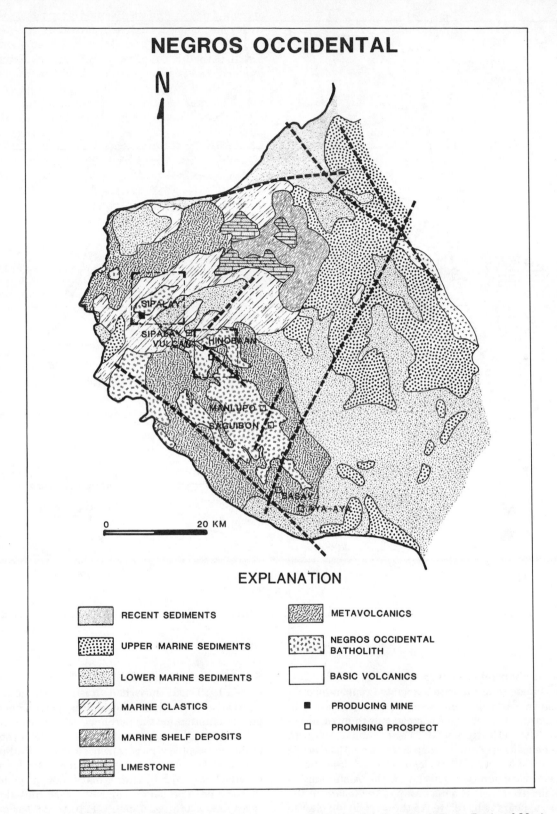

NEGROS OCCIDENTAL

N

SIPALAY

SIPALAY ET
VULCAN HINOBAAN

MANLUPO

SAGUIBON

BASAY
AYA-AYA

0   20 KM

EXPLANATION

RECENT SEDIMENTS	METAVOLCANICS
UPPER MARINE SEDIMENTS	NEGROS OCCIDENTAL BATHOLITH
LOWER MARINE SEDIMENTS	BASIC VOLCANICS
MARINE CLASTICS	■ PRODUCING MINE
MARINE SHELF DEPOSITS	☐ PROMISING PROSPECT
LIMESTONE	

Fig. 19. General Geological Map of Negros Occidental, (after Philippine Bureau of Mines, Regional Map).

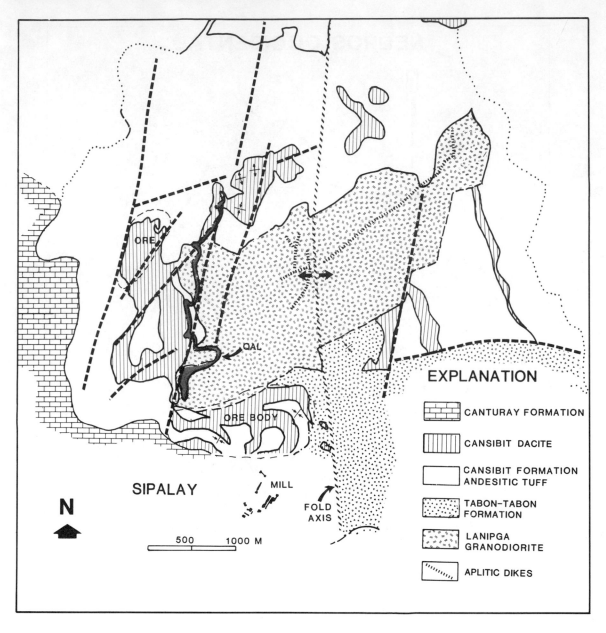

SIPALAY

N

500   1000 M

EXPLANATION

CANTURAY FORMATION

CANSIBIT DACITE

CANSIBIT FORMATION
ANDESITIC TUFF

TABON-TABON
FORMATION

LANIPGA
GRANODIORITE

APLITIC DIKES

Fig. 20.   Geologic map of the Sipalay Mine area.

4. Batanqas-Marinduque area (15–20 m.y. B.P.). This area may include minor diorite associated copper deposits in the eastern Zambales and some porphyry deposits in the Baguio area. This age also corresponds to intrusion at the Frieda River (13–16 m.y. B.P.) and Yanderra (12–14 m.y. B.P.) deposits in New Guinea [*Page and McDougall,* 1972; *Grant and Nielsen,* 1975). *Scholl et al.* [1975] indicate a strong period of igneous activity in the Aleutians between 10 and 16 m.y. B.P. This group of intrusions in the Philippines appears to be related to the west facing Manila Trench. Subduction of the young (at that time) South China

Sea Ridge and basin active 17–32 m.y. B.P. [*Taylor and Hayes,* 1980] could have had unusual tectonic and thermal effects. However, this is not entirely satisfactory as a simple explanation for the porphyry mineralization.

5. Baguio district (4–9 m.y. B.P.). This is the youngest period of porphyry mineralization in the Philippines and may also be represented the Eastern Mindanao area. It may include major hydrothermal events as young as Plio-Pleistocene. The period coincides with metallogenic episodes in Canada, ca. 8 m.y. B.P. [*Christopher and Carter,* 1976], the Andes, ca. 7–10 m.y. B.P. [*Sillitoe,* 1977], the

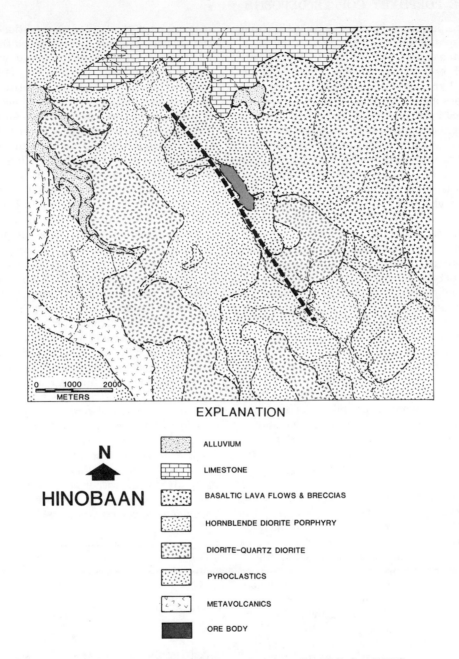

Fig. 21. Geologic map of the Hinbaan prospect (after *Malicdem* [1974]).

New Guinea-Solomon Islands area ca. 1–10 m.y. B.P. [*Page and McDougall*, 1972], and Northern Celebes at Tapadaa [*Lowder and Dow*, 1977].

Petrogenesis of Porphyry Copper Deposits

One of the more puzzling features of porphyry copper related intrusions is that there are few characteristics associated with major or trace element chemistry that distinguish the porphyry intrusions from barren intrusions of similar petrography. The principal contrasts appear to

be the result of hydrothermal activity. It is not entirely clear why some intrusions are mineralized and others are not.

Various proposals have been put forth to explain the formation of porphyry magmas. These include subduction of metaliferous sediments [*Sillitoe*, 1972], level of erosional exposure [*James*, 1971], halogen content of the magma [*Stollery et al.*, 1971], and metallogenic setting of the intrusion [*Noble*, 1974; *Mitchell and Garson*, 1972]. *Oyarzun and Frutos* [1974], *Zohenshain et al.* [1974], and *Clark and*

Foster [1979] have presented subduction based metallogenic models to explain broad geographic zonation in ore deposits. The limited exposures and sampling in the Philippines precluded establishing a regional zonation of chemistry for porphyry deposits. However, some zonation of trace metals does appear to occur in recent lavas as a function of inferred subduction zone depth [*Divis,* 1980]. Although the volcanism appears to be associated with subduction of two separate plates (Philippine Sea and South China Sea), no particular geographical zonation is shown other than that implied to be correlated with subduction zone depth. No distinct correlation with depth appeared to occur in copper and zinc concentrations and most other variations were comparatively subtle. Most major and trace element variations correlated with subduction zone depth could be explained by degree of partial melting and phase stability considerations rather than profound variations in the composition of the subducted slab.

These observations suggest that appealing to anomalous subducted plate or mantle metal concentrations for the derivation of porphyry magmas may not be viable. The question still exists as to whether a 'porphyry magma' has a significantly higher copper content than similar magmas not containing sufficient volatiles for hydrothermal enrichment. In the Baguio district, which has been most extensively studied and sampled, the intra- and post-mineralization sheeted dike swarms and stocks have been interpreted as cosanguinous with the porphyry magma [*Divis and Balce,* 1976; *Divis,* 1977]. These intrusions show slightly higher copper concentrations than those of the older Agno Batholith. Distinctive enrichment trends of Cu and Ag in Type II dikes may be indicative of the early 'magmatic stages' of porphyry formation. Molybdenum does not show a distinct fractionation trend in series. *Putnam and Burnham* [1963] and *Lovering et al.* [1970] have reported a relatively consistent enrichment of whole rock and mineral phase copper in porphyry associated intrusions versus barren intrusions.

Feiss [1978] and *Mason and Feiss* [1979] have demonstrated an interesting correlation between major element chemistry and porphyry mineralization. Utilizing the work of *Burns and Fyfe* [1964], which suggested that the $Al_2O_3/(K_2O + Na_2O + CaO)$ ratio of a magma is proportional to the number of octahedral sites in the melt, they proposed that copper would preferentially fractionate into the liquid in some proportion to this ratio. Thus, the higher the aluminum/alkali ratio, the greater the likelihood of formation of a porphyry deposit due to higher copper concentration of the magma. Their [*Mason and Feiss,* 1979] SiO_2 versus $Al_2O_3/(K_2O + Na_2O + CaO)$ diagrams for deposits and associated intrusives of the southwestern United States and New Guinea-Solomon Island area show a separation into two distinct fields for mineralized and nonmineralized suites. This relationship appears to also hold for the Philippines as shown in Figure 22.

The work of *Holland* [1972] and *Whitney* [1975, 1977] have demonstrated that under an appropriate combination of temperature and pressure conditions a hydrous magma can evolve an aqueous vapor phase capable of extracting metals from an 'average' felsic pluton to produce an ore deposit of signficiant size. *Burnham* [1979] has pointed out that the presence of primary biotite and/or hornblende phenocrysts places a lower limit on magma water contents of 3–4 weight percent. This amount of water would be sufficient to produce a large scale hydrothermal effect compatible with porphyry deposition under the circumstances envisioned by *Whitney* [1975]. Thus, a copper enriched magma may not be required for copper mineralization.

Burnham [1979] has suggested a comprehensive model of porphyry copper formation that is compatible with the geochemical tectonic environment of the Philippines. The model is based on melting of an amphibolitic equivalent of oceanic basalt with only mineralogically bound water present. He points out that given an average thermal gradient between 8.5 and 13.5 °C/km copious production of melt will occur abruptly at a nearly constant pressure of 22–23 Kb (about 75–80 km depth). The amount of melt generated is proportional to the water content of amphibolites. As an example, a 20% melt of an average amphibolitic basalt with 400 ppm sulfur, 87 ppm copper, and 0.6% H_2O can produce a sulfur saturated quartz diorite with 400 ppm copper and approximately 3% H_2O.

Burnham's [1979] discussion of the thermodynamic relations in magmas also provides insight into textural and chemical features of porphyry intrusions. He notes that in hydrous magmas quartz liquidus temperatures increase approximately 20°C to 30°C/kb. Thus, phenocrysts formed from a silica saturated magma will tend to be resorbed as a result of near isothermal or adiabatic depressurization. Such embayed quartz phenocrysts are a typical petrographic feature and 'namesake' of porphyry deposits.

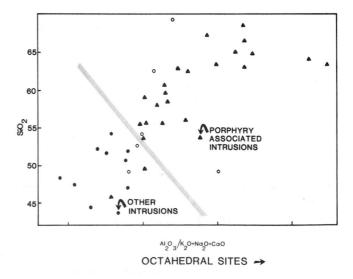

Fig. 22. Alumina/alkali diagram of Philippine porphyry intrusions and barren intrusions. Open circles indicate data from *PBM-JICA* [1976].

The presence of dissolved H_2O may also account for the higher aluminum/alkali ratio previously disucssed. The presence of H_2O tends to lower the crystal-melt equilibrium temperature with increasing pressure, possibly because of the conversion of tetrahedrally coordinated Al-IV to octahedrally coordinated Al-VI. The earliest formed plagioclase then becomes more sodic, raising the aluminum/alkali ratio. Thus, the observations of *Feiss* [1978] and *Mason and Feiss* [1979] may be explained as a consequence of crystal-liquid equilibrium of a hydrous magma. It would appear to be difficult to separate cause and effect in this instance. The formation of a porphyry deposit could require a high aluminum/alkali ratio or the aluminum/alkali ratio could simply be a consequence of the hydrous magma that is required for a hydrothermal system.

Although it is possible to formulate a petrogenic scheme for the evolution of porphyry copper magmas in the Philippines, it is not clear that the details of such a scheme would be unique to porphyry magma evolution as opposed to the formation of barren island arc diorites and quartz diorites. In addition, it is probable that conclusions drawn from the Philippine data may not be generally applicable. In a review of trace element models pertaining to andesite genesis, *Gill* [1978] has pointed out that no satisfactory quantitative trace element models exist for andesitic suites which span a range in excess of 5% difference in silica content (i.e., 55–60%). In the case of porphyry copper magmatic models and trace transition elements, the situation is even less satisfactory.

Primary intrusive compositions may show silica compositional ranges in excess of 10 weight percent. The effects of intense hydrothermal alteration can produce profound fractionation in the trace transition elements of rocks showing relatively minor evidence of mineralogical alteration. The rock suites selected for this study were intentionally screened in an attempt to eliminate such effects. However, the near surface intrusion of hydrous mineral phase bearing magmas is usually associated with the evolution and boiling of a fluid phase. The intrusions are generally porphyritic holocrystalline and do not represent an entirely satisfactory subject for quantitative trace element modeling. The chemical diversity of more readily studied volcanic suites has usually required models involving complex mixing and/or melting processes. Thus, this presentation has been limited to general observations regarding the apparent interrelationship between tectonic and geochemical processes during porphyry magma formation.

Major Element Data

Major element analyses were performed on porphyry associated intrusions from the Baguio Mining District, Dizon, Taysan, Marcopper, Paracale-Polillo, Marcopper, Isao-Pili, Atlas, Sipalay, and Hinobaan mining areas. The most detailed studies were completed in the Baguio area owing to the number of porphyry deposits and the more accessible

exposures. The Kennon Road-Emerald Dike Swarm presented particularly good exposures of a complex of dikes and stocks apparently closely related to porphyry mineralization.

The majority of intrusive rocks mapped and analyzed are hydrous mineral phase bearing andesites, dacites, and plutonic equivalents and diorites and quartz diorites. The most common hydrous mineral phase is pleochroic olive-green–light green hornblende, followed by tan biotite, and chlorite. Pyroxenes are relatively rare in the porphyry associated intrusions of the Philippines.

Typical silica contents ranges from 45% in the basaltic andesites and mafic diorites to 67% in dacites and quartz diorites. The porphyry associated rocks tend to have slightly higher alumina and alumina to alkali ratios than 'barren' rocks. As a result, many are corundum normative. Potassium ranges from 0.25% in trondjemitic intrusions to 4% in quartz diorites and granodiorites. In alkalic rocks of the Cordon deposit potassium may reach 8%.

Rock analyses were performed by X-ray fluorescence spectrometry for SiO_2, Al_2O_3, TiO_2, CaO, K_2O, and P_2O_5 using a Philipps vacuum spectrometer. Na_2O, MgO, and trace element analyses were performed by atomic absorption spectroscopy. Rare earth analyses were completed by isotope dilution or neutron activation. Methodology and precision are summarized in Table 1. Detailed analytical procedures are presented in *Divis* [1976] and *Clark and Viets* [1981].

Isotope Data

Fourteen strontium isotope determinations were made on samples from seven different porphyry localities (Table 7, Figure 23). Relatively minor corrections were made for the post-intrusive evolution of 87-Sr due to the relatively low Rb/Sr ratios of the rocks. The deposits studied ranged in age from approximately 7 to 100 million years in age. Generally, the required correction was less than the analytical uncertainty. The initial 87-Sr/86-Sr ratios ranged from 0.70324 to 0.70404 and averaged 0.70367 with a standard deviation of 0.00024. This coincides well with initial ratios for Philippine volcanic rocks 0.70372 which form a pseudoisochron with an initial 87-Sr/86-Sr ratio of 0.70350 and apparent age of 51.8 m.y. [*Divis,* 1980], Palau — 0.70342 [A. F. Divis, unpublished manuscript, 1982] and ratios for other island arc porphyries summarized by *Kesler et al.* [1975] (Figure 23). Ratios from the Cebu-Negros area appear to be slightly higher than younger samples from Luzon and Marinduque islands, but sampling is insufficient to draw any conclusions.

Trace Element Data

Since one objective of this study was to ascertain the abundance and distribution characteristics of ore forming trace metals, some comparatively low abundance elements (Au, Ag, Sn, Te, and Se) were analyzed with the intention of employing them as tracers. More abundant trace tran-

TABLE 7. Strontium Isotope Data

Measured 87-Sr/86-Sr	Rb/Sr	Assumed Age, m.y.	87-Sr/86-Sr Increase	87-Sr/86-Sr, Initial
Baguio District				
PMB-07 $0.70325 \pm 2 \times 10^{-5}$	0.0485	7	1.4×10^{-5}	0.70324
BAL-37 $0.70375 \pm 3 \times 10^{-5}$	0.0096	25	1.0×10^{-5}	0.70374
BAL-24 $0.70349 \pm 12 \times 10^{-5}$	0.0237	15	1.4×10^{-5}	0.70348
KRL-18 $0.70362 + 18 \times 10^{-5}$	0.0099	7	0.3×10^{-5}	0.70362
BPB-01 $0.70368 \pm 10 \times 10^{-5}$	0.0346	15	2.1×10^{-5}	0.70366
Cordon				
(U. Knittle, manuscript in preparation, 1981)		24.8	isochron	0.70374
Batangas-Marinduque				
MCM-01 $0.70364 \pm 6 \times 10^{-6}$	0.1190	15	7.2×10^{-5}	0.70357
MCM-07 $0.70357 + 5 \times 10^{-5}$	0.0211	15	1.3×10^{-5}	0.70356
MCM-10 $0.70405 \pm 6 \times 10^{-5}$	0.0200	15	1.2×10^{-5}	0.70404
MCM-20 $0.70373 \pm 5 \times 10^{-6}$	0.0385	15	2.3×10^{-5}	0.70371
TML-34 $0.70398 + 8 \times 10^{-5}$	0.0250	15	1.5×10^{-5}	0.70396
PSL-01 $0.70325 + 4 \times 10^{-5}$	0.0333	15	2.0×10^{-5}	0.70323
Negros				
SMN-05 $0.70390 + 5 \times 10^{-5}$	0.0395	30	4.8×10^{-5}	0.70385
Cebu				
AMC-21 $0.70383 + 6 \times 10^{-5}$	0.0287	110	12.7×10^{-5}	0.70370
ILAG $0.70401 + 8 \times 10^{-5}$	0.0240	110	10.6×10^{-5}	0.70390
Cebu [*Walther et al.*, 1981]		107.6	isochron	0.70530
Mean				0.70367
Variance				0.0347
Standard deviation				0.00024

PMB: Santo Tomas II—Philex mine.
BAL: Agno batholith diorites.
KRL: Kennon Road dike swarm.
BPB: Dike—Acupan mine.
MCM: Marcopper mine diorites.
TML: Taysan prospect biotite quartz diorite.
PSL: Paracale quartz diorite.
SMN: Sipalay mine quartz diorite.
AMC: Atlas mine quartz diorite.
ILAG: Quartz diorite outside area of mineralization.

sition metals (Cu, Mn, Cr, Ni, V, and Zn) were also determined. Rb, Sr, and Ba were used in characterizing silicate fractionation processes.

In order to simplify visual presentation, two sets of triangular diagrams have been prepared. The first set presents the interrelationship between K, Rb, Ba, and Sr. The second set considers the fractionation of the elements Ag, Au, Cu, and Zn (Figures 24 and 25). In order to develop a frame of reference for comparison of porphyry magma evolution, a complementary study was made of the geochemical characteristics of recent volcanism in the Philippines [*Divis*, 1980].

One of the most obvious features noted in the examination of the chemical distribution diagrams for volcanic rocks is the very diverse trace element distribution found in hydrous mineral bearing rocks when compared with relatively anhydrous volcanics. This could in part be due to the greater abundance of hydrous phase magmas, but included in this 'hydrous' field are rocks of the Taal Caldera. In Taal, hydrous phases are relatively rare but hydrothermal activity is very obvious within the Caldera lake.

The K-Sr-Rb volcanic diagram (Figure 24a) is readily compared with the porphyry diagram. In the volcanic diagram, an evolutionary path of relatively constant K/Rb ratio with increasing K-Rb abundance is followed by the anhydrous series rocks. The hornblende bearing lavas show an evolutionary path starting with relatively lower strontium and proceeding with a decreasing K/Rb ratio. The porphyry associated intrusions occupy a field very similar to that of the combined volcanic rocks, with a few samples having a slightly higher proportion of strontium than the recent lavas.

The Rb-Ba-Sr diagram (Figure 24b) is interesting in that it appears to show that the hornblende bearing volcanic rocks have a relatively higher Ba-Sr/Rb initial ratios than found in the anhydrous series. Examination of the porphyry diagram (Figure 25b) shows a similar initial abundance distribution to the hornblende bearing volcanics. An additional feature is the apparent evolutionary paths found in the various deposits, suggesting derivation from magmas having distinctly different initial Rb-Ba-Sr compositions.

The Ag-Cu-Au diagrams (Figures 24c and 25c) illustrate

an important contrast between the hornblende volcanic rocks and the hornblende diorite and quartz diorites of the porphyry intrusions. The V shaped distribution pattern of the hornblende volcanics suggests that two different fractionation processes may occur in addition to that represented by the constant Cu/Au ratio distribution occurring in the anhydrous magmas. Surprisingly, the trend toward relative gold enrichment is associated with porphyry deposits (Figure 26). This is partially an anomaly of the normalization technique used in the presentation, but does represent the relative fractionation of the latter two elements. The enrichment factor of copper from intrusion to ore (50 ppm to 0.5%) is approximately 100; for gold (1 ppb to 500 ppb) it is in the range of five times as great. The alkalic rocks of the Marian-Cordon deposit show an unusual scatter from the 'normal' patterns suggesting that several fractionation mechanisms may be occurring side by side.

It is suggested that the Ag to Au enrichment path is a chloride dominated hydrothermal system, while the Ag to Cu path may be related to a sulfide-sulfate system. The volcanoes represented by the Ag-Cu path are generally sulfataric and samples acquired represent their most recent series of flows. It should be emphasized that these diagrams represent an intermediate range of fractionation between purely magmatic and late hydrothermal stages. The scale of presentation is a compromise between illustrating the fine structure of early differentiation versus the later stages of extreme hydrothermal fractionation.

The Ag-Zn-Cu diagrams for the volcanic rocks and porphyry intrusions are very similar, both showing a fractionation toward the copper apex. As in the previous examples, the anhydrous rocks show a limited distribution with a relatively constant Cu/Zn ratio. Figure 27 illustrates the distribution of Cu-Au-Mo in Philippine volcanic rocks and porphyry intrusions compared to porphyry deposits of other island arcs and continents presented by *Titley* [1978]. The relative abundances of Cu, Au, and Mo for unmineralized porphyry associated intrusions, volcanic rocks, and porphyry deposits are very similar, with the points plotting near the copper apex. A few exotic points are probably due to near 'limit of detection' abundances of Mo and Au.

The dikes in the Kennon Road-Emerald Complex of the Baguio district was extensively studied since it presented very good exposures of porphyry associated dikes. Two distinct generations of dikes were noted: an older series of hornblende-plagioclase porphyries (type I) and a younger series of plagioclase porphyries, dacites, and fine grained seriate to porphyritic quartz diorites (type II) (Figure 8). The dikes show log-linear fractionation patterns with respect to a 'solidification index' $= 10(K_2O + Na_2O)/(MgO + FeO + CaO)$, where the compounds are in molecular proportions.

The type I dikes show patterns that would be fairly typical of a 'normal' silicate fractionation trend. Nickel,

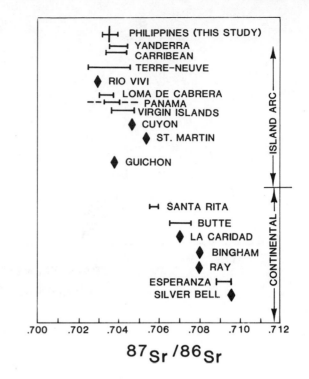

Fig. 23. Strontium isotope diagram for continental and island arc porphyry intrusions (after *Kesler et al.* [1975]).

Zn, Cu, V, and Ag show a fairly consistent decrease with increasing rock differentiation and Au is generally near the limit of detection (0.5 ppb). Rb and Sn show strong enrichment, Sr shows moderate enrichment, and Mn is approximately constant. This pattern would be consistent with the removal of the hornblende and plagioclase phenocryst phases.

The type II dikes show a greater range of fractionation, approximately 45–65% silica versus 45–62% for the type I dikes, but this may only reflect sampling. The type II dikes have at least four times the Au abundance of the type I dikes and show a positive fractionation of Ag and Cu into the more evolved rocks. Type I dikes show higher Cu and Ag abundances at a low 'solidification index' than do type II dikes, but the most fractionated members of the type II series have higher Cu and Ag abundances than the mafic end members of the type I series. Manganese shows a systematic decrease in the type II series. The remaining elements (Rb, Sr, V, Zn, Ni, and Sn) show very similar abundance patterns in both the series.

This could be an example of the octahedral site induced fractionation suggested by *Feiss* [1978] and *Mason and Feiss* [1979], but one would expect to find a similar enrichment in Ni, unless the partitioning of Ni into a crystal phase is significantly greater than Cu and Ag. One possible explanation may lie in the trace sulfide opaque phases found in the Type II dikes. They may represent an exsolved

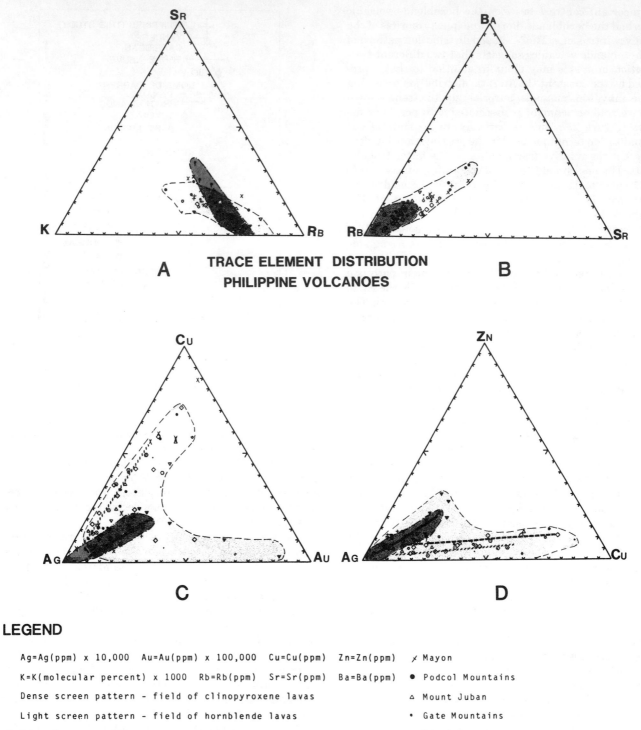

TRACE ELEMENT DISTRIBUTION
PHILIPPINE VOLCANOES

LEGEND

Ag=Ag(ppm) x 10,000   Au=Au(ppm) x 100,000   Cu=Cu(ppm)   Zn=Zn(ppm)   ⟋ Mayon

K=K(molecular percent) x 1000   Rb=Rb(ppm)   Sr=Sr(ppm)   Ba=Ba(ppm)   ● Podcol Mountains

Dense screen pattern - field of clinopyroxene lavas   △ Mount Juban

Light screen pattern - field of hornblende lavas   · Gate Mountains

+ Mount Bacaycay, Balagbag, Labo   ✕ Mount Banoy

❢ Isarog   ⎔ Taal

○ Bulusan   ▽ Mount Ariate

◇ Malinao   ▼ Mount Mandalagan and Mount Canlaon, Negros

Fig. 24.   Trace element distribution diagram for recent central Philippine volcanoes.

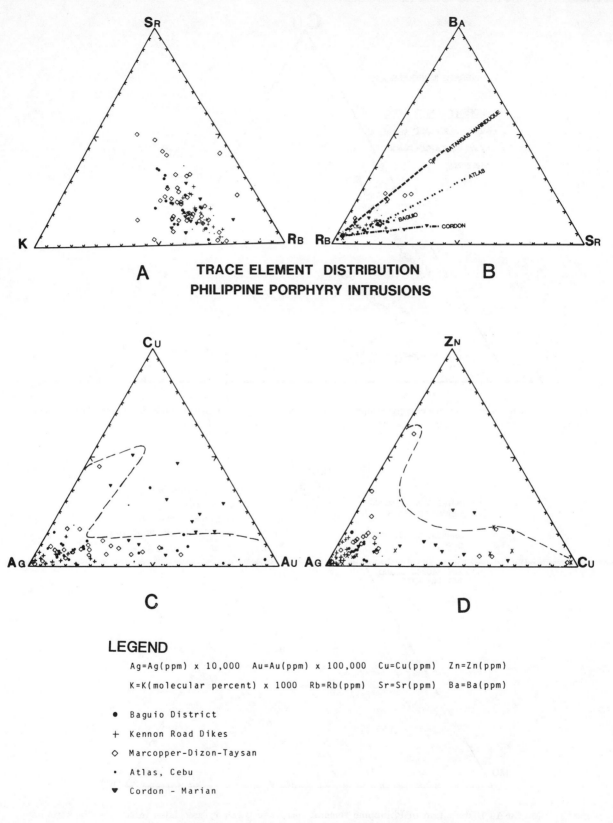

Fig. 25. Trace element distribution diagram for Philippine porphyry copper intrusions.

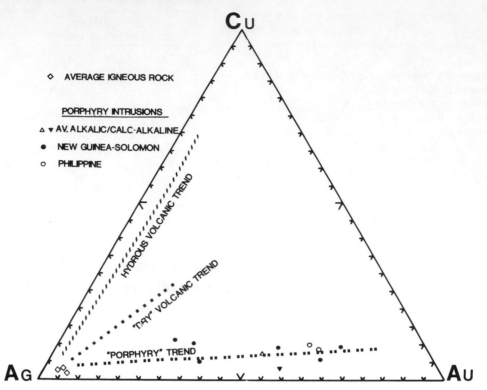

Fig. 26. Ag-Cu-Au diagram for porphyry copper deposits (New Guinea-Solomon Islands data from *Titley* [1978]; average porphyry data from *Sutherland-Brown* [1978]).

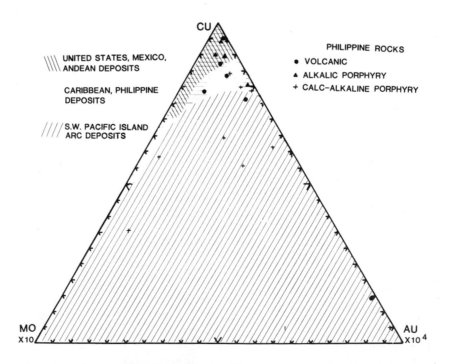

Fig. 27. Mo-Cu-Au Distribution in Philippine volcanic rocks and porphyry associated intrusions. Background fields after *Titley* [1978].

sulfide phase. Unfortunately, the extensive and at times erratic sulfide mineralization in this area can result in disseminated deposition that could be mistaken for primary mineralization. However, it is difficult to resolve this effect with the apparent systematic covariance with major element and Rb and Sr distribution.

This fractionation in the dike complex is very 'preliminary' in terms of the formation of a porphyry deposit. It represents approximately a 5:1 enrichment of Cu and a 2:1 enrichment of Ag. The relative changes in gold are somewhat equivocal in that the abundance is near the limit of detection and a minor increase may be indicated. The abundance variations in rocks showing visible evidence of hydrothermal alteration are generally a factor of ten greater. The separation of a distinct, mobile aqueous phase has a much more profound influence on the formation of an ore deposit than does this early fractionation, but such fractionation may represent an important precursor to such separation.

Rare Earth Elements

Rare earth element (REE) analyses were run on the major porphyry copper intrusions from the Baguio district, the Marcopper Mine, and Atlas Consolidated Mine. In addition, six samples from the Kennon road emerald dike swarm were analyzed (Figure 28). A more detailed summary of rare earth analyses from varied Philippine rock suites will be presented by Goles et al. (manuscript in preparation, 1982). The REE distribution for the two groups of samples is remarkably similar.

Taylor and Fryer [1980] have reported REE abundance patterns for the Bakircay granodiorite porphyry (Turkey) and associated alteration zones. In general, these analyses show much stronger light REE enrichment, approximately 40 times chondrite abundances, a mild Eu enrichment, and heavy REE abundances approximately the same as those of the Philippine rocks (2–8 times chondrites).

The Baguio dikes show slight light REE enrichment relative to the heavy elements but have abundances approximately the same as ocean tholeiites. A point of inflection occurs at Eu and may include a slight positive Eu anomaly. The decrease in relative abundance pattern continues to Tb. From Tb to Lu the pattern is virtually flat.

The porphyry copper quartz diorites show a light REE enrichment only slightly greater than the dike patterns. A point of inflection with a minor positive Eu anomaly is a general feature of the porphyry intrusions. The heavy end of the REE spectrum in these rocks is also flat with an average abundance about five times chondritic abundance. This portion of the REE abundance is significantly less than the average ocean ridge basalt (ORB) abundance.

Geochemical Synthesis

A considerable background of geological and geochemical data suggests that the Philippine porphyry copper deposits formed as near surface subvolcanic intrusions

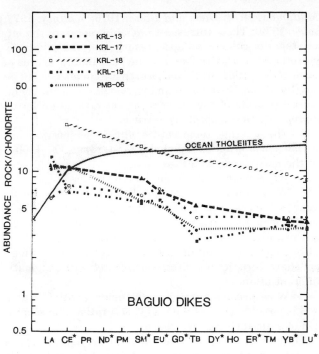

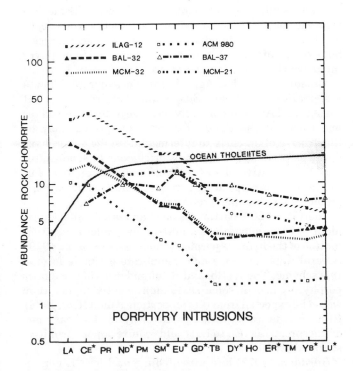

Fig. 28. Chondrite normalized rare earth element patterns for copper porphyry intrusions and Baguio dikes.

[*Madamba,* 1972; *Divis and Balce,* 1976; *Whitney,* 1977; *Balce,* 1979]. These intrusions probably represent the latter stages of caldera collapse, resurgence, and solfataric-hydrothermal activity found in the Taal-Batangas area of Luzon [*Divis,* 1980; *Wolfe and Self,* this volume]. *Gill* [1978] and *Burnham* [1979] have summarized physical and chemical constraints of the latter stages of magma evolution appropriate to the porphyry intrusions:

1. The magma temperature during phenocryst formation and/or eruption is probably in the range 900–1000°C for the hornblende andesites.

2. fO_2 is near or above the Ni-NiO buffer.

3. Water saturation (retrograde boiling) will probably occur in the upper portion of magma chambers and dikes resulting in hydrofracturing, alteration, and hydrothermal mobilization of alkalies and transition metals. This latter effect may result in large scale modification of chemical characteristics not directly associated with magmatic differentiation.

4. Widespread saturation of the magma with CO_2 and/or S is possible even without H_2O saturation and separation of a distinct aqueous phase.

It is obvious that the evolution of the porphyry magmas involves a multi-stage petrologic process that is not readily modeled as a simple system. Some simplification can be accomplished by examining relatively unaltered porphyry associated intrusions. However, considerable variation appears to occur in primary magma characteristics, which is then modified by fractional crystallization. Some correlation appears to occur between subduction zone depth and trace element abundance in Philippine volcanic rocks, but extrapolation of this primary distribution to porphyry intrusive chemistry is unwarranted [*Divis,* 1980].

Vanadium, Cu, Ni, Ag, Ni, and Zn generally show a systematic decrease with SiO_2 in unmineralized rocks, but Cu, Ag, and Au systematically increase in the type II dikes and stocks of the Baguio district. This process may be illustrative of an early to intermediate stage of porphyry development. Copper and Ag enrichment in some suites may be indicative of sulfur associated fractionation, but the lack of corresponding nickel enrichment poses some problems for this interpretation.

The 10% to 20% partial fusion of an amphibolitic equivalent of oceanic basalt is an attractive model for the formation of the hydrous porphyry magmas [*Burnham,* 1979]. Several considerations may complicate a simple form of this scheme. The K, Rb, and Ba abundances of the diorite porphyries are in some cases significantly higher than would be expected from this proportion of melt [*Gill,* 1978]. REE abundance patterns, particularly in the heavy end of the sprectrum (Eu to Lu) could require a more primitive source than ORB.

Burnham [1979] has presented a model for the derivation of porphyry magmas by partial melting of an amphibolite ORB source that is, in most respects, compatible with

chemical and geologic data from Philippine porphyry intrusions:

1. A hydrated equivalent of ORB is subject to abrupt melting at pressures of 22–23 kb (75–80 km depth) and temperatures ranging from 700 to 1050°C. Separation of trondhjemitic to dioritic melt occurs. The presence of residual garnet would tend to result in a decrease in the heavy REE abundance and a steep heavy REE pattern compared to the source. It is therefore difficult to accept significant residual garnet in the source terrain. Extensive residual plagioclase would be expected to result in the formation of a negative Eu anomaly, but moderate amounts of residual plagioclase in the fO_2 range of the Ni-NiO buffer (-8.0 at 1100 C) would produce only a minor anomaly [*Gill,* 1978]. Residual clinopyroxene tends to raise the light REE relative to the heavy REE and prevents large increases in the melt compared to the parent material. Orthopyroxene in the residual material will have the effect of increasing the light REE fairly strongly and heavy REE moderately in the magma [*Arth,* 1976]. Olivine in the residuum will tend to raise the alkalies (Rb, Sr, Ba, K, etc.) and the REE in the magma while reducing Ni in the liquid. Thus, on a qualitative basis, the partial fusion of an ORB composition amphibolite to form a residual refractory assemblage of clinopyroxene, orthopyroxene, and olivine with only minor plagioclase and garnet appears to be a viable model. The principal objections to this model may involve higher than observed abundances of REE and Ni in the hypothetical magma.

This effect is not generally noted in circum-Pacific andesite suites [*Taylor,* 1969; *Kay,* 1977; *Lopez-Escobar et al.,* 1977; *Gill,* 1978]. The more mafic composition indicated for hydrous volcanic rocks suggested in the Rb-Ba-Sr diagram (Figure 24) and relatively primitive REE abundance patterns (Figure 28) may be permissive evidence for this model.

2. Burnham suggests that during the ascent of the magma plagioclase may crystallize without substantial cooling and the magma may become more mafic and hydrous. By contrast, isobaric cooling in a shallow magma chamber would tend to favor precipitation of ferromagnesian minerals. It is suggested that this latter process may be illustrated in the Baguio dikes and other porphyry associated intrusions of the Philippines. Hornblende-plagioclase xenoliths and megacrysts showing comb layering and other cumulate textures are very common in the Baguio area and fractionation of this assemblage would not be incompatible with the evolution of the silicic quartz diorites and mafic granodiorites. This fractionation would probably result in a systematic decrease in Fe, Ni, V, and Cr and increase in silica as observed. Further, it would tend to explain the observed decrease in K/Rb ratios found in hornblende bearing rocks. REE patterns would remain nearly constant or show a slight decrease in heavy REE with a minimum near Dy [*Gill,* 1978]. This is essentially

the trend observed in the REE patterns of the porphyry intrusives and dikes of Figure 28. *Gill* [1978] suggests that a positive correlation between silica contents and Cr/V ratio should be associated with hronblende fractionation. This effect is observed but is not well developed and may be complicated by the onset of magnetite crystallization. A significant drop in Ba/Rb ratio would be expected, as shown in Figure 24.

3. The final stages of porphyry evolution are not readily described in terms of silicate magma-crystal phase equilibrium. Near surface intrusions originating from what may be a much larger subjacent magma chamber form the locus of late magmatic-hydrothermal processes. These intrusions may assume simple stocklike shapes or more complex structures. It is probable that in many cases they are associated with surficial volcanic structures and hydrothermal systems. In the Baguio area, the fluid inclusion work of *Balce* [1979] suggests that the zones of porphyry mineralization occurred within 1500 m of the surface, reasonably near or within the structure of a composite volcanic edifice that may be represented by the Zig Zag and Pugo volcanics. At Santo Nino, *Bryner* [1969] noted that younger andesitic pyroclastics formed a downthrown fault block adjacent to the southwest ore body. Traces of apparently syngenetic copper mineralization occur within the Klondyke series sediments and economic pyrite-chalcopyrite replacement bodies in Pugo-Klondyke equivalent sediments have been mined in the Antamok Mine of the Baguio District. Perhaps one of the most striking examples of volcanic-supergene mineralization is represented by Mount Banoy in the Batangas Peninsula of Luzon. Kuroko type syngenetic sulfide mineralization and sulfide skarns in Pliocene coral reefs occur on the flanks of this structure, which still retains much of its original volcanic topography (J. A. Wolfe et al, manuscript in preparation, 1982).

Thus, the final process of porphyry mineralization appears to be a consequence of the latter stages of hydrous magma crystallization. This model has been presented and refined over the last decade by *Lowell and Guilbert* [1970], *Whitney* [1975, 1977], and *Burnham* [1979]. Porphyry deposits of the Philippines present an interesting series of examples that are 'variations on a theme,' as suggested by *Gustafson and Hunt* [1975]. Hopefully, the data presented herein will provide some useful refinement in the understanding of this basic theme.

Conclusions

Studies of porphyry copper deposits and related intrusive rocks of the Philippines suggest the following conclusions regarding origin and evolution:

1. At least three and probably four major episodes of porphyry intrusion have occurred in the Philippines.

2. The episodes of intrusion show good correlation with regional plate tectonic events, specifically times of rapid plate convergence.

3. Base and precious metal vein mineralization tends to occur on the peripheries of the porphry intrusions.

4. A history of multiple intrusions is generally associated with porphyry deposits and higher aluminum/alkali ratios occur in rock suites related to the porphyries.

5. Anomalously high subducted plate or mantle copper concentrations do not appear to be required for porphyry formation.

6. Evidence for elevated copper concentration in porphyry associated rock suites is equivocal.

7. Hydrous primary mineral phases (biotite and hornblende) are ubiquitous in porphyry associated dike swarms and stocks, suggesting minimum magma water contents of 3–4 weight per cent.

8. The combination of a hydrous, but not necessarily water-saturated, magma may be sufficient to form a porphyry copper deposit if appropriate pressure and temperature conditions exist.

9. Rapid plate convergence and accompanying dilation in the volcanic arc may result in the formation of hydrous magmas in an environment conducive to rapid ascent with minimal dilution of volatiles.

Acknowledgments. This study was carried out at the analytical facilities of the Geology Department of the Colorado School of Mines, the Isotope Geology Branch of the U.S. Geological Survey, and Environmental Science Associates, Denver (ESA). Financial support for a portion of the analytical costs was provided by National Science Foundation grant OCE 76-23381. Support for travel, sample acquisition, and data analysis was provided by ESA. Special appreciation is expressed to the administration and staff of Atlas Consolidated, Sipalay, Marcopper, Santo Nino, Philex, and Benguet Consolidated Mines and the Philippine Bureau of Mines for logistic support, John Wolfe of Pan Asean Technical Services, and Carl Hedge of the U.S. Geological Survey. Gordon Goles of the University of Oregon, School of Volcanology provided neutron activation analyses for REE. Pamela Tarquin, Elaine Anderson, Kent Knudsen, and Terrell York of the ESA staff assisted in the drafting, typing, and preliminary editing of the manuscript. Spencer Titley, John Wolfe, Bob Kamilli, and a anonymous reviewer provided many useful comments and suggestions for improvement of this paper.

References

Agawin, L. R., C. R. Miranda, C. G. DeLeon, and M. E. Colon, Notes on the periods of igneous activity in Luzon as indicated by available K-Ar data, *Tech. Inform. Ser. 26-80,* 8 pp., Bureau of Mines and Geosci., Manila, Philippines, 1980.

Almogela, D. H., Geologic environment and economic possibilities of 'porphyry copper deposits' in the Philippines,

J. Geol. Soc. Philipp., 28, 1–16, 1974.

Apostol, E. L., Porphyry copper deposits in Baguio district Luzon, Philippines, *J. Geol. Soc. Philipp., 18,* 32–43, 1974.

Arth, J. G., Behavior of trace elements during magmatic processes—A summary of theoretical models and their applications, *J. Res. U.S. Geol. Surv., 4,* 41–47, 1976.

Balce, G. R., A. L. Magpantay, and A. S. Zanoria, Tectonic scenarios of the Philippines and Northern Indonesian region, paper presented at the ESCAP CCOP-10C Ad Hoc Working Group Meeting on the Geology and Tectonics of Eastern Indonesia, Bandung, Indonesia, July 9–14, 1979.

Baquiran, G. B., Notes on the geology and exploration of the Marian copper deposit, Cordon, Isabela, *J. Geol. Soc. Philipp., 29,* 1–12, 1975.

Beane, R. E., and S. R. Titley, Porphyry copper deposits, Part II, Hydrothermal alteration and mineralization, *Econ. Geol., 75,* 214–269, 1981.

Bryner, L., Ore deposits of the Philippines-An introduction to their geology, *Econ. Geol., 64,* 644–666, 1969.

Bryner, L., Geology of the sto. nino copper prospect area, *J. Geol. Soc. Philipp., 14,* 33–40, 1970.

Burnham, C. W., Magmas and hydrothermal fluids, in *Geochemistry of Hydrothermal Ore Deposits,* edited by H. L. Barnes, pp. 71–136, John Wiley, New York, 1979.

Burns, R. G., and W. S. Fyfe, Site preference energy and selective uptake of transition-metal ions from a magma, *Science, 144,* 1001–1003, 1964.

Christopher, P. A., and N. C. Carter, Metallogeny and metallogenic epochs for porphyry mineral deposits, in *Porphyry Copper Deposits of the Canadian Cordillera* spec. vol. 15, edited by A. Sutherland-Brown, pp. 64–71, Canadian Institute of Mining and Metallurgy, Ste. Anne de Bellevue, 1976.

Clark, K. F., and C. T. Foster, Compositional variations in Cenozoic mineral deposits located in subduction related magmatic arcs in Western Mexico, *Geol. Soc. Am. Abstr. Programs, 11,* 71, 1979.

Clark, R. J., and J. G. Viets, Multielement extraction system for the determination of 18 trace elements in geochemical samples, *Anal. Chem., 53,* 61–65, 1981.

Corby, G. W., Geology and oil possibilities of the Philippines, *Tech. Bull. 21,* Philipp. Dept. of Agric. and Nat. Resour., 363 pp., Manila, Philippines, 1951.

De Boer, J., L. A. Odum, P. C. Ragland, F. G. Snider, and N. R. Tilford, The Bataan orogene: Eastward subduction, tectonic rotations, and volcanism in the western Pacific (Philippines), *Tectonophysics, 67,* 251–282, 1980.

Diegor, W. G., Geology of the Baguio mineral district, *Tech. Inform. Ser. 13-80,* 12 pp., Bur. of Mines and Geosci., Manila, Philippines, 1980.

Divis, A. F., The geology and geochemistry of the Sierra Madre Mountains, Wyoming, *Q. Colo. Sch. Mines, 71,* 127, 1976.

Divis, A. F., Island arc porphyry copper deposits explora-

tion and evaluation, *Inform. Publ.,* Colo. School of Mines, Golden, Colo., June 1977.

Divis, A. F., The petrology and tectonics of recent volcanism in the central Philippine Islands, in *The Tectonic and Geologic Evolution of Southeast Asian Seas and Islands, Geophys. Monogr. Ser.,* vol. 23, edited by D. E. Hayes, pp. 127–144, AGU, Washington, D. C., 1980.

Divis, A. F., and J. R. Clark, Exploration for blind ore deposits and geothermal reservoirs by lithium isotope thermometry-atomic absorption 'mass spectrometry,' in *Geochemical Exploration 1978, Proceedings of the Seventh International Geochemical Exploration Symposium,* edited by J. R. Watterson and P. K. Theobald, pp. 233–241, Association of Exploration Geochemists, Rexdale, Canada, 1979.

Divis, A. F., and G. R. Palce, Tertiary hypabyssal intrusions of the Philippine Islands, *Geol. Soc. Am. Abstr. Programs, 7,* 1976.

Feiss, G. P., Magmatic sources of copper in porphyry copper deposits, *Econ. Geol., 73,* 397–401, 1978.

Fernandez, H. E., and F. V. Damasco, Gold deposition in the Baguio gold district and its relationship to regional geology, *Econ. Geol., 74,* 1852–1868, 1979.

Fernandez, H. E., F. V. Damasco, and L. A. Sangalang, Gold ore shoot development in the Antamok Mines, Philippines, *Econ. Geol., 74,* 606–627, 1979.

Gervasio, F. C., A study of tectonics of the Philippine archipelago, *Philipp. Geol., 20,* 51–75, 1966.

Gill, J. B., Role of trace element partition coefficients in models of andesite genesis, *Geochim. Cosmochim. Acta, 42,* 709–724, 1978.

Grant, J. N., and R. L. Nielsen, Geology and geochronology of the Yandera porphyry copper deposit, Papua New Guinea, *Econ. Geol., 70,* 1975.

Gustafson, L. B., Some major factors of porphyry copper genesis, *Econ. Geol., 73,* 600–607, 1978.

Gustafson, L. B., and J. P. Hunt, The porphyry copper deposit at El Salvador, Chile, *Econ. Geol., 70,* 857–912, 1975.

Hamilton, W. H., Map of the sedimentary basins of the Indonesian region, *Folio Map 1-875-B,* U.S. Geol. Surv., Reston, Va., 1974.

Hamilton, W., Tectonics of the Indonesian region, *U.S. Geol. Surv. Prof. Pap., 1078,* 345 pp., 1979.

Hawkins, J. W., and C. A. Evans, Geology of the Zambales Range, Luzon, Philippine Islands: Ophiolite derived from an island arc-bach arc pair, this volume.

Hilde, T. W. C., S. Uyeda, and L. Kroenke, Evolution of the western Pacific and its margin, *Tectonophysics, 38,* 145–165, 1977.

Holland, H. D., Granites, solutions, and basemetal deposits, *Econ. Geol., 67,* 281–301, 1972.

Hollister, V. H., *Geology of the Porphyry Copper Deposits of the Western Hemisphere,* 219 pp., Society of Mining Engineers, New York, 1978.

Jackson, E. D., E. A. Silver, and G. B. Dalyrimple, Hawiian-Emperor China and its relation to Cenozoic circumpacific tectonics, *Geol. Soc. Am. Bull., 83,* 601–618, 1972.

James, A. H., Hypothetical diagrams of several porphyry copper deposits, *Econ. Geol., 66,* 43–47, 1971.

Kay, R. W., Geochemical constraints on the origin of Aleutian magmas, *Island Arcs, Deep Sea Trenches and Back-Arc Basins,* Maurice Ewing Ser., vol. 1, edited by M. Talwani and W. C. Pitman III, pp. 229–242, AGU, Washington, D. C., 1977.

Keith, S. B., Possible magmatic and metallogenic products of 120 my to 10 my subduction in southwestern North America, *Geol. Soc. Am. Abstr. Programs, 11,* 47, 1979.

Kesler, S. E., Copper, molybdenum, and gold abundances in porphyry copper deposits, *Econ. Geol., 68,* 106–112, 1973.

Kesler, S. E., L. M. Jones, and R. L. Walker, Intrusive rocks associated with porphyry copper mineralization in island arc areas, *Econ. Geol., 70,* 515–526, 1975.

Kinkel, A. R., Jr., Copper deposits of the Philippines, *Spec. Project Ser. 16,* pp. 1–305, Philippine Bureau of Mines, Manila, 1956.

Larson, R. L., and W. C. Pitman III, World-wide correlation of mesozoic magnetic anomalies and its implications, *Geol. Soc. Am. Bull., 83,* 3645–3662, 1972.

Leith, A., The geology of the Baguio gold district, *Technical Bull. 9,* 70 pp., Dept. of Agric. and Commerce, Manila, Philippines, 1938.

Lewis, S. D., and D. E. Hayes, The structure and evolution of the Central Basin fault, West Philippine Basin, in *The Tectonic and Geologic Evolution of Southeast Asian Seas and Islands, Geophys. Monogr. Ser.,* vol. 23, edited by D. E. Hayes, pp. 78–88, AGU, Washington, D. C., 1980.

Lopez-Escobar, L., F. A. Frey, and M. Vergara, Andesites and high alumina basalts from the central-south Chile High Andes: Geochemical evidence bearing on their petrogenesis, *Contrib. Mineral. Petrol., 63,* 199–228, 1977.

Loudon, A. G., Marcopper porphyry copper deposit, Philippines, *Econ. Geol., 17,* 721–732, 1976.

Lovering, T. G., J. R. Cooper, H. Drewes, and G. C. Cone, Copper in biotite from igneous rocks in southern Arizona as an ore indicator, *U.S. Geol. Survey Prof. Pap. 700 B,* pp. B1–B8, 1970.

Lowder, G. G., and J. A. S. Dow, Porphyry copper mineralization at the Tapadda prospect, Northern Sulawesi, Indonesia, *Trans. Am. Inst. Min. Metall. Pet. Eng., 262,* 1977.

Lowell, J. D., and J. M. Guilbert, Lateral and vertical alteration-mineralization zoning in porphyry ore deposits, *Econ. Geol., 65,* 378–408, 1970.

Madamba, F. A., Geology and mineralization of the atlas copper deposits in Cebu Island, Philippines, *J. Geol. Soc. Philipp., 26,* 13–24, 1972.

Malicdem, D. G., Notes on the geology and exploration of Hinobaan copper project, Hinobaan, Negros Occidental,

J. Geol. Soc. Philipp., 28, 17–25, 1974.

Mason, D. R., and G. P. Feiss, On the relationship between whole rock chemistry and porphyry copper mineralization, *Econ. Geol. 74,* 1506–1510, 1979.

Mason, D. R., and J. A. McDonald, Intrusive rocks and porphyry copper occurrences of the Papua New Guinea-Solomon Islands region: A reconnaissance study, *Econ. Geol., 73,* 857–877, 1978.

Mitchell, A. H. G., and M. S. Garson, Relationships of porphyry copper and circum-Pacific tin deposits to Paleo-Benioff zones, *Trans. Min. Metall. Sec. B, 81,* 10–25, 1972.

Morante, E. M., and C. R. Allen, Displacement on the Philippine fault during the Ragay Gulf earthquake of 17 March 1973, *Geol. Soc. Am. Abstr. Programs, 5,* 744–745, 1974.

Noble, J. A., Metal provinces and metal finding in the western United States, *Mineralium Deposita, 9,* 1–25, 1974.

Oyarzun, J. M., and J. J. Frutos, Porphyry copper and tin bearing porphyries, A discussion of genetic models, *Phys. Earth Planet. Inter., 9,* 259–263, 1974.

Page, R. W., and I. McDougall, Ages of gold and porphyry copper deposits in the New Guinea highlands, *Econ. Geol., 67,* 1034–1064, 1972.

Page, R. W., and R. J. Ryburn, K-Ar ages and geological relations of intrusive rocks in New Britain, *Pac. Geol., 12,* 99–105, 1977.

Pena, R. E., and M. V. Reyes, Sedimentological study of a section of the 'upper Zigzag' Formation along Bued River, Tuba, Benguet, *J. Geol. Soc. Philipp., 14,* 1–19, 1970.

Philippine Bureau of Mines and Japan International Cooperation Agency (PBM-JICA), Report on Geological Survey of Northeastern Luzon, Phase II, Geological and Geochemical Surveys, 141 pp., 1976.

Portacio, J. S., Jr., Porphyry copper: A wealth in rocks, *Mineral. Mag., 8,* 14–29, 1975.

Putnam, G. W., and C. W. Burnham, Trace elements in igneous rocks, northwestern and central Arizona, *Geochim. Cosmochim. Acta, 27,* 53–106, 1963.

Quinto, P. T., Jr., Geology, mineralization and exploration program in Philex mines, Tuba, Benguet, *J. Geol. Soc. Philipp., 14,* 44–46, 1970.

Saegart, W. E., and D. E. Lewis, Characteristics of Philippine porphyry copper deposits and summary of current production and reserves, *Trans. Am. Inst. Min. Metall. Pet. Eng., 262,* 199–208, 1977.

Schafer, P. A., Santo Thomas II deposit in *Copper deposits of the Philippines,* part 1, *SPS 16,* edited by A. R. Kinkel, Jr., et al., Philipp. Bureau of Mines, Manila, 1956.

Scholl, D. W., M. S. Marlow, N. S. MacLeod, and E. C. Buffington, Episodic igneous activity along the Aleutian Ridge: Implications for subduction models, *Geol. Soc. Am. Abstr. Programs, 7,* 370–371, 1975.

Schweller, W. J., D. E. Karig, and S. B. Bachman, Original setting and emplacement history of the Zambales ophiolite, Luzon, Philippines, from stratigraphic evidence, this volume.

Shannon, J. R., Igneous petrology, geochemistry, and fission track ages of a portion of the Baguio mineral district, Northern Luzon, Philippines, Thesis T-2185, Colo. School of Mines, Golden, Colo., 1979.

Shih, T.-C., Marine magnetic anomalies from the western Philippine Sea: Implications for the evolution of marginal basins, *The Tectonic and Geologic Evolution of Southeast Asian Seas and Islands, Geophys. Monogr. Ser.*, vol. 23, edited by D. E. Hayes, AGU, Washington, D. C., 1980.

Sillitoe, R. H., A plate tectonic model for the origin of porphyry copper deposits, *Econ. Geol., 67*, 184–197, 1972.

Sillitoe, R. H., The tops and bottoms of porphyry copper deposits, *Econ. Geol., 68*, 799–815, 1973.

Sillitoe, R. H., Andean mineralization: A model for the metallogeny of convergent plate margins, in *Metallogeny and Plate Tectonics, Spec. Pap. 14,* edited by D. F. Strong, pp. 59–100, Geol. Assoc. Can., 59-100, 1976.

Stollery, G., M. Borcsik, and H. D. Holland, Chlorine in intrusives—A possible prospecting tool, *Econ. Geol., 66*, 361–367, 1971.

Sutherland-Brown, A., *Porphyry Deposits of the Canadian Cordillera,* spec. vol. 15, 510 pp., Canadian Institute of Mining and Metallurgy, Saint Anne de Bellevue, 1976.

Taylor, B., and D. E. Hayes, The tectonic evolution of the south China Basin, *The Tectonic and Geologic Evolution of Southeast Asian Seas and Islands, Geophys. Monogr. Ser.,* vol. 23, edited by D. E. Hayes, pp. 89–104, AGU, Washington, D. C., 1980.

Taylor, R. P., and B. J. Fryer, Multiple-stage hydrothermal alteration in porphyry copper systems in northern Turkey: The temporal interplay of potassic, propolitic and phyllic fluids, *Can. J. Earth Sci., 17*, 901–926, 1980.

Taylor, S. R., Trace element chemistry of andesites and associated calc-alkaline rocks, *Bull. Oreg. Dep. Geol. Miner. Ind., 65*, 43–63, 1969.

Teves, J. S., Philippine structural history and relation with neighboring areas, *Proc. Pac. Sci. Cong. 8th, 2A,* 867–868, 1953.

Titley, S. R., Geological characteristics and environment of some porphyry copper occurrences in the southwestern Pacific, *Econ. Geol., 70,* 499–514, 1975.

Titley, S. R., Copper molybdenum and gold content of some porphyry copper systems of the southwestern and western pacific, *Econ. Geol. , 73,* 977–981, 1978.

Titley, S. R., and R. E. Beane, Porphyry copper deposits, Part I, Geologic settings, petrology, and tectogenesis, *Econ. Geol., 75,* 214–269, 1981.

Titley, S. R., and C. L. Hicks, *Geology of the Porphyry Copper Deposits Southwestern North America,* 287 pp., University of Arizona Press, Tucson, 1966.

Titley, S. R., A. W. Fleming, and T. I. Neale, Tectonic evolution of the porphyry copper system at Yandera, Papua New Guinea, *Econ. Geol., 73,* 810–828, 1978.

Villones, R. D., The Aksitero Formation—Its implication and relationship with respect to the Zambales ophiolite, *Tech. Inform. Ser. 16-80,* 26 pp., Bureau of Mines and Geosciences, Manila, Philippines, 1980.

Walther, H. W., H. Forster, W. Harre, H. Kreuzer, H. Lenz, P. Muller, and H. Raschka, Early Cretaceous porphyry copper mineralization on Cebu Island, Philippines, dated with K-Ar and Rb-Sr methods, *Geol. J., D48,* 21–35, 1981.

Whitney, J. A., Vapor generation in a quartz monzonite magma: a synthetic model with application to porphyry copper deposits, *Econ. Geol., 70,* 346–358, 1975.

Whitney, J. A., A synthetic model for vapor generation in tonalite magmas and its economic ramifications, *Econ. Geol., 72,* 686–690, 1977.

Wolfe, J. A, K-Ar dating in the Philippines, *J. Geol. Soc. Philipp., 26,* 11–12, 1972.

Wolfe, J. A., and S. Self, Structural lineaments and Neogene volcanism in southwestern Luzon, this volume.

Wolfe, J. A., M. S. Manuzon, and A. F. Divis, The Taysan porphyry copper deposit, Southern Luzon Island, Philippines, *Econ. Geol., 73,* 608–617, 1978.

Worley, B. W., Jr., Gold mines of Benguet Consolidated, Inc., *Miner. Eng. Mag., 18,* 1967.

Zohenshain, L. P., M. I. Kuzmin, V. I. Kovalenko, and A. J. Saltykovsky, Mesozoic structural-magmatic patterns and metallogeny of the western part of the Pacific belt, *Earth Planet. Sci. Lett., 22,* 96–109, 1974.

Seismicity Associated With Back Arc Crustal Spreading in the Central Mariana Trough

Donald M. Hussong and John B. Sinton[1]

Hawaii Institute of Geophysics, University of Hawaii, Honolulu, Hawaii 96822

Numerous low-magnitude earthquakes were recorded in the central Mariana Trough by an ocean bottom seismometer (OBS) array deployed during late 1978. Although shallow seismic activity strong enough to be detected on worldwide seismic stations was seldom observed in this back arc basin, on the basis of other geological and geophysical data the basin was thought to be actively spreading. On this assumption, we deployed our OBS array on seafloor structure that has the morphology of a ridge/transform fault/ridge intersection portion of a roughly east-west valley that we named the Pagan fracture zone. Six OBS's recorded an average of 15 local events per day with magnitudes (based on event durations) ranging from 1.5 to 4.0 and low b values of 0.42 or 0.61, depending on the magnitude-duration relationship used. More than 300 hypocenters were determined. An earthquake swarm was located at 17°14′N latitude, 144°55′E longitude, at the base of a bathymetric high at the intersection of the northern spreading center and the transform valley. Hypocenters are concentrated in a zone roughly 15 km wide and 7.5 km deep that trends N30°E between the offset spreading centers, but which does not follow the Pagan fracture zone strike. Hypocenters in the transform zone are deeper than those in or near the crustal spreading areas. The low b values, maximum event magnitude of less than 4.5, and complex bathymetry and hypocenter trends all suggest that spreading in this back arc basin is unstable and is subject to frequent geometric reorientation. The Mariana volcanic arc is a small plate that is tectonically isolated by subduction on its east side and by subduction of the Philippine plate on its west side, producing a highly stressed region under tension even though it lies between major converging plates.

Introduction

Abundant drilling, geological, and bathymetric data acquired in the Mariana Trough have been used to suggest that this extensional back arc basin is active [*Karig*, 1971; *Karig et al.*, 1978; *Bibee et al.*, 1980; *Fryer and Hussong*, 1981; *Hussong and Uyeda*, 1981]. A paucity of earthquake and correlatable marine magnetic anomaly data, however, has made it difficult to prepare a kinematic model for this apparent back arc seafloor spreading.

Several investigators suggest that back arc spreading has some identifiable form of ridge-transform geometry [e.g., *Bibee et al.*, 1980; *Fryer and Hussong*, 1981], but other models include diffuse spreading [*Sclater et al.*, 1972; *Lawver and Hawkins*, 1978] and a central, but incoherent, spreading center [*Karig et al.*, 1978]. These models should

produce distinctly different magnetic anomaly patterns, but in the Mariana Trough these anomalies cannot be distinguished because they have very low amplitudes due to the low latitudes and the hypothesized nearly east-west spreading. Similarly, local seismicity data should be a powerful tool to distinguish among the spreading models. Certainly, the distribution of shallow earthquakes is indicative of mid–ocean basin spreading geometry where earthquakes are localized along transform faults and the axes of spreading. Unfortunately, the few shallow earthquakes detected in the Mariana Trough by the World-Wide Standard Seismograph Network (WWSSN) do not exhibit a pattern that can be related to crustal spreading. We have, however, detected numerous low-magnitude earthquakes in the central Mariana Trough using a small-scale ocean bottom seismometer (OBS) array. In this paper we describe the spatial and temporal distribution of this seismicity as well as the tectonic implications of the earthquakes.

The Mariana Trough is part of the Philippine Sea back

---

[1]Now at Research and Development Department, Conoco, Inc., Ponca City, Oklahoma 74601.

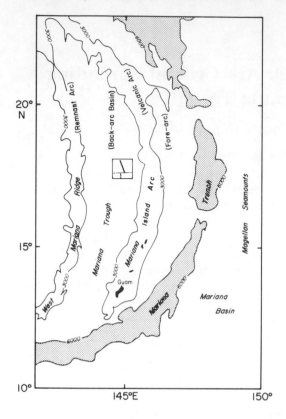

Fig. 1.   Location map showing the regional setting of the Mariana Trench/island arc/back arc basin system. The area of our experiment is indicated by the small boxed region at 18°N latitude.

arc system behind the Mariana Trench. The trough is bounded on the east by the active Mariana Island volcanic arc and on the west by the inactive West Mariana Ridge (Figure 1). Initial opening of the Mariana Trough probably occurred at the end of the Miocene, with subsequent slow spreading continuing to the present at a half-rate of less than 2 cm/yr in the region of 18°N [*Hussong and Uyeda*, 1981].

Available crustal seismic refraction stations are concentrated in the central Mariana Trough near 18°N [*LaTraille and Hussong*, 1980; *Bibee et al.*, 1980; *Ambos and Hussong*, 1982]. In this region the crustal seismic velocity structure is found to be similar to that of the slowly spreading Mid-Atlantic Ridge for ages less than 5 m.y.

Bibee et al. [1980] and *Fryer and Hussong* [1981] also identify and describe an axial graben associated with a regional axial high near 145°E, in which they found fresh basalts and high, variable heat flow. This graben has been suggested as the axis of a symmetrical spreading center within the Mariana Trough because of its general mor-

phology [*Karig et al.*, 1978] and near-axis magnetic anomaly patterns [*Bibee et al.*, 1980].

Karig et al. [1978] identify a fracture zone–like valley trending east-west along 17°35′N, in the area of this study. They note that the trend of the spreading center north of this fracture zone is N30°W, whereas south of the fracture zone the spreading center trends N10°W. From their bathymetry data it was unclear how much spreading center offset exists across the fracture zone or even whether it is an active transform fault, *Fryer and Hussong* [1981], working with a larger data set, interpret the valley as an active transform fault that is unstable and consequently is repeatedly reorienting its strike. They place the orientation of the northern spreading center as N30°W but plot the southern spreading center as nearly north-south. *Fryer and Hussong* [1981] go on to suggest that the character of Mariana Trough spreading may be typical of early rifting in any basin, whether behind an arc or in some other setting.

During the fall of 1978, as part of the 'Studies of East Asian Tectonics and Resources' (SEATAR) program, we deployed an array of ocean bottom seismometers (OBS) centered at 17°35′N by 145°50′E. The configuration of the array was designed to record low-magnitude local seismicity to help define the pattern of tectonic deformation. In particular, we hoped to define the nature of the fracture zone valley discussed above, which we call the Pagan fracture zone (after nearby Pagan Island), as well as the pattern of seismicity within the spreading centers north and south of the valley. At the same time, other seismometers were deployed as land stations on several islands of the active volcanic arc by investigators from Cornell University and as OBS's east of the active island arc near our 18° study areas by investigators from the University of Texas. The analysis of these more regional data will be discussed in another paper.

Data Acquisition and Analysis

The Hawaii Institute of Geophysics (HIG) OBS's used in this study were analog recording instruments containing a vertical and an unoriented horizontal 4.5-Hz seismometer, as well as a hydrophone with similar response. The earthquake data from six of the original 15 OBS's are presented here. A variety of instrumentation and location problems seriously degraded the data from the other OBS's. A map of the array area (Figure 2) shows that three of the useful OBS's (G, S, and E) were deployed within the Pagan fracture zone valley. OBS P was positioned near the southern spreading center valley. OBS L was deployed west of the northern spreading center. OBS J was situated at the base of a bathymetric high on the west side of the intersection of the fracture zone valley and the northern spreading center.

The OBS array was operational from October 29 to No-

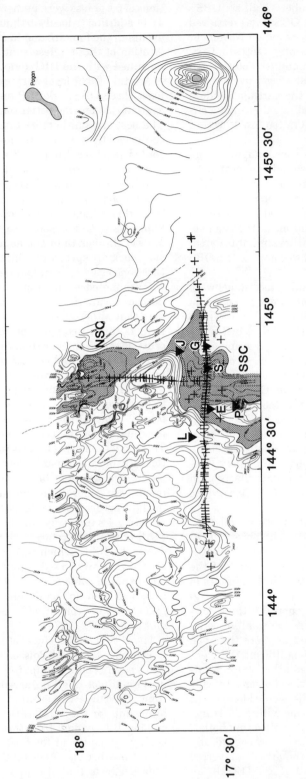

Fig. 2. Positions of receivers, location shots, and refraction lines are indicated by solid triangles (OBS's) and crosses (shots). Corrected bathymetry data from *Fryer and Hussong* [1981]. The stippled region with depths greater than 4000 m indicates the overall crustal spreading geometry. NSC means northern spreading center; SSC means southern spreading center. Contour interval is 200 m.

vember 21, 1978, during which nearly 2000 events were recorded. From October 29 to November 15 all six OBS's were actively recording earthquakes. OBS P was retrieved on November 15, and OBS L was retrieved on November 20. After November 20 the four remaining OBS's (J, G, S, and E) continued to record, but data quality deteriorated as tape speed slowed to the point that after November 21 the data are useless. Some gaps in the earthquake data occur during refraction shooting because gain ranging circuits in the OBS's set their amplifiers too low to record most earthquakes.

The relative locations of the OBS's were computed from travel times of direct water wave signals generated by explosive charges. Eight shots purposely detonated for OBS location as well as shots along two refraction profiles were used in this calculation (Figure 2). The relative error in each location reduced to 0.1 s, or 150 m, in the final solution. A generalized inversion algorithm similar to earthquake location algorithms was used to compute the OBS locations.

After retrieval of the OBS's the analog recordings were digitized for processing. Initial events were picked from the digital data using a computerized event detector. These events were then hand checked, and P and S arrival times and event durations were determined using a graphic display cathode ray tube. The event picker we used compares the average amplitude of one long-term time window with two short-term time windows for discriminating noise from events. The long-term average was computed using a leaky integrator [Gold and Rader, 1969] with a duration of approximately 500 s, while the two back-to-back short-term averages were computed by averaging the amplitude within two 5-s windows immediately in front of the leaky integrator. These three windows provide a particularly good event detector for the local, short-duration (5- to 60-s) events of interest to this study.

The most significant source of site noise on all OBS recordings is tide-generated water current noise [Duennebier et al., 1980]. This noise consists of a monochromatic, amplitude-modulated signal that has a rise time of the order of a few seconds to a few minutes. In the absence of current noise, earthquake events with less than 2:1 signal-to-noise ratios were detected. Generally, these low-amplitude, short-duration events were too small to be located because they were not recorded by more than one OBS. When the current noise became significant, the event detector was adjusted to trigger only on the longer, locatable earthquakes in order to pass by the shorter-term current noise pulses. An example of a well-recorded local event that is typical of earthquakes originating within the transform zone proper is shown in Figure 3. Notice that both P and S phases were recorded at all OBS's, as was generally true for all earthquakes detected by the array. Converted phases (see, for example, Lewis and McClain [1977]) can be identified between the P wave and S wave arrivals on

recordings of OBS's J, S, and G. Small events with durations of 5 s or less were numerous but not locatable (Figure 4). In addition to local earthquakes the array also recorded events in the fore arc and in the trench (Figure 4), but the location of these 'teleseismic' earthquakes cannot be determined with the HIG array alone.

A total of 330 local earthquakes were located using at least three OBS's. Both P and S times were used, when available, to provide better control on the hypocenter depth. The best standard errors in epicenter location and hypocenter depth were 1.5 km for earthquakes that were recorded on at least five OBS's.

An average crustal structure beneath the array was determined from an analysis of the two refraction profiles shot during this experiment (described by Sinton and Hussong [this volume]). They modeled the Pagan fracture zone valley as a region about 15 to 20 km wide that is 1 to 1.5 km thinner than the adjacent crust generated at the slow back arc spreading centers. The transition from fracture zone crust to normal oceanic crust north of the valley was completed within approximately 2.0 km of the valley's northern wall. Sinton and Hussong [this volume] further note that although the crust in this short offset transform fault is thin in comparison to spreading ridge-generated crust, it is still thicker than the 2- to 3-km-thick crust that Fox et al. [1976] and Detrick and Purdy [1980] describe in much longer offset transform systems. Although the crust in the fault valley and the adjoining plates is too complex to be represented by a simple layered model, for the purpose of locating the hypocenters we used a stratified approximation of the crustal velocity structure (Figure 5). This model consists of a two-layer crust underlain by mantle with normal seismic velocities. The model used does not have a sediment layer because in the study area sediments are generally found only in isolated ponds (except in the fracture zone valley where sediments are approximately 0.4 s (two-way time) thick).

All OBS's were referenced, by using delay times, to the depth of OBS L, which was the shallowest and was in an area of virtually no sediments. These delay times provided approximate corrections for the varying sediment thicknesses as well as the different depths of the OBS's. Hypocenters were computed by using program HYPO71 [Lee and Lahr, 1972, 1974].

The P time and S times of all located events were used to estimate the local Poisson's ratio and to constrain the V_p/V_s ratio used in computing earthquake locations. Figure 6 is a plot of all observed S times as a function of P time. The curve represents the S time versus P time relationship for the V_p/V_s ratio of 1.68 used to compute locations. It is evident from this figure that the relationship is slightly nonlinear for P times of less than 2.0 s, where the S times become consistently delayed from the 1.68 line. This suggests that the upper crust has a much higher Poisson's ratio, or higher V_p/V_s ratio, as rays with P times

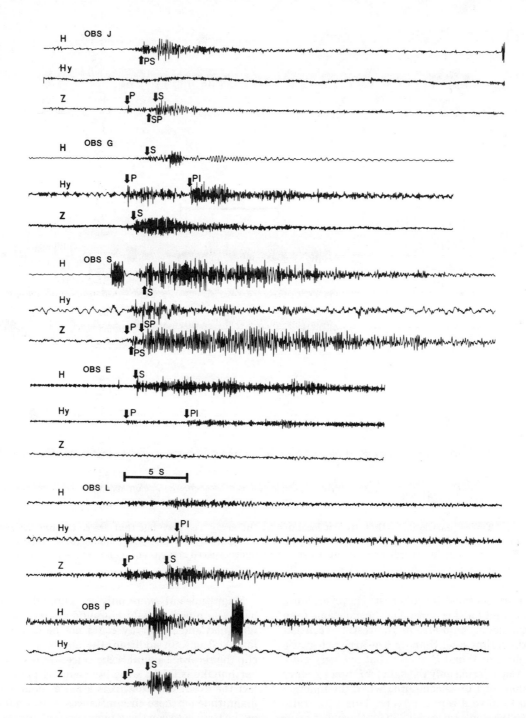

Fig. 3. Sample earthquake recorded on all six OBS's. The magnitude 2.1 earthquake occurred at 2049 UT on November 14 in the axis of the transform zone between OBS J and OBS S. All channels are shown for each OBS, including H (unoriented horizontal), Z (vertical), and Hy (hydrophone). The different phases are labeled P for *P* wave, S for *S* wave, Pl for the water multiple of *P*, and PS or SP for the respective *P* wave/*S* wave conversions.

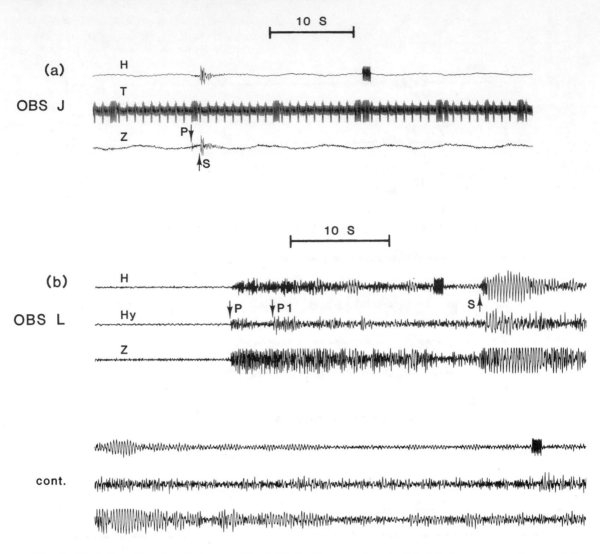

Fig. 4. Typical earthquakes that are not locatable by the array (see text) but that are commonly recorded. Earthquake (a) is a short-duration (less than 10-s) event as recorded only on OBS J. Earthquake (b) is a teleseismic Benioff zone earthquake recorded by our entire array. Our tightly spaced array is equivalent to a single recording point for these longer-range arrivals.

less than 2.0 s are confined to the crust. However, since all but one data point for this portion of the plot came from one OBS (J), the delay could also be the result of consistently picking the S time later than its true arrival because of the interference of P and S waves that are very close to one another. Considering only the data for P times greater than 2.0 s, a slope of 1.68 is computed when the function was not forced to have a zero intercept. This V_p/V_s ratio corresponds to a Poisson's ratio of 0.235 for the upper mantle beneath the array and agrees with the upper mantle value of *Lilwall et al.* [1977] for the Mid-Atlantic Ridge. This low Poisson's ratio at relatively shallow depth is a result of the much thinner crust within the fracture zone valley (Figure 5).

Magnitudes M_L were only computed by event duration because, in part, absolute ground motion could not be determined and the array could not be calibrated against land stations. In addition, the analog instruments often clip the recorded signal if the epicenter is very near or the earthquake is fairly large. As signal clipping does not affect the event duration, it is a more reliable indicator of magnitude for these circumstances. Durations were taken as the time between the P phase arrival and the time when the signal envelope decayed to twice the noise level that existed before the P phase. The empirical magnitude-duration functions of both *Lee and Wetmiller* [1978], $M_L = -0.87 + 2.0 \log(t)$, and *Hyndman and Rogers* [1981], $M_L = -4.4 + 3.2 \log(t)$, were used to compute magni-

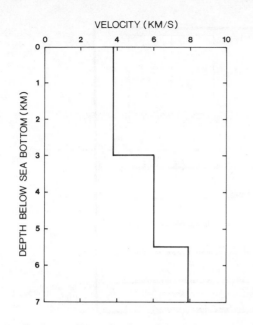

VELOCITY (KM/S)

Fig. 5. *P* wave velocity-depth model used to locate the hypocenters. This represents an average of the model from *Sinton and Hussong* [this volume]. Notice that the crust in this area is abnormally thin in relation to normal ocean crust. The V_p/V_s ratio assumed in the earthquake location calculations was 1.68.

tudes. As with Hyndman and Rogers the epicentral distance correction was negligible because locations were never more than 50 km from any one OBS.

The recent seismicity study by *Hyndman and Rogers* [1981] along the Explorer Ridge spreading center (off Vancouver Island) provides a data set that can be usefully compared with the results of our array, particularly since their study was conducted on a well-defined spreading center. Several aspects of their study are similar to those of this study. Their arrays were of the same dimensions and were deployed in the same position relative to the spreading centers and transform fault. Both studies used OBS's that were similar. Seismicity recorded by their OBS's was favorably compared to that of land stations. Indeed, there is a striking similarity between the earthquakes at the Explorer Ridge and those of the same duration found in the Mariana Trough. The major difference between these two studies is the type of spreading center; the Explorer Ridge is a fast spreading center (5.5 to 6 cm/yr) associated with the boundary of major plates whereas the Mariana Trough contains a slow spreading center bounding small plates.

Results

Temporal Distribution of Earthquakes

Figure 7 shows the temporal distribution of earthquakes recorded by each OBS during the 25 days the array was operating. The histograms are stacked in order of decreas-

ing number of recorded earthquakes. Periods of refraction shooting are noted by dashed lines. Apparently, the top three OBS's either were more sensitive or were closer to centers of activity than the other three OBS's. For OBS's J and P, which recorded the greatest number of earthquakes, the latter was probably true, as both OBS's were located near the intersection of a spreading center and the fracture zone, where the level of local seismicity might be expected to be particularly high. Relative calibration calculations suggest, however, that OBS S was at least twice as sensitive as the other OBS's, although this calibration is not definitive, because it cannot be tied to absolute ground motions as the coupling functions of the OBS's were unknown. The sensitivity of OBS L may be closest to that of OBS S, but because it was located the farthest from the centers of earthquake activity, it recorded fewer earthquakes. OBS G was less sensitive than OBS J, P, and S. The apparent insensitivity of OBS E is worse than we would expect because of the character of the instrument; thus we believe it is influenced by near-receiver propagation effects. The relationship between the position of OBS E and the distribution of epicenters in the vicinity of the OBS is discussed in more detail in the next section.

The other striking features of the histograms in Figure 7 are the two large peaks in activity during November 1 and November 6. Two smaller peaks in activity are also visible during November 11 and 16. The activity peaks stand out against a background level of 2 events per hour.

The character of the larger earthquake swarm during November 1 (Figure 8) is similar to, but more complex than, an aftershock sequence that occurred on the Mid-Atlantic Ridge [*Lilwall et al.*, 1977]. The histogram of Figure 8 was constructed of data from OBS J only, as this OBS was the closest to the swarm area. Four relatively

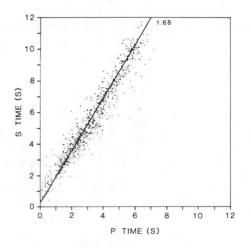

Fig. 6. Variation of Poisson's ratio, or V_p/V_s, as represented by the observed *S* times and *P* times. The curve is a least squares fit to the data which give a $V_p/V_s = 1.68$.

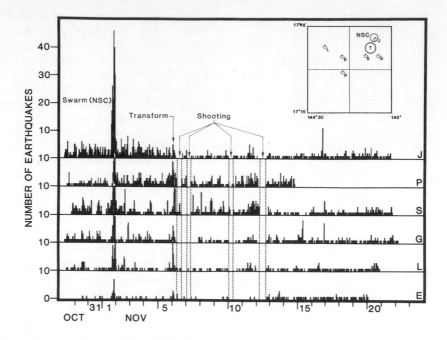

Fig. 7. Seismic activity recorded by each OBS during the 25 days of operation. The histograms are stacked in order of decreasing number of events recorded. Periods of refraction shooting are indicated by the dashed lines. In the inset the circle around OBS J is the area of the northern spreading center (NSC) swarm, and the circle labeled T is the much smaller swarm that later occurred in the transform zone.

large shocks occurred during the 8-hour-long swarm period. Each large (M_L = 2.8–3.2) shock was followed by a peak in seismicity which quickly dropped to the background level before the next large shock (Figure 8). After the large shock at 2237 UT the seismic activity reached a peak of 50 events per hour but dropped to the background

level in 30 min. The progressively greater intensity of seismic activity and increasing magnitude of the large events suggest a gradual buildup of stress during the last 8 hours of November 1. The overall rise in activity above background levels from 1600 to 2000 UT resembles a foreshock sequence, while the smaller-scale activity var-

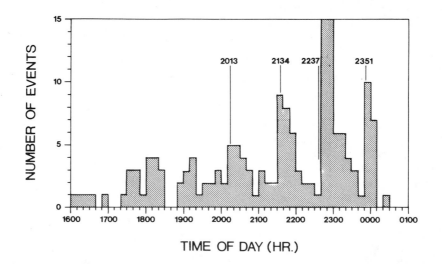

Fig. 8. Histogram showing details of the November 1 northern spreading center (NSC) swarm located near the intersection of the NSC and the transform. Seismic activity reaches a high of 15 events per 10 min. Larger shocks are indicated by their origin times. Each large shock is followed by distinct aftershock activity that reduces to the background seismicity level in 20 to 40 min.

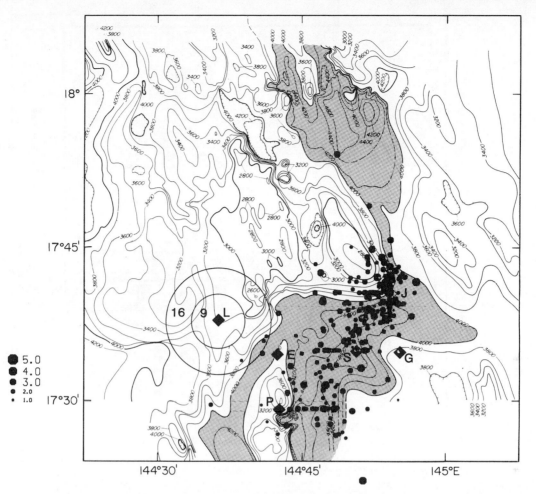

Fig. 9.   Distribution of epicenters. OBS positions are indicated by the solid triangles. Concentric circles about OBS L are ranges of constant $S - P$ time. The numbers inside the circles are the number of events recorded by OBS L within the bands defined by the circles. Other OBS's in the array recorded 4 to 100 times as many nearby events, demonstrating that the Pagan fracture zone is relatively aseismic outside the transform zone.

iations after each major shock are probably aftershock sequences, as evidenced by their characteristic rapid falloff to the background seismicity level. The earthquake at 2237 UT appears to be the main shock, making the last large shock at 2351 UT part of the aftershock sequence. Unfortunately, a recording period of 25 days is precariously short for proper definition of any periodic variations in earthquake activity level.

Epicenter and Hypocenter Distribution

The epicenters in the central Mariana Trough have a distinct distribution that can clearly be correlated with tectonic features. A map of the epicenters of the 330 earthquakes located by our array is shown in Figure 9 along with major bathymetry contour lines. The Pagan fracture zone valley is delineated by the 4000-m contour between 17°30'N and 17°40'N and between 144°43'E and 144°55'E. OBS locations are indicated by the solid diamonds. Three

features stand out in Figure 9: the first is the relatively restricted zone, only 15 km wide and trending N30°E, within which most of the epicenters fall; the second is the major concentration of epicenters about the position of OBS J; the last is the smaller, but still distinct, concentration of epicenters near OBS P.

Within the Pagan fracture zone the apparently active zone of faulting is defined by the N30°E trend of the epicenters in the transform zone between the two spreading centers. This seismically active zone does not follow the trends of either of the adjacent spreading centers or of the fracture zone in the area, yet it encompasses over 98% of the epicenters. Apparently, the transformation of spreading direction and plate motion that must take place in this zone occurs with a complex geometry that is not uniquely resolvable with our data. We can, however, place significant constraints on any tectonic model for seafloor spreading in the area.

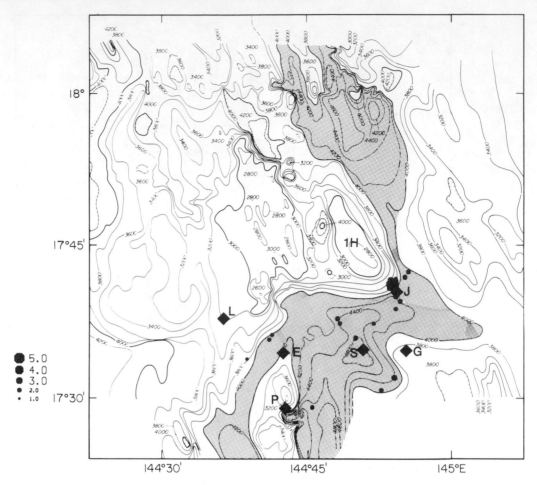

Fig. 10. Epicenters of earthquakes recorded on November 1 during the period of the swarm of Figure 9. The 30 epicenters of the swarm all lie within 2 km of OBS J. The symbol 1H denotes the intersection high.

Surprisingly few earthquakes were located in the spreading center segments near our array. Only three earthquakes were located outside the Pagan fracture zone valley and within the northern spreading center (NSC). The trend of the epicenters within the NSC agrees with the strike of that feature as identified by *Karig et al.* [1978] and *Fryer and Hussong* [1981]. The one epicenter located on the southern spreading center (SSC) does little to help define its trend. Other epicenters located outside of the fracture zone valley area usually lie on or near a major bathymetric scarp.

There are also few earthquakes located east of OBS G, near OBS L, and west of OBS E, although the quiet regions near OBS's G and L are much better defined. OBS L recorded only 25 events with $S - P$ times less than 2 s, while all other OBS's recorded 100 or more such events (Figure 9). OBS L is a sensitive instrument in comparison to the other OBS's, so the apparent lower seismicity near L is not instrument related, indicating that there are few earthquakes west of the SSC. Even at OBS E, on the southern side of the transform valley and only about 5 km west

of the SSC (as positioned by *Fryer and Hussong* [1981]), there were few nearby events. Although OBS E was determined to be relatively insensitive, which is a function of the instrument characteristics as well as coupling and near-receiver attenuation, the very few events detected in its immediate vicinity suggest that there is little seismicity away from the transform zone.

All the epicenters of the November 1 swarm (shown in Figure 8) fell within 2 km of OBS J (Figure 10); indeed, the majority of all locatable earthquakes for the duration of our experiment occurred within 5 to 10 km of OBS J. Epicenters for the smaller November 6 swarm, however, fell farther south, in the transform valley at 17°38′N.

The distribution of hypocenters is shown in Figures 11 and 12 for two orthogonal sets of vertical profiles through the epicenter trend. The standard errors of the hypocenter were at best 1.5 km, and usually closer to 2.5 km. Hypocenter depths clustering at 5 km are artifacts of the locations for which the depth was held fixed or did not change during the piecewise autoregressive hypocenter solution. Figure 11 is a block diagram of east-west profiles that are

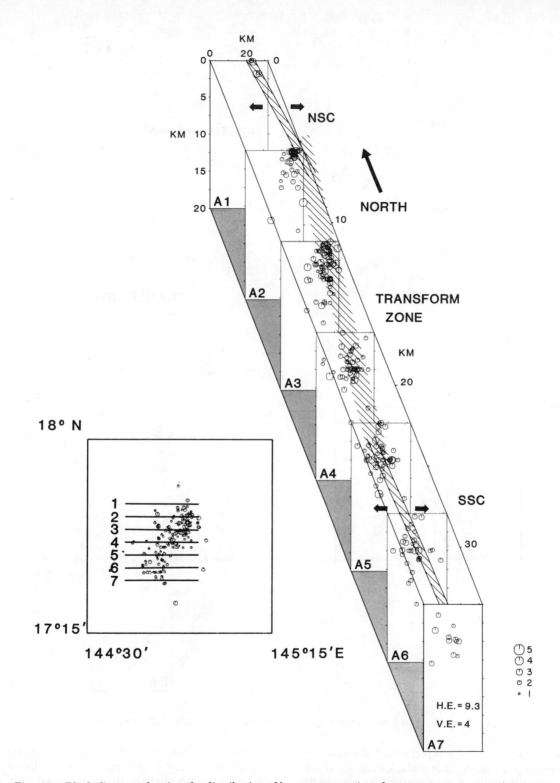

Fig. 11.   Block diagram showing the distribution of hypocenters projected on east-west cross sections approximately parallel to the direction of the Pagan fracture zone. Inset shows the location of the profile lines relative to the epicenter trend.

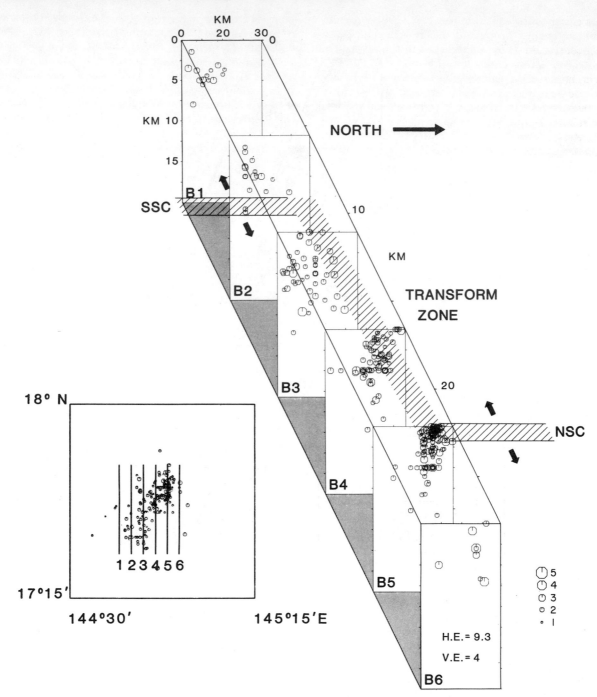

Fig. 12. Block diagram similar to Figure 11 but for north-south cross sections.

approximately parallel to the direction of the back arc spreading trend, and Figure 12 is a block diagram of profiles oriented north-south. In both figures the scale of the isometric axis is exaggerated by approximately 15.

There are subtle changes in the average depth of the hypocenters across the region of the array which we think may be significant, although we realize that the data set is too small to permit definitive interpretation. For instance, it appears that the hypocenters near the more active northern spreading center (profiles A1 and A2, Figure 11, and B5, Figure 12) are shallower than those in the center of the transform zone (A3–A5, Figure 11). The hypocenters closer to the southern spreading center are distributed more evenly throughout the crust in comparison to the concentration in the upper crust in the north. The NSC-transform intersection, transform zone, and SSC-transform intersection can be characterized as having shallow, middle to deep, and shallow to deep crustal earthquakes, respectively. In addition, Figures 11 and 12 show that the seismicity is concentrated in a zone that is only approximately 15 km wide and 7.5 km deep between the NSC-transform intersection and the SSC-transform intersection.

Again, the most obvious feature of Figures 11 and 12 is the shallow depth of the hypocenters in the region of the swarm (profiles A2, A3, and B5). The hypocenters in this region are so densely packed that the true number of events is masked, but approximately 100 of the 330 total located events are shown on profiles A2 and A3 (Figure 11). The mean depth of both the swarm hypocenters and nonswarm activity in the region of the swarm is 0.5 km. This shallow depth contrasts sharply with the distribution of hypocenters in all the other profiles.

Another important feature of the hypocenters is the mostly midcrustal depths determined for earthquakes occurring within the center of the transform zone in the deepest part of the fracture zone valley (profile A4 in Figure 11). The mean depth of these hypocenters is 4.5 km. *Brune* [1968] noted that within a spreading center or transform valley the elastic lithosphere can be considered to be identical to the brittle crust, so that the hypocenter maximum depth defines the elastic thickness. If so, the mean depth of 4.5 km suggests that the elastic layer is thinner within the transform zone than near the SSC-transform intersection and is much thinner than *Cochran*'s [1979] estimate of 10 to 12 km for the thickness of the elastic layer on the Mid-Atlantic Ridge. *Brune* [1968] suggested that the elastic thickness of the Romanche fracture zone is 6.5 km, a value that is not significantly greater than that obtained for the transform zone here. Thus the transform zone appears to be an anomalous region of thin lithosphere similiar to that found by *Detrick and Purdy* [1980] for the Kane fracture zone.

Determination of b Values

The magnitude-recurrence relationships for our Mariana Trough OBS array are shown in Figure 13. The sudden drop in the number of located events below magnitude 2.0 in curve A (based on the recurrence relation of *Lee and Wetmiller* [1978]) and 0.0 in curve B (based on the recurrence relation of *Hyndman and Rodgers* [1981]) represents the threshold magnitudes of the smallest earthquake that can be located with the geometry and receiver spacing of our OBS array. Above these threshold magnitudes the curves in Figure 13 are piecewise linear up to a magnitude of 3.0 (curve A) and 2.2 (curve B). We will call magnitudes 3.0 and 2.2 the 'kink magnitudes,' above

which a single b value cannot fit the observed recurrence relationships well. A maximum likelihood [*Ali*, 1965] b value of 0.61 below $M_L = 3.0$ was calculated for curve A. For curve B the maximum likelihood b value is 0.42 below $M_L = 2.2$. The b values of 0.61 and 0.42 are much lower than normal values, which are nearer to 1.0 [e.g., *Lilwall et al.*, 1977; *Hyndman and Rogers*, 1981]. The fact that the b value for curve B is so much lower than observed elsewhere suggests that curve A is the more accurate recurrence relation; however, the similarities in the equipment, local tectonics, and array geometries of our study to those of Hyndman and Rogers suggest that the recurrence relation of curve B may be the correct one. We cannot choose which curve represents the true recurrence relation for the local seismicity, but both curves point out that stress release for earthquake magnitudes below the kink magnitude is significantly greater than in a region with b values near 1.

Tectonic Implications

Earthquake Distribution

The November 1 earthquake swarm is located at the base of a large bathymetric high at the intersection of the NSC and the transform zone (Figure 14). Thus the swarm is probably associated with faulting rather than with volcanic activity in the NSC rift valley. The scarp associated with the swarm forms the western wall of the NSC rift valley and the northern boundary of the fracture zone valley. The intersection high (Figure 14) is likely due to a combination of the uplift of the western rift valley wall within the NSC and the interaction of the NSC with the transform zone. Intersection highs are a common feature of slow spreading center–transform intersections and can be identified, for example, on the Atlantis, Vema, and

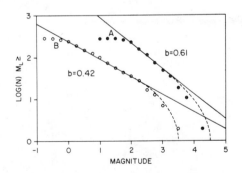

Fig. 13. Magnitude-recurrence relationships for the seismic activity recorded during the 25 days of operation. Curve A was computed using $M_L = 0.87 - 2.0 \log (T)$; curve B was computed using $M_L = 4.4 - 3.2 \log (T)$. T is the event duration in seconds. Each curve has a kink beyond which the b value increases nonlinearly. The location threshold magnitudes are 2.0 for curve A and 0.0 for curve B. The scatter in the curves at the largest magnitudes is due to insufficient sampling.

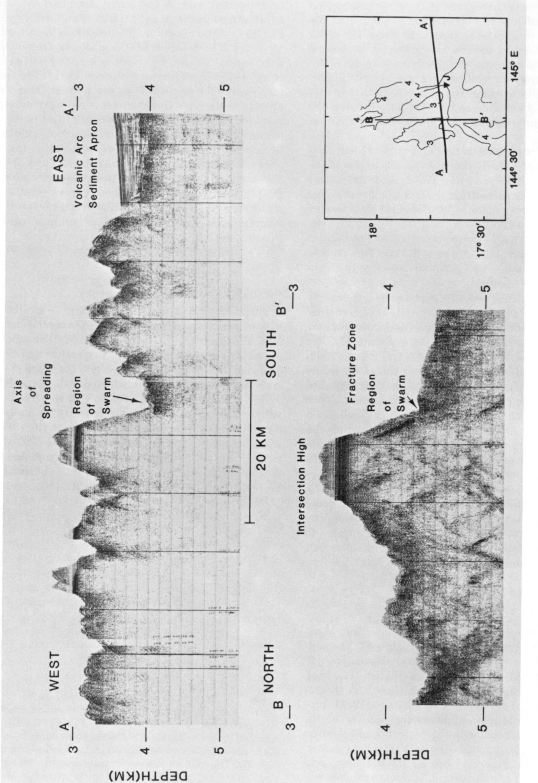

Fig. 14. Reflection seismic profiles showing the scarps that border the intersection high (see text) along its east and south sides. The position of the November 1 swarm is along the base of the intersection high. Contours on the profile location inset are selected 3- and 4-km depths.

Oceanographer fracture zones. For all of these examples the intersection high is located on the transform side (rather than the fracture zone side) of the intersection in the same relative position as shown in Figure 10. Uplift of rift valley walls from a rift valley floor is a dynamic process that is normally attributed to vertical viscous drag of the upwelling magma on the walls of the conduit within the spreading center [*Lachenbruch*, 1973; *Needham and Francheteau*, 1974; *Temple et al.*, 1979] or to viscous head loss of the upwelling magma [*Sleep*, 1975]. Away from the spreading center (defined by the limit of the axial mountains), little seismic activity is expected as ocean floor topography is frozen in [*Needham and Francheteau*, 1974; *McDonald and Luyendyk*, 1977] and the lithospheric plate equilibrates slowly by cooling and subsiding [*Parsons and Sclater*, 1977]. These dynamic processes are complicated by interaction with the transform fault in a way that is still not well understood. *Bonatti* [1978] invoked local vertical movements associated with horizontal thermal contraction of the lithosphere to explain transverse ridges along large fracture zones; this explanation partially accounts for the intersection high.

Regardless of the mechanism by which the intersection high is created, vertical tectonic forces probably generate most of the observed earthquakes. The swarm, and earthquakes located near the swarm, are apparently caused by high-angle dip-slip faulting along the scarps bordering the intersection high in response to dynamic adjustments of the newly formed crust that makes up the intersection high, rather than by strike-slip motion in the transform. Hypocenter distribution (Figures 11 and 12) indicates that the proposed faulting was confined to the top of the crust.

The episodic activity that characterizes the swarm can be modeled by partial release of stress during each major shock/aftershock sequence; in other words the swarm character could be produced by a 'sticky' fault. On this sticky fault there are zones of high strength, or asperities. The weakest asperity fails first, followed by failure of progressively stronger asperities until the faulting stops. This model would explain the observed increase in earthquake magnitude from the first to the last major shock, as well as the aftershock sequences observed after each major shock. Faulting models of this sort have been used to explain the seismicity observed for major transform faults [*Kanamori and Stewart*, 1976] and major subduction zones [*Chung and Kanamori*, 1978]. The only difference between the models for major faults (other than the type of slip) and that proposed here is the size of the asperities involved.

Swarms are commonly observed on spreading center systems. *Reichle and Reid* [1977] identified swarms that displayed episodic seismicity along the east Pacific rise in the Gulf of California, but they were unable to distinguish between a volcanic or a faulting source. *Lilwall et al.* [1977, 1978] observed a swarm that they associated with magmatic intrusion under the rift valley on the Mid-Atlantic Ridge. Their swarm was not episodic and had a temporal

character that differed significantly from the November 1 swarm observed here.

The distribution of epicenters relative to tectonic features as defined by *Fryer and Hussong* [1981] is schematically represented in Figure 15. The hachured region is the present-day zone of transform motion. This motion may be accommodated by a series of short ridge/transform segments as shown in Figure 15. We have no direct evidence for the actual spreading geometry in the transform zone, but P. Lonsdale (personal communication, 1982) has deep-tow survey data in this area that support such a sequence of short spreading segments.

The location of the November 1 earthquake swarm relative to the NSC, transform zone, and intersection high is similar to the locations of epicenters at intersections along the Mid-Atlantic Ridge system, where epicenters are found throughout the corner of the spreading center/transform intersection [*Reid and Macdonald*, 1973; *Francis et al.*, 1978; *Rowlett*, 1981]. In the Pagan fracture zone transform, however, the opposite corner formed by the intersection of the SSC and the transform zone does not display the same epicentral pattern. Instead, seismicity at this intersection is diffuse and shows no preferential trends. The SSC is not as seismically active as the NSC. In contrast to the prominent axial graben in the NSC, there does not seem to be as pronounced a bathymetric expression of a spreading center in the SSC. It is possible that the SSC is not spreading as actively as the NSC. Any pronounced difference in spreading character between the NSC and the SSC can only be a relatively short term phenomenon, or else it would imply impossible rigid plate geometries or misinterpretation of the regional tectonic pattern. Further conclusions require a long-term seismicity study; our 25-day recording period is too short for analysis of long-term tectonics.

Epicenters within the transform zone trend along the minimum distance path between the NSC and the SSC rather than the trend along the roughly east-west Pagan fracture zone trace (Figure 15). The earthquake trend agrees with the instantaneous pole of rotation for the opening of the basin calculated by *Le Pichon et al.* [1975], but we suspect that this correspondence is only coincidental. Unfortunately, the temporal stability of the earthquake trend is in doubt as it is subparallel to the strike of the scarp along the west side of the transform zone valley and is oblique to the scarp along the north side. The earthquake data support the conclusion of *Fryer and Hussong* [1981] that the Pagan fracture zone is unstable.

Many of the transform zone earthquakes are also probably caused by isostatic and thermal adjustment along the adjacent plate boundaries. Six epicenters are located along the west scarp, and many epicenters are concentrated along the north scarp, suggesting that these scarps are still the loci of active vertical motion. These earthquakes would have dip-slip motion, such as that described by *DeLong et al.* [1979] as a product of differential sub-

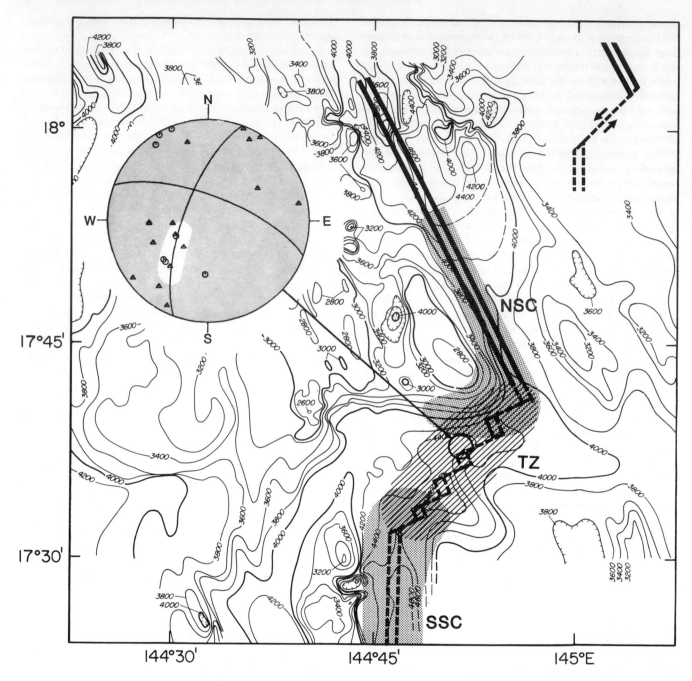

Fig. 15. Summary of possible tectonic features in the transform portion of the Pagan fracture zone. The small transform segments represent a possible spreading geometry that would satisfy the regional structure and epicenter trend. An alternative geometry, shown as an inset at the upper right, is clearly possible and would also fit the crude composite focal mechanism shown. Within the stereonet, motions within the highlighted region in the third quadrant are all small and emergent. All other motions are impulsive.

sidence rates on other major fracture zones.

The character of the scarps that form the northern and western boundaries of the transform valley and the earthquake trend may also represent an episodic variation in

the trend of the transform zone from NE-SW to E-W and back to NE-SW. The present transform zone as represented by earthquakes is more nearly perpendicular to the NSC but oblique to the SSC, whereas the overall east-west trend of the Pagan fracture zone is more perpendicular to the SSC and oblique to the NSC. Thus the predominant east-west trend of the Pagan fracture zone might represent a fossil trend that has rotated, or is rotating, toward agreement with the NSC, whereas the western scarp may be an earlier fossil trend that subsequently rotated to the east-west fracture zone trend. These sequential transform zone rotations from NE-SW to east-west and back to NE-SW occurred over the last few million years, on the basis of the 40-km length of the fracture zone valley and a spreading rate of 1.65 cm/yr. These frequent reorientations are probably typical of the small plate disequilibria that would be expected during initial rifting of an island arc or any other type of crust (including midplate oceanic crust or continental crust). The resultant early rifting crust is thus a welded conglomerate of many small plates, producing a complex bathymetric and magnetic fabric that cannot be correlated except on a very small scale.

The Kurchatov fracture zone, which offsets the Mid-Atlantic Ridge by approximately 40 km [*Searle and Laughton,* 1977], has many of the same characteristics as the Pagan fracture zone transform zone, including the character of the offset of spreading centers, a fault valley approximately 20 km wide, and an irregularly shaped trough outside the active transform fault valley. From an analysis of sidescan sonar data, which provides detailed information on the morphology of small faults, Searle and Laughton modeled the Kurchatov fracture zone as an oblique fault. In this model the fault does not represent a true transform fault. Crust is created within the fault valley so it can be considered as a leaky transform. Within the Kurchatov fracture zone valley, Searle and Laughton identified fault lineaments parallel to a line connecting the two ends of the offset spreading center. These lineaments have the same general geometry relative to the spreading centers as the earthquake trend in the Pagan fracture zone transform region. It follows that the oblique fault model of Searle can fit the transform zone seismicity in the Pagan fracture zone (Figure 15). However, the transform zone of the Pagan fracture zone is probably not an oblique extensional region. *Sinton and Hussong* [this volume] show that the transform zone crust is anomalously thin in comparison to adjoining basin crust and hypothesize that this thickness variation is caused by the thermal effect of the contact of the transform zone with the spreading center. If there is oblique spreading rather than actual transform displacement, the variation in crustal thickness should not occur.

Although limited by an array with only six receivers, we attempted some crude first motion studies using earthquakes having epicenters between OBS J and OPS S. All motions in Figure 15 are impulsive except along the N20°E nodal plane where the P wave motions are all emergent and smaller in amplitude, as would be expected along a nodal plane. This fault plane solution is roughly aligned with the overall trend of the epicenters (Figure 9) and

indicates oblique slip rather than purely strike-slip faulting. This result is significant because all the epicenters from which the stereonet was constructed were located within the transform zone. The partly extensional observed fault slip supports the hypothesis that the Pagan fracture zone does not have a simple transform zone and may not even fit the simple model of short ridge/transform segments in Figure 15.

Values of b

The recurrence relationships (b values) shown in Figure 13 are useful for constraining the interpretation of both the local and regional tectonic geometries. Our data, and the few shallow events located by the WWSSN in the Mariana Trough, suggest that earthquakes in this back arc basin have magnitudes less than 4.5. As long transform faults (length greater than 50 km) have not been identified within the trough, it seems that there is no fault in the Mariana Trough of sufficient dimensions to generate an earthquake of magnitude greater than 4.5. Thus a magnitude 4.5 earthquake may be the maximum possible instantaneous strain release for the Mariana Trough and the Pagan fracture zone transform zone, in the same manner as the earth as a whole has a maximum earthquake magnitude.

In any interpretation of our data we must be cautious because of our short observation time. However, the recurrence relationships in Figure 13 are linear below the well-defined kink magnitudes, indicating that our sampling period is at least adequate for the smaller-magnitude earthquakes. When the 4.5-magnitude restriction is applied to our data, an infinite b value is required at this magnitude, as represented by the dashed curves in Figure 13. The dashed curve recurrence relationship probably represents a transition from the preferred high state of stress below the kink magnitude (constant b value region) to the limiting stress for a 4.5 magnitude. Then the constant b value region (e.g., $b = 0.61$ for curve A) can be interpreted as representing the normal state of stress for the trough.

The local b value curves can be used to estimate the fault lengths associated with the kink and the 4.5 magnitudes. For small earthquakes the magnitude is theoretically thought to be proportional to the cube of the fault length [*Wyss and Brune,* 1968; *Kanamori and Anderson,* 1975], although an empirical linear relationship of the form $M_L = f \log_{10}(L) + g$ has been determined where L is the fault length. For lack of any better relationship we will use the empirical form and the coefficients of *Wyss and Brune* [1968], where $f = 1.9$ and $g = -6.7$ so that L is in centimeters. The coefficients of Wyss and Brune were used for strike-slip faults in southern California, so their validity here might be questioned. These coefficients, however, are from a particularly well-documented data set, and there are no available empirical coefficients specifically calculated from local spreading center–transform earthquakes. At any rate, we will only try to derive an order-of-magnitude estimate of fault length. For $M_L = 4.5$, the maximum fault length is only approximately 8 km, which is half the length of the Pagan fracture zone transform zone. Using the same relationship, the fault length at the kink magnitude is 1.2 km for curve A and 0.5 km for curve B, although the latter estimate was computed from a magnitude that falls below the minimum magnitude for which $M_L = f \log (L) + g$ is thought to be valid. Thus instead of having stress release along faults with lengths of the order of 10 km, the preferred state is to have stress release on shorter faults of the order of 1 km. These faults fit the pattern in Figure 15.

These maximum fault length estimates and their associated magnitudes and b values may be relevant to the dynamics of global plate tectonics. There has long been an intuitive difficulty in relating the apparently tensional tectonics of back arc basin opening to the overall compressional regime one expects in the plate convergence zone at an island arc. In the case of the Mariana region, *Hussong and Uyeda* [1981] note that all the surficial deformation from the trench axis across the arc to the back arc basin appears to be tensional. The low b values suggest that the Mariana Trough is highly stressed. *Chase* [1978] provided a mechanism for this high tensional stress regime by suggesting that all marginal (back arc) basins in the western Pacific are the result of 'trench suction' where the subduction zone retreats from the overlying plate. *Uyeda et al.* [1981] interpret the generation of the Mariana Trough as the result of the Philippine plate being in tension because subduction on its west side pulls it away from the Pacific plate while subduction on its east side anchors the Pacific plate in the asthenosphere, fixing the position of the Mariana Trench. For both those models, the Mariana island arc is a small plate that is decoupled by subduction from both the Pacific and Philippine plates, and from the convergence of the Asia and Pacific plates. This subduction decoupling produces high stress, but isolates the small plate boundaries in the Mariana Trough.

Conclusions

The distribution of epicenters supports the hypothesis that the Mariana Trough is a back arc basin actively opening from a symmetrically spreading ridge-transform-ridge boundary system. The epicenter pattern is also consistent with spreading geometry for the past one million years interpreted from bathymetry, and does not support interpretation of any diffuse spreading or spreading without some form of transform motion. The spreading geometry is, however, unstable over time periods of millions of years and apparently occurs along small ridge segments that undergo frequent reorientation.

Seismicity within the transform zone was greatest near the intersection with the NSC, particularly at the base of the intersection high where two earthquake swarms were recorded. The depth of hypocenters is generally less than

7.5 km and is slightly deeper in the spreading center than in the transform.

Apparently, the distribution of stress in the basin is different from that found along major spreading centers. The maximum earthquake magnitude of 4.5 is much smaller than in the Mid-Atlantic Ridge or the east Pacific rise, and b values in the Mariana Trough are quite low. This is consistent with the short and constantly changing transform segments we have suggested. The unstable geometry of spreading in this back arc basin occurs because it involves small plates and is in a region experiencing strong, and probably variable, tensional stress from the surrounding subduction zones.

Acknowledgments. The authors thank E. Ambos, J. Carter, and P. Milholland for their helpful discussions through the course of this experiment. T. Brocher, F. Duennebier, and R. Cardwell provided valuable reviews of the manuscript. We are grateful for the assistance of Pierre-Luigi Pozzi and the crew of the R/V *Thomas Washington* during deployment of the OBS array. This research was supported under National Science Foundation grant OCE80-12922. Hawaii Institute of Geophysics contribution 1289.

References

Aki, E., Maximum likelihood estimate of b in the formula $\log N = a + bm$ and its confidence limits, *Bull. Earthquake Res. Inst. Univ. Tokyo, 43,* 237–239, 1965.

Ambos, E. L., and D. M. Hussong, Crustal structure of the Mariana Trough, *J. Geophys. Res., 87,* 4003–4018, 1982.

Bibee, L. D., G. G. Shor, and R. S. Lu, Inter-arc spreading in the Mariana Trough, *Mar. Geol., 35,* 183–197, 1980.

Bonatti, E., Vertical tectonism in oceanic fracture zones, *Earth Planet. Sci. Lett., 37,* 369–379, 1978.

Brune, J. N., Seismic moment, seismicity, and rate of slip along major fault zones, *J. Geophys. Res., 73,* 777–784, 1968.

Chase, C. G., Extension behind island arcs and motions relative to hot spots, *J. Geophys. Res., 83,* 5385–5387, 1978.

Chung, W., and H. Kanamori, Subduction process of a fracture zone and aseismic ridges—The focal mechanism and source characteristics of the New Hebrides earthquake of January 19, 1969, and some related events, *Geophys. J. R. Astron. Soc., 54,* 221–240, 1978.

Cochran, J. R., An analysis of isostasy in the world's oceans, 2, Midocean ridge crests, *J. Geophys. Res., 84,* 4713–4729, 1979.

DeLong, S. F., J. F. Dewey, and P. J. Fox, Topographic and geologic evolution of fracture zones, *J. Geol. Soc. London, 136,* 303–310, 1979.

Detrick, R. S., and G. M. Purdy, The crustal structure of the Kane fracture zone from seismic refraction studies, *J. Geophys. Res., 85,* 3759–3777, 1980.

Duennebier, F. K., et al., Lopez island ocean bottom seismometer intercomparison experiment, Final Report,

Rep. HIG-80-4, 272 pp., Hawaii Inst. of Geophys., Honolulu, Hawaii, 1980.

Fox, P. J., E. Schreiber, H. Rowlett, and K. McCamy, The geology of the Oceanographer fracture zone: A model for fracture zones, J. Geophys. Res., 81, 4117–4128, 1976.

Francis, T. J. G., I. T. Porter, and R. C. Lilwall, Microearthquakes near the eastern end of St. Paul's Fracture Zone, Geophys. J. R. Astron. Soc., 53, 201–217, 1978.

Fryer, P., and D. M. Hussong, Seafloor spreading in the Mariana trough: Results of leg 60 drill site selection surveys, Deep Sea Drilling Project, Initial Rep. Deep Sea Drill. Proj., 60, 45–55, 1981.

Gold, B., and C. M. Rader, Digital Processing of Signals, 269 pp., Lincoln Laboratory Publications, McGraw-Hill, New York, 1969.

Hussong, D. M., and S. Uyeda, Tectonic processes and the history of the Mariana arc: A synthesis of the results of deep sea drilling project Leg 60, Initial Rep. Deep Sea Drill. Proj., 60, 909–930, 1981.

Hyndman, R. D., and G. C. Rogers, Seismicity surveys with ocean bottom seismographs off western Canada, J. Geophys. Res., 86, 3867–3880, 1981.

Kanamori, H., and D. L. Anderson, Theoretical basis of some empirical relations in seismology, Bull. Seismol. Soc. Am., 65, 1073–1095, 1975.

Kanamori, H., and G. S. Stewart, Mode of release along the Gibbs fracture zone, Mid-Atlantic Ridge, Phys. Earth Planet. Int., 11, 312–332, 1976.

Karig, D. E., Structural history of the Mariana island arc system, Geol. Soc. Am. Bull., 82, 323–344, 1971.

Karig, D. E., R. N. Anderson, and L. D. Bibee, Characteristics of back arc spreading in the Mariana Trough, J. Geophys. Res., 83, 1213–1226, 1978.

Lachenbruch, A. H., A simple mechanical model for oceanic spreading centers, J. Geophys. Res., 78, 3395–3416, 1973.

LaTraille, S. L., and D. M. Hussong, Crustal structure across the Mariana island arc, in The Tectonic and Geologic Evolution of Southeast Asian Seas and Islands, Geophys. Monogr. Ser., vol. 23, edited by D. B. Hayes, pp. 209–222, AGU, Washington, D. C., 1980.

Lawver, L. A., and J. W. Hawkins, Diffuse magnetic anomalies in marginal basins: Their possible tectonic and petrologic significance, Tectonophysics, 45, 323–339, 1978.

Lee, W. H. K., and J. C. Lahr, HYPO-71: A computer program for determining hypocenter, magnitude, and first motion pattern of local earthquakes, open file report, 100 pp., U.S. Geol. Surv., Menlo Park, Calif., 1972.

Lee, W. H. K., and J. C. Lahr, Revisions of HYPO-71, open file report, 4 pp., U.S. Geol. Surv., Menlo Park, Calif., 1974.

Lee, W. H. K., and R. J. Wetmiller, A survey of practice in determining magnitude of near earthquakes: Summary report for networks in North, Central and South America, open file report, 28 pp., U.S. Geol. Surv., Menlo Park, Calif., 1978.

Le Pichon, X., J. Francheteau, and G. F. Sharman III, Rigid plate accretion in an interarc basin: Mariana Trough, J. Phys. Earth, 23, 251–256, 1975.

Lewis, B. T. R., and J. McClain, Converted shear waves as seen by ocean bottom seismometers and surface buoys, Bull. Seismol. Soc. Am., 67, 1291–1302, 1977.

Lilwall, R. C., T. J. C. Francis, and I. T. Porter, Ocean-bottom seismograph observations on the Mid-Atlantic ridge near 45°N, Geophys. J., 51, 357–370, 1977.

Lilwall, R. C., T. J. C. Francis, and I. T. Porter, Ocean-bottom seismograph observations on the Mid-Atlantic ridge near 45°N—Further results, Geophys. J., 55, 255–262, 1978.

McDonald, K. C., and B. P. Luyendyk, Deep-tow studies of the structure of the Mid-Atlantic Ridge crest near lat. 37°N, Geol. Soc. Am. Bull., 88, 621–636, 1977.

Needham, H. D., and J. Francheteau, Some characteristics of the rift valley in the Atlantic Ocean near 36°48′ north, Earth Planet. Sci. Lett., 22, 29–43, 1974.

Parsons, B., and J. Sclater, An analysis of the variation of oceanic floor bathymetry and heat flow with age, J. Geophys. Res., 82, 803–827, 1977.

Reichle, M., and I. Reid, Detailed study of earthquake swarms from the Gulf of California, Bull. Seismol. Soc. Am., 67, 159–171, 1977.

Reid, I., and K. Macdonald, Microearthquake study of the Mid-Atlantic ridge near 37°N, using sonobuoys, Nature, 246, 88–90, 1973.

Rowlett, H., Seismicity at intersections of spreading centers and transform faults, J. Geophys. Res., 86, 3815–3820, 1981.

Sclater, J. G., J. W. Hawkins, J. Mammerickx, and C. G. Chase, Crustal extension between the Tonga and Lau ridges: Petrologic and geophysical evidence. Geol. Soc. Am. Bull., 83, 505–518, 1972.

Searle, R. C., and A. S. Laughton, Sonar studies of the Mid-Atlantic Ridge and Kurchatov Fracture Zone, J. Geophys. Res., 82, 5313–5328, 1977.

Sinton, J. B., and D. M. Hussong, Crustal structure of a short length transform fault in the central Mariana Trough, this volume.

Sleep, N. H., Formation of oceanic crust: Some thermal constraints, J. Geophys. Res., 80, 4037–4042, 1975.

Temple, D. C., R. B. Scott, and P. A. Rona, Geology of a submarine hydrothermal field, Mid-Atlantic ridge, 26°N latitude, J. Geophys. Res., 84, 7453–7466, 1979.

Uyeda, S., R. McCabe, and N. Sugi, A hypothetical model for the cause of episodic back-arc spreading in the Philippine Sea, paper presented at Seminar of Accretion Tectonics, Oji International, Tomakomai, Japan, Sept. 10–16, 1981.

Wyss, M., and J. N. Brune, Seismic moment, stress, and source dimensions for earthquakes in the California-Nevada region, J. Geophys. Res., 73, 4681–4694, 1968.

Crustal Structure of a Short Length Transform Fault in the Central Mariana Trough

JOHN B. SINTON[1] AND DONALD M. HUSSONG

Hawaii Institute of Geophysics, University of Hawaii, Honolulu, Hawaii 96822

The crustal structure of the Pagan fracture zone, a short length transform fault system in the central Mariana Trough, was determined from a detailed seismic refraction study by using ocean bottom seismometers. The Mariana Trough is a back arc basin with a spreading center opening at a half rate of less than 2 cm/yr. Interpretation of both the seismic travel times and amplitudes suggest that the crust within the transform valley is 1–1.5 km thinner than oceanic crust generated at slow spreading centers away from transform faults but is thicker than fracture zone crust in longer transform offset systems. The crustal differences can be accounted for solely by a thickening, relative to longer transforms, of the lower crust by as much as 3.0 km. Other major differences between the Pagan short transform system and transform systems of greater length are (1) the transition from Pagan transform crust to oceanic crust is only 2.0–4.0 km wide, in comparison with a reported 5–10 km for longer transforms, (2) the 50-mgal free air gravity anomaly over the Pagan transform valley is half that of longer transforms, and (3) whereas bathymetric ridges parallel to longer transform systems are underlain by high-density material, the entire valley of the Pagan transform system is underlain by high-density material. These differences can be explained by the different thermal properties of the third wall at the spreading center/transform intersections, and its affect on crustal accretion. As the third wall becomes colder, the crust becomes thinner, the free air gravity anomaly is larger, the transition from transform to oceanic crust is wider, and the bathymetric relief across the valley is greater. No significant difference was observed between crustal structure in the transform zone in comparison with the fracture zone portion of the Pagan transform system. Use of amplitude modeling with extended WKBJ synthetic seismograms, valid in two-dimensional laterally inhomogeneous media, made it possible to constrain the structural limits of the transform valley crust more closely than would have been possible by using only travel times.

Introduction

We consider a transform fault system to include the active strike slip transform fault offsetting crustal spreading center segments, the inactive fracture zones, and that part of the spreading center that interacts with the transform fault at their intersection. Short length transform systems are distinguished from longer transform systems by the ratio of the spreading center offset to the transform valley width; for shorter transforms this ratio is between 1 and 2, for longer transforms (offsets greater than 50–100 km) this ratio is much greater. Long transforms are used to typify rigid plate tectonics, since they generally lie along small circles about an instantaneous pole of rotation. Short transforms often have geometries that seem to violate the rules of large-scale rigid plate tectonics, although when observed in detail the geometry of even these short transforms becomes more understandable.

More detailed study of transform systems found along spreading axes has recently been made possible by the use of high-resolution geophysical instruments and detailed seismic surveys. An excellent summary of known bathymetric and crustal characteristics of transform systems is given by *Detrick and Purdy* [1980]. Their analysis of seismic refraction data further shows that anomalously thin crust exists within the fracture zone of the Kane transform system. The seismic data also indicate an absence of velocities in the range of oceanic layer 3 [*Fox et*

---

[1]Now at Conoco, Inc., Ponca City, Oklahoma 74601.

al., 1976; *Detrick and Purdy*, 1980] and show that mantle velocities (7.7–8.0 km/s) underly the transform/fracture zone valley crust [*Detrick and Purdy*, 1980]. In fact, fracture zone crustal structure seems remarkably simple; a thin 3- to 4-km/s layer overlying 0–3 km of 6.5-km/s material. Also, a bathymetric high is often found adjacent to the intersection of the spreading center with the transform fault [*Rowlett*, 1981].

A positive gravity anomaly parallel to long transform systems has been explained by a mass excess under the valley walls [*Cochran*, 1973]. Dredge hauls from within transform valleys usually recover basalts and serpentinized ultramafics, indicating that dense material occurs at shallow depth in the crust [*Thompson*, 1973; *Fox et al.*, 1976; *Bonatti*, 1978].

For long transform systems, active strike slip faulting is usually confined to the innermost part of the transform valley [*Searle*, 1979]. This may also be true for short transforms, but *Searle and Laughton* [1977] have used side scan sonar data to show that the strike slip can be oblique to the transform/fracture zone trend for these features.

In spite of the considerable observational data over a few fracture zones, it is not well understood why the crust is often thin in a transform valley, or whether the mechanism for generating this crust is the same for short and long transform systems. Slower magma production [*Detrick and Purdy*, 1980], emplacement of ultramafics by serpentinite diapirism, and crustal thinning by stretching [*Fox et al.*, 1976] have all been used to explain thin transform valley crust. Recent work by *Fox et al.* [1981] shows that the bathymetry at spreading center/transform intersections can be related to the size of the transform offset. By extension, the thickness of transform crust may then be a function of offset since bathymetry and crustal accretion processes are related.

The width of the zone of transition in crustal structure between the transform valley and the adjacent plates is also not well known. Both the origin and extent of the thin transform valley crust is of general interest because these features represent zones of weakness that may influence subduction and crustal accretion process.

A major drawback for fracture zone study has been the distribution of the available data. Existing geophysical surveys using gravity, magnetics, or seismic data have been exclusively collected within long transform systems or fracture zones that are of questionable nature. Except for morphological data from side scan sonar studies, there is a lack of geophysical data for shorter length transforms. It has not been established, for instance, whether short length transforms are also underlain by thin crust.

This study presents seismic refraction and gravity data collected over a short (less than 50-km length) transform system in the Mariana Trough (Figure 1) described by *Karig et al.* [1978] and *Fryer and Hussong* [1981], and named the Pagan fracture zone by *Hussong and Sinton*

[this volume]. The problems addressed by this experiment include the following:

1. What is the velocity structure of the crust within the transform and fracture zone valleys?

2. What is the nature of the crustal transition between the transform valley and adjacent plates?

3. Do shorter transform valleys have the same crustal structure as longer transform valleys?

The Pagan Fracture Zone

The short length transform fault system of the east-west trending Pagan fracture zone (FZ) is located in the central Mariana trough at 17°35′N. The low spreading rate [*Fryer and Hussong*, 1981], seismic crustal structure [*Bibee et al.*, 1980; *LaTraille and Hussong*, 1981; *Ambos and Hussong*, 1982], and general morphology [*Karig et al.*, 1978; *Hussong and Uyeda*, 1981] of the axis of spreading in the Mariana Trough have prompted the authors describing these characteristics to note the similarities of this back arc spreading system to the mid-Atlantic ridge. The active transform fault offsets the spreading center by approximately 20 km between 144°42′E and 144°55′E. The Pagan FZ transform valley is delineated in Figure 2 by the 4000-m contour and its intersections with the north and south spreading centers, NSC and SSC, respectively. Three

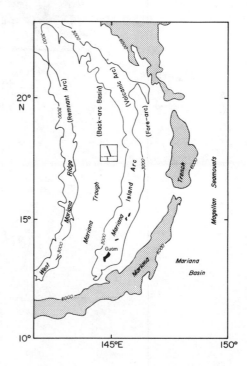

Fig. 1.   Location map showing the regional setting of the Mariana Trench/island arc/back arc basin system. The area of our experiment is indicated by the small boxed region at 18°N latitude.

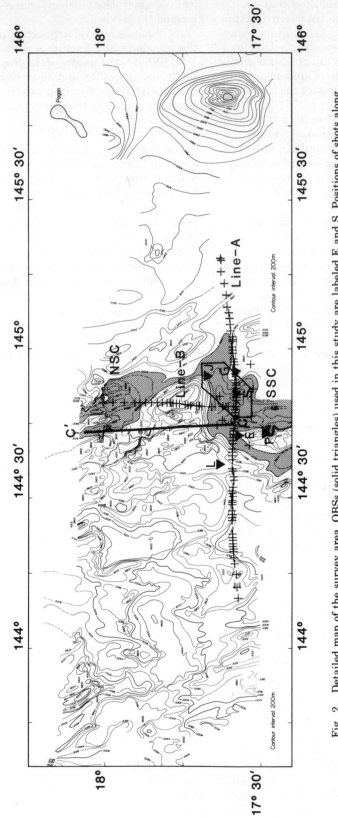

Fig. 2. Detailed map of the survey area. OBSs (solid triangles) used in this study are labeled E and S. Positions of shots along lines A and B are indicated by the crosses. The gravity profile is located along the line labeled C-C'. Water depths greater than 4000 m are shaded along the active spreading center/transform system. The polygonal boxed region within the east-west trending shaded area represents the limits of the present day seismicity concentration associated with the transform fault [Hussong and Sinton, this volume]. The shaded area labeled NSC is the northern limb of the spreading center, while that labeled SSC is the southern spreading center.

large scarps bound the transform valley on the north, east and west. The northern scarp lies parallel to 17°40'N between 144°42'E and 144°55'E. The eastern and western scarps run approximately north-south at 144°55'E and 144°42'E, respectively. The southern limit of the transform valley is bounded by much smaller scarps. The Mariana Trough is spreading at a half rate of less than 2.0 cm/yr [*Bibee et al.*, 1980; *Fryer and Hussong*, 1981; *Hussong and Uyeda*, 1981]. Assuming a half rate of 2.0 cm/yr, this 20 km of offset translates to a maximum age difference of 1.0 m.y. between adjacent crustal blocks.

Although we cannot be certain whether the Pagan FZ transform system studied here is typical, its bathymetry is strikingly similar to that shown by *Searle and Laughton* [1977] for the Kurchatov fracture zone in the mid-Atlantic ridge, and its morphology is similar to fracture zones A and B of the FAMOUS area [e.g., *Laughton and Rusby*, 1975; *Searle*, 1979]. The seismically active zone for the Pagan transform system lies within the deepest part of the transform valley [*Hussong and Sinton*, this volume]. There are no other comparable fracture zones for which such extensive earthquake distribution data sets exist, although the overall concentration of epicenters in the transform zone resembles seismic activity found in long mid-ocean ridge transform systems [*Hussong and Sinton*, this volume]. However, in the Pagan FZ the earthquakes are not confined to the major transform/spreading center intersections. The epicenters trend in a band striking N30°E between the spreading center segments, whereas the overall trend of the Pagan FZ is almost east-west. After noting these characteristics, as well as local bathymetric trends, *Hussong and Sinton* [this volume] conclude that there could be oblique spreading within the Pagan FZ, but that a sequence of very short and unstable transform segments is likely to make up the transform zone.

Data Acquisition

During the fall of 1978 a network of Hawaii Institute of Geophysics ocean bottom seismometers (OBS) was deployed within and around the transform valley shown in Figure 2 by the R/V *Thomas Washington*. While the main purpose of the network was to record the local seismicity in this transform system, two seismic refraction lines were shot to the OBSs in order to determine the local crustal structure for better hypocenter locations.

We present the seismic refraction profiles recorded by OBSs E and S (Figure 2). OBS S was located in the deep part of the transform valley; OBS E was located in the transform valley near the intersection with the southern spreading center. The OBSs continuously recorded three seismic components: a hydrophone, a vertical 4.5-Hz seismometer, and an azimuthally unorientated horizontal 4.5-Hz seismometer. Only the hydrophone data are used here because coupling problems with the seafloor degraded the geophone data. As shown in Figure 2, the shot pattern was an inverted 'T' with a 130-km east-west line running par-

allel to the Pagan FZ valley (line A) and a 50-km line running north from the center of the transform fault portion of the valley (line B). Shot spacing was nominally 1.0 km, but due to rough sea conditions the actual shot spacing varied from 1 to 3 km. Relative shot and receiver locations were determined from a generalized inversion of direct water-wave travel times. This inversion technique is similar to that of the earthquake location problem, but both source and receiver locations are determined. Travel times were converted to slant ranges by using a water velocity of 1.51 km/s; these slant ranges were then used to constrain the relative locations. Use of water velocity structure in this calculation does not significantly improve the results. The final relative locations were determined to within 150 m, which is better than can be achieved with only the shipboard navigation system.

Data Reduction

The seismic amplitudes were corrected for shot size and geometrical divergence by the equation

$$AMP = (\text{Shot Size-kg})^{-0.65} * (\text{Range-km})^{1.2}$$

where AMP is a factor multiplied into the seismogram [*Muller et al.*, 1962]. The affect of instrument automatic gain control (AGC) has also been removed.

Rather than applying topographic corrections to the record sections, lateral variations in bathymetry were included in the model, and a two-dimensional ray trace was used to compute travel time/distance (T-X) plots. The advantage of this procedure over recently used conventional topographic correction techniques [*Whitmarsh*, 1975; *Detrick and Purdy*, 1980] is that there is no inherent error in the topographic contribution to the ray trace because the points of entry and exit of all rays are known. This is particularly important for our study because the bathymetric variations are extreme (as much as 1.5 km). While a correction has been made to move the shots to the ocean surface, no correction has been made to move the OBS to the ocean surface.

Sediments are found only in isolated ponds west of the spreading center and generally thicken toward the active volcanic island arc on the east side of the spreading center. However, the Pagan FZ valley has acted to locally channel sediments coming from the east across the axial graben [*Bibee et al.*, 1980] so it is filled with considerably more sediments. The maximum sediment thickness is found at the eastern end of line A where there are approximately 400 m of material. Within the transform valley there are approximately 200 m of sediments. Since there was no determination of the sediment velocity in this study, the average value of 1.8 km/s determined during DSDP leg 60 drilling [*Hussong and Uyeda*, 1981] in this area was used.

Analysis of the refraction data began by fitting T-X plots, computed from two-dimensional models, to the first arrivals. When the travel times were fit, extended WKBJ (EWKBJ) synthetic seismograms [*Sinton and Frazer*, 1982]

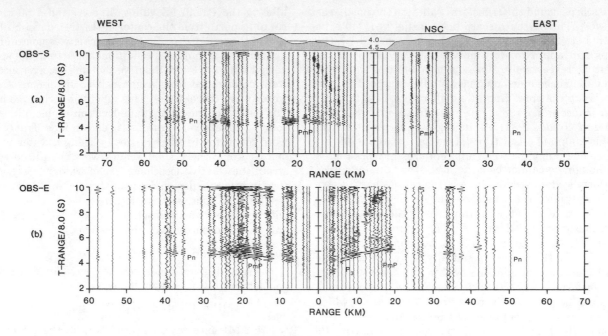

Fig. 3.   Record sections from (a) OBS S and (b) OBS E for shots of line A. These data and all other seismic refraction data shown in this paper are relative amplitude record sections. Bathymetry, in kilometers, is shown at top. P_3 are the crustal phases, PmP is the mantle supercritical reflection, and Pn is the mantle refraction (or head wave).

were used to refine the two-dimensional models by using the amplitudes as constraints. The algorithm discussed by Sinton and Frazer follows from a finite frequency extension of geometrical ray theory discussed by *Frazer and Phinney* [1978]. EWKBJ synthetics are valid for any medium in which the velocity function varies continuously in any direction. No attenuation or S wave propagation is included in the synthetics presented here, although these restrictions are not inherent in the EWKBJ theory.

Refraction data and synthetics were both band pass filtered between 2.0 Hz and 20.0 Hz, although the data of OBS S had a dominant frequency of 8 Hz and that of OBS E had a dominant frequency of 6 Hz. The variations in the dominant frequency are primarily an instrument characteristic related to different analog tape speeds. Ray trace information was computed with sufficient density to define features in the T-X plots with 0.5-km wavelengths for velocities near 4 km/s and with 1.0-km wavelengths for velocities near 8 km/s.

Gravity data used to correlate with the seismic data was taken from the compilation by *Hussong* [1981] along a line shown in Figure 2. Only this north-south profile was used because gravity anomalies within the transform valley are essentially constant.

Refraction Line A: Transform Valley

Split-spread refraction profiles from OBSs E and S are shown in Figure 3. The overall similarity between the two

profiles is apparent, but is complicated by large bathymetric variations. Phases are labeled following the convention of *Spudich* [1979] (see also *Detrick and Purdy* [1980]) for P wave arrivals; P_2 and P_3 are the refracted arrivals from oceanic layers 2 and 3, PmP is the mantle post critical reflection, and Pn is the mantle refraction or head wave. Phases were identified in the refraction profiles by comparison with computed T-X plots.

Note that in both profiles high apparent phase velocities (6.0 km/s and higher) can be found at very short ranges (tentatively identified as phases PmP and Pn). The paucity of secondary arrivals, including S wave and converted arrivals, suggests that there are no major first order velocity discontinuities in the crust, or that the crust is not sufficiently layered to produce such phases with recognizable coherence. The only clear arrivals from the upper crust are the P_3 observed in the eastern half of the profile for OBS E (Figure 3b). These two profiles are very similar in overall character to those of *Detrick and Purdy* [1980] in the Kane fracture zone, although the Pn phase is first observed at a greater range in the Pagan FZ.

There is evidence of near-receiver crustal inhomogeneity for OBS E in Figure 3b. All arrivals east of the OBS E have a consistent 0.2-s delay from corresponding shots on the west side. The delay is almost certainly real because no such delay is observable in the data for OBS S, nor is the delay caused by a timing error or errors in relative shot receiver locations. The ubiquitous nature of this delay

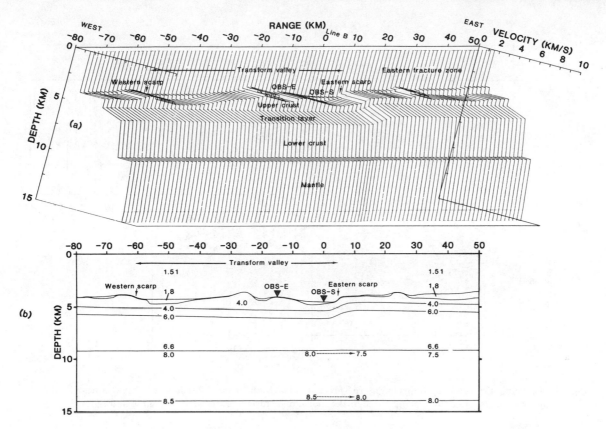

Fig. 4. Two-dimensional crustal model along the axis of the transform/fracture zone valley. Plot (a) is a block diagram of the laterally varying velocity function, while plot (b) is a cross section of the same velocity function along line A. Velocity discontinuities in plot (b) are also shown in plot (a). Velocities between velocity discontinuities are linear functions of depth. Dashed lines in plot (b) show the regions where the velocity changes along the velocity contour directly above or below the dashed lines. All velocities are in km/s.

means that whatever its cause, it must be located very near the OBS and in the upper crust. Possible near surface phenomena were also noted by *Detrick and Purdy* [1980] within the Kane FZ valley, as their data for OBS 5 has a consistently different amplitude character than their data for OBS 4. They could not attribute the different amplitude character of the two OBSs to instrumentation differences and concluded that an unknown feature was affecting their data at OBS 5. It is not surprising that isolated crustal anomalies that significantly affect seismic wave propagation are common in transform or fracture zone valleys. Unfortunately, these crustal anomalies, while important to the formation of the valleys, are impossible to reliably model with our seismic data. Their resolution requires even more detailed seismic field experiments.

Another feature common to both our refraction profiles is the very low amplitude of arrivals from shots on the eastern half of the line. For OBS S this is true for the entire eastern side of the profile, while for OBS E this is only true for shots beyond the 20-km range to the east. This region correlates with the base of a large bathymetric scarp bordering the eastern side of the transform valley

(Figure 2). Within the transform valley, shots west of this scarp have amplitudes at least a factor of 2 higher than the shot amplitudes east of this scarp. We suspect that either the structure of the scarp and the underlying crust causes geometrical defocusing of seismic energy, or the scarp separates regions with differing attenuation characteristics.

On the west side of both split profiles a large increase in arrival amplitude is observed at 20-km range relative to each OBS. Another smaller increase occurs at 40 km. These increases do not correlate with bathymetric features.

The final model for the transform system valley (actually most of the model is in the fracture zone valley outside the transform zone) is shown in Figure 4 and is composed of three crustal layers, excluding the sediments, over a normal mantle. The model fits both refraction profiles. In general, this crustal section is similar to that found in longer transform fracture zone valleys [*Fox et al.*, 1976; *Detrick and Purdy*, 1980; *Detrick et al.*, 1981]. The upper crust is nearly identical to the other fracture zone valleys and the lower crust differs only in the thickness of the 6.0-

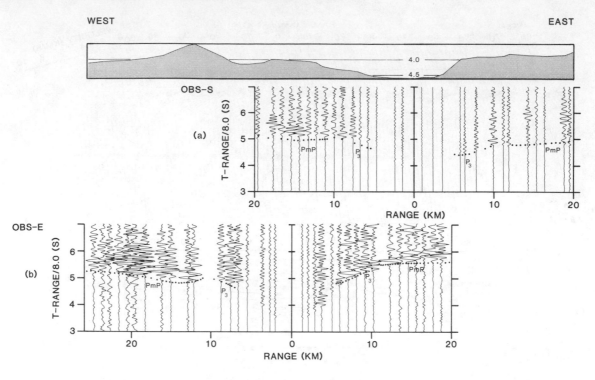

Fig. 5.   Blowups of refraction profiles in Figure 3 for ranges less than 20 km. Solid dots are arrival times calculated by two-dimensional ray tracing through the model of Figure 4. Otherwise, the same as Figure 3.

to 6.6-km/s layer. The crust in the Pagan FZ is 1 to 2 km thicker than that of the Kane FZ and is about 1 to 2 km thinner than young mid-Atlantic crust of comparable age [*Fowler*, 1976]. The uppermost layer has a constant velocity of 4.0 km/s and variable thickness to accommodate the lateral changes in bathymetry. This is underlain by a transition layer where the compressional velocity increases linearly from 4.0 to 6.0 km/s in 0.75 km. The bottom crustal layer is 3.0- to 3.5-km thick with the velocity increasing linearly from 6.0- to 6.6-km/s. The crust is thinnest (4.5 km) underneath OBS S in the deepest part of the active transform valley. The mantle velocity within the transform valley is 8.0 km/s but decreases laterally to 7.5 km/s across the eastern scarp at 144°55′E (Figure 4).

Figure 5 shows the fit of the predicted T-X curve for the model of Figure 4 to the refraction profiles for ranges less than 20 km. In Figure 5, and all other figures detailing line A, the travel times of post-critical reflection branches and the Pn phase predicted by ray tracing are indicated by the solid dots. The 0.2-s delay between the two sides of the profile for OBS E is treated by adding an appropriate amount of extra sediments on the eastern side of the OBS. This sediment infilling treatment, which gives the required 0.2-s delay without adding large near-receiver inhomogeneities, is convenient but is certainly not a unique solution. Arrivals in both profiles have apparent phase velocities ranging between 3.0 and 4.5 km/s for ranges less that 10 km. These phases, labeled P_3, have rays that turn

in the 4.0-km/s layer and in the transition layer. The high-velocity gradient within the transition layer is necessary to minimize second arrivals at about the 10-km range and to keep the P_3 branch as short as possible. All of the variation in apparent phase velocity can be explained by changes in either the sediment thickness or the thickness of the 4.0-km/s layer. Effects of near bottom inhomogeneity are seen in the P_3 arrival in the eastern half of profile OBS S where it crosses the eastern scarp of the transform valley, resulting in an apparent velocity of this arrival that is nonuniform along its length (Figure 5a). In contrast, the P_3 phase in the eastern half of profile OBS E is the least contaminated by lateral inhomogeneities and thus its character varies smoothly (Figure 5b). Similar statements can be made for the P_3 phases along the western halves of each profile. A dip in the transition layer was added below the near bottom inhomogeneities only after all possible bathymetric and sediment affects were determined. The dip was necessary to have a continuous model between the two OBS positions where the refraction profiles are reversed and to fit the observed arrival times of the P_3 phase. It is important to note that the velocities of the 4.0-km/s layer and transition layer do not vary laterally except for the slight dip in the transition layer. Thus almost all of the lateral inhomogeneity in the crust is caused by variations in sediment thickness and variations in bathymetry.

The fault or upwarp in the transition layer beneath the

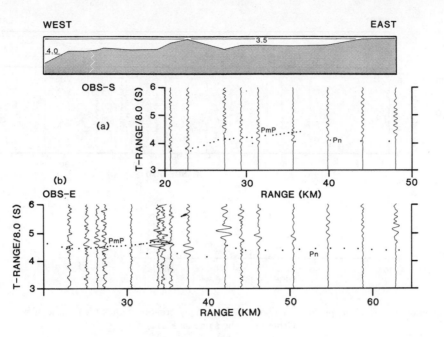

Fig. 6. Blowups of refraction profiles in Figure 3 for ranges greater than 20 km east of each OBS position. Otherwise, the same as Figure 5.

eastern scarp of the transform valley was used to incorporate a relative travel time delay between the P₃ phases of the eastern half of OBS S and all other P₃ phases (Figure 5). This delay is approximately 0.15 s for the eastern half of OBS S. Without the upwarp these P₃ phases arrive too early, even when the change in bathymetry across the scarp is included in the crustal model. By adding a fault in the transition layer that mimics the bathymetry (Figure 4), the observed arrival times of all phases (P₃, PmP, and Pn) for the eastern halves of both OBSs are better matched.

The 0.2-km/s/km gradient in the lower crustal layer is necessary to limit the maximum range of the PmP phase and to minimize the separation of the PmP and Pn phases. Since second arrivals are difficult to identify for ranges less than 25 km, the reflection and refraction branches must be too close to be resolved. Second arrivals are faintly observable in the seismograms of Figures 6a and 7a between the 25- and 30-km range. These low-quality second arrivals were used to limit the maximum range of the PmP phase to approximately 30 km. Both these observations are fit by placing a slight gradient in the lower crust. The most likely magnitude of this gradient is 0.2 km/s/km with a maximum uncertainty of 0.06 km/s/km, as determined by trial and error fitting of T-X plots to the refraction data. Final ray trace T-X plots for the PmP phase are shown in Figures 5, 6, and 7.

Only Pn phases are observed beyond the 35-km range (Figures 6 and 7). Since the largest relative velocity changes occur at the water/ocean bottom interface, it is simplest and geologically plausible to model all short wavelength variations in Pn travel time using lateral changes in either the sediment thickness or bathymetry. Several longer wavelength variations in travel time are a result of either lower crustal or upper mantle inhomogeneities, very large lateral changes in the upper most crust, or inadequate sampling. The first of these travel time variations is associated with the western edge of the transform valley as delineated by a north-south scarp, mentioned above, at 144°42′E (western scarp in Figure 7). Except for the slightly earlier arrival time of shots to the west of this scarp there is no real evidence for a large change in crustal structure across the scarp. In fact, the earlier arrival time of these shots can be completely explained by the shallower bathymetry and lack of sediments west of this scarp. Thus there is no seismic evidence for a deep-seated lateral inhomogeneity beneath the western scarp. In contrast, there must be a marked change in the lower crust or upper mantle across the eastern scarp of the transform valley (Figure 6) similar to that observed for the P₃ phase returning from shallower crustal levels. All Pn arrivals east of this scarp have significantly lower apparent velocities even after the eastward thickening volcanoclastic sediment apron is included in the model. Two models of the crustal structure beneath the eastern scarp were tried: one was a fault at the Moho similar to the fault in the transition layer and the other was a simple horizontal gradient in the mantle. The latter inhomogeneity model is preferred and is shown in Figure 4. The fault model requires too great a throw, amounting to several kilometers, to seem likely. The horizontal gradient used to match the observed Pn travel times in Figure 6 can be no greater than 0.07 km/s/km, which decreases the mantle velocity from 8.0 km/s within the

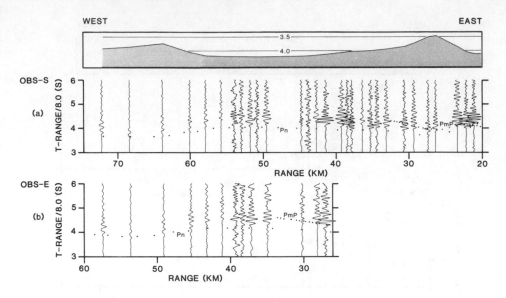

Fig. 7. Blowups of refraction profiles in Figure 3 for ranges greater than 20 km west of each OBS position. Otherwise, the same as Figure 5.

transform valley to 7.5 km/s in the fracture zone to the east. Unfortunately, the eastward thickening volcanoclastic sediment apron buries the possible fracture zone valley so we cannot accurately locate it. *White and Matthews* [1980] also found lower crustal velocities (on the order of 7.5 km/s) in a short Atlantic fracture zone, but they could not determine the crustal thickness.

Using the data of line A, we see no corresponding structural change across the western transition from transform zone to fracture zone valley crust.

We then tried to constrain our model further by interpreting the seismic amplitudes. Because the crustal model for line A is at least two-dimensional, reflectivity synthetics [*Fuchs and Muller*, 1971] cannot be used. Instead, EWKBJ synthetic seismograms [*Sinton and Frazer*, 1982] were computed from the ray trace data. The synthetic seismograms are shown in Figures 8 and 9 along with their respective refraction data.

Before comparing the synthetics and refraction data, two problems concerning EWKBJ synthetics must be addressed. First, although first order discontinuities (FODs) in velocity are not allowable in the EWKBJ theory [*Frazer and Phinney*, 1978], the algorithm used here can be of use for models with FODs. The amplitudes, however, of the synthetics for rays that turn in the region of the FOD will only be an estimate of the true amplitudes, since coupling of P waves and S waves as well as interlayer multiples are not included in the calculations. Fortunately, for refraction lines A and B the only true FODs are the water/sediment and sediment/basement interfaces (labeled P_2 in Figures 8 and 9) that are near the receiver position where no attempt has been made to model amplitudes. Interlayer multiples probably do not make a sig-

nificant contribution to the amplitudes of Figures 8 and 9 because uniformity is not sufficient to support these multiples. Another FOD is modeled at the Moho. In general, seismic amplitudes are used to constrain the character of the Moho [e.g., *Spudich and Orcutt*, 1980; *Kempner and Gettrust*, 1982.] However, whether the Moho boundary is a transition [*Malecek and Clowes*, 1978; *Lewis and Snydsman*, 1979; *Spudich and Orcutt*, 1980; *Kempner and Gettrust*, 1982] or a FOD, as modeled here, is not well constrained by the seismic amplitude data. Therefore, any concern with the inaccuracy of the EWKBJ synthetics near this boundary is overshadowed by the fundamental ambiguity in the nature of the Moho as interpreted from the amplitudes in Figures 8 and 9.

The second problem with EWKBJ synthetics is that rapidly varying two-dimensional features in a model give rise to nongeometrical phases (geometrical phases are those predicted by using geometrical ray theory; nongeometrical phases are not predicted). The nongeometrical phases originate at points along any travel time branch where the derivative of the travel time with respect to range, (dT/dX) changes sign, except at cusps or caustics. At these points the travel time curve is kinked. Normally, these kinks in the travel time curve are smooth, but if they become very sharp (as with rapidly varying lateral inhomogeneities) the nongeometrical phases can be large relative to geometrical phases. The largest source of lateral inhomogeneities is the upper crust, so that wherever the sediment thickness or bathymetry is rapidly varying EWKBJ theory will produce large nongeometrical phases. Such nongeometrical phases are discussed in detail by *Sinton and Frazer* [1982]. Where these phases become large they have either been labeled as Png or have been removed

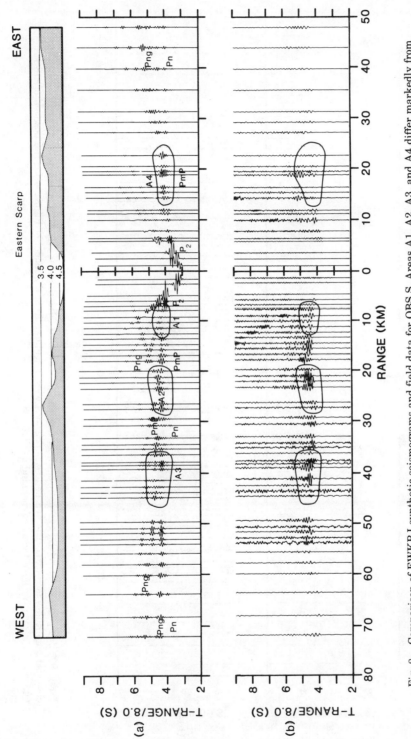
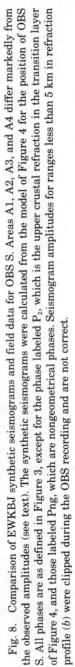

Fig. 8. Comparison of EWKBJ synthetic seismograms and field data for OBS S. Areas A1, A2, A3, and A4 differ markedly from the observed amplitudes (see text). The synthetic seismograms were calculated from the model of Figure 4 for the position of OBS S. All phases are as defined in Figure 3, except for the phase labeled P_2, which is the upper crustal refraction in the transition layer of Figure 4, and those labeled Png, which are nongeometrical phases. Seismogram amplitudes for ranges less than 5 km in refraction profile (b) were clipped during the OBS recording and are not correct.

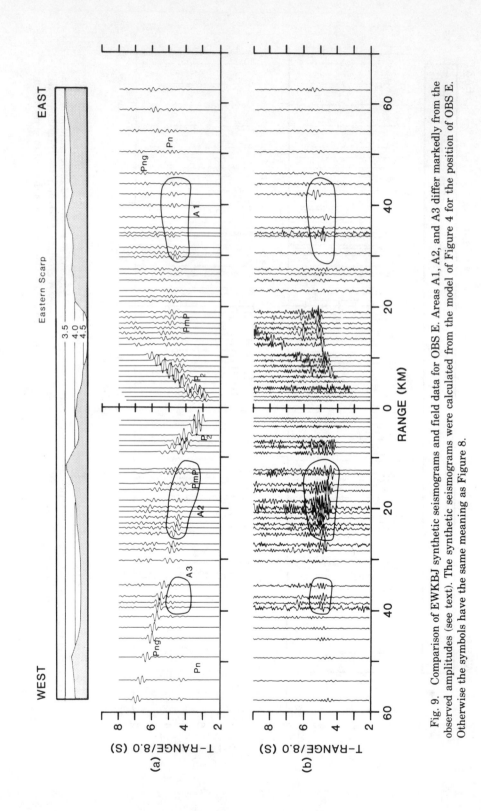

Fig. 9. Comparison of EWKBJ synthetic seismograms and field data for OBS E. Areas A1, A2, and A3 differ markedly from the observed amplitudes (see text). The synthetic seismograms were calculated from the model of Figure 4 for the position of OBS E. Otherwise the symbols have the same meaning as Figure 8.

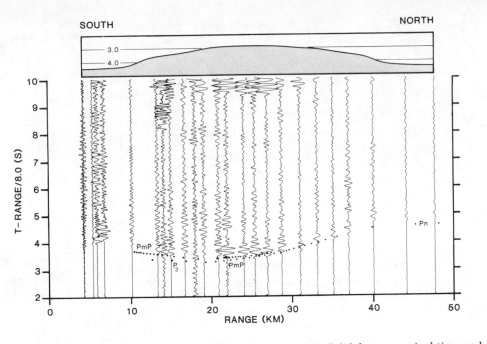

Fig. 10. Seismic refraction profile recorded at OBS S for shots of line B. Solid dots are arrival times calculated using two-dimensional ray tracing from the model of Figure 11. Otherwise the same as Figure 3.

if they are isolated in time from the geometrical phases.

When the synthetics are compared with the refraction data it is obvious that the model of Figure 4 cannot produce all of the small wavelength amplitude variations observed in Figures 8 and 9. However, the relatively lower amplitudes of the seismograms of shots with positions east of the eastern scarp are reproduced in the synthetics (Figures 8a and 9a). This most important amplitude feature must be caused by the scattering effect of the crust underneath the eastern scarp. Other than this point, the EWKBJ synthetics are useful only in estimating how closely the model of Figure 4 represents the true structure.

No attempt was made to improve upon the ray trace model or to match individual seismograms with synthetics. The synthetics do not include the frequency effect of differing charge sizes so that the refraction data beyond the 30-km range are slightly lower frequency than the synthetics. For the profile of OBS S, regions in the synthetics where there are significant deviations from the observed amplitudes are labeled as A1 through A4 (Figure 8), and for OBS E they are labeled A1 through A3 (Figure 9). The most important differences are in the regions A2 and A4 for OBS S and A2 for OBS E where the ray trace model predicts larger second arrivals than are observed. These mismatchs indicate that either the lower crust is more complicated than is modeled in Figure 4, or that off profile affects are important. In both Figures 8b and 9b the increase in arrival amplitude at 40 km, region A3, may be caused by a slight increase in the mantle gradient between 11- and 12-km depth. The best fit for OBS S occurs between 25 and 50 km on the eastern side and between 45 and 65

km on the western side. For OBS E the fit is best for the entire eastern side and beyond 40 km on the western side (Figure 9). It is also apparent that incorrect models result from using only travel time in the analysis of refraction data in regions having large lateral inhomogeneities. Using amplitudes, however, requires that the data density delineates the amplitude variations.

Refraction Line B: Transform-Plate Transition

Line B crosses three structural regions: the transform valley with its thin crust, the intersection high, and the spreading axis which lies under the northernmost 3 shots. The seismic refraction profile for OBS S line B is shown in Figure 10. As for line A, there are no large amplitude second or converted PS wave arrivals, but first arrivals are very distinct. Beyond these similarities in arrival character, line B looks much different from line A, which runs along the transform/fracture zone valley.

OBS S was located in the transform valley. The travel times for shots from the 13- to 30-km range on line B are affected by the intersection high. As would be expected, the first arrivals come in early. Of more interest are the low-amplitude arrivals between 15 and 20 km and the much higher amplitudes between 20 and 30 km. This record section is corrected for shot size, instrument AGC, and geometrical divergence, so the differences in arrival amplitude are real. Another important feature of line B (Figure 10) is the low amplitude and much lower frequency of the last three shots. Although the change in shot size may cause part of this effect, we speculate that it is probably also the effect of a highly attenuative zone.

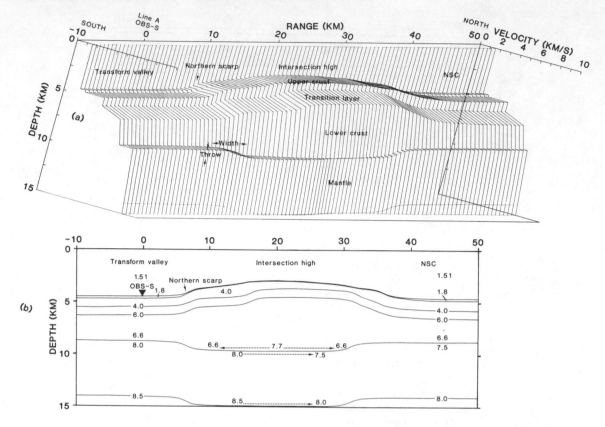

Fig. 11. Two-dimensional crustal model for refraction line B, which is perpendicular to the axis of the transform/ fracture zone valley. Otherwise, labeled as Figure 4.

OBS S was located 4.5 km east of line B, so near receiver crustal structural complexities could not be modeled. For this reason we used only the shots beyond the 10-km range in our analysis and took the upper crustal section directly underneath OBS S from line A. Calculated travel times for rays in the upper crust are consistent with the observed travel times and support the assumption that this section is continuous north from line A up to the scarp terminating the transform valley along 17°40′N (Figure 2). Some error is likely introduced by the assumption of two dimensionality because the actual shooting geometry to OBS S was a fan spread (Figure 2) and the local crustal structure is laterally inhomogeneous. This distortion in the apparent phase velocity is significant near the receiver, but should become negligible beyond 10 km for the geometry of OBS S and shot line B.

In developing the model on Figure 11, the upper crustal model of Figure 4 used beneath OBS S was first adapted for the entire section so that a computed T-X plot would fit the first arrivals of Figure 10. After verifying that the data are satisfied by assuming the transform zone crust is continuous to the valley walls, both the thickness and approximate depth below the ocean bottom of the 4.0 to 6.0 km/s transition zone were held constant beyond the

valley and across the entire model (Figure 11). Thus the upper two layers identified in the transform valley crust retained their relative positions outside the transform valley. Obviously these constraints need not be imposed, but the data require no difference. Furthermore, *Bibee et al.* [1980] found a very similar upper crustal section for a refraction line essentially parallel to and slightly northwest of line B.

After constraining the upper two crustal layers, the thickness and gradient of the lower crustal layer, with velocities between 6.0 and 6.6 km/s, were modified to fit the refraction profile at ranges between 10 and 36 km. The final model shown in Figure 11 has a T-X plot that everywhere fits the refraction first arrivals to within 0.1 s (Figure 10). The best fit is directly over the intersection high between the 15- and 30-km range. We have modeled these first arrivals as supercritical reflections off the crust/mantle interface (PmP). Under the intersection high, the mantle was depressed by 1.0 km and the velocity at the bottom of the lower crust increased to 7.5 km/s in order to fit the observed times. On the northern end of the model the Moho was reinstated to the depth found in the transform valley, although this depth change is not well constrained by the refraction travel times.

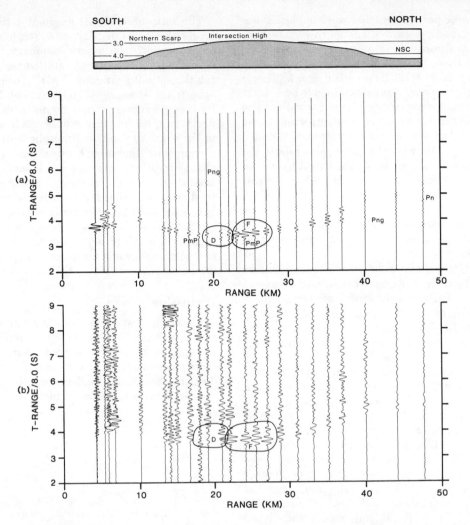

Fig. 12. Comparison of EWKBJ synthetic seismograms and seismic refraction data of Figure 10. Areas D and F are discussed in the text. The northern scarp represents the boundary of the transform fault along 17°42′N in Figure 2. The intersection high is the bathymetric feature created by the interaction of the northern spreading center (NSC) and the transform fault. Otherwise, the same as Figure 8.

To fit the arrival times of the last three shots, a horizontal gradient in velocity was introduced in the mantle so that the velocity under the spreading axis was reduced to 7.5 km/s. This value is typical of 'mantle' velocities found under slow spreading centers [e.g., *Fowler*, 1976]. The fit to the last three shots is adequate considering the large lateral variations in the model of Figure 11.

In contrast to the poor results of using amplitudes in the interpretation of line A, synthetic seismogram modeling of line B was much more successful. In particular, the characteristics of the mantle downwarp, or fault, beneath the northern scarp (Figure 11) are much more sensitive to amplitude than to travel time modeling. The amplitude features over the intersection high that are important to this downwarp are circled in Figure 12*b*, including the defocus (labeled D) between 16 and 20 km,

and the focus (labeled F) between 22 and 28 km. The travel time analysis suggests that all arrivals in this region are the PmP phase, so the amplitude variations must be caused by a lateral inhomogeneity near the Moho. Downwarping the Moho results in a defocus of the synthetic seismograms between 18 and 23 km, and a focus between 24 and 28 km, in the amplitude of PmP phase (Figure 12*a*). The amplitudes and wave forms of the synthetics between 20 and 30 km generally match those of the refraction data. The success of the amplitude modeling of line B is probably due to the more slowly varying structure along the profile in comparison with the high-amplitude, short wavelength bathymetry variations of line A.

Since the amplitude modeling over the intersection high in line B was based entirely on the apparent downwarp, or fault, in the Moho, it is useful to elaborate on this procedure. The three parameters that were important are the throw, angle, and position of the downwarp (Figure 11). The structure shown in Figure 11 represents a compromise between these three parameters. The position of the downwarp can be in error by as much as a kilometer laterally although its center was constrained to be directly beneath the northern scarp of the transform valley. The angle, or width, of the downwarp is more important to the amplitude distribution and cannot vary by more than 0.5 km from its value of 4.0 km. The throw also affects the amplitudes but is most important to the travel times and cannot vary by more than 0.25 km from 1.0 km. Finally, it is important to state that the overlying structure of the transition layer and upper crust in the vicinity of the northern scarp does not affect the amplitude of the PmP phase, since PmP seismic energy does not transverse that region of the crust.

Gravity data collected along the track shown in Figure 2 were modeled by converting the velocities of Figure 11 into densities by using the Ludwig-Nafe-Drake curve [Ludwig et al., 1970] and then computing a synthetic gravity profile (Figure 13). These gravity anomalies are of the same magnitude and overall shape as the anomaly for the Kurchatov fracture zone [Searle, 1979]. The overall fit of the calculated to the observed gravity is everywhere within 2 mgal. Densities within the transition layer, lower crust, and mantle were computed by converting the linear velocity function within each layer to a density function by using the Ludwig-Nafe-Drake curve. Then the mass average of this density function was used for the average specific gravity. The shapes of the transition layer and upper crust were modified to accommodate the difference between the bathymetry under the gravity profile and that of line B. The solid structural boundaries in Figure 13 are those taken directly from the seismic model; the dashed lines are modeled to fit the gravity data.

The major feature in the crust of the intersection high is the inclusion of a body with a slightly higher specific gravity, 2.59 versus 2.54, within the top of the lower crust. Also a root, specific gravity of 2.85 versus 2.75, was placed at the base of the crust, as is suggested by the 7.5-km/s velocity under the intersection high. Both of these features represent building block bodies used to approximate the increase in the velocity gradient under the intersection high. If a constant density of 2.75 g/cc is used for the lower crust under the intersection high, the total gravity anomaly would be 10 mgal too large, while using a density of 2.65 g/cc gives the correct anomaly but is too low for the observed seismic velocities. Thus we see that the 2.59 g/cc body in Figure 13 represents a mass deficiency within the middle of the crust. This mass deficiency could also be placed at the base of the crust, but we found it easier to conceive geologically plausible causes of anomalously low density material at shallower depths. This data set does not provide us with enough insight to speculate on the cause of the mass deficiency beneath the intersection high, although it may be a thermal effect.

The horizontal velocity gradient in the mantle was modeled with a lower density, 3.15 g/cc, to simulate the rising asthenosphere/lithosphere boundary near the northern spreading center. To estimate lithospheric thickness the relationship $9.1 \text{ km/m.y.}^{1/2} \times t^{1/2}$, where t is the lithospheric age, was used [Crough, 1979]. Note that the gravity and refraction profiles do not parallel the axis of the northern spreading center but cross from older lithosphere in the south to younger lithosphere in the north. The asthenosphere/lithosphere boundary in Figure 13 reflects this change in lithospheric age and is consistent with the horizontal gradient in the mantle of Figure 11. The approximate compensation depth for the intersection high using our model is 30 km. The gravity model is most influenced by the decrease in lithospheric age across the intersection high from south to north and the discontinuity between transform valley and intersection high crusts at the northern scarp.

Discussion

Although most crustal formation and alteration in fracture zone/transform systems undoubtedly occurs within the transform valley, the particulars of how and where these processes take place are still obscure. Fox et al. [1976] describe three likely geologic models of crustal formation in long transform systems, including (1) less crustal ac-

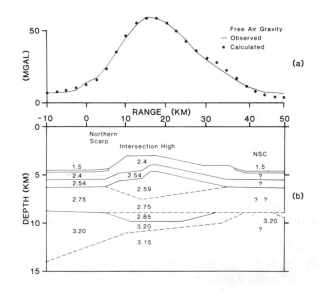

Fig. 13.   Calculated and observed free air gravity anomalies (a) for the crust and upper mantle density distribution (b) of line B. Solid lines are density structures obtained from the refraction seismic model (see text). The dashed structural boundary lines were constructed to obtain the gravity anomaly match (a). This two-dimensional model has to be considered with caution because the true density distribution in the crust and upper mantle is three-dimensional, since the northern spreading center is 5–10 km to the east of line B.

TABLE 1.   Characteristics of Transform Systems for Slow Spreading Rates

Feature	Longer Length	Shorter Length
Offset	50 km and up	Less than 50 km
Bathymetry	(1) Maximum relief: 2–3 km	1–1.5 km
	(2) Valley width: 20–50 km	10–20 km
	(3) Valley shape: asymmetrical	Asymmetrical
Dredge contents	Gabbros, serpentinites, and ultramafics	Same
Hydrothermal	Evidence of active and inactive systems	Same
Gravity	(1) Free air anomalies: 100 mgal peak to trough	50 mgal peak to trough
	(2) Density model: high density material, 3.15 g/cc, at shallow depth on the older side of the FZ	High-density material under entire transform valley
Seismicity	b values probably become lower toward the transform fault center‡	b values remain constant†
Crustal structure		
(A) Fracture zone*	(1) Variable sediments: 0–0.5 km	Same
	(2) 4–5 km/s layer, $h = 1–2$ km	Same
	(3) Not identified	Transition layer
	(4) 6–6.8 km/s layer, $h = 0–3$ km	6.5–6.8 km/s, $h = 3–4$ km
	(5) 7.7–8.0 km/s mantle	7.5–7.7 km/s mantle
	(6) Total thickness 2–3 km	4–6 km
(b) Transform valley	(1) No published data	Variable sediments, 0–0.5 km
	(2) No published data	4–5 km/s layer, $h = 1–2$ km
	(3) No published data	Transition layer
	(4) No published data	6–6.8 km/s layer, $h = 3–3.5$ km
	(5) No published data	8.0-km/s mantle
	(6) No published data	Total thickness 4–5 km
Transition from TV to O crust	10 km wide with crust thickening by 2–3 km	2–4 km wide with crust thickening by 1–1.5 km

*Long transform values from *Fox et al.* [1976], *Detrick and Purdy* [1980].
†From *Hussong and Sinton* [this volume].
‡From *Searle* [1981], *Rowlett* [1981], and *Hyndman and Rodgers* [1981].
TV is transform valley, O is oceanic, and FZ is fracture zone.

cretion at the transform/spreading center intersections, (2) continuous intrusion of basaltic rocks along the entire length of the transform valley, and (3) diapiric intrusion of serpentinized upper mantle material within the transform valley [e.g., *Miyashiro et al.* 1969; *Francis,* 1981]. The last mechanism may continue to be active in the fracture zones, while the first two models are confined to the transform valley. The first model is preferred by *Detrick and Purdy* [1980] and *White and Matthews* [1980] to explain the anomalous seismic crust they found in fracture zones. Both studies imply that the first model is the reasonable choice because it explicitly makes use of rigid plate tectonics, in which crust can only be formed at spreading centers. *Robb and Kane* [1975] and *Bonatti* [1978] prefer serpentinized mantle material as the major constituent of fracture zone crust because of its density, the availability of serpentinizing water, and abundance in rocks recovered from large fracture zones. For the shorter transform system of the Kurchatov FZ, *Searle and Laughton* [1977] propose that oblique spreading is active within the fracture zone valley between the spreading centers. Unfortunately, this model was only based on side scan sonar data and did not include any seismic or gravity data. Thus it is apparent that crust is created within the transform valleys by some mechanism, but may or may not be created or modified within the fracture zones.

Table 1 is a summary of the known characteristics of the short length transform systems, using our data from the Pagan FZ in comparison with longer transform systems for slow spreading centers. The Pagan FZ is clearly associated with a slow spreading center, but it is less clear whether it is a typical short transform system. Regardless, this back arc basin example remains our only short length transform data set.

The seismic zone associated with the short transform system of the Pagan FZ delineates the active transform fault zone shown in Figure 2 [*Hussong and Sinton,* this volume]. The eastern limit of this seismically active zone is bounded by the eastern scarp of the transform valley. There was no significant decrease in number of smaller earthquakes (decrease in b value) at the center of the transform valley. In contrast, the b values of longer transform systems probably decrease along the transform fault away from spreading center/transform intersections. Seismicity within longer transform systems is generally higher near the intersection regions [*Rowlett,* 1981], while large earthquakes occur nearer the center of the transform fault. This appears to be a reasonable assumption because the intersection regions are relatively hot, unable to support large stresses, while the coldest, most competent part of the transform is halfway between the offset spreading centers.

Shorter length transforms are intermediate in crustal thickness between crustal thickness at spreading centers and longer transforms. The transition zone width from transform valley crust to oceanic crust outside the valley increases with length of the transform. The gravimetric and bathymetric expressions across the transform valley of shorter transforms are smaller in amplitude than for longer transforms.

We are aware of no crustal seismic velocity structure data in the transform fault portion of long offset fracture zones. In the Pagan FZ, however, we find very little difference between the structure of the transform zone and the fracture zone to the west. The fracture zone crust to the east is also essentially indistinguishable from the rest of line A, although the Moho velocity is somewhat lower. Considering the resolution of our data, we conclude that there is no difference between the velocity structure of fracture zone crust and transform crust in this short offset transform system. It would be interesting to determine if this consistency holds for longer offset transform systems.

The trends in crustal structure relative to transform offset might be explained by noting that while the crust of shorter transform systems is intermediate in structure between oceanic crust and the crust of longer transform systems, it is initially accreted under similar conditions to the other two environments. We assume that the majority of the crust in the transform valley is formed near the spreading center/transform intersections. The crustal accretion process at a spreading center/transform intersection is modified from the process at spreading centers by the addition of a third boundary, the opposite wall of the transform fault. *Sleep and Biehler* [1970] used this simple third wall model to increase viscoelastic head loss, explaining the relative bathymetric deep at the intersection. Similarly, interaction of upwelling magma with the cooler third wall at the intersection inhibits magma differentiation and upwelling to a degree that is directly related to the difference in temperature, and thus the age of the magma and the cooler wall. The colder the third wall, the less crustal magma emplaced, the deeper the transform valley, the greater the relief across the valley, and the thinner the crust in the valley. This relationship is shown in Figure 14 for transform crust thickness. *Detrick and Purdy* [1980] also note that this interaction of the magma with the colder wall produces thin transform crust, but they did not point out the continuous nature of this interaction as the third wall becomes colder for longer transform offsets. *Fox et al.* [1981] recently proposed a similar relationship between the temperature of the third wall at the intersection and the bathymetric relief across the valley. Unfortunately, with the small and poorly constrained data set on hand we cannot quantify the relationship in Figure 14.

In the Pagan FZ transform system, *Hussong and Sinton* [this volume] suggest that the spreading center may shift across the transform valley by a series of shorter ridge/

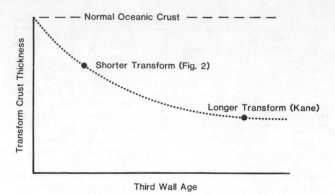

Fig. 14. Schematic representation of the relationship between the temperature of the third wall at the spreading center/ transform intersections and crustal thickness of the transform valley. The age and temperature of the third wall are interchangeable quantities in this relationship. This relationship is asymptotic at longer transform lengths because of the assumption that the third wall temperature approaches an asymptotic value for large ages. Similar relationships probably exist for the bathymetric relief across the valley and the width of the transition from transform valley to oceanic crust.

transform pairs (Figure 15). This would mean that each transform offset would be even smaller, allowing production of transform crust that is more like normal oceanic crust. Obviously, in the Pagan FZ, where the transform/ fracture zone crust is demonstrably thinner than the adjoining back arc basin crust, some appreciable transform offsets must occur.

Conclusions

Our conclusions can be summarized as follows:

1. The crustal structure of the Pagan fracture zone, an example of a short offset length transform fault, differs significantly from longer transform systems. The short transform crust is of intermediate thickness between normal oceanic crust and longer transform crust. The majority of the thickness difference is found in the lower crust, where the shorter transform has a thicker 6.0- to 6.6-km/s layer than longer transforms. Using our seismic data, the crustal structure in the active transform fault portion of the Pagan FZ is indistinguishable from that in the relatively aseismic fracture zone. The high-density crustal material placed on the older side of the fracture zone valley of a longer transform system was not found in the Pagan FZ. The associated gravity anomaly of the short transform is half that of long transforms. Finally, the limited available data suggest that the transition from short offset transform crust to normal crust involves an increase in thickness of 1–1.5 km across a region 2–4 km wide, in comparison with a transition that involves doubling crustal thickness in a zone that is 10 km wide between longer transform crust and normal crust.

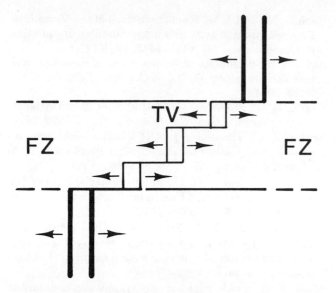

Fig. 15. Model of transform zone for the transform system of Figure 2. This pattern of spreading center-transform pairs would produce the seismicity zone of Figure 2 while still maintaining rigid plate tectonics at the scale of the features within the transform valley (TV). It would also produce the observed thinner transform crust by the relationship of Figure 14.

2.   The use of seismic amplitude information in areas of significant, but slowly varying, lateral crustal inhomogenieties can add much to the interpretation of seismic refraction data. Most of the lateral inhomogeneities are confined to the uppermost crust containing sediments and basaltic basement. The most significant problem of using amplitudes is spatial sampling; it is necessary to ensure that enough data are acquired to define the spacial characteristics of each amplitude variation or else amplitude interpretation in structurally complex environments is very difficult.

3.   Shorter transform systems represent an environment for crustal accretion that is intermediate between a single spreading center and a longer transform system. The major factor in determining the structure of the accreting crust is the thermal character of the lithosphere truncating the spreading center at the transform fault. The longer the transform offset, the cooler will be the truncating lithosphere, and the thinner will be the crust found in the transform valley.

Acknowledgments.   The authors wish to thank E. Ambos for her helpful discussions through the course of this experiment. The reviews of T. Brocher, F. Duennebier, and two anonymous reviewers are also gratefully acknowledged. The OBSs were deployed with the invaluable assistence of Pierre-Luigi Pozzi and the crew of the R/V *Thomas Washington*. This research was supported under National Science Foundation grant OCE80-12922. This paper is contribution no. 1290 of the Hawaii Institute of Geophysics.

References

Ambos, E. L., and D. M. Hussong, Crustal structure of the Mariana trough, *J. Geophys. Res., 87,* 4003, 1982.

Bibee, L. D., G. G. Shor, and R. S. Lu, Inter-arc spreading in the Mariana trough, *Marine Geol., 35,* 183–197, 1980.

Bonatti, E., Vertical tectonism in oceanic fracture zones, *Earth Planet. Sci. Lett., 37,* 369–379, 1978.

Cochran, J. R., Gravity and magnetic investigations in the Guiana basin, western equatorial Atlantic, *Geol. Soc. Am. Bull., 84,* 3249–3268, 1973.

Crough, S. T., Geoid anomalies across fracture zones and the thickness of the lithosphere, *Earth Planet. Sci. Lett., 44,* 224–230, 1979.

Detrick, R. S., and G. M. Purdy, The crustal structure of the Kane fracture zone from seismic refraction studies, *J. Geophys. Res., 85,* 3759–3777, 1980.

Detrick, R. S., M. Cormier, R. A. Prince, and D. W. Forsyth, Seismic constraints on crustal thicknesses at the Vema fracture zone/mid-Atlantic ridge intersection (abstract), *Eos Trans. AGU, 62,* 1050, 1981.

Fowler, C. M. R., Crustal structure of the mid-Atlantic ridge crest at 37°N, *Geophys. J. R. Astron. Soc., 47,* 459–491, 1976.

Fox, P. J., E. Schreiber, H. Rowlett, and K. McCamy, The geology of the oceanographer fracture zone: A model for fracture zones, *J. Geophys. Res., 81,* 4117–4128, 1976.

Fox, P. J., D. G. Galllo, and H. S. Sloan, Ridge-transform intersections: A general model for the bathymetry of fracture zones (abstract), *Eos Trans. AGU, 62,* 1049, 1981.

Frazer, L. N., and R. A. Phinney. The theory of finite frequency body wave synthetic seismograms in inhomogeneous elastic media, *Geophys. J. R. Astron. Soc., 63,* 691–717, 1978.

Francis, T. J. G., Serpentinization faults and their role in the tectonics of slow spreading ridges, *J. Geophys. Res., 86,* 11616–11622, 1981.

Fryer, P., and D. M. Hussong, Seafloor spreading in the Mariana trough: Results of leg 60 drill site selection surveys, deep sea drilling project, *Initial Rep. Deep Sea Drill. Proj., 60,* 45–55, 1981.

Fuchs, K., and G. Muller, Computation of synthetic seismograms with the reflectivity method and comparison with observation, *Geophys. J. R. Astron. Soc., 23,* 417–433, 1971.

Hussong, D. M., Underway geophysics during DSDP leg 60 and related surveys, in *Initial Rep. Deep Sea Drill. Proj., 60,* 71–73, 1981.

Hussong, D. M., and J. B. Sinton, Seismicity associated with back arc crustal spreading in the central Mariana Trough, this volume.

Hussong, D. M., and S. Uyeda, Tectonic processes and the

history of the Mariana arc: A synthesis of the results of deep sea drilling project leg 60, *Initial Rep. Deep Sea Drill. Proj., 60,* 909–930, 1981.

Hyndman, R. D., and G. C. Rogers, Seismicity surveys with ocean bottom seismographs off western Canada, *J. Geophys. Res., 86,* 3867–3880, 1981.

Karig, D. E., R. N. Anderson, and L. D. Bibee, Characteristics of back arc spreading in the Mariana trough, *J. Geophys. Res., 83,* 1213–1226, 1978.

Kempner, W. C., and J. C. Gettrust, Ophiolites, synthetic seismograms, and oceanic crustal structure, 2, A comparison of synthetic seismograms of the Samail ophiolite, Oman, and the ROSE refraction data from the east Pacific Rise, *J. Geophys. Res., 87,* 8463, 1982.

Laughton, A. S., and J. S. M. Rusby, Long-range sonar and photographic studies of the median valley in the FAMOUS area of the mid-Atlantic ridge near 37°N, *Deep Sea Res., 22,* 279–298, 1975.

LaTraille, S. L., and D. M. Hussong, Crustal structure across the Mariana Island arc, in *The Tectonic and Geologic Evolution of Southeast Asian Seas and Islands, Geophys. Monogr. Ser.,* vol 23, edited by D. B. Hayes, pp. 209–222, AGU, Washington, D. C. 1981.

Lewis, B. T. R., and W. E. Snydsman, Fine structure of the lower oceanic crust on the Cocos plate, *Tectonophysics, 55,* 87–105, 1979.

Ludwig, W. J., J. E. Nafe, and C. L. Drake, Seismic refraction, in *The Sea,* vol. 4, Chap. 2, *New Concepts of Sea Floor Evolution,* edited by A. E. Maxwell, 79 pp., John Wiley, New York, 1970.

Malecek, S. J., and R. M. Clowes, Crustal structure near Explorer Ridge from a marine deep seismic sounding survey, *J. Geophys. Res., 83,* 5899–5912, 1978.

Miyashiro, A., F. Shido, and M. Ewing, Composition and origin of serpentinites from the mid-Atlantic ridge near 24° and 30° north latitude, *Contrib. Mineral. Petrol., 23,* 117–127, 1969.

Muller, S., A. Stein, and R. Vees, Seismic scaling laws for explosions on a lake bottom, *Z. Geophys., 28,* 258–280, 1962.

Robb, J. M., and M. F. Kane, Structure of the Vema fracture zone from gravity and magnetic intensity profiles, *J. Geophys. Res., 80,* 4441–4445, 1975.

Rowlett, H., Seismicity at intersections of spreading centers and transform faults, *J. Geophys. Res., 86,* 3815–3820, 1981.

Searle, R. C., Side-scan sonar studies of North Atlantic fracture zones, *J. Geol. Soc. London, 136,* 283–292, 1979.

Searle, R. C., The active part of Charlie-Gibbs fracture zone: A study using sonar and other geophysical techniques, *J. Geophys. Res., 86,* 243–262, 1981.

Searle, R. C., and A. S. Laughton, Sonar studies of the mid-Atlantic ridge and Kurchatov fracture zone, *J. Geophys. Res., 82,* 5312–5328, 1977.

Sinton, J. B., and L. N. Frazer, A method for the computation of finite frequency body wave synthetic seismograms in laterally varying media, *Geophys. J. R. Astron. Soc.,* in press, 1982.

Sleep, N. H., and S. Biehler, Topography and tectonics at the intersection of fracture zones with central rifts, *J. Geophys. Res., 75,* 2748–2752, 1970.

Spudich, P., Oceanic crustal studies using wave form analysis and shear waves, Ph.D. thesis, Univ. of Calif., San Diego, 1979.

Spudich, P., and J. Orcutt, Petrology and porosity of an oceanic crustal site: Results from wave form modelling of seismic refraction data, *J. Geophys. Res., 85,* 1409–1433, 1980.

Thompson, G., Trace element distributions in fractionated oceanic crust rock, gabbros and related rocks, *Chem. Geol., 12,* 99–111, 1973.

White, R. S., and D. H. Matthews, Variations in oceanic upper crustal structure in a small area of the northeastern Atlantic, *Geophys. J. R. Astron. Soc., 61,* 401–435, 1980.

Whitmarsh, R. B., Axial intrusion zone beneath the median valley of the mid-Atlantic ridge at 37°N detected by explosive seismology, *Geophys. J. R. Astron. Soc., 42,* 189–215, 1975.

Seafloor Magnetotelluric Soundings in the Mariana Island Arc Area

J. H. FILLOUX

Scripps Institution of Oceanography, University of California, San Diego, La Jolla, California 92093

Two seafloor magnetotelluric soundings have been performed in the Mariana Island arc and subduction area, the first (station 1) in the Mariana Trough near International Phase of Oceanic Drilling (IPOD) hole 454 (position 18° 01′N, 144° 32′E, depth 3770 m), the second (station 2) in the fore arc basin near IPOD hole 458 (position 18° 06′N, 146° 45′E, depth 3602 m). The electrical conductivity beneath the postulated spreading zone of the Mariana Trough appears to be unexpectedly low in the upper 40 km, increasing slowly and monotonically downward, to 1 S m^{-1} at 700 km. It does not display any significant feature such as lithosphere-asthenosphere or phase transition boundaries. The character of this profile differs considerably from those obtained near the Pacific Rise, suggesting deep as well as shallow structures, generally cooler, and implying less active magmatic processes. Cooler structures may in turn contribute in part to the greater depth versus age of the Mariana Trough compared to that of the main oceanic basins. No indication of the existence of extensive magma concentration of the kind detected on the Pacific Rise at 21°N is recognizable in the magnetic data. This fact, however, may simply result from the distance between station 1 and the spreading axis (~30 km). A cautious speculation on the cause of the implied overall low conductivity values is presented. The MT sounding from the fore arc basin points to (1) very high conductance in the upper zone (0–10 km), accountable for by the sediment blanket, (2) moderate to high conductance in the fore arc upper mantle wedge below, possibly indicative of a moderately high temperature (composition and hydration by water subducted with sediments may also play a role, (3) a large cross section (300–400 km) of unusually little conducting materials, believed to represent the sinking slab and the cooled down environment adjacent to it, and (4) a 20-fold conductivity increase around 420 km depth, sustained over 300 km, followed by a rapid conductivity rise and the crossing of the 1 S m^{-1} conductivity value toward the great depth of 800 km.

Introduction

In one of the final years of project SEATAR (South East Asia Tectronics and Resources) we proposed to carry out a magnetotelluric (MT) traverse at 18.5°N across the exceptionally broad sequence of geological features associated with the Mariana Island arc and subduction zone system. Because of the extensive geophysical research then conducted in this area, we had hoped that a 1000 km or longer MT array extending from the Pacific plate east of the trench to an area well to the west of the Parece Vela Rift could have been possible without excessive interference with the running projects. This ambitious and somewhat hastily conceived program could not be fitted into the later phases of SEATAR and was scaled down to the occupation of two seafloor magnetotelluric (SFMT) sites. The objective was then restricted to a preliminary MT survey that would give the rather novel SFMT techniques

an opportunity to prove themselves by demonstrating experimentally the importance of the initially proposed program. The field work was carried out from the R/V *Thomas Washington* in the fall of 1978. As a result of the limited scope of the project we had ample time as well as a substantial amount of equipment to perform our experimental work in exceptionally favorable conditions. In particular, instrument duplication at both locations provided redundant data for most of the period of observation (65 days). The long duration of the experiment and the duplication of most records thus gives us a justifiably high confidence in the SFMT data obtained during the 1978 Mariana field operations.

Experimental Setting

The two occupied sites were both located along latitude 18°N, the first in the close vicinity of IPOD hole 454 where a known sediment patch could insure better instrumen-

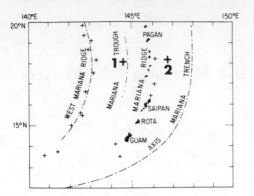

Fig. 1. Location of the two SFMT stations. Station 1 is in the Mariana Trough, position 18° 01′N, 144° 32′E, depth 3770 m. Station 2 is in the fore arc basin, position 18° 06′N, 146° 45′E, depth 3602 m.

tation leveling, the second near IPOD hole 458 in the arc-trench gap (see map in Figure 1). The first location (18° 01′N, 144° 32′E; 3770 m depth) was selected principally because of its proximity to the axis of a ridge-crestlike feature across the Mariana Trough, believed to be the expression of a recently activated spreading center [Karig, 1971a, b]. Inasmuch as such a process implies active magmatism, while MT techniques are particularly sensitive to high conductance features, we assumed that MT data from this area had an enhanced chance to be discriminative and possibly conclusive. The second site was to be located as far to the east as allowable by the available steaming time, hopefully east of the trench on one of the oldest portions of the Pacific Plate. Ship time limitations, however, did not permit us to reach beyond the Mariana fore arc basin. An area near IPOD hole 458, with precisely known bathmetry, was then selected (18° 06′N, 146° 45′E; 3602 m depth).

Two complete sets of EM instruments were deployed at each site, each set including one three-component magnetic variations recorder and one two-component, horizontal electric field recorder. At each station the four instruments were dropped swiftly, with a maximum sea surface separation of 0.75 km at launching time. At recovery they resurfaced with a similar spacing, suggesting the same close spacing on the seafloor. The magnetometers used are of the suspended magnet, optoelectronic feedback type [Filloux, 1980a] with a least count of 0.18 nT. The sampling

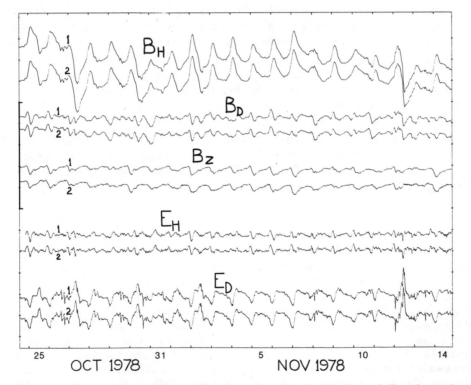

Fig. 2. Illustration of a typical 3-week segment of the data. B_H, B_D, B_Z, E_H, and E_D refer to the magnetic variations to the magnetic north, to the magnetic east and downward, and to the horizontal electric field to the magnetic north and to the magnetic east, respectively. The dark segment along the box border (left) corresponds to 250 nT and 25μV m^{-1}. The numerals 1 and 2 refer to station 1 (Mariana Trough) and station 2 (Mariana fore arc basin). Note the reasonably repetitive daily fluctuations of the solar quiet daily variations and the occasional erratic expressions of solar disturbances (magnetic storms). Note also the more energetic variability of B_H compared to that of B_D and of E_D compared to that of E_H (see additional comments in text).

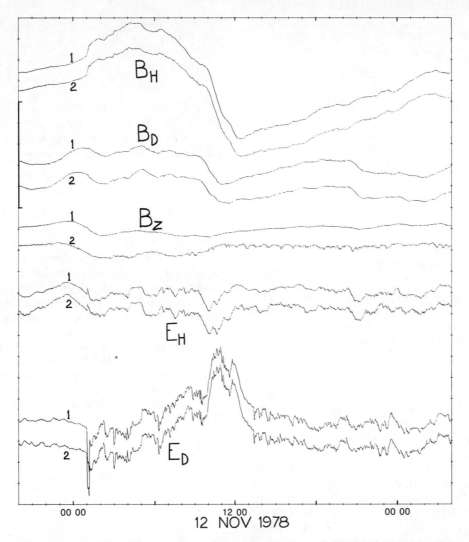

Fig. 3. Illustration of a 32-hour segment of the data selected to cover a magnetically disturbed period (significance of B_H, B_D, B_Z, E_H, and E_D as in Figure 2.) The dark segment along the box border (left) corresponds to 100 nT and $10\mu V\ m^{-1}$. Note the much greater high-frequency activity in the electric field compared to that in the magnetic variations. Note also the high-frequency cutoff in the magnetic signals discussed in the text.

rate was selected at 64 per hour. The electric recorders use salt bridge chopper techniques to reject electrode noise [*Filloux*, 1974]. Their resolution was 0.015 $\mu V\ m^{-1}$ at the sampling rate of 128 per hour. Both types of instruments are self contained and free falling. They return to the surface at a preset time under their own buoyancy, after dropping their tripod ballast.

Data Description and Processing

The data collected are briefly illustrated in Figures 2 and 3, which display two representative data samples. The first of these samples stresses the longer period variations, while the second illustrates, by means of a time scale expanded by a factor of 16, the availability of high-frequency signals up to a high-frequency cutoff where information

in the magnetic variations (though not in the electric records) becomes lost (as noise overcomes a rapidly decreasing signal). This cutoff is an expression of seafloor shielding from ionospheric impulses by the highly conducting oceanic layer. It occurs in this case at around 5–10 cph. In both figures the high correlation between electric and magnetic signals is obvious. A detailed presentation and discussion of a large SFMT data set from another area (Pacific Rise) have been made in an earlier paper [*Filloux*, 1982]. Since the same remarks and similar conclusions would be valid here, we omit such an exhaustive presentation of the Mariana SFMT observations.

Before beginning spectral and cross-correlation analysis of the various EM components, an inspection of all data series was carried out with extreme care, stressing in par-

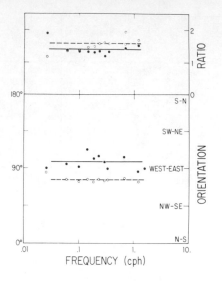

Fig. 4. Frequency dependence of the anisotropy of the seafloor impedances. Solid circles: station 1 (Mariana Trough); open circles: station 2 (Mariana fore arc basin). Lower box: orientation of major polarization axis. Upper box: ratio of major to minor polarization axes.

ticular (1) identity of information between duplicated observations, (2) identification and correction of accidental defects (usually involving single points and known electronic causes), and (3) accuracy of timing and synchronization. The resulting series were then linearly combined, taking into account the trigonometric functions of the appropriate azimuthal and tilt angles, as required for rotation of individual field components to the same reference frame. (Automatic orientation of the instrumentation when reaching the seafloor is only very coarse for the magnetometers, and this adjustment is not performed at all in the electric field recorder system. However, both instruments record orientation and tilt. In this experiment the tilt was negligible in all electric records. The reference frame selected is with respect to magnetic north.) The data processing phase was concluded by plotting all the time series resulting from the experiment with full resolution of individual data points and by conducting a final inspection of these plots.

Data Analysis

Because of the discontinuous spatial coverage of the observations we limit ourselves to a simple one-dimensional and classic MT interpretation. We are aware, of course, of the probable advantage in treating the data within the frame of a two-dimensional structure with east-west variability, as suggested by the generally north-south trend of all distinctive geological features of the area. While this is not possible at this time because of the sparsity of the MT soundings, the present data could nevertheless be incorporated into those of forthcoming experiments.

The approach we use to analyze and to invert the Mariana data is similar to that used, with minor variations, in the handling of comparable SFMT data sets (see details in earlier papers, e.g., *Filloux* [1980b, 1981, 1982]). We therefore give only a brief summary of the various steps.

At each of the two stations the frequency dependant ratio, Z, of horizontal electric field E to horizontal magnetic variations B is estimated from cross correlation of the two component vector fields E and B. The 2×2 tensor Z constitutes the seafloor impedance (or MT tensor) with

$$Z = \begin{vmatrix} Z_{xx} & Z_{xy} \\ Z_{yx} & Z_{yy} \end{vmatrix} \quad (1)$$

in which $Z_{xx}, Z_{xy}, \cdots$ represent the tensor components with respect to a reference frame with x toward the magnetic north and y toward the magnetic east. If certain requirements on Z are satisfied, it is generally found that a meaningful inversion can be accomplished. The important requirements on Z include (1) low anisotropy and, particularly, (2) limited skewness, and of course, (3) sufficiently small error bars. When anisotropy and skewness are substantial but are frequency independent, the data may still be usable. This situation arises when surficial conductivity irregularities modify the flow of electric currents locally, thus also altering the electric field which would otherwise occur in a horizontally homogeneous structure. In such cases, certain correction of Z may be legitimately performed to make it invertible in a useful sense [*Larsen,*

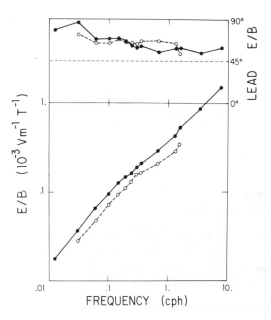

Fig. 5. Seafloor impedances for station 1 (Mariana Trough). Solid circles: for minor polarization axis (of $Z = E/B$). White circles: for minor polarization axis. The higher EM activity in the signals leading to the major axis impedance estimates (black circles) result in a higher coherence and in turn in a wider useful frequency band (more detailed explanation in text).

1975; *Berdichevskii and Dmitriev,* 1976]. At both stations the anisotropy of Z is found to be slight (oriented nearly east-west), and it does not undergo extensive variations with frequency (see Figure 4). The skewness is small and random. The principal axes of Z are those that extremize the ratio $|Z_{x'y'}|/|Z_{y'x'}|$ as Z is rotated to the new reference frame $x'y'$. For station 1 the minimum ratio (for axis x') is 1/1.4 (average value over all frequencies), and it corresponds to an azimuth 8° to the west of due magnetic north. For station 2 the minimum ratio is 1/1.55 for an azimuth 14° to the east of magnetic north. In MT research, such deviations can be considered as favorably low, although this helpful situation is more commonly encountered in the case of oceanic rather than that of continental studies.

The estimated seafloor impedances are plotted over the resolved frequency range in Figure 5 for station 1 and Figure 6 for station 2. Phase as well as amplitude are represented for each one of the principal anisotropy axes. The narrower resolved frequency band for the short polarization axis (x' axis) results from the lower ionospheric signals in the magnetic east-west variations as well as in the correspondingly weaker induced electric field (north-south direction). This situation is readily illustrated by the data plots of Figures 2 and 3, which display much more energetic variations for B to the north and E to the east

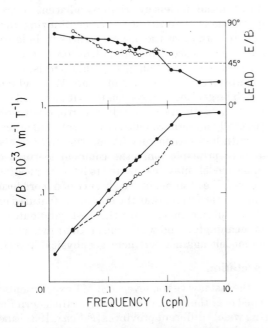

Fig. 6. Seafloor impedances for station 2 (Mariana fore arc basin). The remarks in the caption of Figure 5 hold equally well here. Of immediate significance in this illustration are the rapid change of slope of $|Z|$ toward high frequency and the simultaneous fall of the phase of E with respect to B to values well below 45°; this behavior of T is a clear indication of the existence of a very shallow conducting layer, with a significant conductivity decrease below.

TABLE 1. Seafloor Impedances and Their Uncertainties for Station 1 (Mariana Trough)

f	$|Z|$	ϕ_z	ν	R^2	ϵ	θ
0.0256	0.0365	86.3	32	0.956	0.097	5.5
0.0610	0.0652	68.0	57	0.963	0.065	3.7
0.1016	0.0952	68.8	55	0.966	0.063	3.6
0.1438	0.1237	68.6	74	0.970	0.051	2.9
0.2082	0.1559	64.3	128	0.928	0.060	3.4
0.3183	0.2002	60.1	400	0.885	0.045	2.6
0.6670	0.2981	64.1	250	0.848	0.066	3.8
1.337	0.4904	59.2	420	0.873	0.050	2.9
3.422	0.8892	64.2	90	0.908	0.083	4.8
7.641	1.543	59.8	180	0.590	0.152	6.7

f is frequency (cph), $|Z|$ and ϕ_z are modules and phase of impedances, ν is number of degrees of freedom, R is coherence, ϵ is uncertainty (estimated relative error) on $|Z|$, and θ is uncertainty in ϕ_z, with ϕ_z and θ in degrees.

than for B to the east and E to the north. Thus the estimates of Z which rely on the latter components of B and E are more affected by noise, with the resultant shrinking of the useful frequency band: these estimates correspond roughly to the short polarization axis of Z (x' axis). For this reason our interpretation is based upon the tensor component $Z_{x'y'}$, which corresponds to the long polarization axis (orientation y'). This tensor component contains the most important part of the information in the data, since its estimates cover a much broader frequency range and are more precisely established due to a lesser degradation by noise. However, it is possible that it is more affected by transverse magnetic modal contributions [*Cox et al.,* 1971] than the selected component (transverse electric), as suggested by the nearly north-south trend of geological features. A greater data coverage and data interpretation in terms of two-dimensional models, taking into account both TE and TM modes, would help assess the degree of distortion in the present interpretation. The selected impedance estimates used in optimizing the conductivity models are listed in Tables 1 and 2 together with other important parameters, in particular, relative uncertainties.

TABLE 2. Seafloor Impedances and Their Uncertainties for Station 2 (Mariana Fore Arc Basin)

f	$|Z|$	ϕ_z	ν	R^2	ϵ	θ
0.0256	0.0381	75.0	32	0.941	0.113	6.5
0.0610	0.0764	72.7	57	0.960	0.068	3.9
0.1016	0.1100	73.4	55	0.956	0.072	4.1
0.1438	0.1420	68.9	74	0.958	0.061	3.5
0.2082	0.1936	65.7	130	0.893	0.080	4.6
0.3177	0.2435	62.0	256	0.870	0.065	3.7
0.6670	0.4186	57.8	250	0.832	0.070	4.0
1.376	0.6829	38.3	330	0.805	0.090	5.1
3.422	0.7950	25.0	90	0.763	0.144	8.2
7.641	0.7955	27.9	180	0.517	0.177	10.0

Symbols same as for Table 1.

Conductivity Profiles

Our approach to the inversion of the seafloor impedance estimates into an electrical conductivity function of depth consists of first finding the optimum, isotropic, and horizontally layered model that best fits the data. The fit is judged in terms of the statistical significance of the χ^2 values measuring the least square departure between observed and model impedances normalized with respect to the estimated relative error in the observations, with

$$\chi^2 = \sum_1^N \frac{|Z_n^0 - Z_n^m|^2}{|\epsilon_n Z_n^m|^2} \qquad (2)$$

where N is the number of seafloor impedance estimates, Z_n^0 and Z_n^m are the observed and model impedances, and ϵ_n is uncertainty on Z_n^0 (see, for instance, *Parker* [1980]). If the χ^2 misfit figure is less than that correspondent to the 95% confidence level ($\chi_{95\%}^2 = 2N + 4N^{1/2}$), then one model at least (the current one) fits the data in a statistically acceptable sense. The data are thus invertible within the restrictions assumed for the proposed conductivity structure. If the value of χ^2 approaches its expected value ($\chi_{exp}^2 = 2N$) or falls below, then the significance of the fit decreases or becomes lost altogether as the latitude over which the optimum model may be changed widens. In other words, χ^2 values much lower than χ_{exp}^2 imply model optimizations that rely principally on the noise in the data and not on the EM signals. Even so, model departures from the optimum model generally result (in the case of low noise and invertible data) in a rapid increase of χ^2, without complete loss of the distinctive features of the optimum model as χ^2 reaches significant confidence levels (for instance, $\chi_{95\%}^2$). The problem of finding a systematic way to define the common features of all acceptable models is, to a large extent, unsettled. This subject is investigated in depth by *Parker* [1980] and *Parker and Whaler* [1981]. While providing a formal and conclusive approach to proving or disproving data invertibility, these authors, nevertheless, help us only to perceive the considerable difficulties associated with extracting and with correctly stating the nonambiguous information on conductivity structure hidden in the data. We have attempted to overcome this difficulty in the past in two ways.

First, given an optimized and sufficiently simple layered model, we have estimated the range of conductivity variations in each single layer for which the fit remains acceptable [*Filloux*, 1980b]. We believe that this approach tells us much about which information is significant and which is illusory. Certainly, many features expected from MT sounding clearly emerge from such error bar estimates on conductivity; for instance; (1) the range of depths for which the conductivity estimates are meaningful is intermediate, excluding depths shallower than a minimum or in excess of a maximum penetration depth, (2) MT techniques are selectively sensitive, stressing high conductance layers, (3) the lower bound of the estimated conduc-

tivity of poorly conducting layers is ill defined (because of shielding by the large conductance above), and (4) there is a trade-off between resolution of specific features of the layering geometry and correspondent conductivity resolution.

Second, we have observed so far that the salient features of SFMT conductivity profiles (that is, those features which persist when the number of model parameters is reduced) tend to display a spatial variability that is compatible with their geophysical setting, even when the correspondent profiles overfit the observations somewhat [*Filloux*, 1980b, 1981]. We have therefore taken the position that new SFMT soundings should not be oversimplified from the very beginning, for fear of missing real, and thus important, information, and should be compared with those already available prior to excessive smoothing. Retreating to additional simplification may be proven necessary but can be done afterward. We have adopted this philosophy in our interpretation of the Mariana observations.

For both stations 1 and 2 the number of impedance estimates is $N = 10$ (see Tables 1 and 2). In both cases we started the optimization process using initial models including 10 layers terminated below by a half space. With $\chi_{exp}^2 = 20$ and $\chi_{95\%}^2 = 33$ in this case, then both initial conductivity models overfitted the data, with χ^2 misfit values of 10.5 and 7.5 for stations 1 and 2, respectively. The proposed conductivity distributions (see Figure 7) are derived from these initial models by grouping adjacent layers with nearly the same conductivity and reoptimizing the new layering. In both cases the new structures include six layers plus a half space. This simplification of the profiles has not resulted in excluding any characteristic conductivity feature. The new values of χ^2 are 12.1 and 9.6, not very much larger and still consonant with a model overfit. Therefore the data are definitely invertible with respect to horizontally uniform conductivity models, but other such models with less satisfactory fit do exist and are also acceptable. In principle, only the common features of all acceptable models may be expected to have unquestionable significance. We assume, in the absence of a more formally acceptable alternative, that the dominant features of reasonably simplified models are the best candidates to be proven meaningful, and we decide on their merits by testing the models against pertinent geophysical knowledge.

Interpretation

Over the last few years, several SFMT experiments were conducted over the north central and northeastern Pacific, covering widely different provinces [*Filloux*, 1981] and giving us a background upon which to reference the new Mariana data. In Figure 8 we have plotted together the various conductivity structures resulting from these experiments, including also those from the Mariana, to acquaint the reader with the encountered variability. The ensemble of these results spans, with an irregular and coarse spacing, the entire North Pacific from the Pacific

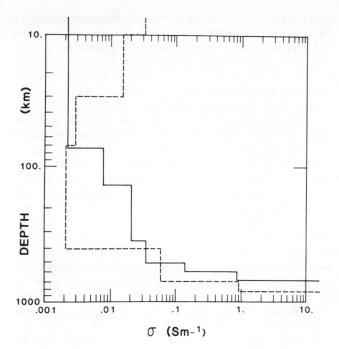

Fig. 7. Proposed conductivity distribution with depth beneath station 1 (Mariana Trough, solid line) and beneath station 2 (Mariana fore arc basin, dashed line). Note the log scales for both depth and conductivity.

Rise spreading center to the Mariana arc and plate subduction system, including areas of intermediate age.

Although more detailed than permitted under the χ^2 criterion referred to earlier, the first three profiles (Pacific Rise at 12°N (12° 06′N, 103° 30′W), Seafloor Revisited (31° 18′N, 128° 20′W), and Aggy III (26° 32′N, 151° 20′W) seem to include a common zone of greatly increased downward conductivity gradient which dips westward. It was interpreted as an expression of the cooling and thickening of the Pacific plate with age [*Filloux* 1980b, 1981]. This distinctive feature persists even if much simplified models are used, such as extremely crude two-layer models. It is also confirmed by other independent and more formal inversions of the same data [*Oldenburg*, 1981]. However, reversal of the conductivity gradient at some depth below, as shown in these profiles, is not supported by the χ^2 criterion, although including it into the models contributes toward an improved fit between data and model predictions.

The conductivity profile proposed for station 1 (Mariana Trough) is shown on Figure 7 (solid line) and also among those of all other SFMT soundings on Figure 8. The depth dependence of σ arrived at in this case is the smoothest of all, with a monotonically increasing conductivity as well as a near-monotonic increase of the conductivity gradient. (In judging the latter, remember the log-log scales in Figures 7 and 8.) If one insists on distinguishing some sig-

nificant features affecting this smoothness, one may suggest the existence of a slight conductivity bulge over the depth range 70–300 km. Even so, the conductivity remains low over this depth interval. This would also mean a sharpening of the conductivity gradient around 80 km followed by a fall back around 180 km. This perturbation, however, is very slight and in no way as definite as some conductivity tongues found in the northeastern Pacific profiles (see Figure 8). While keeping the same standards as those used in interpreting the earlier profiles, no credible hint is provided of the existence of a distinct or even of a diffuse lithosphere-asthenosphere boundary and therefore of a possible depth of occurrence. It is also noteworthy that the conductivity beneath the Mariana trough, at least below 40 km, is systematically lower than at any other station (except at station 2, in the Mariana fore arc basin, for which case there is an acceptable explanation, as discussed later). This fact stresses the considerable difference between the conductivity structure of the lithosphere and

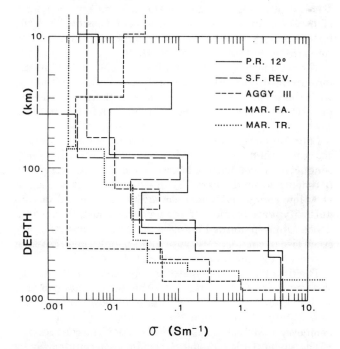

Fig. 8. Proposed conductivity profiles from five SFMT soundings from the North Pacific, including Mariana Trough (MAR. TR) and Mariana fore arc basin (MAR F. A.). The other three soundings are P.R. 12° (Pacific Rise at 12°N), position 12° 06′N, 103° 18′W; S.F. REV. (Seafloor Revisited), position 31° 18′N, 128° 20′W, and Aggy III (Farewell to Aggy, Station III), position 26° 32′N, 151° 20′W. Intercomparison of the various profiles requires simultaneous presentation on a common reference grid. The resulting confusion, difficult to avoid, is greatly reduced if the reader retraces the various profiles in color, using a meaningful code, such as a spectrum color sequence, arranged to match the sequence of station longitudes.

upper mantle of the Mariana trough (a postulated young spreading center) and that of the Pacific Rise, and of course it makes us wonder what might be the explanation. This dilemma is further widened by the moderate to low conductance in the upper 40 km. Here we must proceed with great care. The information from SFMT soundings is quite unspecific for the upper few or few tens of kilometers adjacent to the seafloor. However, high-conductance layers are the most likely to emerge, and an integrated conductivity of 500 S (probably much less) in the upper 20 km should not escape detection. This is particularly true, due to reduced oceanic shielding, in a water depth smaller than the mean oceanic depth (~5000 m), although the actual conductivity distribution may remain quite undefined. Therefore even in the upper region of the Mariana trough profile (say, the upper 40 km) the electrical conductance must be significantly less than in the same depth range of the Pacific Rise profile. Thus the possibility, anticipated at first from current knowledge of the area, that hot and thus highly conducting materials would probably be detected beneath the Mariana trough did not materialize. This is not enough to rule out the existence of some form of magmatism, but it does not appear possible that the latter occurs below the Mariana trough with an intensity comparable to that of the Pacific Rise.

The conductivity profile obtained for station 2, Mariana fore arc basin (see Figures 7 and 8), is definitely the one with the greatest divergence from all others. Yet the implications are more readily explained than in the previous case.

Four important features are discernible in this MT sounding, namely (1) a shallow and relatively thin, very conductive layer with subjacent conductivity decreasing to a depth around 60 km, (2) an extensive cross section of very low conductivity materials (60–420 km), (3) a conductivity jump by more than one decade sustained over a depth range spanning roughly 420–720 km, and (4) a very steep rise in conductivity, with an unusually deep crossing of the 1 S m^{-1} conductivity value at 800 km.

The first feature is unquestionably real, as indicated by the low $|Z| = |E/B|$ ratio and by the decreasing lead of E with respect to B, namely, to phase angles well below 45° (see Figure 6). Although the resolution of the layering geometry of MT is low, as stressed earlier, the conductance of the upper 10 km (~350 S) would be accounted for by 250–400 m of sediments with conductivity ranging from 0.8 to 1.5 S m^{-1}. The sediment accumulation over the fore arc basin, variable and up to 1 km [Latraille and Hussong, 1980], thus could alone explain the nature of the shallow seafloor conduction.

In the 10–60 km depth range the integrated conductivity, also of the order of 300–400 S, is only moderate, but it is not low. The origin of this conduction is not clear. Much water must be subducted with the sediments on the sinking plate, since accretion from plate sediment scraping along the fore arc appears minimal [Uyeda, 1982]. Never-

theless, the importance of water inclusion is supported by the abundance of serpentinized rock samples dredged along the fore arc region at the same latitude [Bloomer and Hawkins, this volume]. However, the closure of pores by pressure should greatly reduce the importance of sea water injection upon rock conduction as depth increases. This reasoning suggests that composition or high temperature or both are responsible for the moderate electrical conductivity of the upper mantle wedge occurring beneath the Mariana fore arc basin.

The exceptionally low conductivity sustained over the extensive depth range 60–420 km must be directly related to the great thickness of the old and cool subducting oceanic plate and also to the cooling it performs on the surrounding environment. Such a feature is conspicuously absent from all other profiles (see Figure 8), yet its magnitude makes it inescapably real. This does not rule out, of course, the existence of conductivity variations with scale sufficiently small, since the conductivity estimate represents an average value. It should be remembered, however, that this 350-km layer ultimately resulted from the lumping together of four adjacent layers of an earlier model with individual conductivity values insignificantly close to each other. Thus any important variability would probably have to be of a scale no larger than a fraction of the thickness of these layers (perhaps 50 km or less).

The 300-km conductivity step below also resulted from the association of two adjacent layers (150 km thick). It is of importance to the optimum fit, but it is not resolved at the 95% confidence level of the χ^2 misfit criterion.

Finally, the rapid conductivity rise and the crossing to conductivities in excess of 1 S m^{-1} at a depth around 800 km constitutes a significant feature of the conductivity dependance with depth. It also approaches the sounding depth limit for the frequency range covered by the data. An area of very rapid conductivity increase at a several hundred kilometer depth seems typical of all SFMT soundings, except that of Aggy III (see Figure 8). This experiment, however, was the first of the series, and its short duration, combined with some instrumental defects not yet detected and corrected at the time of the experiment, resulted in a high noise level at extreme frequencies and, in turn, in a wide uncertainty in the estimated conductivity of the deeper layers (and also of the layers nearest to the seafloor).

Discussion

Before proceeding further in the discussion of the Mariana SFMT soundings, we must first assess the importance of our simplifying assumptions, namely, the use of one-dimension models when the bulk of geophysical evidence points to a two-dimensional structure. The lack of a sufficiently dense east-west data coverage rules out any hope to resolve local and relatively rapid lateral changes. Thus the usefulness of the present results is restricted to information on an average (not well specified) of the conductivity conditions in the seafloor surrounding the individual stations, the averaging spatial scale increasing with depth. In addition, (1) we have shown that the anisotropy of the seafloor impedance was moderate and no larger than for other areas of the North Pacific believed to be tectonically much more uniform (stations Aggy III and S. F. Rev., *Filloux* [1980*b*]) and (2) we have shown that the data are invertible with respect to simple one-dimensional models. Taken together, these facts suggest that it is reasonably safe to interpret the estimated conductivity in a given depth range as a rough lateral average but also as one broader for greater depth.

In the upper 60 km the proposed conductivity structure at station 2 is probably representing reasonably faithfully the fore arc basin conditions. Below, the influences of the bending, dipping, and eventually of the near vertical fall of the slab, well evidenced by other exploratory means [*Katsumata and Sykes,* 1969], may mix to some extent. We assume that the resulting distortion is not so complete as to hide the salient features of the conductivity structure that exist along a vertical below the observation station. A question arises, then, that relates to the considerable thickness of little-conducting, presumably cold materials in the 360 km below the fore arc prism, while estimates of the plate thickness at the 150 m.y. age do not seem to exceed 125 km, crust included [*Leeds,* 1975; *Yoshii et al.,* 1976]. The situation is a repeat of that encountered at S. F. Rev. and at Aggy III stations [*Filloux,* 1980*b*] with the following respective characteristics: plate age, 30 and 72 m.y.; estimated thickness of the low conductivity layer, 80 and 135 km, against estimated plate thickness, 45 and 65 km. Thus MT estimates of the thickness of the low-conductivity layer between seafloor and next sharp conductivity gradient seems to consistently exceed estimates of lithospheric thickness. Since the latter are derived from various investigative techniques considered mutually consistent, it appears that MT soundings do not 'feel' the influence of the asthenosphere before it has reached a certain degree of development. At a small to moderate distance from a spreading center where relatively high temperatures may exist at sufficiently shallow depth (hence at low pressure), the initiation of pore connectivity in a porous matrice filled with magma would greatly increase conductivity [*Shankland and Waff,* 1977]. This initiation of pore connectivity must in turn require a sufficiently high melt fraction threshold and hence a temperature that may be reached only at some depth below the lower plate interface. Whether this mechanism remains plausible at the considerably higher pressure involved in the case of a very old plate area is uncertain.

The rapid conductivity rise at around 420 km depth matches so well that predicted for the olivine to spinel phase transition [*Ringwood and Major,* 1970], that we have wondered whether this explanation could have some merit. We have found, however, no support for this hypothesis in the MT profile for station 1 in spite of its ad-

equacy in terms of both penetration depth and resolution (the error bars in the seafloor impedance estimates are smaller for station 1 than for station 2 (see Tables 1 and 2), thus implying better resolution for the conductivity profile at station 1).

The MT profile for station 1, located near the postulated spreading axis on the center of the Mariana trough, is a surprise because of its prediction of a conductivity relatively low at all depths. Yet magmatism and hydrothermal circulation are supported by solid evidence [*Karig et al.,* 1978; *Hobart et al.,* 1979] and high heat flow values are consistently obtained in the area [*Anderson,* 1975, 1980]. Two explanations may be considered: (1) magma occurs only in disconnected pockets (of unknown size), despite some abundance, and (2) the molten to solid rock proportion is small and the magma supply limited. Because of the predominantly low conductivity in both lithosphere and mantle below the Mariana trough, we tend to prefer the second explanation. We then wonder why the overall conductivity is so unexpectedly low.

It is strongly believed that the island arc and marginal sea system to the west of the Mariana resulted from an extensional process, namely, the progressive increase of distance between the Asian coastline and the subducting zone of the Pacific plate, over a period of perhaps 50 m.y. [*Karig,* 1971*b*; *Hilde et al.,* 1977; *Uyeda,* 1982]. If eastward receding of the subduction zone is a principal component of this extensional motion, then the subduction process must leave in its westward wake a succession of plate fragments. Because of their great thickness and long cooling and remelting time constant, the latter are capable of absorbing heat for a long time from the mantle materials rising to fill the extension-generated gap. This mechanism may explain, through the temperature dependence of conductivity, the low values of the latter over the entire upper mantle area. In turn, it suggests that significant temperature inhomogeneities in the rising replacement mantle may be generated by the irregular distribution, size, and remelting of slab remnants. Thus it is conceivable that hotter mantle portions rising faster may produce locally and for sometime spreading like features. This would account for the spatial scale of large changes between locally coherent high and low heat flow regimes. At the same time the implied lowering of the overall average mantle temperature may contribute to explain, at least partly, the ocean floor depth-age discrepancy discussed by *Louden* [1980] and *Anderson* [1980]. The tectonic heterogeneity of island arc–marginal sea systems, expressed in part in terms of spatial heat flow variability and fossil magmatic features, may constitute the signature of fading out perturbations associated with episodic retreats of the subduction line.

We have pointed out earlier the steepening of the conductivity gradient at great depth with subsequent crossing to conductivities in excess of 1 S m^{-1}. The westward dipping of this transition zone, from 250–300 km near the Pacific

Rise to 800 km beneath the Mariana fore arc basin, must be the expression of some fundamental processes. It is hard not to associate it with the shallower, high gradient zone detected in the upper asthenosphere at almost all stations, which dips similarly westward. Both sloping surfaces suggest a relationship with the streamlines of a thermodynamically driven, creeping or flowing convective motion between rise crest and subduction.

Although obtained within a relatively short distance from the axis of the postulated Mariana Trough spreading ridge (30 km), the high-frequency magnetic variations (see Figure 3) do not include any evidence of a very high frequency enhancement of the kind found above the Pacific Rise at 21°N and interpreted as the EM signature of an extensive shallow magma accumulation spanning the spreading system [*Filloux*, 1982]. A dense array of instruments then provided observations that made the distinctive local variability of the magnetic signals quite obvious. Here, the simultaneous data from station 2 should be adequate to reveal the telltale amplification of the horizontal, high-frequency variations at station 1 if it existed. The negative outcome of this test may simply indicate that the separation between the observation stations and the spreading axis was too large. Clearly, then, observations spanning the postulated Mariana Trough spreading center would be extremely valuable.

The SFMT Mariana experiment may suggest as many questions as it answers. This is not a too surprising result for a relatively novel exploratory method. The proposed conductivity profiles have some unexpected properties, but they do not contain hints of unworkable inconsistencies. The execution of other appropriate regional SFMT investigations would help us to evaluate the scope of the results presented in this article. Among the possible topics for future investigation, S. Uyeda (personal communication) suggests that we include at a high priority the MT exploration of a slow-spreading ridge crest system.

Conclusion

A SFMT experiment limited to two stations cannot, understandably, provide more than a limited contribution to the elucidation of the complex problems associated with an area as vast and as rich in contrasts as a subduction zone–island arc–marginal sea tectonic system. Nevertheless, the soundings discussed here seem to be sufficiently consistent with earlier ones and with many geophysical realities to deserve a cautious but willing credence. Nothing could further enhance their trustworthiness and their significance more than confirmation and extension of the information on conductivity that they have already provided. Supplemental information would be far more than simply additive: it would permit the treatment of all data, past and present, in a much more powerful, two-dimensional perspective. It would be supported by the vast amount of alternate and very complementary knowledge resulting from the SEATAR program.

Acknowledgments. Several constructive criticisms were kindly provided by two reviewers. Valuable comments and suggestions from D. E. Karig, J. W. Hawkins, and J. G. Sclater are acknowledged. The proposal which initiated the present experiment was submitted in the very last years of the SEATAR program when the pressure of not too distant deadlines began to affect many: we are very grateful to the SEATAR team for having unselfishly accepted us. The National Science Foundation supported our research under grants OCE78-16759 and OCE79-18382.

References

Anderson, R. N., Heat flow in the Mariana marginal basin, *J. Geophys. Res., 80,* 4043–4048, 1975.

Anderson, R. N., 1980 Update of heat flow in the East and Southeast Asian Seas, in *The Tectonic and Geologic Evolution of Southeast Asian Seas and Islands, Geophys. Monogr. Ser.,* vol. 23, edited by D. E. Hayes, pp. 319–326, AGU, Washington, D.C., 1980.

Berdichevskii, M. N., and V. I. Dmitriev, Basic principles of interpretation of magnetotelluric sounding curves, in *Geoelectric and Geothermal Studies,* edited by A. Adam, pp. 165–222, Akademia Kiado, Budapest, 1976.

Bloomer, S. K., and J. W. Hawkins, Gabbroic and ultramafic rocks from the Mariana Trench: An island arc ophiolite, this volume.

Cox, C. S., J. H. Filloux, and J. C. Larsen, Electromagnetic studies of ocean currents and electrical conductivity below the ocean floor, in *The Sea,* vol. 4, edited by A. E. Maxwell, pp. 637–691, John Wiley, 1971.

Filloux, J. H., Electric field recording on the sea floor with short span instruments, *J. Geomagn. Geoelectr., 26,* 269–279, 1974.

Filloux, J. H., Observation of very low frequency electromagnetic signals in the ocean, *J. Geomagn. Geoelectr., 32,* Suppl. I, SI1–SI12, 1980a.

Filloux, J. H., Magnetotelluric soundings over the NE Pacific may reveal spatial dependance of depth and conductance of the asthenosphere, *Earth Planet. Sci. Lett., 46,* 244–252, 1980b.

Filloux, J. H., Magnetotelluric exploration of the North Pacific: Progress report and preliminary soundings near a spreading ridge, *Phys. Earth Planet. Inter., 25,* 187–195, 1981.

Filloux, J. H., Magnetotelluric experiment over the ROSE area, *J. Geophys. Res., 87,* 8364–8378, 1982.

Hilde, T. W. C., S. Uyeda, and L. Kroenke, Evolution of the western Pacific and its margins, *Tectonophysics, 38,* 145–165, 1977.

Hobart, M. A., R. N. Anderson, and S. Uyeda, Heat transfer in the Mariana trough (abstract), *EOS Trans. AGU, 60*(18), 383, 1979.

Karig, D. E., Origins and development of marginal basins in the western Pacific, *J. Geophys. Res., 76,* 2542–2561, 76, 1971a.

Karig, D. E., Structural history of the Mariana Island arc system, *Geol. Soc. Am. Bull., 90,* 331–337, 1971*b.*

Karig, D. E., R. N. Anderson, and L. D. Bibee, Characteristics of back arc spreading in the Mariana trough. *J. Geophys. Res., 83,* 1213–1226, 1978.

Katsumata, M., and L. R. Sykes, Seismicity and tectonics of the western Pacific: Izu-Mariana-Caroline and Ryukyar-Taiwan regions, *J. Geophys. Res., 74,* 5923–5948, 1969.

Larsen, J. C., Low frequency (0.1-6 cpd) electromagnetic study of deep mantle electrical conductivity beneath the Hawaiian Islands. *Geophys. J. R. Astro. Soc., 43,* 17–46, 1975.

La Traille, S. L., and D. M. Hussong, Crustal structure across the Mariana Island Arc, in *The Tectonic and Geologic Evolution of Southeast Asian Seas and Islands, Geophys. Monogr. Ser.,* vol. 23, edited by D. E. Hayes, pp. 209–221, AGU, Washington, D.C., 1980.

Leeds, A. R., Lithospheric thickness in the western Pacific, *Phys. Earth Planet. Inter., 11,* 61–64, 1975.

Louden, K. E., The crustal and lithospheric thickness of the Philippine Sea as compared to the Pacific, *Earth Planet. Sci. Lett., 50,* 275–288, 1980.

Oldenburg, D. W., Conductivity structure of oceanic mantle beneath the Pacific plate, *Geophys. J. R. Astron. Soc., 65,* 359–394, 1981.

Parker, R. L., The inverse of E.M. induction: Existence and construction of solutions based upon incomplete data, *J. Geophys. Res., 85,* 4421–4428, 1980.

Parker, R. L., and K. A. Whaler, Numerical methods for establishing solutions of the inverse problem of electromagnetic induction, *J. Geophys. Res., 86,* 9574–9584, 1981.

Ringwood, A. E., and A. Major, The system $Mg_2 S_i O_4$–$Fe_2 S_i O_4$ at high pressures and temperatures, *Phys. Earth Planet. Inter., 3,* 89–108, 1970.

Shankland, T. J., and M. S. Waff, Partial melting and electrical conductivity anomalies in the upper mantle, *J. Geophys. Res., 82,* 5409–5417, 1977.

Uyeda, S., Subduction zones: an introduction to comparative subductology, *Tectonophysics, 81,* 133–159, 1982.

Yoshii, T., Y. Kono, and K. Ito, Thickening of the oceanic lithosphere, in *The Geophysics of the Pacific Ocean Basin and its Margin, Geophys. Monogr. Ser.,* vol. 19, edited by G. H. Sutton et al., pp. 423–430, AGU, Washington, D. C., 1976.

Marine Geology of the Forearc Region, Southern Mariana Island Arc

D. E. KARIG AND BEVERLY RANKEN

Department of Geological Sciences, Cornell University, Ithaca, New York 14853

The Mariana Arc serves as a type example of an oceanic arc system because of its long history without a continental influence and because of the large suite of data collected from that area. The concentration of deep-sea drilling and related survey data near 18°N has been interpreted in support of subsidence and narrowing of the forearc with time as a result of tectonic erosion. On the contrary, interpretation of a lesser concentration of data from the south end of the arc presented here suggests growth and relative uplift of the lower trench slope. Truncation of all forearc elements occurs south of 13°N, probably as a result of strike slip faulting along east-west fractures that define a transform between the back arc spreading ridge and the trench. North of 13°30'N the inner trench slope is ribbed with ridges that trend parallel to or convex toward the trench. These ridges are largest and perhaps most structurally active at the base of the trench slope. Depositional depth of sediments in Deep Sea Drilling Project holes drilled in the upper slope apron, concave upward slopes of this apron, which trap turbidites, and internal arcward fanning of deeper apron strata are cited in support of relative uplift and arcward rotation of the seaward part of the inner slope and of minor absolute uplift of the sediment apron. This pattern of vertical displacement and rotation, coupled with progressive downlap rather than truncation of apron strata, argues against tectonic erosion and subsidence. The conflicting data may be a result of changing response of the arc over time. Forearc volcanism and tectonic disruption of the basement beneath the upper slope apparently ceased by the early Oligocene. Younger features are more compatible with intermittent accretion of oceanic material, possibly tectonically mixed into the arc basement.

Introduction

The Mariana island arc has formed part of the Philippine-Pacific plate boundary (Figure 1) at least since the late Eocene. Because of its long intra-oceanic history this arc has come to be viewed as a type example for oceanic arcs and has been the focus of extensive recent study. Most of these investigations were associated with a transect of drill sites across the arc at 18°N occupied during Deep Sea Drilling Project (DSDP) Leg 60 and have led to the tentative conclusion that at least the recent history of the Mariana forearc has been one including a large component of destruction and subsidence resulting from tectonic erosion [*Hussong and Uyeda,* 1981; *Mrozowski and Hayes,* 1980].

A second concentration of data was generated between 12° and 14°N by earlier DSDP drilling [*Fischer et al.,* 1971] and from random ship tracks in and out of Guam (Figure 2). In this paper we have integrated and interpreted these data in order to investigate the nature of the forearc in the southern section of the arc. The motivation for this study was that an earlier interpretation of the Mariana forearc [*Karig,* 1971a], as growing and rising with time, was contradicted by the conclusions reached recently at 18°N [*Hussong et al.,* 1982]. It seemed appropriate to reassess the earlier data and interpretations from the southern part of the arc with a larger data set as well as to critically evaluate the currently favored erosion-subsidence model with data not considered by *Hussong et al.* [1982].

Regional Tectonic Setting

Subduction along the Mariana Arc has probably been continuous, although not in rate or direction, since at least the early Tertiary [*Jurdy,* 1979]. Vectoral addition of the Philippine-Pacific convergence and a 4 cm/yr separation rate in the Mariana Trough [*Hussong et al.,* 1982] indicate that the Pacific plate is presently being subducted westerly beneath the southern part of the arc at about 8 cm/yr. The seismic zone extends deeper than 500 km beneath most of the arc but shallows rapidly, and is less active, south of Guam [*Katsumata and Sykes,* 1969; *Bracey and Ogden,* 1972]. It dips shallowly beneath most of the forearc but

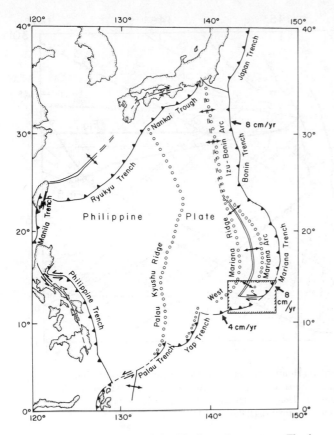

Fig. 1. Tectonic setting of the Mariana Arc system. The ha-chured rectangle outlines the area of investigation. Back arc spreading in the Mariana Trough (double lines) is shown sche-matically as being terminated at the south by a system of trans-form faults. Subduction vectors are calculated from poles of *Karig* [1975], with addition of back arc extension from *Hussong and Uyeda* [1982].

steepens sharply beneath the frontal arc [*Isacks and Bar-azangi*, 1977]. The trench–volcanic arc separation along most of the arc exceeds 200 km, which contrasts markedly with the 90 to 140-km separation observed in much younger oceanic arcs such as the New Hebrides Arc.

The map pattern of the Mariana system describes a very regular arc, but this masks major tectonic complexities south of Guam, in the same area where the seismic activity drops rapidly. South of 13°N the sedimentary apron that defines the gently dipping upper section of the inner trench slope sharply narrows and disappears, leaving only a steep and irregular inner trench slope (Figure 3) [*Karig et al.*, 1978, Figure 7]. This truncation of forearc units was in-terpreted by *Karig et al.* [1978] as resulting from right-lateral displacement along a poorly defined transform sys-tem joining the back arc spreading system to the trench.

Analysis of additional and more recent profiles across this southernmost part of the arc extends the trace of the spreading ridge of the Mariana Trough south of 13°N (Fig-ure 4) and more firmly documents the existence of the zone

with east-west structures. These lineaments appear on re-flection profiles as narrow benches or irregular troughs, generally devoid of sediment (Figure 3). South of 12°20′N the paucity of tracks prevents the mapping of lineations, but the morphology on individual profiles is similar to that between 12°20′ and 13°N. Further southwest, smaller ridge segments of varying trend sit at the top of steep scarps marking the crest of the trench slope. Larger topographic highs near 12°N, 141°30′E may mark yet another segment of the frontal arc.

The frontal arc begins as a continuous feature southwest of Guam. A large topographic block carrying Santa Rosa and Galvez banks is separated from the southern end of the frontal arc by a north-south trending trough at 144°30′E (Figure 4). The trend and position of this trough suggest that separation of these blocks was a result of back arc spreading rather than of the E-W transform faulting.

Northward from about 13°N, the forearc region broadens rapidly and assumes a set of characteristics which appear to persist to at least 16°N (Figure 4). This forearc region can be divided into three major morphotectonic subdivi-sions: (1) the frontal arc, (2) an upper trench slope covered with a large sediment apron, and (3) the steep, irregular lower trench slope. Geological studies of emergent parts of the frontal arc, from Guam to Saipan [*Cloud et al.*, 1956; *Tracey et al.*, 1964], demonstrate that this element has functioned as a mosaic of fault blocks, units of which have oscillated as much as 1 km [*Ingle*, 1975] since the late Eocene. The late Eocene volcanic and bioclastic material from these islands indicates that fairly evolved magmatic centers lay on and to the west of the present frontal arc and that they had reached sea level [*Cloud et al.*, 1956; *Tracey et al.*, 1964].

The acoustically opaque sequence of volcanics and sed-iments forming the frontal arc is bounded on the east by a steep slope that flattens into the upper slope apron [*Ka-rig*, 1971a, Figure 3]. South of 18°N this sediment apron is incompletely delineated because available reflection profiles have not penetrated the entire sediment column, which thickens to 2 km or more near the frontal arc.

The strata of the apron in the southern region form, for the most part, slope-parallel acoustic units that represent dominantly pelagic biogenic and hemipelagic volcanogenic sediments showing various processes of downslope remo-bilization [*Fischer et al.*, 1971; *Karig*, 1971b]. The arcward increase in the thickness of the apron is due, at least in the upper, acoustically resolvable section, primarily to the existence of additional acoustic units at the base of the section closer to the arc. Stated in a different way, younger acoustic units downlap eastward onto basement and ex-tend the apron progressively trenchward with time (Fig-ure 5). Near 14°N, many profiles reveal reflectors beneath the slope-parallel sequence that have very shallow trench-ward dips, which, if corrected for acoustic velocity effects, would be even lower (Figure 5). These reflectors likely represent sediments that overlie and abut a rough acoustic

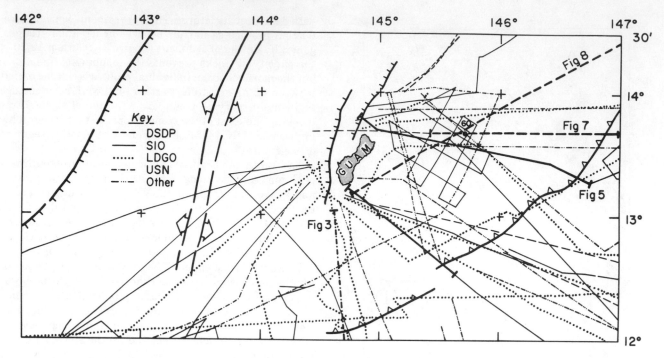

Fig. 2. Ship tracks used in this study of the Mariana forearc. Heavy sections of track show locations of reflection profiles figured in this paper. The thrust fault symbol marks the trench axis, and the normal fault symbol marks the bounding scarps of the Mariana Trough. Abbreviations are Deep Sea Drilling Project (DSDP), Scripps Institution Oceanography (SIO), Lamont-Doherty Geological Observatory (LDGO), and U.S. Navy (USN).

basement. Significantly, no evidence of truncation or removal of the distal edge of the sediment apron has been observed, except along the truncated southern boundary near 13°N. DSDP Hole 60 penetrated the seaward edge of the apron, missing the deeper, subhorizontal reflectors, as well as much of the slope-parallel sequence, and bottomed, probably only slightly above acoustic basement, in noncalcareous ash-bearing clays of late early Miocene age [*Fischer et al.*, 1971].

Acoustic basement beneath the thinner parts of the apron has a local relief of up to 500 m. Denser profile coverage supports the interpretation that this relief defines ridges, trending subparallel to the trench [*Karig*, 1971a]. One northeasterly trending set of ridges, now almost completely buried beneath the apron, is cut off, or capped, by another, more northerly trending set of larger ridges that can be traced southward across the lower trench slope (Figure 4).

Sediments of the apron thin, both southeasterly across the basement ridges and along the troughs between the ridges in a downslope direction. Sediment thickness appears more dependent on distance from the frontal arc that on the geometry of the ridges, as might be expected in a section of predominantly pelagic and hemipelagic origin.

This sediment cover shows that the basement ridges beneath the apron are structurally inactive, but, as at 18°N [*Mrozowski and Hayes*, 1980], the sediment apron is cut

by many small high-angle faults. These faults cannot be correlated between the more widely spaced profiles in the southern area. On individual profiles, faults with displacements generally less than a few hundred meters occur at intervals of several kilometers. Both senses of displacement are seen on these faults, but toward the base of the upper slope apron most faults appear to be downthrown to the east or south.

This smaller-scale faulting may be paralleled by larger-scale displacement of the acoustic basement beneath the trench slope. Rather than descending at a uniform slope from the frontal arc to the trench, this acoustic surface must have a pronounced step beneath the upper slope apron. The forward edge of the step, modified by the apron sediments, defines the trench slope break. The rear edge, although hidden beneath the thickest part of the apron, bounds the steep uppermost slope section. The relief across this rear step in the acoustic basement is generally in excess of 3 km (Figure 5).

Perhaps related to the development of this basement step is the concave upward profile seen in some areas of the forearc, especially between 14°30'N and 17°N (Figure 6) [*Karig*, 1971a, Figure 3]. Pronounced thickening of acoustic units and even ponding of turbidites can occur in the uppermost strata adjacent to the trench slope break in areas where that feature is a topographic high. In these situations the distal part of the apron clearly laps onto the

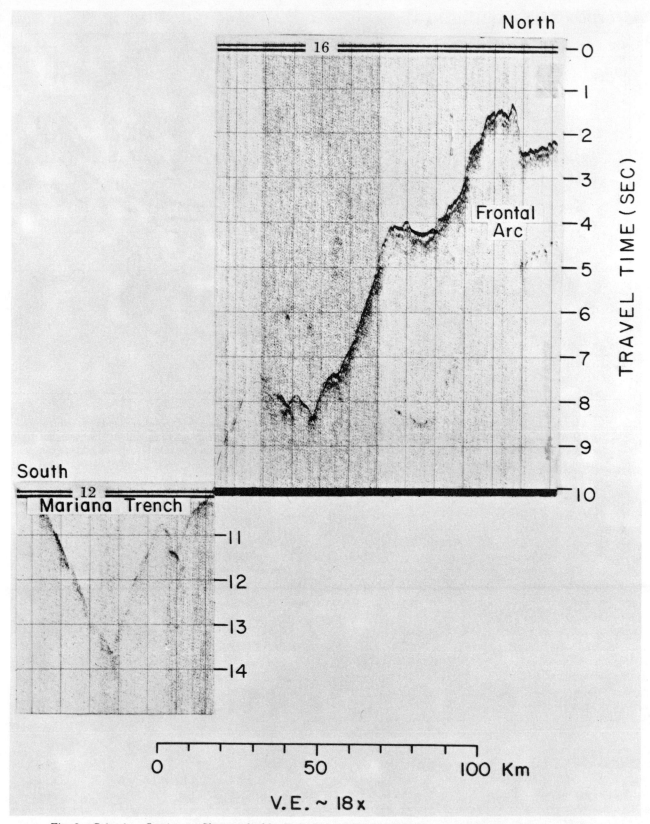

Fig. 3.   Seismic reflection profile (supplied by D. R. Bracey, NAVOCEANO) across the forearc near 144°40′E, illustrating the removal and/or disruption of the upper slope apron. Also shown are unsedimented benches or troughs interpreted to be east-west trending transform faults.

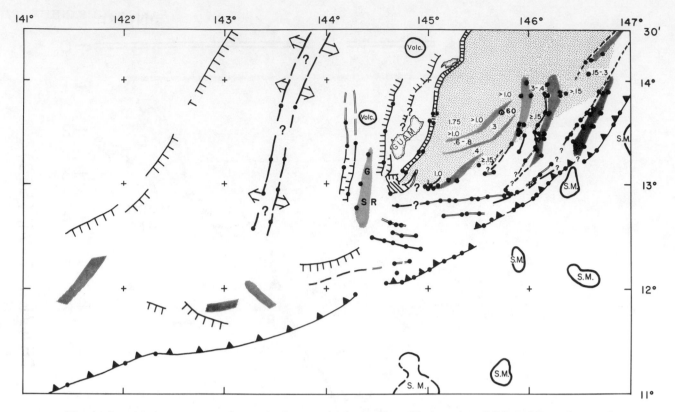

Fig. 4.   Interpretive summary of tectonic elements in the southern Mariana Arc. Solid circles mark control points on tracks crossing ridges except beneath the upper slope apron where line density is greatest. The basement ridges (stippled) on the inner slope may extend northward from this area but cannot be reliably correlated between more widely spaced lines. Solid lines mark troughs or slots. Hachured lines denote scarps, in most cases related to back arc spreading. DSDP site 60 is shown on the apron together with sediment thickness in acoustic travel times. Galvez (G) and Santa Rosa (SR) banks are located on the ridge along 144°30′E longitude.

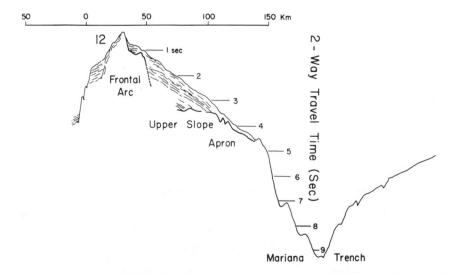

Fig. 5.   Seismic reflection profile (tracing) from Scripps Institution of Oceanography *Circe* cruise across the Mariana forearc showing downlap of upper slope apron sediment on subhorizontal deeper reflectors. Note the lack of truncation of these slope-parallel strata.

270

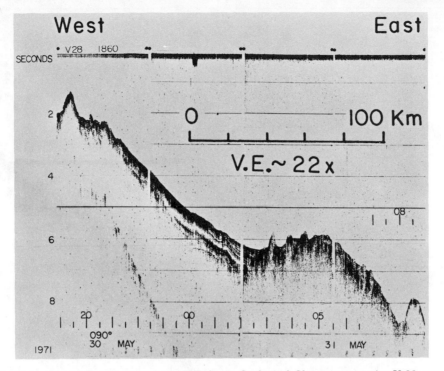

Fig. 6. Seismic reflection profile from Lamont-Doherty Geological Observatory cruise V-28 across the upper slope of the forearc near 14°30′N. This profile illustrates the concave upward slope of the sediment apron and associated thickening of the uppermost acoustic units adjacent to the relatively high trench slope break. One of the large topographic highs of the lower slope, interpreted to be a ridge, is shown at the eastern end of the profile. Deeper, shallower-dipping reflectors of the slope apron are just discernable on this profile.

basement and possibly even onto older sediments covering the trench slope break.

The trench slope break in the Mariana Arc is defined primarily as a regional break in slope. Although on many individual profiles in the southern part of the arc it appears to coincide with one of the basement highs that we have identified as ridges, it is apparent that individual ridges obliquely cross, rather than follow, the trench slope break (Figure 4). The evolution of the trench slope break thus seems to be other than that of a single structural element.

The lower trench slope also displays topographic irregularities, the largest of which can confidently be identified as ridges (Figures 4 and 7). Lower slope ridges generally have greater relief than those of the upper slope, although that relief varies markedly along trend. In plan, these ridges in the southern area appear convex toward the trench and subparallel to it (Figure 4). There is an indication that the ridges increase in size toward the trench. The largest ridge is that closest to the trench, which has a width of 20 km and stands more than 1 km above the regional slope. Both this and the next ridge upslope can be traced for at least 50 km with reasonable confidence. There may, however, be local offsets or amalgamations of smaller units. The large, lowest ridge south of 14°N (Figure 7) parallels the trench to 13°20′N, at which point the trench axis is

offset westward (Figure 4). In this topographically complex area it is possible that the large ridge terminates by intersecting the trench axis.

Behind these large ridges are relative lows or troughs which not only can be traced along the slope but also show monotonically increasing depths as the ridge in front obliquely drops downslope. Deposits with the geometry of turbidites are found in these troughs, up to several hundred meters thick and in some cases far thicker than the deposits in the trench axis (Figure 7). The turbidite channel fills can be several kilometers in width and often show a low but definite arcward tilt. Deeper strata in the best developed example have an even greater arcward tilt.

More recently, D. M. Hussong (written communication, 1980) has gained access to Navy SEABEAM data from this area. These data confirm the presence of ridges on the inner trench slope south of 13°30′N but emphasize irregularities along strike. In particular, subcircular or ovoid seamounts were noted at 13°10′N, 146°02′E and at 13°40′N, 146°E, but such features are not common south of about 14°N. Both these seamounts lie on what are mapped in this study as continuous ridges (Figure 4), on the basis of closely spaced profiles. We suspect that these seamounts are ovoid culminations on basically linear trends. North of 13°30′N, linear ridges are not obvious on the SEABEAM data (D.M. Hussong, written communication, 1980), but

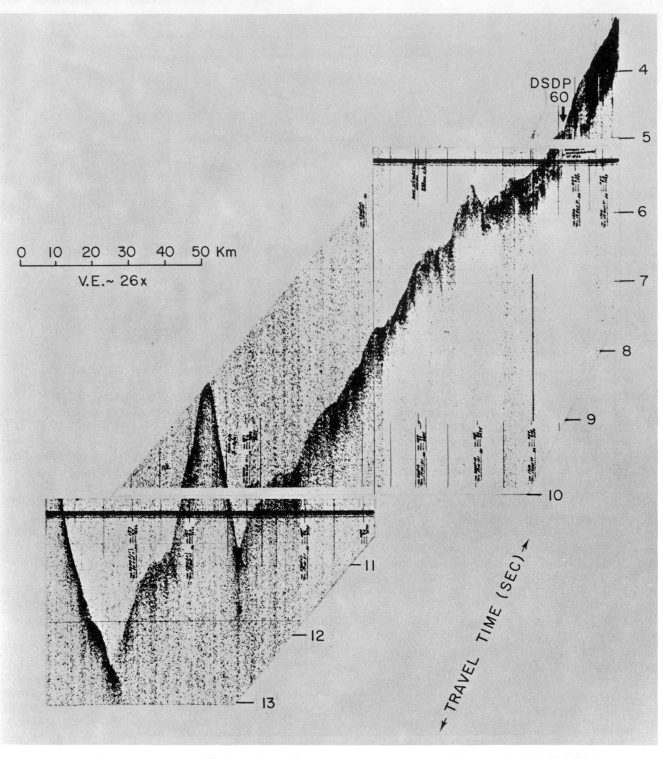

Fig. 7. Seismic reflection profiles across the Mariana forearc east of Guam (from DSDP Leg 20). This profile displays the large basement ridge that can be traced for more than 50 km along the lower trench slope. Behind this ridge is a trough or channel containing several hundred meters of turbidites. Several other ridges are identified upslope, one of which is buried beneath the sediment apron just west of DSDP site 60.

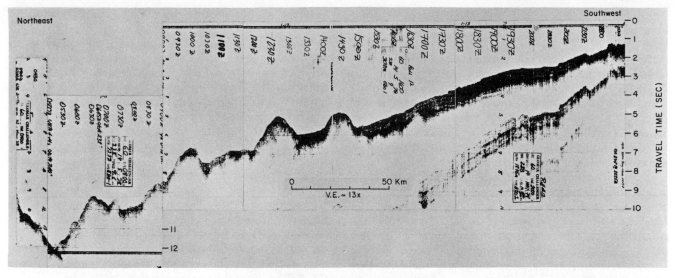

Fig. 8. Seismic reflection profile obliquely crossing the Mariana forearc northeastward from Guam (from DSDP Leg 60). On this profile a series of topographic highs occur, the southernmost of which are reliably identified as ridges. These shallowest discernable ridges are buried and lie inactive beneath the upper slope apron. Also shown on this profile are some of the small high-angle faults in the sediment apron. The descending oceanic plate can be traced on this profile several kilometers beneath the inner trench slope.

having seen a degraded set of these data, I would claim that they are subjected to a range of interpretations, including irregular linear ridges.

The lithologic composition of the lower trench slope south of 18°N is known only through scattered dredge hauls [e.g., *Hawkins et al.,* 1979; *Bloomer and Hawkins,* 1980; *Honza and Kagami,* 1977]. One of these dredges sampled tectonized and serpentinized ultramafic rocks from the seaward side of the large basement ridge near 146°20'E, 13°50'N (Figure 4) [*Honza and Kagami,* 1977]. Ultramafic rocks were also recovered from the ovoid topographic high forming the trench slope break at 18°N, just east of DSDP Site 459 [*Evans and Hawkins,* 1979], suggesting that the ridges on the lower slope are constructed, at least in part, of rocks from the upper mantle.

On the other hand, seismic reflection and refraction studies at 18°N [*Mrozowski et al.,* 1982] outline low-velocity (<4 km/s) material beneath the lowermost slope. This material has sufficient acoustic transparency that the descending oceanic lithosphere can be followed on multichannel reflection profiles at least 15 km landward of the trench axis and beneath DSDP sites 460 and 461 [*Mrozowski et al.,* 1982]. Even on the better single channel profiles to the south the descending crust can be followed beneath the toe of the inner slope for several kilometers (Figure 8).

Discussion

The interpretation of the lower trench slope and its evolution hinges critically on the significance of the topographic highs observed on all profiles across the southern part of the Mariana Arc. We feel, given the persistent appearance of these topographic highs on closely spaced profiles and the coherence of the sedimented troughs behind several of them, that they must in general be ridges rather than subcircular features. We suggest that the subcircular seamounts that are illuminated by the multibeam sounding arrays are parts of these ridges. The dredging of ultramafic rocks and gabbros from at least two of them is a strong argument against interpreting them as volcanic features and favors their identification as units composed of crust and mantle rocks.

The ridges could be interpreted as diapirs of lower density, serpentinized mantle, originating in either the upper or the lower plate. Arguing against this interpretation is the linearity and continuity of these ridges and the apparent lack of a suitable source. If the source were in the upper plate, the density inversion necessary to drive the diapir would have to be generated in an implausibly small vertical distance of only a few kilometers in the case of the lowest ridges. A diapiric source in the descending plate cannot be ruled out, but the linearity of the deeper ridges would then be even more enigmatic. Any diapiric interpretation also has difficulty explaining why the ridges beneath the upper sections of the inner trench slope, where the chance of density inversion is greater, are inactive.

It seems more likely that the ridges have a fault origin and are either mega–slide blocks or thrust slices. Morphologic evidence opposes a slide block origin. The ridges stand well above the regional slope (e.g., Figure 7), which implies a gain rather than a loss in potential energy. Their trenchward convexity and great linear extent are not typical of slump block patterns. Perhaps most difficult to explain by gravitational tectonism is that these ridges underlie both the upper and the lower slopes but are structurally inactive beneath the upper slope apron (Figure 8).

The thrust ridge alternative seems to be consistent with all data from the southern section of the arc. The geometric characteristics cited as incompatible with the slide block model are expected for sequence of thrust sheets. The suggestion of active arcward tilting of those ridges lower on the slope, contrasted with the structural inactivity of ridges beneath the upper slope apron, favors thrusting that is concentrated toward the trench axis.

Thrust disruption of older arc crust and accretion of thrust slices removed from the oceanic plate have both been postulated to occur along convergent margins. The apparent seaward progression of thrust activity, plus the lack of any evidence for truncation or removal of the distal edge of the sediment apron, argues against cumulative tectonic erosion. On the other hand, the likelihood that the large basal ridge south of 14°N merges southward with the trench axis suggests that ridge is a recently accreted slice of ocean crust and upper mantle, the accretion of which led to an eastward jump in the position of the trench axis. The upslope decrease in size of these ridges could reflect continued movement along major slip surfaces, tending to carry the lower ridges into the accretionary prism.

Vertical Displacements

One cannot compare the relative merits of accretion and tectonic erosion without discussing vertical displacements of the lower trench slope. Both 'absolute' displacement (relative to sea level) and displacement of the lower slope relative to the upper slope bear on the mechanics and evolution of the forearc.

Data concerning absolute displacements of the inner trench slope are derived primarily from the results of the Deep Sea Drilling Project holes and consist chiefly of those sediment characteristics that reflect their position relative to the carbonate compensation depth (CCD). The interpretation of vertical displacements requires a knowledge of both temporal and spatial variations in the CCD. The history of the CCD in the westernmost Pacific is not as well known as that further east [e.g., *van Andel*, 1975; *Berger and Winterer*, 1974]. Limited coring and drill hole data place the present CCD near 4.5 km, in agreement with regional estimates of *Berger and Winterer* [1974] and *van Andel* [1975]. Mass transport processes, especially during periods of more rapid sedimentation, probably increase the depth at which carbonate-bearing sediments can be preserved in the forearc. At the latitudes in question (12°–18°N) the CCD has fluctuated around a depth near 4.2 km since the early Oligocene and has generally been shallower than the CCD in the equatorial belt of higher productivity (Figure 9). We must consider not only the temporal variations of the CCD in the forearc sites but also the possibility that these sites have moved northward

since the late Eocene, thus recording latitudinal variations in the carbonate record.

A retrospective reconstruction of the Philippine plate [e.g., *Karig*, 1975] shows that prior to the late Oligocene the southern Mariana Arc lay adjacent to Palau-Kyushu Ridge and the West Philippine Basin. Paleomagnetic data in the West Philippine Basin [*Jarrard and Sasajima*, 1980, and references therein] produce estimates in excess of 20° of northward migration in the past 50 m.y. Such large northward motion of the Mariana Arc is supported by data from DSDP hole 459 [*Bleil*, 1982], but with systematically lower inclination of reversely magnetized units, these values may not pass the reversal test (M. Fuller, personal communication, 1981). Paleomagnetic data from Guam and Saipan [*Larsen et al.*, 1975; M. Fuller, personal communication, 1981] have been interpreted as indicating very little latitudinal shift, although even these more extensive data will permit at least several degrees of northward shift (M. Fuller, personal communication, 1981).

Three DSDP sites have been drilled on the upper trench slope: site 60 of Leg 6 and sites 458 and 459 of Leg 60. The section at DSDP site 60, drilled in 3717 m of water, was sparsely sampled but reveals the downward disap-

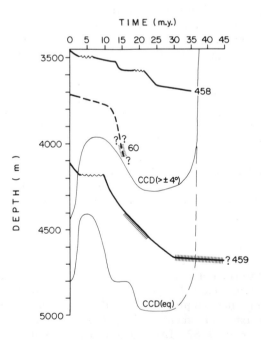

Fig. 9. Paleodepth curves for DSDP sites drilled in the upper slope apron of the Mariana forearc, together with the equatorial and higher-latitude depth history of the CCD from the eastern and central Pacific [from *van Andel*, 1975]. The double-sided hachuring denotes carbonate-free zones; single-sided hachuring denotes zones with very low carbonates and strongly corroded and/or reduced foraminiferal assemblages. If these curves had been corrected for compaction and sediment loading, the older sections would be shifted upward, increasing the likelihood of tectonic uplift of the forearc since the late Eocene.

pearance of foraminifera and nannofossils into a noncalcareous early Miocene ash-rich unit [Fischer et al., 1971] at a total penetration of 348 m (Figure 9). Although the absence of carbonate at the bottom of the hole could reflect rapid ash deposition, the presence of nannofossils in strata of similar lithology and sedimentation rate suggests that the basal strata were deposited below the early Miocene CCD. The present depth of these basal strata should be in the range of 100 m deeper than the original depth, as a result of sediment loading, but even the uncorrected value, and the assumption that the latitude of this site has not changed greatly since the early Miocene, would suggest uplift of several hundred meters.

Site 458 was begun at 3447 m, well above the CCD, and at 250 m entered volcanics that were capped by early Oligocene nannofossil chalk and tuff. This site has clearly remained above the CCD during deposition of the entire section. The paucity of foraminifera throughout the entire section indicates that the sediment surface has remained below the foram lysocline since the early Oligocene. This depth has probably paralleled that of the CCD and has been in excess of 3 km over that period.

Site 459, drilled at a water depth of 4121 m, sampled a much longer and more complicated section. After beginning in calcareous mud the drill passed through a relatively noncalcareous early Miocene section and into nannofossil chalks of late Oligocene age. These overlie a thin (30 m) basal sediment section of noncalcareous clay and ash, spanning the interval from early Oligocene to late Eocene or earlier.

The condensed basal section was most probably deposited below the CCD, which during the early Oligocene dropped to between 4250 m and 4950 m, depending on the latitude of deposition [van Andel, 1975]. The overlying nannofossil chalk-rich section marks the upward passage through the CCD, but the effective lack of forams would place the depth well below the foram lysocline. The paucity of carbonate in the overlying early Miocene turbidites is probably not just an effect of dilution because the more rapidly deposited turbidites that overlie these have a much higher carbonate content and because the carbonate fraction in the early Miocene section still lacks foraminifers. This carbonate fluctuation could be a response to a northward migration of the site through the equator, as suggested in the West Philippine Basin, or to fluctuations in the CCD. Increased carbonate contents and abundance of foraminifera at this site since the early Miocene mark a relative rise of the site with respect to the CCD (Figure 9). Until the paleolatitudinal history of this site can be reliably determined, these carbonate fluctuations cannot be cited as evidence for uplift. However, after restoration of several hundred meters of sediment loading, these data certainly do not permit subsidence of any significance.

Taken together, the sedimentary records from DSDP sites on the forearc apron do not suggest subsidence but rather a small amount of cumulative uplift. Oscillations may be superimposed on this record, but they cannot be demonstrated by the available data.

Data taken from sites 460 and 461, deeper on the trench slope, are both sparse and enigmatic. The sections recovered in the four short holes at these sites contained noncalcareous Quaternary(?) mud overlying Paleogene to Jurassic sediments and igneous rocks that may represent clasts or larger stratal units. Much of these sections is clearly mixed, apparently as a result of slumping, but the shipboard party felt that a 48-m coherent Paleogene section, including carbonate-bearing lithologies, was sampled in hole 460. Arc affinities of some of the igneous clasts, coupled with the lack of a Paleogene section on the westernmost Pacific plate with which to correlate the drilled Paleogene, were considered evidence for very strong subsidence of this section of the forearc.

Any interpretation is speculative with the data now available, but several points can be made. Low core recovery and intense drilling disturbance of these cores preclude any firm conclusion concerning the internal structure or inclination of this supposed Paleogene unit, which appears to lie beneath the layered sequence displayed on the 3.5-kHz profile [Hussong et al., 1982]. There does not seem to be any reason why this block cannot be interpreted as a large slump mass resulting from the same process that produced the reworked material in the adjacent hole (460A) and at site 461. Canyon systems such as the system surveyed near 15°30'N [Karig, 1971a, Figure 10] would be logical conduits for this sort of transport.

Late Jurassic to Cretaceous elements recovered at these two drill sites require that material, up to 3 times the age of the oldest known arc volcanics, was exposed prior to development of the Paleogene section higher on the trench slope. Considering the age and pelagic nature of these rocks, the most logical interpretation is that fragments of the Pacific plate were, by some process, imbedded in the Mariana forearc.

Relative Vertical Motion

Both uplift [Karig, 1971a; Honza and Kagami, 1977] and subsidence of the lower trench slope [Hussong et al., 1982; Mrozowski and Hayes, 1980] with respect to the upper slope apron have been suggested. Strong evidence for recent relative uplift was obtained from the large submarine canyon between 15° and 16°N, in which the channel has been blocked by a topographic high near the trench slope break. Turbidites have been ponded and uplifted behind this blockage [Karig, 1971a]. Several additional reflection profiles crossing this system now reveal that the canyon heads in a number of small tributaries on the sediment apron (Figure 10) rather than near the crest of the frontal arc [Karig, 1971a].

A profile constructed along the lowermost section of the canyon (Figure 10), which trends roughly perpendicular to the trench, shows (1) that there is nearly 500 m of topographic closure behind the topographic blockage, (2)

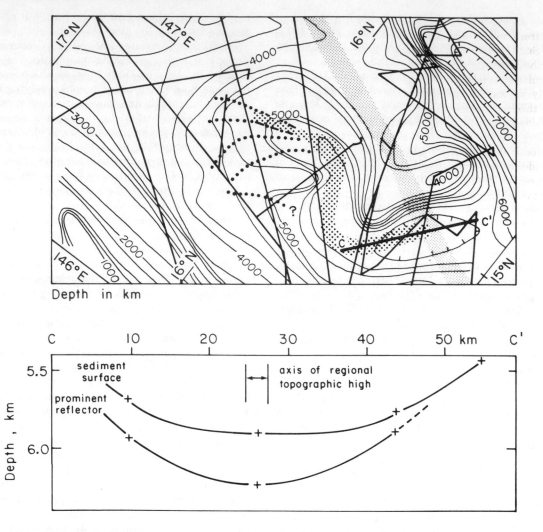

Fig. 10. Bathymetry and axial profile of the large submarine canyon crossing the upper slope apron near 16°N. Only tracks with seismic reflection profiles are plotted; additional bathymetric control is shown by *Karig* [1971a]. The lightly stippled band denotes the topographic crest of the trench slope break. The strong relative uplift of the trenchward part of the slope is shown in the displacements of the surface of the canyon fill and of a prominent reflecting horizon within that fill.

that structural closure, as delineated by a prominent reflector beneath the ponded section, is substantially greater than 500 m, and (3) that the topographic blockage lies 20–30 km trenchward of the axis of the trench slope break as defined by the regional topography. The simplest interpretation of these data remains the relative uplift of the lower trench slope, probably in relatively recent (Neogene?) past.

Relative uplift of the lower slope would also account for the concave upward profiles of the upper slope apron that typify the southern part of the arc. Although such a morphology could be original, with the slope-parallel sediments of the apron mimicking the basement topography, there are a number of profiles in which the deeper reflec-

tors have lower trenchward dips (e.g., Figures 5 and 6) or where turbidites are ponded and thickened behind the trench slope break (Figures 6 and 10). The turbidite section at site 459 may have accumulated behind a similar high along the trench slope break.

Normal faults, both interpreted from reflection profiles along the entire arc and observed on a microscale in the DSDP cores at site 459, are cited as evidence for extension and relative subsidence of the lower slope [*Hussong et al.,* 1982; *Mrozowski and Hayes,* 1980]. Several notes of caution concerning this interpretation are in order. First, the seismically detected faults can be identified only as high angle. Moreover, the implicit assumption that strata have suffered only vertical translation may not be valid. Perhaps these sediments were not originally subhorizontal but had a trenchward slope and were subsequently rotated arcward. Such rotation of originally slope-parallel sediments of the slope apron has just been suggested.

Microfaulting in the cores could either be related to the seismically defined features or represent gravitational sliding of the sediment section. The microfaulting in site 459 seems restricted to the late Oligocene and early Miocene section where slump structures and other evidence of depositional slopes are abundant. Again, we stress the lack of evidence for truncation of the slope apron, and in fact its trenchward progradation, as evidence contrary to significant relative subsidence of the lower trench slope.

The upper slope may also have subsided relative to the frontal arc. Acoustic basement certainly describes a major step beneath the upper slope apron in the southern part of the arc, but the extent to which this is an original feature is unknown. Rounded volcanic clasts and shallow water fossils on Guam and Saipan [Cloud et al., 1956; Tracey et al., 1964] show that some sections of the volcanic chain were at sea level and slightly to the west of the modern frontal arc in the late Eocene. Analysis of benthic foraminifera in the fine-grained components of those strata suggests water depths of several hundred [Garrison et al., 1975] to several thousand [Ingle, 1975] meters over the present positions of Guam and Saipan and thus cumulative uplift of that magnitude. The age of the acoustic basement beneath the thickest part of the apron east of these islands is almost certainly Paleogene and very likely Eocene. The low eastward or possibly even westward dips of the basal strata in this apron reflect arcward rotation, by uplift of the lower slope, by subsidence of the inner edge of the apron, or by a combination of these. If the greater uplift of the frontal arc implied by Ingle's [1975] paleobathymetric interpretation is correct, differential displacement along the rear of the apron is likely. The shallower paleodepth value would be more compatible with interpretation of this step as an original morphologic feature of the young Eocene arc but would require at least several hundred meters of uplift of the lower trench slope. The better quality seismic profiles show relatively little recent displacement along this boundary, suggesting that any displacement occurred early in the arc history.

Further Discussion and Speculations

Both accretion, with uplift of the forearc, and tectonic erosion, leading to subsidence, have been proposed for the Mariana Arc. A first conclusion from our work is that neither model satisfactorily explains all the available data. At present it is easy to enumerate problems with these models but very much more difficult to suggest viable alternatives.

The identification of arc-related igneous rocks beneath at least the upper section of the forearc [Hussong and Uyeda et al., 1982], and beneath the entire inner trench slope by some workers [e.g., Hawkins et al., 1979; Bloomer

and Hawkins, 1980], is perhaps the most perplexing and difficult observation to explain. Eocene arc-related volcanic rocks also occur on the frontal arc from Guam to Saipan [Meijer et al., this volume], where they are interpreted as forming parts of an emergent volcanic chain that lay along and to the west of the modern islands. If these rocks can be interpolated between 15° and 18°N, they would define a band of contemporaneous volcanism nearly 200 km wide during the early stages of development of the Mariana Arc. At least from Saipan south, a typical subaerial volcanic chain lay about 200 km from the modern trench; assumption of tectonic erosion of the forearc would only increase the required original volcanic chain–trench separation.

Such a distribution of arc volcanism is quite unlike that of known modern young oceanic arc. The volcanic chain in these arcs lies 90–150 km behind the trench, and there is, as yet, no evidence for additional volcanism in the intervening region. The possibility, therefore, cannot be overlooked that the initial geometry of the Mariana Arc was quite anomalous.

Because all the igneous rocks drilled beneath the upper slope apron were extrusive, the nature of subjacent basement is unknown. Recovery of ultramafic rocks from the trench slope break at 18°N and the widespread distribution of various crustal to upper mantle lithologies sampled across the lower slope seriously weaken interpretations of the forearc basement as a simple ophiolite sheet. The question remains as to what extent the mixing of lithologies resulted from some tectonic process and from gravitational slumping.

If arc-derived plutonic rocks do form in situ basement beneath the entire inner trench slope, then it is difficult to avoid the inference of massive subsidence, at least relative to the plate interface. We have claimed that such subsidence is not supported by the nature of sediments drilled in the forearc or by the data pertaining to relative vertical displacements within the forearc.

Intermittent accretion of thrust slices consisting of oceanic crust and upper mantle still satisfies the available structural data but is difficult to harmonize with the present petrologic interpretation of dredge hauls from the lower trench slope.

Substantial evidence from the Peru Trench [Kulm et al., 1977] indicates that slices or flakes from the upper part of the oceanic crust can be stripped off the descending plate to form ridges at the base of the trench slope. The conditions attending this process are unknown, but it appears to be favored by a thin sediment cover on the descending plate. It has been suggested that horst blocks produced by normal faulting on the outer trench slope are preferentially involved [Karig and Sharman, 1975], but a contrasting view, that these horst blocks erode the inner trench slope, has also been put forward [Jones et al., 1978]. The fate of these thrust slices, once detached from the descending plate, is equally unclear. Are they accreted to the upper plate at the base of the trench slope, or do they travel some distance along the plate interface to be subcreted or subducted? The preservation of tectonic blocks and slices of basalt in what appear to be very shallow-level accretionary structures in the Sunda Arc [*Moore and Karig,* 1980] and in parts of the Franciscan complex suggests to us that their incorporation into the upper plate at or near the base of the inner slope is quite possible. Additional, but less active, thrusting higher on the inner slope might remove or reduce the ridge topography generated by accretion of large oceanic flakes.

There are several lines of evidence that, together, offer substantial support for the interpretation of the ridges on the Mariana inner slope as accreted masses of oceanic crust and upper mantle. Mesozoic fossils recovered from the DSDP sites 460 and 461, as well as from several recent dredge hauls (S. Bloomer and J. W. Hawkins, personal communication, 1981), are strong evidence for some transfer of material from the Pacific plate to the Philippine plate. If these fossils are reworked into Oligocene and late Eocene strata, then at least some of the transfer occurred early in the history of the arc and to a setting upslope from the drill sites.

The lack of structural activity of these ridges beneath the upper slope apron since the early Miocene or earlier, coupled with the apparent youthfulness and greater size of the basal ridge, fits a pattern of progressive trenchward accretion, although this interpretation is not unique. However, the relative uplift of the lower trench slope and the lack of absolute subsidence in the forearc, for which we have argued, are both more consistent with accretion than with tectonic erosion.

Large-scale accretion might be expected to have produced much greater uplift of at least the seaward part of the forearc than the modest (several hundred meters) figure that we suggest, but analysis of changes of thickness of the upper plate must also take changes in position of the plate interface into account. In the Mariana Arc this plate interface might be depressed not only by the weight of accreted mass but also by an increase in the depth of the descending plate at the trench as a function of its age.

A change in depth of the incoming oceanic basin should result directly in a similar change in the position of the plate interface beneath the inner trench slope and of the sediment surface above. This general notion has been applied to the effect that the passage of a spreading ridge might have on the frontal arc [e.g., *deLong et al.,* 1978] and seems applicable in reverse to the Mariana Arc, where an increase in water depth during most of the Cenozoic would result from the aging of the subducted plate in the reconstruction of *Hilde et al.* [1977].

The extremely large relief of the Mariana inner trench slope would be a logical result of accretion. If it were attributed to tectonic erosion of the base of the trench slope, coupled with relatively little removal of material from the base of the upper plate, and with the acceptance of an arc-

derived basement beneath the entire forearc, the problem of an anomalously wide initial volcanic zone and forearc would be exacerbated. Accretion would also account for the low-velocity material with partial acoustic transparency beneath the toe of the inner trench slope, although clearly there are other explanations.

It is quite possible that, in part, the apparently contradictory evidence from the Mariana Arc results from changes in basic tectonic response since the initiation of the arc system. Recent dredging [*Evans and Hawkins,* 1979; D. M. Hussong, personal communication, 1981] has discounted the interpretation of the forearc seamounts as young volcanoes or accreted seamounts, so that forearc volcanism can now be considered to have ceased by the early Oligocene (>36 m.y. B.P.). Downlap of early Miocene, and probably older, sediments on the basement ridges around DSDP site 60 limit the age of basement morphology beneath the upper slope apron in the southern part of the arc. The sediment-basement relationships at 18°N lead to the similar conclusion that the basement relief on the upper slope developed very early in the history of the arc.

The presence of Jurassic and Cretaceous clasts in presumed Paleogene strata, as well as the gabbroic to ultramafic composition of the seamounts near the trench slope break, would suggest not only that the basement morphology has not been constructed by early arc volcanism but also that a tectonic mixture of arc and oceanic elements was emplaced during the early phase of arc development.

Our interpretation of the vertical displacement data argues against significant subsidence after the early Oligocene and indicates relative uplift of the lower slope during the late Neogene. Other relatively recent and related activity might include the development of the low velocity, and possibly accreted toe of the inner slope, and emplacement of the large ridges on the lower section of the inner trench slope.

Perhaps an early phase of anomalous arc volcanism and structural activity lasting until the early Oligocene was succeeded by a phase or phases of minor accretion, structural mixing of older arc units with the new arrivals, downslope slumping, and arcward rotation of the seaward portion of the forearc.

The dissimilar characteristics of the early Mariana Arc and modern very young arcs raise the possibility that more complicated or unanticipated processes developed an unusual initial plate configuration. For example, a 40 m.y. B.P. reconstruction of the Philippine plate [e.g., *Karig,* 1975] has the actively spreading ridge in the West Philippine Basin intersecting the young Mariana Arc somewhere in its southern half. It is plausible that as a result, very young crust was being created in what soon afterward would be the forearc of the Mariana Arc. A similar plate configuration presently exists in several arc systems and has been called upon to explain the development of some ophiolite sheets [*Karig,* 1982].

Finally, the ambiguities and apparently contradictory

evidence from the Mariana forearc are a warning that it is premature to strongly espouse any model for the evolution of the Mariana Arc and even more unwarranted to extrapolate that model to oceanic arcs in general. An obvious task now is the testing of the disputed interpretations in the Mariana Arc by detailed multibeam bathymetric surveys, bottom-navigated sampling, and near bottom observations.

References

Berger, W. H., and E. L. Winterer, Plate stratigraphy and the fluctuating carbonate line, in *Pelagic Sediments on Land and Under the Sea, Spec. Publ. Int. Assoc. Sedimentol.*, vol 1, pp. 11–48, Blackwell, Oxford, 1974.

Bleil, U., Paleomagnetism of Deep Sea Drilling Project Leg 60 sediments and igneous rocks from the Mariana region, *Initial Rep. Deep Sea Drill. Proj., 60*, 855–873, 1982.

Bloomer, S., and J. W. Hawkins, Arc-derived plutonic and volcanic rocks from the Mariana Trench wall (abstract), *Eos. Trans. AGU, 61*, 1143, 1980.

Bracey, D. R., and T. A. Ogden, Southern Mariana Arc: Geophysical observations and hypothesis of evolution, *Geol. Soc. Am. Bull., 83*, 1509–1522, 1972.

Cloud, P. E., et al., Geology of Saipan, Marianas Islands, *U.S. Geol. Surv. Prof. Pap., 280-A*, 126 pp., 1956.

DeLong, S. E., P. J. Fox, and F. W. McDowell, Subduction of the Kula Ridge at the Aleutian Trench, Geol. Soc. Am. Bull., *89*, 83–95, 1978.

Evans, C., and J. W. Hawkins, Mariana arc-trench system: Petrology of 'seamounts' on the trench-slope break (abstract), *Eos Trans. AGU, 60*, 968, 1979.

Fischer, A. G., et al., *Initial Rep. Deep Sea Drill. Proj., 6*, 1329 pp., 1971.

Garrison, R. E., S. O. Schlanger, and D. Wachs, Petrology and paleographic significance of Tertiary nannoplankton-foraminiferal limestones, Guam, *Palaeogeogr. Palaeoclimatol. Palaeoecol., 17*, 49–64, 1975.

Hawkins, J., S. Bloomer, C. Evans, and J. Melchior, Mariana arc-trench system: Petrology of the inner trench wall (abstract), *Eos Trans. AGU, 60*, 968, 1979.

Hilde, T. W. C., S. Uyeda, and L. Kroenke, Evolution of the western Pacific and its margins, *Tectonophysics, 38*, 145–165, 1977.

Honza, E., and H. Kagami, A possible accretion accompanied by ophiolite in the Mariana Trench, *Chigaka Zasshi, 86*, 8–19, 1977.

Hussong, D. M., and S. Uyeda, Tectonics in the Mariana Arc: Results of recent studies, including DSDP Leg 60, *Oceanol. Acta, 4*, suppl., 203–212, 1981.

Hussong, D. M., and S. Uyeda, Tectonic processes and the history of the Mariana Arc: A synthesis of the results of Deep Sea Drilling Project Leg 60, *Initial Rep. Deep Sea Drill. Proj., 60*, 909–929, 1982.

Hussong, D. M., et al., *Initial Rep. Deep Sea Drill. Proj., 60*, 1982.

Ingle, J. C., Jr., Summary of late Paleogene-Neogene insular stratigraphy, paleobathymetry, and correlations, Philippine Sea and Sea of Japan region, *Initial Rep. Deep Sea Drill. Proj., 31*, 837–855, 1975.

Isacks, B. L., and M. Barazangi, Geometry of Benioff zones: Lateral segmentation and downwards bending of the subducted lithosphere, in *Island Arcs, Deep Sea Trenches and Back-Arc Basins, Maurice Ewing Ser.*, vol. 1, edited by M. Talwani and W. C. Pitman III, pp. 99–114, AGU, Washington, D. C., 1977.

Jarrard, R. D., and S. Sasajima, Paleomagnetic synthesis for southeast Asia: Constraints on plate motions, in *The Tectonic and Geologic Evolution of Southeast Asian Seas and Islands, Geophys. Monogr. Ser.*, vol. 23, edited by D. E. Hayes, pp. 293–317, AGU, Washington, D. C., 1980.

Jones, G. M., T. W. C. Hilde, G. F. Sharman, and D. C. Agnew, Fault patterns in outer trench walls and their tectonic significance, *J. Phys. Earth, 26*, suppl., S85–S101, 1978.

Jurdy, D. M., Relative plate motions and the formation of marginal basins, *J. Geophys. Res., 84*, 6796–6802, 1979.

Karig, D. E., Structural history of the Mariana island arc system, *Geol. Soc. Am. Bull., 82*, 323–344, 1971a.

Karig, D. E., Site surveys in the Mariana area (SCAN IV), Leg 6, Deep Sea Drilling Project, *Initial Rep. Deep Sea Drill. Proj., 6*, 681–690, 1971b.

Karig, D. E., Basin genesis, *Initial Rep. Deep Sea Drill. Proj., 31*, 857–879, 1975.

Karig, D. E., Initiation of subduction zones: Implications for arc evolution and ophiolite development, *Spec. Publ. Geol. Soc. London, 10*, 563–576, 1982.

Karig, D. E., and G. F. Sharman, Subduction and accretion in trenches, *Geol. Soc. Bull., 86*, 377–389, 1975.

Karig, D. E., R. N. Anderson, and L. D. Bibee, Characteristics of back spreading in the Mariana Trough, *J. Geophys. Res., 83*, 1213–1226, 1978.

Katsumata, M., and L. R. Sykes, Seismicity and tectonics of the western Pacific: Izu-Mariana, Caroline, and Ryukyu-Taiwan regions, *J. Geophys. Res., 74*, 5923–5948, 1969.

Kulm, L. D., W. J. Schweller, and A. Masias, A preliminary analysis of the Andean continental margin, 6° to 45°S, in *Island Arcs, Deep Sea Trenches and Back-Arc Basins, Maurice Ewing Ser.*, vol. 1, edited by M. Talwani and W. C. Pittman III, pp. 285–301, AGU, Washington, D. C., 1977.

Larsen, E. E., Jr., et al., Paleomagnetism of Miocene volcanic rocks of Guam and the curvature of the southern Mariana island arc, *Geol. Soc. Am. Bull., 86*, 346–350, 1975.

Meijer, A., M. Reagan, H. Ellis, S. Sutter, M. Shafiqullah, P. Damon, and S. Kling, Chronology of volcanic events in the eastern Philippine Sea, this volume.

Moore, G. F., and D. E. Karig, Structural geology of Nias Island, Indonesia: Implications for subduction zone tectonics, *Am. J. Sci., 280,* 193–223, 1980.

Mrozowski, C. L., and D. E. Hayes, A seismic reflection study of faulting in the Mariana forearc, in *The Tectonic and Geologic Evolution of Southeast Asian Seas and Islands, Geophys. Monogr. Ser.,* vol. 23, edited by D. E. Hayes, pp. 223–234, AGU, Washington, D. C., 1980.

Mrozowski, C. L., D. E. Hayes, and B. Taylor, Multichannel seismic reflection surveys of Leg 60 sites, *Initial Rep. Deep Sea Drill. Proj., 60,* 57–69, 1982.

Tracey, J. I., et al., General geology of Guam, *U.S. Geol. Surv. Prof. Pap., 403-A,* 104 pp., 1964.

van Andel, T. H., Mesozoic/Cenozoic calcite compensation depth and the global distribution of calcareous sediments, *Earth Planet. Sci. Lett., 26,* 187–194, 1975.

Hypothetical Model for the Bending of the Mariana Arc

Robert McCabe

Department of Geological Sciences, University of California, Santa Barbara, California 93017

Seiya Uyeda

Earthquake Research Institute, University of Tokyo, Tokyo, Japan

The southern Mariana Arc has a distinct eastward convex shape which is more pronounced than the typical arcuate structure observed above oceanic subduction zones. The Yap Trench is offset hundreds of kilometers westward from the main Izu-Bonin-Mariana trend. Between the southern Mariana Arc and the Yap Trench, the Mariana Arc has an anomalous east-west orientation and is characterized by a markedly lower seismicity than the main Mariana trend. Situated east of the Yap Trench is the ESE trending Caroline Ridge that geochemically resembles the Hawaiian hot spot trend. Paleomagnetic data from Truk and the generally increasing ages to the WNW trend of the Caroline Ridge suggest that the Caroline Ridge is part of the Pacific plate as suggested by Clague and Jarrard (1973). Pacific plate motion for the Caroline Ridge predicts that the ridge has collided with the Yap Trench during the Tertiary. Other evidence for this collision is observed on the island of Yap and by the fact that this portion of the arc has not had volcanic activity during the Neogene period. Paleomagnetic studies show that since the early Oligocene, Guam has rotated greater than 50° clockwise. During this same period, Saipan has rotated only 35° clockwise. These data, the similar bends of the west Mariana Ridge and the Mariana Ridge, and the orientation of fold axes on Guam and Saipan suggest that the clockwise rotation occurred after the initiation of spreading of the Parece Vela Basin and before the opening of the Mariana Trough. This investigation also suggests that the east-west trending portion of the Mariana Trench is a transform boundary which developed in response to the collision.

Introduction

The process of collision and accretion of various kinds of ridges existing on the downgoing plate against island arcs situated on the upper plate has repeatedly been discussed in the literature to explain the juxtaposition of two allochthonous terranes [*Dewey and Bird,* 1970; *Coney et al.,* 1980; *Nur and Ben-Avraham,* 1978; *McCabe et al.,* 1982; *Matsuda,* 1978]. In most of these studies, the workers are studying structures in Phanerozoic mountain belts and trying with very fragmented pieces of data to reconstruct the tectonic evolution of the area. These studies are beset with numerous difficulties, one major obstacle being the lack of a present-day analogue to these proposed collisions. Thus in order to palinspastically reconstruct these collision areas, we must have a better understanding of the geology of present-day analogues to these collision zones.

One present-day analogue is the southern end of the Mariana Arc. This area is ideal to study, because it is not beset with other structural complications often observed in older collision zones. In this paper, we first look at the paleomagnetic data from the southern portion of the Mariana Arc. Although most of the data are reported elsewhere [*Larson et al.,* 1975; *Fuller et al.,* 1980], new K-Ar age dates [*Meijer et al.,* this volume] from the southern Mariana change the timing of the rotations reported in Guam for these earlier studies. Second, we will look at both the orientation of the ridges in the eastern Philippine Sea and the orientation of fold axes on Guam and Saipan. This will help us time the structural events which resulted from the collision. Although earlier workers suggested that the paleomagnetic rotations of Guam are the results of post-Miocene movements of the arc in response to the opening of the Mariana Trough [*Karig et al.,* 1978; *Larson et al.,* 1975], we suggest that the bending of the arc and the paleomagnetic rotations occurred as a result of the colli-

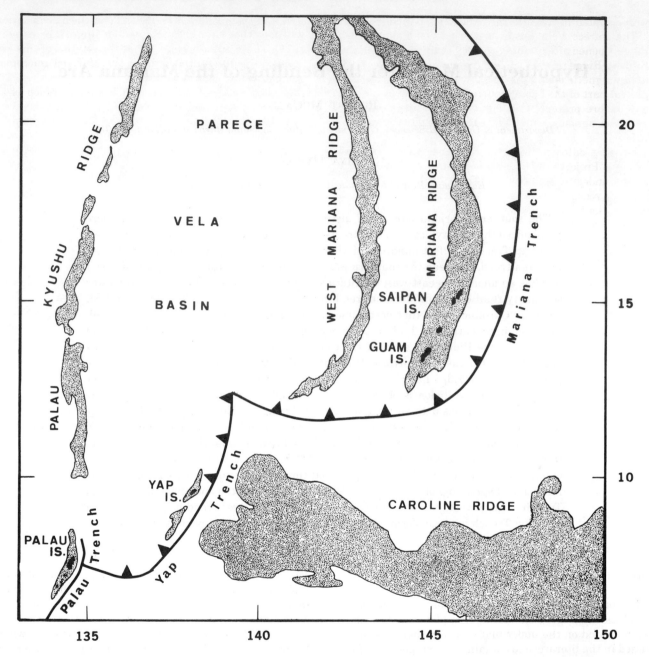

Fig. 1. Outline map of the southeastern portion of the Philippine Sea showing the major bathymetric features.

sion and prior to the opening of the Mariana Trough. In conclusion, we will propose a model for the evolution of the southern Mariana Trench–Yap Trench–Caroline Ridge which may be useful when working in ancient terrane, areas where the evidence is much more fragmented.

Geology of the East Philippine Sea Plate

The Philippine Sea (Figure 1) consists of an early to middle Tertiary western basin and a composite of smaller basins to the east. The eastern and western basins are separated by the N-S oriented Palau-Kyushu Ridge. The eastern composite basin can be subdivided into the larger Parece Vela–Shikoku Basin and the more easterly Mariana Trough.

Drilling of the Palau-Kyushu Ridge (site 448) suggests that the ridge is an old subsided island arc that was erupting island arc tholeiite lavas from 42 m.y. to 29 m.y. before present [*Scott and Kroenke*, 1980; *Meijer*, 1980]. *Meijer* [1980] has shown that these lavas are similar to those that were erupted in Guam during the same period.

The Parece Vela Basin, to the east of Palau-Kyushu Ridge, opened after waning of the volcanism on the Palau-Kyushu Ridge [*Karig*, 1975]. *Mrozowski and Hayes* [1979] used magnetic anomalies to show that the opening of the central part of the basin occurred between 30 m.y. and 17 m.y. before present. This age for the basin is consistent with the age suggested by heat flow studies [*Anderson*, 1980].

Drilling data in the west Mariana Ridge (Deep-Sea Drilling Project (DSDP) site 451) have revealed calc-alkaline andesites and basalts [*Tarney et al.*, 1978]. These volcanic rocks are very similar in composition to Miocene-aged rocks from Guam and Saipan [*Shiraki et al.*, 1978]. Recent work [*Scott and Kroenke*, 1980] has indicated that this ridge was an active area of volcanism between 20 m.y. and 9 m.y. B.P., with activity ceasing after this period.

Karig [1971, 1975] and *Karig et al.* [1978] hypothesized that the Mariana Trough is an area of active back arc spreading. *Hussong et al.* [1978] show that the initiation of spreading in the Mariana Trough started in the late Miocene.

The Mariana Ridge, a concave westward crescent shape that extends from 11° to 25°N, consists of a chain of active volcanoes which are presently erupting material that ranges in composition from basalt to andesite. In the southern portion of the Mariana Ridge, the active volcanic arc is situated just west of an older Eocene-Miocene volcanic arc. This older arc is exposed on the islands of Guam and Saipan, where it is overlain by Miocene and younger limestones.

South of Guam, the shape of the forearc is radically changed from a basically N-S oriented structure to a set of east-west trending ridges and troughs. The forearc itself is abruptly terminated south of 12°30'N. Southwest of this point, a series of east-west trending ridges run from just southwest of Guam to the Mariana-Yap Trench junction. In the southern area, the Mariana interarc basin and the Pacific plate are only separated by an anomalously steep Mariana Trench. In this area, *Katsumata and Sykes* [1969] have shown that the east-west trench is a zone of low seismicity and have obtained a predominantly strike slip focal mechanism solution for one of the larger earthquakes from this area. *Hawkins and Batiza* [1977] reported dredging of highly sheared serpentinized peridotite from a small ridge that runs parallel to the Mariana-Yap Trench intersection (11°15'N, 139°20'E). Near the Yap-Mariana Trench join, *Beccaluva et al.* [1980] have also reported K-Ar dates of 7.8 m.y. B.P. for mid-ocean ridge basalts (MORB) dredged at 11°21'N, 139°00'E.

All of the evidence discussed above suggests that the southern boundary of the Mariana Arc has considerable transform character. *Karig et al.* [1978], *Bracey and Ogden* [1972], and *Larson et al.* [1975] have proposed some degree of right lateral strike slip movement along this southern boundary. Karig et al. and Larson et al. believed that this strike slip component is responsible for the paleomagnetic rotations that have been reported for Guam [*Larson et al.*, 1975]. These authors believe that the rotation of this area is due to a differential drag of the Mariana microplate against the transform boundary.

Located south of the southern Mariana Trench is the Yap Trench. Although this trench has a deep arcuate structure (like the southern portion of the Mariana Trench), it is marked by low seismicity, a volcanic gap, and no Benioff-Wadati zone. The Yap Arc, situated above the Yap Trench, is markedly different than the Mariana Arc. This arc consists largely of metamorphic rocks, whose protoliths appear to be ultramafic and mafic plutonic rocks [*Hawkins and Batiza*, 1977]. Behind the Yap Trench, the Parece Vela Basin is considerably narrower than it is to the north. The equivalents to the west Mariana and the Mariana Trough are not observed.

The seafloor east of the present Yap Trench has a layer number of seamounts and atolls as well as a large mass of thickened oceanic crust which forms the Caroline Ridge; although the origin of the Caroline Ridge is debatable, geochemical studies, age data, and paleomagnetic studies [*Stark and Hay*, 1963; *Fuller et al.*, 1980; *Keating et al.*, 1981a] all support a hot spot origin for the Caroline Ridge as suggested by *Clague and Jarrard* [1973].

Guam

Introduction

Guam (Figure 2), the most southern island of the Mariana Arc, is located at 13.3°N and 144.7°E. The rocks exposed on the island of Guam range in age from Eocene to Recent. Geologic studies [*Tracey et al.*, 1963; *Garrison et al.*, 1975] and K-Ar age dating [*Meijer*, 1980] on volcanic rocks show that Guam is composed of an older Eocene (Facpi and lowermost Alutom formations) to Oligocene (Alutom formation) volcanic core which is overlain by Miocene (Alifan), Plio-Pleistocene (Mariana limestone), and Miocene volcanic rocks (Umatac formation).

Paleomagnetism

Paleomagnetic studies on Guam were carried out by *Larson et al.* [1975]. Because of poor accessibility and intense tropical weathering, Larson et al. only sampled from the southern west portion of the island. The results from Larson's study are shown in Figure 3. From these data, Larson et al. suggested that since the Miocene, either the entire southern Mariana arc-trend system or the southern portion of Guam rotated around 51° clockwise, while suffering no significant latitudinal shift. Most of the sites sampled by Larson et al. were from the Facpi formation (Figure 2). At the time of the study, Larson et al. believed that this section was part of the Miocene-aged Umatac formation, as suggested by *Tracey et al.* [1963]. However, recent radiometric age data reported by *Meijer* [1980] show that the volcanic rocks that make up the Lower Facpi

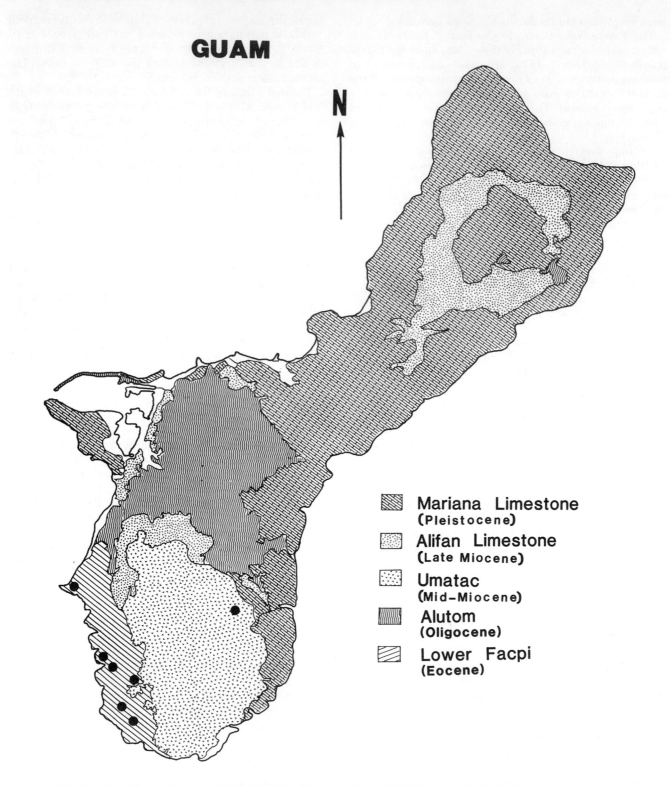

Fig. 2.   Simplified geologic map of Guam [after *Tracey et al.*, 1963]. Solid circles show the location of *Larson et al.* [1975] sites. Note most sites are in Facpi formation.

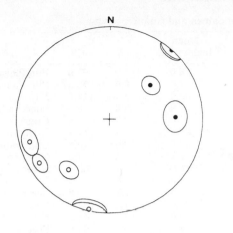

GUAM (Larson et al. 1975)

Fig. 3. Paleomagnetic results from Guam. Results are from *Larson et al.* [1975]. Solid circles are lower hemisphere; open circles are upper hemisphere. Also shown are α_{95} circles of confidence.

member of the Umatac formation (Tracey et al.) are Eocene in age. On the basis of these data, *Meijer et al.* [this volume] assigned the Lower Facpi to the Facpi formation. Since Larson et al. also sampled this older formation, the lower limits of the rotations are post-Eocene and not post-Miocene. At present, the amount of rotation (detectable by paleomagnetism) since the Miocene cannot be evaluated until further studies have been completed. However, by looking at the orientations of folds on Guam and the orientation of bathymetric ridges, a timing of these rotations can be obtained. This timing will be discussed below.

Saipan

Introduction

Saipan (Figure 4), located 200 km NNE of Guam, is composed of an Eocene age rhyolitic core (Sankakuyama formation) which is overlain by Oligocene (Hagman formation) and Miocene (Fina sisu formation) volcanic rocks. These volcanic rocks are overlain by Miocene (Tagpochau) to Pleistocene (Mariana) limestones. K-Ar age dating [*Meijer et al.*, this volume] and paleontological studies [*Tracey et*

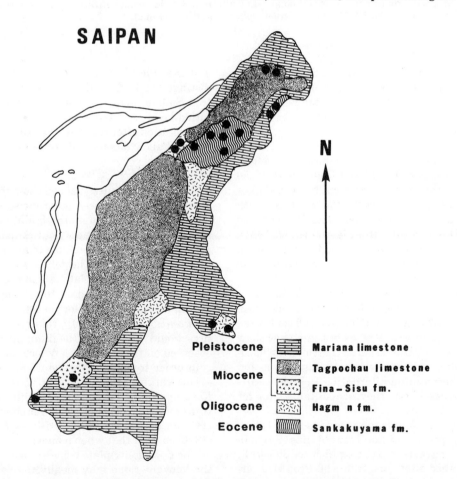

Fig. 4. Simplified geologic map of Saipan [after *Cloud et al.*, 1956]. Circles show the location of sites reported in Table 1.

TABLE 1. Paleomagnetic Results From Saipan and Guam

Formation Name	Location	Number of Samples	Mean Declination	Mean Inclination	α_{95}	K	Age
Mariana limestone	S*	10	358.7	26.0	4.9	97.6	Plio-Pleistocene
Mariana limestone	S	8	354.2	14.0	6.5	73.8	Plio-Pleistocene
Fina sisu (volcanic flow)	S	6	203.0	−39.0	6.4	111.5	Miocene (13 m.y.)
Tagpochau limestone	S	8	207.8	−19.9	20.6	8.2	Miocene (early)
Hagman (volcanic)	S	6	42.6	21.4	22.8	9.6	Oligocene (36 m.y.)
Hagman (volcanic)	S	11	15.4	29.8	11.4	17.1	Oligocene (36 m.y.)
Hagman (volcanic)	S	4	46.6	12.9	10.6	76.8	Oligocene (36 m.y.)
Hagman (sandstone)	S	6	43.7	22.1	35.1	4.6	Oligocene (36 m.y.)
Sankakuyama (volcanic)	S	7	30.0	40.0	6.7	82.9	Eocene (41 m.y.)
Sankakuyama (volcanic)	S	4	48.0	21.5	4.8	364.5	Eocene (41 m.y.)
Sankakuyama (volcanic)	S	4	58.0	22.0	21.1	19.8	Eocene (41 m.y.)
Sankakuyama (volcanic)	S	7	49.0	30.0	4.3	194.1	Eocene (41 m.y.)
Sankakuyama (volcanic)	S	6	49.5	0.0	12.1	41.2	Eocene (41 m.y.)
Sankakuyama (volcanic)	S	6	25.5	24.5	6.7	99.0	Eocene (41 m.y.)
Sankakuyama (volcanic)	S	5	11.0	45.6	4.4	30.5	Eocene (41 m.y.)
Sankakuyama (volcanic)	S	8	357.8	53.3	5.6	100.1	Eocene (41 m.y.)
Sankakuyama and Hagman mean		12 sites	36.6	27.9	11.6	15.1	Eocene-Oligocene
Alutom (volcanic)	G†	5	66.6	11.8	35.3	5.6	Oligocene (35 m.y.)
Alutom (volcanic)	G	7	59.5	21.3	10.9	13.9	Oligocene (35 m.y.)
Lower Facpi (volcanic)	G	5	270.6	−8.4	17.1	20.9	Eocene (42 m.y.)

Directions that are reported represent the values obtained by alternating field demagnetization. Data are from *Fuller et al.* [1980] and Fuller (personal communication, 1982). Ages are from *Meijer et al.* [this volume].
*S is Saipan.
†G is Guam.

al., 1963; *Cloud et al.*, 1956] show that the Sankakuyama, Hagman, and Fina sisu volcanic rocks of Saipan are of the same age as the Facpi, Alutom, and Umatac formations, respectively.

Paleomagnetism

Paleomagnetic studies have been conducted on Saipan [*Fuller et al.*, 1980; *Dunn et al.*, 1979]. Because most of the results discussed in this paper have been presented elsewhere, we will only review these previous studies with respect to the bending of the southern arc. We will also avoid discussing DSDP results from leg 60 [*Bleil*, 1981], in which rotations cannot be detected, and the recent work of *Keating et al.* [1981b], whose studies are still in the preliminary state.

In Saipan the senior author and others [*Fuller et al.*, 1980] collected samples from 15 different sites which represent five different formations ranging in age from Eocene to Pleistocene. Rock magnetic studies of these sites show the following magnetic characteristics: intermediate hardness, low viscous remanent magnetization acquisition, and stability to both thermal and alternating field demagnetization, which suggests that they are suitable sites for paleomagnetic work.

The results from the paleomagnetic study (Table 1, Figure 5) show that the pre–25 million year old sites (Hagman and Sankakuyama) have declinations which are 30°–40°E. The two Miocene-aged sites, one from the Fina sisu formation and the other from the Tagpochau limestone (M. Fuller, personal communication, 1982), are around 25°E,

and the Plio-Pleistocene-aged Mariana limestone is not distinguishable from the present field. These data suggest that Saipan has experienced around 10° of clockwise rotation before the late Oligocene and around 25° of clockwise rotation after the late Oligocene. The Plio-Pleistocene does not show a detectable rotation.

Discussion of Paleomagnetic Results

The studies discussed above show that both Saipan and Guam have experienced a clockwise rotation since early Oligocene. These studies also show that Guam has suffered much more of a rotation than Saipan during the same period. Although it is possible to argue that the clockwise rotation of Saipan may largely be the result of the rotation of the Philippine Sea plate (26° and 23° clockwise for two Miocene sites compared to 10°–18° of clockwise rotation during the past 20 m.y. for the Philippine Sea plate [*Jarrard and Sasajima*, 1980]), this type of entire plate rotation should not produce the differential rotations that we observed on these two closely located islands.

In order to explain the rotation, two possible tectonic events can be hypothesized: (1) *Larson et al.* [1975] and *Karig et al.* [1978] suggest that differential drag along a transform south of Guam could produce the large rotation of Guam, and (2) *Vogt et al.* [1976] and *Kelleher and McCann* [1976] suggest that when a buoyant aseismic ridge located on the downgoing plate comes in contact with the trench, the buoyant ridge may modify the subduction process and partially deform the upper plate margin. The results of such a collision would produce paleomagnetically de-

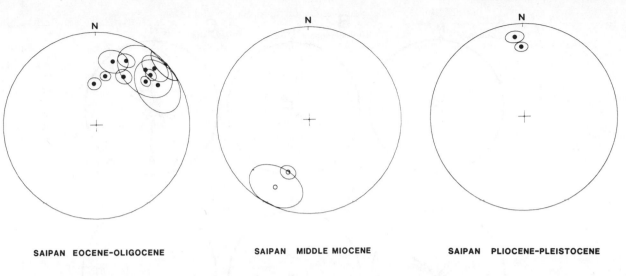

SAIPAN EOCENE-OLIGOCENE   **SAIPAN MIDDLE MIOCENE**   **SAIPAN PLIOCENE-PLEISTOCENE**

Fig. 5.   Paleomagnetic results from Saipan. Solid circles are lower hemisphere; open circles are upper hemisphere. Also shown are α_{95} circles of confidence.

tectable rotations on the upper plate with a sense of motion, which is consistent with the direction of the convergence of the buoyant features [McCabe et al., 1982]. Although the paleomagnetic studies discussed above do not allow one to totally discriminate between these two possible causes for bending of the southern Mariana Arc, a detailed look at the structure from the southern Mariana Arc places additional constraints on the timing of the rotations.

Structural Geology of the Southern Mariana Arc

In structural geology one learns that to understand the major structures of a region, a study of the smaller structures in an area can be used. This concept is easy to apply if the rocks that one is studying have only experienced one structural event, in which the major forces that produced this event are biaxial. Since the tectonic forces that produce major rotations of objects the size of microplates are predominantly caused by horizontal forces, an investigation of the structures of the Mariana Arc was performed to gain insight into the direction of the forces of plate tectonic origin. Since the area of investigation is located on the very edge of the plate, we assume that the minor structures at the boundary of the plate reflect forces that were generated by the interaction of the two plates.

In this study, we investigated two different orders of the structures: the structures of the major ridges in the east Philippine Sea and the orientations of folding in the islands of Guam and Saipan. An examination of the ridges and troughs of the east Philippine Sea shows there is a definite causal relationship between the rotations that we observed paleomagnetically and the orientations of the west Mariana Ridge and Mariana Ridge. Figure 1 shows that the southern end of both the west Mariana Ridge and the Mariana Ridge has a NE orientation, while the Palau-Kyushu Ridge trends north-south.

The next lower division of structure below major ridges is faults and folds. We plotted the strikes of the bedding planes reported by Cloud et al. [1956] for Saipan and by Tracey et al. [1963] for Guam. It should be noted that if fewer than five bedding strikes were reported for a formation; if plunging folds were mapped; if the strikes and the dips reported were along sea cliffs in the young Mariana limestone, and possibly reflecting postdepositional slumping; or if the reported strikes for a formation had less than 5° of dip, we disregarded these formations to avoid events that possibly did not reflect the regional structures of the area. We have also assigned ages to the formations based on new K-Ar age dates of Meijer [1980].

Rose diagrams of the strikes of bedding (on the angular axis) against frequency (on the radial axis) are shown in Figure 6. Although these plots only give a rough estimate of the compressive forces responsible for folding, we can easily see two major orientations. In Guam, the Facpi formation, the Alutom formation, and Maemong members of the Umatac formation show two major trends. A similar case is also observed in Saipan in the older pre-Miocene formations and in the Donni sandstone member which is the stratigraphically lowest member of the Tagpochau limestone. The two trends that are observed in these older (pre–12 m.y. B.P.) formations are a larger NE-SW orientation and a small N-S orientation. This NE-SW orientation is also consistent with (1) the foliations that were reported from Yap [Hawkins and Batiza, 1977], (2) the direction from the paleomagnetic results, and (3) the major trends of the islands and ridges. In the younger Umatac formation, excluding the early Miocene [Garrison et al., 1975] Maemong member and the Tagochau formation and excluding the lower Donni sandstone member, we can see a small or nonexisting NE-SW trend and a much more pronounced N-S trend. All of the post-mid-Miocene for-

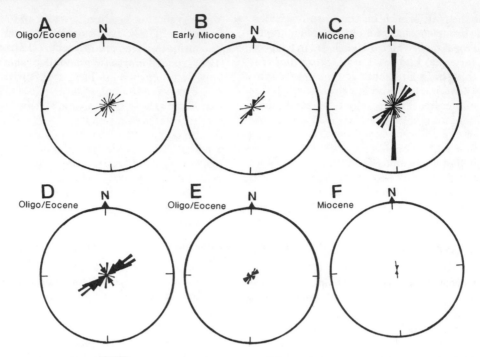

Fig. 6.   Rose diagrams of bedding strikes reported by *Cloud et al.* [1956] for Saipan and *Tracey et al.* [1963] for Guam. The orientation is shown on the angular axis; the frequency that the orientation appears is shown on the radius. Figures 6a, 6b, and 6c are from Saipan and represent (a) late Eocene–early Oligocene Matansa limestone, Denisinyama limestone, Hagmen formation, and the Sankakuyama formation, (b) early Miocene Dunni sandstone member of the Tagpochou limestone, and (c) Miocene Tagpochou limestone. Figures 6d, 6e, and 6f are from Guam, (d and e) from the early Oligocene/late Eocene Alutom formation and Facpi formation, respectively, and (f) from the Miocene Umatac formation (excluding the Maemong limestone member). On both islands, the late Miocene to Recent formations are essentially horizontal. Significance of these plots is discussed in the text.

mations on Saipan and Miocene and younger formations on Guam are predominantly horizontal or dip at a shallow angle (less than 5°).

From the study of the structure of the Mariana region, we suggest the following constraints on the timing of bending of the southern Mariana Arc. As stated earlier, both of the ridges between the Parece Vela Basin and the Pacific plate have a NE orientation. The Palau-Kyushu Ridge trends NS. Assuming that these eastern ridges migrated from the Palau-Kyushu Ridge by back arc spreading [*Karig,* 1971], it can be argued that the bending of the two ridges occurred after the initiation of spreading of the Parece Vela Basin and before the initiation of spreading of the Mariana Trough.

The analysis of fold axes adds additional constraints on the timing of the bending. *Sugimura and Uyeda* [1973] and *Nakamura and Uyeda* [1980] showed that in a volcanic arc, compressional forces are nearly parallel to the direction of convergence. In an arc such as the Marianas, where the convergence direction with the Philippine Sea plate is approximately N70°E [*Uyeda and Kanamori,* 1979], the axes of folds would be predicted to trend approximately N-S. This is indeed the results we see for the Miocene Umatac formation (Figure 6f) on Guam and the Miocene

Tagpochau formation (Figure 6c) from Saipan. An examination of the Dunni sandstone member (Figure 6b), which is the stratigraphically lowest member of the Tagpochau limestone and the Eocene/early Oligocene formations (Figures 6d and 6e) on both Guam and Saipan (Figure 6a), shows a much more pronounced NE-SW trend. Although these data are not totally convincing alone, the fact that they are consistent with the paleomagnetic rotations observed on Guam and Saipan suggests that they result from the same tectonic process. If so, the orientations of the fold axes suggest that the rotations occurred earlier than middle Miocene.

Discussion

An examination of paleomagnetic data and structures in the Mariana Arc shows that the southern portion of the arc has rotated clockwise. Furthermore, the results show that Guam rotated more than Saipan. From these observations and the structures of the southern Mariana Arc we speculate that the differential rotation is a result of a collision between the proto-Mariana-Yap Trench and the Caroline Ridge. This speculation is further supported by the following arguments:

1.   The Truk Islands (7°N, 152°E) are situated about

1600 km east of the Yap Trench on the Caroline Ridge. The Truk Islands group consists of a number of small volcanic islands and coral reefs which are enclosed in a lagoon formed inside a large (65 km) coral atoll. *Stark and Hay* [1963] proposed that the Truk Islands are the eroded remnants of a large Hawaiian type shield volcano, which was active during the Miocene. Two K-Ar age determinations reported by *Fuller et al.* [1980] give ages of 11.2 m.y. and 9.9 m.y. B.P. from two different flows from Truk.

The senior author and others [*Dunn et al.*, 1979; *Fuller et al.*, 1980] conducted a paleomagnetic study of the Truk Islands. The results from Truk show that during the past 10 m.y., Truk has undergone motion that is consistent with the Pacific plate. When these data are combined with the fact that the Truk Islands show an alkali differentiation trend similar to the Hawaiian Islands [*Stark and Hay*, 1963], and the generally increasing ages to the WNW of the Caroline Ridge [*Keating et al.*, 1981a], it strongly supports the suggestion of *Clague and Jarrard* [1973] that the Caroline Ridge is a hot spot trace on the Pacific plate. To this conclusion, we would like to add that the Caroline Ridge is currently moving in a direction that is consistent with Pacific plate motion. This motion would predict that the Caroline Ridge collided with the Philippine Sea plate at the Yap Trench.

2. Located south of the southern Mariana Trench is the Yap Trench. This trench, like the southern end of the Mariana Trench, is also marked by low seismicity and shallow depth earthquakes. The Yap Island group (Figure 1), situated above the Yap Trench, contains a unique basement compared to the other islands located along the Mariana-Palau Arc. The Yap basement complex is composed of regionally metamorphosed rocks whose protoliths are mafic and ultramafic members of rocks possibly formed at an oceanic spreading center [*Hawkins and Batiza*, 1977]. Dredge samples collected by Scripps Institute of Oceanography Expedition INDOPAC, leg 4, show that this complex also extends 65 km to the NNE of Yap.

Hawkins and Batiza have proposed that the Yap Arc formed originally as a volcanic arc in the Eocene or Oligocene times. As subduction continued below Yap, eventually the Caroline Ridge came in contact with the Yap Trench. This collision resulted in eastward thrusting of the seafloor west of Yap over the Eocene-Oligocene volcanic arc and in the termination of subduction in the Yap Trench. The presence of alkalic material in the Yap Trench, the close spacing between the Yap Trench and the Yap Ridge, the low seismicity, and the lack of volcanic activity support their model.

To the above model, we would like to suggest that the easterly directed thrusting and the collision of the Caroline Ridge with the Yap Trench occurred in the late Oligocene to early Miocene period. This age was suggested by two separate lines of evidence: (1) *Cole et al.* [1960] show that the Map formation, which lies in fault contact with the Yap formation, contains late Oligocene to early Miocene smaller foraminifera and mid-Miocene larger foraminifera. These ages suggest that the thrusting of the Yap formation was pre-Miocene. (2) *Mrozowski and Hayes* [1979], using magnetic anomalies, show that the Parece Vela Basin opened 30–17 m.y. B.P. with a spreading rate of 5–6 cm/yr. Although we are unable to predict the actual spreading rate at the latitude of Yap, a similar spreading rate to 18°N would require between 5 and 6 m.y. to open the portion of the Parece Vela Basin behind Yap. This predicts that the spreading ended and the eastward directed thrusting occurred because of a late Oligocene to early Miocene event. It is also of interest that *Mrozowski and Hayes* [1979] needed to assume a much slower spreading rate at 25 m.y. B.P. to make magnetic anomalies in the Parece Vela Basin fit into their spreading model.

Conclusions and Interpretation

In attempting to reconstruct the tectonic evolution of this area, we will draw upon the models of *Vogt* [1973], *Vogt et al.* [1976], and *Kelleher and McCann* [1976]. These models suggest that when a buoyant bathymetric feature (such as Caroline Ridge) interacts with a convergence margin (such as the Mariana Trench), the buoyant ridge will modify the upper plate. These modifications can be expressed as stopping back arc spreading in the area adjacent to the ridge, bending the subduction one, turning off arc volcanism, and changing the nature of seismicity.

In this investigation we have focused on the paleomagnetic and structural aspects of the southern portion of the Mariana Arc. When these observations are combined with geologic and geophysical data from the southern segment of the Mariana Arc, Yap, and the Caroline Ridge (discussed above), they allow for the following five-stage speculative model to be proposed for the present plate geometry in the region.

The presence of Eocene to early Oligocene volcanic rocks on the Palau-Kyushu and Mariana ridges and of Miocene volcanic rocks on the west Mariana Ridge and the Mariana Ridge, the occurrence of two small oceanic basins between these ridges, and DSDP data showing that the easternmost (Mariana Trough) basin is younger than the western (Parece Vela) basin allowed *Karig* [1971] to suggest that this area was formed by rifting of the volcanic arc eastward. Assuming the model of Karig, Figure 7a shows the boundary of the Pacific plate and Philippine Sea plate in the late Eocene to early Oligocene time. During this period, the Pacific plate was being subducted westward beneath the Palau-Kyushu Ridge. The Caroline Ridge, being part of the Pacific plate, moved westward toward the proto-Mariana-Yap Trench. In this figure, question marks are placed on the western edge of the Caroline Ridge. It is impossible to establish the previous existence or nonexistence of such a section of the ridge, because (if it existed) it has since been subducted.

Mrozowski and Hayes [1979] show that rifting began in the Parece Vela Basin in the middle to late Oligocene.

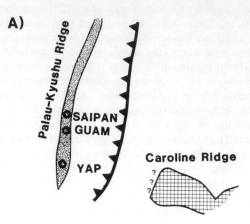

A)

Palau-Kyushu Ridge

SAIPAN
GUAM
YAP

Caroline Ridge

Eocene-Oligocene

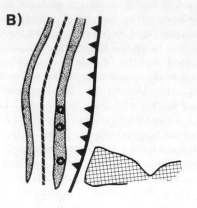

B)

Late Oligocene/Early Miocene

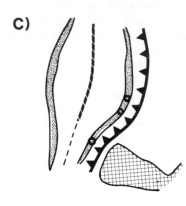

C)

Early-Mid Miocene

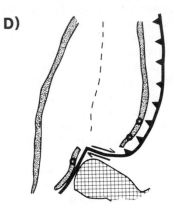

D)

Mid-Late Miocene

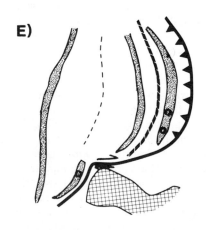

E)

Late Miocene-Recent

Fig. 7. Schematic diagram showing the evolution of the Mariana Arc. See text for discussion.

This rifting (Figure 7b) migrated the proto-Mariana-Yap Trench and the eastern section of the Palau-Kyushu Arc eastward with respect to the remanent section of the Palau-Kyushu Ridge. Although it is not shown in Figure 7b, we suggest that the plate boundary east of Palau acted as a transform boundary along which the Parece Vela spreading center migrated.

The age of the Map formation on Yap and the distance between the Yap Ridge and the Palau-Kyushu Ridge suggest that the collision between the Caroline Ridge and the proto-Mariana Arc occurred around the late Oligocene/early Miocene (Figure 7c). As a result of the collision, spreading stopped behind the Yap section of the arc, the young oceanic crust in the back arc basin behind the Yap thrust over the volcanic arc, and subduction in the Yap Trench was inhibited.

Between the time of the collision and the middle Miocene (Figure 7c), spreading continued in the northern Parece Vela Basin. The bending of the west Mariana Arc occurred sometime between late Oligocene and the middle Miocene (Figures 7c and 7d). The data do not allow enough resolution in the rotations to predict the cause of the bending. At least two possible causes are suggested:

1. Back arc spreading in the northern half of the Parece Vela Basin caused the continued migration of the Mariana section of the proto-Mariana-Yap Trench eastward. As a result of the migration, the southern section of the west Mariana Ridge, still connected to the Yap Ridge, rotated clockwise.

2. As a result of stopping subduction in the Yap Trench, the Philippine Sea plate, immediately behind the Yap region, started moving with Pacific plate motion. This model can be accommodated by either a transform fault along the southern boundary of the entire Philippine Sea plate which is not supported or by the proto–Mariana Arc staying at essentially the same position from Eocene to present while the entire Philippine Sea plate migrated westward by back arc spreading [Ben-Avraham and Uyeda, 1982].

At present, we do not have enough data to differentiate between these two possible causes of bending. In order to resolve this problem, future studies on Guam will be aimed at collecting Oligocene- to Pliocene-aged samples.

Figure 7d shows the configuration of the system in the mid-late Miocene. During this period a transform has developed, allowing the west Mariana Ridge to separate from the Yap Arc.

The present configuration is shown in Figure 7e. Since the bending occurs in both the west Mariana Ridge and the Mariana Ridge, we argue that this bending must have occurred prior to latest Miocene. We also suggest that the transform developed along the southern Mariana Trench. This transform served as a nucleus for the development of spreading in the Mariana Trough.

The relative motion proposed by our model for the Mariana plate suggests that the Mariana plate is moving ENE with respect to the Pacific plate (Figure 7e). It is currently impossible to give a more quantitative description of the motion, because the motion of the Philippine Sea plate with respect to the Pacific plate is currently not well understood. The uncertainties involved with the motions of these plates were previously discussed by both Karig [1975] and Chase [1978]. Second, the motion of the Mariana plate is not known. Karig et al. [1978] suggested that the Mariana plate may not conform to the laws of rigid plate motions.

In this paper we have purposely avoided discussing the possible effects of the collision with the Markus seamounts at the Bonin-Mariana join. Such effects have been suggested by Wu [1978] and Vogt et al. [1976]. Although we do believe that this ridge has an effect on the geometry of the Mariana Arc, not enough data exist to pursue the investigation in this region. It is possible that these seamounts are too far north to produce the local effects that we are concerned with anyway. It is somewhat interesting to note that recent studies by Kodama [1981] found large clockwise rotations, just north of where Markus seamounts intersect the trend of Bonin Island. Also, Nur and Ben-Avraham [1982] have pointed out that there is a volcanic gap in this area, which is similar to the volcanic gap that exists along the Yap Trench.

The results of this study offer an example of how paleomagnetic studies, when combined with geologic studies, can aid in the interpretation of the tectonics of the region. This study also suggests at least two other crucial studies to be carried out in the region. First, future collecting should be aimed at collecting more data from the late Oligocene to Pliocene rocks on Guam. With better data it may be possible to elucidate if either back arc spreading or the stopping of subduction around the Yap Trench is responsible for the bending of the southern Mariana Arc. Also, additional studies are needed from both the Bonin Islands and the southern portion of the Mariana-Yap-Palau region. It is hoped that such studies will show the amount of rotation that is strictly due to the motion of the Philippine Sea plate.

Acknowledgments. We wish to express thanks to the program in East Asian Tectonics and Resources of the United Nations Committee for the Co-ordination of Joint Prospecting for Mineral Resources in Asian Offshore Areas. This group together with the National Science Foundation supplied the financial support for the paleomagnetic data reported in this paper. Also, special thanks are given to M. D. Fuller. Although he is not responsible for the ideas expressed in this paper, some of the ideas expressed in this paper grew out of discussions with him. Also, we want to thank him for use of some of the unpublished data reported in this paper. Thanks are also given to Jim Hawkins and an unidentified reviewer whose numerous comments greatly improved an early draft of this paper. Gratitude is also expressed to the Japanese Government Scholarship Program and to S. Sasajima and the members of his lab at

Kyoto University which allowed the senior author a one-year fellowship in Japan; during this period this manuscript was written. Lastly, thanks are given to Jose Almasco and Elvie McCabe for help in the preparation of this manuscript.

References

Anderson, R. N., 1980 update of heat flow in the east and southeast Asian seas, in *The Tectonic and Geologic Evolution of Southeast Asian Seas and Islands, Geophys. Monogr. Ser.*, vol. 23, pp. 319–326, AGU, Washington, D. C., 1980.

Beccaluva, L., G. Macciotta, C. Savelli, G. Serri, and O. Zeda, Geochemistry and K/Ar ages of volcanics dredged in the Philippine Sea (Mariana, Yap, and Palau trenches and Parece Vela Basin), in *The Tectonic and Geologic Evolution of Southeast Asian Seas and Islands, Geophys. Monogr. Ser.*, vol. 23, pp. 247–268, AGU, Washington, D. C., 1980.

Ben-Avraham, Z., and S. Uyeda, Entrapment origin of marginal seas, *Geodyn. Series*, AGU, Washington, D. C., in press, 1982.

Bleil, U., Paleomagnetism of deep sea drilling project leg 60 sidement and igneous rocks from the Mariana region, *Initial Rep. Deep Sea Drill. Proj., 60*, 855–873, 1981.

Bracey, D. R., and T. A. Ogden, Southern Mariana Arc: Geophysical observations and hypothesis of evolution, *Geol. Soc. Am. Bull., 33*, 1509–1522, 1972.

Chase, C. G., Plate kinematics: The Americas, East Africa and the rest of the world, *Earth Planet. Sci. Lett., 37*, 353–368, 1978.

Clague, D. A., and R. D. Jarrard, Tertiary Pacific plate motion deduced from Hawaiian Emperor Chain, *Geol. Soc. Am. Bull., 84*, 1135–1154, 1973.

Cloud, P. E., R. G. Schmidt, and H. W. Burke, Geology of Saipan, Mariana Islands, 1, General geology, *U.S. Geol. Surv. Prof. Pap., 280*, 126 pp., 1956.

Cole, W. S., R. Todd, and C. G. Johnson, Conflicting age determination suggested by Foraminifera at Yap, Caroline Islands, *Bull. Am. Paleontol., 41*, 77, 1960.

Coney, P. J., D. L. Jones, and J. W. Monger, Cordilleran suspect terranes, *Nature, 288*(27), 329–332, 1980.

Dewey, J. F., and J. M. Bird, Mountain belts and the new global tectonics, *J. Geophys. Res., 75*, 2625–2647, 1970.

Dunn, J. R., M. Fuller, G. Green, R. McCabe, and I. Williams, Preliminary paleomagnetic results from the Philippines, Marianas, Carolines and Solomons (abstract), *Eos Trans. AGU, 60*, 239, 1979.

Fuller, M., J. R. Dunn, G. Green, J.-L. Lin, R. McCabe, K. Toney, and I. Williams, Paleomagnetism of Truk Islands, Eastern Carolines and of Saipan, Marianas, in *The Tectonic and Geologic Evolution of Southeast Asian Seas and Islands, Geophys. Monogr. Ser.*, vol. 23, pp. 235–245, AGU, Washington, D. C., 1980.

Garrison, R. E., S. O. Schlanger, and D. Wachs, Petrology and paleogeographic significance of Tertiary nanno-plankton-foraminiferal limestones, Guam, *Palaeogeogr. Palaeoclimatol. Palaeoecol., 17*, 49–64, 1975.

Hawkins, J., and R. Batiza, Metamorphic rocks of the Yap Arc Trench system, *Earth Planet. Sci. Lett., 37*, 216–229, 1977.

Hussong, D., et al., Near the Philippines, leg 60 ends in Guam, *Geotimes, 23*(10), 19–22, 1978.

Jarrard, R. D., and S. Sasajima, Paleomagnetic synthesis for southeast Asia: Constraints on plate motions, in *The Tectonic and Geologic Evolution of Southeast Asian Seas and Islands, Geophys. Monogr. Ser.*, vol. 23, edited by D. E. Hayes, pp. 293–316, AGU, Washington, D. C., 1980.

Karig, D. E., Structural history of the Mariana Island Arc system, *Geol. Soc. Am. Bull., 81*, 323–344, 1971.

Karig, D. E., Basin genesis in the Philippine Sea, *Initial Rep. Deep Sea Drill. Proj., 31*, 857–859, 1975.

Karig, D. E., R. N. Anderson, and L. D. Bibee, Characteristics of back arc spreading in the Mariana Trough, *J. Geophys. Res., 83*, 1213–1226, 1978.

Katsumata, M., and L. R. Sykes, Seismicity and tectonics of the western Pacific: Izu-Mariana-Caroline and Ryukyu-Taiwan regions, *J. Geophys. Res., 74*, 5923–5948, 1969.

Keating, B., D. Mattey, J. Naughton, D. Epp, and C. E. Helsley, Evidence for a new Pacific hot spot (abstract), *Eos Trans. AGU, 62*, 381–382, 1981a.

Keating, B., C. E. Helsley, K. Kodama, and S. Uyeda, Paleomagnetic of the Mariana Island Arc, paper presented at IAUCEI Symposium on Arc Volcanism, Assoc. of Volcanol. and Chem. of the Earth Interior, Tokyo, 1981b.

Kelleher, J., and W. McCann, Buoyant zones, great earthquakes, and unstable boundaries of subduction, *J. Geophys. Res., 81*, 4885–4896, 1976.

Kodama, K., A paleomagnetic reconnaissance of the Bonin Islands, *Bull. Earthquake Res. Inst. Univ. Tokyo, 56*, 347–365, 1981.

Larson, E. E., Jr., et al., Paleomagnetism of Miocene volcanic rocks of Guam and the curvature of the southern Mariana Island Arc, *Geol. Soc. Am. Bull., 86*, 346–350, 1975.

Matsuda, T., Collision of the Izu-Bonin Arc with central Honshu: Cenozoic tectonics of the fossa magma, Japan, *Adv. Earth Planet. Sci., 6*, 409–422, 1978.

McCabe, R. J., J. Almasco, and W. Diegor, Geologic and paleomagnetic evidence for a possible Miocene collision in western Panay, central Philippines, *Geology, 10*, 325–329, 1982.

Meijer, A., Primitive arc volcanism and a Boninite series: Examples from western Pacific island arcs, in *The Tectonic and Geologic Evolution of Southeast Asian Seas and Islands, Geophys. Monogr. Ser.*, vol. 23, pp. 269–282, AGU, Washington, D. C., 1980.

Meijer, A., M. Reagan, H. Ellis, M. Shafiqullah, S. Sutter, P. Damon, and S. Kling, Chronology of volcanic events in the eastern Philippine Sea, this volume.

Mrozowski, C. L., and D. E. Hayes, The evolution of the Parece Vela Basin, eastern Philippine Sea, *Earth Planet. Sci. Lett., 46,* 49–67, 1979.

Nakamura, K., and S. Uyeda, Stress gradient in arc–back arc regions and plate subduction, *J. Geophys. Res., 85,* 6419–6428, 1980.

Nur, A., and Z. Ben-Avraham, Speculations on mountain building and the lost Pacifica continent, *Adv. Earth Planet. Sci., 6,* 21–39, 1978.

Nur, A., and Z. Ben-Avraham, Volcanic gaps due to oblique consumption and aseismic ridge, *Geodynam. Ser.,* AGU, Washington, D. C., in press, 1982.

Scott, R., and L. Kroenke, Back arc spreading and arc volcanism in the Philippine Sea, in *The Tectonic and Geologic Evolution of Southeast Asian Seas and Islands, Geophys. Monogr. Ser.,* vol. 23, pp. 283–291, AGU, Washington, D. C., 1980.

Shiraki, K., N. Kuroda, S. Mayurama, and H. Urano, Evolution of the Tertiary volcanic rocks in the Izu-Marianas Arc, paper presented at International Geodynamics Project Conference, Tokyo, 1978.

Stark, J. T., and R. L. Hay, Geology and petrology of the volcanic rocks at Truk Islands, East Caroline Islands, *U.S. Geol. Surv. Prof. Pap., 409,* 41 pp., 1963.

Sugimura, A., and S. Uyeda, *Island Arcs, Japan and Its Environs,* 247 pp., Elsevier, New York, 1973.

Tarney, J., N. G. Marsh, and A. D. Sanders, Geochemistry of basalts from West Mariana Transect: POD leg 59 (abstract), *Eos Trans. AGU, 59,* 1188, 1978.

Tracey, J. I., Jr., S. O. Schlanger, J. T. Stark, D. B. Doan, and H. G. May, General geology of Guam, *U.S. Geol. Surv. Prof. Paper, 403a,* 104 pp., 1963.

Uyeda, S., and W. Kanamori, Back arc opening and the mode of subduction, *J. Geophys. Res., 84,* 1049–1061, 1979.

Vogt, P. R., Subduction and aseismic ridges, *Nature, 241,* 189, 1973.

Vogt, P. R., A. Lowrie, D. R. Bracey, and R. N. Hey, Subduction of aseismic ocean ridges: Effects on shape, seismicity, and other characteristics of consuming plate boundaries, *Spec. Pap. Geol. Soc. Am., 172,* 1976.

Wu, F. T., Benioff zones, absolute motion and interarc basin, *Adv. Earth Planet. Sci., 6,* 39–54, 1978.

Gabbroic and Ultramafic Rocks From the Mariana Trench: An Island Arc Ophiolite

S. H. BLOOMER AND J. W. HAWKINS

Geological Research Division A-020, Scripps Institution of Oceanography, University of California, La Jolla, California 92093

Dredge collections from the inner slope of the Mariana Trench include massive and cumulate textured gabbroic and ultramafic rocks and volcanic rocks ranging from basaltic to dacitic in composition. The volcanic rocks are largely products of island arc volcanism and probably represent part of a pre-late Eocene arc complex. This arc complex has been exposed in the inner slope of the trench by tectonic erosion of material by subduction since the Eocene. The ultramafic rocks are largely serpentinized harzburgite; their relict mineralogy is typical of peridotite considered to be the refractory residue of partial melting of the mantle. Some of the ultramafic samples have irregular patches of clinopyroxene that appear to have crystallized from pockets of melt. These resemble rocks termed 'impregnated peridotites.' Cumulate textured ultramafic rocks probably are related to the cumulate gabbros rather than to the residual mantle material. The gabbroic rocks are dominantly cumulate textured PLAG-OPX-CPX rocks. The sequence of crystallization for the cumulate series is OL-(SP)-OPX-CPX-PLAG. The cumulate rock series from the trench slope records a crystallization sequence similar to that of the volcanic rocks. There is an overlap in mineral chemistry between the cumulate rocks and the volcanic rocks, and there is a close match in bulk compositions between the volcanic rocks and melts calculated to have been in equilibrium with the cumulate rocks. We interpret this as an indication that the cumulate rocks were cogenetic with the volcanic rocks and that they both constitute the remnants of an island arc volcanic-plutonic series. This arc-related genesis is supported by chemical characteristics distinct from those of volcanic-plutonic rocks formed at mid-ocean ridges. The Mariana Trench slope samples resemble ophiolite series assemblages characterized by the scarcity of olivine, the abundance and early crystallization of orthopyroxene, Cr-rich spinels (e.g., $Cr/(Cr + Al) > 0.6$) and the association with Si-saturated volcanic rocks depleted in elements such as Ti, Zr, and Y. The Mariana Trench samples would be considered an ophiolite series if exposed on land; the petrologic evidence indicates that it was derived from an island arc rather than from a mid-ocean ridge.

Introduction

Rocks collected from the inner slope of the Mariana Trench (Figure 1) on Scripps Institution of Oceanography Expedition MARIANA include all of the rock types considered typical of ophiolite assemblages. The Mariana Trench and other western Pacific trenches, such as the Tonga, New Hebrides, Philippine, Palau, and Yap trenches, lack evidence for accreted sedimentary prisms, melange or low-temperature (T) and high-pressure (P) metamorphic rocks. These trench slopes are characterized mainly by mafic and ultramafic igneous rocks and to a lesser extent by metamorphic rocks ranging from low-T and low-P to high-T and moderate-P assemblages. The trenches listed are in areas

where there is very minor sediment supply. The igneous rocks exposed on their inner slopes may have been derived from a variety of sources: (1) the initial crust/mantle cross section, ruptured during initiation of subduction; (2) accreted fragments of igneous rock derived from the subducted plate; (3) allochthonous material brought to the trench slope by strike-slip faults; or (4) material exposed by tectonic erosion [*Hussong et al.*, 1978; *Hawkins et al.*, 1979; *Bloomer*, 1982]. There is a striking similarity between the Mariana Trench slope samples and rocks of many ophiolites [*Dietrich et al.*, 1978; *Hussong et al.*, 1978; *Natland and Tarney*, 1981; *Bloomer*, 1982]; there is little doubt that this trench slope assemblage would be considered an ophiolite if exposed on land. We have attempted to deter-

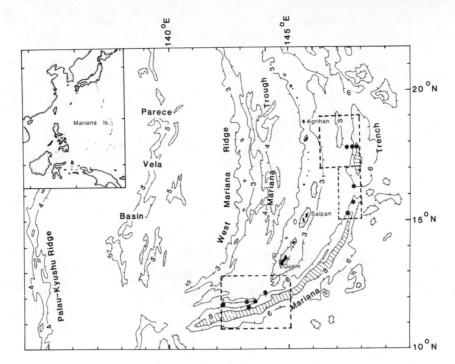

Fig. 1. Location map for Mariana region; dots are *Mariana* dredge locations; areas outlined are areas of maps in Figures 2, 3, and 4.

mine the site of origin of the various rock types collected from the Mariana Trench to see if they represent parts of an originally coherent oceanic crust-mantle series or if they are a random collection of rocks from diverse sources assembled on the trench slope during the subduction process. Understanding the origin of the rocks from the Mariana Trench should offer insight to the origin of some ophiolite suites.

Dredging and drilling in the forearc and inner slope of the Mariana Trench have shown that most of the trench slope is composed of plutonic and volcanic rocks [*Fisher*, 1974; *Hawkins et al.*, 1979; *Dietrich et al.*, 1978; *Hussong et al.*, 1978]. The volcanic rocks include boninites, basalts, andesites, and dacites. Most of these have been interpreted as members of island arc tholeiite or boninite series [*Dietrich et al.*, 1978; *Beccaluva et al.*, 1978; *Hawkins et al.*, 1979; *Bloomer and Hawkins*, 1980; *Meijer*, 1980; *Bloomer*, 1981, 1982].

The presence of plutonic rocks and arc-derived volcanic rocks on the inner trench slope poses problems concerning the petrologic and structural evolution of the trench. Are these plutonic rocks genetically related to the arc-volcanic rocks or do they have another origin? Did the igneous series form on or near the trench wall or are they there because of either tectonic accretion or tectonic erosion? The stratigraphic data from DSDP holes in the forearc show that the volcanic rocks are pre-Late Eocene. This makes them contemporaneous with, or older than, the oldest rocks in the region on the Palau-Kyushu Ridge [*Ingle*, 1975;

Ellis, 1981]. We know of no magma-generating process that would lead to arc volcanism within 50 km of a trench axis in a subduction zone. This, plus the apparent age of the arc-volcanic rocks in the Mariana forearc, supports the interpretation that there has been tectonic erosion of a significant volume of forearc material since Eocene time [*Hussong et al.*, 1978; *Beccaluva et al.*, 1980; *Bloomer*, 1982].

In this discussion we will focus our attention on the petrology of the gabbro and peridotite associated with the arc-volcanic series on the trench slope. The most important problem to be considered is whether or not the gabbro is cogenetic with the volcanic rocks and to determine if the peridotite is genetically related to the gabbroic and arc-volcanic units. If there is a genetic relationship between these rock types, then we may have sampled parts of a disrupted volcanic island arc; this has important implications to the structure of the trench slope and the forearc. If the plutonic and volcanic rocks are related, their chemical characteristics should shed light on shallow-level fractionation processes in the magmatic evolution of arcs.

If the Mariana Trench suite is a random collection of rock types, then we may have a mixture of arc, seamount, and deep-sea floor rocks. The arc-volcanic rocks could be associated with gabbro and peridotite either accreted to the trench wall or representing parts of the crust/mantle substrate upon which the arc rocks were erupted. In one case we could have ophiolite series rocks formed by accretion, obduction, or both or, alternatively, we could have an in situ ophiolite in the sense proposed by *Jakes and*

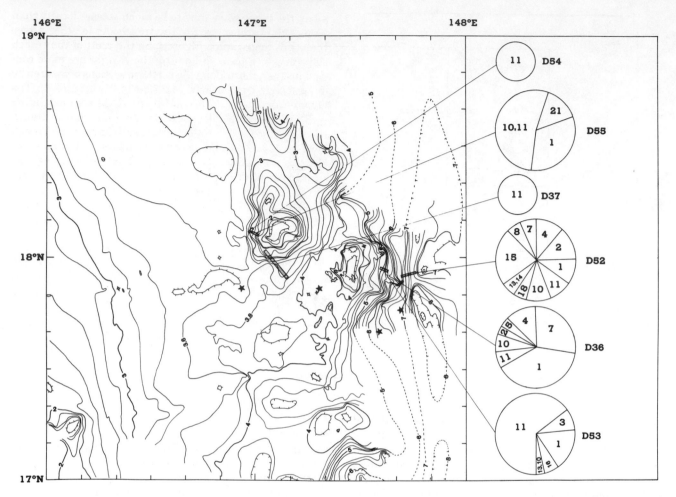

Fig. 2.   Dredge locations and volume percent of rock types, stars are DSDP sites; 1 = serpentinized ultramafic rocks; 2 = gabbros; 3 = diabase; 4 = uniform basalt; 5 = pillow basalt; 6 = glassy basalt; 7 = vesicular basalt; 8 = diorite; 9 = tonalite, granodiorite; 10 = breccia; 11 = siltstone; 12 = volcanic sandstone; 13 = chert; 14 = oolitic limestone; 15 = altered volcanics; 16 = 'melange'; 17 = alteration products; 18 = miscellaneous volcanics; 19 = varved siltstone; 20 = cumulate amphibolite; 21 = Mn-oxide crusts; depth contours in kilometers; dashed contours from unpublished bathymetric chart of Mammerickx.

Gill [1970] and as suggested by *Hawkins and Evans* [1982] for the Zambales, Luzon ophiolite.

Sample Locations

Our samples come from dredge sites on the inner trench slope near 16° and 18°N and from the trench segment that extends to the west of Guam at the southern end of the Mariana Trench. At present, the northern area (16°–18°N) is in a zone where the Pacific plate is being subducted at a high angle to the trench axis. The southern area is located at the zone where the vector of plate convergence is more nearly parallel to the trench and where there is a strong right lateral strike-slip component of motion. Our sample sites are shown in Figures 1–4.

Samples from the lower and mid-slopes of the trench slope are mostly angular blocks and fragments of ultra-

mafic and mafic igneous rocks that range from a centimeter to more than 0.5 m in maximum dimension. They are usually slightly weathered on all sides, sometimes with a thin Mn-oxide coating. Most of these probably came from large talus piles on the trench slope rather than from outcrops. Higher on the trench slope we recovered serpentinite, basalt, and other volcanic rocks, and massive blocks of tuffaceous siltstone. The sedimentary rocks are moderately well indurated light tan to brown volcanoclastic siltstones composed of varying amounts of glass shards, quartz, clinopyroxene, chlorite, and calcium carbonate. Microfossils are rare, but two samples (D37, D24) with discoasters (*Sphenolithus heteromorphus, Cyclic argolithus* (D. Bukry, personal communication, 1981)) indicate mid-Miocene ages. DSDP site 60 found that a large part of the sediment apron east of Guam was composed of sim-

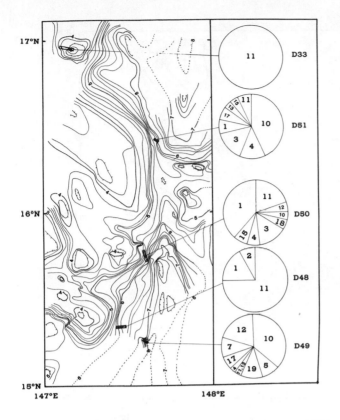

Fig. 3.   Dredge locations and volume percent rock types, 16°N, symbols as in Figure 2.

ilar Miocene volcaniclastics [*Heezen et al.*, 1971]. The glass shards in the siltstones contain about 54–56% SiO_2 based on their refractive indices [*Evans and Hawkins*, 1979].

The northern part of the forearc (15°–18°N, Figures 2 and 3) has a chain of prominent bathymetric highs along the trench slope break that are roughly conical and often slightly elongate subparallel to the trench. They are not continuous features, and, when absent, the trench slope break is marked only by the change in slope. At 18°N there is also a large bathymetric high midway up the inner trench slope. The highs on the trench slope break tend to dam the sediments from the arc. The forearc to the west is a relatively smooth surface, sloping upwards toward the arc, disturbed only by a few shallow depressions and low highs. The trench slope is cut by a number of valleys and canyons roughly perpendicular to the trench axis; these canyons serve as conduits for material that bypasses the trench slope highs and is deposited further down the slope. The trench shoals to depths of nearly 5 km at 16°N where a large seamount chain enters the trench and to 7 km at 18°N where a single seamount blocks the axis.

The southern section of the trench trends westerly, nearly parallel to the plate motion vector, and is dominated today by strike-slip motion. The average depth of the trench axis is significantly deeper here than along the northern section and includes the 10,915 m Challenger Deep. The slopes

along the inner slope tend to be much steeper locally than along the northern part of the trench. As in other areas, there are topographic highs along the crest of the trench slope break (Figure 4). Some of these highs are more continuous then those along the northern sections of the trench; one extends from nearly 142°E to 143°E (Figure 4). The Guam-Saipan frontal arc forms the shoal area extending south at 144°30′E. The west trending part of the trench truncates the forearc, the active arc, and the Mariana Trough back arc basin. As in the northern parts of the trench slope, there are a number of canyons and valleys running perpendicular to the trench axis. However, unlike the northern sections of the trench, there appear to be smaller ridges and depressions subparallel to the trench axis in several parts of the trench slope as well as on the trench slope break (Figure 4).

Gabbros and serpentinized ultramafic rocks occurred in 10 of the 19 dredge collections from the inner slope of the trench (Figures 2–4). Similar rocks have been dredged along the southern section of the trench [*Fisher*, 1974; *Skornyakova and Lipkina*, 1976; *Savelyeva et al.*, 1980; *Dietrich et al.*, 1978]. It has been proposed that some of the ultramafic rocks were emplaced as serpentinite diapirs, which rose along deep faults in the trench slope, and that they dragged slivers of gabbro and volcanic rocks with them [*Evans and Hawkins*, 1979; *Bloomer*, 1981]. Serpentinized ultramafic rocks are an important trench slope component as they are found from the bottom of the trench (8500 m in the north) to depths as shallow as 1200 m. The volumetric proportions of the various rocks types and their stratigraphic relations are shown in Figures 2–4. We distinguished three types of gabbroic or ultramafic rocks in the dredge collections: (1) serpentinized noncumulate ultramafic rocks including both harzburgites and lherzolites; (2) cumulate ultramafic rocks including orthopyroxenites, websterites (OPX-CPX), and harzburgites; (3) cumulate textured gabbros (dominant type) and massive gabbros; their mineralogy is dominated by orthopyroxene-clinopyroxene-plagioclase.

Petrology and Chemistry of Gabbroic and Ultramafic Rocks

Noncumulate Ultramafic Rocks

The majority of the ultramafic rocks in the dredge hauls are serpentinized harzburgite, lherzolite, or dunite with noncumulus textures. Most are serpentinized to some extent, and many are nearly completely replaced by serpentine. We refer to rocks as harzburgites or lherzolites on the basis of their relict mineralogy even when they are serpentinized. In cases of extreme serpentinization, they can be referred to as only as serpentinites.

The harzburgites contain olivine (OL), orthopyroxene (OPX), spinel (SP), serpentine, and secondary Fe-oxides, with minor amounts of talc or tremolite. Modal analyses for some of the least altered samples are listed in Table 1.

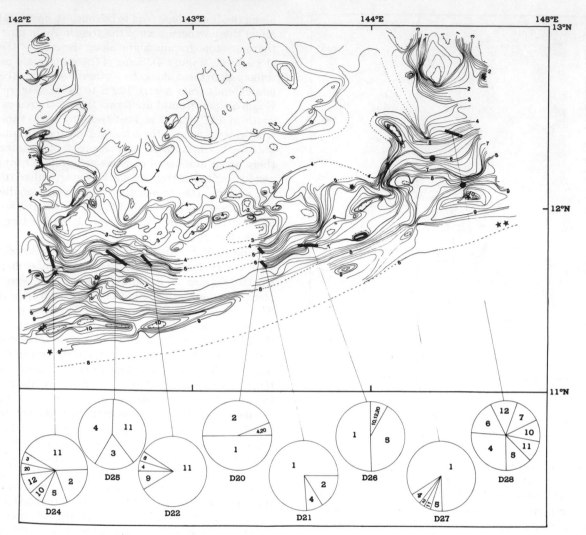

Fig. 4. Dredge locations and volume percent of rocks types, southern section of trench; symbols as in Figure 2 except stars which are dredges by R. L. Fisher, large circles mark dredges by R/V *Mendeleev*.

TABLE 1. Modal Analyses for Ultramafic Rocks From the Inner Slope of the Mariana Trench, Volume Percent

	Olivine	Ortho-pyroxene	Clino-pyroxene	Spinel	Serpentine	Amphibole	Talc	Bastite	Brown/iso-tropic patches	Plagio-clase	Veins	Other
21-1*	—	80.8	0.9	—	—	10.2	—	—	—	1.4	7.9	—
26-4*	6.9	44.0	—	2.0	1.6	26.0	19.3	—	—	—	0.2	—
52-13*	32.6	29.7	8.6	0.2	—	21.3†	—	—	—	3.5	—	—
20-13	4.4	4.9	—	0.3	83.5	—	—	7.0	—	—	—	—
20-25	60.8	—	12.5	1.8	13.1	9.7	.1	—	—	2.1	—	—
26-2A	33.3	5.0	—	0.1	51.6	3.7	4.2	1.3	—	—	0.8	—
26-12	36.1	—	—	0.4	52.9	9.7	0.8	—	—	—	0.1	—
27-1	29.7	14.2	3.8	0.7	44.2	0.3	0.1	0.7	6.4	—	—	—
27-6	12.6	9.2	2.9	0.6	65.8	0.1	—	3.0	5.5	0.4	—	—
27-20	31.7	—	—	0.8	46.9	12.5	5.3	—	—	—	1.4	1.5
36-17A	16.3	0.9	9.8	0.8	65.7	0.3	0.4	0.2	5.7	—	—	—
36-44	16.7	0.1	1.5	0.6	75.6	—	—	0.4	5.1	—	—	—
51-4	—	85.5	7.3	1.4	—	—	—	—	—	—	5.8	—
51-23	9.9	5.2	—	0.5	77.8	1.6	—	3.6	—	—	1.3	—
55-2	22.5	9.2	0.3	1.0	67.0	—	—	—	—	—	—	—

*Cumulate ultramafics.
†6.1 primary brown amphibole, 15.2 secondary amphibole.

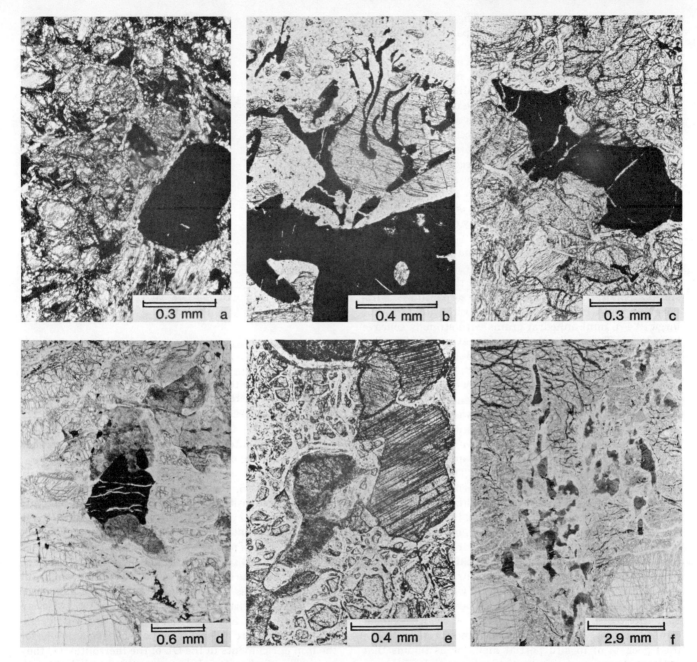

Fig. 5. Spinel habits and textures in ultramafic rocks. (*a*) Rounded spinel, MARA D26-12, serpentinized dunite; (*b*) symplectic spinel intergrown with OPX, MARA D20-13, serpentinite; (*c*) 'holly-leaf' spinel, MARA D27-1, serpentinized harzburgite; (*d*) pull-apart fractures in spinel associated with Ca-Al rich isotropic spot, D27-6, serpentinized CPX harzburgite; (*e*) anhedral CPX with isotropic spot, MARA D36-17, serpentinized lherzolite; (*f*) elongate, subparallel isotropic spots, MARA D27-6.

The serpentine is typically lizardite, less commonly antigorite. The olivine remnants are small grains (0.2–0.3 mm) in a meshwork of serpentine, but optically continous fragments define the boundaries of the original grains that range from 0.2 to 0.3 mm in diameter. In some cases the olivine exhibits a pronounced flattening and may have well-developed kink bands indicative of fairly high tem-

perature deformation [*Carter and AvéLallement,* 1970]. Samples from the southern section of the trench, where convergence is highly oblique, commonly show cataclastic textures and in some cases appear to have been recrystallized to talc-tremolite schists. This extreme deformation and recrystallization is, however, restricted to rocks from the southern boundary. Samples north of 13° show only the high temperature deformation. The orthopyroxene is typically larger than the olivine and is partly or completely replaced by serpentine. The OPX may have thin lamellae of exsolved clinopyroxene (CPX), which are commonly bent or deformed. Where altered, the OPX is replaced by serpentine, talc, and fibrous amphibole. Mildly serpentinized grains tend to have irregular, concave outlines that may be a result of partial melting. Spinels are generally reddish-brown magnesiochromites and occur in many of the habits outlined by *Mercier and Nicolas* [1975] (e.g., euhedral, rounded, holly leaf, and symplectically intergrown with OPX) (Figures 5a–5d). These habits and the variable sizes of the olivine grains may reflect differences in the degree of subsolidus recrystallization of these rocks. Regardless of form, the spinels commonly exhibit pull apart fractures.

Several of the ultramafic rocks from the trench slope include abundant clinopyroxene and can be classified as lherzolites. The OL, OPX, and SP in the lherzolites are similar in most respects to those in the serpentinized harzburgites. The CPX (diopside and diopsidic augite) forms large (0.4–6 mm) anhedral grains with strongly concave boundaries. They typically have large exsolution lamellae of OPX; some may be serpentinized and some have diallage textures (Figure 5e). The CPX seems to be more resistant to alteration and commonly is the only original phase remaining in an otherwise totally serpentinized rock. The CPX occurs as individual grains as well as in glomerocrysts. Most commonly the CPX is associated with elongate, subparallel, irregular dark-brown isotropic patches with high Ca, Al, and Fe contents that appear to be altered plagioclase (Table 4c and Figure 5f). The better crystallized patches contain grains of hydrogarnet. In some cases, small grains of olivine rim these patches, but, more commonly, colorless serpentine rims them. Both the CPX and the isotropic patches occur separately, but typically they are found together (Figure 5e). Similar schlieren of OL, CPX, and PLAG, in which the felsic patches are rodingitized, have been described from the Othris ophiolite [*Menzies and Allen*, 1974]. Segregations of CPX, OPX, PLAG, and, in some cases, spinel have also been found in other ophiolites [*Dick*, 1977; *George*, 1978; C. A. Evans, personal communication, 1981) and have been interpreted as solidified pockets of in situ partial melts or as liquids that infiltrated a harzburgitic host. *Nicolas et al.* [1980] have suggested that these segregations and their host should be referred to as 'impregnated peridotites' and represent an addition to residual and cumulate varieties of ultramafic rocks. Most of the lherzolites from the Mariana Trench slope appear to be examples of 'impregnated peridotites.'

Bulk rock analyses of these rocks are not particularly useful. The high Fe^{+3}/Fe^{+2} values and very high H_2O^+ contents (Table 3, column 11) reflect the extensive serpentinization. The chemistry of the remnant minerals offers the best chance for reconstructing original rock compositions.

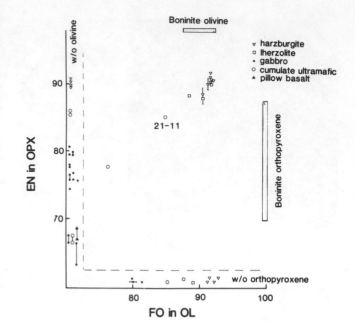

Fig. 6. En composition in orthopyroxene plotted versus Fo in coexisting olivine; stippled bars indicate range of mineral compositions in boninite series rocks from the trench slope; compositions bounded by the dotted line and labeled w/o olivine and w/o orthopyroxene are for samples containing only orthopyroxene or only olivine, respectively.

The noncumulate ultramafic rocks have very restricted ranges of mineral compositions. Olivine ranges from $Fo_{91.2}$ to $Fo_{92.9}$ and orthopyroxene from En_{90} to En_{92}, very rarely to En_{88} (Figure 6, Table 4a). These restricted ranges of refractory compositions are similar to those of tectonite harzburgites in ophiolites and are consistent with the interpretation that these rocks are residual materials left after an episode of partial melting [*Menzies*, 1977; *England and Davies*, 1973]. The CPX in the lherzolites is diopside or diopsidic augite (Figure 7) with Cr_2O_3 contents of about 1% and Na_2O less than 0.5% (Table 4b). The lherzolites tend to have slightly more fayalitic olivine ($Fo_{87.6–90.5}$) than the harzburgites, although there is some overlap in composition. There are at least three possible explanations for the iron enrichments in the OL of the lherzolites: (1) they are not residual materials but are crystal cumulates with interstitial liquid; (2) they represent less depleted mantle than the harzburgites and the CPX is primary; or (3) they were originally similar to the harzburgites, but reequilibration with the trapped liquids has produced slightly more iron-rich compositions. The first explanation might be true for one or two of the samples which may be cumulate dunites but most have textures very similar to the harzburgites and are more likely to be some kind of residual material. The other two possibilities are more difficult to distinguish. The diopsides have similar Cr_2O_3 but lower Na_2O than diopsides in peridotitic xenoliths from kimber-

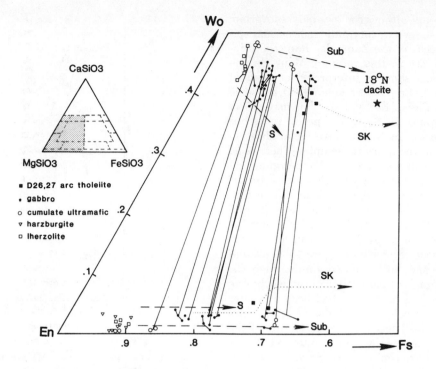

Fig. 7. Pyroxene quadrilateral, short lines connect compositions in individual samples, tie lines connect co-existing low and high Ca pyroxenes; SK is Skaergaard trend from *Nwe* [1976]; S and Sub are solidus and subsolidus trends for gabbros in Marum ophiolite, *Jacques* [1981].

lites and alkali basalts suites (in which Na_2O ranges from 0.7 to 3.3% [*Carswell et al.*, 1979; *Nixon and Boyd*, 1979; *Jagoutz et al.*, 1979]). They are similar to the diopsides in the basal cumulates of the Oman ophiolite (0.8–1.2% Cr_2O_3; 0.12–0.25% Na_2O [*Pallister and Hopson*, 1981]) and to CPX in oceanic peridotites (0.11–0.13% Cr_2O_3; 0.12–0.25% Na_2O [*Hodges and Papike*, 1976]). The CPX in the Mariana lherzolites is usually associated with the altered plagioclase patches, which suggests that the two minerals crystallized together from small pockets of melt. The spinels have a wide range of Cr/Al ratios, and their Al_2O_3 content varies sympathetically with that in the coexisting orthopyroxene (Figure 8). The Al_2O_3 of OPX in spinel peridotites is relatively insensitive to pressure [*Presnall*, 1976; *Danckworth and Newton*, 1978], and this variation in composition may reflect varying degrees of partial melting and depletion of the ultramafic rocks [*Dick*, 1977]. It is important to note that some of the harzburgites have pyroxenes and spinels with Al_2O_3 as high as those in the lherzolites, implying that they are not fundamentally more depleted than the lherzolites. This is consistent with the interpretation of the CPX in the lherzolites as having crystallized from trapped liquid and not being primary CPX in an undepleted host. The most likely explanation for the small iron enrichments in the olivine of the lherzolites is that the CPX- and Ca-rich spots are trapped liquid that has reequilibrated with the host rock.

Nearly all of the noncumulate ultramafic rocks contain some unaltered chrome spinels. These spinels are sensitive indicators of pressure, temperature, and compositional differences in their host rocks [*Irvine*, 1967]. Spinels from both types of noncumulate ultramafic rocks and from the cumulate ultramafic rocks are plotted in Figure 9 (compositions in Table 4c). The spinels in the noncumulates are clearly distinguished from the cumulate spinels by

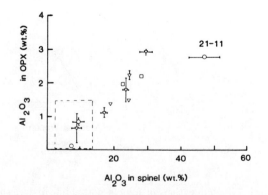

Fig. 8. Al contents in coexisting spinels and OPX, bars indicate range of compositions in individual samples, dashed area encloses boninite compositions, symbols as in Figure 8.

their lower Fe^{+3} contents, although a few analyses overlap. This distinction is also seen in cumulate and residual spinels in ultramafic rocks of the Zambales Range, Philippines [*Hawkins and Evans,* this volume]. Many chrome spinels in the Mariana samples interpreted as residual materials fall within the Zambales residual field, but some also occur to slightly higher Fe^{+3} values. This may reflect a fundamental difference in the two types of rocks but could well be an effect of more extensive reequilibration with Fe-rich liquids in the Mariana samples. There is no clear separation between the spinels of harzburgites and lherzolites. The spinels have an antithetic variation of Cr/Cr + Al and Mg/Mg + Fe^{+2} (Figure 10) similar to that reported for ocean floor basalts and peridotites [*Sigurdsson and Schilling,* 1976; *Dick and Bullen,* 1982], but the Cr/Cr + Al ratios generally are higher than for chrome spinels from ocean floor peridotites (Figure 10) [*Dick and Bullen,* 1982]. Dick and Bullen noted similar high Cr/Cr + Al in spinels from the peridotites of some ophiolite complexes and suggested that there had been a greater degree of melting in the ophiolitic peridotitic suites. Similar Cr-rich spinels have been noted in peridotites of the Zambales Range, Philippines, which is interpreted as a remnant of an island arc complex [*Evans and Hawkins,* 1981; *Hawkins and Evans,* this volume]. If these interpretations are correct, they imply that the noncumulate ultramafic rocks from the Mariana Trench represent very depleted mantle material and that they are similar to ultramafic rocks associated with island arcs.

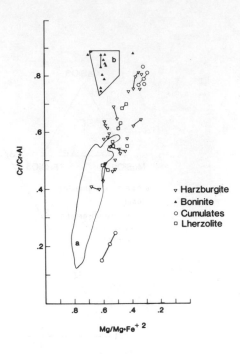

Fig. 10. Cr number versus Mg number for spinels in ultramafic rocks, lines connect compositions in individual samples, field a for spinels in mid-ocean ridge peridotites and field b for boninite series rocks [*Dick and Bullen,* 1982].

Cumulate Gabbros and Ultramafic Rocks

Gabbros were collected at eight sites on the inner slope of the trench (Figures 2–4). We found both massive gabbros with hypidiomorphic-granular textures and cumulate gabbros with heteradcumulate textures. The cumulate gabbros are the much more common of the two types. The massive gabbros are PLAG-CPX rocks; less common are massive PLAG-OPX-CPX gabbros which grade into microgabbros and diabases. Green Ca-amphibole has partially or totally replaced the clinopyroxenes in many samples. These secondary amphiboles are pargasitic or edenitic hornblendes [*Leake,* 1978].

Most of the cumulate gabbros consist of PLAG-CPX-OPX. They have xenomorphic-granular or heteradcumulate textures [*Wager and Brown,* 1967], with a weak to well-developed igneous lamination defined by the orientation of the plagioclase and pyroxene laths (Figure 11a). The orthopyroxene abundance is high and in some cases nearly equal to clinopyroxene (Table 2). Most of the cumulate gabbros are classified as CPX norites or OPX gabbros [*Streckeisen,* 1976]. The OPX forms the largest crystals in most samples and in most cases clearly crystallized before or simultaneously with the CPX and PLAG. The orthopyroxene is primary and has not inverted from pigeonite. Exsolution lamellae are not common. The cumulate gabbros generally are much less altered than the

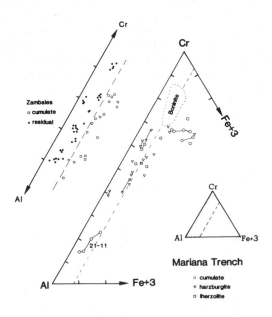

Fig. 9. Trivalent cations in spinels, dotted line divides spinels in cumulate rocks from those in tectonized residual ultramafics in the Zambales Range ophiolite [inset, *Hawkins and Evans,* this volume].

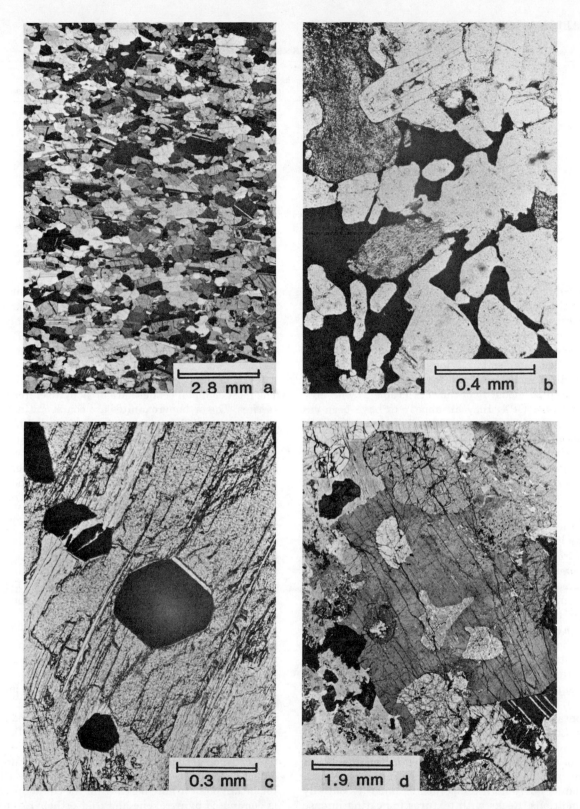

Fig. 11.   Photomicrographs of textures and cumulate rocks and gabbros. (a) Cumulate OPX gabbro, note heteradcumulate textures and orientation of PLAG and CPX, MARA D52-20; (b) Fe-rich gabbro, poikilitic magnetite enclosing PLAG (clear) and CPX (brown), MAR D52-1; (c) cumulate orthopyroxenite, large OPX altered to talc-tremolite, with included Cr-spinel, MARA D26-12; (d) cumulate harzburgite, rounded and irregular resorbed OL poikilitically enclosed in OPX, minor brown amphibole in top left and PLAG in bottom right, MARA D52-13.

TABLE 2. Modal Analyses for Gabbroic Rocks From the Inner Slope, Volume Percent

	Orthopyroxene	Clinopyroxene	Plagioclase	Olivine	Oxides	Secondary Amphiboles	Uralite	Talc/ Amphibole After OPX	Quartz	Alteration	Other
24-45	—	36.5	37.2	23.1	1.1	—	—	—	—	2.1	—
36-5	10.9	34.2	53.1	—	—	—	—	—	—	1.9	—
36-61	30.9	28.2	39.6	—	—	0.7	—	—	—	0.6	—
48-12	6.4	—	61.0	—	—	25.0	—	—	4.2	3.3	—
51-3	5.2	36.1	51.1	0.8	—	—	0.1	—	—	0.2	0.8
51-18m	0.1	48.3	37.0	—	0.1	2.1	—	10.2	—	2.3	—
51-26	16.2	54.5	14.4	—	—	2.5	1.1	—	—	11.3	—
52-1a	—	9.9	67.2	—	8.6	5.0	—	—	—	9.3	—
52-1b	12.0	13.9	55.1	—	0.4	16.6	2.0	—	—	—	—
52-4	12.4	21.7	53.6	—	—	5.7	5.5	—	—	1.1	—
52-14	1.8	15.0	49.4	—	—	24.9	—	6.3	—	6.3	—
52-20	9.6	30.4	48.3	—	—	—	2.8	—	—	8.9	—
52-23	11.7	38.1	47.0	—	1.8	—	0.8	—	—	0.8	—
52-24	11.2	19.6	33.9	—	—	21.5	3.3	—	—	0.5	—
52-38	—	19.0	50.1	28.9	—	—	—	—	—	2.0	—

massive gabbros. Talc and serpentine replace OPX; and CPX is uralitized or partly replaced by amphibole, chlorite, and calcite.

Several varieties of gabbros are represented by only a very few samples. MARA D48-12 is an anorthositic norite. There are five samples of olivine gabbros in the several hundred kilograms of material we collected. Olivine constitutes up to 25 modal % of the rock (Table 2) with 35% PLAG and 35% CPX; they all appear to have been cumulate phases. Magnetite and ilmenite occur as interstitial minerals in two Fe-rich gabbros (Figure 11b). These oxide-bearing samples occur in sharp contact with two pyroxene gabbros, and it is not clear whether they are cumulates or late-stage liquids intrusive into the earlier cumulates. The mineral chemistry suggests the latter.

One of the remarkable characteristics of the cumulate rocks is the general lack of any intercumulus materials. A similar lack of intercumulus phases has been noted in samples from DSDP site 334 on the mid-Atlantic Ridge [Hodges and Papike, 1976], some ophiolites [Jacques, 1981], and layered intrusions [Wager and Brown, 1967]. These textures imply extensive postcumulus growth and equilibrium of the 'trapped' liquid with the overlying magma resevoir presumably through diffusion. It is possible that some of the anhedral textures are due to subsolidus recrystallization, but the persistence of laminated textures suggests that most of it is due to adcumulus growth.

Some samples have fine-scale layering due to rapid changes in grain size between parts of the same rock and sharp contacts between rocks with different mineral proportions. Some of these contacts may be intrusive, but others seem to reflect primary layering.

Many of the gabbroic samples have metamorphic textures and mineral assemblages indicating some recrystallization at fairly high temperatures under essentially static conditions. Rocks with textures indicating intense shearing are rare except in samples from the trench slope southwest of Guam. There, some gabbros have recrystallized cataclastic textures and some have been recrystallized to amphibole-bearing gabbroic gneisses. These strongly sheared textures probably occurred near solidus temperatures possibly during rifting of the West Mariana Ridge. The cataclastic textures may be due to the highly oblique convergence along this section of the trench. Rocks with these textures have not been found in the trench north of Guam.

A few of the ultramafic rocks have cumulate textures. The most common of these are orthopyroxenites and websterites. The orthopyroxenites are coarse grained (Figure 11c), and the OPX commonly have altered clinopyroxene exsolution lamellae. These lamellae are typically slightly warped, and the host grains may be broken and granulated around their margins. Alteration to talc, serpentine, tremolite, and other fibrous amphiboles is common. Orthopyroxenites from the southern section of the trench may be intensely sheared and appear to grade into talc-tremolite schists. Euhedral to slightly rounded magnesiochromites are commonly included in the orthopyroxenes (Figure 11c). The websterites are OPX-CPX cumulates, many of which are severely altered to talc and fibrous amphibole. A cumulate harzburgite (MARA D52-13) has rounded and partly resorbed olivine which is poikilitically enclosed in orthopyroxene, with minor brown pargasitic hornblende (Table 4b) and plagioclase (Figure 11d). Texturally, this intercumulus brown amphibole appears to be one of the few primary igneous amphiboles in the cumulate rocks.

The crystallization sequence inferred for the majority of the cumulate gabbros and cumulate ultramafic rocks is OL, (SP), OPX, CPX, PLAG, oxide (OX), producing cumulate assemblages of OL + SP, SP + OPX, OPX + CPX, OPX + CPX + PLAG ± OX. There are some exceptions to this simple sequence; for example, the rare olivine gabbros record a sequence from OL to PLAG + CPX. The important point to note here is that the cumulate sequence is dominated by pyroxene and that orthopyroxene is volumetrically important and crystallizes early in the sequence indicating that silica saturated magmas were probably the source for these cumulate rocks.

The bulk rock chemistry of the cumulate rocks (Table

TABLE 3. Bulk Rock Chemical Analyses of Selected Samples, Na$_2$O and Cr by Atomic Absorption; Fe$_2$O$_3$ by titration; H$_2$O Gravimetrically; CO$_2$ Manometrically; All Others by X ray flourescence; Methods and Estimates of Accuracy and Precision in *Bloomer et al.* [1982]

	1	2	3	4	5	6	7	8	9	10	11
	27-41	52-1s	24-45	52-38	51-3	48-12	36-5	52-23	52-13	20-8	27-1
SiO$_2$	49.50	47.98	47.29	47.26	48.69	52.68	47.41	49.06	46.60	47.56	42.64
TiO$_2$	0.72	1.46	0.08	0.04	0.63	0.05	0.11	0.22	0.29	0.30	0.08
Al$_2$O$_3$	16.66	17.61	13.41	19.13	15.75	19.00	17.17	14.83	3.74	7.66	2.78
Fe$_2$O$_3$	1.89	8.31	3.60	3.44	1.47	1.03	1.56	2.86	4.17	4.14	3.86
FeO	5.99	6.23	4.35	3.11	3.41	5.79	3.00	6.28	10.64	5.96	4.29
MnO	0.14	0.23	0.14	0.10	0.07	0.12	0.09	0.19	0.25	0.14	0.13
MgO	7.36	4.09	14.96	13.13	11.64	6.72	9.93	10.16	24.68	18.24	36.84
CaO	12.07	8.91	13.48	13.00	14.93	13.02	15.19	14.02	5.97	15.04	2.87
Na$_2$O	3.12	3.52	0.45	0.82	0.82	0.31	1.05	0.99	0.23	0.62	0.10
K$_2$O	0.20	0.38	0.02	0.17	0.14	0.12	0.18	0.15	0.13	0.03	0.01
H$_2$O$^+$	1.78	1.51	2.04	2.93	2.21	0.84	1.60	1.13	1.76	1.12	5.63
CO$_2$	0.04	0.04	0.06	0.02	0.03	0.02	0.03	0.04	0.05	0.14	0.45
P$_2$O$_5$	0.09	0.14	0.04	0.03	0.03	0.04	0.04	0.03	0.03	0.04	0.02
Sum	99.56	100.41	99.92	103.18	99.82	100.28	97.36	99.96	98.54	100.31	100.66
Ni	58	6	424	419	252	109	91	84	855	372	1901
Cr	109	9	1123	351	455	96	232	243	1105	1905	1960
V	243	493	122	49	91	184	110	248	166	—	48
Rb	1	3	1	2	1	1	1	2	1	—	1
Sr	250	218	48	139	106	38	88	406	13	25	82
Ba	38	22	10	4	9	10	—	84	6	13	1
Zr	52	43	6	8	3	16	10	7	25	12	8

Oxides in wt.%, trace elements in ppm.

1, massive gabbro; 2, oxide bearing gabbro; 3,4, olivine gabbro; 5,7,8, two-pyroxene gabbros; 6, anorthositic norite; 9, cumulate harzburgite; 10, olivine websterite; 11, serpentinized harzburgite.

3) shows very low abundances of K$_2$O, P$_2$O$_5$, Rb, Ba, and Zr. This is consistent with the interpretation of these rocks as crystal cumulates with little trapped interstitial liquid. The TiO$_2$ abundance is generally low but varies possibly reflecting changes in the Ti content of the CPX. The variable Zr contents (3–16 ppm in the cumulate gabbros) probably reflect small variations in the amount of intercumulus material. The cumulate rock with the highest Zr (D52-13, 25 ppm) has the largest amount of intercumulus minerals (30% OPX, 6% amphibole).

MARA 27-41, a massive gabbro, and 52-1, a gabbro with interstitial ilmenite, are both near-liquid compositions as opposed to cumulate compositions as evidenced by their high Zr contents. MARA 52-1 is a very fractionated sample similar to arc tholeiites in its low TiO$_2$, low Ni (6 ppm), and low Cr (9 ppm), and high FeO. MARA D27-41 is similar to massive or high-level gabbros commonly found capping the cumulate sequence in ophiolites [*Pallister and Hopson*, 1981; *Salisbury and Christensen*, 1978; *Coleman*, 1977] and probably represents a slightly fractionated liquid composition that crystallized slowly in the upper levels of the magma chamber.

The gabbros as a group show element variations consistent with their postulated cumulate origins. Ni decreases with increasing FeO/MgO ratios, and Ni and Cr decrease sympathetically, supporting the idea of crystallization from an increasingly fractionated magma. The lowest FeO/MgO ratios (0.45) are similar to those in many ocean floor gabbros and to those in the Oman ophiolite

[*Tiezzi and Scott,* 1980; *Smewing,* 1981]. The antithetic behavior of Ni and V is consistent with increasing fractionation in a closed magma chamber and also suggests a tholeiitic magma stem [*Miyashiro and Shido,* 1975]. The cumulate rocks are higher in Mg/Mg + Fe^{+2} and lower in Zr than the volcanic rocks found with them, consistent with their postulated cumulate origin. The massive and Fe-rich gabbros have values of Mg number and Zr in the same range as some of the volcanic rocks.

The cumulate gabbroic and ultramafic rocks have a much wider range in mineral compositions than the noncumulate ultramafic rocks. OL ranges from Fo$_{76}$ to Fo$_{85}$, OPX from En$_{63}$ to En$_{86}$, PLAG from An$_{62}$ to An$_{96}$, and CPX from Wo$_{45}$Fs$_5$En$_{47}$ to Wo$_{45}$Fs$_{15}$En$_{40}$ (Tables 4a, 4b, 4c). There are sympathetic trends of An content in PLAG, Fo in OL, En in OPX, and Mg numbers in CPX (Figures 6, 7, 12, 13), which reflect the same effects of fractionation seen in the variation of Cr, Ni, and V. Olivines of a given Fo value appear to coexist with plagioclase compositions similar to those of mid-ocean ridge gabbros (Figure 12). The trench slope gabbros clearly do not have the high-Fe olivine, high-Ca plagioclase association of the West Mariana Ridge gabbros [*Natland,* 1981]. PLAG compositions vary greatly, even within individual samples, but generally show a correlation with the Mg number of coexisting clinopyroxenes similar to that in gabbros drilled from the Mid-Atlantic Ridge (Figure 13). The massive gabbros have less calcic PLAG with a larger range of compositional zonations than the cumulate PLAG. The anorthositic norite (MARA D48-

TABLE 4a. Representative Analyses of Minerals in Gabbroic and Ultramafic Rocks From the Mariana Trench—Olivine and Orthopyroxene

OLIVINE

	Gabbro 24-45	Cumulate Ultramafic 21-11	52-13	Harzburgite 36-24	51-23	26-2	Lherzolite 27-6	36-9	36-44
SiO_2	39.33	39.44	39.00	40.98	40.77	41.06	41.06	40.40	40.76
TiO_2	.02	0	.04	.02	.04	.01	0	.02	.01
FeO	19.26	14.28	22.64	7.92	8.60	8.27	9.32	11.77	9.49
MnO	.08	.05	.14	.04	.04	.03	.03	.08	.03
MgO	42.99	45.97	40.64	50.65	50.06	51.13	49.86	47.59	49.59
CaO	.01	.04	.08	.04	.04	.04	.03	.04	.04
Cr_2O_3	0	.04	.01	.01	0	0	.05	.03	.01
NiO	.15	.19	.19	.41	.38	.32	.40	.31	.19
SUM	101.84	100.02	102.74	100.07	99.93	100.87	100.76	100.30	100.11

Atoms per 4 oxygens

Si	.989	.988	.988	.996	.995	.992	.997	.997	.996
Ti	0	0	.001	0	.001	0	0	0	0
Fe	.405	.299	.48-	.161	.176	.167	.189	.243	.194
Mn	.002	.001	.003	.001	.001	.001	.001	.002	.001
Mg	1.612	1.717	1.534	1.836	1.822	1.841	1.806	1.751	1.807
Ca	0	.001	.002	.001	.001	.001	.001	.001	.001
Cr	0	.001	0	0	0	0	.001	.001	0
Ni	.003	.003	.004	.008	.007	.006	.008	.006	.004
Fo	79.9	85.2	76.2	91.9	91.2	91.7	90.5	87.8	90.3

ORTHOPYROXENES

	Harzburgites 26-2	36-14	51-23	Lherzolites 27-6	36-44	Cumulate Ultramafics 21-11	48-10	Cumulate 36-61	51-3	gabbros 52-20	48-12
SiO_2	57.65	56.15	56.28	55.25	55.71	54.73	53.91	53.91	55.06	54.53	53.42
TiO_2	0	.01	.02	.24	.09	.02	.08	.14	.14	.14	.06
Al_2O_3	.36	1.52	1.99	2.90	1.98	2.75	1.08	1.25	1.14	1.12	1.08
FeO	5.70	5.20	5.55	6.32	6.29	9.36	20.58	13.54	11.47	13.84	19.86
MnO	.02	.08	.07	.04	.06	.11	.45	.14	.13	.14	.36
MgO	35.90	34.85	34.89	34.02	33.66	31.61	24.90	28.13	30.18	28.70	24.21
CaO	.15	1.26	.65	.65	1.72	.49	1.19	2.13	1.59	1.50	1.36
Cr_2O_3	.12	.66	.54	.68	.59	.18	.14	.12	.07	.05	.04
NiO	.06	.12	.10	.04	.02	.01	.09	.02	.03	.03	.08
SUM	99.96	99.80	100.09	100.13	100.10	99.26	102.44	99.39	99.81	100.05	100.49

Atoms per 6 oxygens

Si	1.981	1.941	1.938	1.910	1.931	1.930	1.948	1.948	1.956	1.954	1.961
Ti	0	0	.001	.006	.002	.001	.002	.004	.004	.004	.002
Al^4	.015	.059	.062	.090	.069	.070	.046	.052	.044	.046	.039
Al^6	0	.003	.019	.028	.012	.044	0	.001	.004	.001	.008
Fe	.164	.150	.160	.183	.182	.276	.622	.409	.341	.415	.610
Mn	.001	.002	.002	.001	.002	.003	.014	.004	.004	.004	.011
Mg	1.839	1.796	1.791	1.753	1.739	1.662	1.341	1.515	1.599	1.533	1.325
Ca	.006	.047	.024	.024	.064	.019	.046	.082	.061	.058	.054
Cr	.003	.018	.015	.019	.016	.005	.004	.003	.002	.001	.001
Ni	.002	.003	.003	.001	.001	0	.003	.001	.001	.001	.002
Wo	.2	2.2	1.2	1.2	3.2	.9	2.3	4.1	3.0	2.8	2.7
Fs	8.2	7.6	8.1	9.3	9.2	14.1	30.9	20.4	17.0	20.7	30.7
En	91.6	90.2	90.7	89.4	87.6	85.0	66.7	75.5	80.0	76.5	66.6

TABLE 4*b*.   Representative Mineral Analyses—Clinopyroxene and Amphibole

| | CLINOPYROXENES | | | | | | | | | | Amphiboles | |
| | l h e r z o l i t e | | | Cumulate Ultramafic | | C u m u l a t e   G a b b r o s | | | | | | |
	36-9	36-17	36-49	48-10	21-11	36-61	51-3	52-20	48-12		52-13	24-15
SiO_2	51.62	52.66	51.22	52.93	52.75	53.59	52.58	52.93	52.65	SiO_2	43.35	53.89
TiO_2	.12	.17	.11	.08	.12	.13	.22	.22	.06	TiO_2	3.59	.25
Al_2O_3	2.84	2.82	2.78	1.04	2.73	.69	1.84	1.61	.82	Al_2O_3	11.06	4.52
FeO	4.14	3.55	3.19	7.69	3.41	5.37	5.92	6.14	9.91	FeO	10.58	7.48
MnO	.16	.03	.13	.24	.10	.23	.21	.14	.27	MnO	.10	.15
MgO	17.77	18.73	16.84	15.60	17.41	16.95	17.99	17.01	14.38	MgO	15.75	20.00
CaO	20.35	21.33	21.40	22.67	23.30	23.10	20.40	22.23	21.07	CaO	11.09	11.48
Na_2O	.14	n.d.	.22	.09	.18	.03	0	.23	.13	Na_2O	2.55	.53
K_2O	.02	n.d.	.01	.01	0	0	.02	.02	0	K_2O	.25	.06
P_2O_5	0	n.d.	.01	0	0	.02	.03	.02	0	P_2O_5	.03	0
NiO	.13	.06	.05	0	.04	.04	.06	.04	.02	NiO	.11	.06
Cr_2O_3	1.12	1.13	1.12	.12	.37	.11	.14	.06	.01	Cr_2O_3	.32	.26
SUM	98.40	100.42	97.09	100.46	100.41	100.26	99.43	100.63	99.33	SUM	98.77	98.61

	Atoms per 6 oxygens										Atoms per 23 oxygens	
Si	1.913	1.906	1.921	1.957	1.918	1.966	1.937	1.938	1.980	Si	6.266	7.489
Ti	.003	.005	.003	.002	.003	.004	.006	.006	.002	Ti	.390	.026
Al^4	.087	.084	.079	.043	.082	.030	.063	.062	.020	Al^4	1.734	.511
Al^6	.037	.036	.043	.002	.035	-	.016	.008	.016	Al^6	.151	.230
Fe	.128	.108	.100	.238	.104	.165	.182	.188	.312	Fe^{+2}	1.199	.805
Mn	.005	.001	.004	.008	.003	.007	.007	.004	.009	Fe^{+3}	.080	.056
Mg	.982	1.012	.942	.860	.944	.927	.988	.929	.806	Mn	.012	.018
Ca	.808	.828	.860	.898	.908	.908	.805	.872	.849	Mg	3.394	4.144
Na	.010	-	.016	.006	.013	.002	0	.016	.009	Ca	1.718	1.709
k	.001	-	0	0	0	0	.001	.001	0	Na	.715	.143
P	0	-	0	0	0	.001	.001	.001	0	K	.046	.011
Ni	.004	.002	.002	0	.001	.001	.002	.001	.001	P	.004	0
Cr	.033	.032	.033	.004	.011	.003	.004	.002	0	Ni	.013	.007
											Cr	.037	.029
Wo	42.2	42.5	45.2	45.0	46.4	45.4	40.8	43.9	43.2			
Fs	6.7	5.5	5.3	11.9	5.3	8.2	9.2	9.4	15.8	$\frac{Mg}{Mg+Fe}^{+2}$	.739	.837
En	51.2	51.9	49.5	43.1	48.3	46.4	50.0	46.7	41.0			

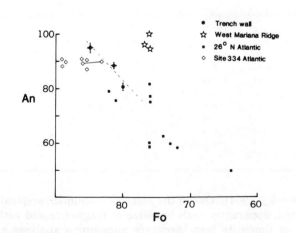

Fig. 12. Plagioclase composition versus composition of coexisting olivines for trench slope, mid-ocean ridge (site 334 [*Hodges and Papike*, 1976]; 26°N [*Tiezzi and Scott* 1980]) and West Mariana Ridge gabbros [*Natland*, 1981].

12) may be more similar to the calcalkaline gabbros found in the West Mariana Ridge and the island of Agrigan; it has An_{96} coexisting with CPX, which has a low Mg number (0.72). Gabbros similar to those at DSDP site 452 on the WMR were found in breccias at DSDP site 461 in the inner trench slope [*Hussong et al.*, 1978; A. Meijer, personal communication, 1981]. When the rocks now exposed on the trench slope were emplaced, some isolated centers in which calcalkaline volcanism and fractionation occurred may have been present in the region.

Pyroxene compositions are plotted in the pyroxene quadrilateral in Figure 7. The CPX and OPX from the cumulate rocks define trends parallel to those defined from other tholeiitic fractionation series (Marum ophiolite and Skaergaard) and plot between the solidus and subsolidus trends defined for the Marum ophiolite [*Jacques*, 1981]. The variation within samples tends to be subparallel to the Wo-En join and may reflect varying degrees of subsolidus reequilibration. None of the pyroxenes are pigeonitic in composition. The apparent clockwise tie-line rotation sug-

TABLE 4c. Representative Mineral Analyses—Spinel, Plagioclase, Fe-Ti Oxide, and Hydrogarnet

SPINELS

	Cumulate Ultramafic		Harzburgite			Lherzolite		
	21-11	26-8	26-2	36-24	51-4	27-6	36-9	36-44
TiO_2	.04	.16	.12	0	.28	.61	.28	.29
Al_2O_3	45.75	6.78	7.24	20.43	15.06	31.18	20.41	23.43
FeO	18.65	23.85	22.74	16.78	19.79	16.17	21.33	18.07
Fe_2O_3	4.47	12.38	16.45	3.20	7.74	5.74	9.10	4.39
MnO	.15	.27	.26	.18	.32	.22	.28	.20
MgO	13.56	6.00	6.65	12.28	9.65	13.76	9.27	11.49
CaO	.02	.02	.05	.02	.01	.04	.03	.02
NiO	.10	.09	.13	.10	.08	.17	.13	.09
Cr_2O_3	18.09	51.83	47.58	49.03	49.06	34.52	41.67	42.85
SUM	100.83	101.37	101.22	102.03	101.99	102.40	102.49	100.83

Atoms per 32 oxygens

Ti	.007	.033	.025	0	.054	.107	.053	.054
Al	12.027	2.188	2.329	5.906	4.536	8.531	6.004	6.791
Fe^{+2}	3.478	5.461	5.189	3.442	4.229	3.139	4.452	3.716
Fe^{+3}	.750	2.550	3.376	.590	1.487	1.002	1.708	.812
Mn	.028	.063	.060	.037	.069	.043	.059	.042
Mg	4.508	2.449	2.705	4.489	3.675	4.761	3.448	4.211
Ca	.005	.006	.015	.005	.003	.010	.008	.005
Ni	.018	.020	.029	.020	.016	.032	.026	.018
Cr	3.190	11.220	10.264	9.508	9.910	6.335	8.222	8.331
$\frac{Cr}{Cr+Al}$	.564	.310	.343	.566	.465	.603	.436	.531
$\frac{Mg}{Mg+Fe+2}$	.210	.837	.815	.617	.686	.426	.578	.551

Isotropic patches/hydrogarnet

	27-6	36-9	36-44
SiO_2	34.23	37.97	29.36
TiO_2	.06	.05	.03
Al_2O_3	21.33	32.74	18.74
FeO	2.52	.35	8.21
MnO	.20	.06	.45
MgO	.60	.76	16.64
CaO	37.35	25.06	10.82
Na_2O	0	.11	0
K_2O	0	0	.02
P_2O_5	.06	0	0
NiO	0	0	0
Cr_2O_3	.06	0	.04
SUM	96.41	97.11	84.31

D52-1, gabbro

	Ilmenite	Magnetite
TiO_2	47.20	2.09
Al_2O_3	.01	1.15
FeO	42.89	29.27
Fe_2O_3	5.91	69.36
MnO	3.71	.28
MgO	.09	.07
CaO	.02	.01
NiO	.01	.01
Cr_2O_3	.05	.04
SUM	99.84	102.01

PLAGIOCLASE

	ultramafic	cumulate		gabbros			massive	gabbros
	21-11	36-61 core	rim	51-3	52-20	48-12	52-1	24-15
SiO_2	43.52	46.40	46.17	45.51	46.65	43.72	51.56	49.47	47.97
Al_2O_3	36.08	33.86	33.57	34.47	33.28	35.34	30.09	32.38	33.49
Na_2O	.36	1.79	1.57	1.67	1.69	.42	4.20	2.99	2.58
CaO	19.34	17.44	17.52	16.94	17.06	19.34	12.05	14.24	15.75
K_2O	.01	.01	0	0	.01	.02	.14	.06	.04
FeO	.11	.51	.57	.41	.56	.18	.58	.71	.23
SUM	99.53	100.01	99.39	99.00	99.26	99.02	98.63	99.84	100.07

Atoms per 32 oxygens

Si	8.096	8.556	8.566	8.462	8.651	8.170	9.488	9.046	8.783
Al	7.914	7.362	7.343	7.557	7.277	7.787	6.529	6.981	7.230
Na	.130	.640	.565	.602	.608	.152	1.499	1.060	.916
Ca	3.855	3.446	3.483	3.375	3.390	3.873	2.376	2.790	3.090
K	.002	-	-	-	.002	.005	.033	.014	.009
Fe	.017	.079	.088	.064	.087	.028	.089	.109	.035
An	96.7	84.3	86.0	84.9	84.8	96.2	61.3	72.4	77.1

gested by the crossing of some of the tie lines has been noted in other gabbroic cumulates and appears to be an effect of subsolidus reequilibration.

The interstitial oxides in some of the late-stage gabbros are ilmenite with magnetite exsolution lamellae (Table 4c). The very pure magnetite lamellae indicate equilibration at very low temperatures (<600°C [*Buddington and Lindsley,* 1964]). One of the samples is submicroscopically zoned, apparently with lamellae of magnetite and rutile in an ilmenite host. Accurate microprobe analyses are difficult, as the lamellae size is on the order of the minimum beam size, 2 μ.

Some of the distinctive features of the Mariana Trench cumulate sequence are more clear in a pseudo-strati-

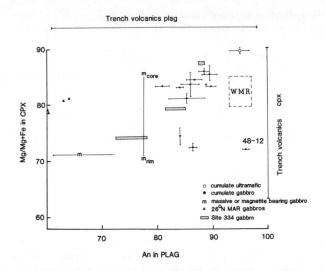

Fig. 13.   Mg number of CPX and An content of coexisting plagioclase in rocks from the inner trench slope; lines indicate range of composition in individual sample, bars at top and right indicate range in mineral compositions in volcanic rocks found with gabbros; site 334 gabbros from *Hodges and Papike* [1976]; WMR range for gabbros from DSDP site 453 [*Natland*, 1981]; 26°N gabbros from *Tiezzi and Scott* [1980].

graphic column (Figure 14). This diagram was constructed by assuming that the more magnesian OPX occurred lower in the section. Mineral compositions were plotted in the vertical sequence based on the Mg number of the coexisting OPX. If no OPX occurred in the sample, the Mg number of the CPX was used to locate the sample in the vertical sequence. This figure is intended to illustrate the range in coexisting mineral compositions and is not meant to imply that the cumulate sections in the Mariana Trench are this simple. The restricted range of olivine crystallization and the very early and persistant appearance of OPX in the cumulate sequence are very clear in this figure. The An content of PLAG and Mg number in CPX vary sympathetically with the En content of OPX, though there are some large deviations particularly in the PLAG compositions. The An contents are fairly constant for much of the section, oscillating around An_{85}.

Parental Magmas and Relation to the Volcanic Rocks

The presence of abundant cumulate gabbros in the trench slope indicates that there are remnants of extensively fractionated magma chambers exposed on the trench slope. This poses two important questions: What was the nature of the magma from which they crystallized, and what was their relation to the volcanic rocks with which they are found? There are two major types of volcanic rocks in the trench slope: (1) boninites and (2) tholeiitic basalts and andesites. Both series have characteristics indicating that they have undergone significant fractional crystallization; the boninitic series was controlled by OL, OPX, CPX, and

PLAG and the tholeiitic series by OL, CPX, PLAG, OPX, and oxides.

The tetrahedral Al contents of the clinopyroxene in the cumulate rocks indicate that they crystallized from a subalkaline magma [*Kuno*, 1960]. The crystallization of two pyroxenes during fractionation is also characteristic of tholeiitic magmas [*Hodges and Papike*, 1976; *Wager and Brown*, 1967; *Pallister and Hopson*, 1981; *Tiezzi and Scott*, 1980]. The V variations in the gabbro bulk rock chemistry are also consistent with their derivation from a tholeiitic magma.

$FeO*/MgO$, Ni, Cr, and TiO_2 contents of the liquid that were in equilibrium with the cumulate olivine and clinopyroxene can be estimated by using partition coefficients. $FeO*/MgO_{solid/liq}$ has been estimated at 0.30 for olivine/liquid [*Roeder and Emslie*, 1970] and that for CPX/liquid at 0.24 [*Stern*, 1979]. There is some debate on how these coefficients change with pressure, temperature, and composition [*Hanson and Langmuir*, 1978; *Bender et al.*, 1978], but for these estimates the average values should suffice. Partition coefficients used for Ni in OL, OPX, and CPX are 10, 4, 2, respectively, and for Cr, 1, 4, 6, respectively [*Irving*, 1978]. Partition coefficients for TiO_2 between mineral and liquid were taken to be 0.3 for CPX and 0.11 for OPX [*Pearce and Norry*, 1978]. The $FeO*/MgO$, Ni, Cr, and TiO_2 values for the liquids calculated to be in equilibrium with the cumulate gabbros and the range for the basaltic and boninitic rocks from the trench slope are presented in Table 5. The $FeO*/MgO$ ratios of the liquids calculated from the OL and CPX are usually in close agreement, indicating they are close to equilibrium compositions. The calculated $FeO/MgO*$ and TiO_2 values indicate that the gabbros crystallized from a range of liquid compositions all of which had low TiO_2. The calculated liquid compositions include most of the range of volcanic rock compositions from the trench slope, although some of the volcanic rocks have lower $FeO*/MgO$ ratios than any

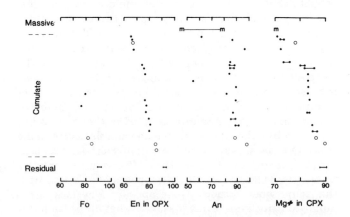

Fig. 14.   'Pseudo-stratigraphic' column for gabbros and ultramafic rocks from the trench slope; compositions of coexisting minerals plotted in vertical sequence based on Mg number of OPX (or of CPX if sample contained no OPX); symbols as in Figure 13.

TABLE 5. Compositions of Liquids Calculated to Have Been in Equilibrium with Gabbros and Residual Ultramafic Rocks From Inner Slope; Also Range in Boninites and Volcanic Rocks From the Inner Slope

	FeO/MgO	Ni (ppm)	Cr (ppm)	TiO$_2$ (wt%)
Liquids in equilibrium with gabbros	0.8–2.05	20–230	90–250	0.37–1.23
Boninites	0.5–1.4	24–380	98–800	0.18–0.4
Forearc volcanics	0.7–3.0	10–150	5–450	0.5–0.9
Liquids in equilibrium with residual ultramafic rocks	0.5–.63	180–260	100–436	0.3–0.5

of the calculated liquids. The FeO*/MgO of those volcanic rocks are, however, similar to compositions calculated to have been in equilibrium with the noncumulate ultramafic rocks. This is consistent with the interpretation of some of the boninites as liquids in equilibrium with an OL-OPX mantle source [*Cameron et al.*, 1978; *Meijer*, 1980; *Bloomer*, 1982; *Hickey and Frey*, 1982].

The boninites are quite fresh, and their compositions can be used for least squares fractional crystallization models. These calculations show that these rocks fractionated from 54 to 68% SiO$_2$ by crystallization of OL, OPX, CPX, and PLAG in approximate proportions of 1:2.5:0.5:0.5. Most of the other volcanic rocks can not be modeled quantitatively because alteration has changed their original chemistry. Despite this alteration, these rocks still show tholeiitic fractionation trends from basalt to dacite that are controlled by the crystallization of OL, CPX, PLAG, and OPX (OPX is less important than in the boninites). The crystallization sequence of the boninites is the same as that deduced for most of the cumulate section, and the abundance of pyroxene as a crystallizing phase in the volcanic rocks suggests a relation to the pyroxene-rich cumulate section.

If the volcanic rocks and gabbroic rocks are genetically related, they should have similar mineral constituents. The ranges in phenocryst mineral compositions in the volcanic rocks are plotted in Figures 6 and 13. The pyroxene and plagioclase compositions for the basaltic rocks overlap the range of mineral compositions in the gabbros. The compositions of OPX in the boninites cover the entire range of those in the gabbros and extend up to those in the noncumulate ultramafic rocks. The olivine of the OL boninites is more magnesian than any sampled in the gabbros but are much like the olivines in the noncumulate ultramafic rocks. Olivine appears to be the liquidus phase in the boninites [*Howard and Stolper*, 1981]. It crystalizes only for a short interval and rapidly disappears by reaction with the liquid when orthopyroxene appears. The lack of cumulates with olivine as magnesian as that in the boninites probably indicates that these very primitive OL boninites underwent only a small amount of fractionation before eruption and thus produced a very small volume of cumulate material.

The similarity between the liquid compositions in equilibrium with the gabbros and the volcanic rocks (including boninites), the overlap in mineral compositions between the volcanic and gabbroic rocks, and the similar crystallization sequences in the two sets of rocks are consistent with the interpretation that the cumulate rocks are cogenetic with the boninitic and basaltic volcanic rocks from the inner trench slope. The relation of the noncumulate ultramafic rocks to the cumulate and volcanic rocks is much less clear, but the data are consistent with an interpretation of these rocks as a residue after an episode of partial melting. The mineral compositions of the cumulate rocks grade into those of the noncumulates, and there are volcanic rocks in the trench slope that have compositions that could have been in equilibrium with the noncumulate ultramafic rocks. These ultramafic rocks are very refractory in composition, as evidenced by the OPX and SP chemistry discussed above, and it is likely that they are the residue from the partial melting event which produced the boninites and volcanic rocks in the trench slope.

Synthesis

A large proportion of the gabbroic rocks dredged from the Mariana Trench had a cumulate origin. There is evidence of extensive fractional crystallization in the associated volcanic rocks, and the bulk rock and mineralogical data for the gabbros are consistent with a cogenetic relation between the two groups of rocks. The depleted ultramafic rocks are most likely unmelted residue from generation of the magma similar to that which produced the cumulate section. The volcanic rocks in the trench slope have island arc type chemistry, and, by implication, the gabbroic plutonic rocks represent the roots of an old (late Eocene) island arc that have been exposed by removal of material from the inner slope of the trench (i.e., by tectonic erosion). The plutonic rocks are remarkably similar over a wide extent of the trench slope, which implies that there may be an extensive layered complex beneath the forearc.

One of the most striking things about the rock samples from the Mariana Trench is their similarity to rocks of ophiolite complexes in that they include residual ultramafic rocks, cumulate ultramafic rocks, cumulate gabbros, diabases, and volcanic rocks [*Coleman*, 1971]. While it is difficult to determine if a dike complex exists, there certainly is abundant diabase, and we interpret the assemblage as a dismembered ophiolite. However, there is strong evidence that the volcanic rocks are island arc volcanic rocks and the gabbroic section appears to be cogenetic with the volcanics. Thus the series is not derived from mid-ocean ridge type crustal rocks.

Miyashiro [1973] suggested that the Troodos ophiolite was formed in an island arc, an idea that generated a great deal of controversy [*Hynes,* 1975; *Miyashiro,* 1975a, b; *Moores,* 1975]. It has become apparent that there are many ophiolites with island-arc-type rocks (including low Ti volcanics resembling boninites) and that arcs must be one of the source regions for ophiolites [*Miyashiro,* 1973; *Beccaluva et al.,* 1978; *Jakes and Gill,* 1970; *Upadhyay and Neale,* 1979; *Hawkins and Batiza,* 1977; *Hawkins,* 1980; *Evans and Hawkins,* 1981; *Hawkins and Evans,* this volume; *Coish and Church,* 1979; *Zouenshain and Kuzmin,* 1978]. The gabbros from the Mariana Trench represent part of an in situ island-arc ophiolite, and a comparison of these gabbros with gabbroic and ultramafic rocks from the ocean floor should help determine how an island arc plutonic complex may differ from a plutonic complex formed at a mid-ocean ridge.

Serpentinized ultramafic rocks have been dredged from many localities in the Atlantic and Indian oceans [*Melson and Thompson,* 1970; *Aumento and Loubat,* 1971; *Engel and Fisher,* 1975; *Bonatti,* 1976; *Bonatti and Honnorez,* 1976; *Bonatti and Hamlyn,* 1978; *Hamlyn and Bonatti,* 1980; *Arai and Fujii,* 1978]. There are far fewer samples from the Pacific [*Krause et al.,* 1964; *Heezen et al.,* 1966; *Dehlinger et al.,* 1970; *Engel and Engel,* 1970; *Anderson and Nishimori,* 1979] partly because of a lack of sampling and partly because of differences in morphology between fast and slow spreading ridges. The ultramafic rocks include serpentinized varieties of harzburgite, dunite, lherzolite, and mylonitized peridotites. The mineral compositions are refractory (Fo_{89-92}, $En_{87-88.8}$, $Wo_{38-49}En_{45-55}Fs_{4-6}$). *Dick and Bullen* [1982], and *Dick* [1978], in studies of peridotites from the Atlantic and Indian oceans, concluded that oceanic peridotites have a higher proportion of lherzolite than do many ophiolitic peridotites, that they had less magnesian olivine and enstatite, and that their spinels had Cr/Cr + Al values less than 0.6, while Cr/Cr + Al in spinels from peridotites in many ophiolite complexes were often greater than 0.6. These differences were particularly striking for circum-Pacific peridotites. They suggested that some of these ophiolites represent fragments of island arcs where remelting of a depleted mantle had left a more refractory residue than is typical beneath mid-ocean ridges.

Massive and cumulate gabbros have been sampled from the deep-sea floor; these include gabbros, norites, ilmenite norites, olivine gabbros, quartz gabbros, and troctolites [*Quon and Ehlers,* 1963; *Bonatti et al.,* 1971; *Melson and Thompson,* 1971; *Bonatti and Honnorez,* 1976; *Clague,* 1976; *Hodges and Papike,* 1976; *Andersen and Nishimori,* 1979;

Tiezzi and Scott, 1981]. Many have greenschist or amphibolite facies metamorphic assemblages. *Engel and Fisher* [1975] have described cumulate rocks from beneath the Indian Ocean Ridge, which include these rock types as well as minor orthopyroxenites. Most of the gabbros from the mid-ocean ridges are CPX + PLAG + OL assemblages. When orthopyroxene does occur it is nearly always as an intercumulus phase [*Hodges and Papike,* 1976; *Engel and Fisher,* 1975]. There are some instances of cumulate OPX in gabbros from 26°N on the mid-Atlantic Ridge, but these are relatively fractionated gabbros (FeO*/MgO as high as 0.84 [*Tiezzi and Scott,* 1981]). OPX is, however, extremely rare in the early stages of crystallization and, in any case, is not a common phase. It is in this absence, or late appearance, of OPX that the ocean floor gabbros are most distinct from the Mariana Trench samples in which two pyroxene gabbros and the early crystallization of OPX are the rule. The lack of OPX as a major phase in the cumulate sequence on the ocean floor may be due to the dominance of open system magma chambers. There is evidence of cyclic units in some of the drilled gabbros [*Hodges and Papike,* 1976] and evidence that the major element compositions of MORB magmas are buffered and effectively uncoupled from the trace elements [*Langmuir et al.,* 1977; *Bryan et al.,* 1976]. This can be explained by an open system [*O'Hara,* 1977] where new batches of magma are frequently injected and mixed into the magma chamber. This keeps the magma from ever reaching the OPX-CPX-PLAG cotectic (Figure 15) and produces a cumulate section that is dominated by OL, CPX, and PLAG. There are two possibilities to explain the abundance of OPX in the Marianas samples: they crystallized from a magma system similar to that producing ocean floor gabbros but are replenished less frequently and thus fractionate to the CPX-OPX-PLAG eutectic; or they crystallized from a melt with an initially higher degree of silica saturation, which crystallized OPX earlier and for a longer interval than do MORB magmas. The second explanation seems most likely because of the large amounts of OPX and the lack of significant volumes of olivine bearing cumulates. The order of crystallization also differs as in the Mariana Trench gabbros OPX generally precedes PLAG (Table 6). The range in liquid compositions calculated to have been in equilibrium with the Mariana Trench gabbros (FeO*/MgO = 0.8–2.05, Table 5) is quite similar to that calculated for ocean floor gabbros (0.88–2.5 [*Hodges and Papike,* 1976; 1.3–3.3, [*Tiezzi and Scott,* 1980]) and for gabbros in the Semail ophiolite (0.8–2.8 [*Pallister and Hopson,* 1981]). All of these gabbros have more restricted ranges of mineral compositions and crystallized from liquids of more restricted FeO*/MgO values than did gabbros from a closed system such as the Skaergaard complex [*Wager and Brown,* 1967]. This suggests that both the gabbros exposed in the Mariana Trench and cumulate gabbros from mid-ocean ridges crystallized in relatively open systems (as compared with a system like the Skaergaard) in which the magma com-

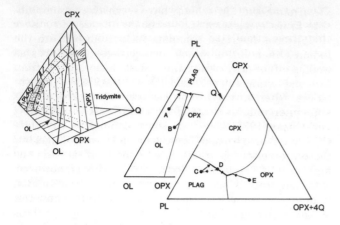

Fig. 15. PL-OL-CPX-OPX phase relations after *Irvine* [1970] indicating possible crystallization sequences; A and C paths for MOR gabbros, producing a crystallization sequence of OL-PLAG-CPX, injections of primitive magma at D prevent liquids from reaching CPX-OPX-PLAG eutectic; paths B and E possible trajectory for magmas like those which produced the gabbros in the Mariana Trench slope, producing a sequence of OL-OPX-CPX-PLAG.

positions were buffered at relatively constant by frequent injections of primitive magma.

In most other respects there are not striking differences between oceanic ridges and the Mariana Trench cumulate sections. Both show iron enrichment trends and both crystallized a large proportion of plagioclase and pyroxene, though the Marianas samples may have slightly more calcic plagioclase. Neither cumulate section has the high-Fe olivine, high-Ca plagioclase association of the active arc or WMR gabbros. This mineral assemblage is considered to reflect high water pressures and the consequent depres-sion of the plagioclase solidus [*Johannes*, 1978; *Stern*, 1979]. The volcanic rocks in the forearc, however, and the bon-inites in particular, tend to have very high water contents (up to 3 wt %) yet the gabbros that are postulated to have fractionated from these magmas have neither abundant hydrous phases nor extremely calcic plagioclase. It would seem that some other factor, perhaps greater depths of crystallization, is required to produce the high-Ca PLAG mineral assemblage.

One might expect island-arc gabbros to have CPX with lower Ti or to have more evolved CPX (lower Mg, Cr) than oceanic gabbros [*Serri*, 1981]. Unfortunately, there have been relatively few detailed studies of the mineral chemistry of oceanic gabbros, so there are few data for comparison. Most of the pyroxene compositions overlap the ocean floor gabbro pyroxene compositions in Figure 16, although many of the Mariana Trench volcanic pyroxenes have significantly lower Ti than MORB CPX at equivalent Mg number. It is apparent that the Ti content of CPX from cumulate gabbros is not a reliable criteria for distinguishing island-arc and mid-ocean ridge cumulate sequences.

Distinguishing between ocean floor plutonic rocks and tholeiitic island-arc plutonic rocks like those exposed in the Mariana Trench is not a straightforward matter. The most striking differences between the two are the more refractory spinels in the Mariana ultramafic rocks and the abundance and early appearance of OPX in the trench slope cumulate section, probably reflecting a greater initial Si saturation in the magma. It has been suggested, based largely on the presence of low TiO_2 volcanic rocks, that there are several ophiolites that may be fragments of crust formed in an island-arc environment [*Upadhyay and Neale*, 1979; *Sun and Nesbitt*, 1978; *Beccaluva et al.*, 1978; *Evans and Hawkins*, 1981]. These include Troodos, parts of the Zambales Range, Philippines; Betts Cove and

TABLE 6. Crystallization Sequence for Mariana Trench Cumulate Section, Mid-Ocean Ridge Gabbros, and Some Ophiolites

Area	Crystallization Sequence	Low Ti Volcanics?	Reference
Mariana Trench Wall gabbros	OL + SP, OPX, CPX, PLAG, OX	yes	
26°N Mid-Atlantic Ridge	OL, PLAG, CPX, OPX	no	*Tiezzi and Scott* [1980]
DSDP 334, Mid-Atlantic Ridge	OL + SP, PLAG, CPX, ± OPX or PIG	no	*Hodges and Papike* [1976]
Canyon Mountain Ophiolite	OL, CPX, PLAG, OPX OL, CPX, OPX, PLAG	—	*Himmelburg and Loney* [1980]
Red Mountain Ophiolite	OL + SP, CPX, PLAG, ± AMPH ± OX	—	*Sinton* [1980]
Zhan Taishir Ophiolite	OL, OPX + CPX, PLAG	yes	*Zouenshain and Kuzmin* [1980]
Papuan Ultramafic Belt	OL, OPX, CPX, PLAG	—	*Davis* [1969]
Vourinos Ophiolite	OL + SP, CPX, PLAG, OPX	—	*Jackson et al.* [1975]
Troodos Ophiolite	OL, CPX, OPX, PLAG	yes	*Greenbaum* [1972]
Appenine Ophiolites	OL + SP, PLAG, CPX, OPX or PIG, OX, AP	—	*Serri* [1980]
Marum Ophiolite	OL + SP, CPX, OPX, PLAG, OX	—	*Jacques* [1981]
Betts Cove Ophiolite	OL, OPX, CPX, PLAG	yes	*Church and Riccio* [1977]
Bay of Islands Ophiolite	OL, PLAG, CPX, OPX	no	*Church and Riccio* [1977]

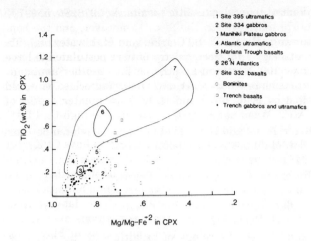

Fig. 16. CPX compositions in rocks from the trench slope compared to clinopyroxenes from other areas; site 395 [*Arai and Fujii,* 1978]; site 334 [*Hodges and Papike,* 1976]; Atlantic ultramafics *Bonatti et al.* [1970]; Mariana Trough (J. Melchior, personal communication, 1981); 26°N Atlantic [*Tiezzi and Scott,* 1980]; site 332 [*Hodges and Papike,* 1977].

Mings Bight, Newfoundland; KhanTaishir, Mongolia; and Vourinos, Greece. Some of the characteristics of these ophiolites, and of some without low Ti basalts, are presented in Table 6. It is apparent that in those ophiolites with abundant low Ti basalts, OPX is usually an important cumulate phase in the gabbros and occurs early in the crystallization sequence. This type of ophiolite represents a lower Ti, higher Mg, more Si saturated magma than that parental to most MORB basalts. While it is conceivable that such low Ti melts could occur on ocean ridges, not even the most primitive MORB compositions found or estimated [*Dungan and Rhodes,* 1978; *Bender et al.,* 1978] approach these compositions, and no basalts even remotely similar have been described from back arc basins [*Hawkins,* 1976, 1980; *Hawkesworth et al.,* 1977; *Melchior and Hawkins,* 1980]. This low Ti, Si-saturated type of magma has in fact been found only in island arc complexes: the Bonin Islands, Mariana forearc, and Papua, New Guinea.

The implication of these data is that ophiolites with low Ti, Si-saturated basalts, abundant cumulus OPX, or refractory spinel compositions do not represent typical mid-ocean ridge derived oceanic crust but in fact are probably some variety of island arc crust. The island arc complex exposed on the inner slope of the Mariana Trench probably was formed during the early stages of island arc activity. The similarities between the rocks from the trench slope and these low-Ti, OPX-rich ophiolites suggests that these ophiolites represent crust formed in the early stages of arc volcanism.

Conclusions

1. The volcanic rocks exposed on the inner slope of the Mariana Trench are island arc volcanics, probably of late Eocene age, and they have been exposed by tectonic erosion of material from the inner slope. The chemistry and mineralogy of cumulate rocks found on the inner slope indicate that these cumulates are genetically related to the volcanic rocks. This implies that these plutonic rocks are part of an island arc complex.

2. The cumulate section of gabbros and ultramafic rocks in the Mariana Trench slope is characterized by the early crystallization, and by persistence through the cumulate sequence, of orthopyroxene. These rocks show a slightly greater range in mineral compositions than do typical MORB gabbros with slightly more calcic plagioclase and a very limited range through which olivine crystallized. They are not calcalkaline gabbros in that they lack the high Fe olivine-high Ca plagioclase association of gabbroic xenoliths found in many island arcs.

3. The noncumulate ultramafic rocks in the trench slope have the chemistry and mineralogy typical of a refractory residue from partial melting. Irregular altered segregations of CPX and altered PLAG are interpreted as trapped pockets of liquid. The Cr spinels in these ultramafic rocks are distinct from those in ocean ridge peridotites in their high Cr/Cr + Al values. The ultramafic rocks may represent the residue from generation of the magma that produced the gabbros and volcanic rocks in the trench slope.

4. The main crystallization sequence in the magmas that produced the cumulate section was OL − (SP) − OPX − CPX − PLAG − (OX). Variations occur indicating slightly different liquid compositions or crystallization paths. The magma chambers were probably open systems that experienced frequent replenishments of primitive liquids.

5. The trench slope assemblage in general is similar to that of an ophiolite. In particular it is similar to ophiolites containing low Ti, Si-saturated volcanic rocks and OPX rich cumulate sections. It is suggested that these ophiolites represent crust formed in the early stages of island-arc volcanism.

Acknowledgments. This work was done as part of the IDOE/SEATAR program and was supported by NSF grants OCE 78-16758, OCE 78-17823, and OCE 80-19016 to J. W. Hawkins. Field work was done on legs 5, 6, and 11 of SIO expedition MARIANA. We thank Capt. C. Johnson of R/V *T. Washington* for his seamanship during the surveying and sampling and especially for his assistance during Typhoon Tess. Chief Marine Technician R. Wilson helped us successfully make 19 deep dredge hauls from the Mariana Trench. We also thank J. Melchior, C. Evans, C. Schimmel, and H. Fujita for extensive help with the lab work. R. Scott and H. Dick deserve special thanks for their careful reviews of the manuscript.

References

Anderson, R. N., and R. K. Nishimori, Gabbro, serpentine and mafic breccia from the East Pacific Rise, *J. Phys. Earth, 27,* 467–480, 1979.

Arai, S., and T. Fujii, Petrology of ultramafic rocks from site 395, in *Initial Rep. Deep Sea Drilling Proj., 45,* 587–594, 1978.

Aumento, F., and H. Loubat, The mid-Atlantic Ridge near 45°N, 16, Serpentinized ultramafic intrusions, *Can. J. Earth Sci., 8,* 631–663, 1971.

Beccaluva, L., G. Macciotta, C. Savelli, G. Serri, and O. Zeda, Geochemistry and K/Ar ages of volcanics dredged in the Philippine Sea [Mariana, Yap, Palau trenches and Parece-Vela Basin], in *The Tectonic and Geologic Evolution of Southeast Asian Seas and Islands, Geophys. Monogr. Ser.,* vol. 23, edited by D. E. Hayes, pp. 247–270, Washington, D. C., 1980.

Beccaluva, L., D. Ohmenstetter, M. Ohmenstetter, and A. Paupy, The Vourinos ophiolitic complex (Greece) has been created in an island arc setting: Petrographic and geochemical evidence (Abstr.), *Ofioliti, 3,* 62–64, 1978.

Bender, J. F., F. N. Hodges, and A. E. Bence, Petrogenesis of basalts from the Project Famous area: Experimental study from 0 to 15 kbars, *Earth Planet Sci. Lett., 41,* 277–302, 1978.

Bloomer, S., Mariana fore-arc ophiolite, structure and petrology (Abstr.), *Eos Trans. AGU, 62,* 1086–1087, 1981.

Bloomer, S., Structure and evolution of the Mariana Trench, Petrologic and geochemical studies, Ph.D. Dissertation, Univ. of California, San Diego, 1982.

Bloomer, S., and J. Hawkins, Arc-derived plutonic and volcanic rocks from the Mariana Trench slope (Abstr.), *Eos Trans. AGU, 61,* 1143, 1980.

Bloomer, S., J. Melchior, C. Evans, and R. D. Francis, Techniques for the chemical analysis of igneous rocks at the Scripps Institution of Oceanography, *Ref. 82-7,* Scripps Inst. of *Oceanogr.,* Univ. of Calif., San Diego, La Jolla, 1982.

Bloomer, S., J. Melchior, R. Poreda, and J. Hawkins, Mariana arc-trench gap studies: Petrology of boninites and evidence for a 'boninite series' (Abstr.), *Eos Trans. AGU, 60,* 968, 1979.

Bonatti, E., Serpentinite protrusions in the oceanic crust, *Earth Planet Sci. Lett., 32,* 107–113, 1976.

Bonatti, E., and P. R. Hamlyn, Mantle uplifted block in the western Indian Ocean, *Science, 201,* 249–251, 1978.

Bonatti, E., and J. Honnorez, Sections of the earth's crust in the equatorial Atlantic, *J. Geophys. Res., 81,* 4105–4116, 1976.

Bonatti, E., J. Honnorez, and G. Ferrara, Periodotite-gabbro-basalt complex from the equatorial mid-Atlantic Ridge, *Phil. Trans. R. Soc. London, Ser. A, 268,* 385–402, 1971.

Bryan, W. B., G. Thompson, F. A. Frey, and J. S. Dickey, Inferred geologic setting and differentiation in basalts from the Deep Sea Drilling Project, *J. Geophys. Res., 81,* 4285–4303, 1976.

Buddington, A. F., and D. H. Lindsley, Iron-titanium oxide minerals and synthetic equivalents, *J. Petrol., 5,* 310–357, 1964.

Cameron, W. E., E. G. Nisbet, and V. J. Dietrich, Boninites,

komatiites and ophiolitic basalts, *Nature, 280,* 550–553, 1979.

Carswell, D. A., D. B. Clarke, and R. H. Mitchell, The petrology and geochemistry of ultramafic nodules from pipe 200, Northern Lesotho, in *The Mantle Sample: Inclusions in Kimberlites and Other Volcanics,* vol. 2, edited by F. R. Boyd and H. O. A. Meyer, pp. 127–144, AGU, Washington, D. C., 1979.

Carter, N. L., and H. G. AvéLallement, High temperature flow of dunite and periodotite, *Geol. Soc. Am. Bull., 81,* 2181–2202, 1970.

Church, W. R., and L. Riccio, Fractionation trends in the Bay of Islands ophiolite of Newfoundland: Polycyclic cumulate sequences in ophiolites and their classification, *Can. J. Earth Sci., 14,* 1156–1165, 1977.

Clague, D. A., Petrology of basaltic and gabbroic rocks dredged from the Danger Island Troughs, Manihiki Plateau, in *Initial Rep. Deep Sea Drilling Proj., 33,* 891–912, 1976.

Coish, R. A., and W. R. Church, Igneous geochemistry of mafic rocks in the Betts Cove Ophiolitic, Newfoundland, *Contrib. Mineral. Petrol., 70,* 20–39, 1979.

Coleman, R. G., *Ophiolites: Ancient Oceanic Lithosphere?,* Springer-Verlag, New York, 1977.

Danckwerth, P. A., and R. C. Newton, Experimental determination of the spinel peridotite to garnet peridotite reaction in the system $MgO-Al_2O_3-SiO_2$ in the range 900°–1100°C and Al_2O_3 isopleths of enstatite in the spinel field, *Contrib. Mineral. Petrol., 66,* 189–201, 1978.

Davies, H. L., Peridotite-gabbro-basalt complex in eastern Papua, *Aust. Bur. Min. Res. Geol. Geophys. Bull., 128,* 1–48, 1969.

Dehlinger, P., P. W. Couch, D. A. McManus, and M. Gemperle, Northeast Pacific structure, 2, *Sea, 4,* 133–189, 1970.

Dick, H. J. B., Partial melting in the Josephine peridotite, 1, The effect of mineral composition and its consequence for geobarometry and geothermometry, *Am. J. Sci., 277,* 801–832, 1977.

Dick, H. J. B., Mineralogy of abyssal peridotites from the far South Atlantic, Caribbean and Equatorial Atlantic, *Geol. Soc. Am. Abstr. Progr., 10,* 388, 1978.

Dick, H. J. B., and T. Bullen, Chromian spinel as a petrogenetic indicator in oceanic environments, submitted to *Geol. Soc. Am. Bull.,* 1982.

Dietrich, V., R. Emmerman, R. Oberhansli, and H. Puchlet, Geochemistry of basaltic and gabbroic rocks from the West Mariana Basin and the Mariana Trench, *Earth Planet Sci. Lett., 39,* 127–144, 1978.

Dungan, M. A., and J. M. Rhodes, Residual glasses and melt inclusions in basalts from DSDP Legs 45 and 46, Evidence for magma mixing, *Contrib. Mineral. Petrol., 67,* 417–431, 1978.

Ellis, C. H., Calcareous nannofossil biostratigraphy—Deep Sea Drilling Project Leg 60, *Initial Rep. Deep Sea Drilling Proj., 60,* 507–536, 1981.

Engel, A. E. J., and C. G. Engel, Mafic and ultramafic rocks, 1, *Sea, 4,* 465–520, 1970.

Engel, C. G., and R. L. Fisher, Granitic to ultramafic rock complexes of the Indian Ocean ridge system, western Indian Ocean, *Geol. Soc. Am. Bull., 86,* 1553–1578, 1975.

England, R. N., and H. L. Davies, Mineralogy of ultramafic cumulates and tectonites from Eastern Papus, *Earth Planet Sci. Lett., 17,* 416–425, 1973.

Evans, C. A., and J. Hawkins, Mariana arc-trench studies: Petrology of 'seamounts' on the trench slope break (Abstr.), *Eos Trans. AGU, 60,* 968, 1979.

Evans, C. A., and J. Hawkins, Island arc and back-arc basin sections within the Zambales ophiolite, Philippines: Differences in the upper mantle peridotites (Abstr.), *Eos Trans. AGU, 62,* 1086, 1981.

Fisher, R. L., Cruise report *Eurydice* 08, Scripps Instit. of Oceanogr., Univ. of Calif., San Diego, La Jolla, 1974.

George, R. P., Structural petrology of the Olympus ultramafic complex in the Troodos ophiolite, Cyprus, *Geol. Soc. Am. Bull., 89,* 845–865, 1978.

Greenbaum, D., Magmatic processes at ocean ridges: Evidence from the Troodos Massis, Cyprus, *Nature, 238,* 18–21, 1972.

Hamlyn, P. R., and E. Bonatti, Petrology of mantle derived ultramafics from the Owen Fracture Zone, northwest Indian Ocean: Implications for the nature of the oceanic upper mantle, *Earth Planet Sci. Lett., 48,* 65–79, 1980.

Hanson, G. N., and C. H. Langmuir, Modelling of major elements in mantle-melt systems using trace element approaches, *Geochim. Cosmochim. Acta, 42,* 725–741, 1978.

Hawkesworth, C. J., R. K. O'Nions, R. J. Pankhurst, P. K. Hamilton, and N. M. Evensen, A geochemical study of island-arc and back arc tholeiites from the Scotia Sea, *Earth Planet Sci. Lett., 36,* 253–262, 1977.

Hawkins, J. W., Petrology and geochemistry of basaltic rocks of the Lau Basin, *Earth Planet Sci. Lett., 28,* 283–297, 1976.

Hawkins, J. W., Petrology of back-arc basins and island arcs: Their possible role in the origin of ophiolites, in *Ophiolites—Proceedings, International Ophiolite Symposium, Cyprus, 1979,* edited by A. Panayiotou, pp. 244–253, Cyprus Geological Survey Department, 1980.

Hawkins, J. W., and R. Batiza, Petrology and geochemistry of an ophiolite complex, Zambales Range, Luzon, Republic of Philippines (Abstr.), *Eos Trans. AGU, 58,* 1244, 1977.

Hawkins, J. W., S. Bloomer, C. Evans, and J. Melchior, Mariana arc-trench system: Petrology of the inner trench slope (Abstr.), *Eos Trans. AGU, 60,* 968, 1979.

Hawkins, J. W., and C. A. Evans, Geology of the Zambales Range, Luzon, Philippine Islands: Ophiolite derived from an island arc–back arc basin pair, this volume.

Heezen, B. C., B. Glass, and H. W. Menard, The Manihiki Plateau, *Deep Sea Res., 13,* 445–458, 1966.

Heezen, B. C., et al., Site 60, in *Initial Rep. Deep Sea Drilling Proj. 6,* 587–630, 1971.

Hickey, R., and F. A. Frey, Geochemical characteristics of boninite series volcanics: implications for their source, *Geochim. Cosmochim. Acta,* in press, 1982.

Himmelburg, G. R., and R. A. Loney, Petrology of ultramafic and gabbroic rocks of the Canyon Mountain ophiolite, Oregon, *Am. J. Sci., 280A,* 232–268, 1980.

Hodges, F. N., and J. J. Papike, DSDP site 334: Magmatic cumulates from oceanic layer 3, *J. Geophys. Res., 81,* 4135–4151, 1976.

Hodges, F. W., and J. J. Papike, Petrology of basalts, gabbros and peridotites from DSDP Leg 37, *Initial Rep. Deep Sea Drilling Proj., 37,* 711–724, 1977.

Howard, A. H., and E. Stolper, Experimental crystallization of boninites from the Mariana Trench, *Eos Trans. AGU, 62,* 1091, 1981.

Hussong, D. M., et al., Leg 60 ends in Guam, *Geotimes, 12,* 19–22, 1978.

Hynes, A., Comment on 'The Troodos ophiolite complex was probably formed in an island arc,' *Earth Planet Sci. Lett., 25,* 213–216, 1975.

Ingle, J. C., Summary of Late Paleogene-Neogene insular stratigraphy, paleobathymetry, and correlations, Philippine Sea and Sea of Japan region, *Initial Rep. of the Deep Sea Drilling Proj., 31,* 837–855, 1975.

Irvine, T. N., Chromian spinel as a petrogenetic indicator, 2, Petrologic applications, *Can. J. Earth. Sci., 4,* 71–103, 1967.

Irvine, T. N., Crystallization sequences in the Muskox Intrusion and other layered intrusions, I, Olivine-pyroxene-plagioclase relations, *Spec. Publ. 1,* pp. 441–476, Geological Society of South Africa, Johannesburg, 1970.

Irving, A. J., A review of experimental studies of crystal/liquid partitioning, *Geochim. Cosmochim. Acta, 42,* 743–770, 1978.

Jacques, A. L., Petrology and petrogenesis of cumulate peridotites and gabbros from the Marum ophiolite complex, northern Papua, New Guinea, *J. Petrol., 22,* 1–40, 1981.

Jagoutz, E., V. Lorenz, and H. Wanke, Major trace elements of Al-augites and Cr-diopsides from ultramafic nodules in European alkali basalts, in *The Mantle Sample: Inclusions in Kimberlites and Other Volcanics,* vol. 2, edited by F. R. Boyd and H. O. A. Meyer, pp. 382–390, AGU, Washington, D. C., 1979.

Jakès, P., and J. Gill, Rare earth elements and the island arc tholeiitic series, *Earth Planet Sci. Lett., 9,* 17–28, 1970.

Johannes, W., Melting of plagioclase in the systems Ab-An-H_2O and Qz-Ab-An-H_2O at P_{H_2O} = 5 kbars, An equilibrium problem, *Contrib. Mineral. Petrol., 66,* 295–303, 1978.

Krause, D. C., H. W. Menard, and S. M. Smith, Topography and lithology of the Mendocino Ridge, *J. Mar. Res., 22,* 236–250, 1964.

Kushiro, I., Si-Al relations in clinopyroxenes from igneous rocks, *Am. J. Sci., 258,* 548–554, 1960.

Langmuir, C. H., J. F. Bender, A. E. Bence, and G. N.

Hanson, Petrogenesis of basalts from the FAMOUS area: Mid-Atlantic Ridge, *Earth Planet Sci. Lett., 36*, 133–156, 1977.

Leake, B. E., Nomenclature of amphiboles, *Am. Mineral., 63*, 1023–1052, 1978.

Lewis, J. F., Petrology of the ejected plutonic blocks of the Soufriére Volcano, St. Vincent, West Indies, *J. Petrol., 14*, 81–114, 1973.

Meijer, A., Primitive arc volcanism and a boninite series: examples from western Pacific island arcs, in *The Tectonic and Geologic Evolution of Southeast Asian Seas and Islands, Geophys. Monogr. Ser.,* vol. 23, edited by D. E. Hayes, pp. 271–282, AGU, Washington, D. C., 1980.

Melchior, J., and J. Hawkins, Petrology of Mariana back-arc basin basalts (Abstr.), *Eos Trans. AGU, 61*, 1980.

Melson, W. G., and G. Thompson, Layered basic complex in oceanic crust, Romanche Fracture, Equatorial North Atlantic, *Science, 168*, 817–820, 1970.

Melson, W. G., and G. Thompson, Petrology of a transform fault zone and adjacent ridge segments, *Phil. Trans. R. Soc. London, Ser. A, 268*, 423–441, 1971.

Menzies, M., Residual alpine lherzolites and harzburgites—Geochemical and isotopic constraints on their origin, *Oregon Dept. Geol. Min. Indus. Bull., 96*, 129–147, 1977.

Menzies, M. A., and C. A. Allen, Plagioclase lherzolite residual mantle relationships within two eastern Mediterranean ophiolites, *Contrib. Mineral. Petrol., 45*, 197–213, 1974.

Mercier, J-C. C., and A. Nicolas, Textures and fabrics of upper mantle peridotites as illustrated by xenoliths from basalts, *J. Petrol., 16*, 454–487, 1975.

Miyashiro, A., The Troodos Ophiolite Complex was probably formed in an island arc, *Earth Planet Sci. Lett., 19*, 218–224, 1973.

Miyashiro, A., Origin of the Troodos and other ophiolites: A reply to Hynes, *Earth Planet Sci. Lett., 25*, 217–222, 1975a.

Miyashiro, A., Origin of the Troodos and other ophiolites: A reply to Moores, *Earth Planet Sci. Lett., 25*, 227–235, 1975b.

Miyashiro, A., and F. Shido, Tholeiitic and calc-alkalic series in relation to the behaviors of Ti, V, Cr and Ni, *Am. J. Sci., 275*, 265–273, 1975.

Moores, E. M., Discussion of 'Origin of Troodos and other ophiolites: A reply to Hynes' by Akiho Miyashiro, *Earth Planet Sci. Lett., 25*, 223–226, 1975.

Natland, J. H., Petrography and mineral compositions of gabbros recovered in Deep Sea Drilling Project Hole 453 on the western side of the Mariana Trough, *Initial Rep. Deep Sea Drilling Proj., 60*, 579–600, 1981.

Natland, J. H., and J. Tarney, Petrologic evolution of the Mariana Arc and back-arc basin system—a synthesis of drilling results in the South Philippine Sea, *Initial Rep. Deep Sea Drilling Proj., 60*, 877–908, 1981.

Nicholls, I. A., Petrology of Santorini Volcano, Cyclades, Greece, *J. Petrol., 12*, 67–120, 1971.

Nicolas, A., F. Boudier, and J. L. Bouchez, Interpretation of peridotitic structures from ophiolitic and oceanic environments, *Am. J. Sci., 280A*, 192–210, 1980.

Nishimori, R. K., The petrology and geochemistry of gabbros from the Peninsular Ranges Batholith, California and a model for their origin, Ph.D. Thesis, Univ. of Calif., San Diego, 1976.

Nixon, P. H., and F. R. Boyd, Garnet bearing lherzolites and discrete nodule suites from the Malaita alnoite, Solomon Islands, southwest Pacific and their bearing on oceanic mantle composition and geotherm, in *The Mantle Sample: Inclusions in Kimberlites and Other Volcanics,* vol. 2, edited by F. R. Boyd and H. O. A. Meyer, pp. 400–423, AGU, Washington, D. C., 1979.

Nwe, Y. Y., Pyroxene crystallization trends in Skaergaard trough bands, *Contrib. Mineral. Petrol., 49*, 285–308, 1975.

O'Hara, M. J., Geochemical evolution during fractional crystallization of a periodically refilled magma chamber, *Nature, 266*, 503–507, 1977.

Pallister, J. S., and C. A. Hopson, Samail Ophiolite plutonic suite: Field relations, cryptic variation and layering, and a model of a spreading ridge magma chamber, *J. Geophys. Res., 86*, 2593–2644, 1981.

Pearce, J. A., and M. J. Norry, Petrogenetic implications of Ti, Zr, Y and Nb variations in volcanic rocks, *Contrib. Mineral. Petrol., 69*, 33–47, 1979.

Presnall, D. C., Alumina content of enstatite as a geobarometer for plagioclase and spinel lherzolites, *Am. Mineral., 61*, 582–588, 1976.

Quon, S. H., and E. G. Ehlers, Rocks of the northern part of the Mid-Atlantic Ridge, *Geol. Soc. Am. Bull., 74*, 1–8, 1963.

Roeder, P. L., and R. F. Emslie, Olivine-liquid equilibrium, *Contrib. Mineral. Petrol., 29*, 275–289, 1970.

Salisbury, M., and N. J. Christensen, The seismic velocity structure of a traverse through the Bay of Islands ophiolite complex, Newfoundland, An exposure of oceanic crust and upper mantle, *J. Geophys. Res., 83*, 803–817, 1978.

Savelyeva, G. N., N. L. Dobrotsov, Yu. G. Lavrentjev, B. G. Luts, F. Dietrich, and R. Coleman, Petrology of ultrabasic rocks, gabbro and metamorphic rocks, in *Geology of the Philippine Sea Floor* (in Russian), A. V. Peive, Publishing Office 'Nauka,' Moscow, 1980.

Serri, G., Chemistry and petrology of gabbroic complexes from the northern Apennine ophiolites, in *Ophiolites—Proceedings, International Ophiolite Symposium, Cyprus 1979,* edited by A. Panayiotou, pp. 296–313, Cyprus Geological Survey Department, Nicosia, 1980.

Serri, G., The petrochemistry of ophiolite gabbroic complexes: A key for the classification of ophiolites into low-Ti and high-Ti types, *Earth Planet Sci. Lett., 52*, 203–212, 1981.

Sigurdsson, H., and J. G. Schilling, Spinels in Mid-Atlantic Ridge basalts, Chemistry and occurance, *Earth Planet Sci. Lett., 29*, 7–20, 1976.

Sinton, J. M., Petrology and evolution of the Red Mountain

Ophiolite Complex, New Zealand, *Am. J. Sci., 280A,* 296–328, 1980.

Skornyakova, N. S., and M. I. Lipkina, Basic and ultrabasic rocks of the Marianas Trench, *Oceanology, 15,* 688–690, 1976.

Smewing, J. D., Mixing characteristics and compositional differences in mantle-derived melts beneath spreading axes, Evidence from cyclically layered rocks in the ophiolite of North Oman, *J. Geophys. Res., 86,* 2495–3134, 1981.

Stern, C., Open and closed system igneous fractionation within two Chilean ophiolites and the tectonic implications, *Contrib. Mineral. Petrol., 68,* 243–258, 1979.

Stern, R. J., On the origin of andesite in the northern Mariana Island arc: Implications from Agrigan, *Contrib. Mineral. Petrol., 68,* 207–219, 1979.

Streckeisen, A., To each plutonic rock its proper name, *Earth Sci. Rev., 12,* 1–33, 1976.

Sun, S., and R. W. Nesbitt, Geochemical regularities and genetic significance of ophiolitic basalt, *Geology, 6,* 689–693, 1978.

Tiezzi, L. J., and R. B. Scott, Crystal fractionation in a cumulate gabbro, Mid-Atlantic Ridge, 26°N, *J. Geophys. Res., 85,* 5438–5454, 1980.

Upadhyay, H. P., and E. R. W. Neale, On the tectonic regimes of ophiolite genesis, *Earth Planet Sci. Lett., 43,* 93–102, 1979.

Wager, L. R., and G. M. Brown, *Layered Igneous Rocks,* W. H. Freeman, San Francisco, Calif., 1967.

Zouenshain, L. P., and M. I. Kuzmin, The Khan-Taishir Ophiolite complex of western Mongolia, Its petrology, origin and comparison with other ophiolitic complexes, *Contrib. Mineral. Petrol., 67,* 95–100, 1978.

Temporal Relationships Between Back Arc Basin Formation and Arc Volcanism With Special Reference to the Philippine Sea

D. E. KARIG

Department of Geological Sciences, Cornell University, Ithaca, New York 14853

The temporal relationship between back arc spreading and arc volcanism is an important element of geodynamic and petrologic models of arc evolution. Recent analyses of this relationship in the Philippine Sea have concluded that back arc spreading occurs or at least is initiated during relative minima in the intensity of arc volcanism. This study demonstrates that, to the contrary, back arc spreading that has been initiated in the worldwide arc systems during the past 5 m.y. is associated with highly active arc volcanism. In addition, a reanalysis of the Philippine Sea data, taking into account the nature of transport of volcanic debris, suggests that pulses of spreading and volcanism are nearly synchronous.

Introduction

The behavior of island arc systems has recently come to be viewed as a reflection of large-scale geodynamic processes. Models have appeared relating tectonic variations among arc systems to such parameters as the age and nature of the descending lithosphere, the motion of the plates involved with respect to each other and to a deeper 'fixed' mantle, and to cyclic thermal and geochemical processes.

A large fraction of the observational data on which these models are based is derived from the kinematic behavior of the back arc region. There are, at present, several models for back arc spreading, including passive upwelling due to retrograde sinking of the descending lithosphere [e.g., *Molnar and Atwater,* 1978] or to rearward absolute motion of the upper plate [*Chase,* 1978; *Uyeda and Kanamori,* 1979], forceful diapirism [*Karig,* 1971; *Oxburgh and Turcotte,* 1971], and secondary convection [e.g., *Hsui and Toksöz,* 1981]. The incidence of back arc spreading, the duration of spreading pulses, and the correlation of these pulses with other arc processes, particularly with arc volcanism, supply important constraints for the resolution of these quite different dynamic models.

In this paper the timing of back arc spreading pulses and their correlation with arc volcanism are explored. The relationship between back arc spreading pulses and arc volcanism is first examined by reviewing modern arc systems in which back arc spreading is presently active. The history of volcanism and spreading in the Philippine Sea is next discussed because, despite extensive investigation, this region has led to quite different ideas concerning this relationship.

Arcs With Actively Spreading Back Arc Basins

As data concerning the kinematic behavior of convergent plate margins have accumulated, it has become progressively clear that not all ocean basins behind arc systems have formed in an identical manner [e.g., *Weissel,* 1981]. Most appear to have originated by back arc spreading behind actively subducting arc systems. Some may be trapped pieces of major ocean basins, which are older than the bounding arc. Others may be formed by extension but not spatially related to a contemporaneously active arc system, whereas yet others may have spread behind an active arc but occur primarily as a result of strike slip motion or of plate edge interactions other than subduction.

Active back arc spreading is most clearly identified behind the Tonga-Kermadec, Mariana, Bonin, Ryukyu, New Hebrides, northern Sunda (Andaman), and Scotia arc systems. Even in this category, however, there are problems in differentiating spreading directly related to subduction from that due primarily to other plate edge effects. Spreading in the Andaman Basin is very nearly parallel to the arc trend and is comparable in magnitude to the component of slip parallel to the convergent margin. The perpendicular component of subduction along the arc is very slow. It has been suggested that spreading is a result of northward motion of the arc with respect to Asia along a dominantly strike-slip boundary [*Curray et al.,* 1979], combined with buckling where the arc-bearing lithospheric strip collides end-on with Asia [*Karig et al.,* 1978].

In several other areas, spreading spatially related to arc systems is even more likely to be the result of transform plate-edge interactions. The Bismark Basin is opening between the West Melanesian and New Britain trenches, but

TABLE 1. Back Arc Spreading and Volcanism in Active Arc Systems

Arc System	Duration of Spreading, m.y.	Reference	Volcanic Activity	Reference
Mariana	0–5	*Hussong* [1982]	10–15 km^3 km^{-1} m.y.$^{-1}$ (very high)	*Sample and Karig* [1982]
Bonin	incipient	*Karig and Moore* [1975a]	high to very high	*Kuno* [1962], *Watts and Weissel* [1975]
Ryukyu	0–4 (?)	*Lee et al.* [1980]	moderate	*Kuno* [1962]
New Hebrides	0–3 (?)	*Carney and MacFarlane* [1977]	high (active since at least 3 m.y. B.P.)	*Carney and MacFarlane* [1977], *Dugas et al.* [1977]
Tonga-Kermadec	0–3.5	*Weissel* [1977]	moderate to high	*Richard* [1962], *Karig* [1970]
Scotia	0–8	*Barker* [1972]	high	*Baker* [1971]
Andaman	0–11	*Curray et al.* [1979]	low to very low	*Weeks et al.* [1967]

extension is approximately parallel to the now defunct West Melanesian Arc and is highly oblique to, as well as being far distant from, the New Britain Arc [*Taylor,* 1979]. The nearby Woodlark Basin is opening parallel to bounding ridges with arc affinities, but neither of these is now an active arc system [*Weissel et al.,* 1982]. Extension in the northwestern corner of Melanesia is quite likely a product of small-plate interactions along this basically transform sector between the Pacific and Indian plates.

Of these, recent (0–5 m.y.) average rates of volcanic production have only been calculated for the Mariana Arc [*Sample and Karig,* 1982], where it is very high in comparison to values calculated in other arcs (references given by *Karig and Kay* [1981]). Average intensities of arc volcanism remain poorly constrained, but my subjective estimate for other arcs with active back arc spreading, based on the volume of cones and volcaniclastic aprons that formed in the actively spreading basins in comparison with the Mariana Arc and other better calibrated arcs [e.g., *McBirney et al.,* 1974], would suggest that high average as well as high historical rates of arc volcanism accompany back arc spreading (Table 1).

Of particular importance is the observation that volcanism in arcs where back arc spreading has most recently been initiated (Bonin, New Hebrides) is very active, based both on the historic activity and on the total volume of volcanic debris. This implies that volcanism has persisted since before initiation of spreading. In these cases it is quite clear that the pulse of volcanism did not lag the back arc spreading episode.

Although the available data will not permit a firm conclusion that highly active arc volcanism and back arc spreading have been in perfect synchroneity during the most recent spreading episode, in all known active basins they are synchronous within the resolution of magnetic anomalies and the volcanic record. Because the active basins represent a range of evolution, from incipient (Bonin) to highly developed (Mariana), it is most logical to assume a model of synchronous spreading and volcanism; certainly there is no support in the recent record for the model of alternating volcanism and back arc spreading.

Back Arc Spreading and Volcanism in the Philippine Sea: Background

The Philippine Sea provides one of the clearest and best studied sequences of back arc basins in the geologic record. For this reason, observations in this region have served as the basis for an evolving series of correlations between back arc spreading and arc volcanism in arcs in general. Early marine geophysical and geological data, combined with the geology of the Mariana Islands and with the results of Deep Sea Drilling Project (DSDP) Leg 6, were used to stress the episodicity of tectonic events in the arc system [*Karig,* 1971]. Pulses of back arc spreading were outlined, each with an implied contemporaneous maximum in arc volcanism. A further implication was that both spreading and volcanism were responses to pulses of subduction. This model was modified and quantified following the drilling of sites during DSDP Leg 31 and the acquisition of additional marine geophysical data throughout the Philippine plate [*Karig,* 1975]. A major revision was the abandonment of the idea that pulses of subduction were necessary for the episodicity in spreading and volcanism.

Recognition of magnetic anomaly patterns in these basins and the extensive data collected for and during the drilling of DSDP legs 58, 59, and 60 led *Scott and Kroenke* [1980, 1981] to postulate a model in which extension alternated with pulses of arc volcanism, or at least was initiated during relative lulls in volcanic activity. Their conclusions have been reiterated, with variations, by *Crawford et al.* [1981] *Rodolfo and Warner* [1980], and *Sharaskin et al.* [1981], among others. *Hussong and Uyeda* [1981], on the other hand, have cited the record of volcanic ash in the Mariana forearc in support of continuous volcanism, with less pronounced fluctuations in production rate. They suggest that some of the apparent decrease in volcanic activity might be attributed to the lack of dispersal of volcanic debris from deep submergence of newly created volcanic arcs associated with initiation of back arc spreading.

In this review, the suite of data resulting from DSDP

legs 59 and 60 is integrated with those of DSDP legs 6 and 31, with island geology, and with marine geologic and geophysical data across the Philippine plate. Particular attention is given to factors governing deposition of volcanic material and to the effects these factors have on the volcanic record at individual sites. The result of this exercise is the reiteration of synchroneity between arc volcanism and back arc spreading in the Philippine Sea and of large fluctuations in the rate of arc volcanism.

Back Arc Spreading Pulses in the Philippine Sea

At least four basins or zones of back arc spreading have been identified in the Philippine Sea: the Mariana Trough, Parece Vela Basin, an incipient zone within the Bonin Arc, and the Shikoku Basin. The ages of these basins are now fairly well known from their magnetic anomaly patterns and from results of deep-sea drilling.

The Mariana Trough, which is actively spreading behind the Mariana Arc, is known from the results of DSDP hole 453 to have been a site of basinal deposition in the earliest Pliocene [*Hussong et al.*, 1981], but most likely it had not begun creating oceanic crust much before 5 m.y. ago. This conclusion is based on the proximity of site 453 to the boundary scarps of the basin, the bottoming of that hole in tectonized deep crustal rocks that suggest a faulted arc setting, and the location of the site west of the western edge of oceanic-type magnetic anomalies (D. M. Hussong, personal communication, 1981).

An incipient or very young zone of extension along the Bonin Arc has been interpreted from bathymetric and seismic reflection profiling, combined with limited geologic data [*Karig and Moore*, 1975a]. This zone cannot have a long period of extension because small closed basins, presumably of extensional origin and from which fresh basalt has been sampled, remain unfilled, despite their proximity to a very active volcanic arc chain.

The Shikoku and Parece Vela basins are linked longitudinally to form a very large, inactive back arc basin behind the pair of actively spreading zones. Combined drilling and magnetic interpretations now indicate that the Parece Vela Basin was active over a slightly older time span (from 30 to 17 or 18 m.y. B.P. [*Mrozowski and Hayes*, 1979]) than was the Shikoku Basin (24 or 25 m.y. to between 17 and 15 m.y. B.P. [*Shih*, 1980a]). However, both basins also show some longitudinal diachronism in the age of initiation.

Pulses of Volcanism in the Philippine Sea

Although arc-related volcanism may have been continuously active at some level in the boundaries around the Philippine Sea since the late Eocene, there do appear to be marked fluctuations in the intensity of volcanic activity over the period during which the back arc basins were generated. These fluctuations may be characteristic of in-

dividual arc systems or may be more regional, even global events [e.g., *Kennett et al.*, 1977]. A brief attempt to quantify the fluctuations along the Mariana Arc is included in the discussion section of this paper.

The volcanic history of the Mariana Arc over its more than 40 m.y. history is the best constrained of the sectors around the Philippine Sea and can be related to the opening of the Mariana Trough and Parece Vela Basin.

Both the historical volcanic record and the average production rate over the past 5 m.y. [*Sample and Karig*, 1982] (Table 1) point to a very high level (≥ 15 km^3 km^{-1} m.y.$^{-1}$) of recent volcanicity in the volcanic chain that developed along the eastern flank of the Mariana Trough. It is more difficult to determine when this high rate of volcanism began and whether there have been fluctuations of intensity during this period. DSDP site 453 revealed that volcanic silts and muds were being rapidly deposited in the Mariana Trough 5 m.y. ago. Although rapidity of deposition may reflect slumping from adjacent topographic highs [*Karig*, 1982], there is no evidence in the sediment section of material reworked from the older arc ridges. In part these high rates could have reflected a restricted area of deposition during early stages of extension [*Karig and Moore*, 1975b], but the fine-grained nature of these sediments suggests a relatively distal setting, which would tend to counteract effects of restricted deposition.

Because most of the products of arc volcanism reside in cones and volcaniclastic aprons (the latter built largely by mass transport processes), the record of recent volcanism outside the Mariana Trough is very subdued or subtle [*Emery*, 1962; *Sample and Karig*, 1982]. Studies of the insoluble fraction of limestones on Guam [*Hathaway and Carroll*, 1964] have revealed periods during which primary volcanic components were supplied to the frontal arc, separated by periods during which only reworked or secondary igneous minerals were available. The youngest group of 'volcanic' limestones is of Plio-Pleistocene age and disconformably overlies the 'nonvolcanic' Janum formation of earliest Pliocene age [*Tracey, et al.*, 1964; *Ingle*, 1975]. The beginning of this period (less than 4.5 m.y. B.P.) is probably a minimum for the upsurge in volcanism because it would probably take several million years before the cones of the new volcanic arc could rise high enough to disperse detritus into shallow water environments.

The forearc slope is largely shielded from input of volcanic mass flow deposits by the frontal arc. Only ash, borne by wind and surface currents, can pass this topographic barrier. Reworking of forearc deposits by biogenic and gravitational processes decreases the resolution of this volcanic record. A qualitative review of ash deposition in DSDP sites 60, 458, and 459 on the forearc slope [*Fischer et al.*, 1971; *Hussong et al.*, 1981] outlines a relatively large ash contribution during much of the past 5 m.y., but the record is degraded by numerous small, short hiatuses and incomplete core recovery, especially in the middle part of this interval.

Prior to 5 m.y. B.P. the volcanic chain of the Mariana

Arc was located on or just west of the West Mariana Ridge. Thus DSDP results from that ridge and from the Parece Vela Basin bear most directly on the Miocene volcanic history of the Mariana Arc. Data from sites 53 and 450 and reflection profiles [e.g., *Karig*, 1971] show a surficial carapace of essentially nonvolcanic clay, representing a lack of significant volcanism along the entire length of the West Mariana Arc after 9 m.y. B.P. A rapid decrease in the depositional rate of volcanic material between 13 or 14 m.y. B.P. and 9 m.y. B.P. in sites 53 and 450 signals a sharp reduction in activity from the preceding volcanic maximum. This reduction in activity is closely mirrored in the limestones of Guam, in which the insoluble components indicate a lack of primary volcanic material between about 4 or 5 m.y. and 13 m.y. B.P. [*Hathaway and Carroll*, 1964].

At DSDP site 451, on the West Mariana Ridge, relatively high rates of volcaniclastic deposition continued until 9 m.y. B.P.[*Kroenke et al.*, 1980] and are cited as evidence to confine the volcanic minimum to the interval between 5 and 9 m.y. [*Scott and Kroenke*, 1980]. These results cannot be quantitatively compared with cores in the Parece Vela Basin, however, because site 451 is very close to the volcanic centers, which would magnify the effects of even minor volcanic activity. Diffuse ash in strata younger than 9 m.y. old at site 451 and a moderate ash content in late Miocene strata on the forearc (site 458) demonstrate that volcanism continued at a low level during this volcanic minimum.

Evidence for the existence of an even stronger mid-Tertiary volcanic pulse is recorded on islands of the Mariana Arc and in the very large volcaniclastic apron that fills the eastern side of the Parece Vela Basin, but the initiation of this period of increased activity has been placed at widely divergent dates. *Karig* [1975] suggested an upsurge in activity about 30 m.y. B.P., at or before the opening of the Parece Vela Basin, whereas *Scott and Kroenke* [1980, 1981] placed this event near 20 m.y. B.P., after most of the basin had been formed.

Volcanic flows and tuffs of the Umatac formation on Guam [*Tracey et al.*, 1964] were dated by *Ingle* [1975] as 30–13 m.y. old based on the observation by *Schlanger* [1964] and by *Garrison et al.* [1975] that the Maemong limestone formed lenses interbedded with the volcanics. *Meijer et al.* [this volume] have reinterpreted this limestone as unconformably underlying the volcanics from which they have obtained radiometric ages of 10–13 m.y. A similar hiatus is also proposed for Saipan. This discrepancy remains to be resolved, but DSDP sites 53, 54, and 450 show that the volcaniclastic apron that formed west of this arc was already very large by the end of the early Miocene (18–20 m.y. B.P.), the age of the oldest dated sediments above basement at these sites. The record of volcanic ash in the forearc suggested to *Hussong and Uyeda* [1982] that, relative to volcanic activity during the previous 7–8 m.y., volcanism was moderately active between 28 and 20 m.y. B.P., after which activity intensified.

Scott and Kroenke [1980] argue for a lack of volcanic activity during the interval between 30 and 20 m.y. B.P., largely based on data from DSDP site 449, in the western Parece Vela Basin. Cores from that site, which penetrated 25-m.y.-old oceanic basement, are dominated by pelagic sediments, with a moderate volcanic component in the early Miocene. However, the lack of rapidly deposited volcaniclastic deposits is not necessarily a result of reduced volcanic activity. Site 449 is located on a prominent topographic high, approximately 500 m above the regional basin depth and almost 150 km from the Palau Kyushu Ridge. Such elevated sites could very likely be protected from deposition of mass-transported volcaniclastics. At distances from the volcanic arc initially in excess of 100 km and increasing with time, deposition as site 449 could easily have been dominated by pelagic mud and calcareous ooze, as is now the situation in an analogous setting in the Lau Basin [*Karig*, 1970; *Griffin et al.*, 1972], where volcanic activity is at a high level.

The ash component in the forearc sites [*Hussong and Uyeda*, 1981] indicates that volcanism was active after 27 m.y., but at perhaps a lower level than after 21 m.y. B.P. The paucity of volcanic ash in the forearc during the late Oligocene could be the result of more tholeiitic [*Natland and Tarney*, 1981], less explosive arc volcanism erupted from great water depth as the new volcanic arc formed along the West Mariana Ridge.

An indirect but strong argument for a high rate of volcanic activity during most or all of the period during which the Parece Vela Basin opened is derived from the size and mode of growth of the volcaniclastic apron built by volcanic centers along the West Mariana Ridge. This apron is constructed of ash and volcanic-lithic detritus carried by currents or by mass transport mechanisms, often after some shallow water reworking on the cones [*Bouma*, 1975; *Rodolfo and Warner*, 1980]. Such aprons grow by outward progradation of the volcaniclastics over the oceanic basin and an older pelagic to hemipelagic cover [e.g., *Karig and Moore*, 1975b]. Seismic reflection profiles in the Parece Vela Basin [e.g., *Mrozowski and Hayes*, 1979] indicate an approximate apron volume of 450 km^3 km^{-1} (250 km^3 dense rock equivalent) along much of the eastern flank, most of which comprise sediments older than the volcaniclastics drilled at DSDP sites 53, 54, and 450.

The total duration of growth of this apron is suggested by a comparison of its volume with that of the smaller but similar apron in the Mariana Trough. Over the past 5-m.y. volcanic maximum the apron in the Mariana Trough grew at a rate of 6–8 km^3 km^{-1} m.y.$^{-1}$ (dense rock equivalent [*Sample and Karig*, 1982]) along the more active central section of the arc, as a product of one of the highest sustained rates of volcanic activity that has yet been measured in an arc system. The much greater volume of the Parece Vela apron would require an average growth rate at least twice this large over the 30- to 13-m.y. interval from initiation of back arc spreading to the decline of volcanism along the West Mariana volcanic arc. If the rate of volcanism were substantially reduced for more than half this interval (from 30 to 20 m.y. B.P.), the rate required after 20 m.y. B.P. to build the apron would imply a sustained level of volcanic activity 4–6 times as great as has yet been recognized.

The upsurge of volcanism that initiated the mid-Tertiary volcanic pulse quite possibly preceded the opening of the Parece Vela Basin. DSDP hole 448 on the Palau Kyushu Ridge recorded rapid volcaniclastic deposition between the base of the sediment column (34 m.y.) and 30 m.y., but this, as we claimed to be the case in site 450, may be a result of proximity to the active centers. However, site 290, drilled into the distal edge of the volcaniclastic apron forming the west flank of the arc, also showed a relatively high rate of ash deposition between 30 and 34 m.y. B.P. [Karig et al., 1975]. The disappearance of volcanic ash from the stratigraphic records of sites on and west of the Palau Kyushu Ridge at 30 m.y. B.P. marked the migration of volcanic centers caused by opening of the Parece Vela Basin and does not necessitate any decline in the level of volcanic activity.

Arc volcanism in the northern sector of the Philippine Sea, which might be associated with the Bonin extensional zone and Shikoku Basin, is much less well constrained than to the south. Cores in DSDP sites 443 and 444 showed rapid, volcanic-rich sedimentation during the past 2 m.y. and prior to the mid middle Miocene. These records are truncated by the age of basement in these Shikoku Basin drill sites. Locally rapid volcanism was occurring along the Palau Kyushu Ridge (site 296) when spreading began in the Shikoku Basin (25 m.y. B.P.) and had been active since the middle or possibly early Oligocene [Karig et al., 1975].

The data from the northern Philippine Sea permit the interpretation that volcanic pulses in the Bonin Arc were synchronous with those in the Mariana Arc, but certainly do not require it. In particular, the present pulse of activity may have begun more recently in the north than in the south.

Conclusions and Implications

That volcanism in arc systems fluctuates is neither an obvious nor a trivial conclusion. Many factors can produce apparent changes in volcanic intensity in the rock record, including site location, both on a local and on a regional scale, and style of volcanism. From the distribution of the products of historic volcanism in the Mariana Arc it is clear that evidence for volcanic activity can be very subtle only a few tens of kilometers away from a volcanic center if volcanism is submarine and the observation site is on a shallow topographic high. Observation points on the volcanic edifice obviously magnify the effects of even minor activity. By far the largest fraction of volcanic detritus from the volcanic arcs is transported by density flow processes, which are very dependent upon topography. For this reason the volcanic record in the forearc depends strongly on the presence or absence of a frontal arc that serves as

a topographic barrier. Another factor to be recognized is that growth of volcaniclastic aprons is by progradation over much more slowly deposited pelagic or hemipelagic sediments that may have a much less obvious volcanic component [Karig and Moore, 1975b]. Thus nannofossil oozes or pelagic muds lying beneath volcaniclastic turbidites cannot simply be interpreted as evidence for a preceding period with a lower rate of volcanic activity.

Calibration using volcanic ash carried by wind and surface currents avoids the biasing created by topography but is subject to dilution, alteration, azimuthal variation in current transport, changes in sedimentation rate, and contamination from sources in other arcs. Recognition of extremely large silicic, ash-producing eruptions [Ninkovitch et al., 1978] raises the question of the dependency of ash content on chemistry and style of volcanic activity, especially at greater distances from the arc. Mafic andesites and arc tholeiites are not likely to produce as much or as far traveled ash as arcs with more silicic, even ignimbritic eruptions.

The most likely setting in which to obtain a representative record of relative volcanic intensity appears to be in the midsection of the volcaniclastic aprons that generally lie along the rear flank of the volcanic centers. The record at such sites is not dominated by activity at a single volcanic center but integrates that of at least several centers. Because the rate of volcanic deposition is moderately high in the midfan setting, variations in total sedimentation rate are large and quantitatively useful. The combination of the good paleontological control generally found in these deposits, combined with simple geometries delineated by seismic reflection profiling, permits quantification of this, a major component of the total volcanic record.

Despite the factors that complicate the interpretation of volcanic activity in arc systems, drilling and seismic reflection profiles clearly outline major fluctuations in that activity around the Philippine Sea, and particularly in the Mariana Arc since the late Eocene. A rough calculation of the volume of the volcaniclastic apron along the eastern flank of the Parece Vela Basin (250 km^3 km^{-1} dense rock equivalent), to which must be added that fraction of the volume of the West Mariana Ridge representing volcanic cones that produced the apron, leads to the conclusion that the mid-Cenozoic pulse produced 5–8 times the volume comprising the late Cenozoic pulse in the Mariana Trough (60 km^3 km^{-1} [Sample and Karig, 1982]). The average rate of mid-Cenozoic volcanism would be at least twice that of the more recent volcanic pulse if activity were continuous during the opening of the Parece Vela Basin and could be more than 5 times as great if the ratio of volcanic material in cones and aprons was about 50:50, as in the Mariana [Sample and Karig, 1982], or if the duration of the mid-Cenozoic pulse were shorter. Volcanic production in the intervening period (13–5 m.y. B.P.) was too low to add significantly to the volcaniclastic apron and must have been near an order of magnitude less than that of the preceding volcanic pulse.

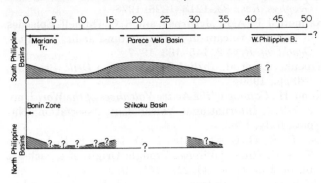

Fig. 1.   Timing of back arc spreading and arc volcanism in the Mariana (upper) and Bonin (lower) arc systems. The relative intensity of arc volcanism is noted by the height of the curve. Spreading half rates are shown for back arc basins where known. The West Philippine Basin did not open behind the Mariana Arc but is added for comparison.

The proposed middle and late Cenozoic volcanic pulses of volcanism in the Mariana Arc appear to have been much more nearly synchronous with back arc spreading episodes (Figure 1) than was claimed by *Scott and Kroenke* [1980]. That the initial phases of presently active back arc basins are associated with volcanism, generally at a moderate to high level relative to contemporaneous volcanic activity in other arcs, also supports synchroneity of spreading and arc volcanism. This conclusion must be tempered by the recognition that not all back arc basins may form from identical processes and that some may not even be related to subduction.

Synchroneity of volcanic fluctuations among arc systems on regional and global scales, the relationship of these fluctuations to changes in subduction rates, and the quantification of the volcanic fluctuations remain extremely important, unresolved questions.

Current geodynamic models lead to a common but usually unstated assumption that intensity of arc volcanism can be correlated with subduction rates. A history of subduction rates cannot be directly determined for the Mariana Arc because the Philippine plate is surrounded by convergent boundaries, but an indirect approach can be used. There is no evidence for large fluctuations in subduction rates between the Pacific and Asian plates since the mid-Tertiary [e.g., *Jurdy*, 1979], so that volcanic fluctuations along the Pacific-Philippine boundary should be anticorrelative with fluctuations along the Philippine-Asian margin if they are dependent upon subduction rates.

Field studies in the northern Philippines [*Balce et al.*, 1979; D. E. Karig, unpublished data, 1980] and the record of volcanic ash at DSDP site 292 [*Donnelly*, 1975] outline volcanic maxima from the late Oligocene into the mid-Miocene and during the past few million years. This remarkably close correlation with the pulses along the Mar-

iana Arc implies that the volcanic fluctuations are not simply the result of subduction rate changes. Of course, once back arc spreading begins, the rate of subduction at one or both of the Philippine plate boundaries must increase by the rate of spreading, which might also intensify the correlation of volcanic activity with spreading.

Volcanic fluctuations around the Philippine plate correlate moderately well with those postulated in other circum-Pacific arcs [*Kennett et al.*, 1977; *Cadet and Fujioka*, 1980]. The late Miocene appears to mark a global relative minimum of volcanic activity, although the transition dates are quite variable. Delineation of volcanic fluctuations in individual arc systems during the Cenozoic is urgently needed to confirm and quantify these variations on a global scale.

References

Baker, P. E., Recent volcanism and magmatic variation in the Scotia Arc, in *Antarctic Geology and Geophysics*, edited by R. J. Aide, p. 57, Oslo Universitets Forlaget, Oslo, 1971.

Balce, G. R., A. L. Magpantay, and A. S. Zanoria, Tectonic scenarios of the Philippines and northern Indonesian region, paper presented at the Ad Hoc Working Group Meeting on the Geology and Tectonics of Eastern Indonesia, U.N. Econ. and Social Comm. for Asia and the Pacific, CCOP, Bandung, July 3–14, 1979.

Barker, P. F., A spreading centre in the east Scotia Sea, *Earth Planet. Sci. Lett.*, 15, 123–132, 1972.

Bouma, A. H., Sedimentary structures of Philippine Sea and Sea of Japan sediments, *Initial Rep. Deep Sea Drill. Proj.*, 31, 471–488, 1975.

Cadet, J.-P., and K. Fujioka, Neogene volcanic ashes and explosive volcanism: Japan Trench transect, Leg 57, DSDP, *Initial Rep. Deep Sea Drill. Proj. 56, 57*, 1027–1041, 1980.

Carney, J. N., and A. MacFarlane, Volcano-tectonic events and pre-Pliocene crustal extension in the New Hebrides, in *International Symposium on Geodynamics in SouthWest Pacific, Noumea*, pp. 91–104, Editions Technip, Paris, 1977.

Chase, C. G., Extension behind island arcs and motions relative to hot spots, *J. Geophys. Res.*, 83, 5385–5387, 1978.

Crawford, A. J., L. Beccaluva, and G. Serri, Tectono-magmatic evolution of the West Philippine-Mariana region and the origin of boninites, *Earth Planet. Sci. Lett.*, 54, 346–356, 1981.

Curray, J. R., D. G. Moore, L. A. Lawver, F. J. Emmel, R. W. Raitt, M. Henry, and R. Kieckhefer, Tectonics of the Andaman Sea and Burma, *Am. Assoc. Pet. Geol. Mem.*, 29, 189–198, 1979.

Donnelly, T. W., Neogene explosive volcanic activity of the western Pacific sites 292 and 296, DSDP Leg 31, *Initial Rep. Deep Sea Drill. Proj.*, 31, 577–598, 1975.

Dugas, F., J. N. Carney, C. Cassignol, P. A. Jezek, and M.

Monzier, Dredged rocks along a cross-section in the southern New Hebrides island arc, and their bearing on the age of the arc, in *International Symposium on Geodynamics in South-West Pacific, Noumea*, pp. 105–116, Editions Technip, Paris, 1977.

Emery, K. O., Marine geology of Guam, *U.S. Geol. Surv. Prof. Pap., 403-B*, 67 pp., 1962.

Fischer, A. G., et al., *Initial Rep. Deep Sea Drill. Proj., 6*, 1971.

Garrison, R. E., S. O. Schlanger, and D. Wachs, Petrology and paleographic significance of Tertiary nannoplankton-foraminiferal limestones, Guam, *Paleogeogr. Paleoclimatol. Paleoecol., 17*, 49–64, 1975.

Griffin, J. J., M. Koide, A. Hohndorf, J. W. Hawkins, and E. D. Goldberg, Sediments of the Lau Basin—Rapidly accumulating volcanic deposits, *Deep Sea Res., 19*, 133–138, 1972.

Hathaway, J. C., and D. Carroll, Petrology of the insoluble residences, *U.S. Geol. Surv. Prof. Pap., 403-D*, 37–44, 1964.

Hsui, A. T., and M. N. Toksöz, Back-arc spreading: Trench migration, continental pull, or induced convection?, *Tectonophysics, 74*, 89–98, 1981.

Hussong, D. M., and S. Uyeda, Tectonic processes and the history of the Mariana Arc: A synthesis of the results of Deep Sea Drilling Project Leg 60, *Initial Rep. Deep Sea Drill. Proj., 60*, 909–929, 1981.

Hussong, D. M., et al., *Initial Rep. Deep Sea Drill. Proj., 60*, 1981.

Ingle, J. C., Jr., Summary of late Paleogene-Neogene insular stratigraphy, paleobathymetry, and correlations, Philippine Sea and Sea of Japan region, *Initial Rep. Deep Sea Drill. Proj., 31*, 837–856, 1975.

Jurdy, D. M., Relative plate motions and the formation of marginal basins, *J. Geophys. Res., 84*, 6796–6802, 1979.

Karig, D. E., Ridges and basins of the Tonga-Kermadec island arc system, *J. Geophys. Res., 75*, 239–255, 1970.

Karig, D. E., Structural history of the Mariana island arc system, *Geol. Soc. Am. Bull., 82*, 323–344, 1971.

Karig, D. E., Basin genesis in the Philippine Sea, *Initial Rep. Deep Sea Drill. Proj., 31*, 857–879, 1975.

Karig, D. E., Initiation of subduction zones: Implications for arc evolution and ophiolite development, *Spec. Publ. Geol. Soc. London, 10*, 563–576, 1982.

Karig, D. E., and R. W. Kay, Fate of sediments on the descending plate at convergent margins, *Philos. Trans. R. Soc. London, Ser. A, 301*, 233–251, 1981.

Karig, D. E., and G. F. Moore, Tectonic complexities in the Bonin island arc system, *Tectonophysics, 27*, 97–118, 1975a.

Karig, D. E., and G. F. Moore, Tectonically controlled sedimentation in marginal basins, *Earth Planet. Sci. Lett., 26*, 233–238, 1975b.

Karig, D. E., et al., *Initial Rep. Deep Sea Drill. Proj., 31*, 926 pp., 1975.

Karig, D. E., R. N. Anderson, and L. D. Bibee, Characteristics of back spreading in the Mariana Trough, *J. Geophys. Res., 83*, 1213–1226, 1978.

Kennett, J. P., A. R. McBirney, and R. C. Thunell, Episodes of Cenozoic volcanism in the circum-Pacific, *Volcanol. Geotherm. Res., 2*, 145–163, 1977.

Kroenke, L., et al., *Initial Rep. Deep Sea Drill. Proj., 59*, 820 pp., 1980.

Kuno, H., *Catalog of the Active Volcanoes of the World*, pp. 245–252, International Volcanology Association, Naples, Italy, 1962.

Lee, C. S., G. G. Shor, Jr., L. D. Bibee, R. S. Lu, and T. W. C. Hilde, Okinawa Trough: Origin of a back-arc basin, *Mar. Geol., 35*, 219–241, 1980.

McBirney, A., J. Sutter, H. Naslund, K. Sutton, and C. White, Episodic volcanism in the Oregon Cascade range, *Geology, 2*, 585–589, 1974.

Meijer, A., M. Reagan, H. Ellis, S. Sutter, M. Shafiqullah, P. Damon, and S. Kling, Chronology of volcanic events in the eastern Philippine Sea, this volume.

Molnar, P., and T. Atwater, Interarc spreading and Cordilleran tectonics as alternates related to the age of subducted oceanic lithosphere, *Earth Planet. Sci. Lett., 41*, 330–340, 1978.

Mrozowski, C. L., and D. E. Hayes, The evolution of the Parece Vela Basin, eastern Philippine Sea, *Earth Planet. Sci. Lett., 46*, 49–67, 1979.

Natland, J. H., and J. Tarney, Petrologic evolution of the Mariana Arc and back-arc system: A synthesis of drilling results in the south Philippine Sea, *Initial Rep. Deep Sea Drill. Proj., 60*, 877–908, 1981.

Ninkovitch, D., R. S. J. Sparks, and M. T. Ledbetter, The exceptional magnitude and intensity of the Toba eruption, Sumatra: An example of the use of deep-sea tephra layers as a geological tool, *Bull. Volcanol., 41*, 1–13, 1978.

Oxburgh, E. R., and D. L. Turcotte, Origin of paired metamorphic belts and crustal dilation in island arc regions, *J. Geophys. Res., 76*, 1315–1327, 1971.

Richard, J. J., *Catalogue of Active Volcanoes of the World Including Solfatara Fields*, part 13, Kermadec, Tonga, and Samoa, International Association of Volcanology, Rome, 1962.

Rodolfo, K. S., and R. J. Warner, Tectonic, volcanic, and sedimentologic significance of volcanic glasses from site 450 in the eastern Parece Vela Basin, Deep Sea Drilling Project Leg 59, *Initial Rep. Deep Sea Drill. Proj., 59*, 603–607, 1980.

Sample, J. C., and D. E. Karig, A volcanic production rate for the Mariana island arc, *J. Volcanol. Geotherm. Res., 13*, 73–82, 1982.

Schlanger, S. O., Petrology of the limestones of Guam, *U.S. Geol. Surv. Prof. Pap. 403-D*, 52 pp., 1964.

Scott, R., and L. Kroenke, Evolution of back arc spreading and arc volcanism in the Philippine Sea: Interpretation of Leg 59 DSDP results, in *The Tectonic and Geologic Evolution of Southeast Asian Seas and Islands, Geophys.*

Monogr. Ser., vol. 23, edited by D. E. Hayes, pp. 283–291, AGU, Washington, D. C., 1980.

Scott, R. B., and L. Kroenke, Periodicity of remnant arcs and back-arc basins of the South Philippine Sea, *Oceanol. Acta, 4,* suppl., 203–212, 1981.

Sharaskin, A. Y., N. A. Bogdanov, and G. S. Zakariadze, Geochemistry and timing of the marginal basin and arc magmatism in the Philippine Sea, *Philos. Trans. R. Soc. London, Ser., A, 300,* 287–297, 1981.

Shih, T., Magnetic lineations in the Shikoka Basin, *Initial Rep. Deep Sea Drill. Proj., 58,* 783–788, 1980a.

Shih, T., Marine magnetic anomalies from the western Philippine Sea: Implications for the evolution of marginal basins, in *The Tectonic and Geologic Evolution of Southeast Asian Seas and Islands, Geophys. Monogr. Ser.,* vol. 23, edited by D. E. Hayes, pp. 49–75, AGU, Washington, D. C., 1980b.

Taylor B., Bismark Sea: Evolution of a back-arc basin, *Geology, 7,* 171–174, 1979.

Tracey, J. I., et al., General geology of Guam, *U.S. Geol. Surv. Prof. Pap., 403-A,* 104 pp., 1964.

Uyeda, S., and H. Kanamori, Back arc openings and the mode of subduction, *J. Geophys. Res., 84,* 1049–1062, 1979.

Watts, A. B., and J. K. Weissel, Tectonic history of the Shikoka marginal basin, *Earth Planet. Sci. Lett., 25,* 239–250, 1975.

Weeks, L. A., R. N. Harbison, and G. Peter, Island arc system in Andaman Sea, *Am. Assoc. Pet. Geol. Bull., 51,* 1803–1816, 1967.

Weissel, J. K., Evolution of the Lau Basin by the growth of small plates, in *Island Arcs, Deep Sea Trenches and Back-Arc Basins, Maurice Ewing Ser.,* vol. 1, edited by M. Talwani and W. C. Pitman III, pp. 429–436, AGU, Washington, D. C., 1977.

Weissel, J. K., Magnetic lineations in marginal basins of the western Pacific, *Philos. Trans. R. Soc. London, Ser. A, 300,* 233–247, 1981.

Weissel, J. K., B. Taylor and G. D. Karner, The opening of the Woodlark Basin, subduction of the Woodlark spreading systems and the evolution of northern Melanesia since mid-Pliocene time, *Tectonophysics,* in press, 1982.

Convergence at the Caroline-Pacific Plate Boundary: Collision and Subduction

K. A. Hegarty

Lamont-Doherty Geological Observatory of Columbia University, Palisades, New York 10964
The Department of Geological Sciences, Columbia University, New York, New York 10027

J. K. Weissel

Lamont-Doherty Geological Observatory of Columbia University, Palisades, New York 10964

D. E. Hayes

Lamont-Doherty Geological Observatory of Columbia University, Palisades, New York 10964
The Department of Geological Sciences, Columbia University, New York, New York 10027

The eastern boundary of the Caroline plate, in the western equatorial Pacific, is composed of three structural provinces distinguished primarily on the basis of morphology. Each province shows evidence for convergence between the Caroline and Pacific plates though the structural style varies considerably between each province. Most notably, the sense of underthrusting appears to change along the boundary at about 3°N. To the south, at the Mussau System, Caroline lithosphere underthrusts beneath the Mussau Ridge (which is part of the Pacific plate), while to the north the Caroline plate appears to overthrust the Pacific plate. Recently collected seismic reflection profiles across each province documents the structural changes along and across strike of the Caroline-Pacific plate boundary. With this information, we estimate that a minimum of approximately 4 km of crustal shortening has occurred at about 5°N due to convergence of the two plates. Further to the south (about 2°N), simple gravity models suggest that about 10 km of Caroline lithosphere lies beneath the present-day Pacific plate. Using a previously determined pole of rotation describing Caroline-Pacific relative motion (Weissel and Anderson, 1978), we grossly estimate the duration of the convergence between these two plates at about one million years. It is suggested that variation in the convergence rate along the plate boundary provides the primary control on the variation of structural deformation observed between provinces; however, favorable thermal conditions are factors that are considered. If the eastern boundary of the Caroline plate is a region of incipient though perhaps transient subduction, as we postulate, then the geophysical and geological evidence presented can constrain models on the initiation of subduction.

Introduction

The Caroline plate is located in the western equatorial Pacific north of New Guinea (Figure 1). The existence of this relatively small plate and the nature of its boundaries have been investigated most recently by *Weissel and Anderson* [1978]. In this study we will look in greater detail at the eastern boundary of the Caroline plate where the subduction process may be in its earliest stages of development.

The eastern portion of the Caroline plate interacts with the Pacific plate along a zone that is characterized by convergence [*Weissel and Anderson*, 1978]. The details of the bathymetric expression of this convergent boundary (Figure 2) have improved with the recent acquisition of over 3000 nautical miles of underway geophysical data (R/V *Vema* 3603). A physiographic interpretation of the available bathymetric and seismic reflection data from this plate boundary segment is shown in Figure 3. Considerable along-strike variation is observed in the style of

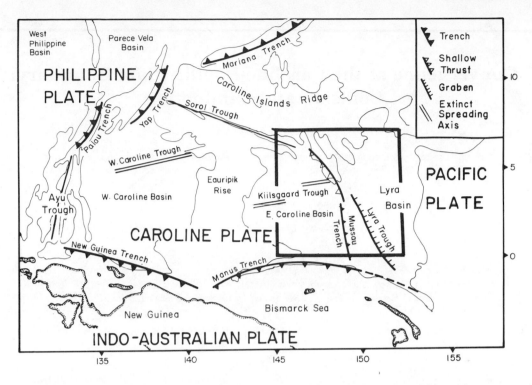

Fig. 1. Tectonic setting of the Caroline plate. The enclosed area delineates the region of this study encompassing the eastern limit of the Caroline plate at the Caroline-Pacific plate boundary.

deformation and amount of shortening across the Caroline-Pacific convergent boundary, and three distinct topographic or structural provinces can be identified (Figure 4). One of the most interesting features of this plate boundary, in addition to its young age, is the change in the sense of underthrusting at about 3°N. The distinctive morphology of the Mussau Trench suggests that the Caroline plate is underthrusting the Pacific plate, while to the north, observed crustal deformation implies that the Caroline plate is attempting to override the Pacific plate.

Geophysical data from the eastern section of the Caroline-Pacific plate boundary may supply constraints that lead to a better understanding of where and how the subduction process begins. There are no obvious present-day analogs on the ocean floor to this early stage in the subduction process which is observed at the Caroline-Pacific boundary. Questions pertaining to the Caroline-Pacific boundary that may have global importance include the following: What is the temporal tectonic relationship, if any, between the three structural provinces, i.e., is one province precursory to another? If the tectonic differences between the three provinces cannot be related to different stages in the development of the subduction process, then what other factors might control the style of deformation at this convergent boundary? Is the relatively small Caroline plate simply responding to forces imposed at its plate boundaries by the much larger plates bounding it, or does

its motion directly reflect large-scale plate driving forces that are active within the mantle?

First, we briefly present the geological and geophysical information that provides constraints on the precise nature of the Caroline-Pacific plate interactions occurring within the structural provinces. General physiographic, bathymetric, seismic reflection, magnetic, gravimetric, heat flow, and seismicity data are summarized. We then focus on those geophysical data that best define the tectonic processes occurring along this young boundary. Following the general descriptions below, our analysis concentrates on seismic reflection and gravity data that can detail aspects of the along-strike tectonic variation at the Caroline-Pacific plate boundary.

General Geological and Geophysical Setting

The bathymetry and structures of the Caroline plate and its boundaries have been previously investigated [*Weissel and Anderson,* 1978; *Mammerickx,* 1978; *Erlandson et al.,* 1976; *Bracey,* 1975; *Bracey and Andrews,* 1974]; however, the past and present tectonics of the boundaries are not well understood. For the purposes of this study, we shall accept the conclusion of *Weissel and Anderson* [1978] that a Caroline plate exists and is composed dominantly of the West Caroline and East Caroline basins (Figure 1). The Caroline plate is bounded to the north and east by the Pacific plate, to the south by the Indo-Australian plate,

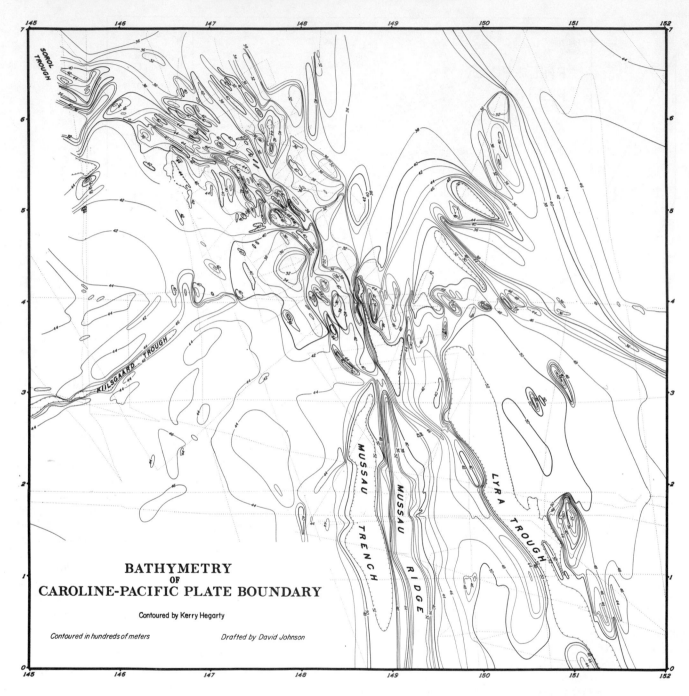

Fig. 2. Bathymetric contours of the Caroline-Pacific plate boundary for the location shown in Figure 1. Contours are in hundreds of meters (corrected) and at 200-m intervals. Depths greater than 5200 m and less than 3200 m across the Mussau Trench and Mussau Ridge are not shown. Track coverage (dotted lines) is sparse to the northeast.

and to the west by the Philippine Sea plate. The physiographic features corresponding to these plate boundaries are labelled in Figure 1. On the basis of bathymetric and seismic profiling data, three structural provinces can be distinguished along the eastern portion of the Caroline plate. Province I is a broad zone of deformation (usually wider than 100 km) characterized by many narrow crustal blocks (usually 2–5 km across) which appear to have been

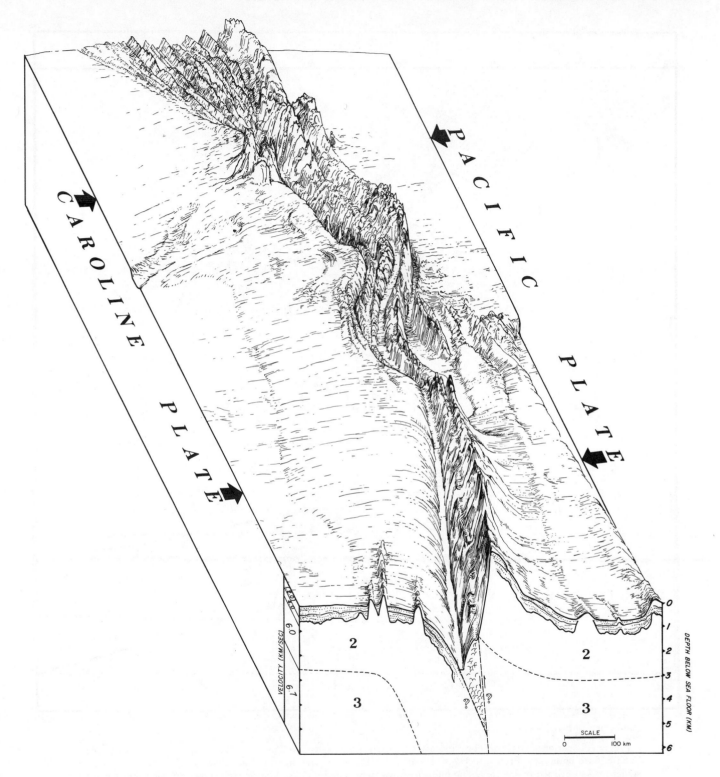

Fig. 3. Physiographic interpretation of the structures associated with the eastern boundary of the Caroline plate. The sense of underthrusting appears to change at the northern limit of the Mussau System. The depths to oceanic layers 2 and 3 are determined from refraction data obtained within the East Caroline Basin. The face of this diagram, constrained by gravity models (Figure 12) discussed in the text, is highly interpretive.

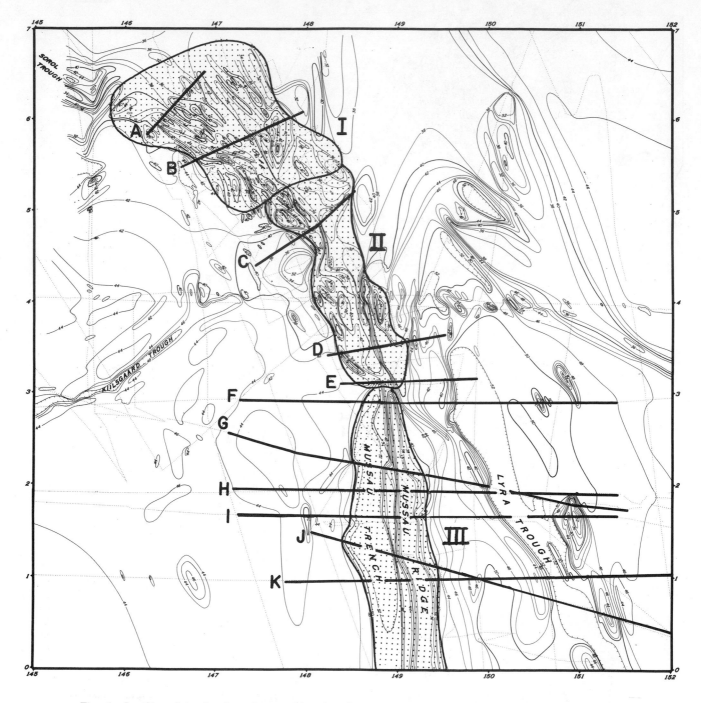

Fig. 4. Location of structural provinces and location of representative seismic reflection profiles at the Caroline-Pacific plate boundary. Profile sections are shown in Figure 5. The provinces are distinguished on the basis of structure as outlined in the text. The main features to be noted include a broad, relatively diffuse zone of deformation associated with province I which narrows and intensifies into province II, but similar structural elements are maintained. Province III is characterized by trenchlike morphology.

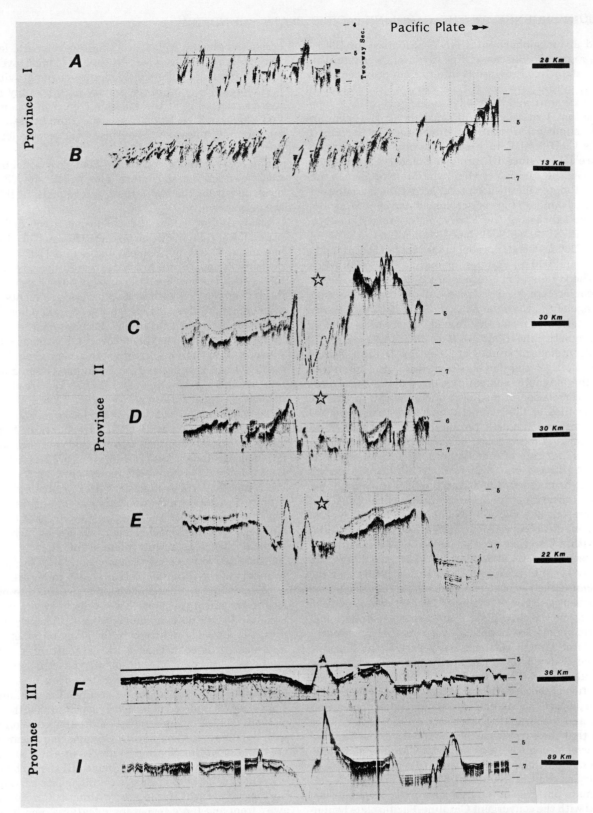

Fig. 5. Seismic profiler records across the three structural provinces located in Figure 4. Note that the distance between profiles E and F is less than 25 km, showing that the trenchlike morphology of province III develops across a very short distance. The star positioned atop profiles C, D, and E identifies the topographic low (generally less than 30 km in width) that characterizes province II. Profile B is used to determine the amount of shortening across province I (Figure 10). Steep scarps on profiles C, D, and E have been highlighted with dashed lines.

rotated about subhorizontal NNW-SSE axes (profiles A and B, Figure 5). Basement dips vary from less than 1° to greater than 15°. Province II differs from province I only slightly, in that province II tapers to a width of about 40 km (maximum) and is always associated with a topographic low. Crustal block dimensions across strike are usually slightly less than the dimensions in province I. Profiles C, D, and E in Figure 5 typify the features of province II. Province III (profiles F and I, Figure 5) encompasses the Mussau System (the Mussau Trench and Mussau Ridge) where shortening across the Caroline-Pacific plate boundary may be accommodated by the Caroline plate underthrusting the Pacific plate [Weissel and Anderson, 1978]. A well-developed trench axis (depths exceeding 7200 m) trends parallel to the Mussau Ridge (depths of less than 2000 m) and lies approximately 30 km to the west. The style and magnitude of the convergence across each province are discussed more fully in the next section.

The age of the Caroline lithosphere in both the eastern and western basins is early to mid-Oligocene as deduced from magnetic lineations [Weissel and Anderson, 1978] and paleontological studies at Deep Sea Drilling Project (DSDP) sites 62 and 63 [Winterer et al., 1971]. Figure 6 shows the magnetic anomalies plotted along ship tracks and the correlation of magnetic anomalies 10 to 13. Projected profiles of the magnetic anomalies are compared with a synthetic magnetic profile (Figure 7) computed using the magnetic time scale of LaBrecque et al. [1977]. A half-spreading rate of 6 cm/yr from 36 to 28.5 m.y. B.P. was used [Weissel and Anderson, 1978]. Careful examination of the magnetic anomalies provides the most direct evidence concerning the age and affinity of the crust east of the present-day Caroline-Pacific plate boundary. Did the present-day observed deformation and shortening originate within Caroline crust or between Caroline and Pacific crust? Magnetic anomalies 9, 10, 11, and 12 (Figure 6) clearly cut into provinces I and II, indicating that the crust comprising the eastern portion of this deformed zone (and perhaps crust even further to the east) was generated as part of the Caroline plate. Furthermore, correlation of magnetic anomalies 12 and 13 suggests that crust north of the Sorol Trough was originally part of the Caroline plate and is now part of the Pacific plate. Track spacing east of about 148.5°E is too large to allow confident determination of the easternmost extent of the Caroline plate; however, based on magnetic, seismic reflection, gravimetric, and topographic observations discussed later, we believe that the current eastern boundary of the Caroline plate developed initially as an intra-Caroline plate phenomenon. We will however continue to refer to the present boundary as the Caroline-Pacific plate boundary.

The main patterns of the free air gravity anomalies associated with the convergent Caroline-Pacific plate boundary include large negative anomalies (−150 mgal) across the Mussau Trench, large positive anomalies (+175 mgal)

across the Mussau Ridge, and small-wavelength, low-amplitude anomalies (between −40 and +40 mgal) over provinces I and II. A more subtle feature of the gravity field is the presence of an outer gravity high possibly related to underthrusting at the Mussau Trench. It lies parallel to the Mussau Trench and has a typical amplitude of < +25 but ranging upward to +40 mgal and a wavelength of about 50–100 km (Figure 8).

Figure 8 shows the location of heat flow, sonobuoy stations, earthquake epicenters, and DSDP site 63. Focal mechanisms for the two largest earthquakes in this area are also shown.

The heat flow values at each station, given in Table 1, are all lower than the value predicted for crust of Oligocene age (about 90 mW/m^2; Parsons and Sclater [1977]). All the measured temperature gradients were linear, so it is unlikely that the low heat flow values are due to hydrothermal circulation within the crust and sediments. Measurements at stations 3 through 6 were taken between or on top of rotated crustal blocks. Perhaps heat is escaping by convective processes along nearby basement outcrops which correspondingly reduces the heat flow in sedimented areas as suggested for the Indian Ocean and Mariana Trough [Anderson et al., 1977; M. Hobart et al., unpublished data, 1982].

Sonobuoy station locations and velocity sections of Hussong [1972] and Den et al. [1971] are shown in Figure 8. Table 2 lists the detailed velocity and layer thicknesses for each sonobuoy. Seismic reflection profiles across the East Caroline Basin and the region between the Mussau Ridge and the Lyra Trough show great acoustic similarity (Figure 5). These observations support the suggestion that crust on each side of the Mussau System is of the same age and that present-day relative motion across this boundary began as an intraplate event.

The relatively sparse seismicity along the eastern boundary of the Caroline plate is shown in Figure 8. Most of the earthquakes are shallow, but too small to provide focal mechanism solutions. Two shallow events near the Caroline-Pacific plate boundary were sufficiently large to provide solutions, and their nodal plane parameters and locations are listed in Table 3. Thrust faulting is indicated by the focal mechanism solution of event I [Bergman and Solomon, 1980]. This solution is based on first-motions (shown in Figure 9a), and the direction of maximum principal stress is inferred to be nearly horizontal with a trend of about 272°. A compressional environment is consistent with the interpretation that the Caroline and Pacific plates are colliding at this position [Weissel and Anderson, 1978]. The lack of seismicity within province III should be noted.

Event II (Figure 8) is clearly an intraplate earthquake. Furthermore, event II is located within the Caroline Basin away from any large topographic features, and hence its focal mechanism solution may define the state of stress within the Caroline plate at this location. The focal mech-

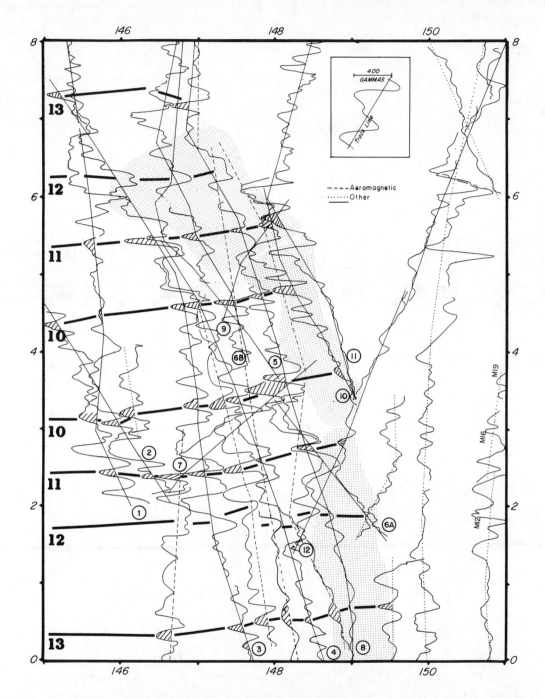

Fig. 6. Magnetic anomalies plotted along ship tracks (solid and dotted lines) and aeromagnetic data (dashed lines) taken from *Bracey* [1975]. The encircled numbers refer to magnetic data along solid lines that are projected onto 000° and shown in Figure 7. Magnetic anomalies 10 through 13 clearly intersect the three structural provinces. Note that northern anomalies 12 and 13 are north of the Sorol Trough. The high fidelity of the magnetic anomalies across the disrupted oceanic crust of provinces I and II suggests that most of the block faulting and rotation has occurred about a northerly, subhorizontal axis and that there has been little, if any, shear or transcurrent motion (see text for discussion). Mesozoic magnetic anomalies have been identified east of the Lyra Trough at about 151°E.

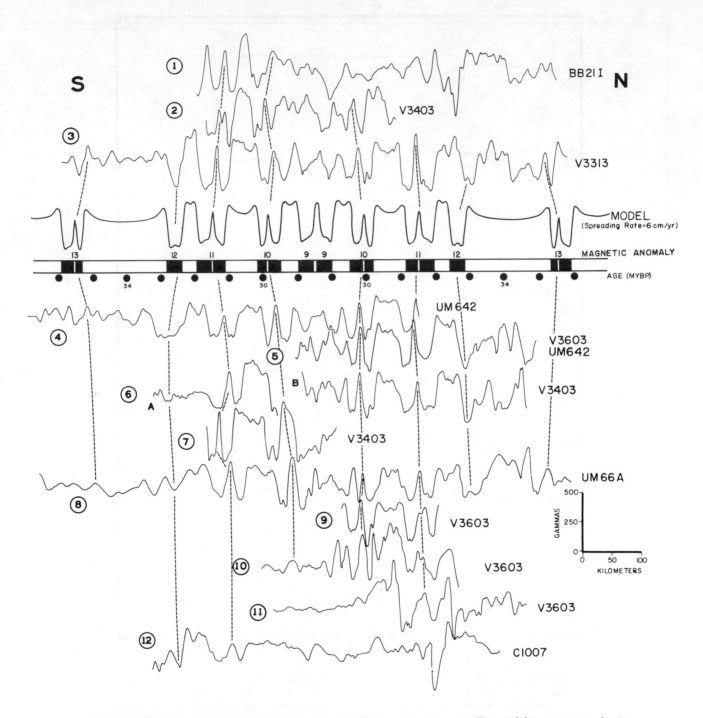

Fig. 7. Projected magnetic anomaly profiles (onto 000°) located in Figure 6. The model was generated using the magnetic time scale of *LaBrecque et al.* [1977] and a half-spreading rate of 6 cm/yr. Anomaly number and age are shown along the block model representing the alternating polarities of the magnetic field.

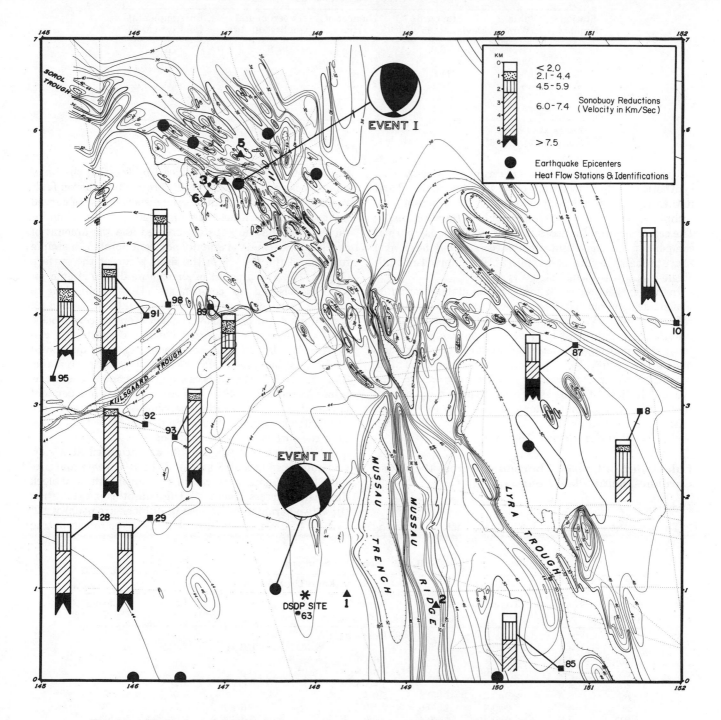

Fig. 8. Additional geophysical information at the Caroline-Pacific plate boundary. Heat flow values and station locations (triangles) are listed in Table 1; sonobuoy solutions (squares) are listed in Table 2; earthquake parameters (circles) and references are listed in Table 3. Note that no seismicity is associated with the Mussau Trench and province II but is observed within province I.

TABLE 1. Heat Flow Stations and Values

Station Number	Value, mW/m^2	Standard Deviation, mW/m^2	Number of Hits	Experimental Quality* (0–10)	Environmental Quality† (A–E)
1	11.1	3.9	7	8	A
2	53.0	17.3	7	8	B
3	53.7	11.0	5	6	D
4	44.2	14.1	5	8	D
5	40.7	9.1	3	7	D
6	45.3	5.1	4	8	D

*Langseth et al. [1966].
†Sclater et al. [1976].

anism solution for event II was determined by comparing synthetic waveforms with the observed first-motion P and naturally rotated SH body waves (Figure 9b) using the technique described by Helmberger and Burdick [1979]. The solution is dominantly strike-slip with a small normal component, and the orientations of the nodal planes (Table 2) are probably accurate to within 5°. The inferred azimuth of the greatest principal stress is 194°.

In the next section we estimate the amount of shortening which has occurred across provinces I and III using seismic reflection and gravity data, respectively. (Estimates across province II are difficult because the apparently more intense deformation, characterized by large basement dips and narrow thrust blocks, prevents the acquisition of high-resolution seismic reflection profiles at standard ship speeds.) In light of these estimates of shortening, we will then examine the implications for the previously determined euler pole describing Caroline-Pacific relative motion [Weissel and Anderson, 1978].

Calculation of Crustal Shortening Across the Caroline-Pacific Plate Boundary

Province I

The amount of apparent horizontal crustal shortening can be estimated across province I by treating the basement blocks as rigid bodies that rotate along curved fault surfaces (with planar faults being a special case of curved fault surfaces). Contact between adjacent blocks is maintained such that no gaps are created and the amount of overlap (h) between originally adjacent points is a simple function of the dip (φ) of the fault plane at the surface (Figure 10). The amount of shortening due to simple rotation consists of two components: (1) shortening due to changes in dip of an originally horizontal surface and (2) shortening due to crustal overlap between adjacent blocks. Thus the amount of apparent horizontal shortening can be directly related to the dip of some originally horizontal surface, e.g., basement, and the attitude of the fault plane. The shortening accrued because of component 1 is determined by measuring the basement dip and is independent of the attitude of the fault plane. Shortening estimates are dominated by component 2 when the dip of the fault plane near the surface is less than about 85°.

Seismic profile B (Figure 10) was acquired at a slow cruising speed (about 5 knots) of the R/V Vema and provides the best data for estimating basement tilts and block dimensions across province I. Uncertainties in estimating these two parameters for each fault block are usually ±2° and ±1 km, respectively, so that the error in estimating

TABLE 2. Sonobuoy Solutions

Station Number	Water Depth, km	Velocity (km/s)/Thickness (km)						
		Layer 1	Layer 2	Layer 3	Layer 4	Layer 5	Layer 6	Layer 7
After Hussong [1972]								
8	4.7	(1.7)/0.3	2.9/0.2		5.4/2.1	6.6/		
10	4.6	(1.7/0.5		4.6/1.4	5.4/2.8			7.7/
85	5.0	(2.0)/0.6		4.6/1.5		6.5/		
87	4.6	(2.0)/0.5		4.9/1.4		6.5/1.3	7.2/	
89	3.8	(2.0)/0.3	3.6/0.9	5.0/1.2		7.0/		
90	4.3	(2.0)/0.2	4.3/0.6	5.1/0.8		6.5/2.1	7.2/	
91	4.3	(2.0)/0.3	4.2/0.4	5.2/0.9		6.3/2.2	7.0/2.5	8.3/
92	4.5	(2.0)/0.4	4.0/0.3		6.0/2.2	6.7/3.1		7.7/
93	4.5	(2.0)/0.4	4.0/0.2		6.0/2.0	6.7/3.8		7.7/
95	4.4	(2.0)/0.4	(4.0)/1.0		5.9/1.4	6.5/0.5	6.9/2.1	8.0/
After Den et al. [1971]								
28	4.5	(2.1)/0.6		5.1/1.4			6.9/3.3	7.8/
29	4.5	(2.1)/0.6		5.1/1.4			7.0/3.5	7.9/

TABLE 3. Earthquake Epicenters and Related Information

Event Number	Date	Location Latitude, °N	Location Longitude, °E	Depth, km	Body Wave Magnitude M_b	Nodal Str/Dip	Planes Str/Dip	P Axis Az/Pl	T Axis Az/Pl	M_s	Moment, dyn cm
I	Aug. 20, 1968	5.43	147.11	28.0*	5.60	34/43SE	340/63SW	272/12	22/56	(5.0)†	· · ·
II	Aug. 30, 1976	1.03	147.56	20.0	5.80	62/85SE	330/69NE	194/12	288/20	5.6	1.5×10^{25}

Str, strike; Az, azimuth; Pl, plunge.
*ISC determination.
†Calculated from the relationship $M_b = 2.5 + 6.3 M_s$.

the shortening for each crustal block is probably less than 100 m. The cumulative error associated with the estimate across the entire section (>20 blocks, Figure 10) is about ±2 km depending on the assumed fault dip. If the fault plane dip at the surface is vertical for each crustal block, then the shortening across profile B is about 1.3 km. For fault plane dip angles of 80°, 70°, and 40°, the shortening is estimated to be 3.9, 7.0, and 23.7 km, respectively. It is assumed that profile B crosses the southwestern and northeastern limits of the deformation.

It is believed that the dip of fault planes across profile B is greater than 40°, but it is difficult to confidently discern the fault plane attitude because of the vertical exaggeration of the seismic reflection profiles (12.5 to 1). True dips of a fault plane between 40° and 90° disrupting basement and sediments would be observed on the seismic reflection record as apparent angles between 84° and 90°, respectively, and therefore difficult to distinguish.

Province III

Shortening within province III is estimated by modelling the observed gravity using simple conceptual models of the elements involved in the initial stages of subduction. A suite of models was generated using different geometries and appropriate densities for the crust and mantle bodies involved in the subduction process. The conclusions of this work that are germane to the question of how much shortening or convergence has occurred between the Caroline and Pacific plates are twofold. First, deflection by a vertical point load of a thin elastic half plate (flexural rigidity of 1×10^{29} dyn cm) overlying a weak fluid can adequately represent the observed surface topography of both the downgoing and the overriding plates (Figure 11). Use of a thin elastic plate in modelling the lithospheric response to loading has been shown to be satisfactory [*Watts et al.*, 1980; *McNutt and Menard*, 1978; *Caldwell et al.*, 1976]. A more complete description of the half-plate approximation to observations at the Caroline-Pacific plate boundary is in preparation (K. A. Hegarty et al., 1982). The three stacked sections in Figure 11 are from east-west projections of profiles G, H, and I. A deflection curve calculated by *Weissel and Anderson* [1978], also using a vertical point load, does not fit the observed topography as well as our new model because they used a continuous plate in their calculations. The fact that the same deflection curve can closely match the observed topography on both sides of the trench axis suggests that each plate edge serves as the load for the other. Furthermore, the same value for rigidity can be used to explain the gross topography on both sides of the trench axis which implies that crust of similar age exists on both sides of the trench axis.

The second conclusion is based on gravity models that are constructed using the elastic plate deflection to define the relief at the water/crust and crust/mantle interfaces. A simple geometric relationship is assumed (Figure 12) to estimate the spatial association of the units involved in

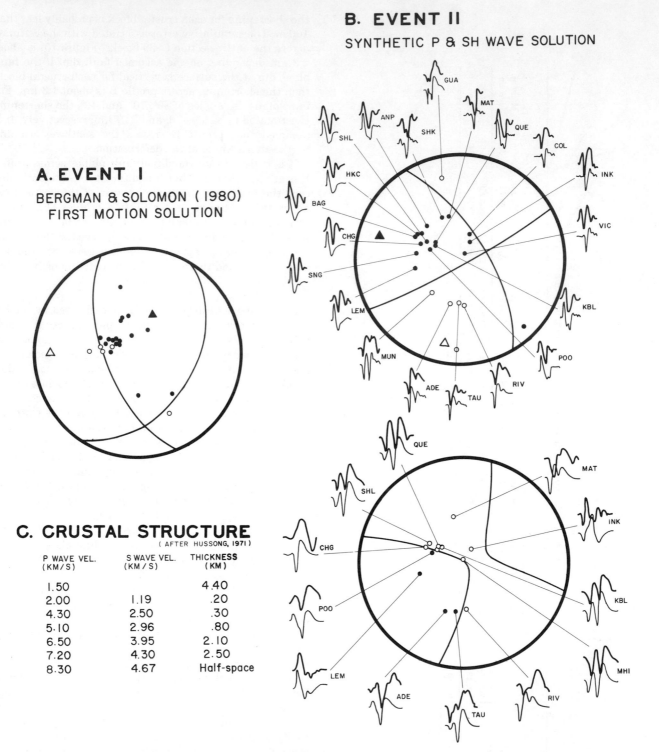

A. EVENT I
BERGMAN & SOLOMON (1980) FIRST MOTION SOLUTION

B. EVENT II
SYNTHETIC P & SH WAVE SOLUTION

C. CRUSTAL STRUCTURE
(AFTER HUSSONG, 1971)

P WAVE VEL. (KM/S)	S WAVE VEL. (KM/S)	THICKNESS (KM)
1.50		4.40
2.00	1.19	.20
4.30	2.50	.30
5.10	2.96	.80
6.50	3.95	2.10
7.20	4.30	2.50
8.30	4.67	Half-space

Fig. 9. (a) Focal mechanism solution for event I shown in Figure 8. *Bergman and Solomon* [1980] identified this seismic event as an intraplate thrust earthquake. (b) Focal mechanism solution for event II as determined in this study. Bold lines are the observed wavelets and fine lines are synthetically derived. See *Helmberger and Burdick* [1979] for a complete description of the technique used to synthesize waveforms. Black and white triangles depict the orientation of the *T* and *P* axes, respectively, for each earthquake. The parameters of each solution are listed in Table 3. (c) Crustal structure used in generating the synthetic *P* and *SH* body wavelets.

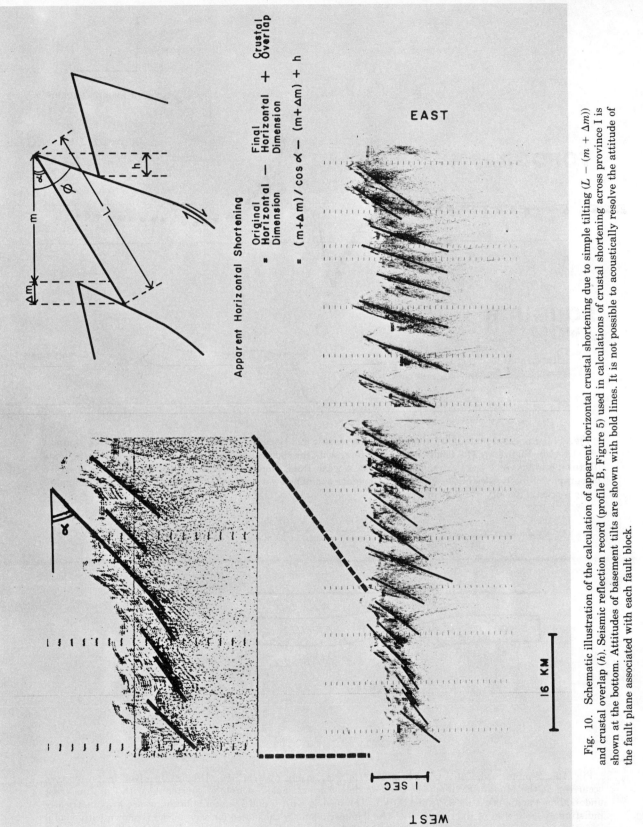

Fig. 10. Schematic illustration of the calculation of apparent horizontal crustal shortening due to simple tilting ($L - (m + \Delta m)$) and crustal overlap (h). Seismic reflection record (profile B, Figure 5) used in calculations of crustal shortening across province I is shown at the bottom. Attitudes of basement tilts are shown with bold lines. It is not possible to acoustically resolve the attitude of the fault plane associated with each fault block.

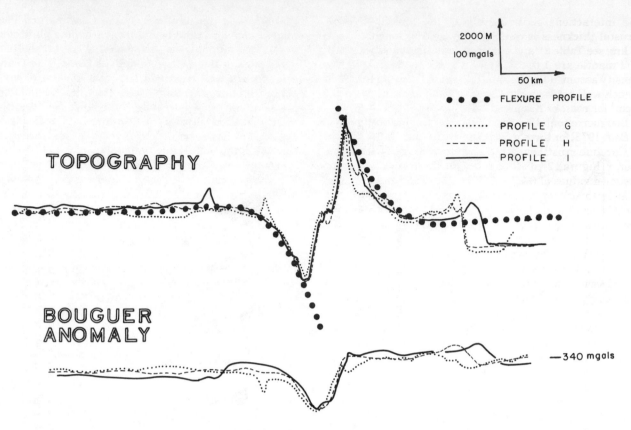

Fig. 11.   Projected (onto 000°) and stacked topographic and Bouguer anomaly profiles for profiles G, H, and I (identified in Figure 4). The boldly dotted line represents the flexural profile of a vertically loaded elastic plate (flexural rigidity is 1×10^{29} dyn cm) overlying a weak fluid. Note the flexure profile matches the observed topography on both sides of the trench axis. Bouguer anomalies were generated from free air gravity anomalies using a crustal density of 2.8 g/cm³.

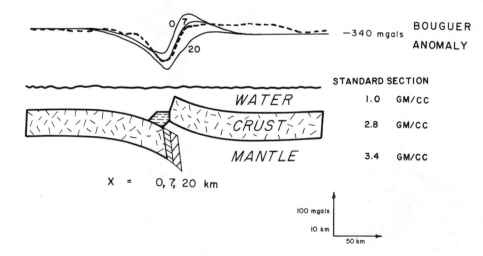

Fig. 12.   Simple model of a possible geometric relationship between the elements involved in the process occurring at the Mussau System. The three models show variable amounts of shortening with Caroline crust underriding Pacific crust by 0, 7, and 20 km. The models with 0 and 7 km of shortening are gravimetrically indistinguishable west of the trench axis. The Bouguer anomaly calculated for each model is shown with a solid line and compared to the average (dashed line) of the observed Bouguer anomalies (stacked in Figure 11). The best fit is obtained when $x \cong 7$ km.

the interactions occurring across the Mussau System. Crustal thickness is controlled by nearby sonobuoy data (6 km; see Table 2), and assumed densities for water, crust, and mantle are 1.0, 2.8, and 3.4 g/cm^3, respectively. The density assumed for the small polygonal body on the inner trench wall between the Caroline and Pacific plates is 2.5 g/cm^3 for reasons discussed later.

Bouguer anomalies were calculated using the method of *Bott* [1973] for three models that differ only by the amount of Caroline crust beneath the Mussau Ridge. The parameter, x (Figure 12), distinguishes the three models and was assigned values of 0, 7, and 20 km, indicating the amount of underthrusting or shortening that has occurred between the Caroline and Pacific plates. The modelled Bouguer anomalies are compared with an average Bouguer profile representing the mean of the observed values along sections G, H, and I. From this comparison (Figure 12) it is concluded that if the assumed density and geometry of the model nearly approximates the current conditions at the Mussau System, then only 7 ± 4 km of Caroline crust lies beneath the Mussau Ridge. It is suspected that more shortening has occurred at the location of profile I than profile G, but these gravity constraints cannot resolve the magnitude of the difference which is predicted to be less than a couple of kilometers.

The bench that is observed on all crossings of the Mussau Trench (profiles F and I, Figure 5) is modelled with a slightly lower density (2.5 g/cm^3) than the value for oceanic crust (2.8 g/cm^3). The bench is probably comprised of either sediments offscraped the downgoing Caroline plate [*Kogan*, 1976] or material downfaulted from the top of the overriding Pacific plate. The origin of the bench material cannot be determined using gravity models because the differences in density may not be significant. However, based on the flexural profile, we favor the hypothesis that the material is downfaulted Pacific crust because the volume of material at the position of the bench approximates the volume delimited by the flexural profile trenchward of the observed leading Pacific edge (Figure 11).

In the above calculations of shortening, we have assumed that the amount of crustal shortening observed at the surface between the Caroline and Pacific plates directly reflects the amount of convergence between these two 'rigid' plates. Further, we have assumed that the shortening can be described with two components: (1) shortening caused by rigid body rotation of fault-bounded blocks and (2) shortening caused by crustal overlap or underthrusting. Shortening caused by deformation of the Caroline crust into long-wavelength (~60 km) undulations is observed but appears to be an order of magnitude less than estimates of shortening from components 1 and 2 above and has been neglected in the calculations.

Discussion

Timing of Deformation

It is possible to determine the duration of the deformation observed within provinces I, II, and III by making the following assumptions: (1) the Caroline and Pacific plates behave as rigid bodies away from the plate boundary; (2) the estimates of the amount of crustal shortening (calculated at the position of profiles B and H, or G and I) are accurate and represent the total amount of convergence that has occurred between the Caroline and Pacific plates at each position; and (3) the euler pole describing relative motion between the Caroline and Pacific plates is known and has remained constant since the inception of the Caroline-Pacific plate boundary.

Weissel and Anderson [1978] estimate the euler pole by vectorially adding poles describing Pacific-Philippine relative motion [*Karig*, 1975] and Philippine-Caroline relative motion [*Weissel and Anderson*, 1978]. The pole position is 13°N and 144°E and has an angular rate of rotation of 0.7°/m.y. There are large uncertainties associated with this relative motion vector because of the large uncertainties in Pacific-Philippine and Philippine-Caroline motions, but from what is presently known about all of the Caroline plate boundaries, we can state that the location of the pole is at least qualitatively consistent with these observations.

The euler pole describing Caroline-Pacific relative motion [*Weissel and Anderson*, 1978] predicts convergence rates of 1.0 and 1.6 cm/yr at the location of profiles B and I, respectively. Therefore if assumption 3 above is correct, then we should expect 1.6 times more shortening at the position of profile I than profile B. The best estimate of the amount of shortening across province III (profiles G, H, and I), based on gravity arguments presented earlier, is about 7 km. Thus we predict that about 4 km of shortening has occurred along profile B, which is the value calculated when all fault planes dip at an angle of 80°. The pole position and rate magnitude are also consistent with values of 7.0 km in the northern province (corresponding to a fault angle of 70°) and about 11 km in province III (where 11 km is within the range of uncertainty in the shortening estimates using gravity models).

Figure 13 shows the relationship between position along the Caroline-Pacific plate boundary and the amount of shortening. The calculations at the two points B and H are seen to be consistent with a pole position at 13°N/144°E although the rate of rotation cannot be constrained by shortening estimates.

If there has been 11 km of shortening between the Caroline and Pacific plates at the position of profile I and convergence occurs at 1.6 cm/yr, then deformation began about 700,000 years ago.

Deformation Structures

The structures described earlier that characterize the deformed zones north of 3°N along the Caroline-Pacific plate boundary are unusual, primarily because of their association with oceanic crust. Similar structural relationships have long been observed in overthrust belts where crystalline basement has been unveiled within continental crust and recently in the equatorial region of the Indian

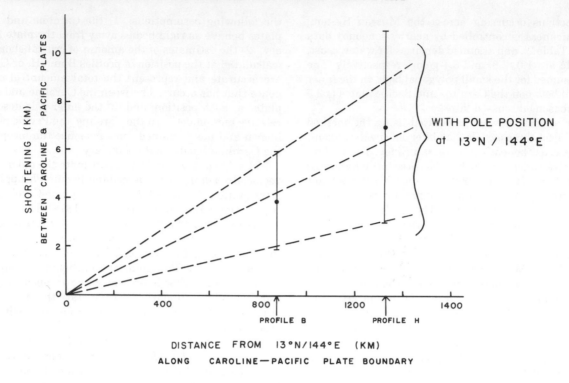

Fig. 13. Observed and predicted relationship between distance along Caroline-Pacific plate boundary from the position 13°N/144°E and the amount of crustal shortening. Shortening and an estimate of error are calculated at two points from seismic reflection data and gravity modelling (see text). A nearly straight line fit through the origin and observed values is required for a euler pole position describing Caroline-Pacific relative motion at 13°N/144°E [Weissel and Anderson, 1978]. Three dashed lines show the best fit and the range of allowed relationships. The rate of relative motion between the Caroline and Pacific plates cannot be constrained by this relationship.

Ocean near the Ninetyeast Ridge [Weissel et al., 1980]. The well-documented Laramide structures of the North American Cordillera [Brewer et al., 1980; Smithson et al., 1979; Sales, 1968; Prucha et al., 1965; Berg, 1961] and the Indian Ocean deformation may serve as appropriate analogs to the structures observed north of the Mussau Trench in provinces I and II. Comparison of seismic reflection records collected from these three areas (Figures 14 and 10) shows that basement-controlled deformation dominates the structures.

According to Brewer et al. [1980], the most intense deformation of the Cordilleran foreland is distinguished by large basement uplifts which are often flanked by curvilinear faults. The deformation seems to have begun with minor upwarping of the basement and sediments followed by fracture of the basement as the compressive stresses build up. The basement appears to have behaved cata-elastically (brittlely) throughout the Laramide deformation in that thrusting was the dominant mode of deformation and folding was secondary.

In the Indian Ocean, reflection records reveal folding of the sediments and high-angle faulting through acoustic basement (oceanic crust) and overlying sediments [Weissel et al., 1980]. Fault block dimensions vary from about 5 to 20 km in width, and the offset across fault planes that disrupt acoustic basement is of the order of hundreds of meters with some evidence from the deformation of the overlying sediments that the faults may be growth structures. Preferred structural asymmetry of the fault blocks (fault blocks generally dip toward the NNW, and associated drape folds verge toward the south) is another parallel to the structures observed within province I. However, the Indo-Australian plate shows deformation of acoustic basement into long-wavelength (100–300 km) undulations with amplitudes of 1–3 km. This scale of deformation structure is not observed at the Caroline-Pacific plate boundary. Deformation of the Indo-Australian plate seems to have begun in the late Miocene and is still active [Weissel et al., 1980].

Laramide structural evolution may parallel the sequence and style of events that are occurring at the Caroline-Pacific boundary north of the Mussau Trench. Once compressive stresses have been established, gentle folds within the sediments are generated followed by basement thrusting (fold thrusts; Berg [1961]) and rotation. In contrast, folds within the deformation field of the Indian Ocean are produced by the faulting. In general, seismic reflection profiles across province I indicate that basement dips in-

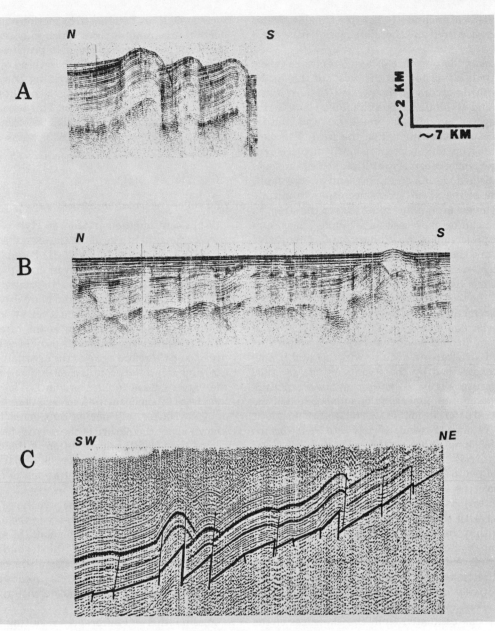

Fig. 14.   Comparison of seismic reflection profiles across deformation features similar to structures at Caroline-Pacific plate boundary within provinces I and II (compare to profiles A, B, C, and D in Figure 5). (*a, b*) Single-channel seismic reflection profiles (R/V *Vema* 3616) over the Bengal Fan in the Indian Ocean (about 3°S, 83°E) showing basement-controlled deformation, constant vergence (SSW) of folded and faulted sediments, and uniform basement dip (toward NNE). Vertical exaggeration at the acoustic basement is approximately 7:1 for both profiles. (*c*) Nonmigrated multichannel seismic reflection profile in the northern Bighorn Basin, Wyoming. Basement-involved block faults and drape fold structures dominate the Laramide deformation. It is not clear if the high-angle faults offset the sedimentary section. Vertical exaggeration at the basement is approximately 3:1. (From *Sacrison* [1977]; reprinted with permission of the author.)

crease toward the east (Figure 10), which suggests two different models describing the development of the deformation.

In model 1, all planes of weakness develop early in the deformational history of the area, and increasing convergence is manifested with greater tilting and displacement along the existing fractures with some fault planes taking up more of the shortening than others. This may explain the central axial low of province II where presumably greater convergence has occurred and therefore more crustal overlap.

Model 2 invokes the presence of a deformational wave front which is migrating toward the west. In this model the earliest faulting occurs at the present eastern limits of provinces I and II. Motion along the fault plane ceases (possibly because of overburden pressures), and another fault plane is established further to the west. This new fault plane takes up the shortening until displacement along a younger, more westerly fault plane requires less force than continued motion on the presently active fault. The magnitude of the shear stress necessary to initiate faulting along a new fault plane must exceed the strength of the rock and can be expressed as a simple linear combination of the projection of the subhorizontal compressive tectonic stress and the pressure imposed by the overburden onto the fault plane. When the overburden or confining pressures become too great because of the amount of overthrusting, the tectonic stress may not be great enough to cause displacement, and failure may occur at a position where the overburden is less.

The amount of displacement that has occurred parallel to the observed structures within provinces I and II may be partially constrained by the fidelity of the recorded magnetic signature. Figure 6 shows that the strength of the magnetic anomalies appears to be independent of the amount of crustal deformation; the correlation of magnetic anomalies 10, 11, and 12 trends clearly into the deformed zones, and the amplitude of the projected magnetic anomalies (Figure 7) shows no obvious diminution which could be attributed to the intense deformation. In order to exactly preserve the strength of the magnetic anomaly, crustal block rotations must occur about an axis which is nearly parallel with the direction of the remanent magnetic vector within the crustal block. Any rotation about an axis normal to the remanent vector would alter the original amplitude and skewness of the magnetic anomaly. The inferred direction of the remanent magnetization vector within positively magnetized blocks of the East Caroline Basin strikes approximately north and lies nearly horizontal ($\pm 2°$) as estimated from the skewness of the observed magnetic anomalies. Thus rotations about a subhorizontal north-south trending axis would have no noticeable effect on the amplitude or skewness of the magnetic anomalies, but rotations about vertical or subhorizontal east-west trending axes would distort the signal creating obvious differences in amplitude between anomalies in the deformed zone and in the undisturbed East Caroline Basin. These differences are not observed. Based on model computations of the effect of changes in magnetic inclination on intensity [Heirtzler et al., 1962], it appears that rotations greater than 15° about an axis normal to the remanent vector would result in amplitude changes of about 15% and observable changes in skewness. Thus it is argued that crustal block rotations within provinces I and II are dominated by motion about subhorizontal north-south trending axes which may imply that there is a greater component of shear in province I than province II due simply to the difference in trend of these two zones. Therefore at the east-west striking Sorol Trough, if this inferred sense of deformation is maintained, we expect almost pure strike slip motion. This is consistent with the suggestion that the Sorol Trough is dominated by an active shear zone [Fornari et al., 1979; Weissel and Anderson, 1978].

Control of Deformation Structures

What controls the variation in style of deformation observed along the strike of the Caroline-Pacific plate boundary within provinces I, II, and III? Possible answers to this queston include (1) variation in crustal age and thermal properties, (2) local variations in strength of the crust or lithosphere, (3) variation in the plate driving forces which may be generating the deformation, (4) amount of crustal shortening, and (5) variation in the rate of convergence. Crust of similar age intersects provinces I, II, and III (Figure 6), and therefore age and the associated thermal effects can probably be discounted as controlling factors, since they are common to each province. There is no evidence for crustal inhomogeneities, so variations in crustal strength as a key factor will not be considered further. Little is known about the nature of the driving forces, but it is suspected that significant variation of the driving forces on the scale of tens of kilometers is unlikely and therefore could not control marked variations in deformation that occur within ~10–30 km.

We see no evidence that the structures observed within provinces I and II are precursory to the structures within province III. We therefore infer that the amount of crustal shortening does not control the tectonic style of deformation. The deformational control imposed by the presence of the Kiilsgaard Trough is unknown, but it is noted that its location does not coincide with the location of the province II-III boundary.

Possible thermal differences resulting from age discontinuities across (as opposed to along) the current Caroline-Pacific plate boundary may control the sense of underthrusting that changes at about 3°N. If the present Caroline-Pacific plate boundary originally served as a transform fault within Caroline lithosphere and the transform offset the easternmost segment of the Caroline crust spreading axis southward of the Kiilsgaard Trough, then the observed sense of underthrusting may be explained by small, but possibly significant, density contrasts across the boundary (Figure 15). Relatively older and therefore colder, denser Caroline lithosphere west of the Mussau Ridge would be expected to be subducted beneath the more buoyant younger lithosphere east of the Massau Ridge. In contrast, to the north where younger, more buoyant lithos-

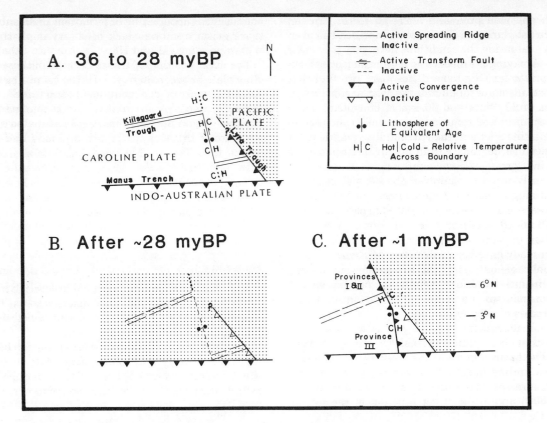

Fig. 15. Simple schematic showing a possible paleoreconstruction of the study area since the early Oligocene. (a) From at least 36 to 28 m.y. B.P., spreading was occurring within the East Caroline Basin and west of the Lyra Trough. During this time the eastern boundary of the Caroline plate is thought to have been the Lyra Trough. The sense of relative motion across the Lyra Trough is unknown. (b) Approximately 28 m.y. B.P., spreading within the basin ceased while the extinct spreading axis continued its approach toward the trench to the south. Relative motion across the Lyra Trough may have ceased. (c) Approximately 1 m.y. B.P., relative motion between the Caroline and Pacific plates began at a zone of weakness delineated by a preexisting fracture zone. The sense of relative motion was convergent. South of about 3°N where older (colder) Caroline lithosphere abutted younger (hotter) Pacific lithosphere, the Mussau System formed such that the Caroline plate was thrust beneath the Pacific plate. North of about 3°N, relatively younger Caroline lithosphere began to override older Pacific lithosphere in the region of provinces I and II.

phere lies west of the plate boundary, the sense of underthrusting by the Pacific plate would be expected to be toward the west, as is observed.

The maximum age difference across the boundary is probably about 9 m.y. (from magnetic anomaly 13 time to anomaly 9 time), which implies a maximum topographic and heat flow difference at some point across the boundary of about 300 m and 11 mW/m^2, respectively [*Parsons and Sclater*, 1977]. Obviously these postulated thermal conditions that may have some control on the development of the deformation cannot be the primary controlling factors, otherwise trenchlike morphology north of 3°N, with the Pacific plate subducting beneath the Caroline plate, should be present. However, if this model is correct in predicting the control on the direction of thrusting, then it is expected

that crust of the same age to each side of the present Caroline-Pacific plate boundary lies at a position near the northern terminus of the Mussau Trench (Figure 15).

We believe that the variation in amount and style of shortening is primarily attributable to changes in the rate of convergence along the Caroline-Pacific plate boundary [*Weissel*, 1980]. Using the pole describing Caroline-Pacific plate relative motion at 13°N and 144°E and 0.7°/m.y. [*Weissel and Anderson*, 1978], the rates of convergence at the transition between provinces I and II and provinces II and III are 1.2 cm/yr and 1.5 cm/yr, respectively. If this is the key factor, a convergence rate greater than 1.5 cm/yr is required for trenchlike morphology to develop in this region.

McKenzie [1977] considered the balance of driving forces

(ridge push and slab pull) and resisting forces (friction along the thrust plane and forces required to bend an elastic plate) in evaluating the conditions necessary to initiate subduction. Applying his model and using an elastic thickness appropriate for Oligocene lithosphere, we find that in order for steady state subduction to occur, shear stresses of the order of 900 bars and 70 km of downgoing plate within the mantle are required. A density contrast between the sinking slab and the surrounding mantle must be maintained in order for slab pull to provide a component in the driving forces. The consumption must occur at a rate sufficiently rapid to maintain the density contrast, and by assuming a thermal time constant of the sinking slab of 10^7 years, convergence is required to occur at a rate of at least $70 \ km/10^7$ years, or 0.7 cm/yr. However, in order for subduction to begin, enough slab to provide a pulling force must already have been subducted. *McKenzie*'s [1977] model is applicable only after the subduction process has already been initiated and slab pull has become an important component in the overall force system that sustains the process of subduction.

In general, it appears from seismic reflection records and topography that deformation intensity increases toward the south. The entire length of the convergent boundary between the Caroline and Pacific plates may be responding to boundary stresses which become more compressive toward the south and are a direct function of the rate of convergence.

If this is true, then why is there no observed seismic activity within provinces II and III? There are two possible explanations. First, aseismic motion may be occurring; however, this explanation seems unreasonable in light of the obvious brittle deformation documented in province II (profiles C and D, Figure 5). A second perhaps more tenable explanation is that small-magnitude seismic energy is being released but escaping detection by the World-Wide Seismic Station Network.

Regional Tectonic Implications

If the duration of deformation (approximately one million years) has been correctly determined, then the regional tectonic implications are as follows:

1. After the cessation of spreading within the Caroline basins (about 28 m.y. B.P.), the Caroline plate probably behaved as part of the Pacific plate.

2. At 10 m.y. B.P. the western Caroline plate boundary was established as a spreading center along the Ayu Trough and as a trench along the Palau and Yap trenches. The Caroline plate remained part of the Pacific plate.

3. At about 1 m.y. B.P. the northern and eastern Caroline plate boundaries were established along the Sorol Trough, province I, province II, and the Mussau System (province III). Fresh basalts and ultramafics dredged from the Sorol Trough suggest a recent age for this feature [*Fornari et al.*, 1979]. Convergence across the eastern

boundary is supported by observations of deformation, but there exists a curious lack of observed seismic activity within provinces II and III.

The exact position of the eastern boundary of the Caroline plate between 36 m.y. B.P. (the beginning of seafloor spreading within the Caroline basins) and the end of spreading remains speculative. As is implied by Figure 15, the Oligocene Caroline-Pacific plate boundary lies eastward of the Mussau Trench and may be represented by the Lyra Trough. Mesozoic magnetic anomalies have been identified [*Taylor*, 1978] just eastward of the Lyra Trough at about 151°E (Figure 6). It is believed that the Lyra Trough has not been active in recent time, but its tectonic role during the Oligocene is unknown.

Summary

In this study we have attempted to describe the location, evolution, style, and control of deformation at the eastern boundary of the Caroline plate. Our conclusions are as follows:

1. Three distinct structural provinces can be identified along the convergent Caroline-Pacific plate boundary. They are distinguished on the basis of their morphology which reflects the style of the deformation occurring in each province.

2. The style of deformation may be dominantly controlled by the rate of convergence between the two plates. Age of the oceanic crust or the amount of shortening accrued across the boundary does not provide the primary control on the style of tectonic shortening; however, the first of these factors may influence the sense of underthrusting. Trench morphology, which characterizes province III, is established when convergence rates are greater than 1.5 cm/yr [*Weissel and Anderson*, 1978].

3. Deformation within provinces I and II may be similar to the Laramide structures observed within the North American Cordillera and intraplate deformational structures in the northern Indian Ocean.

4. Deformation and shortening probably began entirely within the crust of the Caroline plate. Magnetic, bathymetric, seismic reflection, and gravity data and flexure modelling support this conclusion.

5. The amount of shortening across provinces I and III has been determined using seismic reflection profiles and gravity models. These values are consistent with an euler pole describing Caroline-Philippine relative plate motion located at 13°N, 144°E. Utilization of this information indicates that collision between the Caroline and Pacific plates began about a million years ago.

In conclusion, the Caroline basins and associated boundaries appear to have behaved as a discrete 'plate' only as recently as a million years ago. It is believed that the eastern boundary is presently composed of young 'transient' features [*Bracey and Andrews*, 1974; *Weissel and*

Anderson, 1978] that will ultimately develop into a mature subduction zone.

Acknowledgments. We thank L. Burdick, R. Quittmeyer, and G. Karner for use of their seismic wave and gravity and elastic deflection computer programs. K. Crane, G. Karner, R. Leslie, S. Lewis, W. B. F. Ryan, and B. Taylor critically read the manuscript. E. Free and D. Johnson assisted in preparation of the manuscript. This work was supported by National Science Foundation grant OCE 78-19830 and OCE 80-24316. Lamont-Doherty Geological Observatory contribution 3372.

References

Anderson, R. N., M. G. Langseth, and J. G. Sclater, The mechanism of heat transfer through the floor of the Indian Ocean, *J. Geophys. Res., 82,* 3391–3409, 1977.

Berg, R. R., Laramide tectonics of the Wind River Mountains, in *Symposium of Late Cretaceous Rocks, Wyoming and Adjacent Areas, 16th Annu. Field Conf. Guideb.,* pp. 70–80, Wyoming Geological Association, Casper, 1961.

Bergman, E. A., and S. C. Solomon, Oceanic intraplate earthquakes: Implications for local and regional intraplate stress, *J. Geophys. Res., 85,* 5389–5411, 1980.

Bott, M. H. P., Inverse methods in the interpretation of magnetic and gravity anomalies, *Methods Comput. Phys., 13,* 133–162, 1973.

Bracey, D. R., Reconnaissance geophysical survey of the Caroline Basin, *Geol. Soc. Am. Bull., 86,* 775–784, 1975.

Bracey D. R., and J. E. Andrews, Western Caroline Ridge: Relic island arc?, *Mar. Geophys. Res., 2,* 111–125, 1974.

Brewer, J. A., S. B. Smithson, J. E. Oliver, S. Kaufman, and L. D. Brown, The Laramide orogeny: Evidence from COCORP deep crustal seismic profiles in the Wind River Mountains, Wyoming, *Tectonophysics, 62,* 165–189, 1980.

Caldwell, J. G., W. F. Haxby, D. E. Karig, and D. L. Turcotte, On the applicability of a universal elastic trench profile, *Earth Planet. Sci. Lett., 31,* 239–246, 1976.

Den, N., W. J. Ludwig, S. Murauchi, M. Ewing, H. Hotta, T. Asanuma, T. Yoshio, A. Kubotera, and K. Hagiwara, Sediments and structure of the Eauripik–New Guinea Rise, *J. Geophys. Res., 76,* 4711–4723, 1971.

Erlandson, D. L., T. L. Orwig, G. Kiilsgaard, J. H. Mussells, and L. W. Kroenke, Tectonic interpretations of the East Caroline and Lyra basins from reflection-profiling investigations, *Geol. Soc. Am. Bull., 87,* 453–462, 1976.

Fornari, D. J., J. K. Weissel, M. R. Perfit, and R. N. Anderson, Petrochemistry of the Sorol and Ayu troughs: Implications for crustal accretion at the northern and western boundaries of the Caroline plate, *Earth Planet. Sci. Lett., 45,* 1–15, 1979.

Heirtzler, J. R., G. Peter, M. Talwani, and E. G. Zurfluch, Magnetic anomalies caused by two-dimensional structure: Their computation by digital computers and their interpretation, *Tech. Rep. 6,* Lamont-Doherty Geol. Observ., Columbia Univ., Palisades, N. Y., 1962.

Helmberger, D., and L. Burdick, Synthetic seismograms, *Annu. Rev. Earth Planet. Sci., 7,* 417–442, 1979.

Hussong, D. M., Detailed structural interpretations of the Pacific Ocean crust using Asper and ocean-bottom seismometer methods, Ph.D. thesis, 165 pp., Univ. of Haw., Honolulu, 1972.

Karig, D. E., Basin genesis in the Philippine Sea, *Initial Rep. Deep Sea Drill. Proj., 31,* 857–879, 1975.

Kogan, M. G., Gravity anomalies and main tectonic units of the southwest Pacific, *J. Geophys. Res., 81,* 5240–5248, 1976.

LaBreque, J. L., D. V. Kent, and S. C. Cande, Revised magnetic polarity time scale for Late Cretaceous and Cenozoic time, *Geology, 5,* 330–335, 1977.

Langseth, M. G., X. LePichon, and M. Ewing, Crustal structure of the midocean ridges, 5, Heat flow through the Atlantic Ocean Floor and convection currents, *J. Geophys. Res., 71,* 5321–5355, 1966.

Mammerickx, J., Re-evaluation of some geophysical observations in the Caroline basins, *Geol. Soc. Am. Bull., 89,* 192–196, 1978.

McKenzie, D. P., The initiation of trenches: A finite amplitude instability, in *Island Arcs, Deep Sea Trenches and Back-Arc Basins, Maurice Ewing Ser.,* vol. 1, edited by M. Talwani and W. C. Pitman III, pp. 57–62, AGU, Washington, D. C., 1977.

McNutt, M., and H. W. Menard, Lithospheric flexure and uplifted atolls, *J. Geophys. Res., 83,* 1206–1212, 1978.

Parsons, B., and J. G. Sclater, An analysis of the variation of ocean floor bathymetry and heat flow with age, *J. Geophys. Res., 82,* 803–827, 1977.

Prucha, J. J., J. A. Graham, and R. P. Mickelson, Basement controlled deformation in the Wyoming province of the Rocky Mountain foreland, *Am. Assoc. Pet. Geol. Bull., 49,* 966–992, 1965.

Sacrison, W. R., Seismic interpretation of basement block faults and associated deformation, *Mem. Geol. Soc. Am., 151,* 39–49, 1977.

Sales, J. K., Crustal mechanics of cordilleran foreland deformation: A regional and scale-model approach, *Am. Assoc. Pet. Geol. Bull., 52,* 2016–2044, 1968.

Sclater, J. G., J. Crowe, and R. N. Anderson, On the reliability of oceanic heat flow averages, *J. Geophys. Res., 81,* 2997–3006, 1976.

Smithson, S. B., J. A. Brewer, S. Kaufman, J. E. Oliver, and C. A. Hurich, Structure of the Laramide Wind River uplift, Wyoming, from COCORP deep reflection data and gravity data, *J. Geophys. Res., 84,* 5955–5972, 1979.

Taylor, B., Mesozoic magnetic anomalies in the Lyra Basin (abstract), *Eos Trans. AGU, 59,* 320, 1978.

Watts, A. B., J. H. Bodine, and M. S. Steckler, Observations of flexure and the state of stress in the oceanic lithosphere, *J. Geophys. Res., 85,* 6369–6376, 1980.

Weissel, J. K., Pacific/Caroline plate boundary in the eastern Caroline Sea: Preliminary results of the 1979 shipboard program (abstract), *Eos Trans. AGU, 61,* 357, 1980.

Weissel, J. K., and R. N. Anderson, Is there a Caroline plate?, *Earth Planet, Sci. Lett., 41,* 143–158, 1978.

Weissel, J. K., R. N. Anderson, and C. A. Geller, Deformation of the Indo-Australian plate, *Nature, 287,* 1–7, 1980.

Winterer, E. L., W. R. Riedel, R. M. Moberly, Jr., J. M. Resig, L. W. Kroenke, E. L. Gealy, G. R. Heath, P. Brönnimann, E. Martini, T. R. Worsley, Site 63 Shipboard Report, *Initial Rep. Deep Sea Drill. Proj., 7,* 323–340, 1971.

Chronology of Volcanic Events in the Eastern Philippine Sea

Arend Meijer[1], Mark Reagan[1], Howard Ellis[2], Muhammad Shafiqullah[1], John Sutter[3,4], Paul Damon[1], and Stanley Kling[5]

Radiometric and paleontologic ages of samples from chiefly volcanic sections exposed on Guam, Saipan, and in the Palau Islands were determined to provide an improved temporal framework for tectonic and petrologic models for the evolution of the eastern Philippine Sea. The oldest arc related volcanic rocks found in this area are from the Facpi formation on Guam dated at 43.8 ± 1.6 m.y. B.P. (late middle Eocene). Evidence for late Eocene, early Oligocene, and middle Miocene arc volcanism was also found in the Mariana fore arc. The Palau Islands contain volcanic units of late Eocene(?), early Oligocene and early Miocene age. A minimum age of 1.3 ± 0.2 m.y. has been established for the Mariana active arc. Overall, the new data are consistent with Karig's (1971) model for the tectonic evolution of the eastern Philippine Sea. Whether or not arc volcanism and interarc basin spreading can take place at the same time has not been resolved, although no evidence of synchroneity has been found for at least the Parece Vela Basin–South Honshu Ridge arc system.

The Eastern Philippine Sea is composed of a series of volcanic arcs and interarc basins of early Tertiary to Recent age. *Karig* [1971] proposed a tectonic model for the evolution of this region in which volcanic arcs were rifted to produce remnant arcs, frontal arcs, active arcs, and interarc basins. Subsequent studies have largely confirmed the model [e.g., *Karig et al.*, 1975; *Kroenke et al.*, 1980]. Because Philippine Sea arcs were formed in an oceanic setting, they offer a potential prototype for the early evolution of volcanic arcs and, presumably, the initial stages in the development of continental crust. In order to decipher the petrologic evolution of this region, the age relations of the units that compose it must be accurately known. We report paleontologic and radiometric data on selected samples from this region that, when combined with existing data, provide an improved temporal framework for interpretation of its volcanic evolution.

We use the term eastern Philippine Sea for the region east of and including the Palau-Kyushu Ridge (PKR) and west of the Bonin-Mariana-Yap trench system. The Western Philippine Basin lies west of the PKR and east of the Philippine-Taiwan-Ryukyu zone. According to *Karig* [1971],

the proto-PKR developed as a volcanic arc upon oceanic or interarc basin crust in late Eocene to mid-Oligocene time. Part of the arc was rifted to produce the PKR remnant arc, the South Honshu Ridge active arc, and the Parece Vela and Shikoku interarc basins in mid-Oligocene to late Miocene time. The southern portion of the South Honshu Ridge was in turn rifted in late Miocene time to produce the West Mariana Ridge remnant arc, the Mariana frontal arc, the Mariana active arc, and the Mariana Trough interarc basin.

The main sources of geochronologic data for this region (see references below) have been stratigraphic sections on the islands of Guam, Saipan, and Tinian in the Mariana frontal arc, in the Palau Islands on the PKR, and DSDP (Deep-Sea Drilling Project) sites in the area. Most of the published age determinations are based on paleontologic data. Although these ages have provided a general framework for the tectonic evolution of the region, they leave some important questions unresolved, such as (1) the age of the intrusive and basal volcanic units on Guam and Saipan that did not yield fossils, (2) the absolute duration of volcanic activity represented by a given unit, and (3) the extent of reworking of older faunal assemblages into younger deposits resulting in erroneous paleontologic age determinations.

Data Presentation

Our geochronologic studies have concentrated on the islands of Guam, Saipan, and Palau, although several analyses of active arc and interarc basin samples were also completed.

[1] Department of Geosciences, University of Arizona, Tucson, Arizona 85721.

[2] SOHIO Petroleum Company, San Francisco, California 94111.

[3] Department of Geology and Mineralogy, The Ohio State University, Columbus, Ohio 43210.

[4] Now with the U.S. Geological Survey, Reston National Center, Reston, Virginia 22092.

[5] Biostratigraphics, San Diego, California 92111.

Guam

Tracey et al. [1964] studied the geology of Guam and formulated a stratigraphy. According to their descriptions, the Alutom Formation, which outcrops in the central and northern portions of the island, represents the oldest unit on the island and was assigned a lower Oligocene age [*Tracey et al.*, 1964, p. A21]. It is composed of volcanic breccias and tuffs with a few interbedded flows and sills. In central Guam the breccias are unconformably overlain by the Mahlac member of the Alutom formation, a well-bedded fossiliferous shale. The volcanic section exposed in southern Guam was mapped as the Umatac formation, consisting of pillowed lava flows and dikes of the lower Facpi flow member, shallow water limestone of the Maemong limestone member, pillowed and tabular lava flows of the upper Facpi flow member, pyroclastic rocks of the Bolanos pyroclastic member, and flows of the Dandan flow member. Tracey et al. assigned a lower Miocene age to the Umatac formation, based on fauna in the Maemong limestone member. Recognizing that samples from the lower Facpi flow member were chemically more like Alutom formation samples than samples from other members in the Umatac formation, *Meijer* [1974] suggested the need for further investigation of the stratigraphic relationship between the lower Facpi flow member and the Alutom formation and for more definitive age data.

The results of a new mapping effort and petrologic study of southern Guam are presented by *Reagan and Meijer* [1982]. Our radiometric age data are presented in Table 1 and new paleontologic data are summarized in Tables 2*a* and 2*b*. The new results require revision of the stratigraphy formulated by *Tracey et al.* [1964] (see also Figure 2 of *Reagan and Meijer* [1982]). Both the paleontologic and radiometric data suggest the lower Facpi flow member is the oldest unit on Guam with a late middle Eocene age (44 m.y. B.P.). Because it is substantially older than other members of the Umatac formation, as defined by *Tracey et al.* [1964], *Reagan and Meijer* [1982] designated it as a separate formation, the Facpi formation. This formation is paraconformably overlain by the late Eocene to early Oligocene Alutom formation (Tables 1 and 2). The age and areal extent of the Alutom formation remain largely as presented by *Tracey et al.* [1964]. The Alutom formation appears to be composed of a lower section of late Eocene tuffaceous sandstones, shales, limestones and volcanic breccias, and an upper section of early Oligocene coarse breccias interbedded with a few fine grained sedimentary layers and a few pillow lavas (see Tables 1 and 2). Radiometric data on samples collected from many stratigraphic levels of the upper section of the Alutom formation suggest that it was erupted over a relatively short time interval (≥2.0 m.y.) at approximately 35 ± 1 m.y. B.P. (Table 1). The Facpi formation is cut by numerous dikes, originally thought to be the feeders for the pillowed flows within it. The chemistry of some of the dikes [*Reagan and Meijer*, 1982] and an age date of 35.8 ± 0.8 m.y. B.P. obtained on

one of them suggests that a few may have been feeders for magmas of the Alutom formation. The final Oligocene event recorded on Guam was the intrusion of sills into the Alutom formation around 32 m.y. B.P. (Table 1). This age corresponds closely to the estimated age of rifting of the proto-PKR [*Sutter and Snee*, 1980] and suggests that the sills were emplaced during the rifting event.

No volcanic units with ages in the interval from 32 to 13 m.y. B.P. have been found on Guam. During this interval the Maemong limestone member of the Umatac formation accumulated in shallow waters in central and southern Guam, presumably around volcanic highlands composed of Facpi and Alutom formation units [*Tracey et al.*, 1964, pp. A27, A96]. Volcanic activity resumed in middle Miocene time with the eruption of alkalic lavas and breccias of the Umatac formation. The Bolanos pyroclastic member of the Umatac formation overlies the Maemong limestone member where it is present. In other locations it forms the base of the formation. It is composed of interbedded limestone clast-bearing volcanic breccias and calcareous sandstones, with calcareous sandstones being more abundant near its base. A pillow lava flow, which we have designated as the Schroeder flow member (the upper Facpi flow member of *Tracey et al.*, [1964]), is interbedded with the lower units of the Bolanos member near Mt. Schroeder on southern Guam. A clast from the portion of the Bolanos pyrocastic member that overlies the Schroeder flow member yielded a 13.5 ± 0.2 m.y. age (Table 1). This is consistent with a middle Miocene age reported for the Bonya Limestone which overlies the Umatac formation [*Tracey et al.*, 1964, p. A31]. Volcanic activity ceased on Guam after middle Miocene time.

Saipan

The general geology of Saipan has been described by *Cloud et al.* [1956], and *Schmidt* [1957] discussed the petrology of the volcanic rocks. Four volcanogenic formations were mapped on Saipan. From oldest to youngest, these are (1) the Sankakuyama formation, (2) the Hagman formation, (3) the Densinyama formation, and (4) the Fina Sisu formation. The Sankakuyama formation is composed almost entirely of high silica dacite flows and pyroclastic units. Although it yielded no fossils, *Cloud et al.* [1956, p. 40] suggested the Sankakuyama formation was likely to be Eocene in age, based on its stratigraphic position and on 'the presence of tridymite and cristobalite in the groundmass of these rocks.' The results of radiometric dating of two samples of Sankakuyama dacite are presented in Table 1. A late Eocene age (about 41 m.y. B.P.) is confirmed for this formation.

The Hagman formation, which unconformably overlies the Sankakuyama formation [*Cloud et al.*, 1956], is composed of andesitic pyroclastic rocks and lava flows and water-laid volcanogenic sediments. It was assigned a late Eocene paleontologic age by *Cloud et al.* [1956]. We analyzed a clast from the breccia-tuff facies of the formation

TABLE 1. K/Ar Data

Sample Number	Material Dated	K, %, mean	Radiogenic $^{40}Ar \times 10^{-12}$ m/gm mean	Atmospheric ^{40}Ar (mole %), mean	Age, m.y.
Saipan					
Sankakuyama formation					
Dacite S-9	feldspar*	0.1711, 0.1814	13.25, 13.17	33.8, 33.9	41.4 ± 0.9
		0.1733	13.14	34.2	
		0.1745	13.12	33.8	
		0.1810	13.27	34.6	
		0.1868	13.08	33.2	
		0.1878			
		0.1841			
		0.1870			
		0.1874			
Dacite S-12	w.r.†	1.040, 1.040	74.09, 74.09	92.9, 92.9	40.7 ± 1.8
Hagman formation					
Andesite S-57	matrix†	0.601, 0.601	37.55, 37.55	48.3, 48.3	35.7 ± 0.5
Fina Sisu formation					
Basalt S-51	groundmass*	2.019, 2.022	45.25, 45.21	38.0, 38.0	12.0 ± 0.3
		2.017	44.60	38.5	
		2.023	45.90	38.0	
		2.030	45.07	37.6	
		2.018			
Guam					
Facpi formation					
Basaltic andesite GM-F-1	groundmass*	0.2163, 0.2178	16.95, 16.75	78.7, 79.1	43.8 ± 1.60
		0.2195	16.89	78.8	
		0.2177	16.80	78.8	
			16.36	80.1	
Alutom formation					
Andesite GUM-79-2	groundmass*	0.837, 0.849	52.26, 52,98	57.6, 58.4	35.6 ± 0.9
		0.864	52.95	58.9	
		0.850	53.80	58.7	
		0.863			
		0.838			
		0.845			
		0.845			
Basaltic dike GUM-79-6	groundmass*	0.1663, 0.1629	10.08, 10.20	39.5, 39,4	35.8 ± 0.8
		0.1603	10.32	39.3	
		0.1680			
		0.1713			
		0.1571			
		0.1606			
		0.1569			
		0.1608			
		0.1648			
Dacite GUM-80-1b	groundmass*	0.2137, 0.2117	13.12, 13.08	39.8, 40.1	35.3 ± 0.8
		0.2146	13.05	40.2	
		0.2090	13.06	40.2	
		0.2094			
Basaltic andesite G-SR-6	groundmass*	0.3059, 0.3040	18.53, 18.46	21.6, 21.8	34.7 ± 0.7
		0.3026	18.42	21.9	
		0.3044	18.44	21.8	
		0.3049			
		0.3032			
Basaltic andesite clast GM-50	w.r.†	0.502, 0.502	30.15, 30.15	67.1, 67.1	34.3 ± 0.6
Basaltic sill SASA-80-2	groundmass*	0.6684, 0.6666	38.08, 37.58	67.2, 67.6	32.2 ± 1.0
		0.6714	37.08	68.0	
		0.6636			
		0.6631			
Umatac formation					
Basaltic andesite clast GM-63	w.r.†	1.370, 1.370	32.25, 32.25	64.0, 64.0	13.5 ± 0.2

TABLE 1. (continued)

Sample Number	Material Dated	K, %, mean	Radiogenic ^{40}Ar × 10^{-12} m/gm mean	Atmospheric ^{40}Ar (mole %), mean	Age m.y.
Palau					
Aimeliik formation Andesite tuff PAL-16	hornblende*	0.0671, 0.0658 0.0645 0.0658	3.786, 3.718 3.645 3.722	66.0, 66.8 67.6 66.8	32.3 ± 1.1
Ngeremlengui formation Dacite clast PAL-19	groundmass*	0.4226, 0.4206 0.4221 0.4208 0.4201 0.4175	22.24, 22.30 22.22 22.32 22.40	48.7, 48.3 48.2 48.1 48.2	30.3 ± 0.88
Arakabesan formation Andesite clast PAL-3	groundmass*	0.4551, 0.4539 0.4462 0.4589 0.4602 0.4539 0.4526 0.4503	15.80, 15.92 15.84 16.06 15.99	68.0, 67.7 68.0 67.2 67.7	20.1 ± 0.5
Active Arc					
Sarigan Andesite flow SA-137	groundmass*	0.8279, 0.8290 0.8286 0.8303 0.8290	0.755, 0.710 0.698 0.676	98.0, 98.0 97.9 98.1	0.49 ± 0.20
Anatahan Andesite flow ANT-79-24	groundmass*	0.6482, 0.6499 0.6529 0.6514 0.6460 0.6507 0.6459 0.6543	1.426, 1.481 1.605 1.413	26.4, 26.9 27.2 27.0	1.31 ± 0.21
Andesite flow ANT-79-28	groundmass*	0.6557, 0.6554 0.6546 0.6560	0.339, 0.452 0.580	81.2, 88.5 95.1	0.40 ± 0.11
Mariana Trough					
Basalt TSDY-11-D-X	groundmass*	0.0926, 0.0914 0.0924 0.0919 0.0915	0.314, 0.366 0.372 0.343 0.434	97.2, 96.8 96.7 97.0 96.3	2.31 ± 0.47

Constants used: λ_β = 4.963 × 10^{-10} y^{-1}, λ = 5.544 × 10^{-10} y^{-1}, λ_e = 0.581 × 10^{-10} y^{-1}, and ^{40}K/K = 1.167 × 10^{-4} atom/atom.

* Dated at the University of Arizona by methods described in this work and in that of *Baldridge et al.* [1980].

† Dated at the Ohio State University by methods described by *Sutter and Smith* [1979]. K analyses were made using the chemical separation procedure of *Cooper* [1963] and single channel flame photometry on a Zeiss Model PF-5 flame photometer.

and obtained an age of 35.7 ± 0.5 m.y. B.P., which is early Oligocene, not late Eocene. Because the late Eocene paleontologic age appears to be well established [*Cloud et al.*, 1956, p. 47] and because the dated portion of the sample (the matrix) was partly altered, the radiometric age may be too young as a result of Ar loss and/or K gain. We have not analyzed samples from the Densinyama formation.

The Fina Sisu formation was originally thought to be late Oligocene in age based on microfossils [*Cloud et al.*, 1956, p. 62]. Our K/Ar analysis produced an age of 12.9 ± 0.3 m.y. suggesting the formation is middle Miocene in age. This is consistent with a more recent report

of a Miocene fossil age determination by R. Todd, as reported by *Ladd* [1966, p. 8].

Palau Islands

Mason et al. [1956] have discussed the general geology of the Palau Islands and reviewed earlier geologic contributions. They divided the volcanic section of Palau into three formations, which are, in order of decreasing age, the Babelthaup formation, the Aimeliik formation, and the Ngeremlengui formation. Although each of the formations consists dominantly of andesitic to dacitic breccias and tuffs, these workers distinguished the formations on

TABLE 2a. Paleontologic Data for Nannofossils

Sample	Zygrhablithus bijugatus	Triquetrorhabdulus carinatus	Coccolithus miopelagicu	Coccolithus eopelagicus	Coccolithus pelagicus	Cyclicargolithus abisectus	Cyclicargolithus floridanus	'Cyclococcolithus neogammation'	Discoaster delandrei	Sphenolithus moriformis	Bramletteius serraculoides	Coccolithus formosus	Dictyococcites bisectus	Discoaster barbadiensis	Discoaster binodosus	Discoaster saipanensis	Discoaster tanii	Discoaster species	Helicosphaera compacta	Reticulofenestra hillae	Reticulofenestra Sp. cf R. coenura	Reticulofenestra reticulata	Reticulofenestra umbilica	Reticulofenestra sp.	Sphenolithus pseudoradians	Sphenolithus distentus	Sphenolithus spiniger	Sphenolithus predistentus	Calcidiscus Sp.	Calcidiscus kingii	Age
Maemong Limestone GUM-80-33		R	R		F	F	C	F	VR	C			VR																		Early Miocene–?Discoaster deflandrei Subzone (WPN-18b) of the Triquetrorhabdulus carinatus Zone (WPN-18)
GUM-80-77						F	R	F	R	R																					
GUM-80-97			F	R	C	C	R	C	R	R			F															R	R		Late Oligocene Sphenolithus distentus Zone (WPN-16)
GUM-82-15			R		C	C	C	C	C	C	R	F	F															R	R		
GUM-82-10 *(Guam)*		R	R	F	C	C	F	F	F	C		F	C	R?			R	R	R	R		F	F		VR	R		R			Late Eocene Discoaster barbadiensis Zone (WPN-13)
Alutom Formation (basal) GUM-80-58	R		R		C		C		R	R		R	A			R	R	R	R	R		F	F		F						
GUM-80-101			F		C	C	C		F	F		R	C			R	R		R	R	R	R									
GUM-80-107			F	F?	R	R	C		C	C		R	C			C	F		F	F		F	C								
GUM-80-125			F		C		C		F	F		R	R			R	F		R			F	C		F			R			
GUM-82-11																															Late Eocene Discoaster barbadiensis Zone (WPN-13) or Late Middle Eocene Discoaster saipanensis Subzone (WPN-12b) of the Reticulofenestra umbilica Zone (WPN-12)
Facpi Formation (upper) GUM-82-3					C		C		C	C		F	F	R	R									R			R				
GUM-82-9			R		F		F		F	F		R	R	R		R	R		R		R	R					R		R		Late Middle Eocene Reticulofenestra umbilica Zone (WPN-12)
Ngeremlengui Formation PAL-17 *(Palau)*			F		F		F		R	R		F	A	F		R	C		F	C		F	F	F	F			F			Late Eocene Discoaster barbadiensis Zone (WPN-13)

Note: Nannofossil zonation for age determinations is that of Ellis [1981].

TABLE 2b. Paleontologic Data for Radiolaria

	Eusyringium fistuligerum	Lithochitris vespertilio	Lithocyclia ocellus	Lychnocanoma babylonis	Podocyritis papalis	Podocyritis trachodes	Thyrsocyrtis rhizodon	Thyrsocyrtis triacantha	Rhabdolithus pipa	Age
					Guam					
Facpi Formation GM-70	R	R	F	F	F	R	F	R	VR	Late Middle Eocene (Podocyrtis mitra Zone)

the basis of such features as included clast types, presence or absence of veinlets and fractures, and color of matrix.

We have obtained paleontologic and radiometric data on the Aimeliik and Ngeremlengui formations (Tables 1 and 2). Aimeliik formation samples, collected from a well-bedded breccia-ash sequence exposed in a quarry in southern Babelthaup, yielded a late Eocene nannofossil age (Table 2) but an early Oligocene radiometric age (32.3 ± 1.1 m.y. B.P.). The nannofossil age is consistent with an Eocene age determination by *Cole* [1950], based on Foraminifera from the same locality. The radiometric age was obtained on a hornblende separate from a lapilli tuff unit. The matrix of the tuff is highly altered, but the hornblende grains appear unaltered. Microprobe analyses of individual hornblende grains yielded K_2O contents in agreement with the K_2O analysis of the age-dating aliquot (Table 1), suggesting that the K_2O analysis is not in error. Repeat analyses of the Ar content of the hornblende were also consistent. This leads us to suggest the fauna contained in the tuff represents reworked material and that the radiometric age is more reliable.

The Ngeremlengui formation was divided into three members by *Mason et al.* [1956]: the Nghemesed dacitic volcanic breccia, the Medorm andesitic-dacitic volcanic breccia, and the Arakabesan andesitic volcanic breccia, in order of decreasing age. Although no fossils were collected from the formation, it must be younger than the underlying Aimeliik formation and older than the overlying Miocene Palau Limestone. We dated a clast of nearly unaltered dacite from the Medorm member that yielded a middle Oligocene age of 30.3 ± 0.88 m.y. and a clast of andesite from the Arakabesan member which yielded a lower Miocene age of 20.1 ± 0.5 m.y. B.P. Based on these results, we suggest the Ngeremlengui formation should be divided into at least two formations: a middle Oligocene Ngeremlengui formation and a lower Miocene Arakabesan formation.

Mariana Trough

We have dated a single sample of diabasic basalt dredged from the southern Mariana Trough (see Table 3 for sample location). The 2.3 ± 0.5 m.y. age we obtained (Table 1) is consistent with the inferred age of this portion of the trough [*Hussong et al.*, 1981].

Active Arc

In order to place a minimum limit on the age of the inception of volcanism in the Mariana active arc, we dated three samples from what we believed to be the oldest subaerial volcanic sections in the arc. A hornblende andesite sample from Sarigan Island yielded an age of 0.5 ± 0.2 m.y. B.P. The other two samples came from the east and west coasts of Anatahan Island. Geologic mapping suggests that the eastern portion of the island is older than the western portion (A. Meijer et al., unpublished data). The radiometric results bear this out with a 1.3 ± 0.2 m.y. age for the eastern sample and 0.4 ± 0.1 m.y. age for the western sample. These results indicate that the active arc must be least 1.3 ± 0.2 m.y. old.

Discussion

A summary of our age determinations is shown in Figure 1. Also included in this figure are age data from the Bonin Islands, D.S.D.P. legs 59 and 60, and ages inferred for the West Philippine and Parece Vela Basins on the basis of marine magnetics. The oldest arc-related volcanic rocks identified to date in the eastern Philippine Sea are the late middle Eocene flows of the Facpi formation on Guam (Tables 1 and 2). The near coincidence of the age of these lavas (43.8 ± 1.6 m.y.) and the inferred age of the change in Pacific plate spreading direction, as indicated by the bend in the Hawaiian-Emperor Seamount chain (42–44 m.y. B.P., *Clague and Jarrard* [1973]) suggests a causal relationship. Presumably the change in spreading direction

TABLE 3. Sample Localities and Descriptions

Sample	Location	Description
	Saipan	
S-9	Collected in roadcut on Bird Island Bluff, 15°15.60′N, 145°48.49′E	Dacite dome, porphyritic, hyaloophitic texture, phenocrysts 2%, groundmass 90%, vesicles 8%; plagioclase 2%, quartz <1%, opaques <1%, glass 97%; glass is slightly hydrated
S-12	Collected in gully east of road on Bird Island Bluff, 150°15.51′N, 145°48.64′E	Dacite dome, porphyritic, hyalopelitic texture, phenocrysts 5%, groundmass 95%; plagioclase 3%, quartz 2%, opaques <1%, orthopyroxene <1%, glass 94%; orthopyroxene and glass altered to clays and zeolites
S-57	Collected in roadcut along the East Coast Highway, 1 mile north of cross-island highway, 15°12.73′N, 145°46.33′E	Andesite clast, porphyritic, intersertal texture, phenocrysts 28%, groundmass 72%; plagioclase 60%, orthopyroxene 18%, opaques 10%, clinopyroxene 7%, glass 4%, olivine (?) 1%, glass and olivine are altered to clays
S-51	Collected in tunnel in side of gully north of As Perdido Road, 15°08.24′N, 145°42.67′E	Basalt flow, porphyritic, intersertal texture, phenocrysts 4%, groundmass 96%; plagioclase 57%, glass 25%, clinopyroxene 12%, opaques 6%; glass is altered to clays
	Guam	
GM-F-1	Collected in roadcut near Toguan bench mark, southwestern Guam, 13°17.25′N, 144°39.68′E	Basaltic-andesite, porphyritic, hyalopelitic texture, phenocrysts 35%, groundmass 60%, vesicles 5%; plagioclase 60%, clinopyroxene 20%, glass 10%, orthopyroxene 5%, magnetite 5%; glass is altered to clays, vesicles are filled with calcite and clays
GM-50	Collected from a conglomerate in a roadcut on the northwest side of Mt. Santa Rosa, northern Guam, 13°32.40′N, 144°54.42′E	Basaltic andesite clast, porphyritic intersertal texture, phenocrysts 25%, groundmass 75%; plagioclase 50%, glass 33%
GM-70	Collected near Fouha Point, southwestern Guam, 13°18.68′N, 144°39.04′E	Interpillow ooze
GM-63	Collected from a breccia on top of Mt. Jumullong Manglo, southwestern Guam, 13°19.36′N, 144°40.14′E	Basaltic-andesite clast, porphyritic hyalopelitic texture, phenocrysts 18%, groundmass 82%; plagioclase 49%, glass 40%, clinopyroxene 5%, opaques 4%, olivine 2%; glass and olivine are altered to clays
G-SR-6	Collected from a conglomerate in a roadcut on the west side of Mt. Santa Rosa, northern Guam, 13°32.40′N, 144°54.42′E	Basaltic andesite clast, porphyritic, intersertal texture, phenocrysts 28%, groundmass 72%; plagioclase 70%, glass 14%, orthopyroxene 8%, clinopyroxene 5%, opaques 3%, glass is altered to clays
GUM-70-1b	Collected from a breccia unit just below tower at summit of Mt. Alutom, central Guam, 13°25.86′N, 144°42.64′E	Dacite clast, porphyritic, hyaloophitic texture, phenocrysts 39%, groundmass 52%, vesicles 9%; glass 51%, plagioclase 34%, hornblende 6%, orthopyroxene 4%, clinopyroxene 3%, quartz 1%, opaques 1%; glass is partially hydrated
GUM-79-2	Collected just south of radar tower on Mt. Santa Rosa, northern Guam, 13°32.00′N, 144°54.68′E	Andesite pillow lava, porphyritic, hyalopelitic texture, phenocrysts 22%, matrix 73%, vesicles 5%; plagioclase 42%, glass 38%, clinopyroxene 13%, olivine opaques 2%; glass and olivine are altered to clays
GUM-79-6	Collected near Merizo fire station, southwestern Guam, 13°17.49′N, 144°40.35′E	Basaltic dike, nonporphyritic, hyalopelitic texture, vesicles <1%; plagioclase 42%, glass 28%, clinopyroxene 18%, orthopyroxene 8%, olivine 3%, opaques 1%, chromite <1%; glass and olivine are altered to clays
GUM-79-16	Collected on southwest spur of Mt. Schroeder, southwestern Guam, 13°16.60′N, 144°40.35′E	Basaltic pillow lava, porphyritic, hyaloophitic texture, phenocrysts 13%, groundmass 87%; glass 65%, plagioclase 28%, clinopyroxene 6%, opaques 1%, some glass is altered to clays.
GUM-80-33	Collected at northernmost exposure of Maemong limestone, northwest of Mt. Lamlam, southwestern Guam, 13°20.62′N, 144°39.47′E	Limestone; <1% volcanic detritus

TABLE 3. (continued)

Sample	Location	Description
GUM-80-58	Collected near Gaan River, southwestern Guam, 13°22.30′N, 144°39.80′E	Sandy limestone; 15% volcanic detritus
GUM-80-77	Collected southwest of Mt. Lamlam, south-western Guam, 13°20.00′N, 144°39.68′E	Limestone; 1% volcanic detritus
GUM-80-97	Collected on west spur of Mt. Schroeder, southwestern Guam, 13°16.83′N, 144°40.31′E	Sandy limestone; 7% volcanic detritus
GUM-80-101	Collected near water tank above Merizo, southwestern Guam, 13°16.8′N, 144°40.46′E	Sandy limestone; 10% volcanic detritus
GUM-80-107	Collected near road to northeast of Umatac, southwestern Guam, 13°18.08′N, 144°39.87′E	Limestone; <1% volcanic detritus
GUM-80-125	Collected just west of the Agat–Santa Rita school, southwestern Guam, 13°23.56′N, 144°40.38′E	No thin section available
GUM-82-3	Collected just below the Facpi-Alutom contact, approximately 0.8 miles north of Merizo, southwestern Guam, 13°16.47′N, 144°40.29′E	Siliceous limestone matrix of a volcanic breccia
GUM-82-9	Collected just below the top of the Facpi For-mation on a ridge east of Umatac, south-western Guam, 13°17.84′N, 144°40.23′E	Limestone matrix of a volcanic breccia
GUM-82-10	Collected 4 meters above the top of the Facpi Formation of Gum-82-9, southwestern Guam, 13°17.84′N, 144°40.23′E	Limestone; 2% volcanic detritus
GUM-82-11	Collected on a ridge northeast of Mt. Alutom, central Guam, 13°26.06′N, 144°42.51′E	Limestone; 7% volcanic detritus
GUM-82-15	Collected near the Cetti Bay overlook, south-western Guam, 13°19.49′N, 144°39.84′E	Limestone; 7% volcanic detritus
SASA-80-2	Collected in the Sasa River Valley, central Guam, 13°26.57′N, 144°42.03′E	Basaltic sill, nonporphyritic, intersertal texture, vesicles 7%; plagioclase 50%, clinopyroxene 20%, glass 17%, opaques 3%; glass is altered to clays, vesicles are filled with zeolites and clays
	Palau	
PAL-3	Collected from sea cliff near sea plane ramp, Arakabesan Island, 7°20.89′N, 130°29.98′E	Andesite clast, porphyritic, hyalopelitic texture, phenocrysts 36%, groundmass 60%, vesicles 4%; glass and microlites 42%, plagioclase 23%, clino-pyroxene 12%, orthopyroxene 7%, olivine 4%, opaques 2%, olivine is altered to clays
PAL-16	Collected in Ngerusar Quarry, Babelthaup Is-land, 7°21.45′N, 134°32.27′E	Andesite tuff, contains euhedral and angular phen-ocryst fragments 34%, clay and calcite matrix 48%, volcanic rock fragments 6%, and limestone fragments 2%
PAL-17	Collected in Ngerusar Quarry, Babelthaup Is-land, 7°21.45′N, 134°32.27′E	Sandy limestone, 15% volcanic detritus
PAL-19	Collected from cliff near Karamado Bay, Ba-belthaup Island, 7°30.82′N, 134°31.22′E	Dacite clast, porphyritic, hyalopelitic texture, phen-ocrysts 28%, groundmass 72%; plagioclase 46%; glass and microlites 46%, clinopyroxene 3%, or-thopyroxene 2%, opaques 2%, hornblende 1%, quartz <1%; glass is partially altered to clays
	Sarigan	
SA-137	Collected from sea cliff on west side of island, 17°32.45′N, 145°07.50′E	Andesite flow, porphyritic, hyalopelitic texture, phenocrysts 45%, groundmass 55%; plagioclase 60%, glass and microlites 24%, clinopyroxene 5%, opaques 5%, orthopyroxene 3%, hornblende 3%
	Anatahan	
ANT-79-24	Collected from cliff on north coast, 16°22.41′N, 145°42.12′E	Basaltic-andesite flow, porphyritic, hyalopelitic tex-ture, phenocrysts 25%, groundmass 70%, vesicles 5%; glass and microlites 69%, plagioclase 27%, ol-ivine 2%, clinopyroxene 1%, opaques 1%
ANT-79-28	Collected from cliff on west coast, 16°21.94′N, 145°37.66′E	Andesite flow, porphyritic, hyalopelitic texture, phenocrysts 34%, groundmass 61%, vesicles 5%; plagioclase 49%, glass and microlites 40%, clino-pyroxene 5%, orthopyroxene 5%, opaques 2%

TABLE 3. (continued)

Sample	Location	Description
	Mariana Trough	
TSDY-11-D-X	Dredged sample, 14°15′N, 144°06′E	Basalt, nonporphyritic, diabasic texture; plagioclase 51%, clinopyroxene 29%, olivine 13%, glass 5%, opaques 2%

from NNW to WNW caused the initiation of subduction along the proto-PKR, as discussed by *Karig* [1975] and *Scott et al.* [1980]. This implies that the Facpi formation lavas represent a very early stage in the development of oceanic arcs in the eastern Philippine Sea.

Available data suggest that arc volcanism was intermittent along the proto-PKR from late middle Eocene to middle Oligocene time. A late Eocene episode of activity is recorded on Palau(?), Guam, Saipan, Haha-Jima Island in the Bonin Islands, and at D.S.D.P. Site 459B (Figure 1). The presence of volcanic breccias containing clasts of reef limestone in the island exposures indicates this episode was dominated by explosive eruptions in relatively shallow water. On the other hand, the common occurrence of pillows at Site 459B suggests this part of the proto-PKR was at greater depth at this time. The early Oligocene environment was apparently similar, as indicated by fragmental units of this age on Palau (Aimeliik and Ngeremlengui formation), Guam (Alutom formation) and on

the Bonin Islands (Chichi-Jima, *Tsuya* [1937]) and by pillowed flows at Site 458. *Scott et al.* [1980] estimate that the central portion of the proto-PKR arc was rifted between 32 and 29 m.y. B.P. Our data for the Mariana forearc are consistent with this estimate (Figure 1). Furthermore, we suggest that sills in the Alutom formation (32.2 ± 1.0 m.y.) on Guam may have been intruded during the rifting event.

The timing of post–early Oligocene arc activity along the eastward rifted South Honshu Ridge arc is problematic. *Karig* [1975] suggested that this arc was active over the same time interval during which the Parece Vela Basin opened by interarc basin spreading. In contrast, *Scott et al.* [1980] argued that the South Honshu Ridge arc was inactive during most of this time interval. Our dating results do not resolve this issue, although they are consistent with the latter interpretation. The only post–middle Oligocene arc rocks found in the Mariana forearc (Umatac formation on Guam and Fina-Sisu formation on Saipan) have ages that postdate (13 ± 2.0 m.y. B.P.) the cessation

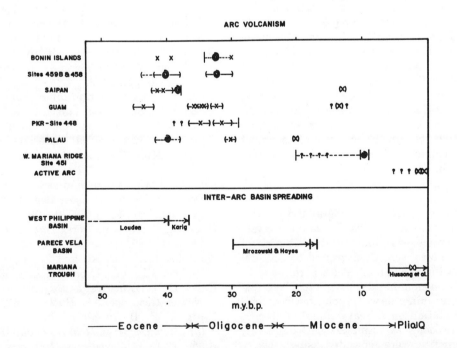

Fig. 1. Summary of age data for volcanic units in arcs and interarc basins of Philippine Sea. Crosses represent K-Ar age determinations. Spirals represent paleontologic ages. Parentheses show approximate error limits for ages. Vertical bars represent estimated age limits for each sequence. Data from *Kaneoka et al.* [1970], *Hussong et al.* [1981], *Cloud et al.* [1956], *Tracey et al.* [1964], *Cole* [1950], *Kroenke et al.* [1980], *Scott et al.* [1980], *Louden* [1976], *Karig* [1975], *Mrozowski and Hayes* [1979].

of spreading in the Parece Vela Basin (Figure 1). Interestingly, the gap in volcanic activity on Palau suggested by our dating (Tables 1 and 2) is of the same age and duration as the gap suggested by *Scott et al.* [1980] for the South Honshu Ridge Arc. Assuming that the Palau gap is real (not an erosional effect), it suggests the gap in arc activity along the South Honshu Ridge arc may not be directly related to the formation of the Parece Vela Basin behind it because an interarc basin was not forming behind the Palau arc during this same time interval.

Conclusions

New age data presented here require revision of the published stratigraphy of Guam. This primarily involves designation of the lower Facpi flow member of the Umatac formation of *Tracey et al,* [1964] as a formation (i.e., Facpi formation) and revising its age assignment from Miocene to late middle Eocene. The Umatac formation, as now defined, is of early to late Miocene age and includes the Maemong limestone member, the Schroeder flow member, the Bolanos pyroclastic member, and the Dandan flow member. The Alutom formation remains as originally defined. Dikes in the Facpi formation appear to be associated in part with eruption of the Alutom formation volcanic rocks.

The Sankakuyama formation on Saipan has been determined to be late Eocene (41 ± 1.0 m.y. B.P.) as suspected by *Cloud et al.* [1956]. Previously undated arc volcanic rocks on Palau have radiometric ages of approximately 30 and 20 m.y. B.P., suggesting either a gap in arc activity or an erosional episode in this time interval. Dating of Mariana active arc samples suggests the arc is at least 1.3 ± 0.2 m.y. old.

Our data are generally consistent with the model of Karig [1971] for evolution of the eastern Philippine Sea, although our data provide a more definitive chronology. The question of whether or not arc volcanism and interarc basin spreading might be synchronous processes cannot be resolved with our data, although no evidence for synchroneity has been found.

Analytical Methods

Potassium-argon determinations made at the University of Arizona were done as follows. Large phenocrysts and xenocrysts were removed by hand picking after initial coarse crushing. Samples were then ground to 150-100 μm. Standard heavy liquid and magnetic separation techniques were used for monomineral separation. Glass, zeolites, clays, and altered minerals were separated from fine-grained rocks by flotation on a heavy liquid of specific gravity 2.50-2.55. K-poor heavy minerals, such as olivine, pyroxene, and magnetite, were separated using a heavy liquid of 2.90-2.95 specific gravity. The remaining concentrate of feldspar and feldspar-rich composite grains were leached with 5% HF for 10 minutes, followed by treatment with a blender to remove any adhering glass, zeolite, clay,

and carbonate. The samples were then resieved to remove the finer grains. The resulting culled groundmass was used for K-Ar dating.

Argon was analyzed using a Nier-type 60°-sector 15.24-cm radius gas source mass spectrometer operated in the static mode. A GAC/SPC 16/55 real time minicomputer changed the magnetic field to focus each beam of ions, in turn, on a Faraday cup collector, then measured and stored the voltages generated. Measurements were time regressed to t_0 (the time gas was introduced), using both linear and parabolic least squares regression routines on individual mass intensities and also on mass intensity ratios. If analyses on different aliquots did not agree within statistical limits, further fusion and analyses were performed until satisfactory results were obtained.

Acknowledgments. We would like to thank K. Mäd for his field assistance and hospitality on Palau. W. R. Riedel kindly evaluated several samples for radiolarians. E. Heyerdahl typed and edited the manuscript. We would also like to thank D. A. Clague for reviewing this manuscript. This work was supported by NSF grant OCE-8018375.

References

Baldridge, W. S., P. E. Damon, M. Shafiquallah, and R. J. Bridwell, Evolution of the central Rio Grande Rift, New Mexico: New potassium-argon ages, *Earth Planet. Sci. Lett., 51,* 309-321, 1980.

Clague, D. A., and R. D. Jarrard, Tertiary Pacific plate motion deduced from the Hawaiian-Emperor chain, *Geol. Soc. Am. Bull., 84,* 1135-1154, 1973.

Cloud, P. E., Jr., R. G. Schmidt, and H. W. Burke, Geology of Saipan, Mariana Islands, *U.S. Geol. Surv. Prof. Pap., 280-A,* 126 pp., 1956.

Cole, W. S., Larger foraminifers from the Palau Islands, *U.S. Geol. Surv. Prof. Pap., 221-B,* 21-26, 1950.

Cooper, J. A., The flame photometric determinations of potassium in geological materials used for potassium-argon dating, *Geochim. Cosmochim. Acta, 27,* 525-546, 1963.

Ellis, C. H., Calcareous nannoplankton biostratigraphy—Deep-Sea Drilling Project Leg 60, *Init. Rep. Deep-Sea Drilling Proj., 60,* 507-536, 1981.

Hussong, D., et al. (editors), *Init. Rep. Deep-Sea Drilling Proj., 60,* 1981.

Kaneoka, I., N. Isshiki, and S. Zashu, K-Ar ages of the Izu-Bonin Islands, *Geochem. J., 4,* 53-60, 1970.

Karig, D. E., Structural history of the Mariana Island arc system, *Geol. Soc. Am. Bull., 82,* 323-344, 1971.

Karig, D. E., Basin genesis in the Philippine Sea area, *Init. Rep. Deep-Sea Drilling Proj., 31,* 857-880, 1975.

Karig, D. E., et al. (editors), *Init. Rep. Deep-Sea Drilling Proj., 31,* 1975.

Kroenke, L., et al. (editors), *Init. Rep. Deep-Sea Drilling Proj., 59,* 1980.

Ladd, H. S., Chitons and gastropods (Haliotidae through

Adeorbidae) from the western Pacific Islands, *U.S. Geol. Surv. Prof. Pap., 531,* 98 pp., 1966.

Louden, K. E., Magnetic anomalies in the West Philippine Basin, in *The Geophysics of the Pacific Ocean Basin and its Margin, Geophys. Monogr. Ser.,* vol. 19, edited by G. H. Sutton et al.., pp. 253–267, AGU, Washington, D. C., 1976.

Mason, A. C., G. Corwin, C. L. Rodgers, O. Elmquist, A. J. Vessel, and R. J. McCracken, Military geology of the Palau Islands, Caroline Islands, report, Intelligence Div., Office of the Eng. Headquarters, U.S. Army Forces Far East and Eighth U.S. Army (rear)/U.S. Geol. Surv., Washington, D. C., 1956.

Meijer, A., A study of the geochemistry of the Mariana Island arc system and its bearing on the genesis and evolution of volcanic arc magmas, Ph.D. dissertation, Univ. of Calif., Santa Barbara, 1974.

Mrozowski, C. L., and D. E. Hayes, The evolution of the Parece Vela Basin, eastern Philippine Sea, *Earth Planet. Sci. Lett., 46,* 49–67, 1979.

Reagan, M., and A. Meijer, Geology and geochemistry of early arc volcanic rocks from Guam, *Geol. Soc. Am. Bull.,* in press, 1982.

Schmidt, R. G., Petrology of the volcanic rocks, in Geology of Saipan, Mariana Islands, part 2, pp. 127–175, U.S. Geol. Surv., Reston, Va., 1957.

Scott, R. B., L. Kroenke, G. Zakariadze, and A. Sharaskin, Evolution of the South Philippine Sea: Deep-Sea Drilling Project Leg 59 results, *Init. Rep. Deep-Sea Drilling Proj., 59,* 803–816, 1980.

Sutter, J. F., and T. E. Smith, ^{40}Ar/^{39}Ar ages of diabase intrusions from Newark Trend basins in Connecticut and Maryland: Initiation of Central Atlantic rifting, *Am. J. Sci., 279,* 808–831, 1979.

Sutter, J. F., and L. W. Snee, K/Ar and ^{40}Ar/^{39}Ar dating of basaltic rocks from Deep Sea Drilling Project Leg 59, *Init. Rep. Deep-Sea Drilling Proj., 59,* 729–734, 1980.

Tracey, J. I., S. O. Schlanger, J. T. Stark, D. B. Doan, and H. G. May, General geology of Guam, *U.S. Geol. Surv. Prof. Pap., 403-A,* 104 pp., 1964.

Tsuya, H., On the volcanism of the Huzi volcanic zone with special reference to the geology and petrology of Izu and the southern islands, *Bull. Earthquake Res. Inst., 15,* 215–357, 1937.

Collision Processes in the Northern Molucca Sea

Gregory F. Moore[1]

Geological Research Division, Scripps Institution of Oceanography, La Jolla, California 92093

Eli A. Silver

Earth Sciences Board, University of California, Santa Cruz, California 95064

The Mindanao-Molucca Sea region is a collision zone between two facing island arc systems. The West Mindanao Arc, which was active from the Cretaceous through the Quaternary, apparently collided in the mid-Tertiary with the East Mindanao Arc, which was active from the Cretaceous through the Oligocene. Following the collision in the Miocene, the Agusan-Davao Trough was filled with coarse clastic sediments. The ophiolite terrane exposed in the Pujada Peninsula was probably emplaced during this collision. In the Molucca Sea region the collision between the Halmahera Arc and the Sangihe Arc is still active. Gravity data and seismic reflection profiles indicate a crustal break between the Talaud Ridge and southern Mindanao. The deformed rocks of the Talaud Islands probably represent the northern limit of the forearc terrane of the Halmahera Arc system. The East Sangihe and West Halmahera thrusts can be followed, via seismic reflection data, northward to the latitude of Talaud. The origin of the Snellius Ridge complex remains obscure, but we interpret it to be the northern extension of the Halmahera Arc terrane. A tectonic reconstruction for this complex region proposes that the collision in the Mindanao region occurred between two intraoceanic arc terranes with very little intervening sediment. The southern end of the eastern Mindanao Arc was connected with the northern end of the Halmahera Arc by a transform fault. We suggest that the thick sediments presently being deformed in the Molucca Sea collision zone were eroded from New Guinea and Halmahera in the south and from the collision zone in Mindanao. A substantial amount of strike-slip motion has probably occurred during the collision in Mindanao. In response to the collision, two new subduction zones, at the Cotabato and Philippine trenches, are propagating southward.

Introduction

The accretion of allochthonous terranes to continental blocks is a fundamental process in the development of orogenic zones. Exotic crustal fragments that have been identified throughout the circum-Pacific mountain belts are often interpreted as the result of collisions between continental or island arc fragments and continental margins. For instance, *Jones et al.* [1977] have suggested that the 'Wrangellian' terranes represent allochthonous blocks incorporated into northwestern North America in the late Mesozoic by collision and strike-slip processes. *Schweickert and Cowan* [1975] have suggested that early Mesozoic rocks of the western Sierra Nevada Mountains of California may represent a collision between opposite-facing subduction

zones. The Sonoma orogeny is believed to be the result of the collision of the Sierran-Klamath Paleozoic arc terrane with the North American margin at the end of the Paleozoic [*Speed*, 1979].

Although the results of collisions have been documented in orogenic belts, the processes by which collisions occur remain largely speculative. This lack of understanding has been due primarily to the paucity of data from upper Cenozoic collisions. Most of our knowledge of the collision process has been derived from the study of completed collisions in very old mountain belts, where the paleoplate motions and boundaries can only be inferred. Moreover, later tectonism may obscure the effects of earlier orogenic events.

An example of a collision zone between two island arc systems occurs in the Mindanao-Molucca Sea region (Figure 1). *Hamilton* [1979] interpreted a large amount of geological and geophysical data from this complicated region and recognized that the collision is nearly complete in

[1] Now at Exploration and Production Research, Cities Service Co., Tulsa, Oklahoma 74102.

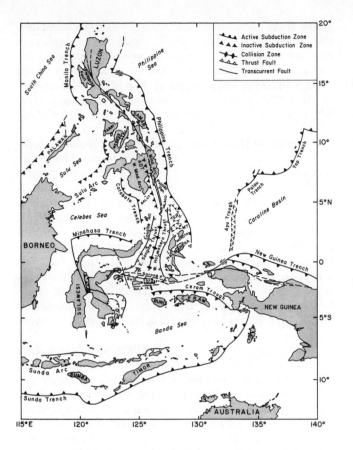

Fig. 1. Tectonic map of the Philippine and eastern Indonesian region modified from *Hamilton* [1979], *Silver* [1981], and *Cardwell et al.*, [1980]. MT = Morotai Trough, SR = Snellius Ridge, and ADT = Agusan-Davao Trough.

Mindanao and is still ongoing in the Molucca Sea. Although the regional tectonic setting and structure of the active Molucca Sea collision zone are known [*Murphy*, 1973; *Hamilton*, 1977, 1979; *Silver and Moore*, 1978; *Cardwell et al.*, 1980; *McCaffrey et al.* 1980], the geology of the Mindanao collision zone is poorly known, and the regional tectonic relations in the transition zone between active and completed collision are uncertain. The purpose of this paper is to present new data from the Molucca Sea/Mindanao region, to interpret the structure and tectonic relationships between the Molucca Sea and southern Mindanao, and to document processes that operate within the collision zone.

Regional Structure

Previous Studies

The Molucca Sea collision zone extends from Mindanao southward to the Sula thrust (Figure 1). The 'Molucca Sea Plate' is presently being subducted along its western margin beneath the Sangihe Arc and along its eastern margin beneath the Halmahera Arc. The Sangihe and Halmahera

arcs are thus colliding, and the southern Molucca Sea represents the collision zone.

A Benioff zone dips westward to greater than 600 km underneath the Sangihe Arc [*Cardwell et al.*, 1980]. The active volcanos extend from northeast Sulawesi through the Sangihe Islands (Figure 1).

The Halmahera Arc is a west-facing island arc system that bounds the eastern side of the Molucca Sea. The volcanic arc of Halmahera has been active only since the early Quaternary [*Sukamto et al.*, 1979]. The active volcanos of Halmahera lie just off the west coast of the island in the southern and central part and just onshore in the northern part. The Quaternary arc was built on an Oligocene-lower Miocene east-facing arc system [*Hamilton*, 1979]. Tuffaceous sandstones interbedded with Eocene limestones have also been reported [*van Bemmelen*, 1949]. The east arms are composed of ultramafic, mafic, and metamorphic rocks, and some schist occurs off the southwest coast. The distribution of rock types on Halmahera seems best explained by westward subduction in the early Cenozoic, followed by a flip to the present sense of eastward subduction in the late Cenozoic [*Hamilton*, 1977; *Katili*, 1978]. The seismic zone dips east from the center of the Molucca Sea to a maximum depth of 229 km beneath Halmahera [*Cardwell et al.*, 1980]. The mantle seismic zone below 100 km extends north northeast of Halmahera under Morotai Island and ends abruptly beneath the Morotai Trough at 3°N. The origin of the Morotai Trough remains unclear. The existence of an east-west regional gravity minimum centered over the basin [*Watts et al.*, 1978] supports the inference of *Cardwell et al.* [1980] that it is the location of a major crustal discontinuity.

The central portion of the southern Molucca Sea collision zone is underlain by an acoustically chaotic mass of low-velocity, low-density deformed sediments [*Silver and Moore*, 1978; *Hamilton*, 1979; *McCaffrey et al.*, 1980]. South of the equator there is a broad gravity low of up to −250 mGal centered over the Molucca Sea melange wedge [*Watts et al.*, 1978; *McCaffrey et al.*, 1980]. In the active collision zone the surface expression of the subduction zones dipping under the Sangihe and Halmahera arcs is obscured by the collision complex, which has been thrust onto the arc flanks with the surface thrusts dipping away from the arcs toward the collision complex. The East Sangihe and West Halmahera thrusts were traced to at least the south end of the Talaud Islands by *Silver and Moore* [1978]. They inferred a northward continuation for both faults on the basis of bathymetry, but our more recent seismic data show that the East Sangihe thrust does not extend north of Talaud (Figure 1). The islands of Mayu and Tifore (Figure 3) are subaerially exposed portions of the collision complex. Mayu contains diabase, porphyrite, gabbro, and harzburgite [*Verbeek*, 1908; E. A. Silver, unpublished field data, 1977] and Tifore consists of serpentinite in the north and intensely folded basalt, quartzo-feldspathic sandstones, and limestones in the south. These islands form a ridge with a 100-mGal gravity high and are thought to represent slices of the Molucca Sea crust that have been thrust into the overlying collision complex [*Silver and Moore*, 1978; *McCaffrey et al.*, 1980].

Northwest-trending belts of melange with ophiolite blocks are in thrust contact with less-deformed Neogene strata on the Talaud Islands [*Sukamto and Suwarno*, 1976; *Soeria Atmadja and Sukamto*, 1980]. This island group has been suggested to be either an uplifted part of the active collision complex [*Hamilton*, 1979] or the southern end of the Samar-East Mindanao Arc [*Cardwell et al.*, 1980].

North of the Talaud Islands, a bathymetric ridge extends toward southeast Mindanao. *Krause* [1966] proposed that the Philippine fault zone continues from the Pujada peninsula southward along the east side of the Talaud Ridge at least as far south as latitude 3°N. He further postulated the existence of a western branch of the fault on the west side of the Talaud Ridge.

Hamilton [1979] suggested that the Molucca Sea melange wedge comes ashore in southeast Mindanao as the Pujada peninsula, where a highly deformed basement complex of ophiolites has been thrust over interbedded graywackes and volcanic rocks [*Melendres and Comsti*, 1951; *Santos-Ynigo et al.*, 1961; *Santos-Ynigo*, 1965].

The deep seismic zone (deeper than 400 km) of the Sangihe Arc continues to the north under western Mindanao, but the shallow seismic zone does not extend north of Davao Gulf, probably because the subducting plate was detached during the collision [*Cardwell et al.*, 1980]. The volcanic arc in Mindanao extends from Balut Island to the north tip of central Mindanao (Figure 1) and is formed of volcanos that were active from the Miocene through historic times. Andesitic tuff, agglomerate and volcaniclastic sedimentary rocks of probable Cretaceous to Paleogene age, and associated basement rocks (peridotites, gabbros and schists) are reported to underlie the young volcanic rocks in central Mindanao [*Casasola*, 1956; *Metal Mining Agency*, 1972], indicating that the Neogene volcanic arc may have been superimposed on a Cretaceous-Paleogene arc. The northern extension of the older arc is unclear. Cretaceous and Paleogene volcanic rocks are exposed on Cebu and Bohol (*Philippine Bureau of Mines*, 1963], but we cannot be certain to which arc they belong.

A north-northwest-trending calc-alkaline volcanic chain developed in southwest Sulawesi during the Paleocene-Eocene [*Hamilton*, 1979]. This volcanic arc may have been continuous with the volcanic arc in Mindanao, as shown by *Hamilton* [1979, Figure 77].

Much of eastern Mindanao is covered with Cretaceous to Oligocene volcanic rocks [*Ranneft et al.*, 1960; *Metal Mining Agency*, 1973; *Matsumaru*, 1974; *Hashimoto*, 1981]. Similar Cretaceous and younger volcanic rocks have been found on Samar and Luzon Islands north of Mindanao [*Reyes and Ordonez*, 1970; *Hashimoto et al.*, 1975; *Hashimoto*, 1981]. These volcanic rocks have been interpreted as a Cretaceous to mid-Oligocene volcanic arc that was active from north of Samar through eastern Mindanao [*Hamilton*, 1979; *Cardwell et al.*, 1980]. The volcanics are overlain by upper Oligocene limestones and Miocene coals and shales, which suggests that volcanism ceased by the upper Oligocene [*Vergara and Spencer*, 1957; *Matsumaru*, 1974]. Serpentinite bodies are common throughout central Mindanao east of the Philippine fault zone [*Teves et al.*, 1951].

The Agusan-Davao Trough of central Mindanao separates the east and west Mindanao arc terranes. The trough contains up to 6 km of Eocene to Recent sedimentary rocks overlying a basement complex [*Ranneft et al.*, 1960]. A period of intense folding is recognized at the end of the mid-Miocene [*Treves et al.*, 1951], and the coarse clastic sedimentary rocks of the upper Miocene to Pliocene Adgoan Formation were derived from an uplifted region, further indicating that the mid-Miocene was a time of tectonic activity in central Mindanao [*Hashimoto*, 1980]. *Cardwell et al.* [1980] propose that the Agusan-Davao Trough might be the forearc basin for the Talaud-East Mindanao Arc and that the suture lies just east of the volcanos in western Mindanao.

A major question addressed in this paper is the nature of the transition from active arc-arc convergence south of 3°N to the zone of completed convergence in Mindanao. Critical evidence is observed in the seismicity and structure of the region. There is a major transition between the northern Molucca Sea and the southern Molucca Sea at approximately 3°N. South of 3°N the bathymetric and gravity trends are northeast, but the trends are north northwest from approximately 3°N to southern Mindanao [*Mammerickx et al.*, 1976; *Watts et al.*, 1978]. Active thrusting of the Molucca Sea melange wedge is occurring south of Talaud. Subduction of at least 700 km of lithosphere is recorded by the Benioff zone extending under the Celebes Sea, and the deep seismicity continues northward under Mindanao, but shallow seismicity is absent under western Mindanao. North of Talaud the thick sediments in Davao Gulf and its southern extension are only moderately deformed, indicating that subduction has stopped or greatly slowed down in this region. In central Mindanao there is presently no active compressional deformation, and large-scale convergence between east and west Mindanao probably ceased in the Miocene [*Hamilton*, 1979].

Because *Cardwell et al.* [1980] interpret convergence between the Talaud Ridge and the Sangihe Arc to have nearly ceased, while convergence still continues between the Halmahera and Sangihe arcs, they require relative motion between the Halmahera Arc and the Talaud Ridge. They suggest that the northern end of the Halmahera Arc may end along a transform fault at approximately 3°N because the active volcanos of the Halmahera Arc and the zone of the intermediate depth earthquakes do not extend north of about 3°N. The transform would strike eastward, possibly along the northern edge of the Morotai Trough. The proposed transform would intersect the Philippine

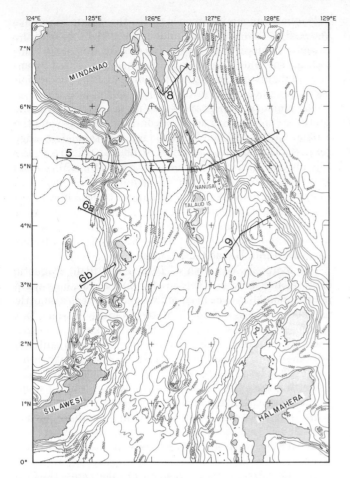

Fig. 2.   Bathymetric chart of the Molucca Sea region, based on all available cruise data, including SIO cruises Indopac, Mariana, and Rama in 1976, 1977, 1979 and 1980 (tracks indicated by ticks at contour crossings). Contour interval is 100 m (uncorrected). Locations of seismic profiles of Figures 5, 6, 7, 8, and 9 are shown with heavy lines.

Trench at about 3°N, where shallow seismic activity along the trench ceases. *McCaffrey* [1982], on the other hand, interprets the seismicity data to indicate a transform in the submerged Molucca Sea plate, but not at the surface. Seismic profiles [*Silver and Moore*, 1978] show no evidence for near-surface faulting between Halmahera and Snellius Ridge.

At the northern end of the Sangihe Arc an arc polarity reversal is now underway in response to cessation of west-directed underthrusting [*Hamilton*, 1979]. The Cotabato Trench is defined by bathymetric and gravity minima south of Mindanao [*Mammerickx et al.*, 1976; *Watts et al.*, 1978]. Shallow seismicity at the trench suggests that there is active subduction occurring [*Stewart and Cohn*, 1977]. The lack of an active volcanic arc, along with the short length of the inclined seismic zone under Mindanao, suggest that

subduction along the Cotabato Trench began relatively recently [*Cardwell et al.*, 1980].

The Philippine Trench extends along the continental slope east of Mindanao southward to at least the north end of Halmahera (Figure 1). The small accretionary prism and shallowing of seismicity to the south led *Hamilton* [1979] and *Cardwell et al.* [1980] to infer that the Philippine Trench is a very young feature that is propagating southward. The youth of the southern Philippine Trench, the remnants of a volcaniclastic apron east of that trench [*Karig*, 1975], and the abundance of island arc volcanic rocks along the Talaud-East Mindanao ridge are explained by *Cardwell et al.* [1980] through interpretation of the ridge as part of a west-facing island arc.

New Data

Geophysical maps. Our new bathymetric, free-air gravity, and magnetic anomaly maps of the Molucca Sea

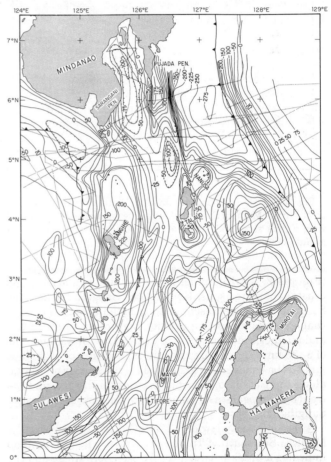

Fig. 3.   Free-air gravity anomaly map of the Molucca Sea region updated from *Watts et al.* [1978] with new SIO data (ship tracks with gravity data are indicated with dots). Contour interval is 25 mGal.

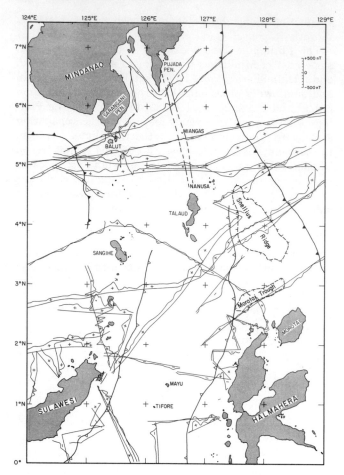

Fig. 4.   Residual magnetic anomaly data plotted along ship tracks in the Molucca Sea region. Positive anomalies are indicated with a plus sign, and negative anomalies are indicated with a minus sign.

(Figures 2, 3, 4) help delineate tectonic boundaries and crustal characteristics. North of the Sangihe Islands, the Sangihe Arc is represented by a broad submarine ridge that exhibits a positive gravity high of up to 225 mGal and magnetic anomalies of up to 800 nT with wavelengths of 5–35 km (Figure 4), suggesting the existence of shallow igneous basement. The magnitude of the gravity minimum along the Cotabato Trench decreases southward, and the trench disappears as a morphologic feature southwest of Sangihe Island.

A major gravity low of −175 mGal is centered over the Molucca Sea south of Talaud (Figure 3). The gravity field indicates a break between the Mayu-Tifore Ridge and the Talaud Ridge. The regional structural grain and gravity field of the southern Molucca Sea trend northeast (Figures 2, 3). A detailed survey south of Tifore, however, shows that the small-scale structures within the ridge trend northwest.

Magnetic anomalies measured in the Molucca Sea are generally of low amplitude and long wavelength, with the exception of local 200–250 nT, 10–25 km wavelength anomalies south of Talaud and Tifore islands. These anomalies probably reflect the location of ophiolites and ophiolite fragments as seen in the geology of the islands.

The major free-air gravity minimum centered over the southern Molucca Sea does not continue north beyond the Talaud Islands (Figure 3). Although a bathymetric ridge connects Talaud with southeast Mindanao (Figure 2), the free-air gravity high continues southward from Mindanao only to about 5°N. The gravity field north of Talaud is very complex but indicates that the Talaud crustal block does not continue northward. A free-air minimum extends west of Talaud from approximately 3°N to 5°N. This belt of negative anomalies is offset to the west at 5°N and trends northward into Davao Gulf (Figure 3).

The Molucca Sea north of Talaud has little magnetic expression (Figure 4). There are low-amplitude, long-wavelength anomalies over most of the area. There is also little magnetic expression north of the Nanusa Islands, but southeast of Nanusa there is a 100-nT, 10-km-wavelength anomaly (Figure 4). A −275 mGal gravity low is centered over the trench slope north of Talaud. The gravity minimum is displaced 25–30 km landward of the trench axis, indicating the existence of low-density material.

The composition and structure of Snellius Ridge are obscure, but short-wavelength (7–10 km), high-amplitude (170–280 nT) magnetic anomalies and a 150-mGal gravity high over the ridge are compatible with an island arc origin.

Seismic reflection data.   Seismic profiles across the Cotabato Trench (e.g., Figure 5) illustrate deformation of thick, layered sediments in the Celebes Sea. The zone of deformation is about 40 km wide, and a frontal thrust has recently propagated west of the main accretionary prism. The width of the accretionary prism decreases rapidly southward to less than 10 km in line 7 (Figure 6a), and deformation is not apparent in line 11 (Figure 6b). The latter profile indicates deformation of the arc slope, but unlike the lines to the north the deeper layers in the Celebes Sea do not dip eastward toward the arc. Farther south, several north-trending ridges occur near the base of the western arc slope. All profiles across the base of the slope show signs of vertical motion, but no clear indications of thrusting are found south of 3°20′N. Significantly, the southern extent of the Cotabato Trench corresponds fairly closely to the northern extent of the East Sangihe thrust.

Northeast of the Sangihe Islands, the East Sangihe thrust is absent, and a trough 10–40 km wide is filled with at least 2 km of nearly undeformed sediments (Figure 7). North of approximately 5°N, a ridge (at 35 km in Figure 7) separates the sediments in the basin into two sub-basins. This ridge is the southern continuation of the Pujada Peninsula (Figure 2). Deep strata in the basin appear to be deformed by this ridge, but younger strata lap onto the

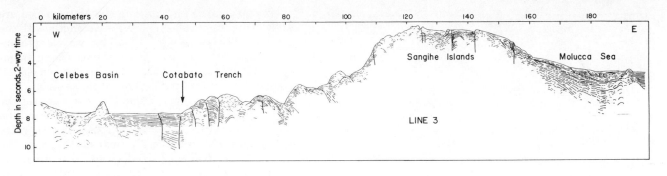

Fig. 5. Line drawing of single-channel seismic reflection profile across the northern Molucca Sea, Sangihe Arc, and Cotabato Trench. Location of profile is shown in Figure 2.

ridge and are not deformed by it. In the profile of Figure 5, sediments in the eastern sub-basin are deeper than sediments in the western basin. This ridge is acoustically opaque (Figure 7) and exhibits a free-air gravity high (Figure 3) but does not have any appreciable magnetic expression (Figure 4), indicating that it probably comprises deformed sedimentary rocks. The basin west of the ridge continues northward into the Davao Gulf, where sediments are broadly folded into an anticline at the base of the slope off the Sarangani Peninsula [see *Cardwell et al.,* 1980, Figure 8]. Seismic reflection profiles collected by AMOCO International Oil Company across Agusan-Davao Trough in central Mindanao clearly show that the basement of the West Mindanao Arc dips eastward under Agusan-Davao Trough. A distinctive Eocene limestone is onlapped by younger sediments that are relatively undeformed, except on the eastern side of the basin where they are truncated by the Philippine fault zone. Seismic profiles

collected across the southern portion of Agusan-Davao Trough show low-amplitude anticlinal folds formed over west-dipping thrust faults, indicating recent compression across the south part of the basin.

The ridge north of Talaud is a horst block between two fault traces (Figures 1, 2). This horst block does not appear to be a very young feature because sediments in the basin to the west lap onto the western portion of the ridge (Figure 7). The fault strands can be identified in seismic profiles (Figures 7, 8) and in the regional bathymetry and gravity (Figures 2, 3) from north of Talaud to southeast of the Pujada Peninsula, where they probably connect with strands of the Philippine fault zone.

The eastern boundary of the southern Molucca Sea collision zone is the West Halmahera thrust (Figure 1), along which the collision complex is being thrust eastward onto the Halmahera Arc apron [*Silver and Moore,* 1978]. The western margin of the Snellius Ridge dips westward at

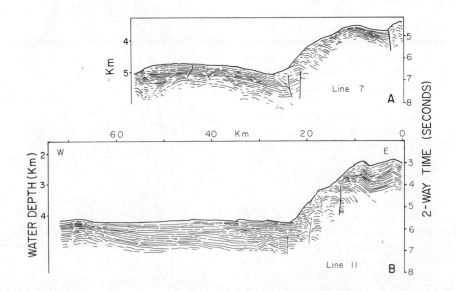

Fig. 6. Seismic reflection profiles across the Cotobato Trench. See Figure 2 for locations.

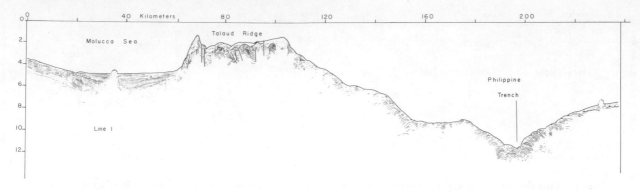

Fig. 7. Line drawing of single-channel analog reflection profile across the Molucca Sea north of Talaud. See Figure 2 for location.

least 20 km underneath the collision complex (Figure 9), and there are indications of westward-dipping imbricate structures within the collision complex (Figure 9). Our new seismic data show that the Halmahera thrust can be traced northward to approximately 4°30'N, but the thrust is not visible on a seismic profile at 4°55'N (Figure 7), although the structure on this line is complex.

The Snellius Ridge has a flat-topped summit as shallow as 500 m below sea level and is tilted to the west (Figure 9). The eastern portion of the ridge has up to 1 s of sediments covering acoustic basement.

Seismic reflection profiles that cross the landward slope of the Philippine Trench between Talaud and Mindanao show a very steep lower slope and benchlike topography between 5000 m and 2500 m (Figure 7). The origin of this topography is unclear, but it may be due in part to strike-slip faulting.

Geological field data. On the Talaud Islands, mid-Miocene to Pliocene marine strata crop out in approximately north-south trending belts and overlie melange and ophiolite slabs [*Moore et al., 1981a*]. The ophiolite bodies are large blocks incorporated into the melange, but our mapping does not constrain their thickness. Benthic Foraminifera in the Neogene strata indicate that the island has been uplifted from lower bathyal (<2000 m) depths since the mid-Miocene. Miocene island arc volcanic rocks (two-pyroxene andesites) crop out in east central Talaud, on the Nanusa Islands east of Talaud, and on Miangas Island on the ridge north of Talaud [*Sukamto and Suwarno, 1976; Sukamto et al., 1980; Evans et al., 1982*]. The Talaud Islands are thought to represent the forearc region of a Neogene west-facing arc system [*Moore et al., 1981a*].

Our field mapping [*Moore et al., 1981b*] in eastern Mindanao has not yet revealed the existence of melange terranes. We observed deformed Eocene and older graywackes in long-wavelength folds north and east of the Pujada Peninsula. These Eocene graywackes, derived from andesitic volcanic rocks, are interbedded with volcanic rocks along the northern edge of Pujada Peninsula. They are

separated from the basement rocks of the Pujada Peninsula by a strand of the Philippine fault zone. The Pujada Peninsula comprises a northwest-trending imbricate terrane of ophiolite slabs thrust over chlorite-actnolite schists. Amphibolite facies metabasites and ultramafic rocks occur as lenses aligned on major north-northwest-trending west-dipping thrusts [*Hawkins et al., 1981*]. The ophiolites consist of sheared and serpentinized ultramafic rocks, gabbro, diabase, and basalt. The igneous and metamorphic rocks are overlain by mid-Miocene to Pliocene conglomerates containing fragments of the older rocks, indicating pre-mid-Miocene ophiolite emplacement.

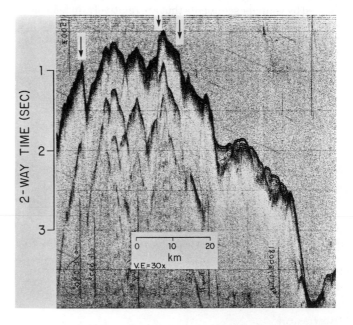

Fig. 8. Photograph of a single-channel analog seismic reflection profile south of the Pujada Peninsula showing the traces of the Philippine fault zone. Location shown in Figure 2.

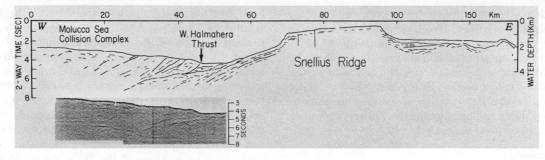

Fig. 9. Multichannel seismic reflection profile across the Snellius Ridge and West Halmahera Thrust. See Figure 2 for location.

Our new field data from eastern Mindanao confirm that the volcanic rocks exposed over large areas [*Metal Mining Agency*, 1973] are island arc tholeiites [*Wright et al.*, 1981]. The volcanics are interbedded with Cretaceous limestones, but we cannot yet determine the facing direction of the arc. The volcanic rocks are overlain by upper Oligocene limestones that have no volcanic detritus, indicating that volcanism in eastern Mindanao had ceased prior to late Oligocene time.

The Philippine Fault is an important, but poorly known, tectonic feature in Mindanao. Although *Hamilton*'s [1979] examination of airphotos provided no evidence for strike-slip motion on the fault, *Cardwell et al.* [1980] inferred significant strike-slip motion on the basis of earthquake data. Our field work along the northeast coast of Pujada Peninsula (Figure 1) delineated a series of en echelon folds and faults adjacent to the Philippine Fault that are consistent with left-lateral wrenching (G. F. Moore, unpublished field data, 1981). In addition, we observed one possible stream offset on an aerial photograph from north of Pujada Peninsula. We therefore tentatively interpret the Philippine Fault in Mindanao as being a left-lateral strike-slip fault. Considerable additional field work is necessary to define the amount and rate of motion along the fault.

Plio-Pleistocene volcanic activity has occurred in the Masara area east of the head of Davao Gulf [*Malcidem and Pena*, 1967]. There are many active hot springs in this area, and we observed steam eminating from the region around one of the young intrusions as overburden was removed during a mining operation. We cannot presently determine whether this young volcanism is related to the initiation of a volcanic chain associated with the young Philippine Trench or whether it might be related to activity along the Philippine fault zone.

Discussion

Structural Relations

The major structural complication in this region is in the central portion of the Molucca Sea. The Mayu-Tifore Ridge trends north northeast, whereas the Talaud-Mindanao Ridge trends north northwest. The transition between these two areas appears to be near the Talaud Islands. The Philippine fault zone does not continue south of Talaud, and the East Sangihe thrust continues northward only to the latitude of Talaud. The West Halmahera thrust has not been followed north of Talaud, but the structural control for that feature is less certain in the northern region. Although *Cardwell et al.* [1980] and *Moore et al.* [1981a] interpreted Talaud as the southern extension of the East Mindanao Arc, we now favor a different interpretation. The gravity field north of Talaud is not continuous with that over the ridge south of Mindanao, the structural vergence on Talaud is different from the vergence in southeast Mindanao, and the Cretaceous and Paleogene volcaniclastic sedimentary rocks of Mindanao are absent from Talaud. We therefore suggest that the Talaud Islands are not part of the East Mindanao Arc. Talaud and East Mindanao appear to be structurally continuous now only because of the recent movement along the Philippine fault zone, which has propagated south into the Talaud block. We suggest that Talaud may be the forearc terrane of the east-dipping Halamhera arc system and has overridden the Sangihe subduction complex during the collision. Convergence between Talaud and Sangihe has ceased since initiation of the Philippine and Cotabato trenches.

The origin of the Snellius Ridge complex is still unclear. Because of its gravity, magnetic, and seismic character, we infer that the Snellius Ridge represents an eroded island arc. According to *Silver and Moore* [1978], the Halmahera thrust represents convergence between the Halmahera Arc and its forearc to the west, although it is now difficult to distinguish Halmahera forearc from Sangihe forearc and either from material added to and deformed in the collision zone after collision initiated. The importance of this view is that Talaud and Snellius Ridge, although presently separated by the Halmahera thrust, may have originated as part of the same arc-forearc system.

The origin of the Morotai Basin is also problematical. It could represent a transform fault between the Snellius Ridge and the rest of the Halmahera Arc. Although seismicity data strongly indicate a transform fault cutting the lower plate [McCaffrey, 1982], reflection data have so far failed to reveal evidence of surface faulting.

Seismic reflection profiles collected and interpreted by Mobil Oil Company geologists across the southern part of the Philippine Trench east of Halmahera show evidence of subduction-related deformation of trench sediments as far south as 1°40'N, indicating that the Philippine Trench has propagated at least to this latitude.

Collision Processes

From our study and previous studies of the Molucca Sea region, it appears that the history of arc-arc collision has been diachronous, proceeding from north to south [Hamilton, 1979]. Collision occurred in Mindanao during the Miocene and appears inactive at present. It is presently active in the southern Molucca Sea [Silver and Moore, 1978; McCaffrey et al., 1980; McCaffrey, 1982] and much less active seismically and structurally in the vicinity of northern Talaud. This pattern may be thought of as a southward scissoring against the Sangihe Arc by either a long, continuous Mindanao-Halmahera Arc [e.g., Hamilton, 1979], a double Mindanao-Talaud and Halmahera arc system [Cardwell et al., 1980], a double Mindanao and Talaud-Halmahera arc system (our preferred scenario), or a series of small, unconnected arc segments (B. Taylor, personal communication, 1982).

Several periods of arc polarity reversal have occurred in response to the collision of various crustal blocks in the Molucca Sea region. The Samar-East Mindanao Arc apparently was active during the Paleogene and has been interpreted as a west-facing arc [Cardwell et al., 1980]. Following the mid-Tertiary collision of this arc with the central Mindanao arc, subduction ceased and remained inactive during much of the Miocene [Karig, 1975]. A new subduction zone has recently started at the Philippine Trench, converting this arc into an east-facing arc system. The active subduction zone at the Philippine Trench is presently propagating southward beyond Halmahera [Cardwell et al., 1980].

In response to the collision in central Mindanao, westward subduction beneath western Mindanao ceased [Cardwell et al., 1980]. The present deep seismic activity under west Mindanao may be due to sinking of a detached slab. The Quaternary volcanism in central Mindanao may also be an effect of this detached slab. The cessation of subduction may be propagating southward toward the north tip of Sulawesi. In order to take up the Molucca Sea-Celebes Sea convergence, a new subduction zone is presently developing southwest of Mindanao (Cotobato Trench). The Cotobato Trench is also propagating southward parallel to the inactivation of the Sangihe Trench.

Strike-slip faulting is also very important in the development of this collision zone. The active Philippine fault in eastern Mindanao is a left lateral feature that has been superimposed on the mid-Tertiary collision zone and can be traced from Luzon southward through the Philippine Islands to Mindanao [Allen, 1962]. We believe that the present location of the Philippine Fault in Mindanao is along the old suture zone, which represented a zone of weakness, and that the left lateral motion is probably in response to oblique convergence between the two arc systems [Cardwell et al., 1980]. We have traced the fault zone south from Mindanao to the Talaud Islands. It is likely that a strand of the fault also passes between the Talaud and Nanusa islands. The Snellius Ridge and Nanusa Islands may be portions of the Halmahera block that are currently being transported northward by motion along the southern continuation of the Philippine fault zone.

Neogene Plate Interaction

Several attempts have been made to reconstruct the Neogene tectonic history of the northern Indonesia/southern Philippines region. Roeder [1977] proposed that this region may be characterized as a rotating convergence system between the Halmahera and Sangihe arcs and that the arc-arc collision has been followed by the west arc overriding the east arc in Luzon and the east arc overriding the west arc in the Molucca Sea region. The data presented by Silver and Moore [1978] demonstrated that the crustal structure assumed by Roeder [1977] is incorrect and that the collision is not characterized by one arc overriding another.

Hamilton [1979, Figure 77] has presented a reconstruction in which the west-facing Halmahera Arc was continuous with the East Mindanao Arc from 20 m.y. to 5 m.y. In his reconstruction the southern boundary of the Molucca Sea plate is a transform fault connecting the Sunda Arc with the arc-arc collision in northern New Guinea. The Sangihe Arc was continuous with the Sunda Arc at 20 m.y. The collision occurred first in Mindanao and progressed southward with time. No oblique convergence or strike-slip faulting for the collision in Mindanao is shown in Hamilton's [1979] scenario. He infers that the Sula microcontinent was ripped from the north part of New Guinea and translated westward along a strike-slip fault until it collided with Sulawesi.

Cardwell et al. [1980], on the other hand, infer considerable strike-slip motion associated with the collision in Mindanao. Their reconstruction places the Halmahera Arc on a plate distinct from the East Mindanao Arc and requires the existence of a transform fault between northern Halmahera and the Palau Trench. Cardwell et al. [1980] and Moore et al. [1981] interpreted Talaud as the southern portion of the East Mindanao Arc.

We propose a new tectonic reconstruction for the eastern Indonesia/southern Philippines region (Figure 10). Ours is somewhat similar to those discussed above, with modifications based on the data and interpretations presented in this paper. The reconstruction is presented as an attempt to synthesize our ideas about the tectonic processes by which this complicated arc-arc collision zone has de-

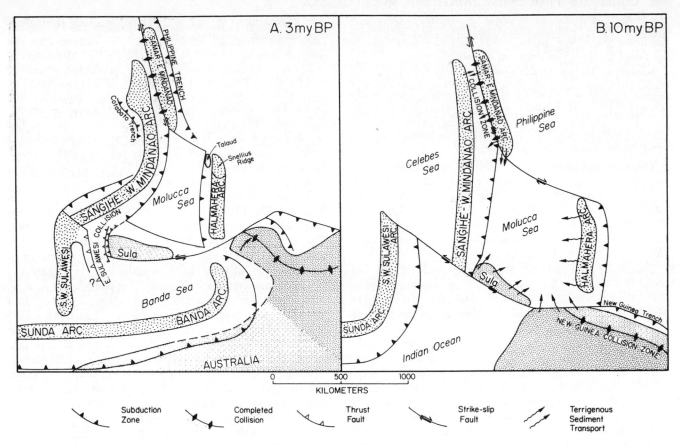

Fig. 10. Schematic tectonic reconstructions of the southern Philippine/northeastern Indonesia region at 10 m.y. (late Miocene) and 3 m.y. (mid-Pliocene). The Zamboanga Peninsula of west Mindanao and the Sulu Arc are left out to simplify the reconstructions. See text for discussion.

veloped. Because convergence rates are unknown, the separations between the arc systems shown in Figure 10 and the implied rates of convergence are estimates and are shown only to indicate the relative positions of crustal blocks and the timing of events. Also, the reconstruction neglects the evolution of the Banda Sea because, though it is extremely significant, insufficient data are available to distinguish between major alternatives.

The Sangihe-West Mindanao Arc has been an east-facing arc system from the Miocene to the present. Because the geology of the region north of Mindanao is poorly known, we have drawn the northern termination of the arc just north of Mindanao, although the arc could have continued farther north. The east Mindanao Arc is shown as west facing, although this is not required by the present data. We have chosen the southeast corner of Mindanao as the southern termination of the East Mindanao Arc because the volcanic rocks do not appear to continue farther south.

The reconstruction shown for 10 m.y. (Figure 10) implies 300 km of strike-slip displacement along the Philippine fault zone. Although we have no geological constraints for

the rate of displacement along the Philippine Fault, the implied rate (3 cm/yr) is of the order of magnitude of displacement along other strike-slip faults. Halmahera has been moved back in time 800 km southeast, based on the length of the present seismic zones underneath Halmahera and the Celebes Sea. The Halmahera block has been drawn with the assumption that the Snellius Ridge is part of the Halmahera Arc and that the arc extended southward to the present location of the Sorong fault zone (Figure 1). Talaud represents the forearc terrane of the Snellius Ridge.

Our assumption of continuity between Snellius Ridge and Halmahera is based on their present continuity, the lack of defined faults between them, the geophysical signatures of Snellius Ridge supporting (though not exclusively so) the interpretation of the ridge being an eroded island arc, and simplicity. Arguments against this continuity are the seismicity data, indicating a transform fault in the lower (Molucca Sea) plate [*Cardwell et al.*, 1980; *McCaffrey*, 1982], and the presence of the Morotoi Trough, a bathymetric and gravity low, separating the two features. Because we have no evidence for surface faulting in the trough, it makes most sense to view the earthquake activity as originating wholly within the subducted plate. We are not distressed by the presence of a basin within an island arc system. In general, arcs are not continuous ridges built to sea level, but rather they are a series of ridges separated by basins of various sizes. The gravity low associated with the basin is unusual, though not unique. If it does indicate earlier tectonic complications between ridge and Halmahera, we do not have the geologic control to evaluate their potential magnitude.

The West Mindanao and East Mindanao arcs were probably intraoceanic arcs and therefore had little trench sediment fill. This explains the lack of a thick mass of deformed sediments in the Miocene collision zone of Mindanao. The collision in Mindanao was complete by 10 m.y. B.P., and coarse clastic sediments were being shed into the Agusan-Davao Trough (Figure 10).

Coarse clastic sediments were being shed into the Molucca Sea as well. We suggest this for several reasons. First, the enormous thickness of material in the southern Molucca Sea [McCaffrey et al., 1980] contrasts sharply with the very small amount of collision complex observed in southern Mindanao. Offscraping is enhanced by the presence of voluminous, young unlithified sediments [Moore, 1975]. We have dredged volcaniclastic rocks from the Mayu ridge [Silver and Moore, 1978] and have sampled a large outcrop of coarse arenite (nearly pure quartz-feldspar sandstone) on Tifore (E. A. Silver, unpublished field data, 1977). The volcaniclastics probably had local sources (such as Mindanao or Halmahera), but this is unlikely for the arenite. Its most likely source is the Sula Island platform to the south or the Australia-New Guinea region at an early stage of its collision with the Indonesian arc system (see Figure 10).

Our reconstruction differs from others in that the East Mindanao Arc is connected to the Halmahera Arc by a transform fault rather than being one continuous arc, as shown by Hamilton [1979], or being completely unrelated arcs, as shown by Cardwell et al. [1980]. The Halmahera Arc reversed polarity approximately 10 m.y. ago, possibly as a result of the collision between the East and West Mindanao arcs, the buoyant crust colliding with the Halmahera Arc from the east, or a combination of such factors. The timing of events on the Talaud Islands [Moore et al., 1981a] suggest that the polarity reversal may have started first in the north and propagated south. After reversal of Halmahera's polarity, the Molucca Sea closed at a rapid rate.

In the southern part of the Molucca Sea, collision between the Sula Islands and the Sulawesi-Sangihe Arc began approximately 10 m.y. ago [Hamilton, 1979]. This collision involved ophiolite emplacement and appears to have resulted in a significant clockwise rotation of the northern arm of Sulawesi, with convergence being taken up at the Minahasa Trench north of Sulawesi (Figure 1). Otofuji et al., [1981] present paleomagnetic evidence for the amount and approximate timing of this rotation. That collision is the subject of a separate study.

At 3 m.y. (mid-Pliocene), the Halmahera and Sangihe arcs were nearly colliding (Figure 10). Terrigenous clastic sediments, eroded from the arcs and from the continental mass of Australia-New Guinea, were deposited in the Molucca Sea and were beginning to be deformed in the collision zone to form the Molucca Sea collision complex. The northern end of the Halmahera forearc was being thrust over the Sangihe forearc, uplifting the ophiolite/melange terrane of the Talaud Islands. The Philippine and Cotabato trenches were propagating southward by this time.

The present regional tectonic configuration is shown in Figure 1. The Talaud Islands have been uplifted above sea level, and convergence in this area has nearly stopped. The Philippine Trench has propagated southward to the latitude of Halmahera, and the Cotabato Trench has propagated to the latitude of the Sangihe Islands (Figure 1).

The reconstruction in Figure 10 appears to us to most simply and easily explain the observations, however other more complex scenarios are possible. One relates Halmahera with the Mariana-Yap system to the north at an early stage, removing the need for the long Halmahera-Mindanao transform fault. Another views East Mindanao, Snellius Ridge, Talaud, and Halmahera as separate, unrelated blocks that fortuitously amalgamate to their present alignments. Perhaps with careful study of Halmahera and Snellius Ridge a more complex scenario will prove necessary. At the present time we can explain the geology and geophysics without resorting to unconstrained complexity.

Significance for Interpretation of Ancient Mountain Belts

The tectonic development of the arc-arc collision zone in the Molucca Sea region has implications for the development of collision complexes in ancient mountain belts. For instance, it is obvious that this area has developed mostly by collision of small crustal fragments, both island arc terranes and slivers of continental crust. Most of the collisions have not occurred by subduction alone but rather have proceeded with an unknown, but probably considerable, component of strike-slip motion.

Due to the arc-arc collision in the Mindanao-Molucca Sea region, the subduction zones in Mindanao have ceased, and new subduction zones have been initiated at the Philippine and Cotabato trenches. In this collision zone the arcs have not overridden one another, nor have the subduction zones 'flipped,' as shown by Roeder [1975, 1977]. Convergence ceased after the collision of the two arcs, but before the intervening sedimentary basin (Agusan-Davao Trough) was strongly deformed. In response to continued plate motion between the Philippine Sea and southeast Asia, new subduction zones have broken at preexisting continental margins and are propagating southward. It appears that it is mechanically easier for a new trench to break behind the volcanic arcs than for the crustal blocks to continue converging as in a continent-continent collision. There was no sudden flip of the subduction zones, but rather the new subduction zones have propagated from the more complete to the less complete parts of the collision zone. This change in subduction pattern has caused diachronous structural relations along the collision zone.

If observed in a mountain belt, this collision zone would be characterized by a narrow 'suture zone' with ophiolites along it and a large, nearly undeformed 'molasse' trough adjacent to it. Detailed and regional structural and stratigraphic studies would probably identify the diachronous nature of the collision but might fail to recognize the significant amount of strike-slip displacement. Final suturing in this region will not occur until Australia collides with the Asian continent, further deforming and obscuring the structural relations of the Molucca Sea collision zone in the process.

Acknowledgements. C. Mrozowski, C. von der Borch, W. Hamilton, S. Lewis and R. Cardwell offered valuable suggestions on the manuscript. A particularly thorough review by Brian Taylor led to significant improvement of this paper. This study was supported by the National Science Foundation, grants EAR 80-07429 (SIO) and OCE 78-08693 (UCSC). We thank Amoco Production Co. (International) for providing access to their Mindanao seismic data and for partial support of our Mindanao field work.

References

Allen, C. R., Circum-Pacific faulting in the Philippines-Taiwan region, *J. Geophys. Res., 67,* 4795–4812, 1962.

Cardwell, R. K., B. L. Isacks, and D. E. Karig, The spatial distribution of earthquakes, focal mechanism solutions, and subducted lithosphere in the Philippine and northeast Indonesian Islands, in *The Tectonic and Geologic Evolution of Southeast Asian Seas and Islands,* edited by D. E. Hayes, *Geophys. Monog. Ser.,* vol 23, pp. 1–35, AGU, Washington, D. C., 1980.

Casasola, A. G., Petroleum and geological field investigation of western Davao, *Philipp. Geol., 10,* 76–88, 1956.

Evans, C. A., J. W. Hawkins, and G. F. Moore, Petrology and geochemistry of ophiolitic and associated volcanic rocks on the Talaud Islands, Molucca Sea collision zone, N.E. Indonesia, in *Geodyn. Ser.,* edited by T. Hilde, in press, AGU, Washington, D. C., 1982.

Hamilton, W., Subduction in the Indonesian region, in *Island Arcs, Deep Sea Trenches, and Back-Arc Basins,* edited by M. Talwani and W. C. Pitman III, *Maurice Ewing Ser.,* vol. 1, pp. 15–31, AGU, Washington, D. C., 1977.

Hamilton, W., Tectonic map of the Indonesian region, scale 1:5,000,000, *Misc. Invest. Ser., Map I-875-D,* U.S. Geol. Surv., Reston, Va., 1978.

Hamilton, W., Tectonics of the Indonesian region, *Geol. Surv. Prof. Pap.* (U.S.) *1078,* 345pp., 1979.

Hashimoto, W., Geologic development of the Philippines, *Geol. Paleontol. Southeast Asia, 21,* 83–170, 1981.

Hashimoto, W., E. Aliate, N. Aoki, G. Balce, T. Ishibashi, N. Kitamura, T. Matsumoto, M. Tamura, and J. Yanagida, Cretaceous system of southeast Asia, *Geol. Paleontol. Southeast Asia, 15,* 219–280, 1975.

Hawkins, J. W., Petrology of back-arc basins and island arcs: Their possible role in the origin of ophiolites, in *Ophiolites, Proceedings, International Ophiolites Symposium, Cyprus, 1979,* edited by A. Panayioutou, pp. 244–254, Geological Survey Department, Nicosia, Cyprus, 1980.

Hawkins, J. W., C. A. Evans, E. F. Wright, and G. F. Moore, East Mindanao ophiolite belt: Petrology of metamorphosed basal cumulate rocks (abstract), *Eos Trans. AGU, 62,* 1086, 1981.

Jones, D. L., N. J. Silberling, and J. Hillhouse, Wrangellia—A displaced terrane in northwestern North America, *Can. J. Earth Sci., 14,* 2565–2577, 1977.

Karig, D. E., Basin genesis in the Philippine Sea, *Initial Rep. Deep Sea Drill. Proj., 31,* 857–879, 1975.

Katili, J. A., Past and present geotectonic position of Sulawesi, Indonesia, *Tectonophysics, 45,* 289–322, 1978.

Krause, D. C., Tectonics, marine geology and bathymetry of the Celebes Sea-Sulu Sea region, *Geol. Soc. Am. Bull., 77,* 813–832.

Malicdem, D. G., and R. Pena, Geology of copper-gold deposits of the Masara Mine area, Mabini, Davao, *Philipp. Geol., 21,* 16, 1967.

Mammerickx, J., R. L. Fisher, F. J. Emmel, and S. M. Smith, Bathymetry of the east and southeast Asian seas, *Map Chart Ser., MC-17,* Geol. Soc. Am., Boulder, Colo., 1976.

Matsumaru, K., Larger Foraminifera from east Mindanao, Philippines, *Geol. Paleontol. Southeast Asia, 14,* 101–115, 1974.

McCaffrey, R., Lithospheric deformation within the Molucca Sea arc-arc collision: Evidence from shallow and intermediate earthquake activity, *J. Geophys. Res., 87,* 3663–3678, 1982.

McCaffrey, R., E. A. Silver, and R. W. Raitt, Crustal structure of the Molucca Sea collision zone Indonesia, in *The Tectonic and Geologic Evolution of Southeast Asian Seas and Islands,* edited by D. E. Hayes, *Geophys. Monogr. Serv.,* vol 23, pp. 161–177, AGU, Washington, D. C., 1980.

Melendres, M. M., and F. Comsti, Reconnaissance geology of southeastern Davao, *Philipp. Geol., 5,* 38–46, 1951.

Metal Mining Agency, Report of geological survey of eastern Mindanao in the first phase, Overseas Tech. Coop. Assoc., Tokyo, 1972.

Metal Mining Agency, Report of geological survey of eastern Mindanao in the second phase, Overseas Tech. Coop. Assoc., Tokyo, 1973.

Moore, G. F., C. A. Evans, J. W. Hawkins, and D. Kadarisman, Geology of the Talaud Islands, Molucca Sea collision zone, N.E. Indonesia, *J. Struct. Geol., 3,* 467–475, 1981a.

Moore, G. F., J. W. Hawkins, and R. Villamor, Geology of

southern Mindanao, Philippines (abstract), *Eos Trans. AGU, 62,* 1086, 1981*b.*

Moore, J. C., Selective subduction, *Geology, 3,* 530–532, 1975.

Murphy, R. W., Diversity of island arcs: Japan, Philippines, northern Moluccas, *Australian Petrol. Explor. Assoc. J., 11,* 19–25, 1973.

Otofuji, Y., S. Sasajima, S. Nishimura, A. Dharma, and F. Hehuwat, Paleomagnetic evidence for clockwise rotation of the northern arm of Sulawesi, Indonesia, *Earth Planet. Sci. Lett., 54,* 272–280, 1981.

Philippine Bureau of Mines, Geological map of the Philippines, scale 1:1,000,000, 9 sheets, Manila, 1963.

Ranneft, J. S. M., R. M. Hopkins Jr., A. J. Froelich, and J. W. Gwinn, Reconnaissance geology and oil possibilities of Mindanao, *Am. Assoc. Geol. Bull., 44,* 529–568, 1960.

Reyes, M. V. and E. P. Ordonez, Philippine Cretaceous smaller foraminifera, *J. Geol. Soc. Phil., 24,* 1–47, 1970.

Rceder, D. H., Tectonic effects of dip changes in subduction zones, *Am. J. Sci., 275,* 252–264, 1975.

Roeder, D., Philippine arc system—Collision of a flipped subduction zone?, *Geology, 5,* 203–206, 1977.

Santos-Ynigo, L., Distribution of iron, alumina, and silica in the Pujada laterite of Mati, Davao Province, Mindanao Island, Philippines, *Philipp. Geol., 19,* 97–110, 1965.

Santos-Ynigo, L., M. R. Lucas and R. De Guzman, Geology, structure, and origin of the magnesite deposits of Piso Point, municipality of Lupon, Davao Province, *Philipp. Geol., 15,* 108–133, 1966.

Schweikert, R. A., and D. S. Cowan, Early Mesozoic tectonic evolution of the Western Sierra Nevada, *Geol. Soc. Am. Bull., 86,* 1329–1336, 1975.

Silver, E. A., A new tectonic map of eastern Indonesia, in *Geology and Tectonics of Eastern Indonesia,* edited by A. J. Barber and S. Tjokrosapoetro, Geol. Res. Dev. Centre, Bandung, 1981.

Silver, E. A., and J. C. Moore, The Molucca Sea collision zone, Indonesia, *J. Geophys. Res., 83,* 1681–1691, 1978.

Soeria Atmadja, R., and R. Sukamto, Ophiolite rock association on Talaud Islands, East Indonesia, *Bull. Geol. Res. Dev. Centre, 1,* 17–35, 1980.

Speed, R. C., Collided Paleozoic microplate in the western United States, *J. Geol., 87,* 279–292, 1979.

Stewart, G. S., and S. N. Cohn, The August 16, 1976, Mindanao, Philippine earthquake (M_s = 7.8)—Evidence for a subsection zone south of Mindanao (abstract), *Eos Trans. AGU, 58,* 1194, 1977.

Sukamto, R., Tectonic significance of melange on the Talaud Islands, northeastern Indonesia, in *Geology and Tectonics of Eastern Indonesia,* edited by A. J. Barber and S. Tjokrosapoetro, Geol. Res. Develop. Centre, Bandung, 1981.

Sukamto, R., and H. Suwarno, Melange di daerah Kepulauan Talaud, paper presented at Indonesian Association of Geologists meeting, [Direktorat Geologi, Bandung], December 1976.

Sukamto, R., T. Apandi, S. Supriatna, and A. Yasin, The geology and tectonics of Halmahera Island and its surroundings, paper presented at Ad Hoc Working Group Meeting on the Geology and Tectonics of Eastern Indonesia, July 9–14, Bandung, Indonesia, 1979.

Sukamto, R., N. Suwarna, J. Yusup, and M. Monoarfa, Geologic map of Talaud Islands, scale 1:250,000: Geol. Res. Develop. Centre, Bandung, Indonesia, 1980.

Teves, J. S., A. Vergara, and N. Badillo, Reconnaissance geology of Agusan, *Philipp. Geol., 5,* 24–40, 1951.

van Bemmelen, R. W., *The Geology of Indonesia,* 793 pp., Printing Office, The Hague, 1949.

Verbeek, H., Molukken-Verslag, *Jaarb. Mijnwezen Ned. Oost-Indie, 37,* 826 pp., 1908.

Vergara, J. V., and F. D. Spencer, Geology and coal resources of Bislig-Lingig region, Surigao, *Phil. Bur. Mines Spec. Proj. Ser. 14,* 1–63, 1957.

Watts, A. B., J. H. Bodine, and C. O. Bowin, Free air gravity field, Geophysical atlas of east and southeast Asian seas, *Map Chart Ser., MC-25,* Geol. Soc. Amer., Boulder, Colo., 1978.

Wright, E., J. W. Hawkins, C. A. Evans, and G. F. Moore, East Mindanao ophiolite belt: Petrology of volcanic series rocks (abstract), *Eos Trans. AGU, 62,* 1086, 1981.

The Halmahera Island Arc, Molucca Sea Collision Zone, Indonesia: A Geochemical Survey

J. D. Morris,[1] P. A. Jezek,[2] S. R. Hart,[1] and J. B. Gill[3]

The Halmahera island arc, northeastern Indonesia, is the east flank of the Molucca Sea collision zone which is the site of an active arc-arc collision. One unique aspect of the arc is the vast thickness of marine sediments outboard of the trench, a result of ~1000–1500 km of closure in the Molucca Sea basin. The Halmahera arc is underlain by a 45° east dipping Benioff zone, which is present to depths of 230 km. The volcanoes form a single front which lies ~100 km above the top of the slab. The arc can be separated into three regions, on the basis of tectonic setting and chemistry. Most volcanoes are part of the normal calc-alkaline oceanic arc segment. Lavas here are basalts through dacites, with basaltic andesites and andesites dominant. Suites are medium-K and show little to moderate Fe enrichment. Abundances of Al_2O_3, the alkali elements, compatible elements, and the high field strength elements are typical of calc-alkaline island arc lavas. $^{87}Sr/^{86}Sr$ ratios are 0.70357–0.70438; the average value for all volcanic centers is almost the same, but most centers show a large range of real variation around that average. Pb isotopic compositions are $^{206}Pb/^{204}Pb = 18.55–18.62$, $^{207}Pb/^{204}Pb = 15.55–15.63$, $^{208}Pb/^{204}Pb = 38.48–38.67$. On a Pb-Pb diagram, they form a linear cluster of steep slope, between oceanic sediments and the less radiogenic end of the mantle array. Pb and Sr isotopic compositions for the oceanic segment can be used to test models for the origin of arc lavas. Both isotope systems can be satisfied by a three-component mixing model where normal oceanic island basalt-type magma is mixed with MORB and contaminated by heterogeneous sediments. A two-component model, where OIB-type magmas (which are heterogeneous with respect to both Sr contents and $^{87}Sr/^{86}Sr$ ratios) are contaminated with sediments, can also explain the data. This generates isotopically heterogeneous suites, and much of this heterogeneity is preserved through the eruption process. A continental suite occurs on the island of Bacan, where the arc intersects a continental fragment. This interaction is reflected in high $^{87}Sr/^{86}Sr$ (up to 0.724) and high $^{207}Pb/^{204}Pb$ (≈ 15.80) ratios. Further south, an alkaline suite occurs on the inactive volcanic islands which are spatially associated with the Sorong fault zone, a transform plate boundary. These lavas have more radiogenic isotopic compositions ($^{87}Sr/^{86}Sr = 0.7058–0.7087$; $^{206}Pb/^{204}Pb \approx 18.77$) than the oceanic, calc-alkaline segment of the arc.

Introduction

A number of arcs occupy unusual tectonic settings, a fact which is often reflected in the geochemistry of the arc lavas. The Sunda Arc, which changes from continental to oceanic along strike, and the Banda Arc, which is the site of an ongoing arc-continent collision, are examples. The Halmahera Arc, which forms the east margin of the Molucca Sea collision complex in northeastern Indonesia, is another. The Halmahera Arc is actively colliding with the Sangihe-Sulawesi Arc. Presently, the two arcs are separated by ~250 km; tectonic models suggest that the arcs were 1000–1500 km apart in mid-Pliocene time. Since one effect of this collision is to confine marine sediments into a narrow basin where they reach depths of 10–15 km, one might expect to see the effects of sediment subduction reflected in the chemistry of the arc magmas.

[1]Center for Geoalchemy, Department of Earth and Planetary Sciences, Massachusetts Institute of Technology, Cambridge, Massachusetts 02139.

[2]Stone and Webster Engineering Corp., Boston, Massachusetts 02107.

[3]Earth Sciences Board, University of California, Santa Cruz, California 95064.

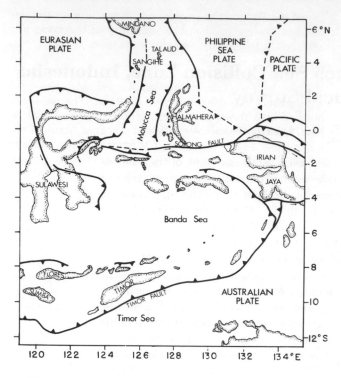

Fig. 1. Tectonic map of the Molucca Sea region, following *McCaffrey et al.* [1980]. Map shows complex interaction of Eurasian, Pacific, Philippine Sea, and India-Australian plates. Sulawesi-Sangihe and Halmahera arcs are overriding Molucca Sea plate. Teeth are drawn on Molucca Sea plate to indicate partial obduction of Molucca Sea sediments onto arc apron.

The controversial question of sediment subduction was originally proposed to explain Pb isotopic characteristics of volcanics from the Lesser Antilles [*Armstrong and Cooper*, 1971]. It came under fire from several directions. Some argued against it as a physically unlikely process [*Karig and Sharman*, 1975]. Several geochemical studies [*Meijer*, 1976; *Sinha and Hart*, 1972; *Oversby and Ewart*, 1972] used Pb isotopes to document the absence of a pelagic sediment component in the Mariana and Tonga arcs; the Pb isotope characteristics were then used as evidence against sediment subduction as a process in island arcs. Nevertheless, many island arcs have isotopic compositions most easily explained by sediment involvement. Furthermore, new geophysical studies [*Sharman*, 1980] indicate mechanisms whereby sediments could be carried to depth in a subduction zone.

The focus of this work is an attempt to identify the source materials which contribute to the magmas of the Halmahera Arc. In this paper we present results of a reconnaissance geochemical survey of the arc using major and trace element and Sr and Pb isotope analyses. This report, along with those for the Sangihe-Sulawesi Arc on the west side of the collision complex [*Jezek et al.*, 1982; *Morrice et*

al., 1982] provides a complete set of regional geochemical data for the arcs of the Molucca Sea.

Tectonic Setting

The Molucca Sea collision zone of northeastern Indonesia is the site of an active arc-arc collision between the Halmahera Arc on the east and the Sangihe-Sulawesi Arc on the west (see Figure 1 for tectonic map). The collision started earliest in the north and is complete in the area north of Talaud Island [*Cardwell et al.*, 1980]. South of Talaud, convergence is being taken up along the two subduction zones which flank the collision complex and which dip away from the center of the Molucca Sea.

There are several unique aspects to the Molucca Sea as a result of the collision. The oceanic basement is overlain by 10–15 km of deformed Tertiary limestones, sandstones, and siltstones [*McCaffrey et al.*, 1980]. Possible sources of the sandstones and siltstones, which show varying ratios of volcaniclastic to continental debris, include Irian Jaya [*Hamilton*, 1978] and the Philippines. Along the central axis of the Molucca Sea, the sediments are associated with mafic and ultramafic rocks including peridotite, serpentinite, and gabbro. *Silver and Moore* [1978] and *McCaffrey* [1981] have interpreted these as slivers of Molucca Sea oceanic crust which have been thrust to the surface during the collision. This is supported by the large number of shallow earthquakes with thrust mechanisms noted for the central ridge [*Fitch*, 1970; *Cardwell et al.*, 1980]. In addition, geochemical data suggest that the ophiolites exposed on Talaud are best modeled as pieces of oceanic crust [*Evans et al.*, 1982].

The Halmahera Arc, on the east side of the collision complex, is underlain by a seismic zone which dips ~45° to the east to depths of ~230 km [*Cardwell et al.*, 1980]. The associated trough is only ~3 km deep [*U.S. Geological Survey*, 1974], being largely obscured by sediments of the collision complex [*Silver and Moore*, 1978]. The Halmahera subduction zone terminates just north of Halmahera Island; north of this, west dipping subduction occurs in the Philippine Trench. To the south, the arc-trench system appears to end at the Sorong fault, a major zone of left lateral strike slip movement.

The Halmahera Arc is shown in more detail in Figure 2. Triangles indicate Quaternary volcanoes and active volcanoes are shown by stars. The volcanoes from Ibu to Kajoa have erupted through Tertiary volcanic rocks. To the south the oceanic arc intersects a continental fragment; on Bacan, several volcanoes are built on the Paleozoic (?) rocks of the Sibela metamorphic complex [*Hamilton*, 1979; G. Russman, personal communication, 1981]. Southeast of the active arc, a number of small extinct volcanic islands (e.g., Pisang, Woka, Kekik) are distributed along the Sorong fault zone, a transform plate boundary. The age of volcanism is thought to be late Tertiary, although no dates are available. Their relationship to the active Halmahera Arc is unclear, but, because of their spatial association

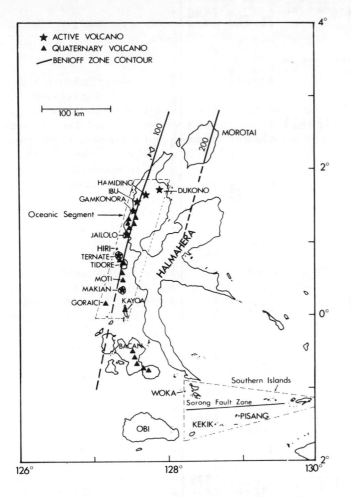

Fig. 2.   Map of the Halmahera island arc, indicating Quaternary (triangles) and active (stars) volcanoes. Contours to the Benioff zone are from *Cardwell et al.* [1980].

with the arc, these volcanics will be discussed along with those which are clearly arc related.

The Quaternary volcanoes form a chain about 300 km long. Active volcanoes are separated by uniform distances of ~60 km. As can be seen from the Benioff zone contours in Figure 2, the Halmahera volcanoes form a single well-defined volcanic front. Depths to the seismic zone vary only from ~90 to ~125 km.

In summary, although the Molucca Sea is a tectonically complex area, the tectonic setting of the Quaternary Halmahera Arc is relatively straightforward. The active arc is associated with a shallow, well-defined Benioff zone, and the volcanoes are all about 100 km above the top of the downgoing slab. In most cases, the lavas erupt through Tertiary volcanics which overlie oceanic crust. On southern Bacan the lavas pass through a continental fragment. The inactive volcanic islands south of Halmahera (of unknown relationship to the active arc) are built on oceanic crust in the region of the Sorong fault, a transform boundary of the Halmahera microplate. These varying tectonic settings are reflected in the chemistry of the lavas and will form a framework for later discussion.

Results

The volcanoes of the Halmahera Arc are composite cones with craters 700–1700 m above sea level. The cones are built of lava flows, pyroclastic debris and occasional lahars. Locally, flows have pillowed textures. Many of the cones show a suite from basalt through basaltic andesite to andesite and occasionally dacite. Other volcanic centers, such as Tidore, are composed of andesite only.

The petrographic features of the Halmahera volcanics are those of typical calc-alkaline arc lavas. Most samples are moderately crystalline with 30–50% phenocrysts. In the more basic members, phenocryst assemblages are typically plagioclase, olivine, clinopyroxene, and magnetite. In slightly more acidic samples, olivine is rimmed with orthopyroxene and opx appears as a microphenocryst phase. In some of the more siliceous andesites on Ternate, the phenocryst assemblage is plagioclase, clinopyroxene, hornblende, and magnetite. Phenocrysts are usually <2 mm in length and often less than 1 mm. Groundmass textures cover the entire range from glassy to granular. Pillowed lavas are occasionally oxidized, but most samples show little alteration. Glassy groundmasses are variably devitrified and plagioclase crystals are occasionally cloudy. Rarely, olivine crystals show slight alteration to iddingsite. Only the freshest samples, with no petrographic evidence for alteration, were chosen for isotopic analysis.

Major and Trace Element Chemistry

Representative major and trace element analyses for the Halmahera Arc lavas are presented in Table 1. For clarity, we will discuss the data region by region, from north to south.

The active arc from Kayoa north (see Figure 2 for island names and locations) appears to be 'normal' oceanic arc. It has many of the characteristics considered typical for island arcs. Basaltic andesites (53–57% SiO_2) and andesites (57–63% SiO_2) constitute ~75% of the 81 analyzed samples from this segment, with basalt the next most common rock type (21%). Al_2O_3 contents are typically high, between 17% and 20%. TiO_2 contents are almost always less than 1%. K_2O contents in the andesites are generally 1.0–1.9% and they are medium-K suites according to the classification of *Gill* [1978].

The analyses are plotted on a FeO*/MgO versus SiO_2 diagram in Figure 3. Virtually all islands have samples which plot on both sides of the calc-alkaline/tholeiitic boundary line. Some suites, such as the suite of Jailolo, have very steep slopes, indicating no iron enrichment. Others, such as Moti and Ibu, form arrays which are best fit by a line coincident with the TH-CA boundary line. The lavas thus show small to moderate degrees of iron enrichment, variable from island to island.

TABLE 1. Major Element, Trace Element, and Isotope Chemistry of the Halmahera Arc Lavas

	Ibu							Jailolo				Hiri		
	NH 45	NH 46	NH 48	NH 43	NH 44	NH 10	NH 12	NH 5	NH 11	NH 7	H 9	H 1	H 2	H 5
SiO_2	52.45	53.67	55.22	55.24	61.10	50.56	50.54	53.08	55.80	58.86	51.55	51.84	51.85	61.89
Al_2O_3	19.39	18.27	17.78	20.29	16.81	20.48	19.49	18.44	18.50	18.00	19.02	18.91	19.04	17.85
FeO	8.43	8.56	8.19	6.98	7.22	8.71	8.92	9.29	6.88	6.04	9.38	9.57	9.77	5.44
MgO	4.77	4.84	4.58	3.23	2.58	4.34	5.72	5.00	4.71	4.15	5.07	4.93	4.78	2.68
CaO	10.13	9.27	8.54	9.14	5.97	12.14	11.92	9.98	9.20	8.22	10.43	9.70	9.78	6.53
Na_2O	3.03	2.65	2.81	3.13	3.68	2.23	2.14	2.38	2.85	3.23	2.87	3.00	3.01	3.80
K_2O	0.67	0.81	1.48	1.04	1.86	0.86	0.50	0.61	1.00	0.79	0.56	0.65	0.66	1.10
TiO_2	0.84	0.81	0.79	0.74	0.91	0.82	0.82	0.81	0.69	0.60	0.90	0.99	1.00	0.47
P_2O_5	0.13	0.18	0.20	0.16	0.20	0.11	0.10	0.09	0.09	0.11				
Total	99.81	99.37	99.59	99.95	100.33	100.25	100.15	99.68	99.72	100.00	99.78	99.64	99.89	99.76
K	5318	9399	12293	8671	15326	7144	4152	5065	8337	6599	4623	5436	5461	9172
Rb	8.24	19.54	28.39	12.94	34.67	9.92	10.65	11.13	19.91	14.5	9.95	10.70	10.76	20.0
Cs	.31	1.28	1.58	0.92	2.37	0.87	0.61	0.80	1.26	0.92	0.65	0.35	0.38	—
Ba	132.2	214.7	254.9	278.7	336.2	129.9	104.5	126.5	175.9	165.9	115.0	133.8	134.1	259.1
Sr	338.2	403.1	394.5	528.7	393.5	427.3	338.9	356.3	405.9	491.0	338	376.2	376.6	364.9
Y	—	—	24.8	—	30.4	17.4	14.4	18.1	—	—	—	25.6	25.0	24.0
Zr	—	—	92.8	—	110.2	60.8	52.7	68.9	—	—	—	76.0	79.1	126.7
Nb	—	—	3.6	—	3.3	3.4	3.4	3.4	—	—	—	4.0	3.3	3.3
V	—	—	235.3	—	181	285.4	205.3	291.2	—	—	—	270.0	273.5	92.7
Cr	—	—	27	—	5.4	17.7	32.2	27.8	—	—	36	19.1	19.4	15.6
Ni	—	—	18.9	—	3.8	14.7	14.4	14.0	—	—	—	12.2	13.9	8
Cu	—	—	117.1	—	24.6	108.5	61.8	91.8	—	—	—	80.4	87.3	24.2
Zn	—	—	75.	—	84.8	66.7	52.8	75.4	—	—	—	61.8	63.6	71.3
Rb/Sr	0.0249	0.0485	0.0720	0.0245	0.0881	0.0232	0.0314	0.0312	0.0491	0.0296	0.0294	0.0284	0.0286	0.0548
K/Rb	645	481	433	670	442	720	399	455	429	455	465	508	508	458
K/Cs	17377	7343	770	9425	6467	8240	6806	6331	6775	7173	7080	15532	14221	—
K/Ba	40.2	43.8	48.2	31.1	45.6	55.0	39.7	40.0	48.5	39.8	40.2	40.6	40.7	35.4
K/Sr	16.1	22.3	31.2	16.4	38.9	16.7	12.3	14.2	21.0	13.4	13.6	14.5	15.5	25.1
Ba/Sr	0.399	0.533	0.646	0.527	0.854	.304	0.308	0.355	0.433	0.338	0.340	0.356	0.356	0.710
$^{87}Sr/^{86}Sr$	0.70414±6	0.70357±4	0.70420±5	0.70396±4	0.70414±5	0.70417±7	0.70428±5	0.70414±5	0.70380±6	0.70363±4	0.70389±6	0.70389±6	0.70391±5	0.70416
$^{206}Pb/^{204}Pb$	18.564	18.611	18.569		18.569	18.619	18.598			18.589		18.627		18.564
$^{207}Pb/^{204}Pb$	15.583	15.634	15.589		15.589	15.603	15.591			15.563		15.630		15.589
$^{208}Pb/^{204}Pb$	38.523	38.668	38.554		38.554	38.644	38.591			38.438		38.724		38.553

Total Fe as FeO. All Sr ratios normalized to $^{86}Sr/^{88}Sr = 0.1194$ and reported relative to E&A $SrCO_3 = 0.70800$. Uncertainties are 2σ of the mean. Pb isotopic compositions corrected for fractionation. Reproducibilities are: $^{206}Pb/^{204}Pb$, ±0.10%; $^{207}Pb/^{204}Pb$, ±0.15%; $^{208}Pb/^{204}Pb$, ±0.20%.

TABLE 1. (continued)

	Ternate				Tidore				Moti			
	T 7	T 10	T 2	T 1	TD 14	TD 4	TD 13	TD 5	MT 3	MT 11	MT 7	MT 1
SiO_2	53.72	56.73	57.23	61.65	54.37	56.64	58.10	61.75	50.61	52.84	55.95	60.71
Al_2O_3	18.66	17.57	17.72	16.69	18.98	18.73	18.54	19.50	19.41	18.93	19.29	19.26
FeO	9.01	8.53	7.84	6.88	8.21	7.56	6.87	4.67	8.29	7.67	7.55	5.17
MgO	4.54	3.27	3.62	2.04	4.16	3.67	3.37	1.51	6.15	5.34	3.24	1.79
CaO	8.78	7.43	7.62	5.13	8.94	7.88	7.33	6.51	10.56	10.10	8.41	6.85
Na_2O	2.92	3.45	3.26	4.08	3.10	3.59	3.30	4.08	3.17	3.44	3.71	4.15
K_2O	1.11	1.55	1.60	2.22	1.03	1.25	1.72	1.74	0.67	0.79	0.98	1.10
TiO_2	0.98	1.02	0.86	0.93	0.82	0.85	0.74	0.59	0.94	0.97	0.68	0.43
P_2O_5	0.09	0.17	0.13	0.22	0.16	0.13	0.16	0.21	0.14	0.11	0.14	0.20
Total	99.81	99.72	99.88	99.84	99.77	100.30	100.13	100.56	99.92	100.19	99.95	99.66
K	9178	12859	13254	18423	8523	10367	14284	14430	5581	6523	8172	9106
Rb	24.91	36.33	41.56	57.38	22.34	29.59	37.49	40.81	14.79	12.97	25.67	30.33
Cs	1.69	2.50	2.77	3.94	1.34	1.99	2.16	2.68	0.22	0.44	1.56	1.86
Ba	210.8	314.9	295.9	415.6	172.7	219.1	245.8	249.7	147.0	127.9	181.9	235.9
Sr	356.7	362.9	338.1	384.2	436.0	317.3	320.4	353.1	361.2	415.3	374.3	464.3
Y	26.0	33.9	25.8	41.8	23.7	27.8	26.8	—	20.7	—	25.3	—
Zr	97.7	123.1	124.7	169.4	93.9	118.4	129.0	—	86.6	—	109.1	—
Nb	5.1	2.3	6.1	4.9	3.0	2.2	4.1	—	5.5	—	3.8	—
V	256.4	254.2	195.4	113	202.6	195.8	167.1	—	242.6	—	131.7	—
Cr	16.6	5.8	53.1	5	6.6	7.6	12.1	—	54.1	86.0	10.4	—
Ni	16.5	7.6	24.2	4	8.4	9.1	11.9	—	31.6	—	10.6	—
Cu	125.2	142.3	33.4	52.8	43.0	54	38	—	31.4	—	39.0	—
Zn	79.5	81.4	57.5	76.5	53.8	69.2	58.1	—	59.8	—	68.9	—
Rb/Sr	0.0698	0.1001	0.1229	0.1494	0.0512	0.0933	0.1170	0.1156	0.0409	0.0312	0.0686	0.0653
K/Rb	369	354	319	321	382	350	381	354	377	503	318	300
K/Cs	5431	5139	4778	4677	6347	5212	6601	5395	25370	14993	5249	4896
K/Ba	43.5	40.8	44.8	44.3	49.4	47.3	58.1	57.8	38.0	51.0	44.9	38.6
K/Sr	25.7	35.4	39.2	48.0	19.6	32.7	44.6	40.9	15.5	15.7	21.8	19.6
Ba/Sr	0.590	0.868	0.875	1.08	0.396	0.691	0.767	0.707	0.407	0.310	0.486	0.508
$^{87}Sr/^{86}Sr$	0.70415±4	0.70419±4	0.70410±4	0.70424±5	0.70416±4	0.70422±5	0.70417±4	0.70424±6	0.70427±6	0.70383±4	0.70409±5	0.70408±4
$^{206}Pb/^{204}Pb$	18.500		18.588	18.610		18.619		18.598	18.545			18.544
$^{207}Pb/^{204}Pb$	15.604		15.597	15.620		15.619		15.625	15.5833			15.604
$^{208}Pb/^{204}Pb$	38.629		38.612	38.680		38.729		38.683	38.532			38.594

TABLE 1. (continued)

	Makian				Goraici Group		Kajoa		Bacan				Pisang			Woka	Kubi	Kekik
	MK 10	MK 3	MK 2	MK 8	GG 2	G11	KJ 7	KJ 2	BC 35	BC 15	BC 3	BC 6	P 4	P 7	P 3	W 1	K 1	KK 1
SiO_2	50.70	52.94	59.41	60.17	51.35	58.61	54.22	57.66	51.81	57.81	65.64	67.46	52.84	55.37	61.91	53.21	54.84	52.68
Al_2O_3	19.57	19.76	18.64	18.65	20.06	18.37	17.44	16.57	18.98	17.49	15.71	16.08	18.69	18.30	17.39	17.63	17.89	16.24
FeO	8.78	8.60	6.27	5.93	8.29	6.77	10.23	6.86	9.34	6.39	4.79	3.89	6.75	5.59	3.48	6.35	6.57	7.87
MgO	5.05	4.73	2.69	2.71	5.22	2.82	3.36	4.62	5.63	4.72	2.51	1.88	5.41	4.88	2.80		6.42	7.11
CaO	10.81	9.53	7.22	6.77	10.82	7.54	6.47	7.97	8.04	8.69	5.33	4.74	9.36	7.39	4.97	9.35	7.46	8.60
Na_2O	2.56	2.92	4.02	3.93	2.34	3.60	4.07	3.28	3.59	2.31	2.86	2.94	3.31	3.89	4.76		3.25	2.89
K_2O	1.28	0.86	1.29	1.32	0.62	1.41	2.46	1.51	1.03	1.28	2.57	2.67	2.08	2.29	3.44	2.88	3.13	2.60
TiO_2	1.04	0.79	0.55	0.58	0.84	0.62	0.99	0.82	0.85	0.59	0.55	0.45	1.01	0.89	0.54	0.97	0.73	1.25
P_2O_5	0.08	0.07	0.12	0.14	0.19	0.19	0.15	—	0.15	0.14	0.11	0.10	0.27	0.22	0.27	0.23	—	0.27
Total	99.87	100.20	100.21	100.40	99.54	99.93	99.35	99.51	99.42	99.42	100.07	100.21	99.72	98.82	99.56	100.08	100.29	99.51
K	10611	7691	9933	10984	5164	11698	20446	12573	8572	10626	21350	22148	17293	19010	28557	23908	26017	21584
Rb	22.88	23.58	34.65	34.44	4.82	25.09	44.83	29.98	14.41	15.36	100.16	118.4	46.61	62.21	85.84	58.02	71.61	48.52
Cs	1.07	1.40	.814	2.27	0.208	0.482	0.838	1.27	0.29	0.89	4.96	6.75	1.85	2.43	3.14	1.22	3.03	0.245
Ba	224.4	162.6	237.8	264.6	138.7	200.7	374.3	173.0	85.4	150.9	403.2	425.5	957.8	1030.0	1089.1	194.8	776.8	803.0
Sr	563.1	352.8	331.7	327.4	438.5	368.3	348.9	722.3	336.3	436.0	289.5	260.2	1080.7	1068.6	1081.8	648.7	649.4	832.2
Y	25.5	24.3	25.9	—	22.7	24.5	30.3	42.5	19.1	—	—	—	17.5	16.8	12.1	18.7	—	13.1
Zr	96.4	98.2	129.3	—	78.3	89.4	102.5	133.1	62.8	—	—	—	212.0	194.7	175.5	179.9	—	194.3
Nb	3.7	3.3	3.9	—	4.6	1.4	3.6	4.6	2.1	—	—	—	25.3	21.2	21.1	8.3	—	21.7
V	293.3	185.1	98.4	—	290.3	177.8	311.1	235.0	329.5	—	—	—	228.0	174.9	96.2	222.4	—	194.8
Cr	21.6	11.1	7.3	—	22.8	7.2	15.2	21.7	20.9	—	—	—	17.8	74.8	47.5	195	—	309
Ni	18.7	13.3	7.1	—	26.7	6.9	14.1	12.7	23.1	—	—	—	32.4	54.8	33.3	62.1	—	113.8
Cu	144.4	69.1	40.0	—	124.3	33.6	286.6	187.5	154	—	—	—	114.7	49.9	20.7	75.7	—	63.3
Zn	69.3	72.8	60.0	—	67.3	61.9	96.8	64.0	84.5	—	—	—	62.5	59.4	42.4	60.6	—	70.5
Rb/Sr	0.0406	0.0668	0.1045	0.1052	0.0110	0.0681	0.1285	0.0415	0.0429	0.0352	0.3460	0.455	0.0431	0.0582	0.0793	0.0894	0.1103	0.0583
K/Rb	464	326	256	319	1071	466	456	419	595	692	213	187	371	306	333	412	363	445
K/Cs	9963	5494	12203	4839	24825	24270	24399	9877	29357	11939	4302	3281	9358	7823	9095	19597	8578	88098
K/Ba	47.3	47.3	41.8	41.5	37.2	58.3	54.6	72.7	100.4	70.4	53.0	52.1	18.1	18.5	26.2	122.7	33.5	26.9
K/Sr	18.8	21.8	29.9	33.6	11.8	31.8	57.6	17.4	25.5	24.37	73.7	85.1	16.0	17.8	26.4	36.9	40.1	25.9
Ba/Sr	0.398	0.461	0.717	0.808	0.316	0.545	1.05	0.240	0.254	0.346	1.39	1.63	0.886	0.964	1.00	.300	1.20	0.965
$^{87}Sr/^{86}Sr$	0.70371±5	0.70435±4	0.70408±5	0.70438±4	0.70393±5	0.70358±5	0.70397±7	0.70403±6	0.70380±3	0.70399±5	0.71987±6	0.72397±5	0.70642±7	0.70615±6	0.70606±5	0.70697±6	0.70870±4	0.70568±5
$^{206}Pb/^{204}Pb$	18.554			18.578	18.524	18.544			18.337		19.884	19.942	18.802	18.757				
$^{207}Pb/^{204}Pb$	15.571			15.605	15.580	15.550			15.526		15.800	15.831	15.657	15.642				
$^{208}Pb/^{204}Pb$	38.484			38.623	38.434	38.353			38.127		40.170	40.331	38.866	38.799				

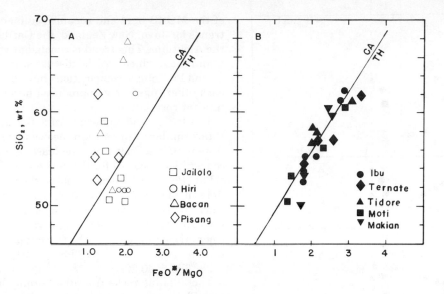

Fig. 3.   FeO*/MgO versus SiO₂ for Halmahera lavas. Suites in Figure 3a show no Fe enrichment; those in Figure 3b lie along the TH-CA boundary, indicating moderate Fe enrichment.

Incompatible element concentrations vary substantially within a suite, largely as a function of silica content. This is shown in Figure 4, where Cs, Ba, and K_2O are plotted against SiO_2. The concentrations of alkali elements, Ba, and Sr are typical for oceanic island arcs [Taylor, 1981; Gill, 1981; Bailey, 1981]. Samples from a given volcano often form smooth trends on these variation diagrams, despite the fact that only two suites, Ternate (solid diamonds) and Tidore (solid triangles), are isotopically homogeneous. Sr contents, not plotted, remain nearly constant through the entire range of silica content, indicating that bulk $D^{Sr} \approx 1$ [Gill, 1978].

Incompatible element ratios are listed in Table 1 and plotted in Figure 5. Some of the basalts have fairly high K/Rb (640–720) and K/Cs (up to 35 K) for arc lavas, although these values are still substantially lower than those reported for MORB (average K/Rb ~ 1100, average K/Cs ~ 80 K). K/Rb and K/Cs often decrease very rapidly with increasing SiO_2 up to ~53% SiO_2 at which point the curves flatten out and incompatible element ratios decrease only slightly with further increase in SiO_2.

The lavas from the oceanic segment show the characteristic depletion in high field strength and compatible elements associated with arc rocks [Gill, 1981]. For example, basalts with ~50.5% SiO_2 and 5.7–6.2% MgO contain only 14–32 ppm Ni.

The basalt (BC35) and andesite (BC15) from the northern part of Bacan Island are quite similar to those from the oceanic segment. On the southern part of the island, however, lavas that erupted through continental crust are largely dacitic with relatively high concentrations of Rb (up to 118 ppm) and Cs (6.75 ppm). Trends on alkali-element versus SiO_2 variation diagrams (Figure 4) are slightly steeper than those for the oceanic arc. Ba is typically higher and Sr lower in the dacites, relative to extrapolated trends for the oceanic volcanoes. It is worth noting that in this case of severe continental contamination ($^{87}Sr/^{86}Sr$ ratios for the dacites are as high as 0.724), alkali element contents are only slightly elevated above extrapolated trends for the rest of the arc.

The alkaline lavas from the extinct volcanoes along the Sorong fault zone are different from the oceanic arc samples in some respects but are similar in others. Andesite is the dominant rock type, with subequal amounts of basalt, basaltic andesite, and dacite. Al_2O_3 is high, TiO_2 is low, and there is no trend toward Fe enrichment with fractionation. K_2O plus Na_2O is greater than 5%, with Na_2O often less than K_2O. Alkali contents are generally higher than in lavas from the oceanic arc, and Ba (195–1090 ppm) and Sr (650–1082 ppm) show dramatic enrichment. Incompatible element ratios are generally similar to those from the oceanic arc.

It is possible to briefly characterize the three regions. The oceanic arc segment closely corresponds to the 'typical' island arc suite. In many characteristics, it is very similar to the Sangihe-Sulawesi Arc [Morrice et al., 1982]. In Bacan, interaction with the crust is reflected in higher alkali and Ba contents for most samples. The islands south of the active arc have many characteristics similar to shoshonites [Morrison, 1980], although they do not display the expected depletion in Zr and Nb. The association is unclear; they may be part of either the Quaternary or Tertiary Halmahera arcs or they may be related to the transform plate margin.

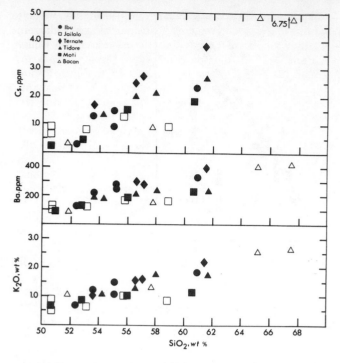

Fig. 4. Variations in K_2O, Ba, and Cs as a function of SiO_2. Indicated symbols are the same as in Figure 3. Only two suites, Ternate (solid diamonds) and Tidore (solid triangles) are isotopically homogeneous.

Sr and Pb Isotopes

Sr and Pb isotopic compositions for selected Halmahera Arc samples are listed in Table 1 and plotted in Figures 6 and 7. As with major and trace element data, we will discuss the data region by region.

Figure 6 shows $^{87}Sr/^{86}Sr$ ratios plotted as a function of geographic location within the arc. For the oceanic segment, volcanoes are listed from north (Ibu) to south (Kayoa). Most volcanic centers are isotopically heterogeneous, and many centers show a range nearly as great as that demonstrated by the entire arc. The isotopic compositions for the oceanic part of the arc (0.70357–0.70438) are similar to those from many island arcs. Isotopic heterogeneity within a single volcanic center, similar to that shown here, has been reported for a number of island arcs including the Lesser Antilles [Hawkesworth and Powell, 1980; Hawkesworth et al., 1979], the Aleutians [Morris and Hart, 1981; Kay et al., 1978], the Banda Arc [Whitford and Jezek, 1979], the Scotia Arc [Hawkesworth et al., 1977], and the Sunda Arc [Whitford, 1975].

The Pb isotope data are plotted on a $^{206}Pb/^{204}Pb$ (α) versus $^{207}Pb/^{204}Pb$ (β) and $^{208}Pb/^{204}Pb$ (γ) diagram in Figure 7. Fields for eight island arcs, MORB, and oceanic sediments are shown for comparison. Solid circles are data points for samples from the oceanic segment of the Halmahera Arc. They form a steep array between the high $^{206}Pb/^{204}Pb$ end

of the MORB field and oceanic sediments, similar to the trends for Java, New Zealand, the Caribbean islands, and the Aleutians. This trend is consistent with sediment contamination, which will be discussed in more detail later. Sr and Pb isotopic compositions have been plotted against each other (figure not reproduced here) and show no significant correlation.

The basalt and andesite from Bacan have Sr isotope ratios similar to those from the oceanic segment. A basalt from northern Bacan has the least radiogenic Pb isotopic composition reported for the entire arc, $^{206}Pb/^{204}Pb = 18.34$, $^{207}Pb/^{204}Pb = 15.53$, which plots on the margin of the MORB field in Figure 7. The dacites, which erupt through continental crust, have very radiogenic Sr isotopic compositions (0.720–0.724), requiring a very large degree of continental contamination. The Pb isotope systematics reflect this contamination: $^{206}Pb/^{204}Pb \approx 19.91$, $^{207}Pb/^{204}Pb \approx 15.80$, $^{208}Pb/^{204}Pb \approx 40.20$.

The alkaline rocks from the Sorong fault zone have Sr and Pb isotopic characteristics which are distinct from both the oceanic arc and Bacan. $^{87}Sr/^{86}Sr$ ratios are 0.70568–0.70870 and Pb ratios are more radiogenic than those from the oceanic segment. The Sr ratios are quite high for island arcs. Anomalously radiogenic oceanic islands such as Kerguelen and Gough have Sr isotopic compositions which fall at the low end of the range reported here [Dosso et al.,

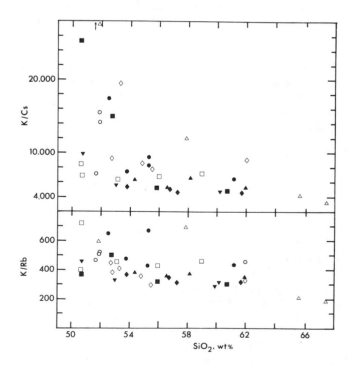

Fig. 5. K/Rb and K/Cs ratios plotted against SiO_2 for suites from the oceanic segment, the continental fragment of Bacan, and the alkaline islands southeast of the active arc. Symbols as in Figure 3.

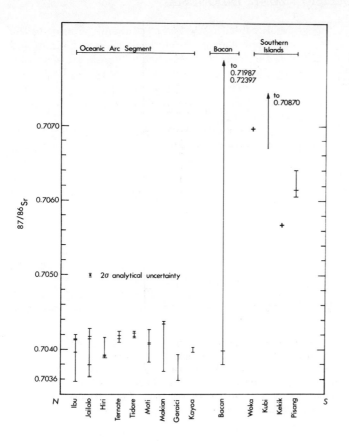

Fig. 6. Sr isotopic compositions as a function of geographic location within the arc. Volcanoes are listed from north to south and separated into an oceanic segment, Bacan, and the alkaline islands southeast of the active arc.

1979; *Gast et al.,* 1964] and Pb isotopic compositions which are nearly as radiogenic as those reported here [*Sun,* 1980; *Dosso and Murthy,* 1980].

Discussion

Models for Magma Genesis: Pb Isotope Evidence

The Pb and Sr isotope data for the Halmahera Arc provide a basis for testing models which have been proposed to explain the origin of island arc lavas. For example, the Pb isotope trend for the Halmahera volcanics, shown in Figure 7, rules out pure MORB as the parental magma. Concentrations of Pb in seawater are very low [*Schaule and Patterson,* 1981], and so remelting of variably altered slab could not generate the observed trend either.

Other models, which are consistent with the Pb isotope data for the Halmahera Arc, are illustrated schematically in Figure 8. Model 1 corresponds to mixing within the mantle wedge of slab-derived and mantle-derived melts. The solid line within the MORB-OIB array is a model mixing line for remelted MORB and magmas similar to those of normal oceanic island basalt. The Halmahera data trend is completely distinct from this mixing line. The

other model illustrated under 'mantle mixing' involves mixing of a slab-derived melt with anomalously radiogenic magmas similar to the basalts of the oceanic islands Gough, Kerguelen, and Reunion [*Sun,* 1980; *Dosso and Murthy,* 1980]. The field labelled 'anomalous OIB' in Figure 8 shows the $^{207}Pb/^{204}Pb$ and $^{206}Pb/^{204}Pb$ ratios required of this mantle source, if it is to be a suitable end-member for the Halmahera array. Figure 9 shows the Halmahera Pb compositions plotted on a Pb isotope correlation diagram with fields indicated for Gough, Reunion, and Kerguelen. The Halmahera data extend to higher $^{207}Pb/^{204}Pb$ ratios than either Gough or Kerguelen, suggesting that anomalously radiogenic OIB-type magmas such as these are not a suitable end-member either. However, because the difference in $^{207}Pb/^{204}Pb$ between the upper end of the Halmahera array and the Gough field is small, we cannot rule this model out strictly on the basis of the Pb isotopes. The Sr isotope systematics applicable to this scenario will be discussed later.

Model 2 in Figure 8 illustrates mixing of an isotopically homogeneous mafic end-member, of composition M with a homogeneous sediment of composition S to generate the Halmahera trend. Model 3 shows mixing of a mafic magma M with isotopically heterogeneous sediments, thereby generating the Halmahera array. The mixing lines for this model are drawn with a common, isotopically homogeneous mafic end-member, although this is not actually required by the data, as will be shown later.

The Pb data permit us to rule out pure MORB remelts and MORB plus-normal OIB mixes as the parental magma for the Halmahera lavas. Unfortunately, they don't allow us to distinguish between the three remaining models sketched in Figure 8. Neither do they provide a basis for identification of the mafic end-member M. The Sr isotope systematics, presented in the next section, allow us to address some of these unanswered questions.

Sr Isotope Constraints

The Pb and Sr isotope ratios of the Halmahera lavas do not correlate with each other (cloud not reproduced here). However, in general, any model which explains the Pb data must also explain the Sr isotope systematics, and these twin constraints allow us to evaluate more critically the models illustrated in Figure 8.

Sr isotope ratios for four isotopically heterogeneous suites from the calc-alkaline oceanic segment of the Halmahera Arc are plotted in Figure 10. This figure is a linearized hyperbolic mixing diagram; simple two-component end-member mixing is represented by a straight line (as is combined assimilation-fractional crystallization if bulk $D^{Sr} = 1$).

The Halmahera data, taken as a whole, form a broad, poorly defined ellipse of positive slope. If, however, one focusses on the data island by island, some trends begin to emerge. The majority of analyses for any island can be

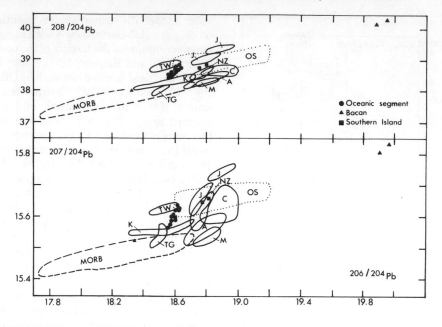

Fig. 7. $^{206}Pb/^{204}Pb$ (α) versus $^{207}Pb/^{204}Pb$ (β) and $^{208}Pb/^{204}Pb$ (γ) for Halmahera Arc lavas, with symbols as indicated. Analytical uncertainty (reproducibility) less than twice symbol size. Fields for island arcs for comparison. Abbreviations and data sources as follows: TG, Tonga; K, Kermadec [*Oversby and Ewart*, 1972]; M, Mariana Arc [*Meijer*, 1976]; A, Aleutians [*Kay et al.*, 1978]; C, Caribbean; NZ, New Zealand [*Armstrong and Cooper*, 1971]; J, Java [*Whitford*, 1975]; TW, Taiwan [*Sun*, 1980]. MORB field following *Sun* [1980]. Oceanic sediment field includes sediments measured outboard of Mariana, Tonga, Ryukyu, and Aleutian trenches and the Cascades [*Church*, 1976] as well as open ocean sediments [*Sun*, 1980].

fit well by the straight lines which are drawn through the data points and labelled with the appropriate symbol in parentheses. The straight lines project toward high $^{87}Sr/^{86}Sr$ ratios and low Sr contents, characteristics associated with pelagic sediments. At the lower end, the mixing arrays project toward mantle values; note that the mixing trends do not converge on a single homogeneous mafic end-member. The data suggest that four isotopically distinct

mafic magmas are sampled in the four different volcanic centers.

Sr isotope ratios and Sr concentrations for the four mafic end-members discussed in Figure 8 (N-type MORB, E-type MORB, OIB, and anomalous OIB as represented by Kerguelon) are also plotted on Figure 10. Data sources for these magma types are included in the figure caption. Alkali basalts from oceanic islands typically have high Sr

SCHEMATIC Pb MIXING MODELS

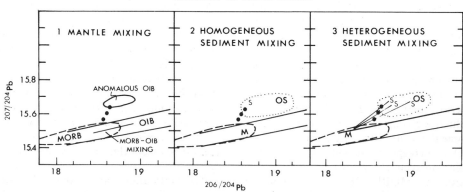

Fig. 8. Schematic Pb mixing models for the Halmahera Arc lavas. Solid circles represent the trend for the oceanic segment of the Halmahera Arc. MORB and normal OIB fields [*Sun*, 1980] are the same in all diagrams.

Pb Isotope Correlation Diagram

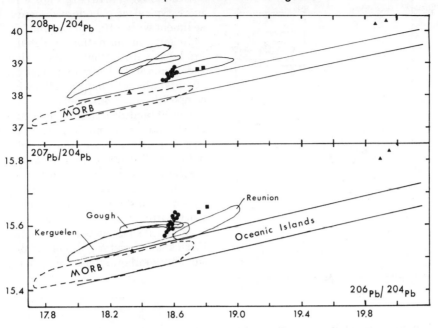

Fig. 9.   Pb isotope correlation diagram for the Halmahera Arc, with fields for the anomalously radiogenic islands of Gough [*Gast et al.*, 1964; *Sun,* 1980], Kerguelen [*Dosso and Murthy,* 1980], and Reunion [*Oversby,* 1972] shown for comparison. Normal OIB trend following *Sun* [1980].

contents; the anomalously radiogenic islands are characterized by high $^{87}Sr/^{86}Sr$ ratios as well. This is illustrated by the star labelled Kerguelen, which plots far from either end of the Halmahera mixing arrays. Thus the Sr data also suggest that mixing of slab-derived melts with anomalously radiogenic alkali basalts such as those from Kerguelen (model 1 in Figure 8) is not a viable mechanism for generation of the Halmahera Arc lavas. Magmas with lower Sr contents but derived from a similar source could conceivably be acceptable end-members.

The primary mixing arrays in Figure 10 diverge at the more radiogenic end as well, suggesting that the sediment end-member is also heterogeneous. This is especially true if one considers the data points which fall off the primary mixing trends. Because the isotopically homogeneous suites show little variation in Sr content, it is unlikely that these points were shifted off the primary arrays by fractionation. One explanation for these data is that they reflect mixing of a mafic magma with sediments which are variable even within a single source region. The lines passing through these anomalous data points are connected with the primary mixing arrays to illustrate this explanation; the lines are joined to indicate a single homogeneous mafic end-member for each center except Ibu.

The heterogeneity in the sedimentary end-member indicated by the Sr isotopes rule out model 2 (Figure 8) as the explanation for the Halmahera lavas. Model 3, mixing of heterogeneous sediments with multiple, isotopically dis-

tinct, mafic end-members, is the only model which satisfies both the Pb and Sr constraints.

Unfortunately, it is more difficult to identify uniquely the source of the mafic end-members. Because the mixing arrays project toward the region between average OIB-type magmas and E-type MORB melts, it is tempting to suggest that the mafic end-members are formed by mixing of melts derived from enriched MORB with magmas from the overlying mantle. This would imply mafic end-members which range from 100% slab derived to 100% mantle derived.

It is possible, however, to generate these mafic end-members in at least two other ways. Basalts from oceanic islands show a wide range of variation in both isotopic composition and Sr concentration; pure mantle-derived magmas could be parental to the Halmahera Arc lavas if they had initial Sr contents which varied from approximately 400 to 800 ppm. Alternatively, mixing melts derived from normal MORB with OIB-type magmas can also yield compositions which plot on the Halmahera mixing arrays. This is illustrated in Figure 10 by the straight line joining average OIB with remelted N-type MORB; numbered tick marks indicate proportion of slab-derived magma in the mix. If the mantle beneath the arc has a 'plum pudding' structure, with depleted and undepleted regions intermixed, the mafic end-members could be generated by mixing, in the mantle, of MORB and OIB sources. Finally, fluid phase transfer from the slab to an OIB-type mantle

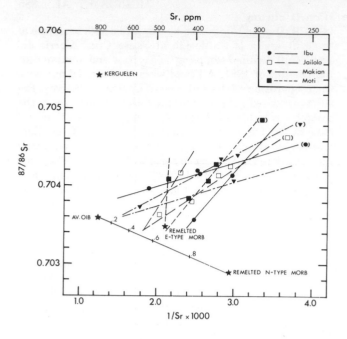

Fig. 10.   Linearized hyperbolic Sr mixing diagram for Halmahera volcanics with symbols as indicated. See text for discussion. Average OIB, N-type MORB, and E-type MORB Sr contents from *Sun* [1980]. Sr isotopic compositions averaged from published analyses. Average value for Kerguelen taken from *Dosso et al.* [1979]. Sr contents for remelted N-type and E-type MORB calculated on the basis of 30% partial melting of water undersaturated quartz eclogite with garnet and clinopyroxene in the residue, following *Gill* [1974] and *Apted* [1981].

could be invoked, although it is not clear that systematic Sr-$^{87}Sr/^{86}Sr$ relationships would be maintained.

Estimates of the proportion of sediment contaminant in the mafic magmas are hard to make as the mass balance calculations are dependent on the exact Pb and Sr contents and isotopic compositions chosen for the sediments. In Halmahera, contaminating sediments range from volcaniclastic with 350–450 ppm Sr (J. D. Morris, unpublished data, 1982) to pelagic with higher $^{87}Sr/^{86}Sr$ ratios and much lower Sr contents [e.g., *Faure*, 1977].

If we assume that volcaniclastic sediments have ~450 ppm Sr and $^{87}Sr/^{86}Sr \sim 0.7063$ (unpublished data for Molucca Sea volcaniclastic sediments), then up to ~20% sediments can be accommodated in the magmas. This proportion of sediment to mafic magma will change somewhat if other values are chosen for the sediment contaminant or the mafic melt. Even still, the percentage of sediment contaminant is likely to remain higher than the 2–5% sediment quoted as upper limits for other arcs [*Kay et al.*, 1978; *Meijer*, 1976]. Note that volcaniclastic sediments will have chemical compositions quite similar to their island arc sources, and so admixtures of volcaniclastic sediments with mafic magmas need not produce arc volcanics with anomalous major or trace element chemistry.

Other mixing arrays in Figure 10 project toward values more typical of pelagic sediments. For a model pelagic sediment with ~100 ppm Sr and $^{87}Sr/^{86}Sr \sim 0.709$, 15–20% sediment contamination of a mafic magma will produce arc lavas with isotopic compositions similar to the more radiogenic of the Halmahera volcanics. Again, proportions of contaminant to mafic magma will change if end-member compositions are changed.

The Pb and Sr data are both consistent with sediment contamination of a mafic magma. This automatically gives rise to questions regarding the site of sediment contamination. Sediments could be incorporated either by melting of subducted sediments at the top of the slab, by fluid phase transfer from the slab to the overlying mantle, or by assimilation of shallow water sediments deposited on the arc crust during evolution of the arc. Two lines of reasoning suggest that subducted sediments are the contaminant. First, shallow water sediments from the Molucca Sea (presumably similar to those deposited on the arc, but probably

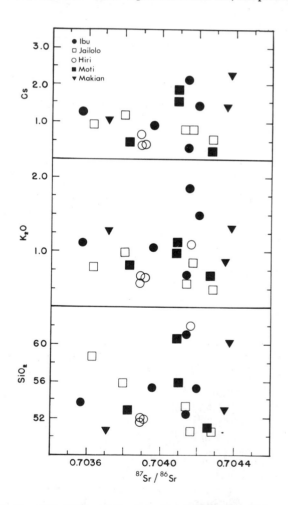

Fig. 11. SiO_2, K_2O, and Cs plotted against $^{87}Sr/^{86}Sr$ for isotopically heterogeneous suites from the oceanic segment. Symbols as in preceding figure.

different from deep sediments which would be subducted) have Pb isotope ratios (unpublished data) which overlap those of the volcanics. Consequently, they cannot be an end-member for the Halmahera trend. Second, progressive fractionation and assimilation of sediments should generate positive correlations between elements such as Si, K, and Cs and $^{87}Sr/^{86}Sr$ ratios. Figure 11 indicates that no such correlations exist. Consequently, the subduction zone seems the most likely site for sediment contamination of these arc lavas. Unfortunately, this does not allow us to uniquely distinguish between low-temperature melting of sediments and fluid phase transfer.

Conclusions

The Halmahera island arc can be divided into three regions, each area associated with a distinct chemistry and tectonic setting. The origin of the alkaline rocks on the inactive volcanic islands along the Sorong fault zone, a transform plate margin, is unclear. The rocks have some chemical characteristics of shoshonites and may be arc-related rocks which have been displaced by motion along the Sorong fault zone. Alternatively, they may have erupted in place and represent mantle magmas erupted to the surface along a transform plate boundary.

The continental suite on Bacan reflects the intersection of an oceanic arc with a continental fragment. Dacites have isotopic compositions which require very large amounts of a continental component. Despite this severe contamination, the major and trace element contents are only slightly different from those expected of uncontaminated dacites.

Most of the volcanoes of the Halmahera arc are part of the calc-alkaline oceanic segment. The lavas have major and trace element and isotopic compositions which are similar to a number of other arcs. The Pb and Sr isotopes rule out the possibility that the Halmahera arc parental magmas were generated by remelting of the slab alone or by tapping of a MORB-like mantle wedge. Rather, the isotopes are best matched by a three-component mixing model where MORB melts mix with magmas from the overlying wedge and are contaminated with oceanic sediments of variable composition. A two-component model, where magmas with variable Sr contents, derived from an OIB-type mantle wedge, are mixed with oceanic sediments is also consistent with the data. Both models indicate that the mafic end-members for the Halmahera volcanics are not chemically identical, even over the limited distance from Ibu to Makian, about 200 km. This suggests that models for the origin of island arc lavas which call upon a single, homogeneous mafic parental magma may not be applicable and may lead to inconsistent or inconclusive results.

Appendix: Methods

Field work was done by P. A. Jezek and J. Morris. Jezek collected reconnaissance samples from all volcanic centers in 1979. During this period he was a research associate at the University of California at Santa Cruz. Morris did additional detailed sampling on Ternate and Makian during the fall of 1981. All field work was carried out with the cooperation of the Indonesian Geological Survey. Results presented in this paper are based on analyses of the samples collected by Jezek. All sample preparation was done at the University of California at Santa Cruz. Representative samples were split and crushed in a jaw crusher then powdered in a tungsten carbide shatter box. The same powder aliquots were used for all analytical techniques.

Major element compositions were determined at the Smithsonian Institution by microprobe analysis of fused glasses. The technique is described in the work of *Jezek et al.* [1978]. All Fe is reported as FeO. Analytical uncertainties are typically <5%, except for TiO_2, which may vary by ~13%. K_2O contents are from isotope dilution. K, Rb, Cs, Ba, and Sr were determined by isotope dilution. All other trace elements were measured on pressed powder pellets, using the automated Phillips X ray fluorescence spectrometer at Woods Hole Oceanographic Institution. Analytical uncertainties are <5%, based on numerous replicate analyses of U.S. Geological Survey whole-rock standards [*Schroeder et al.,* 1980].

All isotope analyses were done on the 9-inch 60° mass spectrometer in Stan Hart's lab at Massachusetts Institute of Technology. The technique for Sr separation and analysis are as reported in the work of *Hart and Brooks* [1977]. All isotopic compositions are normalized to $^{86}Sr/^{88}Sr = 0.1194$ and reported relative to E&A $SrCO_3 = 0.70800$. Uncertainties (reproducibilities) in concentration data are ≈1% for all elements except Cs which is ≤5%. Uncertainties reported for isotopic compositions are 2σ of the mean. Procedures for Pb isotope analyses follow the Paris technique [*Manhes et al.,* 1978] and will be presented by J. D. Morris and S. R. Hart (work in preparation). Total blanks for the procedure are ~2 ng total Pb. Analyses were corrected for fractionation on the basis of multiple analyses of NBS SRM 981, the common lead standard. Reproducibilities for reported values are ±0.05%/mass unit (i.e., $^{206}Pb/^{204}Pb = ±0.10\%$, $^{207}Pb/^{204}Pb = ±0.15\%$) based on numerous analyses of an in-house whole-rock silicate standard.

Acknowledgments. The field work in Indonesia was carried out in cooperation with the Indonesian Geological Research and Development Centre. Our thanks to M. Hartono, Director, and to M. Monoarfa and K. Hardjadinata for assistance in the field under occasionally trying circumstances. We want to thank Claude Allegre and his colleagues for their advice regarding techniques for Pb separation and analysis, and also B. J. Pegram, who developed the Pb system here at Massachusetts Institute of Technology. Data collection and analysis were begun through funding from the International Decade of Oceanic

Exploration (IDOE), National Science Foundation grant OCE-7920999. Subsequent work has been supported by National Science Foundation grant EAR-8024041. Thanks to Alan Zindler and Alan Divis for constructive reviews. Our especial thanks to Donna Hall and Theresa Miele for manuscript preparation.

References

Apted, M. J., Rare earth element systematics of hydrous liquids from partial melting of basaltic eclogite: A re-evaluation, Earth Planet. Sci. Lett., 52, 171–182, 1981.

Armstrong, R. L., and J. A. Cooper, Lead isotopes in island arcs, Bull. Volcanol., 35, 27–63, 1971.

Bailey, J. C., Geochemical criteria for a refined tectonic discrimination of orogenic andesites, Chem. Geol., 32, 139–154, 1981.

Cardwell, R. K., B. L. Isacks, and D. E. Karig, The spatial distribution of earthquakes, focal mechanism solutions and subducted lithosphere in the Philippine and north-eastern Indonesian islands, in The Tectonic and Geologic Evolution of Southeast Asian Seas and Islands, Geophys. Monogr. Ser., vol. 23, edited by D. E. Hayes, pp. 1–36, AGU, Washington, D.C., 1980.

Church, S. E., The Cascades Mountains revisited: A re-evaluation in light of new lead isotopic data, Earth Planet. Sci. Lett., 29, 175–188, 1976.

Dosso, L., and V. R. Murthy, A Nd isotopic study of the Kerguelen Islands: Inferences on enriched oceanic mantle source, Earth Planet. Sci. Lett., 48, 268–276, 1980.

Dosso, L., P. Vidal, J. M. Cantagrel, J. Lameyre, A. Marot, and S. Zinine, 'Kerguelen: Continental fragment or oceanic island?': Petrology and isotopic geochemistry evidence, Earth Planet. Sci. Lett., 43, 46–60, 1979.

Evans, C. A., J. W. Hawkins, and G. F. Moore, Petrology and geochemistry of ophiolite and associated volcanic rocks on the Talaud islands, Molucca Sea collision zone, northeast Indonesia, J. Geodyn., in press, 1982.

Faure, G., Principles of Isotope Geology, John Wiley, New York, 1977.

Fitch, T. J., Earthquake mechanisms and island arc tectonics in the Indonesian-Philippine regions, Bull. Seismol. Soc. Am., 60, 565–591, 1970.

Gast, P. W., G. R. Tilton, and C. Hedge, Isotopic composition of lead and strontium from Ascension and Gough islands, Science, 145, 1181–1185, 1964.

Gill, J. B., Role of underthrust oceanic crust in the genesis of a Fijian calc-alkaline suite, Contrib. Mineral. Petrol., 43, 29–45, 1974.

Gill, J. B., Role of trace element partition coefficients in models of andesite genesis, Geochim. Cosmochim. Acta, 42, 709–724, 1978.

Gill, J. B., Orogenic Andesites and Plate Tectonics, Springer-Verlag, New York, 1981.

Hamilton, W., Tectonic map of the Indonesian region, Map I-875-D, Misc. Invest. Ser., scale 1:500,000, U.S. Geological Survey, Reston, Va., 1978.

Hamilton, W., Tectonics of the Indonesian region, U.S. Geol. Surv. Prof. Pap., 1078, 1979.

Hart, S. R., and C. Brooks, The geochemistry and evolution of Early Precambrian mantle, Contrib. Mineral. Petrol., 61, 109–128, 1977.

Hawkesworth, C. J., and M. Powell, Magma genesis in the Lesser Antilles island arc, Earth Planet. Sci. Lett., 51, 297–308, 1980.

Hawkesworth, C. J., R. K. O'Nions, R. J. Pankhurst, P. J. Hamilton, and N. M. Evensen, A geochemical study of island-arc and back-arc tholeiites from the Scotia Sea, Earth Planet. Sci. Lett., 36, 253–262, 1977.

Hawkesworth, C. J., R. K. O'Nions, and R. J. Arculus, Nd and Sr isotope geochemistry of island arc volcanics, Grenada Lesser Antilles, Earth Planet. Sci. Lett., 45, 237–248, 1979.

Jezek, P. A., J. M. Sinton, E. Jaroseqich, and C. R. Obermyer, Fusion of rock and mineral powders for electron microprobe analysis, Smithson. Contrib. Earth Sci., 46–52, 1978.

Jezek, P. A., D. J. Whitford, and J. B. Gill, Geochemistry of recent lavas from the Sangihe-Sulawesi Arc, Indonesia, Geology, in press, 1982.

Karig, D. E., and G. F. Sharman III, Subduction and accretion in trenches, Geol. Soc. Am. Bull., 86, 377–389, 1975.

Kay, R. W., S. S. Sun, and C.-N. Lee-Hu, Pb and Sr isotopes in volcanic rocks from the Aleutian Islands and Pribilof Islands, Alaska, Geochim. Cosmochim. Acta, 42, 263–273, 1978.

Manhes, G., J. F. Minster, and C. J. Allegre, Comparative uranium-thorium-lead and rubidium-strontium of St. Severin amphoterite: Consequences for early solar system chronology, Earth Planet. Sci. Lett., 45, 14–24, 1978.

McCaffrey, R., Crustal structure and tectonics of the Molucca Sea collision zone, Indonesia, Ph.D. thesis, Univ. of Calif. at Santa Cruz, Santa Cruz, 1981.

McCaffrey, R., E. A. Silver, and R. W. Raitt, Crustal structure of the Molucca Sea collision zone, Indonesia, in The Tectonic and Geologic Evolution of Southeast Asian Seas and Islands, Geophys. Monogr. Ser., vol. 23, edited by D. Hayes, pp. 161–179, AGU, Washington, D.C., 1980.

Meijer, A., Pb and Sr isotopic data bearing on the origin of lavas from the Mariana arc system, Geol. Soc. Am. Bull., 87, 1358–1369, 1976.

Morrice, M. G., P. A. Jezek, J. B. Gill, D. J. Whitford, and M. Monoarfa, An introduction to the Sangihe Arc: Volcanism accompanying arc-arc collision in the Molucca Sea, Indonesia, J. Volcanol. Geotherm. Res., in press, 1982.

Morris, J. D., and S. R. Hart, Transverse geochemical variations across the Aleutian Arc, Cold Bay to Amak (abstract), Eos Trans. AGU, 61, 400, 1981.

Morrison, G. W., Characteristics and tectonic setting of the shoshonite rock association, Lithos, 13, 97–108, 1980.

Oversby, V. M., Genetic relationships among the volcanic

rocks of Reunion: Chemical and lead isotopic evidence, *Geochim. Cosmochim. Acta, 36,* 1167–1179, 1972.

Oversby, V. M., and A. Ewart, Lead isotopic compositions of Tonga-Kermedec volcanics and their petrogenetic significance, *Contrib. Mineral. Petrol., 37,* 181–210, 1972.

Schaule, B. K., and C. C. Patterson, Lead concentrations in the northeast Pacific: evidence for global anthropogenic perturbations, *Earth Planet. Sci. Lett., 54,* 97–262, 1981.

Schroeder, B., G. Thompson, M. Sulanowska, and J. N. Ludden, Analysis of geologic materials using an automated x-ray fluorescence system, *X Ray Spectrom., 9,* 198–205, Oct. 1980.

Sharman, G. F., The accretion/subduction balance (abstract), *Eos Trans. AGU, 61,* 1121, 1980.

Silver, E. A., and J. C. Moore, The Molucca Sea collision zone, Indonesia, *J. Geophys. Res., 83,* 1681–1691, 1978.

Sinha, A. K., and S. R. Hart, A geochemical test of the subduction hypothesis for generation of island arc magmas, *Year Book Carnegie Inst. Washington, 71,* 309–312, 1972.

Sun, S.-S., Lead isotopic study of young volcanic rocks from mid-ocean ridges, ocean islands and island arcs, *Philos. Trans. R. Soc. London Ser. A, 297,* 409–445, 1980.

Taylor, S. R., Island arc basalts, in *Basaltic Volcanism on the Terrestrial Planets,* Pergamon, New York, 1981.

U.S. Geological Survey, Bathymetric map of the Indonesian region, *Map I-875-A, Misc. Invest. Ser.,* Reston, Va., 1974.

Whitford, D. J., Geochemistry and petrology of volcanic rocks from the Sunda arc, Indonesia, Ph.D. thesis, Australian Natl. Univ., Canberra, 1975.

Whitford, D. J., and P. A. Jezek, Origin of late-Cenozoic lavas from the Banda Arc, Indonesia: Trace element and Sr isotope evidence, *Contrib. Mineral. Petrol., 68,* 141–150, 1979.

Paleomagnetism and Age Determination of Cretaceous Rocks From Gyeongsang Basin, Korean Peninsula

Yo-ichiro Otofuji,[1] Jin Yong Oh,[2] Takao Hirajima,[1] Kyung Duck Min,[2] and Sadao Sasajima[1]

Paleomagnetic measurements and age determinations have been made on samples of sediments and lavas of Cretaceous age from the Gyeongsang Basin, South Korea. The K-Ar age of 68.1 ± 3.4 m.y. for basalts from the Hagbong Volcanic Member confirms the Cretaceous period of Gyeongsang Supergroup. Paleomagnetic directions are fairly grouped with a mean confidence interval of declination of 26.6°, inclination of 62.3°, and α_{95} = 8.3°. The reliability of the direction is ascertained through thermal and alternating field demagnetization techniques and comparison of the remanent directions of sedimentary rocks and a lava flow. The South Korean pole obtained (191°E, 69°N) is in excellent agreement with the poles from neighboring regions of North Korea, China, Kolyma, and the Siberian Platform. From this concordance it is inferred that since the Late Cretaceous the Korean Peninsula has been a part of the Asian continent and was not subjected to any tectonic rotation relative to the Asian continent. This high stability of the Korean Peninsula against tectonic disturbances is explained by the presence of a thick and stiff lid within the lithosphere, inferred from the comparison between the character of the lithosphere beneath the Korean Peninsula and Japan.

Introduction

Paleomagnetism, marine magnetic anomaly, and seismological studies document large-scale relative displacements or rotations of peninsulas to their host continents. The Indian Peninsula has been colliding with the Asian continent since the Paleogene [*Molnar and Tapponnier,* 1975, 1977], and the Arabian Peninsula is now colliding at the Zagros region of Iran [*Bird et al.,* 1975]. The Iberian, Italian, and Malay peninsulas have all been subjected to tectonic rotations [*Van der Voo,* 1969; *Lowrie and Alvarez,* 1974; *McElhinny et al.,* 1974]. Since peninsulas are situated at the edge of a continent, where crust is thin [*Cummings and Shiller,* 1971], the interactions among plates may easily cause the relative motion of the peninsula with respect to the host continent.

The Korean Peninsula is one of the largest peninsulas in the eastern margin of the Asian continent. The work presented here is an attempt to obtain the relative movement between the Korean Peninsula and the Asian continent from the viewpoint of paleomagnetism.

The Cretaceous strata of Gyeongsang Basin in South Korea are very suitable for paleomagnetic study, because the strata consist of many kinds of rocks (lava, tuff, shale, and sandstone) with different kinds of natural remanent magnetization (NRM) and their bedding planes of strata are clearly recognized. The present paleomagnetic work could therefore provide more reliable paleomagnetic directions than the previous works from igneous rocks [*Kienzle and Scharon,* 1966; *Ito and Tokieda,* 1980].

Geology and Sampling

Gyeongsang Basin is located in the southeastern part of the Korean Peninsula (Figure 1). This basin consists of a thick sequence of fluviolacustrine sediments intercalated with lavas and volcanoclastic rocks of Cretaceous age. These strata constitute the Gyeongsang Supergroup. According to the definition of *Chang* [1975] the Gyeongsang Supergroup is divided into three groups: the Sindong, Hayang, and Yuchon groups in ascending order (see Table 1). The Sindong Group consists of sediments of the prevolcanic phase, the Hayang Group comprises nonvolcanic sediments with some volcanic horizons, and the Yuchon Group consists of volcanic formations.

The Sindong Group, the lowest of the Gyeongsang Supergroup, is extensively fossiliferous, and it is documented as lowest Cretaceous in age [*Yabe,* 1905; *Kawasaki,* 1928; *Um and Reedman,* 1975]. The K-Ar data for granitic plutons intruding the Gyeongsang Supergroup range from 62 to 88 m.y. [*Chang,* 1975; *Lee,* 1980], which puts the end of the Cretaceous as the upper limit of the supergroup. These data indicate that the Gyeongsang Supergroup was formed during the Cretaceous.

More than 120 samples were obtained from 10 widely

[1] Department of Geology and Mineralogy, Faculty of Science, Kyoto University, Kyoto, Japan.

[2] Department of Geology, Yonsei University, Seoul, Korea.

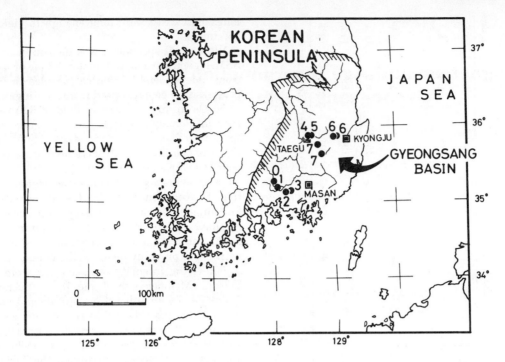

Fig. 1. Sampling localities in Gyeongsang Basin: site 0, Nagdong Formation; site 1, Hasandong Formation; site 2, Jinju Formation; site 3, Chilgog Formation; sites 4 and 5, Hagbong volcanic members; site 6, Chaeyagsan Formation; and site 7, Yuchon Group. Stratigraphic horizons of sampling sites are given in Table 1.

spaced locations in the Gyeongsang Basin. Figure 1 shows the distribution of sampling sites. The samples at sites 0–2 are from the Sindong Group, those at sites 3–6 are from the Hayang Group, and those at site 7 are from the Yuchon Group. The bedding planes were clearly recognizable at all sampling localities except for site 5 (of basaltic lavas). Detailed stratigraphic horizons of sampling sites and the rock types are given in Tables 1 and 2. The basaltic rock

samples were collected for K-Ar age determination from the Hagbong Volcanic Member of site 4, in the Hayang Group.

Paleomagnetism

The block samples collected were drilled into 2.4-cm-diameter cores and cut into individual specimens 2.4 cm

TABLE 1. Stratigraphy of Gyeongsang Supergroup in Gyeongsang Basin and the Stratigraphic Horizons of Sampling Sites

Group	Subdivision	Site
Yuchon Group		7
Hayang Group Jindong Subgroup	Geoncheonri Formation Chaeyagsan Formation Songnaedong Formation Banyaweol Formation	6
Uiryong Subgroup	Haman Formation Hagbong Volcanic Member Silla Conglomerate	5 and 4
Chilgog Subgroup	Chilgog Formation	3
Sindong Group	Jinju Formation Hasandong Formation Nagdong Formation	2 1 0

Data after *Chang* [1975].

TABLE 2. Results of Paleomagnetic Measurement

Division	Locality Latitude, °N	Locality Longitude, °E	Rock Type	N	Level of Demagnetization	D, deg	I, deg	α₉₅, deg	K	Virtual Geomagnetic Pole (North) Latitude, °N	Virtual Geomagnetic Pole (North) Longitude, °E	δp, deg	δm, deg
Site 7 Yuchon Group	35.72	128.74	andesite	22	200 Oe	20.8	57.7	19.4	3.5	73.2	203.6	20.9	28.5
Site 4 Hagbong Volcanic Member	35.89	129.03	basalt	6	300 Oe	6.6	66.3	5.1	142.8	76.3	147.3	8.4	8.4
Site 3 Chilgog Formation	35.17	128.16	shale	27	200 Oe	27.8	58.7	1.9	210.8	67.5	198.8	2.1	2.8
Site 2 Jinju Formation	35.16	128.10	sandstone	7 24	150°C 300 Oe	53.1	60.1	7.5	12.9	48.4	193.2	8.6	11.4
Site 1 Hasandong Formation	35.17	128.03	sandstone	20	300 Oe	18.8	58.8	7.5	19.8	74.5	196.1	8.3	11.2
Mean	37.	128.		5		26.6	62.3	8.3	85.0	68.9	191.2	10.1	12.9

Parameters are as follows: N, number of specimens; D, declination; I, inclination; α_{95}, radius of 95% confidence circle; K, precision parameter; and δp and δm, semiaxes of ovals of 95% confidence.

long; most cores were long enough to yield two or three specimens.

NRM of all specimens was measured on a Schonstedt SSM-1A spinner magnetometer. Three or more specimens from independently oriented samples in each site were progressively demagnetized in alternating field (af) in steps of 50 Oe up to a maximum field of 600 Oe. Progressive thermal demagnetization was also undertaken on three specimens from each site in steps of 100° or 150°C to the Curie temperature, with increment steps reduced to 50°C near the Curie temperature.

Thermal demagnetization experiments (Figure 2) show that the specimens from the Hasandong (site 1), Jinju (site 2), Chilgog (site 3), and Chaeyagsan (site 6) formations retained a large fraction of NRM after being heated to 650°C and lost their magnetization at 700°C. Their magnetizations probably reside in the hematite. After heating to 600°C the intensity of specimens from the Hagbong Volcanic Member (sites 4 and 5) and Yuchon Group (site 7) dropped by an order of magnitude and became random in direction, indicating that the magnetization resides in magnetite. The remanent directions from all sites differed only slightly from the NRM direction before treatment during thermal cleaning up to near the Curie temperature. Thermal cleaning experiments demonstrate that the specimens have a single and stable component of magnetization.

There is little difference in the direction changes after alternating field and thermal demagnetization for pilot specimens from the Hasandong Formation, Chilgok Formation, Hagbong Volcanic Member, and Yuchon Group (Figure 2). The optimum demagnetization was confined to the af demagnetization method; the optimum demagnetization level for each site was chosen as the value for which more than three directions of pilots were most clustered. The NRM's after cleaning by optimum demagnetization fields are drawn in Figure 2.

Remanent magnetizations of specimens from the Jinju Formation (site 2) showed a different behavior in response to af and thermal demagnetization. Specimens were cleaned either by af demagnetization in a peak field of 300 Oe or by thermal demagnetization at 150°C. The site mean was calculated from both af and thermal demagnetization data, because there was little difference in direction between the two data sets.

Specimens from the Nagdong Formation (site 0) had a remanent intensity of less than 10^{-8} emu/g. These specimens were rejected to avoid the introduction of errors through instrumental noise on measurements. Pilot specimens from the Chaeyagsan Formation (site 6) showed a stable remanent magnetization in response to both af and thermal demagnetization. However, the precision parameter k [*Fisher*, 1953] of three pilot specimens was less than 2.0 at any demagnetization level. The directions of all specimens ($N = 12$) demagnetized after 300 Oe were also highly scattered with a precision parameter of only 1.3. It

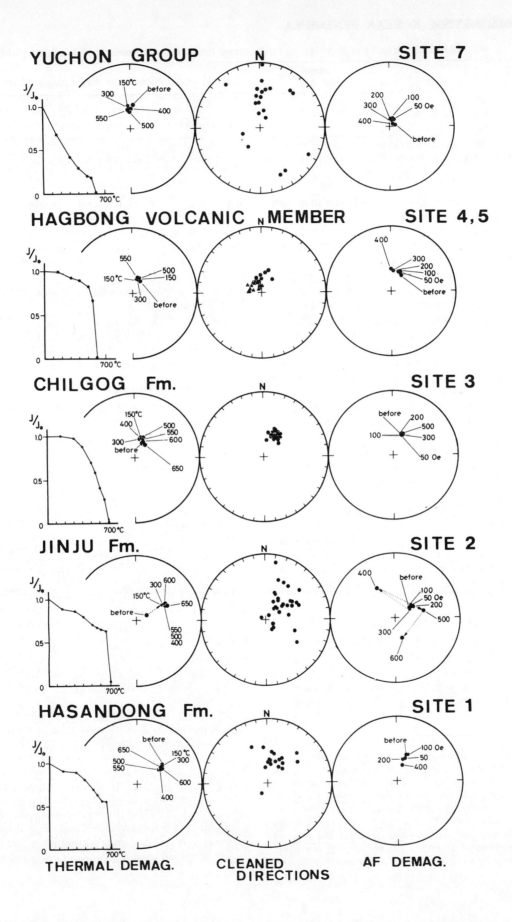

YUCHON GROUP — N — SITE 7

HAGBONG VOLCANIC MEMBER — N — SITE 4,5

CHILGOG Fm. — N — SITE 3

JINJU Fm. — N — SITE 2

HASANDONG Fm. — N — SITE 1

THERMAL DEMAG.　　CLEANED DIRECTIONS　　AF DEMAG.

TABLE 3. Results of K-Ar Dating for Volcanics From Hagbong Volcanic Members

Sample	Locality	Isotope Age, m.y.	Scc ^{40}Ar/g $\times$ 10^{-5}	^{40}Ar, percent	K, percent
PG 35	129.03°E, 35.90°N	68.1 ± 3.4	0.295	74.8	1.10
			0.298	69.8	1.10

The constants for the age calculation are $\lambda_\beta = 4.962 \times 10^{-10}$ yr^{-1}, $\lambda_\epsilon = 0.581 \times 10^{-10}$ yr^{-1}, and ^{40}K $= 1.167 \times 10^{-4}$ atom per atom of natural potassium [*Steiger and Jager*, 1977]. The error indicated for the age consists of a summation of all analytic errors (5%).

is possible that the block samples from this site may have been subjected individually to arbitrary rotation or movement after they had acquired the stable remanent magnetization. Data from this site were therefore omitted from further paleomagnetic considerations.

Table 2 represents a summary of the paleomagnetic results after correction for geological dip. The largest dip correction was 15°; most were in the range of 11°. The characteristic NRM direction of specimens from site 5 of the Hagbong Volcanic Member ($D = -19.0$, $I = 789.9$, and $\alpha_{95} = 6.9$) deviates by 22.2° from the mean of the Gyeongsang Supergroup, while the deviation of the NRM direction of site 4, from the same volcanic member, is only 9.9° from the mean. The large deviation of site 5 is ascribed to the lack of the dip correction for direction. This result documents the importance of the dip correction for obtaining reliable paleomagnetic directions.

The mean declination and inclination values were calculated from five directions into which the directions at five sites in the Gyeongsang Basin were converted at the representative point of the Korean Peninsula (128°E, 37°N). The Gyeongsang Supergroup yields a mean direction of $D = 26.6°$, $I = 62.3°$, and $\alpha_{95} = 8.3°$ at the representative point. The reliability of the direction is ascertained through thermal demagnetization, dip correction, and the agreement of directions between lava flow and sediments, although no specimens with reversed polarity were observed.

K-Ar Age of Basalts From the Hagbong Member

K-Ar whole rock age of basalt from the Hagbong Volcanic Member in the Hayang Group was determined by the Geochronometry section of Teledyne Isotope. Prior to the request for K-Ar dating, thin sections of the samples were examined. The rock is composed of phenocrysts of olivine and clinopyroxene and groundmass of clinopyroxene, plagioclase, opaque mineral, biotite, and less than 3% glass. Although all olivine crystals and some glass are altered to talc (sometimes iddingsite) and chlorite, respectively, the other constituents of rock, particularly interstitial phases, remain unaltered. According to criteria discussed by *Mankinen and Dalrymple* [1972] this rock is relatively suitable for dating.

Results of K-Ar dating are given in Table 3. The age obtained, 68.1 ± 3.4 m.y., is consistent with the geological estimation [*Chang*, 1975] that the Gyeongsang Supergroup was formed during the Cretaceous.

Significance of Results

The most striking result from this work is the great predominance of normally magnetized rocks. The question arises whether this predominance of normal polarity is primary or due to overprinting by the later heat of the intrusion of the plutons known as the Bulgugsa Granite Series [*Um and Reedman*, 1975]. The thermal demagnetization tests strongly suggest that no significant thermal overprinting has occurred and the magnetization is primary. The radiometric data support this implication: Since the age determination shows that the Gyeongsang Supergroup was formed during the Cretaceous, rocks could acquire magnetization with normal polarity, which was dominant in the middle to later Cretaceous age [*Larson and Hilde*, 1975; *Labrecque et al.*, 1977; *Lowrie and Alvarez*, 1981].

No systematic movements are observed in the distribution of virtual geomagnetic pole positions for five acceptable sites as shown in Figure 3. Radiometric data and the stratigraphic sequence suggest that the rock units sampled covered nearly the entire range of Cretaceous time. It seems reasonable therefore to assume that the mean pole position obtained (191°E, 69°N) averages over enough time to eliminate secular variation effects and gives a reliable estimate of the Cretaceous paleomagnetic pole position.

Comparison of the pole position with other Cretaceous pole positions from the Korean Peninsula is made in Figure 3. Overlapping circles of 95% confidence indicate that the Gyeongsang Supergroup pole and other poles from volcanics [*Kienzle and Scharon*, 1966], granites [*Ito and Tokieda*, 1980], and North Korean sediments [*Gurarii et al.*, 1966] are not significantly different and probably repre-

Fig. 2. (Opposite) Equal-area projections showing cleaned directions (center) of remanent magnetization together with examples of thermal (left) and alternating field (right) demagnetization of specimens. Solid symbols are lower hemisphere plots. Optimum temperature and field strength of the demagnetization are listed in Table 2. The remanent directions without tilting correction in Hagbong volcanic members are shown as triangles. The decay of remanent magnetization intensity is also plotted versus temperature during thermal demagnetization: J_0 is the initial intensity.

(A)

(B)

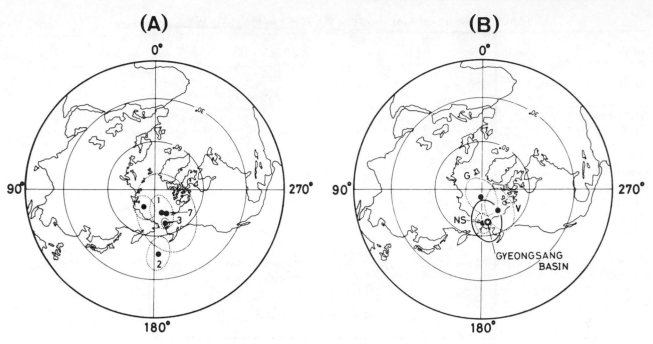

Fig. 3.   (a) Pole positions and 95% cones of confidence for Gyeongsang Basin. Site numbers are as in Table 2. (b) Cretaceous pole positions and 95% cones of confidence for the Korean Peninsula: pole V, volcanic rocks [*Kienzle and Scharon,* 1966]; pole NS, North Korean sediments [*Gurarii et al.,* 1966]; and pole G, granitic rocks [*Ito and Tokieda,* 1980].

sent much the same position of the Cretaceous pole. While the Gyeongsang Basin pole is in excellent agreement with the pole from the sediments of North Korea, the poles from the volcanic and granitic rocks are rather far removed from the Gyeongsang Basin pole. The cause of these deviations may be attributed, in part, to the lack of dip correction for data from the volcanic and granitic samples, because the strata of the Gyeongsang Supergroup which the granites intrude and the volcanics overlie have a southward dip with low angles of about 10°–30°. Thus it seems likely that the paleomagnetic pole (longitude 190°E, latitude 69°N) calculated from data of the Gyeongsang Supergroup and North Korean sediments represents a true Cretaceous pole.

Figure 4 shows the Cretaceous poles for neighboring regions of the Korean Peninsula [*McElhinny,* 1973; *Sasajima,* 1981; *Nishida et al.,* 1980; *Khramov et al.,* 1981]. The Korean pole obtained is in good agreement with poles from China, Kolyma, and the Siberian Platform: The polar distances between the Korean pole and the other poles are less than 13°. However, two circles of 95% confidence for the Korean pole and the southwest Japan pole fail to overlap. The polar distance between them is larger than 23° This separation is explained by the southward drifting accompanied by the clockwise rotation of southwest Japan from the Asian continent [*Hilde and Wageman,* 1973; *Kobayashi and Isezaki,* 1976]. The agreement between the Korean pole and poles from the neighboring regions ex-

cepting southwest Japan, on the contrary, suggests that the Korean Peninsula has been a part of the Asian continent at least since the Cretaceous and that the peninsula was not subjected to rotational movement relative to the continent.

Discussion

The Korean Peninsula is flanked by the Sea of Japan, which is a typical marginal basin floored by young oceanic crust. The origin of the Sea of Japan has been explained by back-arc spreading since Cretaceous time, based on the geophysical evidence [*Matsuda and Uyeda,* 1971; *Hilde and Wageman,* 1973; *Uyeda and Miyashiro,* 1974; *Isezaki,* 1975]. The spreading may be due to an upwelling thermal diapir [*Karig,* 1971] or evolution of the convective process in the asthenosphere [*Sleep and Toksöz,* 1971; *Andrews and Sleep,* 1974]. The large-scale extension must have been induced to the edge of the Asian continent just before the spreading. The Chugarian rift valley, which extends from Seoul to Wansan, is possibly the remnant of the extension region in the Korean Peninsula [*O. J. Kim,* 1975; *S. G. Kim,* 1980]. The Korean Peninsula, however, did not drift off from the Asian continent as Japan did, as is concluded in the previous section. This implies that the Korean Peninsula is strong enough to resist the active extension.

The resistance to extension is probably attributed to the stiffness of either the crust or the lithosphere beneath the

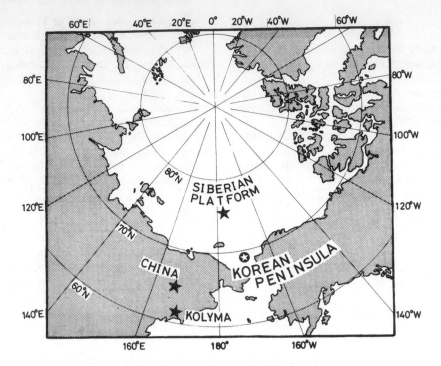

Fig. 4.   Cretaceous pole positions for the Korean Peninsula and neighboring regions. The Korean Peninsula pole calculated from data of the present study and North Korean sediments is 189.5°E, 68.9°N. Poles of neighboring regions are a compilation from data of *McElhinny* [1973], *Sasajima* [1981], *Khramov et al.* [1981], and *Nishida et al.* [1981]: Siberian Platform pole ($N = 10$), 180.9°E, 75.4°N; Kolyma pole ($N = 7$), 166.1°E, 61.3°N; China pole ($N = 8$), 164.1°E, 64.0°N; southwest Japan pole (not shown) ($N = 11$), 196.1°E, 45.9°N. N is the number of poles averaged for the compilation.

Korean Peninsula. The crust beneath the Korean Peninsula is rather thin, less than 30 km [*Cummings and Shiller,* 1971], while the thickness beneath Honshu (Japan) is about 40 km [*Yoshii,* 1979; *Hashizume et al.,* 1981; *Hurukawa,* 1981]. This indicates that the resistance of the landmass against the extension hardly depends on the thickness of the crust.

Honshu has thin lithosphere of about 40 km, and the crust beneath Honshu comes directly into contact with underlying asthenosphere [*Yoshii,* 1979]. On the other hand, the Korean Peninsula has a stiff lithosphere under the crust (lid): The propagation of S_n, a short-period seismic shear wave traveling in the uppermost part of the mantle, is very efficient and shows a much higher Q value than below Honshu [*Molnar and Oliver,* 1969]. Because the Korean Peninsula is part of the Sino-Korean Precambrian shield [*Um and Reedman,* 1975], the thick and stiff lid is also expected from thickening plate theory [*Parker and Oldenburg,* 1973; *Oldenburg,* 1975; *Kono and Amano,* 1978]. This evidence strongly suggests that the stiffness of the lid within the lithosphere controls the occurrence of the isolation of the landmass from the continent. There remains, however, another possible explanation of the stable condition of the Korean Peninsula, that the most active extension area was far from the peninsula.

Conclusion

Five reliable sites from volcanic and sedimentary rocks in the Gyeongsang Basin, in the southeastern part of the Korean Peninsula, have yielded an average Cretaceous pole of 191.2°E, 68.9°N. The representative Korean pole of Cretaceous age, 190°E, 69°N, calculated from Gyeongsang Basin and North Korean sediments is in good agreement with poles from China, Kolyma, and the Siberian Platform excepting southwest Japan. This implies that the Korean Peninsula has been a part of the Asian continent at least since the Cretaceous and that during this time it has not been subjected to any rotational motion relative to the continent. This strong resistance to tectonic disturbances of the Korean Peninsula is probably attributed to the presence of a thick and stiff lid within the lithosphere beneath the peninsula.

Acknowledgments.   We are grateful to the two reviewers for their constructive comments on the manuscript and for significant improvements of the English. Special thanks are due to M. Torii for the use of the thermal demagnetization unit.

References

Andrews, D. J., and N. H. Sleep, Numerical modelling of tectonic flow behind island arcs, *Geophys. J. R. Astron. Soc., 38,* 237–251, 1974.

Bird, P., M. N. Toksöz, and N. H. Sleep, Thermal and mechanical model of continent-continent convergence zones, *J. Geophys. Res., 80,* 4405–4416, 1975.

Chang, K. H., Cretaceous stratigraphy of southeast Korea, *J. Geol. Soc. Korea, 11,* 1–23, 1975.

Cummings, D., and G. I. Shiller, Isopach of the earth's crust, *Earth Sci. Rev., 7,* 97–125, 1971.

Fisher, R., Dispersion on a sphere, *Proc. R. Soc. London, Ser. A, 217,* 295–305, 1953.

Gurarii, G. Z., P. N. Kropotkin, M. A. Pevzner, R. S. Von, and V. M. Trubikhin, Laboratory evaluation of the usefulness of North Korean sedimentary rocks for paleomagnetic studies, *Izv. Acad. Sci. USSR Phys. Solid Earth.,* Engl. Transl., *11,* 128–136, 1966.

Hashizume, M., K. Ito, and T. Yoshii, Crustal structure of south-western Honshu, Japan and nature of the Mohorovičić discontinuity, *Geophys. J. R. Astron. Soc., 66,* 157–168, 1981.

Hilde, T. W. C., and J. M. Wageman, Structure and origin of the Japan Sea, in *The Western Pacific,* edited by P. J. Coleman, pp. 415–434, University of Western Australia Press, Perth, 1973.

Hurukawa, N., Normal faulting microearthquakes occurring near the Moho discontinuity in the northeastern Kinki district, Japan, *J. Phys. Earth, 29,* 519–535, 1981.

Isezaki, N., Possible spreading centers in the Japan Sea, *Mar. Geophys. Res., 2,* 265–277, 1975.

Ito, H., and K. Tokieda, An interpretation of paleomagnetic results from Cretaceous granites in South Korea, *J. Geomagn. Geoelectr., 32,* 275–284, 1980.

Karig, D. K., Origin and development of marginal basins in the western pacific, *J. Geophys. Res., 76,* 2542–2561, 1971.

Kawasaki, S., General geological map of Chosen (1:1,000,000), Geol. Surv. Chosen, Seoul, 1928.

Khramov, A. N., G. N. Petrova, and D. M. Pechersky, Paleomagnetism of the Soviet Union, in *Paleoreconstruction of the Continents, Geodyn. Ser.,* vol 2, edited by M. W. McElhinny and D. A. Valencio, pp. 177–194, AGU, Washington, D. C., 1981.

Kienzle, J., and L. Scharon, Paleomagnetic comparison of Cretaceous rocks from South Korea and late Paleozoic and Mesozoic rocks of Japan, *J. Geomagn. Geoelectr., 18,* 413–416, 1966.

Kim, O. J., Granites and tectonics of South Korea, *J. Korean Inst. Min. Geol., 8,* 223–230, 1975.

Kim, S. G., Seismicity of the Korean Peninsula and its vicinity, *J. Korean Inst. Min. Geol., 13,* 51–63, 1980.

Kobayashi, K., and N. Isezaki, Magnetic anomalies in the Sea of Japan and the Shikoku Basin: Possible tectonic implications, in *The Geophysics of the Pacific Ocean Basin and Its Margins, Geophys. Monogr. Ser.,* vol. 19, edited by G. H. Sutton, M. H. Manghnani, and R. Moberly, pp. 235–251, AGU, Washington, D. C., 1976.

Kono, Y., and M. Amano, Thickening model of the continental lithosphere, *Geophys. J. R. Astron. Soc., 54,* 405–416, 1978.

Labrecque, J. L., D. V. Kent, and S. C. Cande, Revised magnetic polarity time scale for Late Cretaceous and Cenozoic time, *Geology, 5,* 330–335, 1977.

Larson, R. L., and T. W. C. Hilde, A revised time scale of magnetic reversals for the Early Cretaceous and Late Jurassic, *J. Geophys. Res., 80,* 2586–2594, 1975.

Lee, Y. J., Granitic rocks from the southern Gyeongsang Basin, southeastern Korea, I, General geology and K-Ar ages of granitic rocks, *J. Jpn. Assoc. Mineral. Petrol. Econ. Geol., 75,* 105–116, 1980.

Lowrie, W., and W. Alvarez, Rotation of the Italian Peninsula, *Nature, 251,* 285–288, 1974.

Lowrie, W., and W. Alvarez, One hundred million years of geomagnetic polarity history, *Geology, 9,* 392–397, 1981.

Mankinen, E. A., and G. B. Dalrymple, Electron microscope evaluation of terrestrial basalts for whole-rock K-Ar dating, *Earth Planet. Sci. Lett., 17,* 89–94, 1972.

Matsuda, T., and S. Uyeda, On the Pacific-type orogeny and its model: Extension of the paired belts concept and possible origin of marginal seas, *Tectonophysics, 11,* 5–27, 1971.

McElhinny, M. W., *Paleomagnetism and Plate Tectonics,* 358 pp., Cambridge University Press, New York, 1973.

McElhinny, M. W., N. S. Haile, and A. R. Crawford, Paleomagnetic evidence shows Malay Peninsula was not a part of Gondwanaland, *Nature, 252,* 641–645, 1974.

Molnar, P., and J. Oliver, Lateral variations of attenuation in the upper mantle and discontinuities in the lithosphere, *J. Geophys. Res., 74,* 2648–2682, 1969.

Molnar, P., and P. Tapponnier, Cenozoic tectonics of Asia: Effects of a continental collision, *Science, 189,* 419–426, 1975.

Molnar, P., and P. Tapponnier, The collision between India and Eurasia, *Sci. Am., 236,* 30–41, 1977.

Nishida, J., C. L. So, K. Maenaka, T. Ikeda, O. Tamada, S. Sasajima, and T. Yokoyama, Paleomagnetic study of Jurassic acid igneous rocks and Cretaceous red bed in Hong Kong, *Tsukumo Earth Sci., 15,* 27–37, 1980.

Oldenburg, D. W., A physical model for the creation of the lithosphere, *Geophys. J. R. Astron. Soc., 43,* 425–451, 1975.

Parker, R. L., and D. W. Oldenburg, Thermal model of ocean ridges, *Nature Phys. Sci., 242,* 137–139, 1973.

Sasajima, S., Pre-Neogene paleomagnetism of Japanese Islands (and vicinities), in *Paleoreconstruction of the Continents, Geodyn. Ser.,* vol 2, edited by M. W. McElhinny and D. A. Valencio, pp. 115–128, AGU, Washington, D. C., 1981.

Sleep, N. H., and N. Toksöz, Evolution of marginal basin, *Nature, 233,* 548, 1971.

Steiger, R. H., and E. Jager, Subcommission on Geochronology: Convention on the use of decay constants in geo and cosmochronology, *Earth Planet. Sci. Lett., 36,* 359–362, 1977.

Um, S. H., and A. J. Reedman, *The Geology of Korea,* 139 pp., Geological and Mineral Institute of Korea, Seoul, 1975.

Uyeda, S., and A. Miyashiro, Plate tectonics and the Japanese Island: A synthesis, *Geol. Soc. Am. Bull., 85,* 1159–1170, 1974.

Van der Voo, R., Paleomagnetic evidence for the rotation of the Iberian Peninsula, *Tectonophysics, 7,* 5–56, 1969.

Yabe, H., Mesozoic plants from Korea, *J. Sci. Coll. Imp. Univ. Tokyo., 20,* article 8, 1905.

Yoshii, T., A detailed cross-section of the deep seismic zone beneath northern Honshu, Japan, *Tectonophysics, 55,* 349–360, 1979.